WORLD SOYBEAN RESEARCH CONFERENCE II: PROCEEDINGS

Other Titles of Interest

New Agricultural Crops, edited by Gary A. Ritchie

Successful Seed Programs: A Planning and Management Guide, Johnson E. Douglas

Animals, Feed, Food and People: An Analysis of the Role of Animals in Food Production, edited by R. L. Baldwin

Pest Control: Cultural and Environmental Aspects, edited by David Pimentel and John H. Perkins

Pest Management in Transition: With a Regional Focus on the Interior West, Pieter de Jong, project coordinator

Small-Scale Processing and Storage of Tropical Root Crops, edited by Donald L. Plucknett

The Mineralogy, Chemistry, and Physics of Tropical Soils with Variable Charge Colloids, Goro Uehara and Gavin P. Gillman

Azolla as an Aquatic Green Manure: Use and Management in Crop Production, Thomas A. Lumpkin and Donald L. Plucknett

Irrigated Rice Production Systems: Design Procedures, Jaw-Kai Wang and Ross Ellis Hagan

Managing Pastures and Cattle Under Coconuts, Donald L. Plucknett

Rice in the Tropics: A Guide to the Development of National Programs, Robert F. Chandler, Jr.

About the Book and Editor

World Soybean Research Conference II: Proceedings

edited by Frederick T. Corbin

The result of strong international interest in the soybean, the World Soybean Research Conference II was held March 26–29, 1979, at North Carolina State University. This volume contains seventy-four of the invited papers presented at that meeting. The authors, international authorities in their fields, represent sixteen areas of professional specialization. In this overview of major soybean research conducted since 1975, they describe specific accomplishments and advances in entomology, pathology, engineering, economics, food science, genetics, physiology, and agronomy. In addition, various aspects of research on soybean production, marketing, and utilization are examined. Also available from Westview Press is *World Soybean Research Conference II: Abstracts,* which contains summaries of more than two hundred papers presented at the conference.

Frederick T. Corbin is professor of crop science at North Carolina State University in Raleigh.

Published in cooperation with
North Carolina State University

WORLD SOYBEAN RESEARCH CONFERENCE II
March 26–29, 1979

CONFERENCE SPONSORS

Agency for International Development
American Soybean Association Research Foundation
INTSOY/University of Illinois
National Soybean Crop Improvement Council
North Carolina Agricultural Extension Service
North Carolina Agricultural Research Service
North Carolina Crop Improvement Association
North Carolina Foundation Seed, Inc.
North Carolina Soybean Producers Association
School of Agriculture and Life Sciences, NCSU
Science and Education Administration, USDA

B.E. Caldwell, Conference Chairman

PUBLICATIONS COMMITTEE

F. T. Corbin, Department of Crop Science, NCSU
H. T. Daniel, Department of Economics, NCSU
D. P. Schmitt, Department of Plant Pathology, NCSU
J. W. Van Duyn, Department of Entomology, NCSU
P. A. Wilson, PawPrint Graphics and Typesetting

WORLD SOYBEAN RESEARCH CONFERENCE II:

PROCEEDINGS

edited by

Frederick T. Corbin

Routledge
Taylor & Francis Group
LONDON AND NEW YORK

First published 1980 by Westview Press

Published 2018 by Routledge
52 Vanderbilt Avenue, New York, NY 10017
2 Park Square, Milton Park, Abingdon, Oxon OX14 4RN

Routledge is an imprint of the Taylor & Francis Group, an informa business

Library of Congress Cataloging in Publication Data
World Soybean Research Conference, 2d, North Carolina State University, 1979.
World Soybean Research Conference II.
Includes index.
1. Soybean—Congresses. 2. Soybean products—Congresses. I. Corbin, Frederick T.
SB205.S7W67 1979b 633.3'4 80-10785

ISBN 13: 978-0-367-21400-5 (hbk)

TABLE OF CONTENTS

FOREWORD v

PREFACE vii

KEYNOTE ADDRESSES

Desirable Qualities of Soybeans—Japanese Viewpoints
H. Nakamura 1

Sharing Agricultural Technology
P. Findley 11

MINERAL NUTRITION

Cation Nutrition and Ion Balance
J. E. Leggett and D. B. Egli 19

Response of Soybeans to Liming on Acid Tropical Soils
F. Abruña 35

Mineral Nutrition and Nodulation
D. N. Munns. 47

NITROGEN FIXATION

Nitrogen Input with Emphasis on N_2 Fixation in Soybeans
R. W. F. Hardy, U. D. Havelka, and P. G. Heytler 57

Rhizogenetics of Soybeans
R. W. Zobel 73

Rhizobium Inoculation and Soybean Production
J. C. Burton 89

PHYSIOLOGY

Amino Acid Loading and Transport in Phloem
L. E. Schrader, T. L. Housley, and J. C. Servaites. 101

Nitrogen Translocation in the Xylem of Soybeans
D. W. Israel and P. R. McClure 111

Coupling Between Photosynthesis and Nitrogen Fixation
J. G. Streeter, H. J. Mederski, and R. A. Ahmad 129

Regulation of Senescence
L. D. Noodén 139

Effect of Pod Filling on Leaf Photosynthesis in Soybeans
W. A. Brun and T. L. Setter 153

Postponement of Severe Water Stress in Soybeans by Rooting Modifications: A Progress Report
H. M. Taylor 161

BREEDING

Role of Classical Breeding Procedure in Improvement of Self-Pollinated Crops
V. A. Johnson and J. W. Schmidt 179

Population Improvement in Self-Pollinated Crops
D. F. Matzinger and E. A. Wernsman 191

Long- and Short-Term Recurrent Selection in Finite Populations—Choice of Population Size
J. O. Rawlings 201

Strategies for Introgressing Exotic Germplasm in Breeding Programs
W. J. Kenworthy 217

Role of Physiology in Soybean Breeding
D. N. Moss 225

Accomplishments and Priorities in Plant Breeding
G. F. Sprague 233

Mobilization, Conservation and Utilization of Soybean Germplasm in the USSR
N. I. Korsakov 241

ENTOMOLOGY

Systems Approach to Pest Management in Soybeans
J. L. Stimac and C. S. Barfield 249

Interactions of Control Tactics in Soybeans
L. D. Newsom 261

Pest Problems of Soybeans and Control in Nigeria
M. I. Ezueh and S. O. Dina 275

Soybean Insects and Their Natural Enemies in Brazil—A Comparison with the Southern United States
S. G. Turnipseed 285

Soybean Insect Problems in India
A. K. Bhattacharya and Y. S. Rathore 291
Insect Problems of Soybeans in the United States
M. Kogan 303
Entomopathogens for Control of Insect Pests of Soybeans
C. M. Ignoffo 327
Insect Pest Management in North Carolina Soybeans
J. R. Bradley, Jr. and J. W. Van Duyn 343

PLANT PATHOLOGY

Role of *Phomopsis* Sp. in the Soybean Seed Rot Problem
A. F. Schmitthenner and K. T. Kmetz 355
Research on Phytophthora Root and Stem Rot: Isolation, Testing Procedures, and Seven New Physiologic Races
B. L. Keeling 367

WEED CONTROL

Ecological Factors Affecting Weed Competition in Soybeans
C. G. McWhorter and D. T. Patterson 371
Weed Control Systems in the Corn Belt States
F. W. Slife 393
Weed Control Systems in Southern U.S.
R. Frans 399

PRODUCTION

Foliar Fertilization of Soybeans During Seed-Filling
J. J. Hanway 409
Factors Affecting the Response of Soybeans to Molybdenum Application
F. C. Boswell 417
Soybean Harvesting Equipment: Recent Innovations and Current Status
W. R. Nave 433
Critical Factors in Soybean Seedling Emergence
H. D. Bowen and J. W. Hummel 451

ENGINEERING

Mechanization Alternatives for Small Acreages in Less-Developed Countries
M. L. Esmay and M. Hoki 471

Soil Conservation Practices in Soybean Production
J. C. Siemens 477

Equipment for Narrow-Row and Solid Plant Soybeans
R. I. Throckmorton 485

Soybean Damage Detection
M. R. Paulsen 493

Drying Methods and the Effect on Soybean Quality
G. M. White, I. J. Ross, and D. B. Egli 501

Soybean Seed Storage Under Controlled and Ambient Conditions in Tropical Environments
E. J. Ravalo, E. D. Rodda, F. D. Tenne, and J. B. Sinclair 519

MODELING SOYBEAN SYSTEMS

Simulation for Research and Crop Management
D. N. Baker 533

Simulation of Insect Damage to Soybeans
W. G. Rudd 547

Simulation of the Vegetative and Reproductive Growth of Soybeans
R. B. Curry, G. E. Meyer, J. G. Streeter, and H. J. Mederski. 557

Application of Modeling to Irrigation Management of Soybeans
J. W. Jones and A. G. Smajstrla 571

A Mathematical Model for the Threshing and Separating of Soybeans
V. M. Huynh, T. E. Powell, and J. N. Siddall. 601

Simulation Applications in Soybean Drying
J. H. Young and R. N. Misra 623

RESEARCH TECHNIQUES

Application of Von Bertalanffy's Equation to Soybean Growth Modeling
R. Guarisma 637

Simulation of Soil Compaction Under Tractor Wheels
H. D. Bowen and H. Jaafari 647

UTILIZATION—OILS

Raw Material and Soybean Oil Quality
T. L. Mounts 659

Nutritional Aspects of Soybean Oil Utilization
E. A. Emken. 667

Food and Industrial Uses of Soybean Lecithin
B. F. Szuhaj 681

Industrial Uses of Soybean Oil
K. T. Zilch 693

UTILIZATION—PROTEIN

Anti-Nutritional Factors as Determinants of Soybean Quality
I. E. Liener. 703

Soy Protein Product Characteristics
M. F. Campbell 713

Analysis of Soy Protein in Meat Products
A. C. Eldridge and W. J. Wolf. 721

Oriental Soybean Foods
D. Fukushima and H. Hashimoto 729

Foods from Whole Soybeans
A. I. Nelson, L. S. Wei, and M. P. Steinberg 745

PROTEIN AND OIL

Soybeans in World Trade
R. E. Bell. 763

Soybean Proteins in Human Nutrition
W. J. Wolf 767

Pricing Soybeans on the Basis of Chemical Constituents
N. J. Updaw and T. E. Nichols, Jr. 781

AGRIBUSINESS

Protectionism in Trade—Expansion in Oilseeds
S. J. Hauck. 801

Demand for Soybeans and Soybean Meal in the European Common Market
H. C. Knipscheer 807

Improving Grades and Standards for Soybeans
L. D. Hill 823

Role of Soy Protein Products in National Development
F. H. Schwarz and J. K. Allwood 831

MARKETING, TRANSPORT AND STORAGE

Principal Determinants of Variations in Soybean Prices
R. McFall Lamm, Jr. 839

Soy Oil Utilization—Current Situation and Potential
D. R. Erickson and R. A. Falb 851

Risk Management in Marketing Soybeans
D. E. Kenyon 859

Competitive Position of New Oilseed Sunflowers with Soybeans
H. O. Doty, Jr. 871

Exporting Soybeans Through the Mississippi River Gulf Port
W. M. Gauthier, G. R. Hadder, and H. D. Traylor. 883

Soybean Processing Industry—U.S. and Foreign
J. J. Mogush 889

AUTHOR INDEX 897

FOREWORD

Worldwide interest in soybeans continues to increase. In the past several decades, soybeans have risen from a minor crop to a major world source of oil and protein. With this growth in importance, research interest has also increased, and the number of people involved has grown rapidly.

The First World Soybean Conference was held at Urbana, Illinois, in 1974. The conference was well attended with 600 participants from about 50 countries. At the conclusion of the conference, the general consensus was that a second conference should be held in three to five years. The location and exact date were not determined. With the encouragement of industry leaders such as Mr. Robert Judd, National Soybean Crop Improvement Association, and Dr. Robert Howell, Chairman of the 1974 conference, North Carolina State University agreed to host the Second World Soybean Conference.

World Soybean Conference II was held at the Jane S. McKimmon Center for Continuing Education, N. C. State University, Raleigh, N. C. This volume contains 74 of the 83 invited papers presented at the conference. Abstracts of all papers were published separately and distributed at registration. There were a total of 220 papers presented with 29% of these from foreign participants.

Our goals from the conference were to improve communication among soybean research and extension specialists, to share current information on the various disciplines, to provide a forum for multidisciplinary discussions, and to stimulate new research ideas. We feel that we accomplished these goals. Many participants indicated that they gained a better understanding of their own area as well as a familiarity with the current research in other areas. Several new cooperative endeavors were established and others updated and expanded.

Because of the number of papers and the array of disciplines, concurrent sessions were necessary. However, an attempt was made to minimize conflicts and maximize multidisciplinary exchanges. Plenary sessions were used for topics of general interest to all participants.

In an effort to make the conference international and to assist participation from soybean specialists from developing countries, a USAID grant was obtained. Also USAID funds plus USDA-SEA funds were used to assist with the publication of Abstracts and Proceedings. Others listed in this volume contributed to conference support. This support was critical to the conduct of the conference.

We want to thank all those who contributed to the conference—our sponsors, the participants, the faculty and their spouses who made World Soybean Conference II a success. My special thanks to the faculty who provided leadership for the various committees and to Dr. Fred Corbin and Mrs. Pam Wilson for thier efforts in the timely completion of this volume.

—Billy E. Caldwell
Conference Chairman

PREFACE

The *World Soybean Research Conference II: Proceedings* should serve as an excellent resource for scientists involved currently in various phases of soybean research. Furthermore, this volume will be a valuable aid to students by introducing recent concepts and literature relevant to soybeans. The book is divided into sixteen sections of particular professional specializations. Specific accomplishments and recent advances within each section are reported by international authorities.

The chapter titles are identical to the titles listed in the Abstracts which were distributed at the conference. The names and addresses of authors of contributed papers are also printed in the Abstracts.

I am grateful to Dr. Billy Caldwell for his support and leadership as conference chairman, to Dr. H. D. Gross for program organization and assistance in providing a timetable for manuscript receipt, to the publications committee for their valuable suggestions in book production and manuscript review, to Mrs. Julia Bender for editorial review, and to Dr. C. D. Raper for graphic design.

I am especially grateful to Mrs. Pam Wilson for her assistance in technical production. In addition to the typesetting, Mrs. Wilson coordinated the mailing and final proofing of the manuscripts. Her suggestions were invaluable in the publication of both the Abstracts and the Proceedings.

— Frederick T. Corbin
Editor

DESIRABLE QUALITIES OF SOYBEANS—JAPANESE VIEWPOINTS

H. Nakamura

It is my great honor and privilege to be invited to the second World Soyben Research Conference and to make a keynote address on the desirable qualities of soybeans for the world market.

I think it is historically significant for this conference to be held in North Carolina, because in the early period of United States soybean production at the beginning of this century, soybeans were first grown here and the domestically grown soybeans were processed into oil and meal in 1915 by the Elizabeth City Oil and Fertilizer Company at Elizabeth City in this State. Also, several cottonseed oil plants in North Carolina began processing soybeans for oil and meal, but only a small quantity of soybeans was available for crushing due mainly to the high prices of the seed which was in strong demand for planting as hay during those early years. In the past fifty years since its small beginning in North Carolina, American soybean production has increased at an astonishing pace and today the United States has become by far the largest producer and processor of soybeans in the world.

The growth and present prosperity of the American soybean industry has been brought about by sincere efforts of countless numbers of people in all segments of the American soybean industry, including growers, processors, scientists, handlers, and all others who are connected in some way with the production, utilization and marketing of soybeans in this country. The U.S. government has also played an important role in the growth of the soybean industry.

On the basis of a glorious history of success of the American soybean industry in the past, I think it is most appropriate for us to be gathered here in this historic soybean state of North Carolina and to look into the future of soybean production and utilization in the world. For this purpose I think it is important at this time to review the characteristics of soybean demand in the world and to recognize the preferences of the consumers of soybeans in the world market.

With the above in mind, today I will first explain the nature of soybean demand in the Japanese market and then mention the desirable types of soybeans from the standpoint of soybean consumers in Japan.

The Japanese soybean market may be described as one of the most diversified in the world along with the soybean markets in other Asian countries. As you all know, soybeans have been consumed as food in Asia for hundreds of years before they were used as raw material for the processing industry.

In Japan, recorded history shows that soybeans were already used as an ingredient in various types of daily food in the sixth century. Since time immemorial soybeans have been an important part of the Japanese diet, together with fish and rice. Through the centuries different types of soybean foods have been developed and some have become traditional and popular daily food items, such as tofu or soybean curd, miso or fermented soybean soup, natto or fermented soybeans, kinako or roasted soybean flour and soy sauce. An average Japanese cannot help eating some of these soybean foods every day. Indeed, without tofu, miso, and soy sauce most Japanese dishes are without taste and flavor.

For the purpose of manufacturing the traditional soybean foods, Japan consumes approximately 800,000 tons or about 29 million bushels of whole soybeans each year. In addition, about 300,000 tons of soybean meal are used for manufacturing soy sauce and other soybean foods. Thus, when you combine whole soybeans and soybean meal and express their total amounts in terms of whole soybeans, then 1.2 million tons are consumed annually for food purposes. On a per capita basis, this is about 10 kg or 22 lb of soybean consumption for direct food use per year.

The Japanese soybean food manufacturers have long been accustomed to using domestic or imported Chinese soybeans in their manufacturing process. Since they use whole soybeans directly as food they are very much concerned with their quality. They prefer soybeans of a clear yellow color, uniform size, and high protein content with no foreign materials. When American soybeans were first used by the Japanese soybean food industry in the laste 1950's because of the lack of Chinese soybeans, many complaints were raised against American soybeans regarding the presence of foreign materials, percentage of green beans, irregularity of size and lower protein content. As it was necessary to clean American soybeans, various cleaning devices have been developed to get rid of the foreign materials and dust as well as broken particles of soybeans. Thus, through mechanical cleaning operations, food manufacturers obtain dockage-free American soybeans of uniform size.

With regard to the level of protein concent, most Japanese food manufacturers have discovered that soybeans grown in the central and northern parts of Illinois, Indiana, and Ohio are generally high in protein and suitable for their soybean products. Therefore, they pay some premiums for soybeans from these areas. However, miso manufacturers still prefer Chinese soybeans and they say they can obtain a better product through the fermentation process when they use Chinese beans as the raw material. The reasons for these preferences are not well defined scientifically, but are said to be related to the softness of beans when boiled, and the thickness of hulls of American soybeans.

Tofu or soybean curd manufacturers prefer Indiana-Ohio beans, because they can make a better product by using high protein American soybeans grown in these areas. In terms of quantity of tofu, people annually consume half a million tons or about 18 million bushels of U.S. soybeans.

The general quality requirements of the traditional Japanese soybean food manufacturers are clean yellow soybeans with a high protein content. Depending upon different products they have a definite preference for a type of soybeans to be used as raw material, and when they find their favorite type they are willing to pay a higher price for the suitable beans. They say the tastes of their products are related to the types of beans they use. Under such circumstances there is an opportunity for plant breeders to develop a high protein and a high yielding soybean variety suitable for various types of soybean foods in the Japanese and other Asian markets. Moreover, if and when more soybeans are used directly as human food in the world, there will arise a greater need for developing a suitable soybean variety for food uses. Seed and hilum color, size of seed, cooking time, swelling, taste, and texture will all become important factors for developing such a food variety. Preferences of consumers must be the basis for developing a new soybean variety for direct food uses.

In the calendar year 1978 Japanese soybean consumption for all purposes amounted to approximately 4.1 million tons or about 150 million bushels, of which 80% or 3.3 million tons were used by the soybean processors to make oil and meal and 20% or 800,000 tons were used by food manufacturers. In terms of current world consumption it is estimated that more than 90% of the total soybean demand is for crushing and only 7 to 8% is used for food products, mostly in Asian countries.

Now let me speak on preferred types of soybeans by the Japanese processors. As a matter of fact, desirable quality requirements of soybean processors may be universal throughout the world, because processors are first interested in high oil content beans, which will enable them to get a higher yield of oil per bushel in their extraction process. Since the price of oil per pound is generally 2.5 to 3 times higher than that of meal, a higher oil yield means a higher products value per bushel of soybeans crushed, and naturally a better margin for processors.

Of course, processors are also concerned with the protein content of soybeans, as soybean meal is traded on protein basis. A protein level high enough to make 44% protein meal without dehulling is a minimum requirement.

With respect to quality of soybeans for processing, it must also be mentioned that even for large-scale processing operations it is desirable, as in the case of soybean food manufacture, that the level of foreign materials be as low as possible. Harvesting and handling equipment must be so improved as to reduce foreign materials in harvesting and handling operations. Also, it is necessary to control weeds in the soybean fields in order to reduce admixtures of weed seeds in soybeans.

One characteristic of the Japanese processing industry is that all the processors refine their crude oil production and sell refined oil under their brand names in the domestic market. Now, let me briefly explain the edible liquid vegetable oil market in Japan. The annual domestic production of soybean oil in 1978 was about 600,000 tons, rapeseed oil production 350,000 tons, corn oil and other vegetable oil production combined amounted to about 200,000 tons. In addition, Japan imported about 45,000 tons of cottonseed oil and other vegetable oils in 1978. The total of all these edible liquid vegetable oils in 1978 was 1,200,000 tons or about 2.65 billion lb. These figures do not include palm oil and coconut oil.

Of the total supply of 1.2 million tons of edible seed oils, about one-third was sold by processors in consumer packs in liquid form for household consumption; one-third went for restaurant and other institutional uses; and another one-third was sold in bulk to manufacturers of margarine, shortening, mayonnaise, and other food items.

In the category of household market for edible vegetable oils, rapeseed oil accounts for 70 to 80% of the consumption, and use of soybean oil is quite limited due to its color and flavor reversion problem.

Oils for restaurant and other institutional uses are sold in a five-gallon can and consist mostly of refined soybean oil. In this category oil is consumed in a relatively short period of time and color and taste reversion is not much of a problem.

Oils used by the food fats and oils processing industry are sold in bulk and both refined soybean oil and rapeseed oil are used extensively. Since the Japanese oilseed processors generally refine all their crude oil production and sell refined oil under their brand, they are very much interested in the quality of crude oil.

Polyunsaturated oils are preferred for health reasons. For household consumption oils are sold mostly in glass bottles or plastic containers. Because of the flavor and color reversion tendency of soybean oil, rapeseed oil is definitely preferred in consumer packs in the Japanese market. In a reversed sense, this means that if the stability of soybean oil against light, heat, and oxidization is enhanced, demand for soybean oil, hence for soybeans, will be greatly increased. Such a change in the nature of soybean oil could be attained possibly only through breeding a new variety of soybeans having a new composition of fatty acids.

In this connection, the Canadian success of rapeseed breeding activities may be cited as an example. As you know already, the concerted efforts of the government and research institutions in Canada have culminated in developing new rapeseed varieties with low erucic acid and low glucosinolate content. The low erucic acid characteristics of Canadian rapeseed have certainly contributed to a greater acceptance of rapeseed oil in the world market. Also, low glucosinolate rapeseed meal may be accepted more readily by feed manufacturers than before. Thus, the basic changes in the nature of Canadian rapeseed through breeding have greatly enhanced the prospects of demand for Canadian rapeseed and products.

As in the case of Canadian rapeseed, if the fatty acid composition of soybeans could be changed to make soybean oil more stable by developing new varieties, the market for soybean oil would be further expanded. Of course, yields of new varieties must be as high as the present varieties to make them an economical crop.

Now let me review next the soybean meal market in Japan. In the calendar year 1978 Japan produced 2.54 million tons of soybean meal from the 3.3 million tons of soybeans processed. In addition, Japan imported 350,000 tons of soybean meal in 1978, of which 300,000 tons were from the United States, and most of the balance from Brazil. Thus, the total supply of soybean meal was about 2.9 million tons. Subtracting the 300,000 tons of meal used for manufacturing traditional soybean foods such as soy sauce, the balance of 2.6 million tons was used to manufacture 21 million tons of mixed feeds in 1978.

The demand for soybean meal for feeds in Japan has increased at a very rapid rate in the past two decades along with the increasing production of mixed feeds for the fast growing livestock industry. For instance, the quantity of soybean meal used for the manufacture of feeds in 1960 was only 160,000 tons, which in a period of 18 yr increased nearly 15 times to a level of 2.6 million tons in 1978. During the same period the production of mixed feeds increased seven times from 2.9 million tons to 21 million tons. From these figures it can be noted that the demand for soybean meal has increased at a rate much greater than that of production increase of mixed feeds due to the fact that the percentage of soybean meal mixture in the total ingredients of mixed feeds has increased from 5.4% in 1960 to 11.3% in 1978. This increase in the mixing rate of soybean meal has been brought about by the recognition by the Japanese feed manufacturers of the excellent feeding value of soybean meal as a source of protein at a reasonable cost. Also, dependable and increasing supplies of soybeans from the United States have provided a background for the greater use of soybean meal by the feed industry in Japan.

Since soybean meal is used by the feed industry as a high protein ingredient, the price of meal is determined by protein content. A higher protein level commands a higher price. So, given the oil content, processors prefer soybeans with a high protein level. Thus, it is very clear from the standpoint of soybean processors that soybeans with high oil and high protein content are most desirable, and superior quality soybeans in this respect could command a premium in the market. In the light of the negative correlation between the level of oil and protein, it may be very difficult to satisfy high oil and high protein content requirements in one variety, but such a combination is clearly most desirable from a processor's standpoint.

After satisfying the requirements for oil and protein content, soybean buyers prefer soybeans with the least amount of foreign materials. The presence of foreign materials incurs cleaning costs. For overseas buyers additional ocean freight is also incurred. For manufacturers of traditional soybean foods in Japan, soybeans must be completely free of foreign materials when soybean foods are made directly from soybeans.

With regard to the nature of soybean protein for feeding purposes, the amino acid composition is well balanced except for the low level of methionine. The low level of methionine is supplemented by synthetic methionine or fish meal. Therefore, if a new soybean variety containing more methionine could be developed, the value of soybean meal would be enhanced and the market for soybean meal would be further expanded. In this connection, the development of high-lysine corn in the United States may be recalled as a case in point. A corn variety with a high protein level has a better feeding value and it is only natural to make efforts to develop such a variety.

Now let us consider the use of soybeans directly as human food. Although in the minds of the American people soybeans are defined as an oilseed which is used by soybean processors to make oil and meal for industrail uses, in Japan and Asia soybeans have long been considered a food grain like rice or wheat and used in various types of daily foods as a source of protein. Soybeans are properly called the "meat of the fields" and constitute an important part of the Japanese diet along with rice, fish, meat, and other livestock products. In addition to protein and oil, soybeans contain Vitamins A, B, D, and E, and also lecithin, calcium, phosphoric acid, and potassium, which are all necessary for the maintenance of the human body.

In terms of calories, 100 g of soybeans have 400 to 430 calories compared with 350 calories in 100 g of rice and other food grains. In terms of protein, rice has a protein content of about 6.2%, corn 8.2%, and wheat 10.5%, while soybeans have a protein content of 38%. Such protein-rich and high-calorie soybeans, if used directly as food, can help alleviate the shortage of protein food in the world, and for this purpose various recipes utilizing soybeans must be studied. It would be of interest in this respect to review how soybeans are consumed in different ways as food in the Oriental countries, and how some of the soybean foods or their modifications may be introduced in the world market.

Of course, the scope is limited for incorporating Oriental soybean foods into Western types of dishes because the traditional soybean foods are generally most suitable for cooked rice. But recently new freeze-dry technology has made it possible to produce an instant soup mix out of a traditional wet-type Miso product. Also, traditional soy sauce is modified into a modern-type barbecue sauce for roasted meat. Tofu or soybean curd is deep-fried in soybean oil and then dried and used in some instant type foods. Soy milk is also a nutritious soybean product. These are only a few examples of how traditional soybean foods are utilized in the present-day Japanese diet. Nutritionally, they all serve as a source of vegetable protein. It seems that vegetable oils and vegetable proteins have the virtue of avoiding obesity among the Japanese people and contribute to making the average life expectancy in Japan one of the longest in the world. At any rate, I think from the standpoint of economy, nutrition, and health it is worthwhile to make efforts for better utilization of soybean protein directly as human food.

Recently, in addition to utilizing whole soybeans directly for human food, efforts have been renewed to develop new uses of soybean meal for edible purposes. However, at the present stage of development, the quantity used for new protein products is yet very limited. The future potential, though, could be great in this area when research efforts are concentrated in developing new soy-protein products.

Now let us briefly look at soybean production in the world. In 1977-1978 the total world soybean production was estimated at about 77 million tons or 2.8 billion bushels, of which the U.S. production was 48 million tons, Chinese production 12 million tons, and Brazilian production 10 million tons. The recent trend is that although the United States is the biggest producer of soybeans in the world, the American share is gradually declining due to the increasing production in South America.

From the standpoint of balancing the world supply and demand of soybeans and to maintain a reasonable price level for both producers and consumers in the world market, it is necessary for the producing countries to increase soybean production steadily each year in the future. When we consider intercrop competition among soybeans and other crops, high-yielding soybean varieties with higher oil and protein content must be developed to strengthen the competitive position of soybeans. This may be a difficult task for plant breeders, but if any one of the desirable improvements is achieved, both profitability and competitiveness of soybeans will be greatly enhanced to the mutual benefit of both growers and consumers of soybeans in the world. I think both government and industry should do their best for such a purpose.

In addition to the development of new and better varieties, it is also important to improve cultural practices and production technology in order to increase soybean yields. Preparation of fields, selection of good seed and seeding times, weed and pest control, proper fertilizer application, and careful harvesting are all important factors in determining yields, provided of course the weather is normal. In this connection, it may be pointed out, for example, that in the United States there are various levels of yields in different regions, generally high in the north and low in the south, but if the average yields are raised by one bushel per acre in the United States by better production practices, the American soybean production will be increased by more than 60 million bushels. Of course, raising the soybean yield all over the world by one bushel per acre would mean a tremendous addition to current production. For such a purpose, scientists gathered at this conference can make a real contribution in their respective fields.

Next, let me speak on the trade practices of soybeans with regard to quality. At present, oil and protein content is not specified in the quality standards of soybeans, partly because there are not enough kinds of soybeans for different uses and also because it is difficult to ascertain oil and protein content accurately in a simple manner at different levels of marketing channels. But if different varieties of soybeans with different levels of oil and protein content become available in the future, both oil and protein content should be incorporated in the quality standards of soybeans. I think efforts should be made to develop different varieties for different uses, so that soybeans could be traded on an oil or a protein basis in the future. Low-cost testing equipment for determining oil and protein content, easy to operate by any person, must be developed for the fore-mentioned purpose.

Agricultural scientists in the United States have accomplished in the past 50 yr the tremendous job of turning an Oriental curiosity crop, "soybeans", into a modern "Cinderella" golden crop of American agriculture through breeding and other research activities. These scientific efforts have been concentrated mainly on increasing efficiency of soybean production and protection of the soybean crop from various hazards, such as plant diseases, insect damage, weeds, and other environmental conditions. Of course, such efforts must be continued in the future in order to maintain and strengthen further the competitive position of soybeans in world agriculture. High oil, high protein, high-yielding, and highly disease-resistant varieties are always desirable. In addition, I now suggest that the time has come to make new efforts to improve the quality of soybeans in terms of desirable characteristics of oil and protein for better acceptance by the market. Soybean quality may also be improved for utilization directly as human food.

In summary, I have explained to you the diversified nature of the soybean market in Japan and have mentioned the desirable quality characteristics from the standpoint of both the traditional soybean food industry and the soybean processing industry.

Soybean demand in the world is expected to grow steadily in the future, because both soybean oil and protein are an important source of human food. As long as the world economy grows and the consumers' income keeps increasing, soybean requirements will certainly expand at a rate greater than that of the world population increase. In this connection, it may also be mentioned that the world-wide market development activities for soybean products by the American Soybean Association are contributing to the expansion of the soybean market in the world.

Soybeans, rich in protein and oil, are an excellent food, and in fact, are utilized extensively in various types of daily food in Asian countries. Food uses of soybeans can contribute to alleviating the protein shortage of the world, and ways and means of utilizing soybeans directly for human food may be further studied.

Consumers of soybeans in the world always look for increasing and stable supply of soybeans, and in order to attain such a goal this World Soybean Research Conference can be an ideal place for discussion and exchange of frank opinions on the production, utilization and marketing of soybeans in the world. It is indeed gratifying to note from the conference program that the various problems which I have referred to in my address are going to be discussed in the individual sessions. I wish for the great success of this conference.

NOTES

Hiroshi Nakamura, Hohnen Oil Co., Ltd., Tokyo, Japan.

SHARING AGRICULTURAL TECHNOLOGY

P. Findley

They tell me there is nothing like springtime in North Carolina and I can believe it. I certainly hope we have seen the end of winter. I'm sure you all heard about our terrible winter in Washington. It got so cold that the politicians were putting their hands in their *own* pockets!

When Dean Legates of the School of Agriculture and Life Sciences here invited me to participate in your conference, I jumped at the chance because it gives me the opportunity to do some missionary work.

I am here tonight to preach the gospel—the gospel of salvation. You might call it the Gospel According to Paul. That's Paul Findley, not St. Paul, I hasten to add.

Seriously, the gospel I preach tonight is the gospel of salvation from pestilence, from famine, from malnutrition.

It is a call to duty, a call to action for a cause that embodies the greatest and deepest humanitarian commitments of Christianity, Judaism, and the Moslem faith—feeding the hungry, helping the downtrodden, bringing light where there is darkness, hope where there is despair.

All this gospel needs is converts, witnesses, and evangelism. And the result will be an end to famine, an end to malnutrition, an end to hunger and starvation.

If that sounds like a tall order coming from a Congressman, hear me out.

The gospel I preach is famine prevention. And the way to this glorious destination is to be found in Title XII of the Foreign Assistance Act, also known as the Famine Prevention Act.

Title XII offers food to a hungry world. Here is how it works, plain and simple, with all the bureaucratic language stripped away. Title XII leads to systems of education for the small farmer struggling for survival in distant fields. It holds promise that these farmers will soon begin to receive on a regular basis technical information and guidance that will enable them to grow more food and thus have a better life for themselves and their families.

That's what Title XII is all about—establishing a system of farmer education in each developing country which is interested, through which small farmers will receive information that will help them get better production from the resources they have at hand.

That is the essence, the key, the heart of Title XII—a system through which the small farmer will receive information helpful in producing food.

Under Title XII, the system of education will be local—and adapted, of course, to local needs. It will be established by the local government in cooperation with U.S. university specialists who are experienced in establishing and operating the marvelous system through which U.S. farmers get information helpful in producing food. It is self-help in the best sense of the term. The U.S. universities will help the local government build a farmer-education program durable and flexible enough to meet whatever challenges the future may bring.

The gospel of salvation that I preach teaches us that foreign assistance should emphasize *dignity* rather than the *dole.*

This gospel teaches us that it is far better for the developing countries—the hungry nations—if we help them help themselves, instead of creating their dependency on others.

And that gospel is the essence of Title XII which was authored by the late Senator Hubert Humphrey and me and signed into law in late 1975.

I remember the first time that I heard Senator Humphrey speak . . . he was in the second hour . . . of a five-minute talk. It is true that Senator Humphrey's oratory is legendary. I remember someone once clocked one of his speeches at 350 words a minute . . . with gusts up to 700.

Believe me when I say that I am not taking liberties with the memory of that very remarkable man. He had a great sense of humor, but he also had a great feeling of compassion for the hungry and malnourished of the world.

Hubert Humphrey was one of the most inspiring persons I have ever known and he reached a new height of eloquence at the Famine Prevention Symposium a year ago December. He was literally a dying man and three weeks later he was gone. Despite his frailty, he spoke with great eloquence, fervor, and passion of the need for the Title XII Famine Prevention Program.

This program brings American know-how to the agriculture of developing nations, upon which three-fourths of mankind depends. It is a remarkable cooperative venture that harnesses into a working team our university system and government. It recognizes the great contributions that agricultural colleges and universities—especially the land-grant institutions—have made in developing America's bountiful agricultural output. Through this team work, it offers the vast educational, extension, and research resources of these institutions to help develop agriculture in lesser-developed countries.

Title XII set up the Board for International Food and Agricultural Development, generally referred to by its acronym as BIFAD, to organize this gigantic job in coordination with the Agency for International Development of the U.S. State Department.

Titles XII is not a handout program. It does not dole out food to poor countries. However laudable a dole approach to food aid might be in the short term, in the long run it tends to make the recipient country subservient to the donor nation and undermines efforts by the recipient country to improve its own food production.

Title XII is designed to help developing nations help themselves to become self-sufficient, solve their own problems, and thereby strengthen their economies and give them national pride.

The key is the establishment of a system through which useful technology and know-how can be transferred to the farmers of developing nations. I don't have to tell you that soybeans and soybean products loom large in the picture of upgrading diets in those countries. Soybean meal figures prominently in high protein animal feeds. And soybean products provide protein-rich food for human consumption. Both are essential to improving diets and inevitably will be prominent in the development of agriculture through Title XII.

Our goal, our great humanitarian opportunity, is to provide developing nations a system of agricultural education similar to that provided our own farmers. In most countries the system must necessarily be elementary and basic, at the same level as were the training programs in our own land grant institutions when we were developing from subsistence-level to productive-level farming. Our goal is to teach, to demonstrate, and thereby enable the farmer to grow two blades of grass where one grew before.

It is a process which could be described by farmers back home as "getting the hay down where the calves can get at it."

Such training, in addition to formal instruction, must include demonstration-type extension programs and adaptive research extension programs modeled to the agriculture of the particular country and its production potential.

Through the Title XII program we can share our land-grant resource with a hungry world and share it broadly.

I would be less than honest with you, however, if I did not point out that there are grave uncertainties that could threaten the future of the Title XII program. While the program was enacted three and a half-years ago, it has made only a small beginning toward the enormous goal of helping developing nations help themselves to become self-sufficient.

The time has come to give this program a much higher priority than it has been given in the Agency for International Development (AID). Human life is literally hanging in the balance and each day's delay translates into countless deaths and debility among the world's starving and malnourished peopled.

There are those in the U.S. Senate who want to take an axe to the funding of this program, ignoring the fact that the Title XII program is a low-cost approach to achieving a long-lasting solution to the problem of world hunger. It has a highly favorable cost-benefit ratio and is indeed notable in that respect.

Then there is the uncertainty caused by the vacuum of administration leadership resulting from the recent resignation of AID Administrator John Gilligan. My fear is that the program will be allowed to languish without the impetus that a strong and sympathetic administrator could give it.

And there is the uncertainty of whether our land grant universities and their boards of trustees really understand the program well enough to give it the full commitment it must have to fulfill its mission.

We must develop a sense of urgency about this important program. If that sense of urgency were linked in inverse proportion to the growing despair of the world's hungry, there is no question in my mind that the Title XII program would have the highest national priority—both in government and on the campuses of our nation.

While I have mentioned the concern over possible Title XII budget-cutting attempts in the Senate, I am happy to say that Title XII is not without friends on that side of Capitol Hill. Senator Frank Church, Foreign Relations Committee Chairman, supports Title XII and has asked Senator George McGovern, who serves on both Foreign Relations and Agriculture Committees, to take a special oversight role in regard to Title XII's progress. I am most happy to say that Senator McGovern has given me every assurance that he is vitally interested in safeguarding the mission of Title XII.

I must frankly confess that my fervert hopes for this program have been dashed by frustrations over the time that it has taken to get this far. I am appalled by the slow pace of bureaucratic endeavors.

I have deep concern about the future of Title XII.

More emphasis must be given to the extension approach provided by the program—an approach that will help that small farmer in a developing country help himself and his family to survive. Extension gets useful information right to the farmer in the field. That is not to say that I object to research programs encompassed by the Title XII program. There is a glamor to conducting research and its results are identifiable products in the eyes of university boards of trustees. The establishment of a local system for teaching poor farmers in another country to do a better job of producing foods is a slow, painstaking process—not glamorous, not a process which often lends itself to doctoral treatises.

Yet I submit to you that it is extension training and education which puts research to work, and in my mind, research is useless unless its benefits are channeled to the farmer in the field.

I am concerned also about the natural tendency to use the consortium approach in assigning university responsibility to carrying out Title XII programs. I strongly urge in every case that a single university be given the prime contractor role with other universities available in what you could call subcontractor roles. The consortium approach tends to diffuse responsibility and accountability, a discipline that is basic and essential to the success of any Title XII project.

I am also troubled by what I perceive will be a year of budget-cutting. Don't misunderstand me, I am not against budget-cutting—not when budgets are pruned with prudence. This is a period of budget retrenchment, and I urge you to help me make known the fact that the Title XII Program is already a low-cost way to achieve a long-lasting solution to the problem of world hunger. It is a foreign-aid program that yields far more than the meager investment in it.

It is encouraging that the food and nutrition budgetary resources allocated to Title XII have increased and I have every hope that this trend will not be slowed. In Fiscal Year 1976, Title XII types of activities amounted to about $100 million or 17% of the Agency for International Development's total investment in food and nutrition programs. The Title XII investment has increased each year to an estimated $250 million or 32% of AID's total food and nutrition program in the current fiscal year. It is projected that in Fiscal 1980 the Title XII program will be about $400 million or 56% of AID's food and nutrition program.

As of mid-March 1979, eighteen country projects whose life-of-project costs will be about $64 million have been shaped and implemented by the review and contractor selection process carried out by the Board for International Food and Agricultural Development and the Agency for International Development.

Fifteen other country projects, with ultimate costs of $69 million, have been approved. The process of selecting the U.S. contracting institutions is underway. Still another 19 projects, costing about $183 million, are in an advanced planning stage and will soon be fully approved.

Let me cite a few examples to give you the flavor of Title XII:

— The University of Wisconsin will continue its long-term work in Indonesia through a Title XII project designed to assist Bogor Agricultural University in increasing its capacity to provide agricultural leaders to meet the increasing demand for well trained people in that nation's agricultural sector.

— Also, in Indonesia, Iowa State University is engaged in a collaborative effort to assist the Indonesia Department of Agriculture in upgrading its planning and administrative capacities.

— In the Yemen Arab Republic, the University of Arizona will take the lead in a broad-based agricultural development project by introducing the U.S. land grant college concept of integrated teaching, research, and extension. Initial activities will be development of Yemen's first agricultural training institution and establishment of a nationwide seed program. Two additional sub-activities of farm water management and soil fertility will be started also.

— In the African country of Botswana, South Dakota State University is assisting the government in the creation of a locally staffed agricultural training institution at the basic and intermediate skill levels in agriculture, animal health, and community development. The immediate beneficiaries will be the 47 Botswanans trained at the school; the secondary beneficiaries will be the 298 students enrolled annually at the school; and the ultimate beneficiaries will be the rural population served by the people trained at the school as agricultural demonstrators, veterinary assistants, and assistant community development directors.

— In Tanzania, the University of West Virginia is working with the agricultural faculty of the University of Dar-es-Salaam to improve the faculty's ability to train students in agricultural teaching, extension, management, and technical skills. About one million farm families are presently served in one way or another by agricultural extension agents. The people trained in this project will be the supervisors and trainers of extension personnel who assist farmers and villages in increasing their production and income levels.

This is but a sample of the agricultural development projects now underway and many, many more need to be in place and active if we are to win the war against hunger.

If the farms of the developing nations are to become and remain effective and efficient contributors to the continued well-being of their societies and to the world, they require service by local institutions which will assure sustained modernization. This will happen only when each of these nations, in some way, creates this local capacity to provide directly to small farmers the continuous stream of essential information and adapted technology required to improve levels of agricultural production.

Dr. Clifton Wharton, chancellor of the State University of New York, is the very able chairman of the Board for International Food and Agricultural Development. In his recent testimony before the Senate Foreign Relations Committee he said:

"In structuring any program to attack world hunger, certain basics must be kept 'front and center.' One of these is that the productivity of resources employed on the farms of the developing countries must be increased. The farmers and their farms are the focus of ultimate action. It is only through quantum increases in the productivity of land, labor, capital, and management on these farms that the food needs of rapidly expanding populations can be met. It is only with such increases in productivity that the difficult equity problems facing much of the world's population might be approached in any meaningful way."

To that I say a fervent: "AMEN!"

The goal is not to finance agricultural research in some distant corner of the globe. The goal is to provide the world's small farmers with the know-how to rise above subsistence level production and thereby alleviate the world's hunger pains.

Believe me when I say that this approach is *not* dramatic. Education is a good investment but it takes time. It requires a long-term commitment . . . by the recipient nation as well as those who provide the assistance.

Keeping the Title XII program on track is not easy. Senator George McGovern and I are joined in encouragement and oversight of the Title XII program. We need your understanding and support to insure that the program will move along.

If you want to help, now is the time to "hit the sawdust trail" and preach the gospel of salvation for the world's hungry and malnourished.

Preach it in the offices of the Agency for International Development. Preach it in the halls of Congress with your Senators and Representatives. Preach it in your own state capitals and university board rooms. Preach the need for the university funds it takes to match people to the program. Preach the need of adequate federal funds to keep it going. And for our visitors here tonight from other countries, go back and preach the need for this Title XII program to your own government officials and institutions of learning.

And when you preach this gospel of salvation, remember for whom you are preaching—for hungry millions of the world.

For the sake of humanity, I wish you well.

NOTES

The Honorable Paul Findley, U.S. House of Representatives, 20th District, State of Illinois.

CATION NUTRITION AND ION BALANCE

J. E. Leggett and D. B. Egli

Plant tissue will usually have a charge imbalance because of the difference in equivalent charge concentration between inorganic cations and inorganic anions. The cell adjusts the differences in equivalent charge concentration by increasing or decreasing the organic anions. Ion balance in plant tissue pertains to the maintenance of electro-neutrality between the inorganic cations and the sum of inorganic anions and organic anions. Ion balance will be considered as it relates to the nitrogen source available for accumulation and the cation accumulation by the soybean plant.

The nutrient status of a plant is the result of many interactions which occur in the growth medium during ion absorption and translocation, and during the metabolism of some nutrient ions within the plant. These interactions are not only evident among ions of similar charge, but also among ions of dissimilar charge. Cations and anions may be absorbed independently and they may be absorbed in unequal quantities, but within limits electroneutrality must be maintained both within the plant and in the nutrient medium. Hence, ion balance within a plant is of considerable importance in plant nutrition. The ion ratios and the nitrogen source in the nutrient medium, as well as changes in the ionic form within the plant, will have a marked effect on metabolism and on the accumulation of other nutrients.

Our present understanding of ion interactions was developed primarily from experiments using single salt solutions and low salt excised roots. Investigations utilizing short term kinetic studies on intact plants to bridge the gap between observations on excised roots and those collected in field experiments are becoming more popular. Studies with intact plants are necessary to determine the influence of anions especially those changing ionic form, on cation accumulation.

The expression of ion interactions during the initial ion entry process, as measured with excised roots, is minimized because of the complexities of the metabolic reactions during plant growth. The use of intact plants has provided a better understanding of ion accumulation and how ion balances at critical growth stages influence yield.

Ion balance in the plant cell may result from cation-cation, anion-anion, or anion-cation interactions. Interactions between or among ions of similar charge are most prominent in the processes involved in the initial entry processes in the root. The inhibitory interactions between chemically similar ions are usually of a competitive nature, however in a few cases inhibitory ion interactions are not competitive and may be actually one of activation, i.e. the Viets Effect. Cation-anion interactions are not only important during entry into the root, but also in cellular processes that occur after entry into the plant. Although interactions occur after ion entry, we usually assume that they are similar to the interaction during the entry process.

The interactions among ions after entry are probably more important than those interactions at the root cell plasmalemma in a plant growing in the field. Ion interactions after ion entry into the plant are influenced by the growth rate of the plant, change in the ionic form of the nutrient, synthesis of organic anions, binding of ions, and retranslocation of ions. In terms of growth and yield, these interactions are probably more important in plant nutrition than ion entry across the root plasmalemma.

NUTRIENT COMPOSITION OF PLANTS

The ion composition and ionic balance in most plants are dominated by six macronutrient elements. The contribution of micronutrient elements to the total ion composition and ionic balance is negligible and will not be considered in this discussion. The six macronutrients, their ionic form, and their concentration in vegetative tissue of soybeans *[Glycine max* (L.) Merrill] are listed in Table 1. Of these six nutrients, nitrogen has the greatest influence upon the ion balance because (a) it is accumulated in large quantities, (b) it may enter the plant as an anion, a cation or as a neutral element via nitrogen fixation, and (c) it may change ionic form and be incorporated into organic compounds.

Leguminosae, in most situations, accumulate nitrogen predominantly as reduced N via nitrogen-fixation or as NO_3. Independent of the mode of nitrogen accumulation, the plants contain higher concentrations of inorganic cations that inorganic anions. This is caused by low accumulation of anions, other than NO_3, and by the conversion of NO_3 to NH_4 and into organic compounds.

Excess accumulation of cations by plants has been related to a change in organic acid levels to maintain electro-neutrality in the plant tissue (1-3,

Table 1. **Macronutrient composition of soybeans.**

Nutrient	Ionic Form	% in Tissue[a]
Calcium	Ca^{2+}	0.68
Magnesium	Mg^{2+}	0.26
Nitrogen	N_2, NO_3^-, NH_4^+	2.44
Phosphorus	H_2PO_4, HOP_4^{2-}	0.19
Potassium	K^+	1.56
Sulfur	SO_4^{2-}	--

[a]Henderson and Kamparth (13). Three year average for total above ground portion of the plant during vegetative growth; cultivar—Lee.

14,15,17-20,27,33,35). The concentration of cations, anions, and organic acids will be influenced by plant species, plant growth stage, and the composition of the nutrient medium.

An indication of the ionic balance in five plant species during vegetative growth is provided by the data of Pierce and Appleman in Table 2. The plants were grown in sand culture with an inorganic source of nitrogen. Total cation concentration exceeded total anion concentration because of NO_3 reduction in the plant and the relatively low level of accumulation of SO_4 and H_2PO_4. The organic acid levels increased to balance the excess cation levels in the tissues. This pattern was consistent across the five species included in Table 2.

NUTRIENT ACCUMULATION

Excised Roots

Cation interactions in excised roots occur both at the membrane and intracellular levels. Interactions at the membrane would be expected to be exhibited during the initial ion entry to low salt roots, essentially for time periods of less than 1 hr. During longer accumulation periods the interactions among intracellular ions would be expected to dominate the resultant ion levels.

Interactions of ions in short-term studies have been investigated primarily using excised barley *(Hordeum vulgare)* roots. However, differences between plant species would not be expected to be large in a qualitative sense. Excellent discussions on ion interactions and proposed models for the observations may be found in the literature (6,12,13,17,28,29,32).

Reports of investigations on ion accumulation by excised soybean roots are very limited in the literature. This is probably because it is much easier to obtain uniform excised root segments of barley than soybeans. Reproducible results can be obtained with soybeans, similar to those obtained with

Table 2. **Concentrations of cations, anions, and organic acids in five plant species (27).**

Plant	Tissue	Cation	Anion	Excess cation	Organic acids
		– *meq/100 g dry wt* –			
Glycine max (Mer..)	Leaf	276	63	213	249
	Stem	243	79	164	177
Phaseolus limensis (Macf.)	Leaf	287	60	227	250
	Stem	249	57	192	171
Pisum sativum (L.)	Leaf	231	69	162	213
	Stem	214	77	137	181
Medicago sativa (L.)	Leaf	244	78	166	186
	Stem	206	67	139	141
Triticum aestivium (L.)	Leaf	237	143	94	122

excised barley roots, by using roots representing a uniform length from the root apex of soybean plants. This procedure is time consuming relative to the ease in obtaining excised barley roots.

Cation accumulation by excised soybean roots does not agree with ideas developed with barley roots. Excised soybean roots from plants grown on a dilute $CaSO_4$ solution to obtain low salt roots, as in barley, are quite different when compared to barley roots. The K concentration is approximately 40 to 50 meq/g fr wt for soybeans relative to 15 to 20 meq/g fr wt for barley. This can be of importance since the maximum K accumulation is near 100 meq/g for both roots, hence the net K accumulation will be greater for barley than for soybean roots. The Mg and Ca content is similar for both barley and soybeans grown on a $CaSO_4$ solution. The Viets Effect (34) where Ca increases ion accumulation can be observed in both species. However, soybeans appear to be more responsive to the presence of Ca in the nutrient solution. The effect of Ca has been related to maintenance of the reduced ion efflux (25).

Potassium accumulation by excised barley roots is linear with time for approximately 6 hr, whereas linear accumulation of K by soybean roots is observed for 24 hr. This difference between barley and soybean roots is not understood, but may reside in the relative rate of K accumulation of the root cells and the rate of exudation of K out the cut end of roots.

The most interesting aspect of cation accumulation in soybean roots is the interaction among the cations (22). Inhibition of Mg accumulation by either K or Ca alone was negligible. A decrease in Mg accumulation was evident only when the nutrient solution included both K and Ca. Magnesium accumulation from a solution of 2.5 mM $MgCl_2$ + 5 mM KCl was reduced

50% by addition of 0.2 mM $CaCl_2$. The inhibitory reaction of K on Mg in soybean roots depends on the presence of Ca in the external solution. Although the mode of action is not understood, the interaction would appear to be one of Ca modifying the membrane, a subsequent reduction in K efflux and hence a more favorable competitive position for inhibition of Mg.

Acceleration of ion accumulation by divalent cations was maximum when both Ca and Mg were present in the external solution. The interaction appears to be at the membrane-solution interface since K and Cl accumulation were maximum and Ca + Mg accumulation was relatively low (22). A reduced K efflux, suggested by Mengel and Helal (25), appears to be a satisfactory explanation because of the high K accumulation, whereas Mg accumulation at high levels was associated with a decrease in both K and Ca. The change in efflux and/or influx rates implies a specificity modification of the membrane resulting in a larger accumulation of K rather than accumulation of Mg + Ca.

The influence of Ca and Mg on K accumulation by an intact plant would be expected to be similar as in excised roots. However, on the basis of the Ca content of excised roots, one would expect a relatively low Ca concentration in the top of an intact plant. This is not the case, which leads to the conclusion that the root tissue is saturated with a low Ca concentration (ca. 10 to 15 meq Ca/g fr wt) compared to a higher Ca concentration in the leaf tissue (ca. 20 to 35 meq Ca/g fr wt) for 7 day old plants (Table 3). Therefore, Ca appears to move through or around the root cells into the xylem in an amount not expected from the excised root data. Moore, et al. (26) reported a similar finding in excised barley roots.

Excised root experiments may be useful in establishing ion interactions at the root cell, but may not represent the amounts of an ion transported to

Table 3. **Time course of ion accumulation by *Glycine max* (L.) Merr., cv. Hawkeye. Nutrient solution: 0.5 X Hoagland (modified) containing either 5 mM NO_3 or Cl, pH 5.**

		NO_3				Cl			
Time		K	Mg	Ca	NO_3	K	Mg	Ca	Cl
– hr –					*– meq/g fr wt –*				
Shoots	1	63	13	20	27	67	15	24	4
	7	74	15	25	43	77	17	25	2
	24	96	20	32	62	96	17	20	4
	33	131	22	23	231	106	16	15	7
Roots	1	28	5	14	1	19	2	8	10
	7	56	6	15	32	65	5	13	19
	24	76	8	16	31	66	5	14	25
	33	91	8	12	67	84	7	10	31

the shoot. Therefore, use of intact plants is necessary in order to gain a better understanding of the cation interactions both for short-term and long-term experiments. In addition, the final evaluation probably must be made across growth stages—from vegetative to reproductive.

Intact Plants—Short Term

The relative accumulation rate of the anion from the medium will determine the accumulation rate of a cation such as K particularly at the higher solution concentrations (6,14). In this respect soybeans are similar to excised barley roots. Potassium accumulation from solutions of KCl, K_2SO_4 or KNO_3 increased in the order $Cl<SO_4<NO_3$ for *Glycine max*, but was in the order $SO_4<Cl<NO_3$ for *Vigna sinensis* and *Pisum sativum* (Table 4) (23). A greater K accumulation from a NO_3 solution was expected, however the reverse relationship of $Cl<SO_4$ of *Glycine max* from the other two plants was not expected. This is another example of differences among species and can be accounted for by a reduced accumulation of Cl in *Glycine max* relative to *Pisum sativum* and *Vigna sinensis*. On the basis of these results, K accumulation by plants obtaining their N from fixation would be expected to be reduced. This aspect of cation accumulation will be expanded in a later part of this discussion.

Data on the accumulation of ions from a single or simple salt solution are useful for obtaining initial maximum rates of uptake and for studying ion interactions during the entry process. Nevertheless, at some point in understanding plant nutrition, one should advance toward the nutrient medium and growth conditions of a "normally" developing plant. Investigations on ion accumulation by intact low salt plants from complete nutrient solutions for short time periods reduce the effect of growth on ion interactions during the accumulation processes.

Soybean plants grown on a dilute $CaSO_4$ solution, then transferred to a complete nutrient solution, accumulate K and NO_3 primarily (Table 4). The concentration of NO_3 is that present in the tissue at the time of sampling and does not represent total NO_3 accumulation because of reduction of NO_3 to other forms of N. The relatively large increase in NO_3 between 24 and 33 hr must result from a lowered rate of NO_3 reduction in the tissue as the system readjusts from the low salt status. Changes in Mg and Ca concentration were essentially negligible during the experiment. Although Cl accumulation was much less than NO_3, the K accumulation was not reduced to the same extent. In the presence of Cl as the dominant anion, K must be entering the plant by exchange for H, and in association with the SO_4 and H_2PO_4. The trend appears to be for a reduction in K accumulation in the Cl system with the longer time periods.

In soybeans, as for other plants, the major changes in ion concentration will be a result of K, NO_3 and Cl accumulation. Major shifts in ion balance

Table 4. Potassium accumulation in plants from solutions of three K salts (23) (K = 10 meq/1, 24 hr).

Plant	Shoot			Root		
	NO_3	SO_4	Cl	NO_3	SO_4	Cl
			– net k, meq/g fr wt –			
Glycine max	22	11	4	40	27	21
Pisum sativum	42	24	38	32	16	26
Vigna sinensis	55	9	23	52	36	48

will not be associated with Mg or Ca as a result of their small changes in concentration. Likewise, the relatively slow rates and lower concentration levels of S and P will not be a major factor in rapid shifts of ion balance within the tissue.

INTACT PLANTS--LONG TERM

General Characteristics

Although short-term kinetic studies on cation accumulation in soybeans are limited, the nutrient composition of field grown plants has been documented adequately (10,12,13,21). These observations during the growth and development of soybean plants provide an overview of total nutrient accumulation as well as the redistribution of nutrients during reproductive growth.

The growth curve of a soybean community in the field consists of a period of relatively slow growth from emergence to approximately 40 days after emergence, followed by a period with a rapid constant growth rate that extends from 40 days after emergence to near maturity (4,13). The accumulation of most of the nutrients parallels the dry matter accumulation rate during vegetative growth (13). Accumulation as a percent of the total for nitrogen and phosphorus is slightly less than the accumulation of dry matter for several days from 40 days after emergence to near maturity. Magnesium accumulation as a percent of the total is slightly less than the accumulation for dry matter during the time period between 40 days after emergence and maturity. Potassium accumulation as a percent of the total exceeded dry matter accumulation between 40 and 80 days after emergence, then K accumulation was less than dry matter accumulation between 80 days and maturity. The accumulation of calcium was similar to dry matter accumulation between 40 and 80 days after plant emergence, then Ca exceeded dry matter accumulation between 80 days and maturity.

At the beginning of reproductive growth, vegetative growth ceases, and leaves and petioles are lost from the plant during reproductive growth. The change in the size of the individual plant parts across the growth stages is represented in the data of Hanway and Weber (9). Considerable redistribution of nutrients from vegetative to reproductive plant plants occurs during reproductive growth to the point that little or no accumulation may occur during this period.

The major nutrient element redistributed from the soybean leaf is nitrogen. This element has been described as being the most limiting of the nutrient elements to yield. In addition, Sinclair and de Wit (30,31) characterized the soybean plant as being self-destructive as a result of N redistribution from vegetable tissue to the seed. Although redistribution of N from vegetative to reproductive plant parts influences plant senescence, Egli et al. (5) suggested that other factors were also involved in leaf senescence.

Redistribution of cations from vegetative tissue to the seed occurs for K (8,13) to a very limited extent for Ca while Mg exhibits either no redistribution (8) or redistribution plus accumulation from the soil (13).

Cation Accumulation

The soybean plant is involved in nutrient redistribution at both ends of its life cycle. Nutrients are translocated from the cotyledons to the root and hypocotyl during germination and emergence. The cation content of a Hawkeye soybean seed will be approximately 92, 32, and 8 meq of K, Mg and Ca, respectively. A 7 day old plant from this seed, grown in $CaSO_4$, will produce roots having a cation concentration of 37, 2, and 15 meq/g fr wt of K, Mg, and Ca, respectively. The shoot will have a K, Mg and Ca concentration (meq/g fr wt) of 57, 15, and 45, respectively.

A soybean seed planted in soil will redistribute nutrients from cotyledons and will accumulate nutrients from the soil as the root radicle develops. The plant depends upon NO_3 from the soil as the sole N source until nodule development for N-fixation. Hence, cation entry should be at the maximum potential rate, assuming a sufficient NO_3 concentration in the soil solution.

The primary source of nitrogen for the plant is usually from N-fixation by the nodules. Development of the nodules to the point that they can make a sufficient contribution of N to the plant is somewhat delayed, but increases rapidly near flowering (11). The maximum rate of N-fixation is usually attained early in the grain filling period and then it decreases rapidly as the plant approaches maturity. The sequence of N sources available to the plant, i.e. first primarily NO_3, then primarily from N-fixation, followed by low levels of NO_3 and N-fixation during grain filling should influence the accumulation of cations.

The potential for cation accumulation should be maximum early in the growing season when soil NO_3 is at the highest level. Less cation accumulation would be expected as the soil NO_3 level drops and N-fixation by the plant becomes more dominant. In a field situation cation accumulation would not appear to be rate limiting to plant growth and yield. Nevertheless, it is important to understand the influence of N sources on cation accumulation as this condition applies to soybean yield.

The yield of soybeans is influenced greatly by environmental conditions during flowering and pod set when seed number is established. Factors which may influence yield after seed number is fixed include the environment, redistribution of nutrients from vegetative to reproductive parts, and the influence of a degrading root system on nutrient accumulation. Since the early reproductive stage is important in establishing seed number, this growth stage may also be important in cation accumulation and ion balance as they influence yield. Cation accumulation during the early reproductive stage will be discussed relative to ion balance and N sources.

A model illustrating the interrelationships of the processes involved in ion accumulation by roots and export to the shoots has been recently proposed by Israel and Jackson (17). This model attempts to illustrate our present understanding of ion accumulation processes and considers how the plant controls or modifies ion accumulation by the internal and external pH. The interactions are due primarily to differences in plant species and sources of nitrogen.

Soybean plants (cv Ransom) dependent on N_2 fixation as the sole nitrogen source were utilized to investigate cation and anion accumulation and the organic acid levels (17). The observations were made at 65 days after planting (near end of flowering) and at 105 days. The inorganic ion concentration (meq/plant) was 94 and 278 for cations and 9 and 20 for the anions at 65 and 105 days, respectively. The difference between inorganic cations and inorganic anions (C-A) increased from 38 to 96 meq-plant during the 40-day period, thus the cations increased more than did the anions. However, on a plant dry wt basis the (C-A) value decreased slightly from 1.08 to 1.01 meq/g. The large excess of inorganic cation accumulation over anion accumulation suggests that the roots synthesized relatively large amounts of organic anions to maintain electro-neutrality.

The influence of NO_3 and urea on the exudate composition of nodulated soybean plants was also studied by Israel and Jackson (17). Xylem sap was collected from nodulated plants at the late flowering stage of growth. Nitrate in the nutrient medium increased the inorganic anion concentration in the exudate, but the cation concentration in the exudate was similar for all treatments, zero N, NO_3, and urea. Likewise the (C-A) for the NO_3 treatment was less than for the other treatments. These results indicate that the N source does influence the organic anion concentration in the xylem. The long term effect on growth and yield is not known.

The ratio of C/A for nodulated plants was 10.4 and 13.9 for 65- and 105-day old soybean plants, respectively. A C/A value of 5.9 was observed in the exudate from nodulated plants. The larger value for C/A in the intact plants compared to the C/A value in xylem exudate may be evidence for a recycling of organic acid in the intact plant. Although the C/A value in the exudate differs from the C/A of the intact plant, exudate composition has merit for comparing various treatment solutions on ion transport through the root system.

The investigations of Israel and Jackson (17) provide clear evidence of cation movement and ion balance by organic anions in the root system. The question arises whether the root system can maintain its ability to accumulate and transport cations over a major part of the reproductive stage. The studies of Vincent (personal communication) on the influence of plant age on the root characteristics, and the ability of the root system to accumulate cations provide information on this question. Fiskeby V soybeans grown in 0.5 X Hoagland's solution (16) were harvested at growth stages R1, R5, R6, and R8. The measured parameters of the root system are listed in Table 5. The dry wt and root length/plant were similar from R1 to R8. However, the number of root tips decreased rapidly from R1 and were essentially absent at R8. These data on dry wt and root length may be evidence that in solution culture the root system would retain a constant area for cation entry during the reproductive stages.

Rubidium accumulation was determined for 24-hr periods on Fiskeby V soybeans grown in solution culture at developmental stages R1, R5, R6, and R8 (7). The accumulation of Rb was greatest at the first observation, i.e. R1 and the concentration of Rb in both the leaves and roots decreased with maturation of the plant. Rubidium concentration in the roots decreased from 11 to 1.5 meq/g fresh wt from R1 to R8. During these developmental stages, Rb concentration in the leaves decreased from 50 to 5 meq/g dry wt. However, the total Rb accumulated by the tops during this time was only reduced from 125 to 90 meq/plant. Although the concentration in the tissue and the total Rb in the top decreased, the plant maintained a significant potential for cation entry through the root system.

Exudate from stems, after removal of tops, was measured for 1 hr following the 24-hr exposure to Rb. The volume of exudate and Rb in the exudate decreased from the growth stage R1 to R8. At R1 the exudate volume was 0.7 ml/hr and the Rb in the exudate was 1.4 meq/hr, and this decreased to 0.1 ml/hr and 0.2 meq Rb/hr at R8. Although the volume of exudate decreased, the Rb concentration in the exudate remained constant for the two growth stages. Total Rb moving through the soybean root system to the tops at R8 was approximately 75% of the amount observed at R1. Hence the ability of the root system to absorb ions over the reproductive stage has remained relatively high. This conclusion does not support the idea of the roots being greatly limited by transport of photosynthate from the tops as suggested by Sinclair and de Wit (30,31).

Table 5. Characteristics of Fiskeby V soybean root system of four growth stages[a]

Parameter/Plant	Growth Stage R1	R5	R6	R8
Dry Weight (mg)	678.2 ± 181.9	1003.3 ± 317.9	591.3 ± 129.4	681.5 ± 209.2
Length (m)	120.5 ± 52.9	183.9 ± 57.5	135.0 ± 35.8	165.5 ± 49.3
Root Tips (no.)	1678 ± 918	165 ± 133	223 ± 206	0

[a]Vincent (personal communication).

In an attempt to determine the long term influence of N sources on cation composition, soybeans (cv Fiskeby V) were grown from growth stage R2 to R6 with NO_3, NH_4, urea and N_2 fixation as the sole N source. All plants were grown on a 0.5 X Hoagland solution to growth stage R2 in a gravel culture system in the greenhouse. There were 5 plants/pot and each treatment was replicated at planting and additional inoculum was added 5 days after emergence. Nitrate, NH_4, and urea were added to a minus N 0.5 X Hoagland solution at equivalent N concentration of 7.5 meq/1. The solutions were renewed weekly or more often when necessary to control pH.

Plants were harvested at growth stages R2, R4, and R6, separated into leaves, stems + petioles, and pods, dried at 60 C, and ground to pass through a 40 mesh screen. Cations were extracted by a $HNO_3/HClO_4$ procedure and NO_3 extracted by hot water. Analysis for cations was by atomic absorption and NO_3 by the enzyme procedure of Lowe and Hamilton (24).

The concentrations of K, Mg, Ca and NO_3 in the leaves and the concentrations of K, Mg, and Ca in the stems + petioles of plants from the four nitrogen treatments are listed in Table 6. The concentration of NO_3 in the leaves was low and less than 5% of the total cations. Nitrate concentration in the stem + petiole sample was too low to measure with any degree of accuracy, hence the values are not reported. The concentrations of P and S were not measured, but would not be expected to exceed a value of 0.3 meq/g expressed as H_2PO_4 and SO_4. Hence the total inorganic anion concentration would not exceed 20 to 25% of the total inorganic cation concentration. In which case, the change in total cation concentration will reflect a change in organic anion levels as influenced by the treatments.

The amount of dry matter in the leaves and stems + petioles were not significantly different for the growth stages and they were not affected by the nitrogen treatments. Since growth rates were similar for all treatments, cation concentration in the tissue was not influenced by dry matter production. Changes in cation concentration must then be related to the treatment and/or the growth stage.

Table 6. Cation concentration in soybean leaves and stem + petioles as influenced by the nitrogen source.

Treatment	Dry Wt		Leaf				Stem + Petiole		
	Leaf	Stem	K	Mg	Ca	NO_3	K	Mg	Ca
	– g –					– meq/g –			
R2	1.16	0.53	1.19	0.33	1.12	0.120	1.30	0.48	1.02
R4 Nod	2.78	1.22	1.00	0.38	1.21	0.058	0.84	0.46	0.78
R4 NO_3	2.38	1.02	0.93	0.30	1.12	0.052	0.94	0.45	0.92
R4 Urea	2.61	1.05	0.84	0.33	0.92	0.041	0.63	0.45	0.66
R4 NH_4	2.64	1.01	0.84	0.39	0.87	0.061	0.74	0.40	0.66
R6 Nod	2.58	1.16	0.86	0.36	1.20	0.017	0.67	0.47	0.75
R6 NO_3	2.90	1.22	0.95	0.34	1.30	0.045	0.76	0.40	0.74
R6 Urea	2.77	1.20	0.74	0.32	0.90	0.024	0.44	0.39	0.49
R6 NH_4	2.29	0.93	0.76	0.37	0.92	0.036	0.61	0.44	0.56
Growth Stage	NS	NS	**	NS	+	**	**	NS	**
Nut	NS	NS	**	**	**	**	**	NS	**
Nut X GS	NS	NS	+	**	NS	NS	NS	NS	NS

**Significant at $\alpha = .01$

+Significant at $\alpha = .10$

The K concentration in leaves and stems + petioles was reduced by urea and NH_4 at R4 and R6. Generally, K in the tissue was higher for the NO_3 treatment than for nodulated plants, except at R4 in the leaves. The K concentration in the leaf and stem + petiole tissue was lower at growth stage R6 compared to R4 for all treatments.

The treatments had a significant effect on the Mg concentration in the leaves and stems + petioles. However, the changes in Mg concentration were not great and did not greatly influence the total cation concentration. There was no significant effect of growth stages on the tissue Mg concentration.

The Ca concentration in the tissue was reduced by urea and NH_4 for both plant parts relative to the nodulated and NO_3 treatments. Differences between the NO_3 treatments and the nodulated plants were not consistent. A decrease in Ca concentration in the leaves and stems + petioles for plants in the urea and NH_4 treatment was a result of a decreased accumulation and translocation of Ca from leaves and stems + petioles to the pods. Inhibition of accumulation was probably because of the absence of a mobile anion in the external solution for Ca entry. The influence of urea and NH_4 on H concentration changes within or near the root surface is not known, but could be a factor in cation accumulation.

The results from the N treatments provide evidence that the cation concentration in the soybean plant may change during the reproductive stage. Potassium and Ca were particularly decreased in the urea and NH_4 treatments. Primarily, these changes in concentration resulted from redistribution because of a reduced accumulation by the total plant. The NO_3 treated plants were slightly higher in K and Ca than were the nodulated plants at R6. This evidence suggests that over a longer period of time, i.e. from the vegetative to the reproductive stage, the differences between nodulated plants and the NO_3 treated plants would be even greater.

SUMMARY

An understanding of cation accumulation and ion balance may be attained finally when we consider all possible types of investigations. It should be obvious that studies on excised roots, short-term intact plants, and field experiments provide only a general understanding. It may also be a possibility that a concern with cation accumulation and ion balance will not contribute to a potential yield increase in soybeans.

Cation accumulation is related to the anion composition of the medium, i.e. anions of greater accumulation will increase cation accumulation. It is in this context that the N source available to the plant influences cation accumulation and organic acid synthesis. With NH_4 as the source of N, it will compete with other cations, especially K, in accumulation processes and lead to an imbalance of cations-anions in the plant tissue.

The rate of dry matter production will be the primary determinant of cation accumulation. In short term experiments the shift in (C-A) may be observed readily. However, with long term growth periods the plant adjusts its growth rate with the result being essentially a constant (C-A) concentration in the tissue.

Finally, it is not known presently whether (C-A) is an important parameter influencing soybean yield. The influence of (C-A) on vegetative yield over short time periods is not a good indicator of potential yield. The question remains whether (C-A) influences node number, hence a possible change in pod and seed number. Any factor reducing seed number/ha will necessarily lead to a reduction in yield.

NOTES

J. E. Leggett and D. B. Egli, Department of Agronomy, University of Kentucky, Lexington, Kentucky 40506.

REFERENCES

1. Dijkshoorn, W. 1962. Metabolic regulation of the alkaline effect of nitrate utilization in plants. Nature 194:165-167.
2. Dijkshoorn, W. 1971. Partition of ionic constituents between organs, pp. 447-476. In R. M. Samish (Ed.) Recent advances in plant nutrition, Vol. 2. Gordon and Breach Scientific Publishers, New York.
3. Dijkshoorn, W. 1973. Organic acids and their role in ion uptake, pp. 163-188. In G. W. Butler and R. W. Bailey (Eds.) Chemistry and biochemistry of herbage, Vol. 2. Academic Press, New York.
4. Egli, D. B. and J. E. Leggett. 1973. Dry matter accumulation patterns in determinate and indeterminate soybeans. Crop Sci. 13:220-222.
5. Egli, D. B., J. E. Leggett and W. G. Duncan. 1978. Influence of N stress on leaf senescence and N redistribution in soybeans. Agron. J. 70:43-47.
6. Epstein, E., D. W. Rains and O. E. Elzam. 1963. Resolution of dual mechanisms of potassium absorption by barley roots. Proc. Nat. Acad. Sci. 49:685-92.
7. Fehr, W. R. and C. E. Caviness. 1977. Stages of soybean development. Special Report 80. Iowa State Univ. CODEN: 1WSRBC (80) 1-12.
8. Hammond, L. C., C. A. Black and A. B. Norman. 1951. Nutrient uptake by soybeans on two Iowa Soils. Agr. Expt. Sta. Bul. 384.
9. Hanway, J. J. and C. R. Weber. 1971. Dry matter accumulation in soybean [*Glycine max* (L.) Merrill] plants as influenced by N, P, and K fertilization. Agron. J. 63:264-266.
10. Hanway, J. J. and C. R. Weber. 1971. N, P, and K percentages in soybean [*Glycine max* (L.) Merrill] plant parts. Agron. J. 63:286-290.
11. Hardy, R. W. F., R. D. Holsten, E. J. Jackson and R. C. Burns. 1968. The acetylene-ethylene assay for N_2 fixation: laboratory and field evaluation. Plant Physiol. 43:1185-1207.
12. Harper, J. E. 1971. Seasonal nutrient uptake and accumulation patterns in soybeans. Crop Sci. 11:347-350.
13. Henderson, J. B. and E. J. Kamparth. 1970. Nutrient and dry matter accumulation in soybeans. North Carolina Expt. Sta. Tech. Bul. No. 197.

14. Hiatt, A. J. 1968. Electrostatic association and Donnan phenomena as mechanisms of ion accumulation. Plant Physiol. 43:893-901.
15. Hiatt, A. J. and J. E. Leggett. 1974. Ionic interactions and antagonisms in plants, pp. 101-134. In E. W. Carson (Ed.) The plant root and its environment. Univ. Press of Virginia, Charlottesville.
16. Hoagland, D. R. and D. I. Arnon. 1950. The water culture method for growing plants without soil. Calif. Ag. Expt. Sta. Cir. 347.
17. Israel, D. W. and W. A. Jackson. 1978. The influence of nitrogen nutrition on ion uptake and translocation by leguminous plants, pp. 113-129. In C. S. Andrew and E. J. Kamparth (Eds.) Mineral nutrition of legumes in tropical and sub-tropical soils. Commonwealth Scientific and Research Organization, Melbourne, Australia.
18. Jacobson, I. and L. Ordin. 1954. Organic acid and metabolism and ion absorption in roots. Plant Physiol. 29:70-75.
19. Kirkby, E. A. and A. H. Knight. 1977. Influence of the level of nitrate nutrition on ion uptake and assimilation, organic acid accumulation, and cation-anion balance in whole tomato plants. Plant Physiol. 60:349-353.
20. Kirkby, E. A. and K. Mengel. 1967. Ionic balance in different tissues of the tomato plant in relation to nitrate, urea, or ammonium nitrition. Plant Physiol. 42:6-14.
21. Leggett, J. E. and M. H. Frere. 1971. Growth and nutrient uptake by soybean plants in nutrient solutions of graded concentrations. Plant Physiol. 48:457-460.
22. Leggett, J. E. and W. A. Gilbert. 1969. Magnesium uptake by soybeans. Plant Physiol. 44:1182-1186.
23. Leggett, J. E. and A. J. Hiatt. 1971. Interaction of cations and anions during salt uptake by plants, pp. 97-103. In E. Broda, A. Lacker, H. Springer-Lederer (Eds.) First European biophysics congress, Baden, Austria.
24. Lowe, R. H. and J. L. Hamilton. 1967. Rapid method for determination of nitrate in plant and soil extracts. J. Ag. Food Chem. 15:359-361.
25. Mengel, K. and M. Helal. 1967. Der einfluss des austauschbaren Ca^{2+} junger gerstenwurzeln auf den flux von K^+ und phosphat-eine interpretation des Viets Effektes. Z. Pflanzenern., Dung., Bodenk 57:223-234.
26. Moore, D. P., B. J. Mason and E. V. Maas. 1965. Accumulation of calcium in exudate of individual barley roots. Plant Physiol. 40:641-644.
27. Pierce, E. C. and C. O. Appleman. 1943. Role of ether soluble organic acids in the cation-anion balance in plants. Plant Physiol. 18:224-237.
28. Pitman, M. G. 1967. Ion uptake by plant roots, pp. 95-128. In U. Luttge and M. G. Pitman (Eds.) Transport in plants II, Part B. Tissue and organs. Encyclopedia of Plant Physiology, Springer-Verlag, Berlin.
29. Raven, J. A. and F. A. Smith. 1976. Nitrogen assimilation and transport in vascular land plants in relation to intracellular pH regulation. New Phytol. 76:415-431.
30. Sinclair, T. R. and C. T. de Wit. 1975. Comparative analysis of photosynthate and nitrogen requirements in the production of seeds by various crops. Science 565-567.
31. Sinclair, T. R. and C. T. de Wit. 1976. Analysis of the carbon and nitrogen limitations to soybean yield. Agron. J. 68:319-324.
32. Smith, F. A. and J. A. Raven. 1976. H^+ transport and regulation of cell pH, pp. 317-346. In U. Luttge and M. G. Pitman (Eds.) Transport in plants II, Part A, Cells. Encyclopedia of Plant Physiology, Springer-Verlag, Berlin.
33. Ulrich, A. 1941. Metabolism of non-volatile organic acids in excised barley roots as related to cation-anion balance during salt accumulation. Am. J. Botany 28:526-537.
34. Viets, F. C., Jr. 1944. Calcium and other polyvalent cations as accelerators of ion accumulation by excised barley roots. Plant Physiol. 19:466-80.

35. de Wit, C. T., W. Dijkshoorn and J. C. Noggle. 1963. Ionic balance and growth of plants. Versl. Landbouwk Onderz 69:15.

RESPONSE OF SOYBEANS TO LIMING ON ACID TROPICAL SOILS

F. Abruña

Millions of hectares of Ultisols and Oxisols occur in South and Central America, Africa, and Asia. These soils have been devoted to native forests, extensive uses such as rangeland for cattle, or subsistance farming based on shifting cultivation. The leached, acidic, highly weathered conditions of these soils make intensive farming possible only if judicious intensive management systems are used.

Although erratic weather and outbreaks of insects and diseases are generally limiting factors, soil acidity factors such as toxic concentrations of aluminum (Al) and/or manganese (Mn) commonly limit crop yields. Most cultivated crops respond to liming on these soils, but the magnitude of the response is determined largely by the relative tolerance of the species, or even of the variety, to high levels of Al or Mn in the soil solution. For example, coffee is tolerant to high concentrations of Al but is sensitive even to moderately high concentrations of soluble Mn in the soils (1). Similarly, sweet potatoes are quite tolerant (3) but tobacco (2) is very sensitive to both exchangeable Al and soluble Mn in the soil.

Soybean [*Glycine max* (L.) Merr.] production is expanding rapidly in the humid tropics, becoming a major low-cost source of protein which is the main deficiency in the diet of millions of people in these areas. Considerable emphasis has been placed during recent years on research to determine the lime requirements of soybeans for maximum production in acidic tropical soils. Most of the research work has been carried out in the Campos Cerrado region of Brazil and the Llanos Orientales of Colombia, since the most extensive areas of Ultisols and Oxisols in South America occur there. Mikkelson et al. (12) found while working with Latosols and Regosols in the Campos

Cerrado area, that soybeans were less responsive to liming than corn. Latosols in the Brazilian classification scheme correspond to Oxisols in the U.S. taxonomy. Freitas et al. (7) significantly increased soybean yields by appling 2 mt $CaCO_3$/ha to a Humic Latosol from the Campos Cerrado with an initial pH of 4.5. In another study by Freitas et al. (8), soybeans responded strongly to liming on Latosols from the Campos Cerrado area, where initial pH values ranged from 4.3 to 5.1. Martini et al. (11). working with Oxisols from Río Grande Do Sul in Brazil, reported that Bragg soybeans responded significantly to liming. Highest yields were obtained when Al saturation was around 5%. Liming increased the Ca and reduced the Mn content of the leaves. Soares et al. (15) found that the application of 5 mt $CaCO_3$/ha to Dark Latosols from Campos Cerrado reduced the Al saturation from 86 to less than 10%, but soybean yields increased only 7%. Foster (6), working with soils in Uganda (probably Oxisols), reported significant increases in soybean yields due to liming even though the initial soil pH was 5.86. It seems that the response in this case was to Ca as a nutrient rather than to the effect of liming on acidity factors. Mascarenhas et al. (10) reported an increase of 30% in soybean yields when only 1.6 mt/ha of limestone was applied to a Red Latosol from Brazil with a pH of 5.5 and only traces of exchangeable Al. The increase in yields was sustained for 2 consecutive yr. Again, it appears that this soil had a low effective cation exchange capacity (CEC), and the response was to CA as a nutrient, since higher rates did not increase yields further. In another report, however, Mascarenhas et al. (9) found that soybeans on a recently cleared Red Latosol with a pH of 4.8 did not respond to liming over a 2-yr period.

The use of soybean varieties tolerant to high Al or Mn concentrations in acid soils has been gaining favor as an alternative to liming. Armiger et al. (5) found considerable variability in tolerance to Al toxicity among soybean varieties representing the 10 maturity groups. Sartain and Kamprath (13) found that the 'Lee', 'Lee 68', 'Odgen', and 'Dare' soybeans were relatively tolerant, as measured by root elongation, to Al toxicity. The 'Lee', 'Bragg', 'Pickett 71', and 'York' varieties were the most tolerant in terms of top growth.

The objectives of the experiments represented in this chapter were (a) to determine the response of 'Hardee' soybeans to liming on Utisols and Oxisols, and (b) to evaluate the effect of soil acidity factors on yields and foliar composition.

MATERIALS AND METHODS

Experiments were conducted on one Oxisol and two Utisols of Puerto Rico. Each experiment consisted of a series of 4 m^2 plots arranged in a completely randomized design using 30 on Corozal clay (Aquic Tropudults),

60 on Humatas clay (Typic Tropohumults), and 40 on Coto sandy clay (Tropeptic Haplorthox), to which variable increments of hydrated lime were applied, resulting in a wide range of acidity in the upper 20 cm of soil.

Soil samples were taken from each plot 6 months after liming and analyzed for pH, exchangeable bases, exchangeable Al and exchangeable Mn, following standard procedures. Exchangeable bases were extracted with normal neutral ammonium acetate and exchangeable Al with normal KCl. Calcium and Mg were determined by titration with Versenate, K by flame photometry and Al by titration with NaOH.

'Hardee' soybeans were planted at 8-cm intervals in rows 45 cm apart. All plots received 30 kg P, 50 kg K, 30 kg Mg, and 30 kg of a minor element mixture containing B, Zn, Mo, and Cu/ha. At the preblooming stage, the fourth and fifth pairs of leaves from plants in the central row of each plot were analyzed for Ca and Mg by the Versenate method, for K by flame photometry, for N by the Kjeldahl method, and for P and Mn colorimetrically. Yields of clean beans (14% moisture) were determined in all plots and related to the various soil acidity factors by regression analysis.

RESULTS

Corozal Clay (Aquic Tropudults)

Table 1 shows that highest yields were obtain at pH 5.60, corresponding to 3% Al saturation and an Al/base ratio of 0.03. Yields decreased 37% when Al saturation increased to 13% and pH dropped to 5.0. When Al saturation increased to 26%, yields dropped to 28% of maximum. Yields were almost identical in the 4.6 to 4.8 pH range, although Al saturation continued to increase up to 44%. This might have been due to less active forms of Al at these pH values or to other factors affecting yields on specific plots. At extreme soil acidity levels yields were very low, averaging only 62 kg/ha. These data clearly showed that 'Hardee' soybeans are quite sensitive to exchangeable Al.

Soil acidity factors strongly affected the content of some elements in the soybean leaves, but did not affect others. Calcium and N contents decreased as percent Al saturation increased, as pH decreased, and as the Al/base ratio increased. Other leaf constituents were not affected consistently, indicating that the response of soybeans to liming in the Corozal soil was linked to Ca uptake and N fixation. There was a drastic reduction in nodulation with increasing Al saturation of the soil, as Table 1 shows. The number of nodules/plant dropped from 75 at the 3% Al saturation value to only 4 at the 44% saturation level. No nodulation occurred at extreme soil acidity. The Ca/Mn ratio in the leaves decreased with increasing Al saturation of the soil, largely because the Ca content decreased as the Mn content remained constant.

Table 1. The effect of soil acidity factors on yield and foliar composition of soybeans, Hardee variety, grown on two Ultisols and one Oxisol from Puerto Rico.

	Soil Factors		Yield		Foliar Composition							
pH	% Al[a] sat	Ratio[b] Al/Bases	Actual kg/ha	Relative %	Ca	Mg	N	P	K	Mn	Ratio Ca/Mn[c]	Nodules/ Plant
							– % –			ppm		
					Corozal Clay – Aquic Tropudults							
5.60	3	0.03	2081	100	1.71	0.23	4.49	0.30	1.63	152	155	75
5.00	13	.13	1311	63	1.49	.20	3.47	.26	1.56	152	135	62
4.80	26	.36	585	28	1.39	.16	3.35	.25	1.54	150	127	44
4.70	33	.49	558	27	1.26	.16	3.03	.24	1.48	152	114	13
4.60	44	.78	532	26	1.02	.14	2.81	.24	1.47	152	92	4
4.50	56	1.28	64	3	.79	.17	2.17	.21	2.09	151	72	0
4.30	67	2.03	62	3	.80	.17	2.16	.26	2.10	150	73	0
					Humatas Clay – Typic Tropohumults							
5.60	2	0.02	915	100	1.78	0.24	5.14	0.21	2.19	60	400	--
4.90	15	.17	906	99	1.63	.26	4.67	.21	2.30	64	350	--
4.80	22	.28	875	96	1.58	.22	4.83	.20	2.22	54	400	--
4.60	36	.56	602	66	1.57	.23	4.54	.24	2.20	68	317	--
4.50	45	.81	475	52	1.47	.24	4.63	.24	2.38	99	204	--
4.40	55	1.20	430	47	1.47	.27	4.50	.24	2.40	120	168	--
4.20	65	1.84	392	43	1.42	.27	4.37	.24	2.33	105	186	--
3.90	17	3.33	61	7	1.04	.25	4.07	.26	2.50	125	115	--
					Coto Sandy Clay – Tropeptic Haplorthox							
5.30	5	0.05	3555	100	2.00	0.15	4.78	0.16	1.31	367	75	--
4.70	16	.19	2997	84	1.96	.19	4.84	.18	1.32	468	58	--
4.50	25	.33	2755	78	1.86	.19	4.84	.18	1.39	514	50	--
4.30	36	.56	2540	71	1.87	.18	4.81	.20	1.37	528	49	--

[a]Al saturation percentage = (Exchangeable Al)/(Exchangeable Ca + Mg + K + Al)
[b]Ratio Al/bases = (Exchangeable Al)/(Exchangeable Ca + Mg + K)
[c]Ca/Mn = (Chemical Ca equivalents)/(Chemical Mn equivalents)

Table 2 shows that there were significant correlations between soybean yields and the various soil acidity factors with r values exceeding 0.90. Figure 1 shows the close correlation between the percent Al saturation of the soil and grain yields. Similarly, leaf Ca and N contents were significantly correlated through regression analysis with soil acidity factors as well as with yields. The best correlation was obtain between N content of the leaves and grain yields, as shown in Figure 2. A 5% N content was associated with maximum yield.

Humatas Clay (Typic Tropohumults)

Yields on this soil were lower than those obtained on the other soils (Table 1) due to constant strong winds, excessive rainfall, and insect damage. The response of soybeans to liming on this soil was not as strong as on the Corozal soil, probably because the above mentioned factors limited yields.

The Humatas soil had a wider range of pH, of Al/base ratios, and of percent Al saturation than the Corozal soil. Although absolute yields were lower, soybeans produced relatively higher yields than on the corozal soil at comparable pH, Al/base ratio or percent Al saturation values. For example, at pH 4.4 and 55% Al saturation, grain yields were 47% of maximum on Humatas soil, as compared with only 3% of maximum in the Corozal soil. Yields remained essentially constant up to 22% Al saturation and then decreased consistently as Al saturation and Al/base ratio increased and pH decreased.

The effect of liming on foliar composition was similar to that on the Corozal soil. Calcium content decreased as soil acidity increased. Nitrogen content also decreased, but intermediate values did not vary so strikingly as those on Corozal soil did. Magnesium, P, and K values were not affected by variations in soil acidity. The Mn content remained fairly constant up to pH 4.6 and then increased as pH dropped and percent Al saturation increased. The Ca/Mn ratio remained fairly constant up to pH 4.8 and then decreased consistently as pH decreased and percent Al saturation increased.

Table 2 shows that acidity factors and leaf Ca and N contents were significantly correlated with soybean yields on this soil, although the r-values were not as large as those on the Corozal soil. Leaf Ca content was significantly correlated with soil acidity factors, but in contrast with the Corozal soil, N content was not. The response of soybeans to liming on this soil is apparently related primarily to an increase in Ca uptake induced by inactivation of the Al in the soil solution and on the exchange complex. The Ca/Mn ratio of the leaves was correlated significantly with yields, but the r-value was only 0.41.

Coto Sandy Clay (Tropeptic Haplorthox)

This Oxisol is fairly low in total exchangeable Al compared with the two Ultisols; the cation exchange capacity is much lower at comparable pH levels.

Table 2. Relationships and correlation coefficients between soil acidity factors, yields, and foliar composition of 'Hardee' soybeans.

	pH	% Al sat	Al/Base	% Ca	% N	Ca/Mn
			Corozal Clay			
	r = .94**	r = .93**	r = .92**	r = .75**	r = .94**	
Yield	$Y=15269.44+5116.04x-359.88x^2$	$Y=72.59+2233.3(.954^x)$	$Y=254.72+2034.35(.104^x)$	$Y=1055.47+1653.25x$	$Y=1667.10+806.87x$	Nonsignificant
	r= .84**	r= .82**	r= .83**			
% Ca	$Y=14.31+5.64x-.494x^2$	$Y=1.72+.014x$	$Y=1.699-.99x+.252x^2$	–	–	–
	r=.90**	r= .84**	r= .83**			
% N	$Y=14.64+5.63x-.393x^2$	$Y=4.458-.037x$	$Y=4.41-2.709x+.739x^2$	–	–	–
			Humatas Clay			
	r= .77**	r= .78**	r= .77**	r= .67**	r= .63**	r= .41**
Yield	$Y=6997.7+2735.3x-234.0x^2$	$Y=1493.1-531.6(1.012^x)$	$Y=914.3-432.7x+52.5x^2$	$Y=508.3+706.8x$	$Y=1812.5+513.8x$	$Y=315.1+1.59x$
	r= .60**	r= .59**	r= .65**			
% Ca	$Y=.207+.367x$	$Y=1.82-.008x$	$Y=2.24-.336x$	–	–	–
% N	Nonsignificant	Nonsignificant	Nonsignificant	–	–	–
			Coto Sandy Clay			
	r= .55**	r= .59**	r= .57**			r= .45**
Yield	$Y=547;66+744.89x$	$Y=3438.89-25.99x$	$Y=3456.75-2625.02x+1654.21x^2$	Nonsignificant	Nonsignificant	$Y=1970.7+2288x$
	r= .34*	r= .33*	r= .37*			
% Ca	$Y=1.274+1.39$	$Y=2.25-.01x$	$Y=2.24-.336x$	–	–	–
% N	Nonsignificant	Nonsignificant	Nonsignificant			

*Significant at 5% level (Duncan's new multiple range test).
**Significant at 1% level (Duncan's new multiple range test).

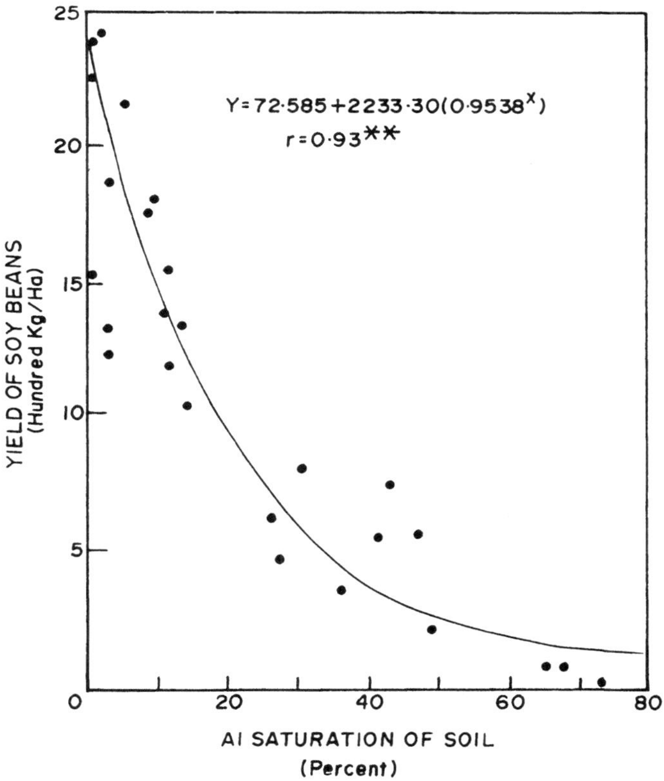

Figure 1. **Influence of the aluminum saturation percentage of Corozal clay soil on soybean yields.**

This soil also has a lower percent Al saturation than the Ultisols at corresponding pH levels (Table 1) but is higher in both exchangeable and easily reducible Mn. Soybeans produced higher yields on this soil than on the Ultisols. Maximum yields of 3555 kg/ha were produced at 3% Al saturation and lowest yields of 2540 kg/ha at 36% Al saturation. This soil responded much less to liming than the Ultisols. As Table 1 shows, at the highest Al saturation value of 36%, soybeans produced 71% of maximum yield, as compared with only 27% at corresponding percent Al saturation level on Corozal clay.

Soil acidity affected some of the foliar constituents, but not as strikingly as on the Ultisols. Calcium content of the leaves decreased and Mn content increased as percent Al saturation and Al/base ratio increased. The other leaf constituents were unaffected by soil acidity.

Table 2 and Figure 3 show that soybeans responded in yield to liming, but not as strongly as on the Ultisols. Yields were signficiantly correlated with all soil acidity factors: pH, percent Al saturation, and Al/base ratio.

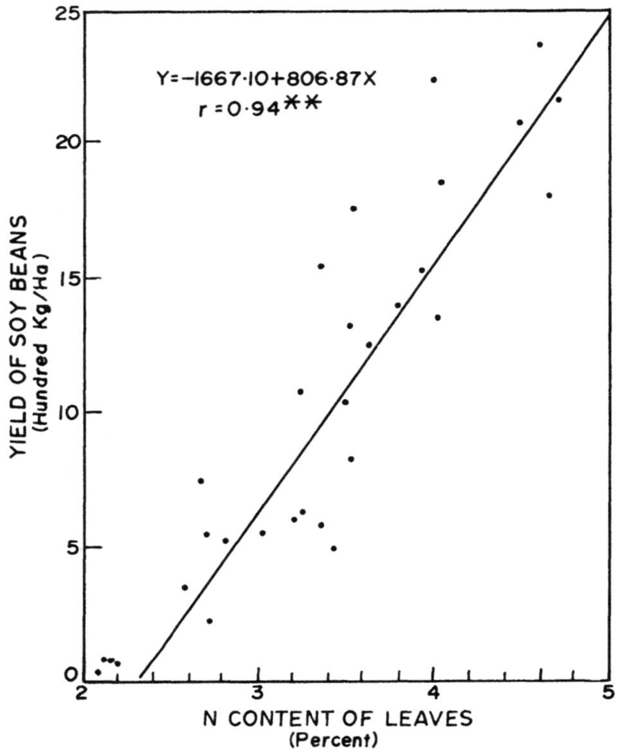

Figure 2. **Relationships between grain yields and the nitrogen content of the leaves of soybean grown on Corozal clay soil.**

Calcium content of the leaves was correlated with soil acidity factors, although the correlations were barely significant at the 5% level. Nitrogen content of the leaves was not correlated with the soil acidity or yields. However, the correlation between the Ca/Mn ratio in the leaves and grain yield was significant.

DISCUSSION AND CONCLUSIONS

The data presented show that soybeans respond to liming on both Oxisols and Ultisols, but they respond more strongly on the Ultisols. Among the Ultisols, a stronger response was evident in the less weathered Corozal soil than on the Humatas soil, probably because of other factors that limited yields on the Hamatas soil. The response seemed to be linked to N and Ca nutrition. On Corozal clay, nodulation was curtailed severely as soil acidity increased. Although nodulation was not measured on the Hamatas and Coto soils, it was probably not affected as severely, since the N content of the

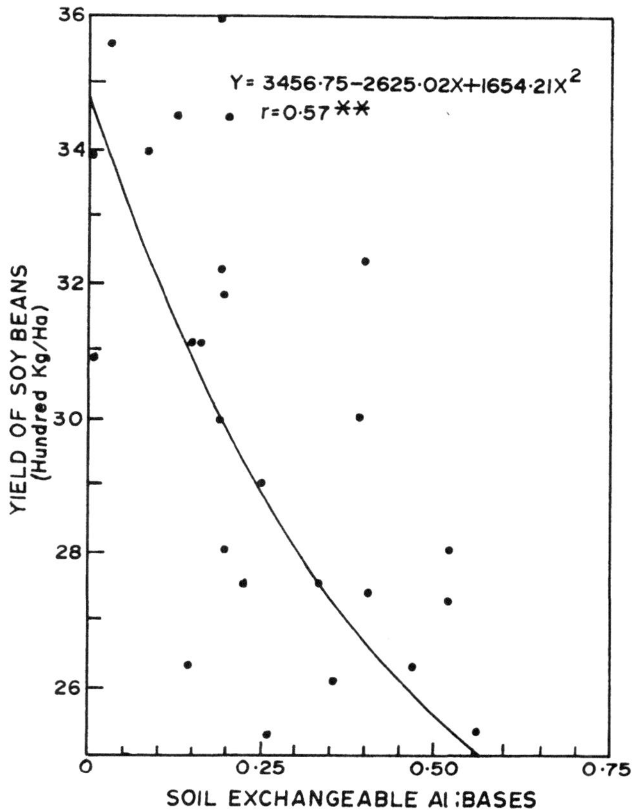

Figure 3. **Influence of the aluminum/base ratio of Coto sandy clay on soybean yields.**

leaves on these soils was not significantly affected by changes in pH, percent Al saturation, or Al/base ratios.

The Ca content of the leaves was modified in a similar way as N by the intensity of the soil acidity factors. The most striking variations in the Ca content were recorded on the Corozal clay; this soil also showed the strongest response to liming. The other two soils showed variations in the Ca content that were roughly similar to the pattern of response to liming. The least variation in the Ca content was recorded on the Coto soil; this soil also responded least to liming. It seems that Ca uptake is the main factor involved in the 'Hardee' soybeans' response to liming since it was the only foliar constituent correlated with acidity factors on the three soils.

The main factor responsible for both N nutrition and Ca uptake seems to be the Al ion, although Mn may be a contributing factor in Oxisols. Table 3, taken from Pearson et al. (14), shows the Al and Mn molar activities in the soil solution for one Ultisol (Humatas) and two Oxisols (Coto and Catalina).

Table 3. Soil solution cation composition of Humatas and Coto soils (14).

Soil	Solution pH	Soil Solution Composition					Al Activity	Mn Activity
		Ca	Mg	K	Al	Mn		
				– mM –			– μM –	
Humatas clay	6.5	2.6	1.7	.1	0	0	0	0
	5.0	12.2	2.2	1.2	0	.3	0	125
	4.7	7.0	1.0	1.3	.1	.4	20	200
	4.1	13.6	4.7	.2	.4	.6	85	229
	3.8	8.3	1.9	.3	.7	.5	179	289
	3.7	11.2	3.7	.3	1.4	.6	331	309
	3.6	13.9	4.8	1.0	2.6	.8	500	339
Coto clay	6.7	3.2	2.2	.1	0	0	0	0
	5.8	1.8	.9	.1	0	0	0	0
	4.5	5.1	.6	.1	.2	1.0	56	578
	4.3	3.2	.5	.3	.3	.8	91	702
	4.2	5.4	.5	.1	.3	1.3	101	717
	4.1	8.4	1.2	.5	.8	1.9	196	901
Catalina clay	6.4	2.2	1.0	.1	0	0	0	0
	5.0	8.4	2.8	.1	0	.3	0	148
	4.4	26.0	2.3	.5	.2	22.3	24	7777
	4.3	16.5	4.6	.5	.3	20.6	46	7526
	4.2	15.2	2.7	.5	.4	19.8	59	7441
	4.1	18.8	1.8	.8	.5	20.8	74	8355

The Al activity, although closely related to pH, varies considerably among soils. It was highest in the Humatas soil and lowest in the Catalina soil. [Catalina (Tropeptic Haplorthox) is more weathered and leached than Humatas or Coto.] The Al activity corresponds with the degree of weathering of each soil. Although there are no data available, Al activity in the Corozal soil is probably higher than in the Humatas soil.

Manganese activity is many times greater in Oxisols than in Ultisols. This strong activity of the Mn ion in Oxisols apparently reduces the Al activity, depressing it in such a way that crops sensitive to Al toxicity but tolerant to Mn do not respond to liming, even at low pH values and fairly high percent Al saturation values. The predominance of the Mn ion in the soil solution of Puerto Rican Oxisols at pH values under 5 is a distinct characteristic of this group of soils.

At comparable Al saturation values, soybeans responded more strikingly to liming on Corozal than on Humatas or Coto soils. Calcium uptake seems to be the determining factor in lime response, and the Al ion seems to be the main factor governing Ca uptake. On Oxisols, Mn seems to replace Al as the main factor controlling Ca uptake. Since Mn is also absorbed differentially by plants, the ratio of Ca/Mn in the leaves seems to be an appropriate index for measuring soybean response to liming on Oxisols. Significant correlations between yields and Ca/mn ratio were obtained in Humatas and Coto soil (Table 2), although they were not as striking as those obtained with snap-beans (4).

We concluded that 'Hardee' soybeans respond strongly to liming in soils where the percent Al saturation is the main cause of infertility (Ultisols) but to a lesser extent on Oxisols, where soluble Mn predominates in the soil solution at pH levels lower than 5.0. The response of soybeans to liming seems to be a direct result of the inactivation of Al and the consequent increase in Ca uptake and N fixation.

NOTES

Fernando Abruña, Science and Education Administration, Agricultural Research, U.S. Department of Agriculture, Rio Piedras, PR 00923.

REFERENCES

1. Abruña, F., J. Vicente-Chandler, L. A. Becerra, and R. Bosque-Lugo. 1965. Effects of liming and fertilization on yields and foliar composition of high-yielding sun-grown coffee in Puerto Rico, J. Agr. Univ. P.R. 49(4) 413-428.
2. Abruña, F., J. Vicente-Chandler, R. W. Pearson, and S. Silva. 1970. Crop response to soil acidity factors in Ultisols and Oxisols: I. Tobacco. Soil Sci. Soc. Am. Proc. 34(4) 629-635.
3. Abruña, F., J. Vicente-Chandler, J. Rodríguez, J. Badillo, and S. Silva. 1979. Crop response to soil acidity factors in Ultisols and Oxisols. V. Sweet potato. J. Agr. Univ. P. R. (In press).
4. Abruña, F., R. W. Pearson, and R. Pérez-Escolar. 1974. Lime response of corn and beans in typical Ultisols and Oxisols of Puerto Rico. Proc. Soil Mgmt. in Tropical America. Univ. Consortium of Soils of the Tropics, N. C. State Univ., Raleigh, NC.
5. Armiger, W. H., C. D. Foy, A. L. Fleming, and B. E. Caldwell. 1968. Differential tolerance of soybean varieties to an acid soil high in exchangeable aluminum, Agron. J. 60:67-70.
6. Foster, H. L. 1970. Limin continuously cultivated soils in Uganda East Afr. Agr. For. J. 36:58-69.
7. Freitas, L. M. M. de, A. C. McClung, and W. L. Lott. 1960. Field studies on fertility problems of two Brazilian campo cerrados soils. IBEC Research Inst. Bull. 21.
8. Freitas, L. M. M. de, E. Labato, and W. V. Soares. 1971. Experimentos de calagen e abubacao em solos sob vegetacao de Cerrado do Distrito Federal. Pesq. Agropec. Bras. Ser. Agron. 6:81-89.
9. Mascarenhas, H. A. A., S. Miyasaka, and T. Igue. 1968. Abubacao de soja. VII. Efeito de doses crescentes de calcário, fósforo e potássio en solo Latosolo Roxo con vegetacao de Cerrado recem desbravado. Bragantia 27:279-289.
10. Mascarenhas, H. A. A., S. Miyasaka, T. Igue, E. S. Freire, and G. Di Sordi. 1969. Respostas da soja a calagen e a abubaccao minerais com fósforo e potássio em solo Latosolo Roxo. Bragantia 28:17-21.
11. Martini, J. A., R. A. Kochhann, O. J. Siqueira, and C. M. Barkert. 1974. Response of soybeans to liming as related to soil acidity, Al and Mn toxicities and P in some Oxisols of Brazil. Soil Sci. Soc. Am. Proc. 38:616-620.
12. Mikkelson, D. S., L. M. M. de Freitas, and A. C. McClung. 1963. Effects of liming and fertilizing cotton, corn, and soybeans on Camp Cerrado soils, State of Sao Paulo, Brazil. IRI Research Inst. Bull. 29.

13. Sartain, J. B. and E. J. Kamprath. 1978. Aluminum tolerance of soybean cultivars based on root elongation in solution culture compared with growth in acid soils. Agr. J. 70:17-20.
14. Pearson, R. W., R. Pérez-Escolar, F. Abruña, Z. F. Lund, and E. J. Brenes. 1977. Comparative response of three crop species to liming several soils of the Southeastern United States and of Puerto Rico. J. Agr. Univ. P. R. 61(3) 361-382.
15. Soares, W. V., E. Labato, E. González, and G. C. Naderman. 1974. Liming of soils of the Brazilian Cerrado. Proc. Soil Mgmt in Tropical America. Univ. Consortium on Soils on the Tropics, N.C. State Univ., Raleigh, NC.

MINERAL NUTRITION AND NODULATION

D. N. Munns

The purpose of this chapter is to describe some aspects of mineral nutrition that specifically afect symbioses between rhizobia and legumes, soybeans in particular. Most of the information comes from research with other legumes. Extrapolating it to soybeans is uncertain, because legume symbioses vary in response to mineral nutrition, even at the level of rhizobial strain or host plant variety. This variation is a hindrance to easy generalization. More important, it may allow selection of plants and rhizobia for useful tolerance of nutritional stresses.

"Nodulation", for present purposes, will comprise the events and processes that lead to effective symbiotic nitrogen fixation. It includes rhizobial growth and persistence in soil, rhizobial colonization of new root surfaces, infection, nodule development, and maintenance of effective nodule function throughout the plant's development. Each of these may be influenced by a nutritional disorder. The plant's dependence on symbiotic nitrogen fixation can in some respects increase its sensitivity to nutritional stress.

ASSESSMENT OF NODULATION

Nodulation is assessed frequently by measurements of nodule number, nodule size, nodule color, or nitrogenase acitivty (rate of reduction of acetylene or rate of assimilation of $^{15}N_2$). These measure only part of the whole, part of the time. Though useful, they do not substitute for measurements of the yield of symbiotically acquired nitrogen and its consequences, if any, to yield and quality of plant product.

Figure 1 provides a case in point. Wide variation in nodule number (and mass in this case) due to differences in rhizobial strain or soil pH had little

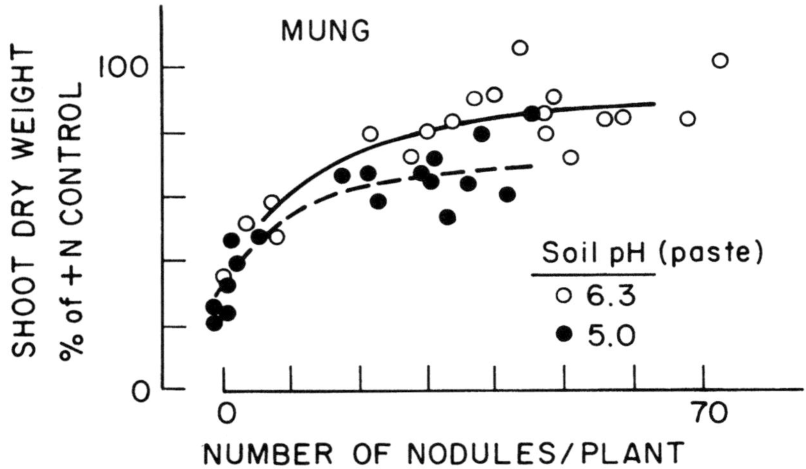

Figure 1. **Relationship between growth and nodule number as influenced by soil pH and rhizobial strain in mung plants dependent on N_2 (31).**

effect on plant growth except at very low nodule numbers, and accounts for only a part of the adverse effect of acidity on growth. Indeed, some nutritional stresses such as deficiency of S or Mo can severely inhibit nitrogen fixation and growth while increasing nodule number (28,37).

EFFECTS OF NODULATION ON MINERAL NUTRITION

Nodulation does not just supply N to the plant: it has at least the following three effects on mineral nutrition.

Increased Requirement for Certain Nutrients

Root nodules are rich in Mo, Co, Fe, Zn, P, S, and N (28,37). The enrichment must reflect the high bacteroidal concentrations of nucleotides, cobalamin, and protein, including Fe-, S-, and Mo-protein, and the presence of Fe in leghemoglobin. However, the Fe, Zn, S, P and N in nodules normally comprise little of the plant's total content, so that deficiencies of these elements possible have no inhibitory effects on nodulation except as a consequence of poor plant growth. By contrast deficiencies of Mo or Co in legumes are due primarily to symbiotic failure, in N-deficient media. If combined N is not deficient, much less Mo and Co is needed (28,37).

Poor Root Development and Nutrient Uptake

Nodules consume photosynthate that could otherwise go to roots (34). In addition, they alter levels of growth substances (25). Therefore, nodulation should influence root growth and activity. In fact, soybean and cowpea plants that were symbiotic (dependent on nodulation) have been found to have substantially smaller root systems and root:shoot ratios than otherwise comparable nonnodulated plants supplied with nitrate (10). Inhibition of root growth should impair uptake of phosphate and other nutrients whose mobility in the soil is low. In a field trial in a P-deficient Ultisol, soybeans that were almost completely symbiotic needed about 20% more added superphosphate for 90% maximal yield than did sparsely nodulated plants growing with nonlimiting N-fertilizer rates, although the latter plants yielded better (10).

Acidification of the Soil

Plants alter ionic concentrations and pH in the root medium, especially in the rhizosphere region immediately next to the root (13,15,17,18,32). If there is little NO_3, then growth supported by N_2, seed reserves, or NH_4 will result in excess uptake of cations and net proton efflux, acidifying the soil. The magnitude of the acidification by symbiotic legume growth has been calculated from plant composition and measured directly in soil (17,18,32). It is comparable to, though less than, the effect of NH_4 fertilization. Among other effects, acidification should aggravate Al toxicity; and species differences in Al-tolerance have been attributed to differences in estimated acidifying effect (13,17,18).

Acidification, and depletion of Ca and P, in the rhizosphere of legume plants in low-NO_3 soil should interfere with rhizobial colonization and infection (18,29). Glass microelectrode measurements on seedlings of soybeans, mung and cowpea confirm that the effects may be substantial. Soybean plants, for example, induced a decline of at least 0.3 pH units next to 1-day-old roots in a low-NO_3 soil and a 1-unit increase in a high-NO_3 soil (P. W. Vonich and D. N. Munns, unpublished).

EFFECTS OF MINERAL NUTRITION ON NODULATION

Nitrogen fixation and growth of the legume plant tend to vary together under the influence of nutritional and other factors. The relationship is commonly linear (19). It can reflect dependence of growth on nodulation or dependence of nodulation on growth. If a stress inhibits growth because it inhibits nodulation, the plant will be more sensitive to the stress when symbiotic (dependent on N_2) than when it is getting ample combined N from the soil (Figure 2a). This is the case when Mo deficiency specifically limits synthesis of nitrogenase or soil acidity specifically prevents formation of nodules. If a stress inhibits nodulation only because it inhibits growth of the plant, the plant will be equally sensitive whether it is symbiotic or not (Figure 2b). This case may be the more common, e.g. as the normal result of deficiencies of P, S, and K; nodulation being limited either by root growth (28) or supply of photosynthate (4,7,34).

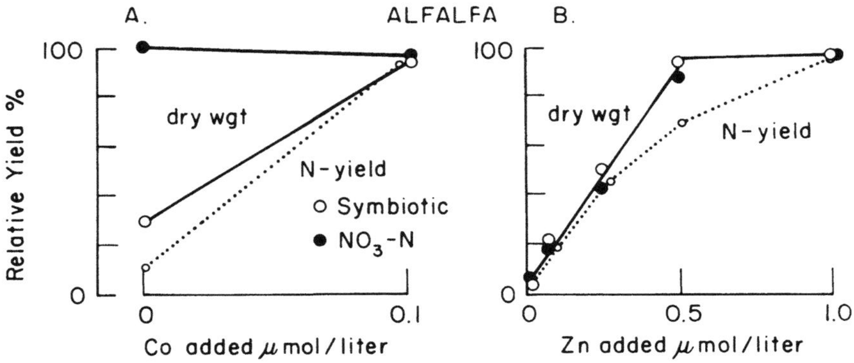

Figure 2. **Contrasting responses: (A) Symbiotic plants more Co-responsive than N-fertilized plants (11); and (B) Symbiotic and nonsymbiotic plants equally Zn-responsive (26). Solid lines indicate plant dry wt. Dotted lines indicate N-yield of symbiotic plants.**

Attempts to distinguish these interdependencies and explain their mechanisms have been discussed at length by mineral nutritionists (9,36,37). Though interesting, the distinctions would hardly matter in practice if the only remedy for poor plant nutrition were improvement of the soil. But another remedy is improvement of plant tolerance by selection. To select rationally one needs to know whether the symbiosis is more sensitive than the nonsymbiotic plant. If it is, selection should be addressed to plants growing symbiotically, dependent on N_2; and the possiblity arises that rhizobia are the weak partner. Rhizobia are easier to select than plants.

Effects Arising from Poor Plant Nutrition

When nodulation is impaired by deficiencies of P, S, K, Zn, Fe or Mn, or by Mn toxicity, the poor nodulation probably results from reduction in root growth, N metabolism, or photosynthetic energy supply. In general, these disorders affect the growth of the legume similarly whether the plant is symbiotic or not.

Nevertheless, it is impossible to argue with confidence that the effects on nodulation are always solely and simply due to disorders in the plant. There may be more direct effects on nodulation at particular levels or combinations of stress, or with particularly sensitive rhizobia or plants. For example, severe S deficiency in clovers reduces plant growth and consequently nitrogen fixation by limiting protein synthesis (28,37), but mild S deficiency in *Stylosanthes humilis* may inhibit nodule activity as its prime effect (20). Frequently, maximal nitrogen fixation and protein yield need higher levels of fertility than maximal dry matter accumulation (1,14) (Figure 2). Very high rates of P combined with high K increased the numbers of nodules in field-grown soybeans (12). One case has been observed where symbiotic plants needed higher rates of P fertilization than N-fertilized plants (10). In P-deficient soils, mycorrhizal innoculation can improve nodule formation and raise plant N concentration; an effect attributable to improve P nutrition in most, though not all, cases (30).

Phosphate requirements of rhizobia are not known—a symptom of our ignorance about mineral nutrition of these bacteria (37). Phosphate becomes so rapidly and severely depleted around young roots that an organism capable of colonizing the rhizosphere must be efficient at accumulating P from external solutions of low concentration. K. L. Clarkin and Munns (unpublished) have results with the cowpea *Rhizobium* strain CB756 indicating that the external P concentration it needs for maximal growth rate is no higher than 0.2 μM (0.006 ppm). There is evidence that rhizobial strains vary in ability to grow in low-P media (21,22).

Direct Nutritional Effects on Nodulation

Deficiencies of Mo, Co and B. The common consequence of these micronutrient deficiences in legumes growing in N-deficient media is nodulation failure and N-starvation, correctable by addition of combined N (28,37). Molybdenum and Co have clearly defined special roles (respectively as a component of nitrogenase and as a nutrient for rhizobia (28,37). Boron deficiency prevents meristematic and vascular development in nodules, as elsewhere (28,37); perhaps there is poorer translocation of the element to nodules than to other meristems in the plant. Molybdenum deficiency is most common in acid soils and can be relieved by liming (28). Its misdiagnosis as an acidity problem could lead to inflated estimates of legume lime requirement. Molybdenum fertilization is cheap, and can be satisfactorily incorporated in seed coatings or pelleting materials compatibly with peat inoculation (28).

Salinity. A few definitive studies show that nodulation is more sensitive to NaCl than ordinary growth of the host plant, but not in all legume species. Alfalfa rhizobia can grow at very high salt concentration, e.g. 200 mM NaCl (16,42). Additions of salt to soil, sufficient to halve the growth of alfalfa, reduced or delayed nodule formation negligibly (M. G. Zaroug and D. N. Munns, unpublished). Symbiotic and N-fertilized alfalfa plants have been found to respond equally to NaCl (3) (Figure 3). Likewise, in several tropical pasture legumes, salt treatments injurious to growth induced no clear case of faulty nodulation (39). By contrast, soybean rhizobia and other slow-growers are less salt-tolerant (16,44), and soybeans when symbiotic are more sensitive than when N-fertilized (3) (Figure 3). *Glycine wightii* behaves similarly. In detailed studies with this species (44), addition of NaCl at 75 or 150 mM to nodulated plants without N caused almost immediate cessation of N_2 fixation and plant growth, with a drop in leaf %N (44). When salt was removed from the growth medium, nodules recovered effectiveness.

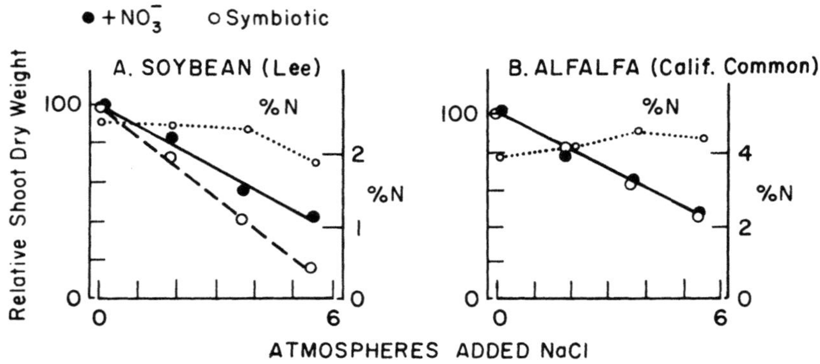

Figure 3. **Contrasting salt responses (3): Symbiotic plants more sensitive than N-fertilized plants in soybean (A) but not in alfalfa (B). Solid lines indicate plant dry wt. Dotted lines indicate % N in shoots.**

Chickpea (*Cicer arietinum*) also becomes more salt sensitive when symbiotically dependent. Although its rhizobia can grow at salt concentrations up to 150 or 300 mM, NaCl at about 50 to 100 mM delays the formation and lowers the activity of nodules. Some strains of chickpea rhizobia form more tolerant symbioses than others (D. L. Lauter, K. L. Clarkin, J. S. Hohenberg and D. N. Munns, unpublished).

Soil Acidity Factors. Low pH, low Ca, high Al and high Mn have received much attention as inhibitors of legume nodulation. The story has become a little less simple. A few years ago (28), adverse effects of soil acidity on nodulation could be attributed mainly to interactive effects of low Ca and low pH. Effects of Al were not apparent: most of the work had been done with legumes sensitive to H^+ itself. Whether or not Mn toxicity had important effects other than through plant growth was doubtful, and still is. Moderate acid stress can be partly offset by increasing rhizobial numbers, i.e., size of inoculum. But slightly more severe acidity had been shown to specifically inhibit nodule initiation, regardless of rhizobial numbers. Rhizobial infection appeared to be more sensitive than rhizobial growth, root growth, the function of nodules once established, or the nonsymbiotic growth of the host. Rhizobial strains were known to differ in ability to grow in acid media (Figure 4). Soybean rhizobia and other slow-growing strains tend to grow at lower pH than the fast growers. This agrees with data on their relative distribution and performance in acid soils. There was evidence that alfalfa rhizobia could be selected for improved ability to form nodules in acid soil.

New conclusions and complexities have been added by recent research, concentrated on more acid-tolerant host species and their slow-growing rhizobia. Calcium deficiency, and perhaps acidity, can inhibit nodule function as well as nodule formation (5,29,37). The evidence includes an experiment

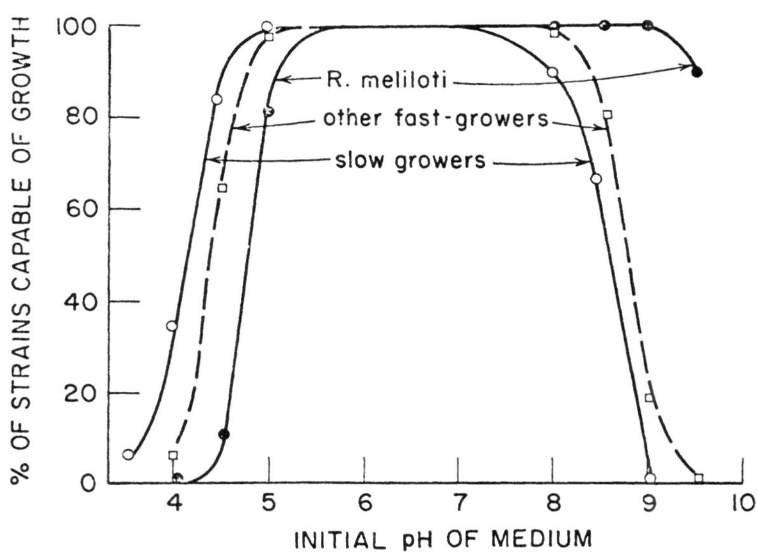

Figure 4. **Variation in pH requirements for growth in culture medium, among different strains and species of rhizobia (16).**

in which soybean nodules lost acetylene-reducing activity when Ca in the nutrient solution was dropped from 500 to 50 μM (5).

Aluminum, like H^+ or Ca-deficiency, can interfere with nodule initiation. This has been shown clearly in controlled media, with very acid-tolerant species of *Stylosanthes* (9). The effective Al concentrations were in the range of 50 to 120 μM, characteristic of concentrations in solution in soils of pH 4.5 or lower. Similar Al concentrations slow the growth of acid-tolerant strains of rhizobia effective with cowpea, mung, peanut, soybean and *Stylosanthes,* completely stopping some of them (22). Tolerance of rhizobial growth to low Ca (e.g., 50 μM) and to high Mn (e.g., 200 μM) has been confirmed with a large number of strains of soybean and cowpea rhizobia (23).

Rhizobia of the cowpea group differ in ability to nodulate cowpea, mung and peanut in acid soils. Some strains cannot nodulate effectively under acid and Al stress even though they can grow fairly well under these conditions (24,31,36). Rhizobia can be selected for symbiotic acid tolerance combined with high effectiveness.

Varieties of legume differ in tolerance to soil acidity. This has been demonstrated in greenhouse and field trials with alfalfa, bean, soybean, and cowpea, in most cases with adequate combined N present (6,8,13,29,37,41). The provision of high N-fertility in this type of screening has been criticized, since it is the symbiosis that is sensitive to soil acidity, tolerance screening should be done on symbiotically-dependent plants (28).

Soybeans may be an exception, no more sensitive when symbiotic than when nonsymbiotic, at least to certain kinds of soil acidity. There is little doubt that in very acid artificial media (28) or organic soils that are probably not Al-toxic (27), nodulation can fail and the plant becomes N-starved. But in acid mineral soils the response might be different (Figure 5). A group of current strains of soybean rhizobia were screened for tolerance of acidity in a California Ultisol, using varieties Evans and Williams as hosts. Unlike cowpea rhizobia in previous trials, all the strains behaved alike, More important, inoculated and NH_4NO_3 control plants responded alike to lime (Figure 5A). The inoculated plants made fewer nodules at low pH, but probably only because the plants grew little (Figure 5B). The nodules they did produce were enough to keep the plants high in N (Figure 5C). Unlike "normal" symbiotic legumes under acid stress—small, unnodulated, low in N and yellow—these soybeans were small, but nodulated, high in N and dark green (too dark, as if Al-toxic). This seems to be a common appearance for soybeans in too-acid soil in the field (36,38,43, and K. Cassman, A. A. Franco, R. S. Smith, personal communication). Perhaps this implies that an important point of weakness in the soybean symbiosis is the aluminum sensitivity of the host plant, exceeding that of the rhizobia or the nodulation process. The absence of rhizobia during selection for Al-tolerance might, for soybeans, be excusable.

NOTES

D. N. Munns, University of California, Davis, CA 95616.

REFERENCES

1. Andrew, C. S. 1977. Nutritional restraints on the legume symbiosis, p. 253-274. In J. M. Vincent, A. S. Whitney, J. Bose (eds.) Exploiting the legume-Rhizobium symbiosis in tropical agriculture. Univ. Hawaii, Coll. Trop. Agr. Misc. Publ 145.

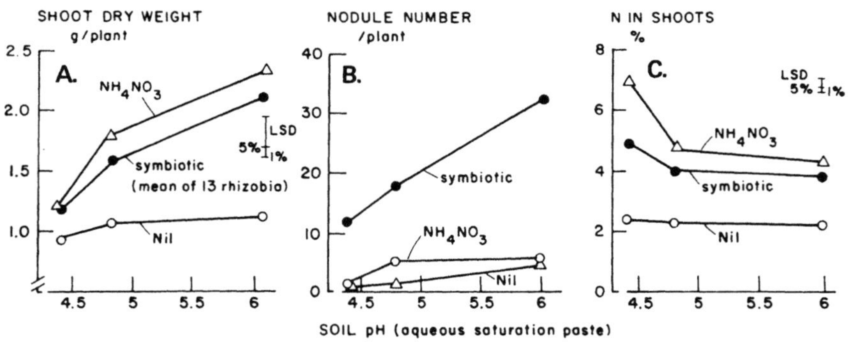

Figure 5. **Response of symbiotic and non-symbiotic soybeans to acidity in subsoil material from an Ultisol. Data are means for two varieties and 13 rhizobial strains which behaved similarly.**

2. Andrew, C. S. 1978. Mineral characterization of tropical forage legumes, p. 93-112. In C. S. Andrew and E. J. Kamprath (eds.) Mineral mutrition of legumes in tropical soils. CSIRO, Melbourne, Australia.
3. Bernstein, L. and G. Ogata. 1966. Effects of salinity on nodulation, nitrogen fixation and growth of soybean and alfalfa. Agron. J. 68:201-203.
4. Bethlenfalvay, G. J. and D. A. Phillips. 1978. Interactions between symbiotic N_2 fixation, combined N applications and photosynthesis in *Pisum sativum.* Physiol. Plant 42:119-123.
5. Blevins, D. G., N. M. Barnett, and F. J. Bottino. 1977. Effects of Ca ion and ionphore A1338 on nodulation of soybean. Physiol. Plant 41:235-239.
6. Brown, J. C. and W. E. Jones. 1977. Fitting plants nutritionally to soil. I. Soybeans. Agron. J. 69:399-404.
7. Brun, W. A. 1976. Relation of N_2 fixation to photosynthesis, p. 135-143. In L. Hill (ed.) World soybean research, Interstate Printers, Inc., Danville, IL.
8. Buss, G. R., J. A. Lutz, and G. W. Hawkins. 1975. Yield response of alfalfa cultivars and clones to several pH levels in Tatum subsoil. Agron. J. 67:331-334.
9. Carvalho, M. M. de. 1978. Comparative study of responses of six *Stylosanthes* species to acid soil factors. Ph.D. Thesis, University of Queensland.
10. Cassman, K. 1979. The P nutrition of two grain legumes as affected by mode of N nutrition. Ph.D. Thesis, University of Hawaii.
11. Delwiche, C. C., C. M. Johnson, and H. M. Reisenauer. 1961. Influence of cobalt on nitrogen fixation by *Medicago.* Plant Physiol. 36:73-78.
12. DeMooy, C. J. and J. Pesek. 1966. Nodulation responses of soybeans to added P, K, and Ca salts. Agron. J. 58:275-280.
13. Devine, T. E., C. D. Foy, A. L. Fleming, C. H. Hanson, T. A. Campbell, J. E. McMurtrey, and J. W. Schwartz. 1976. Development of alfalfa strains with differential tolerance to aluminum toxicity. Plant Soil 44:73-79.
14. Eirz, P. A., D. L. Almeida, and W. C. Silva. 1972. Factors nutricionais limitantes do desenvuovimento de tres leguminsas forrageiras em um solo podzolico vermelho-amarelo. Pesq. Agrpec. Bras. 7:185-192.
15. Foy, C. D., A. L. Fleming, W. H. Armiger. 1969. Al tolerance of soybean varieties in relation to Ca nutrition. Agron. J. 61:505-511.
16. Graham, P. H. and C. A. Parker. 1964. Diagnostic features in characterization of root nodule bacteria of legumes. Plant Soil 20:383-396.
17. Helyar, K. R. 1978. Effects of Al and Mn toxicities on legume growth, p. 207-232. In C. S. Andrew and E. J. Kamprath (eds.) Mineral nutrition of legumes in tropical soils, CSIRO Press, Melbourne.
18. Israel, D. W. and W. A. Jackson. 1978. Influence of N nutrition on ion uptake and translocation by leguminous plants, p. 113-130. In C. S. Andrew and E. J. Kamprath (eds.) Mineral nutrition of legumes in tropical soils, CSIRO Press, Melbourne.
19. Jones, R. J. 1972. The place of legumes in tropical pastures. Food and Fert. Technol. Center (Taipei) Tech. Bull. 9.
20. Jones, R. K., P. J. Robinson, K. P. Haydoch, and R. G. Megarrity. 1971. Sulphur-nitrogen relationships in the tropical legume *Stylosanthes humilis.* Aust. J. Agr. Res. 22:885-894.
21. Kamata, E. 1962. Morphological and physiological studies on nodule formation in leguminous crops. Proc. Crop Sci. Soc. Japan 31:78-89.
22. Keyser, H. H. and D. N. Munns. 1979a. Tolerance of rhizobia to acidity, Al, and P. Soil Sci. Soc. Am. J. 43:519-523.
23. Keyser, H. H. and D. N. Munns. 1979b. Effects of Ca, Mn and Al on growth of rhizobia in acid media. Soil Sci. Soc. Am. J. 43:500-503.

24. Keyser, H. H., D. N. Munns, and J. S. Hohenberg. 1979c. Acid tolerance of rhizobia in culture and in symbiosis with cowpea. Soil Sci. Soc. Am. J. 43 (In press).
25. Libbenga, R. R. and R. S. Bogers. 1974. Root nodule morphogenesis, p. 430-472. In A. Quispel (ed.) The biology of N_2 fixation. North-Holland Publishing Co., Amsterdam.
26. Lo, S. Y. and H. M. Reisenauer. 1968. Zn nutrition of alfalfa. Agron. J. 60:464-466.
27. Menzel, D. B. and E. J. Kamprath. 1978. Effect of soil pH and liming on growth and nodulation of soybean in Histosols. Agron. J. 20:959-963.
28. Munns, D. N. 1977. Mineral nutrition and the legume symbiosis, p. 211-236. In R. W. F. Hardy and A. H. Gibson (eds.) A treatise on dinitrogen fixation. IV. Agronomy and Ecology, Wiley, New York.
29. Munns, D. N. 1978. Soil acidity and nodulation, p. 247-264. In C. S. Andrew and E. J. Kamprath (eds.) Mineral nutrition of legumes in tropical soils. CSIRO Press, Melbourne.
30. Munns, D. N. and B. Mosse. 1979. Mineral nutrition of legume crops. In A. J. Bunting (ed.) Advances in legume science. Oxford Univ. Press (In press).
31. Munns, D. N., H. H. Keyser, V. W. Fogle, J. S. Hohenberg, T. L. Righetti, D. L. Lauter, M. G. Zaroug, K. L. Clarkin, and K. A. Whitacre. 1979. Tolerance of soil acidity in symbioses of rhizobia with mung bean. Agron. J. 71:256-260.
32. Nyatsanga, T. and W. H. Pierre. 1973. Effect of N fixation by legumes in soil acidity. Agron. J. 65:936-940.
33. Parker, M. B. and H. B. Harris. 1977. Yield and leaf N of nodulating and non-nodulating soybean as affected by N and Mo. Agron. J. 69:551-554.
34. Pate, J. S. 1976. Nutrient mobilization and cycling: Case studies of C and N in organization of a legume, p. 447-462. In I. F. Wardlaw and J. B. Passioura (eds.) Transport and transfer processes in plants. Academic Press, New York.
35. Phillips, D. A., K. D. Newell, S. A. Hassell, and C. E. Felling. 1976. Effect of CO_2 enrichment on root nodule development in symbiotic N_2 fixation in *Pisum sativum*. Am. J. Bot. 63:356-362.
36. Rerkasem, B. 1977. Differential sensitivity to soil acidity of legume-*Rhizobium* symbioses. Ph.D. thesis, Univ. of W. Australia.
37. Robson, A. D. 1978. Mineral nutrients limiting N_2 fixation in legumes, p. 277-294. In C. S. Andrew and E. J. Kamprath (eds.) Mineral nutrition of legumes in tropical soils, CSIRO Press, Melbourne.
38. Ruschel, A. P. and P. A. da Eira. 1969. Fixcao simbiotica do nitrogenio na soja: Influencia da adicao de calcio ao solo e molibdenio ao revistimento da semente. Pesq. Agropec. Bras. 4:103-106.
39. Russell, J. S. 1976. Comparative salt tolerance of some tropical and temperate legumes and tropical grasses. Aust. J. Exp. Agr. An. Husb. 16:103-109.
40. Sartain, J. B. and E. J. Kamprath. 1975. Effects of liming a highly Al-saturated soil on top and root growth and nodulation of soybean. Agron. J. 67:507-511.
41. Tropical soils research program, annual report for 1975, p. 40-65. Soil Science Dept., North Carolina State Univ., Raleigh, NC.
42. Steinborn, J. and R. J. Roughley. 1975. Toxicity of sodium and chloride ion to *Rhizobium* Spp. in broth and peat culture. J. Appl. Bact. 39:133-138.
43. Vidor, C. and R. J. R. Freire. 1972. Controle da toxidez de aluminio e magnanes em *Glycine max* pelo calagem e adubacao fosfatada. Agron. Salriogr. 8:73-87.
44. Wilson, J. R. 1970. Response to salinity in *Glycine*. VI. Some effects of a range of short-term salt stress on growth, nodulation, and N fixation of *G. wightii*. Aust. J. Agric. Res. 21:571-582.

NITROGEN INPUT WITH EMPHASIS ON N_2 FIXATION IN SOYBEANS

R. W. F. Hardy, U. D. Havelka, and P. G. Heytler

Nitrogen is more important quantitatively and nitrogen input is more complex for soybeans than for any other major world crop. About 100 kg N is used by the soybean crop for each 1000 kg grain produced. Nitrogen input occurs by both uptake of combined N and biological fixation of atmospheric N. Our contribution will consider the following aspects of N input with emphasis on N_2 fixation: (a) Estimates of world soybean production demands, N needs and general approaches to meet those needs; (b) Status of knowledge on biological N_2 fixation viewed on a "what's wrong" with the natural process for soybean production and thereby identifying the opportunities for technological improvements; (c) Critical subsets of the N_2 fixation system in soybeans, e.g., photosynthesis, translocation, nodular production of energy, reductant and carbon skeletons, N_2 fixation including concomitant H_2 production, H_2 utilization, and other reactions. The cost of N_2 fixation by soybeans measured by a comparison of respiration by nodulated and non-nodulated isolines will be reported; and (d) Possible future technologies to provide N by biological N_2 fixation or other routines.

ESTIMATED SOYBEAN WORLD PRODUCTION DEMANDS, NITROGEN NEEDS AND GENERAL APPROACHES TO MEET THOSE NEEDS

World soybean production of about 70 Mt in 1975 constituted slightly more than 50% of total world grain legume production (Table 1). The average yield was 1470 kg/ha which is an unsatisfactory 0.1% conversion of solar energy to harvested grain. The world soybean crop in 1975 used 7 Mt fixed N assuming 2/3 fixed N in grain and 1/3 in non-grain parts at harvest. About 90% of this nitrogen came from biological N_2 fixation and combined N in

Table 1. World 1975 grain legumes—production, area and yield (4).

Legume	Production	Area	Yield
	– 10^6 *Mt* –	– 10^6 *ha* –	– *kg/ha* –
Soybeans	68.4	46.5	1471
Groundnuts	19.1	19.4	986
Dry beans	13.3	24.7	535
Dry peas	10.6	10.6	999
Chick peas	5.7	9.7	590
Broad beans	6.4	5.4	1166
Pigeon peas	2.0	2.7	699
Vetches	1.6	1.6	997
Lentils	1.2	1.9	640
Cow peas	1.1	5.2	212
Lupins	0.6	0.9	647
Other pulses	3.7	6.8	543
Total or average	133.7	135.4	990

the soil with only about 10% from fertilizer N applied directly based on the application of 15 kg fertilizer N/ha of soybeans in the U.S. (39) However, some of the 90% part comes from carryover of fertilizer N applied to prior crops, such as corn.

World demand for soybeans as for other grain legumes may be projected to quadruple to about 275 Mt by 2000 A.D. (27). The driving force for this increase will be the demand for protein-rich animal feeds to satisfy needs created by a combination of increased population and altered dietary habits in which increasing affluence strongly favors an increase in the ratio of animal to plant foods. The established importance of soybean meal in meeting this need is documented by the fact that 85 to 90% of the about 40 Mt increase from 42 to 85 Mt in total world meal production from 1965 to 1979 was provided by the about 36 Mt increase from 18 to 54 Mt in world soybean meal production (Figure 1).

This projected demand for 275 Mt soybeans in 2000 A.D. will use about 27 Mt fixed N annually or an increase of 20 Mt over that used in 1975. Although the absolute amount is quantitatively impressive, it may be even more meaningful to translate this additional N to its current value of $3 billion as fertilizer N.

General approaches to meet this challenging quantitative need and economic cost are enhancing biological N_2 fixation, development of an efficacious fertilizer N system or a combination of both. Much research emphasis is being placed on enhancement of biological N_2 fixation. Recent books (15,20,33,38,43,56,57) popular review articles (13,28), and this session are devoted exclusively to this approach although no practical solutions for high

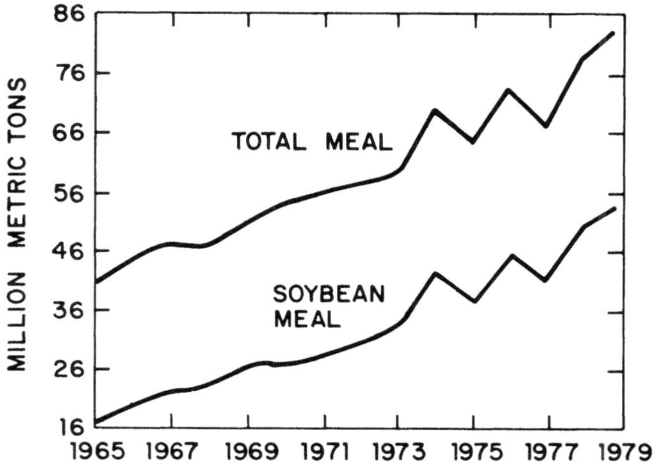

Figure 1. Soybean meal and total world meal production from 1965 to 1979 (44). The total is composed of fish, peanuts, sunflower seeds and rapeseed meals in addition to soybean meal. The absolute amount of each of fish, peanuts, sunflower seeds and rapeseeds have not changed much in recent years with almost all increases attributable to soybeans.

yield soybeans have been found. Minimal attention is directed towards an efficacious fertilizer N system documented by a single presentation at this meeting (26) rather than a session on this approach. Although multiple foliar applications of a mixture of N-P-K-S during pod fill have failed to reproduce consistently the substantial yield increase observed initially, the approach is attractive and worthy of much effort to discover a useful system. Since a combination of technologies for enhancing biological N_2 fixation and an efficacious use of fertilizer N may be the best solution, joint consideration of both approaches is needed (3,10,25,40,65).

STATUS OF KNOWLEDGE ON BIOLOGICAL N_2 FIXATION BASED ON A WHAT'S WRONG WITH THE NATURAL PROCESS APPROACH

Knowledge of biological N_2 fixation at the biochemical, genetic, biological and agronomic level has expanded rapidly during the past two decades and current activity suggests that this trend will continue for at least the next several years. Somewhat surprisingly, the knowledge for the agriculturally important *Rhizobium*-legume system is less developed than that for the relatively agriculturally insignificant *Clostridium, Azotobacter,* and *Klebsiella.* However, the existence of a single nitrogenase with only small variations may support extension of some of the fundamental information to *Rhizobium* from that for the naturally free-living diazotrophs, especially microaerphiles.

During recent years we have tabulated the knowledge (27,29,37) on N_2 fixation and CO_2 fixation in terms of limitations or what's wrong with the naturally evolved system for maximum crop productivity. Chemical, genetic, physical or cultural corrections of the most significant of these "what's wrongs" should enable the contribution of N_2 fixation for crop production to increase. In Table 2 a tabulation is provided of 81 "what's wrongs" for N_2 fixation by soybeans at the technical level—mathematic, chemical, biochemical, genetic, biological and agronomic—and two "what's wrong's" at the policy level—proprietariness and quality. The composition of this list especially in the technical area will change as the knowledge base from which it is derived expands. The significance of the individual "what's wrongs" varies from highly significant to insignificant. It is important that the most significant opportunities receive greatest attention. The possibility of providing solutions to the "what's wrongs" also varies from highly probable to improbable based on currently available solution-type techniques, and constraints of practicality, safety, and implementability.

Some general comments will be made on selected "what's wrongs." It would seem desirable to develop mathematical models for steps in the inoculation, rhizosphere colonization, infection, and nodule development for *Rhizobium*—legume symbioses as well as for others such as *Azospirillum*-grass association. An example of such an approach is that used recently for the *Rhizobium*-cereal root association in which a modified Langmuir absorption isotherm fitted the data (63). Chemical effectors of the *Rhizobium*-soybean

Table 2. What's wrong with biological N_2 fixation in soybeans.

I. Technical

A. Mathematical

No treatment for *Rhizobium*-legume although one exists for *Rhizobium*-cereal

B. Chemical

No chemical effectors

C. Biochemical

Minimal variation of nitrogenase from all sources examined although variable complementariness of parts; large enzymes; two component enzyme—Fe (nitrogenase reductase) and Mo-Fe protein (nitrogenase); complex Fe-Mo cofactor; processing of precursor proteins and insertion of Fe-Mo cofactor; optimum component ratio; allosteric component ratio; O_2 lability of each component; O_2 inhibition of reaction; special O_2-handling molecules and systems, e.g., LHb; temperature instability; biphasic arrhenius plot with high apparent activation energy at $<$ 18 to 20 C; Mo, Fe, and S content; systems for uptake and storage of Mo and Fe; low turnover; intermediates—dinitrogen hydride(?); electrons for reduction; special electron donors—Fd and Fld; source of electrons—isocitric dehydrogenase for bacteroids(?); high redox potential (NADPH/

Table 2. (Continued)

NADP > 100); high direct ATP requirement (4 ATP's/2e); ADP inhibition; high ATP/ADP ratio > 10; high ATP-generating capacity; substrate promiscuity in H_2O-H^+ reduction; uptake hydrogenase needed; regulating molecules for nitrogenase and process; inhibition by NO_3^- and/or NO_3^- product; ammonia exporting system; special NH_3 incorporating system—Gln, Glu, Ala; multiple forms of glutamine synthetase and role in regulation; allantoin formation; associated molecules for development of symbioses—lectins, cell wall polysaccharides; accumulation of β-hydroxybutyrate polymer; CH_2O substrate used by bacteroid(?); and CO_2 refixation by PEP carboxylase—significance.

D. Genetic

No useful variation in genes for structural proteins; large number of genes for *nif* (14); large size of above genes (30 Kbases); prokaryotic limitation; plasmid or chromosomal location of *nif* in *Rhizobium;* other genes in addition to *nif* for symbioses; both host (e.g., LHb) and bacteroid genes; communication between bacteroid and host(?); higher DNA content in bacteroids than bacterium; multiple copies of cDNA for LHb; mRNA for LHb larger than required; only part of *nif* genes inserted into plasmid; circular linkage map in *Rhizobium* but not for *nif;* similarity of *Agrobacterium* and *Rhizobium* with movement of Ti plasmic from *A.* to *R.;* regulation of nitrogenase expression—glutamine synthetase(?); minimal information on the identification and location of genes for symbioses; optimum matching of host and bacteroid genes; and claims of super mutants(?)

E. Biological and Agronomic Limitations

Low activity of isolated bacteroids; asymbiotic fixation by only some *Rhizobium* and fixation is usually much less than in symbiosis; low pO_2 required for asymbiotic fixation but adaptability from > 5% O_2 for nodulated plant; membrane to separate host and *Rhizobium;* fast growers mutate to slow growers(?); fixed N (NO_3^-, NH_4^+) inhibits infection, nodulation and induces senescence; denitrification in *Rhizobium,* Bacteroids (?); role of capsular, lipo- and exopolysaccharide in specification; essentiality of lectins (plant) for infection(?); cellulase (host) and pectinase (bacteria) for cell wall dissolution; evolution of H_2 by several *Rhizobium;* high and inefficient use of carbohydrate (10 kg CH_2O/kg N_2 fixed); photosynthetic inefficiency of soybeans; time profile of N_2 fixation does not match need; premature senescence; inadequate amount for high yield soybeans; inhibition by fixed N as occurs in high fertility soils; high cost for synthesis and maintenance of nodules; need for a multiplicity of strains dependent on cultivar, soil, climate; problem of manufacture, storage, handling, and application of labile *Rhizobium;* competition between applied and endogenous *Rhizobium;* emphasis on greater competitiveness may be undesirable; and measurement techniques—indirect vs. direct—kinetic vs. integrative.

II. Policy

A. Proprietariness

Inadequate proprietariness to encourage adequate investment in exploration, development, and implementation of solutions to overcome the key limitations.

B. Quality

Quality control of product and use.

symbiosis have not been reported. They could provide useful probes as well as possible beneficial plant growth regulators.

Thirty-six "what's wrongs" are listed for the biochemical level. Unfortunately, there is only minimal variation in nitrogenase from different sources. All nitrogenases are composed of an Fe protein (nitrogenase reductase) and an Mo-Fe protein (nitrogenase or dinitrogen reductase (24) (Figure 2) with a common general mechanism. This information suggests little useful genetic variability at least for the structural components of nitrogenase. The excessive and unnecessary energy requirement of nitrogenase is a major biochemical "what's wrong." The substrate promiscuity in which protons are reduced concomitantly to H_2 along with N_2 to $2NH_3$ worsens further the energy waste and necessitates another enzyme system for uptake and reutilization of H_2 to conserve some of this waste. These aspects of biological energy cost and H_2 uptake will be discussed below. Both large and small molecules that regulate synthesis, proteolysis, and activity of nitrogenase and associated reactions are understood inadequately (47-49,58, 62). Furthermore, NO_3^- or a product of its reduction inhibit nitrogenase activity.

Eighteen "what's wrongs" are identified based on information of the genetics. Knowledge of this area is in an exponential phase and undoubtedly the limitations will expand in step with the knowledge. Almost all of the molecular genetic information is based on *Klebsiella* (17,46,50,62) with only recent studies initiated on *Rhizobium* (6-9,12,45,49,55) so that the juvenility

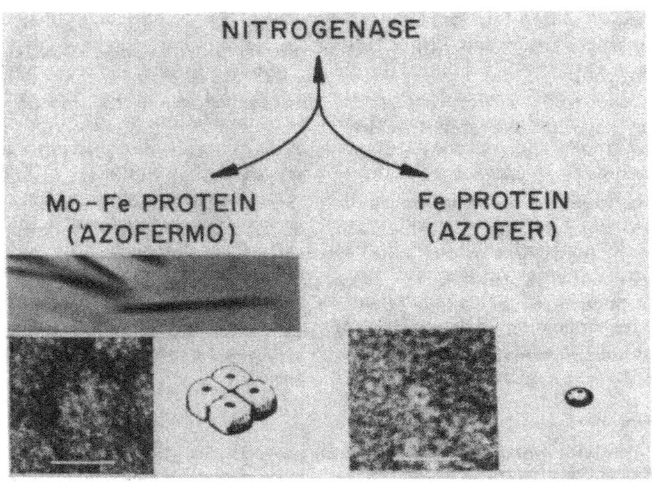

Figure 2. **Nitrogenase and its component proteins—the Fe protein and the Mo-Fe protein showing light micrographs of crystalline *Azotobacter* Mo-Fe protein and electron micrographs and models of the Mo-Fe and Fe proteins (34).**

of the information on *Rhizobium* is a major limitation. Undoubtedly, this will be changed by World Soybean Research Conference III. Recent research has identified 14 *nif*, or N_2 fixation associated genes occupying a continuous area of 30 kbases of the *Klebsiella* chromosome (Figure 3) (59).

Three genes are related to structural proteins—one for the Fe protein subunit and two for the two types of subunits of the Mo-Fe protein. Three are related to the Fe-Mo cofactor. Two or three may be involved in processing the precursor proteins for the structural proteins of nitrogenase. Up to five genes may be involved in regulation suggesting a complex regulatory system for N_2 fixation. No useful variation is anticipated in the genes for the structural proteins based on biochemical information. Regulation and other genes could vary and there will be several *Rhizobium* and soybean genes involved in the steps of specification, infection, and development of the N_2-fixing symbiosis (Figure 4) in which up to 10,000 microsymbionts containing nitrogenase will occur in each host cell.

Twenty-three "what's wrongs" are listed for the biological and agronomic area. Asymbiotic N_2 fixation by *Rhizobium* is at this stage limited to only some *Rhizobium (Rhizobium japonicum* asymbiotically fix N_2 under specific culture conditions) and is usually of low activity so that a useful quantitative evaluation of a *Rhizobium* can only be made by formation of the *Rhizobium*-whole legume plant symbiosis (1,20,31,42,69) and a meaningful evaluation will require an integrated measurement over the whole season in

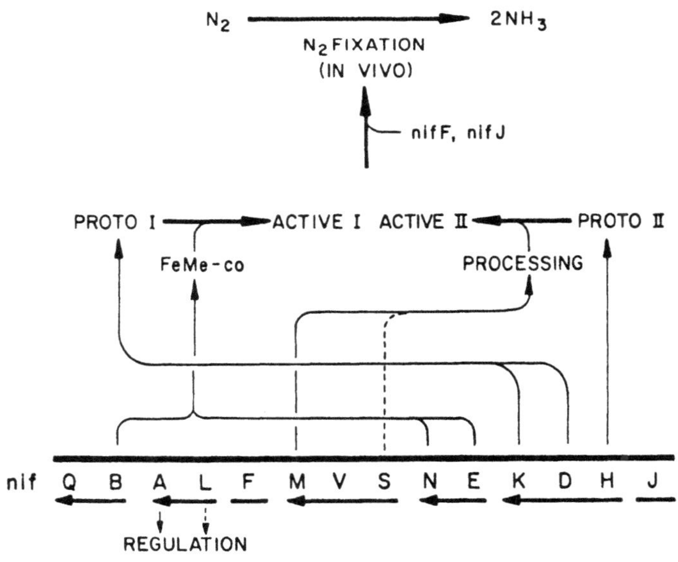

Figure 3. The 14 *nif* genes of *Klebsiella* (59).

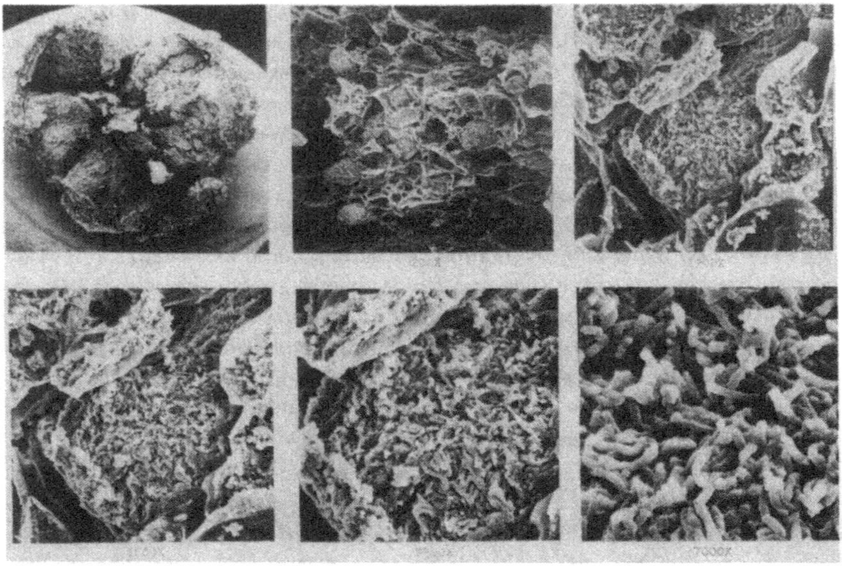

Figure 4. **Scanning electron micrograph of a freeze-cleaved soybean nodule at a low magnification showing one-half of the nodule to high magnifications which which show the up to 10,000 *Rhizobium* bacteroids within each host cell. Within these bacteroids large amounts of energy are consumed and N_2 is reduced to NH_3 as well as protons to H_2 (27).**

a field situation rather than claims for supermutants made on limited or one-time growth room measurements (53). High levels of combined N especially NO_3^- such as occur in high fertility soils inhibit the biological N_2 fixation process.

The *Rhizobium*-soybean symbiosis requires about 10 kg carbohydrate/kg of N_2 fixed documenting the high energy requirement of this process. This high energy requirement for N_2 fixation in legumes such as soybeans is coupled unfortunately with a C_3 or photosynthetically inefficient system. The time-course of nitrogen input by N_2 fixation does not match the N need of the soybean during later pod fill since N fixation often declines prematurely about mid pod fil (35). Optimization of *Rhizobium* for a variety of conditions—host cultivars and environmental factors of temperature and soil such as pH and ions (54) will require a multiplicity of strains. The problems of manufacture, distribution, and application of a labile microorganism further complicate the effective use of *Rhizobium.* Competitiveness is an important factor since endogenous soil strains often outcompete applied *Rhizobium* in nodule formation (16). A solution to the competitiveness problem has not been found. It may be desirable to develop *Rhizobium* with built-in lethality coupled with elimination of endogenous *Rhizobium* so that improved applied

Rhizobium will not have to compete with endogenous ones that have been selected because of their increasing competitiveness. Alternatively, highly competitive endogenous *Rhizobium* may be upgraded genetically so as to improve them, such as incorporation of genes for H_2 reutilization (67).

There are policy as well as technical limitations for the *Rhizobium*-soybean symbiosis. There is inadequate proprietariness to justify the substantial private sector research required to produce order of magnitude improvements for some of the "what's wrongs" by genetic approaches. Although composition of matter patents may become a reality in the U.S., e.g., the Bergey selected strain for improved antibiotic production and the Chakrabarty strain genetically engineered by plasmid modification for oil degradation, the proprietariness for a much improved soil microorganism such as *Rhizobium* would probably be of little effectiveness. Quality of *Rhizobium* inoculum is also a policy problem. Establishment of a repository for *Rhizobium* by the USDA at Beltsville is a positive policy step to preserve and use existing germ plasm. Evaluation of each *Rhizobium* is important.

CRITICAL SUBSETS OF THE N_2 FIXATION SYSTEM

Photosynthesis, translocation, nodular production of ATP, reductant and carbon skeletons, N_2 fixation, H_2 reutilization, and other aspects are critical subsets of the N_2 fixation system. Substantial and convincing data have accumulated to document the importance of photosynthesis for N_2 fixation by soybeans (35,66). Decreasing photorespiration by altered CO_2/O_2 ratio of the aerial atmosphere increases N_2 fixation within 24 hr when the ratio is increased above atmosphere by CO_2 enrichment or O_2 depletion while N_2 fixation is decreased by elevated O_2 (18,32). Field experiments utilizing CO_2 enrichment increase substantially N_2 fixation by soybeans and produce up to a 98% increase in yield (Table 3) (19,35,36). In addition to photorespiration, other reactions including starch synthesis and remobilization and regulatory aspects may offer opportunities to increase photosynthesis for increased N_2 fixation (11,22,64,68).

Table 3. Yield of field-grown soybeans grown under ambient and 1500 ppm CO_2 (37).

	Dry Matter Yield (kg/ha)		
Component	**Ambient**	**1500 ppm CO_2**	**% Increase**
Total dry matter[a]	9,373	15,504	68
Seed	3,003	5,946	98

[a]Leaves, stems, pods, seeds, and roots.

The requirement for ATP and a reductant equivalent to the H_2 electrode at pH 7 by isolated nitrogenase suggested several years ago that N_2 fixation is energy intense. A number of approaches have been used in recent years to estimate the actual cost of the *Rhizobium*-legume symbiosis (5,14,23,51,52, 60) and values from 5 to 20 kg of carbohydrate/kg N_2 fixed have been reported. Recently, we have used a comparison of root or nodulated root respiration by a nodulated and a non-nodulating Clark isoline of soybean (41). The respiration rate as measured by CO_2 production per mass for nodulated roots was 2 to 4 times that for non-nodulated roots. The specific nodulated root respiration varied with N_2-fixing activity so that the calculated cost of N_2 fixation decreased from a high of 100 kg glucose/kg N_2 fixed at high rates of N_2 fixation. These data suggest that variability in the biological cost of N_2 fixation reported by different groups may be attributed at least in part to the rate of N_2 fixation. A corollary is that the incremental biological cost in fixing N_2 at a high basal rate of activity is much less than at a lower basal rate.

The above data suggest the use of 280 g or 1.55 moles of glucose to produce about 60 ATP equivalents to fix 28 g or 1 mole of N_2 in the most efficient case. In vitro measurements on nitrogenase suggest that 28 of these 60 ATP equivalents are used for N_2 fixation and concomitant H^+ reduction to H_2, e.g.,

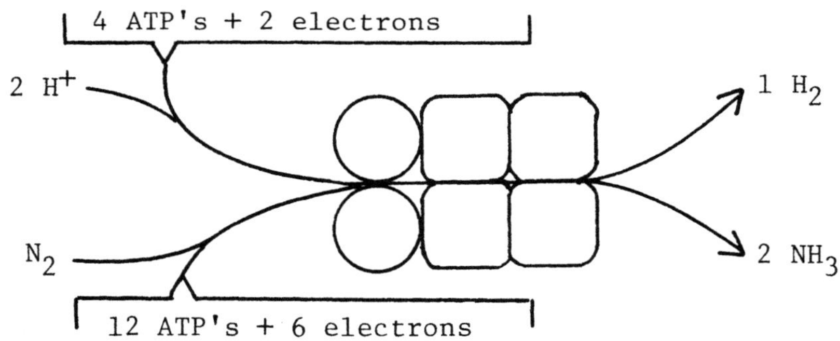

with each pair of electrons being equivalent to 3 ATP's. No approaches have been discovered that reduce or eliminate this wasteful reduction of protons although its elimination should increase N_2 fixation by 30% (36). This objective should still be pursued. The less effective solution based on net energy use is to reutilize the H_2 through a hydrogenase variously referred to as an uptake or reutilization or "after burner" type which enables oxidation of H_2 to H_2O with the generation of possibly 2 to 3 ATP's. Note that 7 ATP equivalents were consumed in production of the H_2 so that most of the energy is lost. Considerable attention has been focused on H_2 uptake (2,21, 61). Increases in dry wt and N content of soybeans nodulated with *Rhizobium* that possess H_2 uptake vs. no H_2 uptake capacities have been found (Table 4).

These increases were obtained in N-free greenhouse experiments while those so far observed in field situations may be about 5%. These results suggest that all *Rhizobium* should be assessed for uptake hydrogenase and uptake hydrogenase capabilities should be incorporated in all *Rhizobium* (67). In the case of some soybean *Rhizobium* this may have been done in many cases on an empirical basis since those *Rhizobium* already selected as better ones often possess uptake hydrogenase activity.

Of the 60 ATP equivalents used to fix 28 g N_2, about 30 are used for synthesis and maintenance of the system as well as import and export of nodular substrates and products. The efficiency of the N_2-fixing system may be improved by economization in this area which is understood only poorly at this time.

POSSIBLE FUTURE TECHNOLOGIES

A tabulation proceeding from the near term most probable technologies for N input into soybeans to the long-term technologies is presented in Table 5. On the short term, multiple cropping of legumes and nonlegumes, effective rhizobial nodulation technology, *Rhizobium* with H_2 uptake genes, N-fertilizer responsive systems, and *Rhizobium* and/or soybeans insensitive to fixed nitrogen are reasonable possibilities. In the intermediate term, increased photosynthate available to the nodule and use of regulatory chemicals are possibilities. In the longer term, improved nodular efficiency such as elimination of H_2 evolution, nitrogen fertilizer manufacture by a process

Table 4. Significance of hydrogenase in dry matter accumulation and N_2 fixation by nodulated soybeans grown on N-free media (2).

R. japonicum	Rel. Eff.	Plant Dry Wt	Total Nitrogen
		– g –	*– g –*
Expt. 1[a]			
H_2-Uptake + ve	0.99	29.5	.94
H_2-Uptake – ve	0.69	25.5	.74
Exp. 2[b]			
H_2-Uptake + ve	0.99	13.6	.45
H_2-Uptake – ve	0.81	10.3	.30

[a] Expt. 1 – H_2-Uptake + ve: USDA 136, 100, 122; 311b-143; 311b-6; H_2-Uptake – ve: USDA 16, 135, 120, 117, 3. Grown on N-free media and harvested at 70 days.

[b] Expt. 2 – H_2-Uptake + ve: USDA 122 (DES), SR; H_2-Uptake – ve: SRI, SR2, SR3. Grown on N-free media and harvested at 50 days.

Table 5. Possible future technologies for providing nitrogen for soybeans.

Multiple cropping of legumes and nonlegumes.
Effective technology for Rhizobial inoculation.
Nitrogen fertilizer responsive soybeans.
Nitrogen fertilizer or system for soybeans.
Plant growth regulators to improve compatibility of high N fertilizer and N_2 fixation.
Rhizobium with H_2 uptake genes.
Rhizobial mutants and/or soybeans in which nodulation and subsequent steps are insensitive to fixed nitrogen.
Increased photosynthate available to nodule, e.g., plant growth regulator to inhibit photorespiration, deregulate photosynthesis, or alter assimilate partitioning or soybean mutants with similar characteristics.
Improved efficiency of nodule.
Synthetic nitrogen fertilizers by zero-direct energy input process.
Incorporation of *nif* genes in soybeans.
Synthetic genes that code for small, stable (O_2 and temperature insensitive), high turnover, absolute substrate specificity (No H^+ reduction), zero-direct energy requiring (No ATP) N_2-fixing enzyme with appropriate repression by fixed nitrogen.

requiring zero direct-energy input and possibly transfer of *nif* genes to soybeans may be possible. In the longest term, the construction of synthetic genes that code for a N_2-fixing enzyme without the "what's wrongs" of the naturally evolved nitrogenase may be achieved.

SUMMARY

Annual world soybean production is projected to increase to 275 Mt by 2000 A.D. from 70 Mt in 1975 with a concomitant increase to 27 Mt from 7 Mt in use of nitrogen. This nitrogen will be provided by a combination of enhanced biological N_2 fixation and use of N fertilizer. Eighty-one technical "what's wrongs" and two policy "what's wrongs" are identified for N_2 fixation by the *Rhizobium*-soybean symbiosis with several suggested to represent significant opportunities to improve the process. Critical subsets of the N_2 fixation system which include photosynthesis, translocation, nodular production of energy, reductant and carbon skeletons, N_2 fixation including concomitant H_2 production, H_2 reutilization and other reactions are described. Alteration of photosynthetic rate by alteration of aerial CO_2/O_2 ratio produces parallel changes in N_2 fixation. The biological cost of N_2 fixation by comparison of nodulated and non-nodulated isolines varies from 100 to 10 kg glucose/kg N_2 fixed with the highest efficiency occuring at the highest rates of N_2 fixation. This overall cost of 10 kg glucose/kg N_2 fixed is about double the in vitro cost of nitrogenase. Elimination of H_2 production by nitrogenase is most desirable but incorporation of H_2 reutilization capabilities into *Rhizobium* appears to be of some benefit until the former can be accomplished.

NOTES

R. W. F. Hardy, U. D. Havelka, and P. G. Heytler, Central Research and Development Department, Experimental Station, E. I. du Pont de Nemours and Company, Wilmington, Delaware 19898.

REFERENCES

1. Abu-Shakra, S. S., D. A. Phillips, and R. C. Huffaker. 1978. Nitrogen fixation and delayed leaf senescence in soybeans. Science 199:973-974.
2. Albrecht, S. L., R. L. Maier, F. J. Hanus, S. A. Russell, D. W. Emerich, and H. J. Evans. 1979. Hydrogenase in *Rhizobium japonicum* increases nitrogen fixation by nodulated soybeans. Science 203:1255-1257.
3. Anderson, J. B., J. E. Harper, and R. H. Hageman. 1979. Contribution of nitrate and dinitrogen to total plant nitrogen and seed protein in soybeans, p. 55. In F. T. Corbin, (Ed.), Abstracts World Soybean Res. Conf.—II, N.C. State Univ., Raleigh, NC.
4. Anonymous. 1976. 1975 Production Yearbook, Vo. 29, Food and Agricultural Organization of the United Nations, Rome, Italy.
5. Atkins, C. A., D. F. Herridge, and J. S. Pate. 1978. The economy of carbon and nitrogen in nitrogen-fixing annual legumes: Experimental observations and theoretical considerations, pp. 211-242. In Isotopes in Biological Dinitrogen Fixation, International Atomic Energy Agency, Vienna, Austria.
6. Beringer, J. E., J. L. Beynon, A. V. Buchanan-Wollaston, and A. W. B. Johnston. 1978. Transfer of drug-resistance transposon Tn5 to *Rhizobium*. Nature 276: 633-634.
7. Beringer, J. E., S. A. Hoggan, and A. W. B. Johnston. 1978. Linkage mapping in *Rhizobium leguminosarum* by means of R palsmid-mediated recombination. J. Gen. Microbiol. 104:201-207.
8. Beringer, J. E., and D. A. Hopwood. 1976. Chromosomal recombination and mapping in *Rhizobium leguminosarum*. Nature 264:291-293.
9. Beringer, J. E. and A. W. B. Johnston. 1978. The genetics of the *Rhizobium*-legume symbiosis, pp. 27-39. In Isotopes in Biological Dinitrogen Fixation, International Atmoic Energy Agency, Vienna, Asutria.
10. Bethlenfalvay, G. J., S. S. Abu-Shakra, and D. A. Phillips. 1978. Interdependence of nitrogen nutrition and photosynthesis in *Pisum sativum* L. Plant Physiol. 62: 127-120.
11. Bethlenfalvay, G. J., and D. A. Phillips. 1977. Ontogenetic interactions between photosynthesis and symbiotic nitrogen fixation in legumes. Plant Physiol. 60: 419-421.
12. Bishop, P. E., F. B. Dazzo, E. R. Appelbaum, R. J. Maier, and W. J. Brill. Intergeneric transfer of genes involved in the *Rhizobium*-legume symbiosis. Science 198:938-940.
13. Brill, W. J. 1977. Biological nitrogen fixation. Scientific American, March, pp. 68-81.
14. Broughton, W. J. 1979. Effect of light intensity on net assimilation rates of nitrate-supplied on nitrogen-fixing legumes. In H. Clysters, R. Marcelle, and M. van Poucke (Eds.) Photosynthesis and Plant Development. Dr. W. Junk, The Haag, The Netherlands (In press).
15. Burns, R. C., and R. W. F. Hardy. 1975. Nitrogen fixation in bacteria and higher plants. Springer Verlag, New York, NY.

16. Burton, J. C. 1980. Rhizobia and soybean production, pp. 89-100. In F. T. Corbin (Ed.) World Soybean Research Conf.—II, Proceedings, Westview Press, Boulder, CO.
17. Connor, F. C., G. E. Riedel, and F. M. Ausubel. 1977. Recombinant plasmid that carries part of the nitrogen fixation (*nif*) gene cluster of *Klebsiella* pneumoniae. Proc. Natl. Acad. Sci. U.S. 74:2953-2967.
18. Criswell, J. G., and R. W. F. Hardy, unpublished results.
19. Havelka, U. D., and R. W. F. Hardy. 1976. $N_2[C_2H_2]$ fixation growth, and yield response of field-grown peanut (*Arachis hypogaea* L.) when grown under ambient and 1500 ppm CO_2 in the foliar canopy. Agron. Abstr., p. 72.
20. Devine, T. E., and D. F. Weber. 1977. Genetic specificity of nodulation. Euphytica 26:527-535.
21. Evans, H. J., T. Ruiz-Argueso, N. Jennings, and J. Hanus. 1977. Loss of energy during the fixation of atmospheric nitrogen by nodulated legumes, pp. 61-76. In Report on the public meeting on genetic engineering for nitrogen fixation, U.S. Government Printing Office, Washington, D.C.
22. Giaquinta, R. T. 1977. Possible role of pH gradient and membrane ATPase in the loading of sucrose into the sieve tubes. Nature 267:369-370.
23. Gibson, A. H. 1966. The carbohydrate requirement for symbiotic nitrogen fixation: A "wholeplant" growth analysis approach. Aust. J. Biol. Sci. 19:499-515.
24. Hageman, R. V., and R. H. Burris. 1978. Nitrogenase and nitrogenase reductase associate and dissociate with each catalytic cycle. Proc. Natl. Acad. Sci. U.S. 75:2699-2702.
25. Ham, G. E. 1978. Use of ^{15}N in evaluating symbiotic N_2 fixation of field-grown soybeans, pp. 151-162. In Isotopes in biological dinitrogen fixation, International Atomic Energy Agency, Vienna, Austria.
26. Hanway, J. J. 1980. Foliar fertilization of soybeans, pp. 409-416. In F. T. Corbin (Ed.) World Soybean Research Conf.—II, Proceedings, Westview Press, Boulder, CO.
27. Hardy, R. W. F. 1977. Increasing crop productivity:agronomic and economic considerations on the role of biological nitrogen fixation, pp. 77-106. In Report of the public meeting on genetic engineering for nitrogen fixation, U.S. Government Printing Office, Washington, D.C.
28. Hardy, R. W. F. 1978. Food, famine, and nitrogen fixation, pp. 224-237. In 1979 Yearbook of Science and the Future, Encyclopaedia Britannica, Inc., Chicago, IL.
29. Hardy, R. W. F. 1979. Chemical plant growth regulation in world agriculture, pp. 165-206. In T. K. Scott (Ed.) Plant Regulation and World Agriculture, Plenum Press, New York, NY.
30. Hardy, R. W. F., F. Bottomley, and R. C. Burns (Eds.). 1979. A treatise on dinitrogen fixation. Inorganic and physical chemistry and biochemistry, John Wiley & Sons, Inc., New York, NY.
31. Hardy, R. W. F., R. C. Burns, and R. D. Holsten. 1973. Applications of the acetylene-ethylene assay for measurement of nitrogen fixation. Soil Biol. Biochem. 5: 47-81.
32. Hardy, R. W. F., J. G. Criswell, and U. D. Havelka. 1977. Investigations of possible limitations of nitrogen fixation by legumes: (1) Methodology, (2) Identification, and (3) Assessment of significance, pp. 451-467. In W. Newton, J. R. Postgate, and C. Rodriguez Barrueco (Eds.) Recent developments in nitrogen fixation, Academic Press, London.
33. Hardy, R. W. F., and A. H. Gibson (Eds.). 1977. A treatise on dinitrogen fixation. Agronomy and ecology, John Wiley & Sons, Inc., New York, NY.

34. Hardy, R. W. F., and U. D. Havelka. 1975. Nitrogen fixation research: a key to world food? Science 188:633-643.
35. Hardy, R. W. F., and U. D. Havelka. 1976. Photosynthate as a major factor in limiting nitrogen fixation by field-grown legumes with emphasis on soybeans, pp. 421-439. In P. S. Nutman (Ed.) Symbiotic nitrogen fixation in plants, Cambridge University Press, London.
36. Hardy, R. W. F., and U. D. Havelka. 1977. Possible routes to increase the conversion of solar energy to food and feed by grain legumes and cereal grains (crop production): CO_2 and N_2 fixation, foliar fertilization, and assimilate partitioning, pp. 299-322. In A. Mitsui, S. Miyachi, A. San Pietro, and S. Tamura (Eds.) Biological Solar Energy Conversion, Academic Press, New York, NY.
37. Hardy, R. W. F., U. D. Havelka, and B. Quebedeaux. 1978. The opportunity for and significance of alteration of ribulose 1,5-bisphosphate carboxylase activity in crop production, pp. 165-178. In H. W. Siegelman, and G. Hind (Eds.) Photosynthetic carbon assimilation, Plenum Press, New York, NY
38. Hardy, R. W. F., and W. S. Silver (Eds.). 1977. A treatise on dinitrogen fixation. Biology. John Wiley & Sons, Inc., New York, NY.
39. Havelka, U. D. and R. W. F. Hardy. 1977. Research on nitrogen and carbon input to increase domestic crop protein production, pp. 204-235. In M. Milner, N. S. Scrimshaw, and D. I. C. Wang (Eds.) Protein resources and technology: status and research needs, AVI Publishing Co., Inc., Westport, CN.
40. Havelka, U. D., and R. W. F. Hardy. 1977. Response of nodulating and non-nodulating isolines of clark soybeans as affected by CO_2 enrichment and/or soil or foliar nitrogen fertilization. Agron. Abstr., p. 86.
41. Heytler, P. G., and R. W. F. Hardy. 1979. Energy requirements for nitrogen fixation by rhizobial nodules in soybeans. Plant Physiol. Abstr. (In press).
42. Holl, F. B. 1975. Host plant control of the inheritance of dinitrogen fixation in the *Pisum-Rhizobium* symbiosis. Euphytica 24:762-770.
43. Hollaender, A., and R. H. Burris, P. R. Day, R. W. F. Hardy, D. R. Helinski, M. R. Lamborg, L. Lowens, and R. C. Valentine. 1977. Genetic engineering for nitrogen fixation. Plenum Press, New York, NY.
44. Holtz, A. 1979. World oilseed and meal output to be up in 1979, but rate of gain will be less. Foreign Agriculture, Jan. 8, pp. 2-4.
45. Hooykass, P. J. J., P. M. Klapwijk, M. P. Nuti, R. A. Schelperoort, and A. Rorsch. 1977. Transfer of the *Agrobacterium tumefaciens* T1 plasmid to avirulent agrobacteria and to *Rhizobium ex planta.* J. Gen. Microbiol. 98:477-484.
46. Kennedy, C. 1977. Linkage map of the nitrogen fixation (*nif*) genes in *Klebsiella pneumoniae.* Mol. & Gen. Genet. 157:199-204.
47. Lim, S. T., H. H. Hennecke, and D. B. Scott. 1979. Regulation of symbiotic nitrogen fixation: Guanosine-3',5'-cyclic monophosphate as a modulator of nitrogenase biosynthesis in *Rhizobium japonicum*. (In press).
48. Lim, S. T., and K. T. Shanmugam. 1979. Regulation of hydrogen utilization in *Rhizobium japonicum* by cyclic AMP. Biochem. Biophys. Acta. (In press).
49. Ludwig, R. A., and E. R. Signer. 1977. Glutamine synthetase and control of nitrogen fixation in *Rhizobium.* Nature 267:245-247.
50. MacNeil, D., T. MacNeil, and W. J. Brill. 1978. Genetic modifications of N_2-fixing systems. BioScience 28:576-579.
51. Mahon, J. D. 1977. Respiration and the energy requirement for nitrogen fixation in nodulated pea roots. Plant Physiol. 60:817-821.
52. Mahon, J. D. 1977. Root and nodule respiration in relation to acetylene reduction in intact nodulated peas. Plant Physiol. 60:812-816.

53. Maier, R. J., and W. J. Brill. 1978. Mutant strains of *Rhizobium japonicum* with increased ability to fix nitrogen for soybean. Science 201:448-450.
54. Munns, D. N. 1980. Mineral nutrition and nodulation, pp. 47-56. In F. T. Corbin (Ed.) World Soybean Research—II, Proceedings, Westview Press, Boulder, CO.
55. Nuti, M. P., A. M. Ledeboer, A. A. Lepidi, and R. A. Schelperoort. 1977. Large plasmids in different *Rhizobium* species. J. Gen. Microbiol. 100:241-248.
56. Nutman, P. S. (Ed.) 1976. Symbiotic Nitrogen Fixation in Plants, Cambridge University Press, London.
57. Newton, W., J. R. Postgate, and C. Rodriguez-Barruenco (Eds). 1977. Recent Developments in Nitrogen Fixation, Academic Press, London.
58. O'Gara, F., and K. T. Shanmugam. Regulation of nitrogen fixation in *Rhizobium* spp. Isolation of nutants of *Rhizobium trifolii* which induce nitrogenase activity. Biochem. Biophys. Acta 500:277-290.
59. Roberts, G. P., T. MacNeil, D. MacNeil, and W. J. Brill. 1978. Regulation and characterization of protein products coded by the *nif* (nitrogen fixation) genes of *Klebsiella pneumoniae.* J. Bacteriol. 136:267-279.
60. Ryle, G. J. A., C. E. Powell, and A. J. Gordon. 1978. Effect of source of nitrogen on the growth of Fiskeby soya bean: The carbon economy of whole plants. Ann. Bot. 42:637-648.
61. Schubert, K. R., and H. J. Evans. 1976. Hydrogen evolution: A major factor affecting the efficiency of nitrogen fixation in nodulated symbionts. Proc. Natl. Acad. Sci. U.S. 73:1207-1211.
62. Shanmugam, K. T., and R. C. Valentine. 1975. Molecular biology of nitrogen fixation. Science 187:919-924.
63. Shimshick, E. J., and R. R. Hebert. 1978. Adsorption of rhizobia to cereal roots. Biochem. Biophys. Res. Commun. 84:736-742.
64. Siegelman, H. W., and G. Hind (Eds.) 1978. Photosynthetic carbon assimilation, Plenum Press, New York, NY.
65. Sorensen, R. C., and E. J. Penas. 1978. Nitrogen fertilization of soybeans. Agron. J. 70:213-216.
66. Streeter, J. G., H. J. Mederski, and R. A. Ahmad. 1980. Photosynthesis and N_2 fixation, pp. 129-138. In F. T. Corbin (Ed.) World Soybean Research Conf.—II, Proceedings, Westview Press, Boulder, CO.
67. Valentine, R. C. Personal communication.
68. Zelitch, I. 1979. Photosynthesis and plant productivity. Chem. and Eng. News 57(6):28-48.
69. Zobel, R. W. 1980. Rhizogenetics of soybeans, pp. 73-88. In F. T. Corbin (Ed.) World Soybean Research Conf.—II, Proceedings, Westview Press, Boulder, CO.

RHIZOGENETICS OF SOYBEANS

R. W. Zobel

"Rhizogenetics" is the genetic study of below ground plant organs, i.e., roots and nodules. There is an abject paucity of literature on this subject, especially in reference to soybean. Several laboratories around the world are currently attempting to rectify this situation, with a considerable amount of success. A paper by H. M. Taylor in these proceedings will discuss several aspects of rhizogenetics with emphasis on roots, especially tap roots. There have been several studies attempting to describe the structure and pattern of soybean root systems (13,9,16). Unfortunately, these studies have utilized very few differing varieties, and therefore cannot be used to generalize for the species.

Nodulation and nitrogen fixation are not significantly better off than roots. Although many studies have been initiated to explore the physiology and biochemistry of nodulation and nitrogen fixation from the plants point of view, very few have compared different varieties, therefore severely weakening generalizations which have been attempted. In addition, genetic studies of soybean nitrogen fixation pale when compared to the work of Nutman (10) with clover or even more recently that of Holl (7) and his co-workers with field peas. 'Detailed' genetic studies have been presented for the rj (non-nodulating) mutant series, but these do not give any real insight to the role of the plant in symbiosis. Recent work at the USDA nitrogen fixation laboratory (3) is remedying some of this, but much more is needed, and with host variants of other types.

Contrasted to the situation with the plant host is the scientific base for information on the bacterial symbiont. Where host research is hampered by length of growth cycle, the availability of high precision genetic tools, field, laboratory, and greenhouse space, and most important personnel deficiencies, that of the rhizobium is certainly stimulated. The tools of recombinant DNA, coupled with standard microbiological techniques allow this phase of the research to progress with comparative speed. Bacteriological results demonstrate that most of the characteristics of nodulation and nitrogen fixation are controlled at least indirectly by the bacterial symbiont. One example is the evolution of hydrogen which is controlled apparently by the presence of a hydrogenase in the bacteriods (14).

Is the situation bleak, or are we nearing a point of rapid advancement in soybean rhizogenetics? To get an answer, we need to look at what we know physiologically and genetically, and where this information leads us.

PATTERNS OF DEVELOPMENT

Physiological

Several laboratories have demonstrated the presence of an interesting pattern in nitrogen assimilation: as nitrogen fixation begins, nitrate reduction drops off rapidly. Thibodeau and Jaworski (17) confirmed this phenomena and demonstrated that in Missouri the cultivar Wayne, when grown under field conditions, apparently stops nitrate reduction and simultaneously begins nitrogen fixation during early podfill. The timing of this phenomena agrees fairly closely with that of Harper and Hageman (5) and also with the nitrogen fixation data of Hardy et al. (5). Thibodeau and Jaworski used the same cultivar as Hardy who studied field grown material in Delaware (cv. Wayne). Harper and Hageman, on the other hand, used cv. Beeson in a modified field/hydroponics environment. The pedigrees of these two lines are unfortunately very similar.

Skrdleta et al. (15) followed the course of nitrogen fixation with the cultivar Altona and found that nitrogen fixation peaked at late flowering rather than early podfill. Their study was, however, a greenhouse experiment and may not be directly comparable. One aspect the first two papers did not account for is the significance of nitrate reduction in the roots and nodules, and its timing in relation to nitrogen fixation and shoot nitrate reduction. Radin (11) indicated that as much as 25% of the nitrate reduced by soybean plants may be reduced in the root system while Randall et al. (12) implicated nodules for similar levels. It is possible that the pattern is different for root nodule nitrate reduction than for leaf, and therefore needs quite close testing and confirmation.

Morphological

Hardy, in the experiments mentioned previously, used only one variety in his field studies, cv. Wayne. In this study he demonstrated an apparent absence of nodulation on plants until about the time of flowering. This coincided with the timing of the first measurable acetylene reduction. No data as to numbers of nodules are presented, but nodule wt per sample ranged up to nearly 5 g. Thibodeau and Jaworski, using the same variety, describe a similar pattern for nodule development, with a maximum nodule wt of about 2 g/ plant. Figure 1 presents a developmental plot of nodule wt for the variety Wayne in the field of Missouri, Delaware, and our New York studies (number and diameter). The nodule development pattern is approximately the same in all 3 locations, except for much earlier initial nodulation in New York. Nodule wt used in the Missouri and Delaware studies was in actuality a product of nodule number and diameter. A change in number over a season indicates new nodule initiations, and increases in diameter imply increases in the amount of nitrogen fixed per nodule. Thus the amount of nitrogen fixed per plant can vary due to increases in 1) nodule number, 2) nodule size, and 3) nodule efficiency. Theoretically any one of these three can change over time without any concomittant changes in others. Since nodule diameter continues to increase throughout the season, this continued growth represents growth of the nodules coupled with senescence of the core. This will result in increasing nodule volume, with relatively smaller increases in nodule mass, and a less rapid increase or even a decrease in the amount of nodular tissue fixing nitrogen actively. In reference to the cultivar Wayne, there appears to be general agreement between the 3 sets of experiments on nodule development, though timing may differ from site to site. Nodulation was well started by 2 wk in New York, and 3 wk in Missouri, while the Delaware plants apparently did not nodulate until onset of flowering (9 to 10 wk). These differences may be due to environmental factors such as the amount of residual nitrogen in the soil, temperature, etc.

Harper and Hageman's study with 'Beeson' gives a slightly different plot, with maximal nodule wt occuring at about mid podfill. This coincides with the achieving of maximum rate of nitrogen fixation/plant. Lawn and Brun (8) studied 2 varieties in detail, and both gave different patterns. Chippewa 64 peaked in specific activity very early in the season (at about 6 to 7 wk after planting) but total activity did not peak until after flowering finished (about 12 wk). Total nodule fresh wt also peaked at this later point. Nodule number, however, appeared to plateau at about 8 wk, after nodules had reached maximum specific activity. The second variety, 'Clay' demonstrated a less clear plateau in specific activity at about the same time as Chippewa 64, and a plateauing of nodule number at the same time in the season as Chippewa 64. Clay, however, peaked in total nodule activity prior to the end of flowering and reached a plateau in nodule fresh wt at about the same time. Both specific activity and total nodule activity dropped precipitously at the end of flowering in Clay, about a month before significant drops occurred in

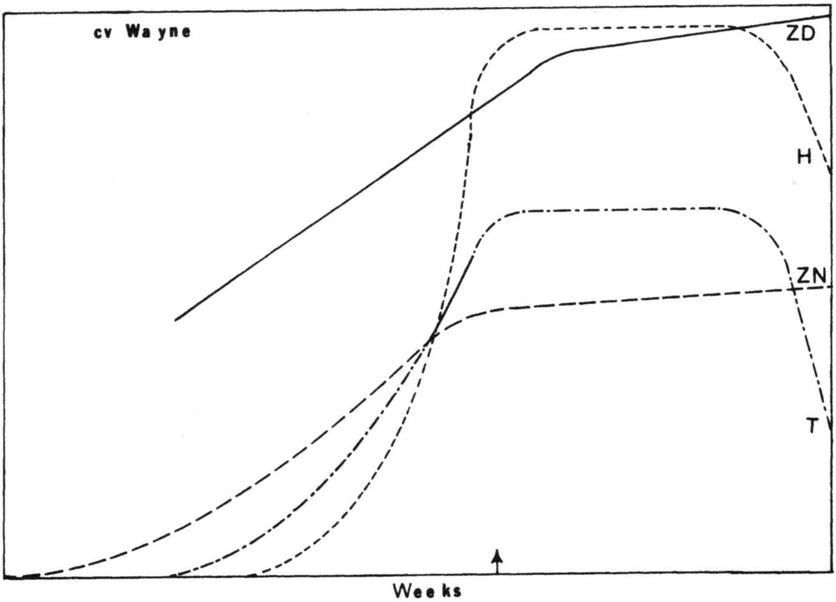

Figure 1. Idealized plots of nodule development in cv. 'Wayne.' X-axis represents time in weeks, and Y-axis represents nodule wt in g/sample or plant, or nodule number, or nodule diameter in mm, as indicated for each plot. The plot 'H' is nodule biomass data taken from Hardy et al. (5); 'T' is nodule biomass/plant data taken from Thibodeau and Jaworski (17); 'ZD' and 'ZN' are nodule diameter and number, respectively, taken from 1978 field plots in New York (averages of three replicates, three samples/replicate/date). The arrow represents time of 100% flowering.

nodule number, or nodule wt. With Chippewa 64, all four occurred simultaneously, and during early podfill.

Skrdleta et al. (15) in their study of 'Altona' soybeans, found an additional pattern. Altona increased nodule numbers up to 45 days after sowing, just after flowering, then remained constant for the duration of the experiment. Nodule wt, on the other hand, increased to the end of flowering (75 days) and then began to drop. Total nodule activity peaked before the end of flowering and then dropped rapidly. Although their work was done in the greenhouse in pots, the pattern and timing of nodule number increases correspond very closely to the results of our 1978 field experiments in New York. Figure 2 is a composite drawn from the results of the research on these 5 varieties (Wayne, Chippewa 64, Clay, Beeson, and Altona). There are obvious and dramatic differences between varieties, but as discussed, separate experiments carried out at very different locations demonstrate virtually identical variety dependent patterns.

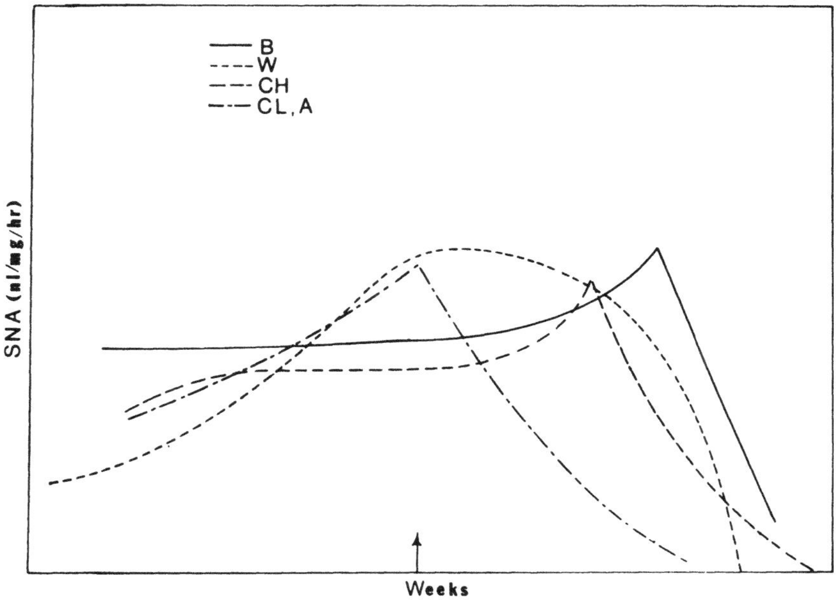

Figure 2. **Idealized plots of Specific Nodule Activity (SNA) for five varieties of soybean, plotted against time. 'Beeson' (B) [data taken from Harper and Hageman (6)], 'Wayne' (W) [data taken from Thibodeau and Jaworski (17)], 'Chippewa-64' (CH) [data taken from Lawn and Brun (8)], 'Clay' (CL) [data taken from Lawn and Brun (8)], and 'Altona' (A) [data taken from Skrdleta et al. (15)]. The arrow represents 100% flowering.**

MEASUREMENT OF NITROGEN FIXATION IN THE FIELD

The measurement of rates of nitrogen fixation by plants is the bottom line when any genetic or physiological study is made on fixation. There are a multitude of different methods, but generally they can be separated into four classes: 1) ^{15}N, 2) measurement of actual uptake over time, 3) measurement of hydrogen evolution, and 4) acetylene reduction. Most nitrogen fixation research uses a combination of these rather than only one, and much of the research reported in the literature is greenhouse, growth chamber, or modified field research. Very little research has been done on field-grown plants where standard agronomic practices were used. Unfortunately, greenhouse and growth chamber research results often are not applicable to field-grown plants.

Applications of ^{15}N can provide perhaps the best measure of nitrogen fixation if little differential uptake of isotopes is assumed. With the technology available for differentiation between the two isotopes of nitrogen (^{14}N and ^{15}N) it is possible to determine just exactly how much nitrogen has been fixed over a given period of time, or a season. Because of cost, effort, and other constraints involved in the use of ^{15}N it is really not applicable for use under normal field conditions.

A recent study in our laboratory (Spaeth, M.S. Thesis, 1978) demonstrated that nitrogen uptake over time is a cultivar-dependent phenomena. Figure 3 demonstrates some of the differences between cultivars in their nitrogen uptake patterns. When total N uptake is observed in isolation, it is impossible to determine the components, or their role in overall N assimilation. Measurements of total N uptake must be compared with estimates of nitrate reduced, and nitrogen fixed to provide any meaningful information about relative roles of these two processes in total plant yields.

Hydrogen evolution has been shown to be a suitable measure of nitrogen fixation under certain conditions (14). This measure of nitrogen fixation has the advantage that there is no need to apply a substrate to the plant for production to occur. On the other hand, under field conditions, the hydrogen observed as evolved in situ probably does not represent amounts produced within the nodule. This is because of the presence of microorganisms in the rhizosphere which can metabolize hydrogen and therefore reduce final amounts liberated to the atmosphere, and subsequently measured. Strains of *Rhizobium* evolve differing amounts of hydrogen from nodules. Unless field soil is sterilized or has never seen soybean rhizobia, there is little control over specific strains which actually induce a given nodule. The situation is, however, even more complex. During a given growing season pea nodules produce hydrogen at rates which do not parallel acetylene reduction patterns (1). Although there is no evidence that a similar situation may exist in soybean, the actual germplasm surveyed for hydrogen evolution is exceedingly small, leaving open the question of similarity of acetylene reduction and hydrogen evolution patterns in soybean as a species. There is evidence that a further problem does exist in the use of hydrogen evolution as a measure of nitrogen fixation. Carter et al. (2) have demonstrated that there are plant varietal differences in the acetylene/hydrogen ratio in soybean. If the plant genotype conditions this ratio, then surely the physiological status of the plant can modify the ratio, thus severely impairing its reliable use as a measure of nitrogen fixation, per se.

Hardy et al. (5) established the suitability of using acetylene reduction as a measure of rates of nitrogen fixation in soybean. This method of measurement is being used currently in many forms, some of which appear to be more applicable than others. The technique of stripping nodules off roots and measuring acetylene reduction on the nodules alone is perhaps the poorest technique since stripping of nodules frequently has been demonstrated to severely reduce measured rates of fixation. Using roots with nodules attached has proven more suitable since it does not involve the wounding of the nodule itself or the root in the immediate vicinity of the nodule. Such

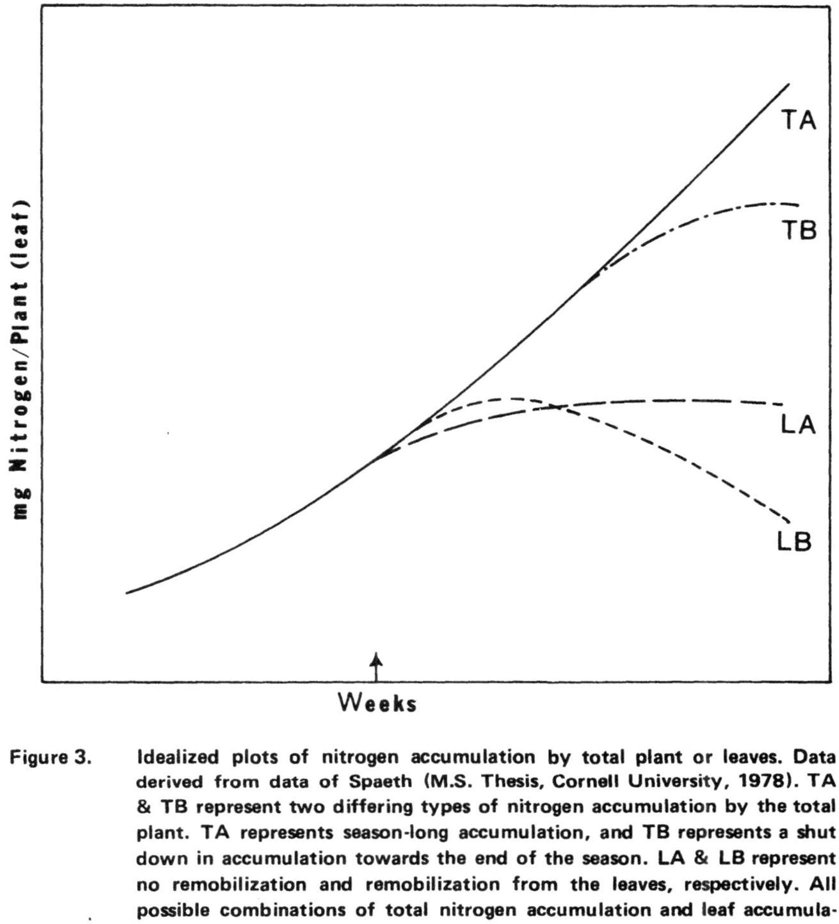

Figure 3. **Idealized plots of nitrogen accumulation by total plant or leaves. Data derived from data of Spaeth (M.S. Thesis, Cornell University, 1978). TA & TB represent two differing types of nitrogen accumulation by the total plant. TA represents season-long accumulation, and TB represents a shut down in accumulation towards the end of the season. LA & LB represent no remobilization and remobilization from the leaves, respectively. All possible combinations of total nitrogen accumulation and leaf accumulation were observed in differing cultivars.**

damage to the root results, however, in the immediate modification of hormonal patterns, and disruption of normal physiological and biochemical processes, not to mention the disruption of even localized nutrient movement. Whole root systems, from which the tops have been severed are one step closer to intact plants, but not far removed from use of nodulated roots. There is some potential for movement of nutrients between differing roots within the intact root system, but this is probably a negligible advantage in this case. Both intact root systems and nodulated roots suffer the disadvantage of not maintaining normal translocation found in intact plants. One modification of the intact root system method described by Hardy et al. (1967) has several additional disadvantages. This method measures acetylene reduction of an intact soil core which contains the central portion of the root system, and the included nodules essentially undisturbed. This would appear to be an advantage, but their results demonstrate no difference from nodulated roots in 50% of their experiments, with reduced rates in 38% and increased rates in 12%. The soil core method was not clearly superior to the other methods, and was far more inconsistent in the results obtained. The explanation offered was that diffusion was restricted; but this hypothesis suffers from the observation that 50% of the experiments were identical in values to nodulated roots, demonstrating no diffusion resistance. Other possibilities include differing amounts of 'wounding' and, therefore, differing amounts of disruption of normal patterns, or secondly that the variability they observed was real, and the apparent uniformity of the nodulated root method was artifactual. None of these points can be resolved finally until intact plants/roots/nodules are measured in the field, in situ (that is undisturbed), and these results compared to methods used by others.

A study to accomplish this task is underway in our laboratory. It uses open-ended cylinders buried in the soil at the time the soybean crop is sown. Acetylene/air mixtures are injected into the cylinders from the bottom, and then removed from the top of the cylinder after passing over the roots and nodules at a given rate. This allows measurement of acetylene reduction rates while the plant is allowed to continue its normal growth and development undisturbed. The acetylene and ethylene concentrations are below those which would be inhibitory to normal development. Preliminary results demonstrate variability on the level of those from the soil cores of Hardy et al. (5) even though tests demonstrate no significant inhibition of gas flow, nor channeling through the soil (Denison, M.S. Thesis, 1979). Several reports, including that of Hardy et al. (5) discuss diurnal patterns, and seasonal patterns in nitrogen fixation. With the use of in situ measurement methods, it is possible to look at each of these with much greater precision, and less likelihood of artifacts. In addition, this system provides an excellent method of comparing cultivars under similar environments, without the possibility of differential responses to pre-testing manipulations.

GENETIC VARIABILITY

Plant breeders and applied plant geneticists often grow large numbers of cultivars and PI lines or other accessions, in a single year, at a single location, and often unreplicated, for the purpose of surveying the genetic variability available, and identifying lines of potential use in breeding/genetic programs. Though this approach lacks the precision found in experiments replicated over years and locations, it provides an estimate of the probable maximum values to be found in a similar population grown under those conditions. Growth of large numbers of cultivars or lines under 'one environment' reduces most of the variability to that which is due to the genetic variability between lines rather than environmental variability. The larger the population and the broader the germplasm base, the greater the proportion of genetic variation to environmental variation. The greatest advantage of this procedure is that it provides the experimentor with an opportunity to select lines at the extreme of populations for further use, and provides an initial analysis of the probable genetic complexity of a character, and genetic diversity of a species.

Further use of lines selected from mass screens, by planting out populations derived from intercrosses into experiments replicated over time and locations, allows the partitioning of environmental and genetic characteristics. Further, genetic correlations between characters can be estimated from these latter studies, giving insight into the feasibility of use of associative breeding techniques for a given character or characters. Associative breeding relies on genetic correlation between a character measured easily, and a difficult to measure character which is of importance. Strong correlation allows the selection of the easy to measure character with the expectation that the desired though unmeasured character will be carried along. This procedure has been used successfully by animal breeders (4). There are numerous, though indirect, examples of this in plant breeding. For example, soybean cultivars derived from Harosoy usually have high rates of photosynthesis (50 to 60 $mg/dm^2/hr$) while those from Lincoln have relatively low rates of photosynthesis (35 to 45 $mg/dm^2/hr$ (Lugg and Sinclair, in press).

In addition to screening large numbers of lines for given characteristics, additional information can be derived from frequent sampling at one location of cultivars which have been selected for their apparent genetic differences. This experimentation reveals differences in developmental patterns which would not be visable with one point in time measurements, nor infrequent sampling. The differences between two specific lines is characterized not only by quantitative differences at a given point in time, but also developmental timing. An example is presented in Figure 2 which demonstrates the difference in timing with which several differing soybean cultivars reach maximum rates of nitrogen fixation.

Recently, Zobel (unpublished) has concluded several experiments of the above two types, and the results are very interesting in their implications. He screened 342 named varieties and breeding lines for root and nodule characteristics in 1977 and grew 31 selections out in a replicated trial at a site in central New York (yield trials in an adjoining plot gave average yields of 40 bushels in 1977 and 31 bushels in 1978). In 1978 500 PI lines also were screened for root and nodule characters. The results of these experiments are summarized in Figures 4 to 6, and Table 1.

From the graphs in Figure 4, it is obvious that there is a relatively continuous increase in diameter of soybean roots and nodules during the season. Size of rhizosphere organs appears to be little affected by

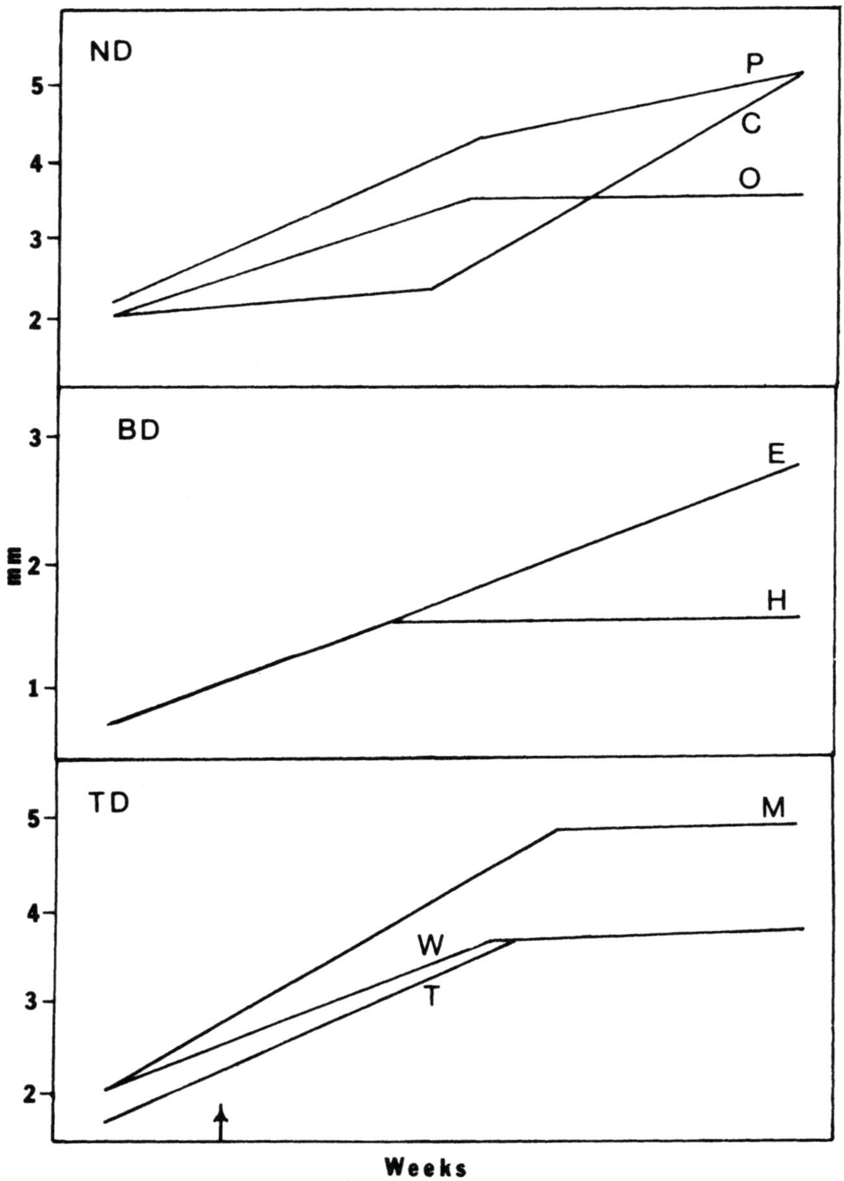

Figure 4. Idealized plots of nodule diameter (ND), basal diameter (BD), and tap root diameter (TD). X-axis is time with the arrow represening 100% flowering. The Y-axis is diameter in mm. P is 'Provar', C is 'Comet', 0 is '052-903', E is 'Elton', H is 'Harosoy', M is 'Macoupin', W is 'Wilkin', and T is 'T-85'.

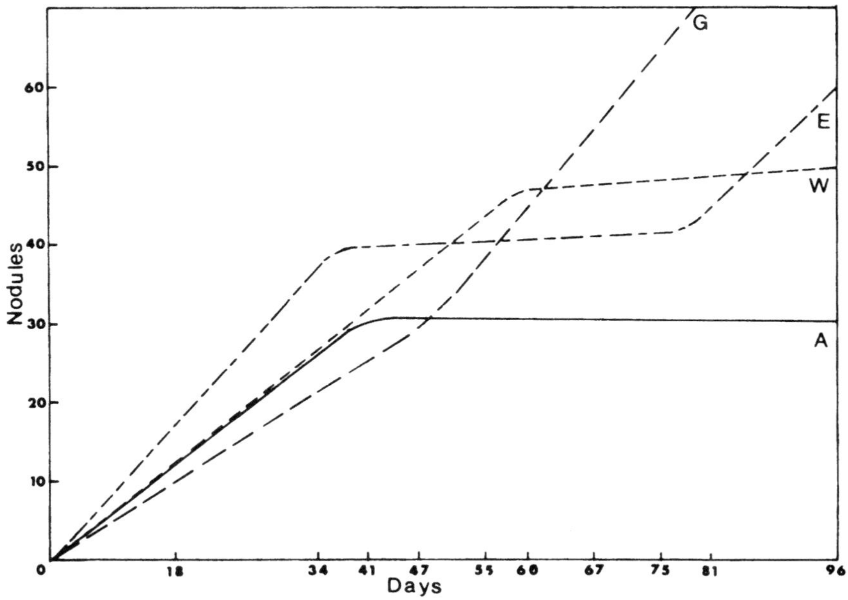

Figure 5. **Idealized plots of nodule number against time for four cultivars (G is 'Guelph', E is 'Elton', W is 'Wayne', A is Altona'). The data are from the 1978 field experiments in New York, and are the average of three samples/replicate, and three replicates/variety/date. Dates shown are sampling dates after sowing.**

developmental cycles within the plant. On the other hand, rates with which the organs increase in size differ between cultivars, and are apparently insensitive to developmental stage. Other characteristics are very much dependent upon stage of development, witness the changes in rates of increase in nodule number. These 'sensitive characters', (nodule number, plant ht, plant biomass, root biomass, basal, tertiary, and quaternary root numbers) all change with stage of development, and demonstrate different patterns in different cultivars.

Figure 6 and Table 1 demonstrate the extent of variability to be found in PI material and cultivars. Table 1 demonstrates that most of the characters scored in the screen of 500 PI lines are correlated, though the correlation coefficients are so low as to be questionable. This statistical correlation with a low coefficient must be due to the extreme variability observed, and number of individuals scored. This probably also represents a correlation between each character and total plant biomass. In addition to the extensive amount of variability observed for every character, it was observed that

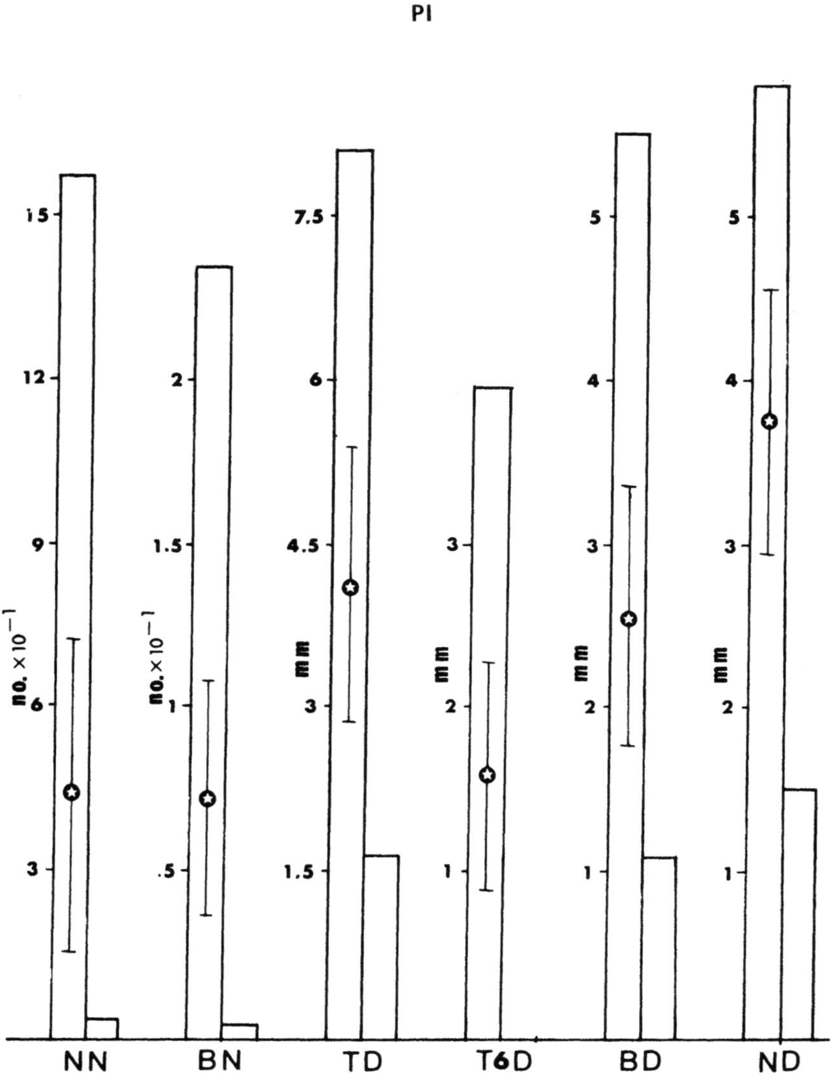

Figure 6. Bar graphs demonstrating the maximum and minimum PI line means plus the population mean and standard deviation for six rhizosphere characters scored in a planting of 500 plant introductions in New York in 1978. The figure '✪' represents the population mean. 'NN' is nodule number, 'BN' basal root number, 'TD' tap root diameter, 'T6D' tap root diameter 6 cm below the hypocotyl, 'BD' basal diameter, and 'ND' nodule diameter.

Table 1. Correlation coefficients and probability of significance of R for six rhizosphere characters: nodule number, nodule diameter, basal root (major secondaries) number, basal root diameter (measured at the tap root), tap root diameter (measured just below the hypocotyl), tap-6 diameter (measured 6 cm below the hypocotyl). Data taken from the 1978 planting of 500 PI lines in New York.

		Nodule Diameter	Basal Number	Basal Diameter	Tap Root Diameter	Tap-6 Diameter
Nodule	R	0.187	0.235	0.129	0.079	0.107
Number	P	0.0001	0.001	0.0001	0.0034	0.0001
Nodule	R		0.185	0.105	0.088	0.115
Diameter	P		0.0001	0.0001	0.0007	0.0001
Basal	R			0.389	0.290	0.118
Number	P			0.0001	0.0001	0.0001
Basal	R				0.465	0.387
Diameter	P				0.0001	0.0001
Tap Root	R					0.565
Diameter	P					0.0001

10 PI lines have novel pear-shaped nodules. The cause of the difference in shape is not yet known, but tests currently in progress will determine their effectiveness, as well as relation to host genetics.

As discussed earlier, where similar varieties have been studied at other laboratories, under different environmental conditions, the same developmental patterns are obtained. The timing of initiation of nodulation, for example, varies with location, but the timing of maximum nodule number remains consistent at given developmental stages, as does acetylene reduction. Severe stress does, of course, disturb normal patterns, but in its absence, data derived at one location may be applicable at most other locations. This stability of pattern argues in favor of the use of associative breeding techniques to modify these characters.

The presence of a wealth of variability in rhizosphere characters indicates that there is ample room for breeding and genetic manipulation of rhizosphere organs, especially nodules. If it can be determined which characters need to be modified in what direction, there will be a relatively short wait for soybean lines with improved genetic capacity for nodulation and nitrogen fixation, as well as other rhizosphere characteristics.

COUPLED HOST/SYMBIONT RESEARCH

One major aspect of rhizogenetics not discussed above is the genetic control of host/symbiont interactions per se. This control can occur at several levels, recognition of the host or symbiont by each other, initial growth of the infection thread, initial stages of proliferation of the cells and bacteriods which make up the nodules, and production of the final physiological, biochemical and morphological framework within which nitrogen is fixed. It is likely that there is equivalent genetic variability for these characteristics as with the morphological ones discussed previously. One set of characters which has been studied intensely and is currently the primary thrust of an interesting project in the USDA is the *rj* mutant series, the non-nodulating soybean lines (3).

This project is studying the host/symbiont interaction to identify strains of rhizobia that can infect and nodulate *rj* lines, resulting in normal fixation. The discovery of such strains will allow the use of specific strains of bacteria in a field of a given cultivar containing the *rj* gene, with confidence that no wild strains of bacteria will infect the plants and reduce nitrogen yields. This technique will allow the tailoring of plant and bacterium such that maximum rates of fixation will be observed in the field.

CONCLUSIONS

Rhizogenetics, though only in its embryonic stages as a field of research, is demonstrating extensive genetic variability in plant rhizosphere organ development and physiology. Physiological/developmental data published in the past can only indicate some of the potential patterns in nitrogen fixation and assimilation. Because of the wealth of variability in overall patterns, and in specific efficiencies, it is presumptive, if not inaccurate, to make generalizations from the results of experiments on a few cultivars. Much more research is needed with many differing cultivars, and it needs to be coupled tightly with research on other physiological/developmental genetic aspects of soybean, such as photosynthesis, mineral assimilation, water assimilation, and morphology. There is a long way to go, and much to do, but it is obvious that many options will be opened up in terms of genetic variability.

NOTES

R. W. Zobel, Assistant Professor, Departments of Agronomy, and Plant Breeding and Biometry, Cornell University, Ithaca, New York 14853.

REFERENCES

1. Bethlenfalvay, G. J. and D. A. Phillips. 1977. Ontogenetic interactions between photosynthesis and symbiotic nitrogen fixation in legumes. Plant Phys. 60:419-421.
2. Carter, K. R., N. T. Jennings, J. Hanus, and H. J. Evans. 1978. Hydrogen evolution and uptake by nodules of soybeans inoculated with different strains of *Rhizobium japonicum.* Can. J. Microbiol. 24:307-311.

3. Devine, T. E. and D. F. Weber. 1977. Genetic specificity of nodulation. Euphytica. 26:527-535.
4. Falconer, D. S. 1961. Quantitative genetics. The Ronald Press, New York, p. 312-329.
5. Hardy, R. W. F., R. D. Holsten, E. K. Jackson, and R. C. Burns. 1968. The actylene-ethylene assay for N_2 fixation: Laboratory and field evaluation. Plant Phys. 43: 1185-1207.
6. Harper, J. E. and R. H. Hageman. 1972. Canopy and seasonal profiles of nitrate reductase in soybeans. Plant Phys. 49:146-154.
7. Holl, F. B. 1975. Host plant control of the inheritance of dinitrogen fixation in the *Pisum-Rhizobium* symbiosis. Euphytica 24:767-770.
8. Lawn, R. J. and W. A. Brun. 1974. Symbiotic nitrogen fixation in soybeans. I. Effect of photosynthetic source-sink manipulations. Crop Science 14:11-16.
9. Mitchell, R. L. and W. J. Russell. 1971. Root development and rooting patterns of soybean [*Glycine max* (L.) Merrill] evaluated under field conditions. Agronomy Journal 63:313-416.
10. Nutman, P. S. 1969. Genetics of symbiosis and nitrogen fixation in legumes. Proc. Roy. Soc. B. 172:417-437.
11. Radin, J. R. 1978. A physiological basis for the division of nitrate assimilation between roots and leaves. Plant Science Letters 13:21-25.
12. Randall, D. D., W. J. Russel, and D. R. Johnson. 1978. Nodule nitrate reductase as a source of reduced nitrogen in soybean (*Glycine max.*). Phys. Plant. 44:325-328.
13. Raper, C. D., Jr. and S. A. Barber. 1970. Rooting systems of soybeans. I. Differences in root morphology among varieties. Agron. J. 62:581-584.
14. Schubert, K. R., J. A. Engelke, S. A. Russel, and H. J. Evans. 1977. Hydrogen reactions of nodulated leguminous plants. I. Effect of rhizobial strain and plant age. Plant Physiol. 60:651-654.
15. Skrdleta, V., V. Vasinec, A. Hyndrakova, and M. Nemcova. 1978. Dinitrogen fixation-acetylene reduction in soybeans during the reproductive growth period. Biol. Plant 20:210-216.
16. Tanaka, N. 1977. Studies on the growth of root systems in leguminous crops. Agricultural Bulletin of Sage University (Japan) 43:1-82.
17. Thibodeau, P. S. and E. G. Jaworski. 1975. Patterns of nitrogen utilization in the soybean. Planta. 127:133-147.

RHIZOBIUM INOCULATION AND SOYBEAN PRODUCTION

J. C. Burton

The soybean, *Glycine max* L. Mer., appears to have no geographical or environmental limitations which cannot be overcome with a few years of study by plant scientists. With more than 40 million ha planted to soybeans in 1978 and projected increases in the decade ahead, the soybean is truly a world crop supplying ever increasing amounts of nutritious protein food for man and animal. Indeed the latest news releases suggest that the soybean may soon prove itself in outer space. Nonetheless, in our research conference this week we should perhaps restrict ourselves to the planet we know best.

Historians tell us that the soybean is native to eastern Asia and that it was domesticated in the eastern half of North China about 1100 B.C. (16). From China, the soybean spread to Korea and Japan sometime between 200 B.C. and 200 A.D. and into Europe early in the 18th century. Culture of soybeans in the U.S. was first mentioned early in the 19th century (14). Mease grew soybeans in Pennsylvania around 1804. However, development of the soybean as a major grain legume in the U.S. has occurred during the past 40 yr.

The origin and spreading of the root nodule bacteria of the soybean, *Rhizobium japonicum,* are even more obscure than that of the host. We can conclude, however, that somehow some of these bacteria were carried along with the seed or plants because without rhizobia to form nodules and supply N, the soybean could never have become such an important world crop. Since the important function of rhizobia in the legume nodule was not discovered until 1888 (6) the soil microbiologist cannot claim credit for the increase in popularity of the soybean during the first 19 centuries of its culture. Albeit, the rhizobiologist has played an important role in increasing grain production by the soybean over the past 40 yr.

The task of the rhizobiologist has been to select effective efficient strains of *Rhizobium japonicum,* culture them in the laboratory and develop delivery systems to provide an abundance of the selected rhizobia to assure nodulation and a dependable nitrogen supply. Like plant breeders (12), physiologists, and other scientists working with the soybean, rhizobiologists take pride in the progress soybeans have made.

RESEARCH OBJECTIVES

The problems and research objectives of the soybean rhizobiologist have increased in complexity as soybean culture has expanded to diverse soils, climates and various levels of farm management. Soybean breeders continue to develop new improved genotypes for various stress factors: A) increased resistance to disease and insects, B) better adaptation to narrow row and solid seeding, C) greater efficiency in hot and cold climates, D) tolerance to low pH and high concentrations of soluble aluminum and manganese, E) better adaptation to alkaline soils, F) insensitivity to photoperiodism, or G) simply higher yields as a result of less shattering and plant lodging. Modification of the host plant can affect N_2 fixation and symbiosis with different strains of rhizobia.

The soybean rhizobiologist must keep abreast with plant breeders and screen promising new genotypes with the strains of rhizobia which are likely to be used on them at the first opportunity. The importance of discreet matching of symbionts cannot be overlooked. Simple tests in the greenhouse or growth chamber can provide valuable clues regarding the need for further testing under field conditions.

Response of New Cultivars to Various Strains of Rhizobia

The response of a dwarf type soybean, Elf, and one of its parents are shown in Table 1.

Table 1. Nitrogen fixed by a dwarf cultivar of soybean, Elf, and Williams, one of its parents.

Inoculation treatment	Cultivar: Elf	Cultivar: Williams
(Strain)	*mg N/10 plants*	
None	113	91
A96	357	298
A101	414	334
A118	422	297
A124	388	400
A133	389	275
A148	356	259
A149	374	326
A150	441	304

Williams gave a highly effective response to only about 4 whereas Elf worked effectively with all 8 strains of rhizobia tested. The indications are that the Elf has retained or improved its symbiotic capabilities through the developmental phases.

Responses of 2 night-length insensitive cultivars of soybeans to 7 strains of rhizobia are given in Table 2. The 7 strains were selected for their effectiveness on many of the older cultivars. All strains of rhizobia were effective nitrogen fixers on both cultivars, but strains A121 and A122 were much more effective on Kino than they were on Rillito.

Developing More Effective Strains of Rhizobia

In addition to testing new genotypes of the host for response to proven strains of *R. japonicum,* the rhizobiologist must search continuously for more efficient strains of *R. japonicum* for currently productive commercial soybean cultivars as well as new ones. This is achieved by isolating and screening numerous wild type strains of the microorganism or by genetic manipulation and screening of numerous mutants (24). Nitrogen fixation by two mutants and the parent wild strain on Hark soybeans is shown in Figure 1. Strains SM31 and SM35 were derived from strain 61A76 after treatment with a chemical mutagen (13). On Hark soybeans, dry wt and nitrogen content of the soybeans inoculated with the mutants did not differ from that of soybeans inoculated with the wild strain either at 21 or 35 days age. Under field conditions, these mutants have produced increased yields both in Nigeria and Israel. (Personal communication W. J. Brill.)

In order to demonstrate a high N_2 fixing potential under field conditions rhizobia must be competitive as well as have good N_2 fixing abilities. They must be able to nodulate their host in the presence of numerous infective native rhizobia. The competitiveness of 12 strains of soybean rhizobia were tested in a growth chamber experiment using a technique described by

Table 2. Nitrogen fixation by night-length insensitive cultivars of soybeans.

Inoculation treatment	Cultivar	
	Kino	Rillito
(Strain)	*mg N/10 plants*	
None	108	86
A108	419	359
A110	327	298
A118	387	383
A121	410	318
A122	383	291
A124	369	348
A133	306	324

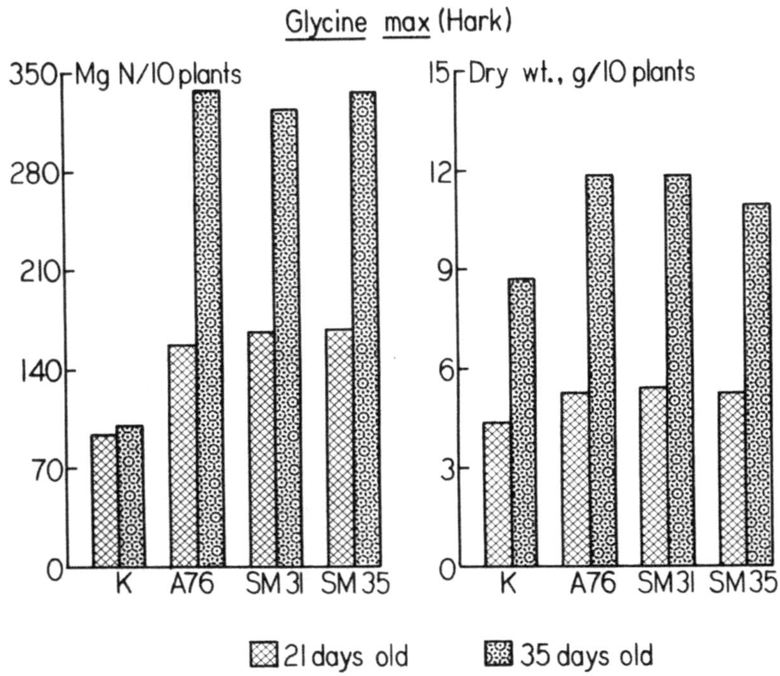

Figure 1. **Nitrogen fixation by Hark soybeans inoculated with two mutants, strains SM31 and SM35 and the wild strain A76.**

Jones, Burton, and Vaughn for sub clover, *Trifolium subterraneum* (9). A peat-base inoculum prepared with the test strain is applied to the seed. The inoculated seed are planted in two series of Leonard jars—one containing only sterile sand and a second series in which the substrate sand is impregnated with infective strains which fix little or no nitrogen. Growth and N_2 fixation obtained by the test strains under these two conditions give a good measure of the test strain's competitiveness. The results (Fig. 2) show that certain strains, A124 and A134, fixed more nitrogen than the other 10 strains in the sterile sand, but with ineffective rhizobia in the sand the amounts of N fixed did not differ greatly. It appears that all the strains included in this test were good competitors. The results may have been different with more agressive ineffective strains in the substrate.

Changing the Susceptibility of the Host Plant to Infection

A major problem in the soybean producing areas of the midwest and Mississippi delta where soybeans have been grown for many years is providing

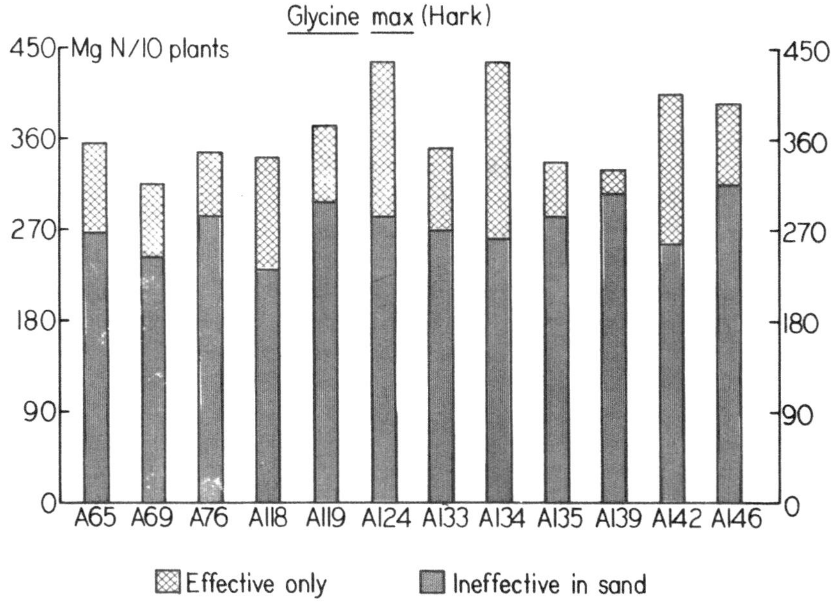

Figure 2. **Comparative N fixation of Hark soybeans grown in pure sand and in sand containing ineffective rhizobia.**

sufficient inoculum bacteria to offset the great number of carry-over rhizobia in the soil. In Minnesota, Hicks and Ham (7) had to apply from 800 to 1000 times the usual inoculum to get as many as 50% of the nodules produced by the applied rhizobia. Even then, there were no increases in yield because the left-over rhizobia were apparently quite effective. Yields were not limited by inadequate nitrogen. Similar results were reported in Ontario, Canada, by Quashie-Sam (17), in Indiana (15) and in Illinois (10). Van Rensburg, Strijdom and Kriel (23) reported different results in South Africa. A survey of 98 soybean fields on 43 farms revealed that inoculated soybeans were generally better nodulated than non-inoculated beans grown on soils where soybeans had been grown previously.

Attention is now being focused on developing soybean genotypes which are resistant to infection by the common carry-over soybean rhizobia in the soil, but susceptible to nodulation by specific selected strains which may be applied. Devine and Weber (5) reported that rhizobia strains have been identified which are both effective and efficient with a host genotype (rj1, rj1) which excludes most indigenous strains. In screening around 1600 genotypes, Kvien, Ham, and Lambert (11) reported finding lines which would not nodulate well with native strains in Minnesota, but yields were increased up to 625 kg/ha when these soybeans were inoculated with other selected strains of rhizobia.

The cowpea and soybean rhizobia are similar in many ways. Cowpeas may be nodulated by *R. japonicum* in some instances and soybeans may be nodulated by cowpea rhizobia. Since cowpea rhizobia are prevalent in many of the soils of Asia and Africa (23) attention is being given to developing soybeans which will be effectively nodulated by native cowpea rhizobia in order to avoid inoculating the seed. Also, the cowpea rhizobia are more efficient users of energy because hydrogen is recycled and used rather than being released (18,19,20). The wild soybean, *Glycine ussuriensis,* grows wild in these countries. Seeds are smaller and resemble cowpeas. However, *G. ussuriensis* or *Glycine soja* is effectively nodulated by *R. japonicum* strains effective on *Glycine max.* The forage species, *Glycine wightii,* is nodulated both by soybean and cowpea rhizobia but the cowpea rhizobia are generally more effective.

IMPROVED DELIVERY SYSTEMS

It is generally recognized now that it takes a large inoculum to nodulate soybeans effective in old fields and in new fields. In 1978, Bezdicek et al. (1) studied soybean inoculation in a soil free of *R. japonicum* and low in N. Soybeans were inoculated in the field with peat-base inocula applied with gum arabic and by using a granular peat inoculum. Higher yields were generally obtained with the granular inoculum because it was possible to apply larger numbers of rhizobia. Application rates varied from 4 x 10^6 to 115 x 10^7/m. It was shown that soybeans fixed more than 300 kg/ha N, an amount considerably higher than that reported anywhere else. The proportion of plant N derived from N fixation ranged from 71 to 80%. It was concluded that the micro-symbiont is not a limiting factor in soybean yield where effective rhizobial strains are present. Limitations in soil moisture, nutrients, light, CO_2 and plant genotype may be more important.

The problem of supplying adequate numbers of rhizobia to bring about effective nodulation becomes more critical as soybean culture shifts from the cool temperate climates to subtropical and tropical conditions. Soils tend to become less fertile, have a lower pH and higher concentrations of soluble aluminum and manganese which are toxic to some rhizobia. In an Oxisol in Puerto Rico, Smith et al. (22) found that 3.9 x 10^5 rhizobia/cm was required to achieve maximum numbers of tap root nodules. Soybeans in rows receiving 3900 rhizobia/cm produced no nodules.

Soil temperature at planting time is often high in tropical areas. Scudder (21) states that soil temperatures at seed depth are often 42 C and can go higher when soil moisture is low. A temperature of 50 C for 30 min or longer can be lethal to soybean rhizobia. The survival of *Rhizobium japonicum* in a peat-base inoculum sealed in a polyethylene package stored at different temperatures is shown in Figure 3.

The death rate in soil at 42 C would undoubtedly be greater than that in a polyethylene package with a moist peat base and no drying. On the other hand, the temperature of a package of inoculum exposed to the sun or inoculated seeds in a metal seed hopper can reach lethal temperature in a very short time. Most of the rhizobia could be killed before the seeds are planted in the soil.

The hazards to rhizobia or soybean seeds placed in subtropical soils were pointed out by Hinson (8) in 1969. Two alternative methods of inoculating soybeans were compared to seed inoculation. In one treatment, peat-base inoculum was mixed with moist builders' sand using up to 10 times the normal amount of inoculum and then drilled at different depths in rows before planting. This proved very satisfactory.

A second method consisted of early broadcast drilling of 100 kg inoculated soybean seed to produce a green manure treatment before planting the production crop. Plants allowed to grow until they had 5 or 6 nodules/plant, about 6 wk, before turning under as a green manure, provided sufficient rhizobia for effective nodulation of the subsequent crop. When soybeans were disced under earlier there were not adequate rhizobia for effective nodulation.

The so-called bulk inoculants, granular and liquids which are applied separately from the seeds but usually in the seed furrow at the time of planting, permit application of large numbers of rhizobia and placing them where their chances of survival are greatly improved. Also, when these forms of inoculants are used, fungicides and insecticides may be used on the seeds without the risk of killing excessive numbers of the rhizobia.

The merits of granular and liquid inocula applied separately from the seeds are demonstrated dramatically in Scudder's experiments in Florida (21). Bragg soybeans inoculated with granules 5.6 kg/ha produced 2920 kg/ha beans as compared to 2300 kg/ha for the seed applied inoculum. With the liquid inoculum which provided 2.9×10^8 viable rhizobia/m of row, the same as was provided by the granules, the yield was 2740 kg/ha soybeans. There was no significant difference between the yields from the granular and the liquid inocula as long as they supplied the same number of viable rhizobia. The same strains of rhizobia were in both inocula. Results with the Hardee cultivar were equally dramatic.

One of the key problems of soybean production in the tropics is obtaining effective nodulation. The granular and liquid inoculants which can be adjusted to provide the required number of viable rhizobia are methods of overcoming this problem.

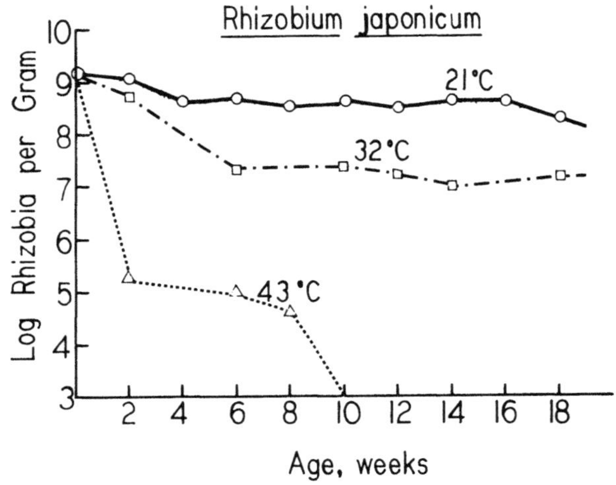

Figure 3. **Survival of *Rhizobium japonicum* in peat-base sealed in moisture-proof polyethylene bags kept at various temperatures.**

COMPATIBILITY OF *R. JAPONICUM* WITH PESTICIDES

Treatment of seeds with chemical fungicides is often necessary to obtain good stands in tropical countries. Hark soybean seeds were inoculated with a peat-base inoculum slurry containing a chemical fungicide which was added immediately before applying to the seeds. The seeds were allowed to dry for 1 hr. A sample of the seeds was then planted to determine the number of viable rhizobia. At the same time some of the seeds was planted in sterile Leonard jars where they were allowed to germinate and grow for 14 days before harvesting and counting the tap root nodules. Rhizobia survival is shown in the left graph and tap root nodules are shown in the right graph Figure 4. The inoculant and chemicals were applied at the rates recommended by the respective manufacturer (3,4):

THIRAM (75% wettable) 0.6 g/kg seed, mixed with 4.4 g peat-base inoculant
CAPTAN (75% wettable) 0.8 g/kg seed, mixed with 4.4 g peat-base inoculant
CARBOXIN (75% wettable) 1.1 g/kg seed, mixed with 4.4 g peat-base inoculant
PCNB (20%) 0.9 g/kg seed mixed with 4.4 g peat-base inoculant.

Fungicides and inoculum were mixed in the proper portions with water to form a slurry. The slurry was applied to the seed at the rate of 13 ml/kg.

Thiram was not toxic to soybean rhizobia as measured by numbers of viable rhizobia or tap root nodulation at 14 days age. Captan and Carboxin did not reduce drastically the number of viable cells on the seed, but nodulation was reduced greatly even 1 hr after treatment. PCNB was highly toxic to

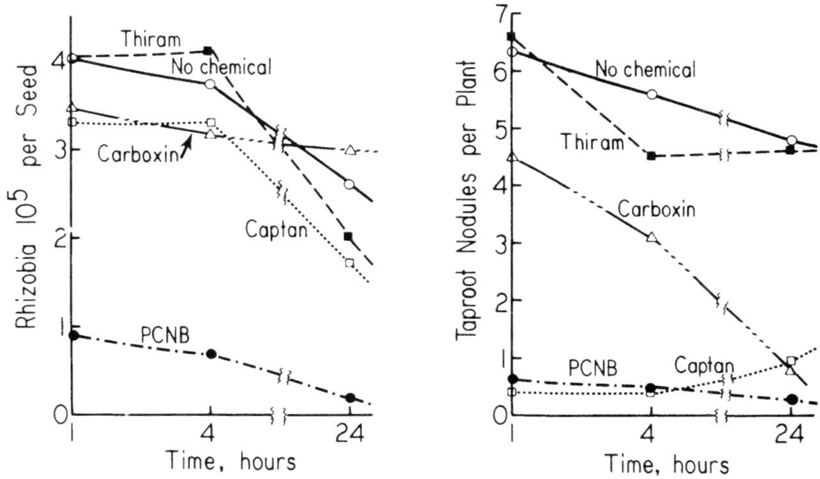

Figure 4. **Survival of *R. japonicum* on soybean seeds treated with various fungicides. Thiram (Methythiuramdisulfide); Captan (N-trichloromethylthio-4-cyclohene-1,2-dicarboximide); Carboxin (5,6 dihydro-2-methyl-1,4-oxathin-3-carboxanilide) and PCNB (pentachloronitrobenzene).**

the rhizobia on the seeds, and plants were almost free of nodules even when planted 1 hr after treatment. It should be pointed out here that these tests do not measure the respective values of the chemicals as protectants against soil fungi. Nonetheless, when soybeans are planted on soils which need rhizobia badly, one should either use one of the more compatible fungicides or use the granular inoculum. Fungicides can destroy rhizobia on seeds even when planted immediately after inoculating. Seeds treated with fungicides have no adverse effect on the rhizobia when placed in the seed furrow with granular or liquid inocula.

RHIZOBIA COMPATIBILITY WITH SYSTEMIC INSECTICIDES

Insecticides are needed frequently to protect seeds and seedlings from the ravages of worms and insects which flourish under tropical conditions. Systemic chemical insecticides absorbed in a granular carrier (clay or corn cobs) often are drilled into the seed furrow along with the seeds at planting. Also granular or liquid inoculants may be planted in the furrow with the seed. Insecticides and inoculants usually are drilled from different hoppers or sections of hoppers, but when only one applicator is available, they may be mixed and drilled from the same hopper.

Field conditions were simulated to study the effects of systemic chemical insecticides on rhizobia planted in the same furrow and also to study survival of rhizobia on the inoculant:insecticide mixtures. The insecticides reduced tap root nodulation to some extent at 14 days (Table 3), but plants with 3 or more tap root nodules at this age are considered nodulated effectively. Differences in nodulation disappear as the plants get older. Number and mass of nodules increase as the plants approach maturity. The treatments are considered compatible and safe should both treatments be needed.

In another experiment, the Soil Implant or granular inoculum was mixed with two granular systemic insecticides in the proportions recommended by their respective manufacturers. At various intervals the inoculum-insecticide mixtures were planted in furrows in moist sand with non-inoculated soybean seeds. After 2 wk growth the plants were harvested and the nodules counted (Table 4). There was no indication of inadequate nodulation under ideal planting conditions such as these. With hot dry storage of seed, the rhizobia kill could have been drastic. The safest system is to drill inoculum and insecticide from different compartments.

Table 3. Effect of systemic insecticides on nodulation of soybeans when inoculum and insecticide are placed in the same furrow.

Treatment	Tap Root Nodules/Plant at 14 Days Age
Inoculum, 55 g/100 m of row	8.0
Inoculum, 55 g/100 m of row FURADAN 110 g/100 m*	4.5
Inoculum, 55 g/100 m of row TEMIK 69 g/100 m*	3.2
Inoculum, 55 g/100 m of row THIMET 73 g/100 m*	7.0

*All systemic insecticides were used at the rates recommended by their respective manufacturers:

"FURADAN" - 2,3 dihydro-2,2 dimethyl benzofuranyl-7-N-Methyl carbonate
"TEMIK" - 2 Methyl-3-Methylthiopropionaldehyde 0-Methylcarbamoyl oxime
"THIMET" - 0-0-diethyl-S-(ethyl thiomethyl) phosphorodioate

Table 4. Nodulation of soybeans planted with mixtures of inoculum and insecticide.

Time In Mixture	Inoculum Only 55 g/100 m	Inoculum, 55/g/100 m FURADAN 110 g/100 m	Inoculum, 55 g/100 m TEMIK 70 g/100 m
		Tap root nodules/plant at 14 days	
1 hr	8.4	6.0	7.9
2 hr		5.5	8.8
4 hr		5.6	6.0
8 hr	9.6	5.8	6.2

SUMMARY AND OUTLOOK

Our knowledge on culture of soybeans continues to grow, but numerous challenging problems remain for the inquisitive rhizobiologist, plant physiologist, plant breeder, soils specialist, pathologist and the entomologist. Selection of *Rhizobium japonicum* strains for tropical conditions has just begun, and much more work is needed. Undoubtedly, diverse strains will be needed for diverse soils and plant cultivars. It is very unlikely that a single "super" strain will do the job.

The indications are strong that conventional delivery systems are very inadequate to provide sufficient rhizobia to effectively inoculate plants the first year they are grown. This loss on first year production is very important and can be avoided.

The question of competition between chemical nitrogen and biologically-fixed nitrogen in soybeans continues to cause frustration. Responses to N fertilizer at any time during the plants' growth are rare and unpredictable. Since nodules become senescent early during pod filling, one would certainly expect foliar N to have a sparing action on nodule and root nitrogen and possibly prolong the N_2 fixation period, but that does not seem to be true (2). Foliar feeding of N to soybeans has been just as disappointing as any other system of supplying it.

If photosynthate is the limiting factor to soybean production, varieties or cultivars selected for improved photosynthetic ability should be the highest yielders but progress in this area of research has been slow. Gains in photosynthetic capacity apparently are offset by losses in other areas.

NOTES

Joe C. Burton, Research Department, The Nitragin Company, Inc., Milwaukee, Wisconsin.

REFERENCES

1. Bezdicek, D. F., D. W. Evans, B. Abede, and R. E. Witters. 1978. Evaluation of peat and granular inoculum for soybean yield and N fixation under irrigation. Agron. J. 70:865-868.
2. Boote, K. J., R. N. Gallaher, W. K. Robertson, K. Hinson, and L. C. Hammond. 1978. Effect of foliar fertilization on photosynthesis, leaf nutrition, and yield of soybeans. Agron. J. 70:787-791.
3. Burton, J. C. 1976. Some practical aspects of legume inoculation. In Proceedings, Winter Meeting Mississippi Section, American Society of Agronomy, p. 51-68.
4. Burton, J. C. and R. L. Curley. 1975. Compatibility of *Rhizboium japonicum* with chemical seed protectants. Agron. J. 67:807-808.
5. Devine, T. E. and D. F. Weber. 1977. Genetic specificity of nodulation. Euphytica 26:527-535.
6. Hellriegel, H. and H. Wilfarth. 1888. Untersuchungen uber die stickstoffnahrung der gramineen und leguminosen. Beilageheft zu der Ztschu ver Rubenzuker Industries, Deutschen Reichs, p. 234.

7. Hicks, D. R. and G. E. Ham. 1975. Soybean inoculation. In Crop News 30. Agron. and Plant Genetics, University of Minnesota.
8. Hinson, K. 1969. Alternatives to seed-packet inoculation of soybeans with *Rhizobium japonicum*. Agron. J. 61:683-686.
9. Jones, M. B., J. C. Burton, and C. E. Vaughn. 1978. Role of inoculation in establishing sub clover on California annual grassland. Agron. J. 70:1081-1085.
10. Kapusta, G. and D. L. Rouwenhorst. 1973. Influence of inoculum size on *Rhizobium japonicum* serogroup distribution frequency in soybean nodules. Agron. J. 65:916-919.
11. Kvien, C., G. E. Ham, and J. W. Lambert. 1978. Improved recovery of introduced *Rhizobium japonicum* strains by field-grown soybeans. Agron. Abstracts, p. 142.
12. Luedders, V. D. 1977. Genetic improvement in yield of soybeans. Crop Science 17:971-972.
13. Maier, R. J. and W. J. Brill. 1978. Mutant strains of *Rhizobium japonicum* with increased ability to fix nitrogen for soybeans. Science 201:448-450.
14. Mease, J. 1804. Willich's domestic encyclopedia, 1st Amer. Ed., Vol. 5:13.
15. Nelson, D. W., M. L. Swearingin, and L. S. Beckham. 1978. Response of soybeans to commercial soil-applied inoculants. Agron. J. 70:517-518.
16. Probst, A. H., and R. W. Judd. 1973. Origin, U.S. history and development and world distribution. In B. E. Caldwell (ed.) Soybeans, improvement, production and uses, Amer. Soc. Agron., Madison, Wisc., p. 1-12.
17. Quashie-Sam, S. J. 1978. Ecological studies on *Rhizobium japonicum* and inoculation of soybeans in the soybean growing areas of Ontario. Ph.D. Thesis, University of Guelph, Guelph, Ontario, Canada.
18. Schubert, K. R. and H. J. Evans. 1976. Hydrogen evoluation: A major factor affecting efficiency of nitrogen fixation in nodulated symbionts. Proc. National Academy Sci. USA Bot. 73:1207-1211.
19. Schubert, K. R., J. A. Engelke, S. A. Russell, and H. J. Evans. 1977. Hydrogen reactions of nodulated leguminous plants. I. Effect of rhizobial strain and plant age. Plant Physiol. 60:651-654.
20. Schubert, K. R. and H. J. Evans. 1977. The relation of hydrogen reactions to nitrogen fixation in nodulated symbionts. In W. E. Newton, J. R. Postgate, and C. Rodriquez-Barrueco (eds.) Recent developments in nitrogen fixation, Academic Press, New York, p. 469-485.
21. Scudder, W. T. 1975. *Rhizobium* inoculation of soybeans for sub-tropical and tropical soils. I. Initial field trials, Proc. Soil and Crop Science Soc. of Florida 34:79-82.
22. Smith, R. S., M. A. Ellis, and R. E. Smith. 1978. Effect of *Rhizobium japonicum* inoculation rates on nodulation establishment in a tropical soil. Agron. Abstracts, p. 147.
23. Van Rensburg, H. J., B. W. Strijdom, and M. M. Kriel. 1976. Necessity for seed inoculation of soybeans in South Africa, Phytophylactica 8:91-96.
24. Wacek, T. J. and W. J. Brill. 1976. Simple rapid assay for screening nitrogen-fixing ability in soybeans. Crop Science 16:519-522.

AMINO ACID LOADING AND TRANSPORT IN PHLOEM

L. E. Schrader, T. L. Housley, and J. C. Servaites

Transport of nitrogen in soybeans occurs in both the phloem and xylem. Nitrate-N and much of the nitrogen arising from nitrogen fixation in the nodules are transported in the xylem. Israel (8) has discussed the translocation of nitrogen in the xylem. Nitrogen exported from leaves to other plant parts is transported in the phloem. Most of this nitrogen is in the form of amino acids derived either from the reduction of nitrate in the leaves, remobilization of protein stored in the leaves, or from the assimilation of ureides, amides, or amino acids produced in the root and nodules during nitrogen fixation and/or nitrate reduction before transport via the xylem to leaves.

Sucrose is the principal photosynthetic product in the phloem. It is actively and selectively loaded into the phloem (2), mediated by a specific carrier (11). Phloem loading of sucrose from the free space is sensitive to metabolic (2,9,11) and sulfhydryl group inhibitors (3), supporting the premise that loading is a metabolically-dependent process. Much less is known about loading and transport of other compounds (e.g., amino acids) found in the phloem. In this paper, we report evidence for a metabolically-dependent mechanism for amino acid loading that is distinct from that for sucrose loading.

EVIDENCE FOR PHLOEM TRANSPORT

Phloem transport of amino acids from leaves was demonstrated by heat girdling the phloem of a soybean petiole by passing an electrical current through a nichrome wire (6). The relative water content of the leaf was monitored until the next day by Beta-gauging to insure that the xylem was not disrupted by the heat treatment. The leaf attached to the girdled petiole was offered $^{14}CO_2$ for 2 hr, but no transport of [^{14}C] from the leaf was detected. This finding indicated that the transport of both sugars and amino acids was stopped by the heat-girdling, suggesting that both are transported from the leaf in the phloem (6).

PARTITIONING OF ASSIMILATES

A fully-expanded leaf (source leaf) on a soybean plant was fed $^{14}CO_2$ in a steady-state labeling experiment for 2 hr, the plant was divided into several fractions, and the plant parts were extracted in hot ethanol. When the ethanol-soluble fractions were fractionated on ion-exchange resins, about 70% of the [^{14}C] recovered from the source leaf was in the neutral fraction (sugars), whereas about 8 to 17% (mean of 14%) was in the basic fraction (amino acids), and 13 to 20% (mean of 17%) was in the acidic fractions (mostly organic acids) (7). However, in the path (petiole of fed leaf and stems) about 92% of the [^{14}C] was recovered in sugars and 2 to 6% was recovered in both amino acids and organic acids. The marked changes in distribution of radioactivity among these compounds in the path as compared to the source leaf suggest that (a) some barrier exists to the loading of amino and organic acids, (b) sucrose is loaded preferentially as compared to amino and organic acids, or (c) amino and organic acids are compartmentalized (e.g., in the vacuole) so that they are not readily available for loading and transport.

Sucrose contained almost all of the [^{14}C] recovered in the sugar fraction. Only a limited number of amino acids were labeled (Table 1). Three amino acids (alanine, serine, and glutamate) contained about 75% of the radioactivity recovered in the amino acid fraction from the source leaf (Table 1). Those same amino acids were highest in concentration in the leaf, as shown by the amino acid analysis which detected both [^{14}C] and [^{12}C] amino acids (Table 1). In the petiole of the fed leaf, five amino acids contained most of the recovered radioactivity. In addition to the three that were predominant in the fed leaf, aspartate and gamma-aminobutyrate (GABA) were heavily labeled in the petiole (Table 1). The amino acid analyses confirmed that GABA and aspartate concentrations were higher in the petiole than in the fed leaf. Perhaps amino acids are loaded selectively, compartments exist in the leaf that supply the conducting channels directly, or some conversions occur during, or after, the loading of certain amino acids. An earlier review (13) indicated that GABA transaminase is present in the phloem of some species. Glutamic acid decarboxylase and the transaminase were both present in soybean callus cells (12).

Table 1. **Distribution of selected amino acids (g/100 g of amino acids), and the percentage distribution of radioactivity in the amino acid fraction from the fed leaf and its petiole of a nodulated soybean plant after a 2 hr steady-state labeling with $^{14}CO_2$.**

	Amino Acid Distribution		^{14}C Distribution	
Amino Acids	Fed Leaf	Petiole	Fed Leaf	Petiole
Aspartate	3	18	4	13
Serine	18	11	33	12
Glutamate	26	18	15	22
Alanine	18	7	26	11
γ-Aminobutyrate	2	11	1	15
Phenylalanine	1	1	5	4
Asparagine	3	8	1	0
Glutamine	1	4	1	2

SELECTIVITY OF AMINO ACID LOADING

We earlier compared the loading and transport of serine, leucine, lysine, and sucrose (6). Serine was selected as an amino acid normally present in large amounts in the phloem whereas leucine and lysine were barely detectable. Sucrose and [^{14}C] amino acids were applied to an abraded spot on a soybean leaf as described earlier (6). Transport of the [^{14}C] to a sink leaf was monitored. Transport velocities for the four compounds were similar (Table 2) and compared favorably with the velocity for [^{14}C] assimilates formed and transported from a comparable source leaf fed $^{14}CO_2$. Hence it appears as if the amino acids and sugars applied exogenously to the abraded spot were loaded in much the same manner as the assimilates formed in the chloroplasts of the mesophyll cells. These results indicated that no selective barrier exists to prevent entry of leucine or lysine into the phloem from the apoplast. The small amounts of leucine or lysine normally transported may be due to low production of these amino acids in the source leaf or due to a barrier to export of these compounds from the mesophyll cells into the apoplast.

In later experiments (10), transport velocities and mass transfer rates were compared for three compounds that differ in their functional groups. Leucine, a neutral amino acid, was compared with glutamate, a dicarboxylic amino acid, and GABA, which lacks an alpha-amino group. Mass transfer rates for the three compounds were similar, suggesting that similar amounts of each are loaded if present in equal concentrations at the site of loading (Table 2). Transport velocities for these three compounds compared favorably with those reported by Housley et al. (6). The experiment with GABA indicated that loading does not discriminate against a compound

Table 2. **Transport velocities for [^{14}C] photosynthate and applied sucrose and amino acids [Experiment I from (6); Experiment II from (10)].**

Radioactive compound fed[a]	Transport velocity (Expt. I)	Transport velocity (Expt. II)	Mass transfer rate (Expt. II)
	cm/min	*cm/min*	*nmol/cm^2·hr*
Sucrose	1.06		
	0.75		
Lysine	0.77		
	0.83		
Serine	0.90		
	0.70		
Leucine	0.99	1.3	22
	0.81		
$^{14}CO_2$-assimilates	0.97		
Glutamate		1.2	24
γ-Aminobutyrate		1.0	28

[a]Expt. I: For sucrose and amino acids, 10 μCi of [^{14}C] compound were applied to an braded spot on the source leaf. The $^{14}CO_2$ was offered as a 10 min pulse, generated from 50 μCi of [^{14}C] bicarbonate, and followed with a 110 min chase period in air. Arrival of the [^{14}C] in the sink leaf was monitored with a GM tube. Expt. II: [^{14}C] amino acids (500 nmol; 10 μCi) were applied to an abraded spot. After 2 hr, the leaf was sacrificed and the amount of label exported was determined to estimate mass transfer rates.

lacking the alpha-amino group. Hence the GABA found in the petiole (Table 1) could have been loaded from the leaf rather than being synthesized after entering the phloem. Although these experiments are consistent with the earlier report (6) that no barrier exists to the loading of these compounds, it is not evident whether there are specific carriers for each amino acid or carriers which can load several or all amino acids. Additional competition experiments must be completed in which more than one amino acid is applied to the abraded spot concurrently before definitive statements can be made about the selectivity of the loading mechanism for amino acids.

Berlin and Mutert (1) reported the existence of distinct transport systems for uptake of L-amino acids by cultured tobacco cells. Distinct transport systems for neutral, acidic, and basic amino acids were kinetically characterized. However, it is not known whether the same specificity exists for export of amino acids from soybean mesophyll cells into the apoplast, or if specific carriers are present for loading amino acids into companion and/or sieve tubes of the phloem.

CONCENTRATION DEPENDENCE OF LOADING MECHANISM

The effect of leucine and sucrose concentrations applied to the abraded spot on a leaf was examined to ascertain whether their concentrations in the free space could be limiting translocation. A biphasic concentration response was observed for sucrose (Fig. 1A), and was similar to that reported for sugarbeet (11). The first phase extended from 1 to 110 mM sucrose and had a K_m of about 35 mM, which is similar to that reported for sugarbeet leaves (11) and leaf discs (3). The second phase extended from 110 to 400 mM with a K_m of 169 mM.

Leucine translocation showed a triphasic saturation response to concentration (Fig. 1B). The first phase was from 0 to 4 mM, the second from 4 to 40 mM, and the third from 40 to 100 mM. The latter concentration approaches the limits of solubility of leucine in water at 25 C. The apparent K_m's for the three phases were 3, 21, and 52 mM, respectively. The presence of many saturated phases in the concentration response curves is evidence that transport of both sucrose and amino acids is mediated by carriers, but the differences in kinetics suggest that different carrier systems are involved in sucrose and amino acid loading.

Further evidence for distinct carrier systems for sucrose and amino acids is provided by other data. The [^{14}C] sucrose and [^{3}H] leucine were combined so that the sucrose concentration was always four times higher than that of leucine. When various concentrations of this mixture were applied to the abraded spot on the leaf, the presence of sucrose did not decrease the amount of leucine transported (represented by the open circles in Fig. 1B). In another experiment, exogenous sucrose at 100 mM had no effect on translocation of 10 mM leucine (data not shown). Both experiments indicate that sucrose and leucine were not competing for the same carrier site.

METABOLIC DEPENDENCE OF LOADING MECHANISM

Metabolic inhibitors were applied to the abraded areas of some source leaves 30 min before, during, and after the addition of labeled sucrose and/or leucine. The uncouplers, 2,4-dinitrophenol (DNP) and carbonylcyanide-m-chlorophenylhydrazone (CCCP), effectively inhibited the translocation of 30 and 100 mM leucine and sucrose, but sucrose translocation was more sensitive than leucine to inhibition by DNP (Table 3 and Fig. 1). The different phases of the saturation curves (Fig. 1) were eliminated by 5 mM DNP, but the linear response observed in the presence of DNP suggests that some leucine and sucrose diffused into the phloem.

p-Chloromercuribenzenesulfonic acid (PCMBS) complexes with sulfhydryl groups and has been shown to be a specific and potent inhibitor of sucrose phloem loading and translocation (3). PCMBS at 1 mM inhibited

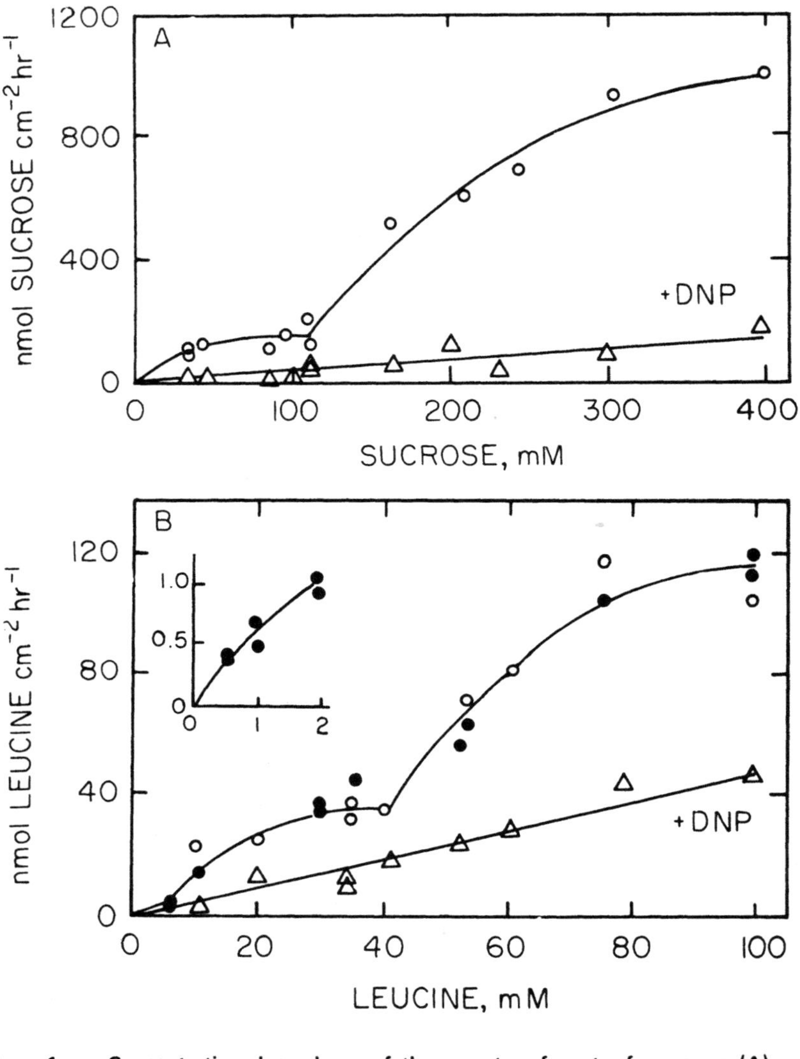

Figure 1. Concentration dependence of the mass transfer rates for sucrose (A) and leucine (B) applied exogenously to leaves. The [^{14}C] sucrose and [^{3}H] leucine were mixed so that sucrose was four times higher in concentration. Transport of sucrose and leucine was monitored simultaneously by applying an aliquot of this mixture to an abraded area on a soybean leaf. The amount transported from the leaf after 2 hr was determined (—○— in both figures). Another group of plants was treated similarly except that 50 μl of 5 mM DNP were added as a pretreatment for 30 min before adding the mixture (—△— in both figures). A third group of plants had no sucrose applied with leucine (—●— in Fig. 1B).

Table 3. Effect of inhibitors on the mass transfer rates of exogenous leucine and sucrose applied to abraded spots on soybean leaves. Modified from Servaites et al. (10).

Treatment	Experiment	Leucine		Sucrose	
		30 mM	100 mM	30 mM	100 mM
		Percent of Control[a]			
DNP, 5 mM	1	40		5	
	2	22		11	
	3		56		12
CCCP, 0.1 mM	1	51		47	
PCMBS, 1 mM	1	23		10	
	2	37		11	
	3		39		7
KCl, 0.1 M	1	51		46	
Sorbitol, 0.8 M	1	222		74	
	2	117		80	

[a]Control rates (nmol cm^{-2} hr^{-1}) for leucine, expts. 1-3 were: 42, 51, and 91; for sucrose, expts. 1-3 were: 91, 104, and 248.

sucrose translocation in soybeans by about 90% and leucine translocation by about 65% (Table 3). Giaquinta (3) concluded that sucrose was actively accumulated in the phloem from the apoplast and that membrane sulfhydryl groups were involved in the process. Because PCMBS inhibited both sucrose and leucine transport in soybeans, sulfhydryl groups may be important for amino acid loading as well. These data support a recent model (5) for phloem loading of sucrose in which the loading is driven by an ATPase-generated proton gradient across the sieve tube plasmalemma.

High concentrations (100 mM) of KCl inhibited both sucrose and leucine translocation (Table 3) by about one-half in soybeans. High K^+ concentrations also inhibited sucrose phloem loading in sugarbeet (4) and castor bean (9), presumably by depolarizing the proton gradient between the phloem and free space.

The application of 0.8 M mannitol to sugarbeet leaves resulted in partial plasmolysis of mesophyll cells without disturbing translocation of exogenous sucrose (11). We found that 0.8 M sorbitol inhibited sucrose translocation slightly, but stimulated leucine translocation (Table 3). The reason for this stimulation is not known, but perhaps the high osmoticum plasmolyzed mesophyll cells permitting a faster penetration of the leucine into the free space. Regardless of the reason for this stimulation, these data support the role of the apoplast as the primary route for amino acid loading.

In general, sucrose and leucine translocation responded similarly to inhibitors, suggesting that both are metabolically dependent. These inhibitors could affect amino acid translocation by inhibiting some process common to loading of both amino acids and sucrose, or they could inhibit amino acid transport indirectly by inhibiting sucrose translocation directly. In the latter case, amino acid transport could be dependent on, or limited by, sucrose loading into veins to provide the driving force for solute movement in the phloem. However, if amino acid transport were solely dependent on sucrose transport, then some stimulation of amino acid transport should have been apparent in Figure 1B when high concentrations of sucrose were added with leucine. It seems likely that amino acid availability and/or an amino acid loading system regulate amino acid transport in soybeans.

CONCLUSIONS

1. Transport of amino acids and sucrose from a soybean source leaf was halted by heat girdling the phloem of the petiole of that source leaf.
2. Partitioning of [^{14}C] assimilates from current photosynthate differed in the source leaf and transport path. The [^{14}C] amino acids comprised 13% of the recovered radioactivity in the source leaf but only 4% in the petiole.
3. Three amino acids (serine, alanine, and glutamate) were predominant in the source leaf. These three plus aspartate and GABA were high in the petiole.
4. No barrier seems to exist to the loading of other amino acids into the phloem from the apoplast.
5. Transport velocities for amino acids were similar to that for sucrose, but mass transfer rates were lower for the amino acids.
6. Loading of leucine and sucrose into the phloem were concentration-dependent. Leucine loading showed a triphasic saturation response whereas sucrose showed a biphasic saturation response.
7. Loading of leucine and sucrose was inhibited by metabolic inhibitors, suggesting a dependence on one or more metabolic processes.

NOTES

L. E. Schrader, Department of Agronomy, University of Wisconsin—Madison. T. L. Housley, Department of Agronomy, Purdue University, West Lafayette, Indiana (formerly postdoctoral fellow at the University of Wisconsin—Madison). J. C. Servaites, USDA/SEA, Light and Plant Growth Lab, BARC-West, Beltsville, Maryland (formerly postdoctoral fellow at the University of Wisconsin—Madison).

The authors gratefully acknowledge that Dr. J. A. Weybrew, Department of Crop Science, North Carolina State University, Raleigh, provided the amino acid analyses shown in Table 1, and that Marna Miller provided expert technical assistance. This research was supported by the College of Agricultural and Life Sciences, University of Wisconsin—Madison, by American Soybean Association Research Foundation Grant 75-ASARF-208-3, and USDA-CSRS Grant 616-15-72.

REFERENCES

1. Berlin, J., and U. Mutert. 1978. Evidence of distinct amino acid transport systems in cultured tobacco cells. Z. Naturforsch. 33:641-645.
2. Geiger, D. R., and S. A. Sovonick. 1975. Effects of temperature, anoxia, and other metabolic inhibitors to translocation. p. 256-286. In. A. Pirson and M. H. Zimmerman (ed.) Encyclopedia of plant physiology, new series, vol. I. Springer, Berlin.
3. Giaquinta, R. T. 1976. Evidence for phloem loading from the apoplast. Plant Physiol. 57:872-875.
4. Giaquinta, R. T. 1977. Phloem loading of sucrose. Plant Physiol. 59:750-755.
5. Giaquinta, R. T. 1977. Possible role of pH gradient and membrane ATPase in the loading of sucrose into the sieve tubes. Nature 267:369-370.
6. Housley, T. L., D. M. Peterson, and L. E. Schrader. 1977. Long distance translocation of sucrose, serine, leucine, lysine, and CO_2 assimilates. I. Soybean. Plant Physiol. 59:217-220.
7. Housley, T. L., L. E. Schrader, M. Miller, and T. L. Setter. 1979. Partitioning of $[^{14}C]$ photosynthate, and long distance translocation of amino acids in preflowering and flowering, nodulated and non-nodulated soybeans. Plant Physiol. (In press.)
8. Israel, D.W., and P. R. McClure. 1979. Nitrogen translocation in xylem of soybeans. This volume.
9. Komor, E. 1977. Sucrose uptake by cotyledons of *Ricinus communis* L.: Characteristics, mechanism, and regulation. Planta 137:119-131.
10. Servaites, J. C., L. E. Schrader, and D. M. Jung. 1979. Energy-dependent loading of amino acids and sucrose into the phloem of soybeans. Plant Physiol. (In press.)
11. Sovonick, S. A., D. R. Geiger, and R. J. Fellows. 1974. Evidence for active phloem loading in the minor veins of sugar beet. Plant Physiol. 54:886-891.
12. Tokunaga, M., Y. Nakano, and S. Kitaoka. 1976. The GABA shunt in the callus cells derived from soybean cotyledon. Agr. Biol. Chem. 40:115-120.
13. Ziegler, H. 1974. Biochemical aspects of phloem transport. Soc. Expt. Biol. Symp. 28:43-62.

NITROGEN TRANSLOCATION IN THE XYLEM OF SOYBEANS

D. W. Israel and P. R. McClure

The predominate form of nitrogen transported from the root to shoot in the xylem varies widely among higher plants. Nitrate, amino acids, amides and ureides (allantoic acid and allantoin) (1,22,23,25) have been shown to be predominate forms of N transported in various plants. Recently, it has become apparent that the predominate form of N transported in the xylem of soybean plants depends upon the source of N that is assimilated for growth. Studies from several laboratories (3,15,18,31) have demonstrated that soybean plants, primarily dependent upon N_2-fixation, transport predominately allantoic acid and allantoin in the xylem; and plants that are primarily dependent upon assimilation of NO_3 transport predominately NO_3, asparagine and glutamine.

The objectives of this chapter are to (a) summarize experiments which established the connection between the form of N assimilated by soybean plants and the form of N transported in the xylem, (b) examine how major N transport forms may be synthesized in the root system and utilized upon delivery to the shoot, (c) discuss physiological and biochemical implications that result from measurement of the N composition of xylem sap, and (d) examine practical implications of the association between the source of N nutrition and predominate N transport forms. As these ideas are explored, attempts will be made to identify gaps in current understanding and experimental approaches which may hold promise for filling in some of the gaps.

NITROGEN TRANSPORT FORMS AS INFLUENCED BY NITROGEN NUTRITION

Kushizaki et al. (12,13) observed that the ratio of ureide-N (allantoin and allantoic acid) to amino acid-N in the soluble fraction of various above-ground tissues was influenced by the application of N fertilizer, and thus were the first to suggest that ureides might have an important role in the N metabolism of the soybean plant. They observed (14) that about 80% of the total soluble N in the hot H_2O soluble fraction of upper stem tissue was ureide-N raised the possibility that ureides were the predominate forms of N delivered to the shoot of soybean plants in the xylem stream. It was reported subsequently (15) that ureides were the predominate forms of N in xylem sap collected over a 24-hr period from the stems of nodulated, N_2-fixing soybean variety, A62-1, decapitated just above the cotyledonary node. A non-nodulating variety, A62-2, assimilating NO_3 accumulated low levels of ureide in the upper stems, and xylem sap collected during the first 12 hr after decapitation from this variety contained very low concentrations of ureide (15). These studies suggested a relation between the source of N assimilated and the form of N transported in the xylem sap.

In contrast to the results of the Japanese workers, Streeter (29) reported that asparagine was the principle N compound in the xylem sap of nodulated soybean plants growing in the field. The problem with this study was that the importance of asparagine as a transport form was assessed by determining its relative abundance as a percentage of the sum of the sap NO_3-N and amino acid-N concentration rather than as a percentage of the total sap N concentration. Streeter (31) has subsequently reanalyzed sap samples (stored frozen) from this study and has reported that ureide-N concentrations during reproductive development were two to six times greater than amino acid-N plus NO_3-N concentrations.

The aforementioned studies on xylem sap composition were either limited to one time during vegetative growth, based on analyses of sap samples collected over long periods of time after decapitation (12 to 24 hr), incomplete in terms of a detailed sap N composition, or conducted with plants growing on poorly defined sources of N supply. For these reasons a series of experiments was initiated in our laboratory to characterize the influence of (a) seasonal development, (b) source of N nutrition, and (c) rhizobial strain and plant genotype on the form of N transported in the xylem of soybean plants.

That changes in rate of water flow and in flux of solutes to the xylem upon decapitation of plants do not significantly alter the relative abundance of N compounds in the xylem sap is an assumption (22,29) used in the interpretation of data in this report. A time course experiment in which the relative distribution of N compounds in sap changed only to a small extent during the first 20 min after decapitation of nodulated plants (18) suggested that this assumption was reasonably valid. In all experiments, collection periods of 20 min or less were used.

In this report total sap nitrogen refers either to a summation of ureide-N, total amino acid-N (ASN, GLN and common α amino acid), NO_3-N and NH_4-N or to total N determined by a modified Kjeldahl procedure in which NO_3 was reduced quantitatively to NH_4^+ (18). Total sap N concentration determined by the summation procedure was always verified by analysis of separate sample aliquots by a modified Kjeldahl procedure (18).

The effect of development from vegetative to reproductive growth stages on the N composition of xylem sap of soybean plants (Ransom cultivar) exposed to single sources of N and to changes from one source of N to another was examined in a greenhouse culture system (18). Nitrogen composition of sap from plants dependent either on nodular N_2 fixation or assimilation of NO_3 was essentially constant from early vegetative to mid podfill growth stages. Sap from plants dependent on nodular N_2 fixation contained seasonal averages of 78 and 20% of the total nitrogen as ureide-N and total amino acid-N, respectively. Sap from non-nodulated plants assimilating NO_3 contained seasonal averages of 6, 36, and 58% of the total N as ureide-N, total amino acid-N and NO_3-N, respectively. When well nodulated plants were supplied with 20 mM KNO_3 nutrient solution beginning at the late flowering growth stage, the ureide-N content declined to about 12% of the total sap N while the NO_3-N and total amino acid-N content each increased from 40 to 45% of total sap N over the following three or four weeks. This study demonstrated clearly that the predominate form of N in the xylem was dependent upon the source of N assimilated and that the predominate transport forms do not vary appreciably as plants progress from the vegetative to the reproductive growth stage provided the source of nitrogen is adequate and does not change.

Data in Table 1 illustrate in more detail the effect of decreasing the proportion of total plant N demand met by nodular N_2-fixation on the N composition of the xylem sap. As the NO_3 level increased, nodule development and N_2 fixation activity were restricted severely (Table 2). As NO_3 supplied to inoculated plants was increased to 20 mM, the mean sap ureide-N declined from 86 to 8.7% of the total N while mean total amino acid-N and NO_3-N increased from 13.68 and 0% to 57 and 34%, respectively (Table 1). The sap nitrogen composition of inoculated plants supplied 20 mM NO_3 (+, 20 mM NO_3 treatment) was not appreciably different from that of non-inoculated plants receiving the same level of NO_3 (-, 20 mM NO_3 treatment). Allantoic acid comprised greater than 85% of the total ureide-N in xylem sap of plants from all treatments (data not shown).

Table 1. Effect of source of N on the N composition of xylem sap of soybean plants.[a] Davis soybean plants were cultured as described previously with the exception that deionized H_2O rather than tap H_2O was used for irrigation (18). One group of plants was inoculated at transplanting with *Rhizobium japonicum* strain USDA 110 and during growth received nutrient solutions containing 0, 5, 10 and 20 mM KNO_3 and another group of plants was not inoculated and received nutrient solution containing 20 mM KNO_3. Since plants were cultured in late fall, natural light was supplemented for 12 hr each day with metal halide lamps which provided 400 $\mu Em^{-2}s^{-1}$ of PAR light at bench level. Flowering was prevented by extending the photoperiod to 16 hr with low intensity light from incandescent lamps. At 48 days after transplanting, sap was collected and analyzed subsequently for ureides, amides, amino acids, NH_4^+ and NO_3 as described previously (18).

N-Source		Total[b] Sap N (μg/ml)	% Total Sap N			
Nodules	NO_3-N mM		Ureide[c] N	Total Amino Acid-N	NH_4-N	NO_3-N
+	0	446	86.29	13.68	0.03	0
+	5	342	59.25	29.78	0.03	10.93
+	10	450	14.32	41.81	0.07	43.81
+	20	485	8.74	57.51	0.10	33.65
-	20	466	5.86	50.65	0.08	43.40
LSD .05		103	7.26	5.68	0.03	8.17

[a] Values represent means of four replicates.
[b] Summation of all N compounds measured.
[c] Allantoic acid + allantoin. Values determined with an amino acid analyzer were verified by analysis of separate aliquots by the Young and Conway method (35).

The change in total amino acid-N content of the sap in response to increased NO_3 supply was due almost entirely to an increase in the relative abundance of asparagine (Table 3). Mean asparagine-N increased from 7.6 to 48.6% of the total sap N while the relative abundance of glutamine and α amino acids was about the same for all treatments.

Since about 56% of the total N in sap of non-nodulated plants assimilating NO_3 (-, 20 mM NO_3 treatment, Table 1) was in reduced forms, it may be inferred that appreciable NO_3 reduction occured in the root system. Substantial root NO_3 reduction is also suggested by similar increases in asparagine and NO_3 sap content in response to increased external NO_3 supply (Tables 1 and 3). Because of the possibility of reduced N compounds, asparagine in particular, being transported from the shoot to the root system in the phloem and being exchanged into the xylem, the exact extent of root NO_3 reduction cannot be determined from the relative proportions of reduced N and NO_3-N

Table 2. Relation of xylem sap ureide content of N_2-fixation activity. Experimental conditions are described in the legend of Table 1.[a]

N-Source Nodules	NO$_3$-N mM	% Total Sap N As Ureide	Estimate[c] % Total N Input by N_2-Fixation	Nodule Dry Wt g/Plant	C_2H_2 Red. Act. µM C_2H_4/hr/plant
+	0	86.29	100.00	1.22	106.7
+	5	59.25	73.51	1.00	117.3
+	10	14.32	12.26	0.17	16.2
+	20	8.74	1.72	0.06	2.8
-	20	5.86	0	0	0.5
r^2[b]		--	0.964	0.923	0.850

[a]Values listed represent the means of four replicates.
[b]Percent total sap N as ureide-N was correlated with the other three parameters listed in the table. Correlations are based on values for each replicate of each treatment.
[c]Derived from measurement of C_2H_2 reduction activity at the time of sap collection and from total plant N accumulated during the growth period.

Table 3. Effect of source of N assimilated on the distribution of N in the amino acid fraction of xylem sap of soybean plants.[a] Experimental conditions and total sap N concentrations are identical to those cited for Table 1.

N-Source		% Total Sap N As		
Nodules	NO$_3$-N mM	ASN	GLN	α Amino Acids
+	0	7.64	2.53	3.51
+	5	20.21	3.27	6.30
+	10	33.41	2.32	6.08
+	20	48.65	2.83	6.03
-	20	41.71	3.26	5.68
LSD .05		4.55	0.92	1.40

[a]Values represent the means of four replicates.

in the sap. This problem could be clarified to some extent with experiments in which $^{15}NO_3$ is supplied to roots of non-nodulated plants at the same time as decapitation of the shoot. Appearance of ^{15}N label in the reduced N fraction of the sap would provide support for root NO_3 reduction. The rate at which the atom percent excess ^{15}N increased in the reduced N fraction of the xylem sap would permit inferences about the extent of recycling of reduced N from the shoot.

Recently we have examined the influence of rhizobial strain and host plant genotype on sap N composition. Three varieties (Essex, Davis and Ransom) of early, mid and late maturity, respectively, under North Carolina conditions were nodulated by *Rhizobium japonicum* strains USDA 110 and 31. Strains 110 and 31 have been characterized previously as "efficient" and "inefficient" N_2-fixing strains, respectively, based on differences in the amount of H_2 gas evolved during nodular N_2 fixation (26). The three varieties growing in low N greenhouse culture system for a full season were shown to produce twice as much seed and to accumulate twice as much total N when nodulated by strain 110 than when nodulated by strain 31 (unpublished data from same experiment).

Sap collected from plants nodulated by strain 110 during early pod development contained significantly more total nitrogen than sap from plants nodulated by strain 31 (Table 4).

For all rhizobial strain-host plant genotype combinations ureide-N predominated the sap N composition, containing 75 to 89% of the total N (Table 4). However, there were subtle, but statistically significant, changes in the relative sap N composition in response to nodulation by the different strains. The relative ureide-N content was 9 to 13% higher in sap from plants nodulated by strain 31 than in sap from plants nodulated by strain 110, while the relative total amino acid-N was 9 to 13% lower. More detailed examination of the sap N composition indicates that allantoic acid-N increased and allantoin-N, asparagine-N and glutamine-N decreased in relative abundance in response to nodulation by strain 31 (Table 5). Although the relative abundance of allantoic acid-N was higher, its absolute concentration (product of total N concentration and proportion of total N) as well as that of the other compounds was lower in sap from plants nodulated by strain 31 (Tables 4 and 5). Furthermore, multiplication of exudation rate during the collection period by the concentration of specific N compounds indicated that nodulation by strain 31 caused a greater decrease in the flux of all compounds than in their concentrations (Table 6). These changes in response to rhizobial strain were consistent among the varieties (Tables 4, 5, and 6).

The lower absolute concentrations and lower fluxes of specific N compounds and of total N in the sap of plants nodulated by strain 31 are consistent with the observations that strain 31 has a lower N_2-fixation capability

Table 4. **Influence of rhizobial strain and host plant genotype on the N composition of xylem sap.[a] Plants of three cultivars (Essex, Davis and Ransom) were inoculated at transplanting with pure cultures of *Rhizobium japonicum* strains USDA 110 and USDA 31. Plants were supplied minus N nutrient solution throughout growth (18) and cultured in a greenhouse under natural sunlight and daylength conditions from May 20 until early to mid August. Sap was collected during early pod development (71 days, 92 days and 93 days after transplanting for Essex, Davis and Ransom varieties, respectively). All other experimental conditions were as described previously (18).**

			% Total Sap N As			
Maturity Group	Genotype	Rhizobial Strain	Total[b] Sap N μg/ml	Ureide-N[c]	Total Amino Acid-N	NH_4-N
V	Essex	USDA 110	316	75.75	24.16	0.08
		USDA 31	154	84.43	15.22	0.35
VI	Davis	USDA 110	254	73.01	26.89	0.20
		USDA 31	104	86.28	13.51	0.20
VII	Ransom	USDA 110	254	80.54	19.33	0.12
		USDA 31	63	89.13	10.48	0.39
LSD .05			60	3.46	3.49	0.13

[a]Values represent means of four replicates.
[b]Summation of all N compounds measured.
[c]Allantoic acid + allantoin.

Table 5. **Influence of rhizobial strain and host plant genotype on the relative concentrations of specific N compounds in the xylem sap.[a] Experimental conditions and total sap N concentrations are identical to those cited in Table 4.**

		% Total Sap N As			
Genotype	Rhizobial Strain	Allantoic Acid[b]	Allantoin[b]	ASN	GLN
Essex	USDA 110	57.68	18.07	15.45	4.32
	USDA 31	78.48	5.95	7.67	1.10
Davis	USDA 110	52.97	20.05	16.97	4.17
	USDA 31	83.76	2.46	5.12	0.85
Ransom	USDA 110	69.11	11.43	13.00	2.69
	USDA 31	86.10	3.12	3.34	0.47
LSD .05		6.10	5.33	2.39	1.07

[a]Values represent means of four replicates.
[b]Values obtained from an amino acid analyzer were verified by analysis of separate sample aliquots by the method of Young and Conway (35).

Table 6. **Influence of rhizobial strain and host plant genotype on flux of total N and of specific N compounds in the xylem sap.[a] Experimental conditions are described in legend of Table 4.**

Genotype	Rhizobial Strain	Exudation Rate µl/min/plant	N Flux (µg N/min/plant) As Total Nitrogen	Allantoic Acid	Allantoin	ASN	GLN
Essex	USDA 110	84.6	27.08	15.71	4.92	4.10	1.15
	USDA 31	47.6	7.21	5.67	0.42	0.55	0.08
Davis	USDA 110	163.9	43.12	22.14	9.41	7.14	1.94
	USDA 31	152.9	18.10	14.94	0.66	0.96	0.16
Ransom	USDA 110	144.6	35.11	24.30	4.00	4.57	0.95
	USDA 31	113.4	7.08	6.06	0.24	0.26	0.03

[a] Values represent means of four replicates.

than strain 110 (26, unpublished results). The increase in relative abundance of allantoic acid and decrease in relative abundances of allantoin, asparagine and glutamine suggest that the sizes of pools accessible for loading into the xylem were decreased to differing extents in response to the decrease in N_2-fixation rate.

SITE OF UREIDE SYNTHESIS

The close correlation between parameters related to nodule development and function and percent of total sap N as ureide-N suggest that the nodule is the site of synthesis of the greater part of ureide that is transported and accumulated in nodulated soybean plants (Table 2.) That roots have some capacity to synthesize ureides is suggested by the occurrence of a low level of ureide in the sap of non-nodulated plants supplied NO_3 (Table 2). This is also supported by the observation that sap from a non-nodulating soybean variety, A62-2, contained low concentrations of ureide (15) and by the observation that ureides are predominate N transport forms in the xylem sap of non-inoculated plants of *Phaseolus vulgaris* supplied high levels of NO_3 (34).

The close association between ureide transport in the xylem sap and nodule development and function could result from (a) ureide synthesis in root tissue from substances delivered to them from the developing nodules or (b) ureide synthesis within the nodules. To examine the function of nodules in ureide production, Fujihara and Yamaguchi (5) fed nodulated and nodule-detached soybean plants $^{15}NH_4$ (50 ppm, 10% atom excess) for one week and then measured the ^{15}N excess in the N fraction of different plant parts. The exogenous NH_4^+ concentration (50 ppm) was not enough to alter N_2-fixation activity of the nodules. The rationale was that if a given N compound was produced mainly in the roots high incorporation of ^{15}N into this substance would be detected and, alternatively, if it was produced in the nodule in conjunction with assimilation of N fixed via N_2 fixation the ^{15}N abundance would be relatively low. The atom percent excess of the applied NH_4^+ indicated that about 80% of the N in this fraction of roots and stems of nodulated plants was derived from the NH_4^+ supplied during the treatment period. While the ureide-N fraction of roots and nodules contain ^{15}N, the enrichment was low in comparison to that of the amino acid-N fraction. The ^{15}N enrichment of the amino acid-N and ureide-N fractions of nodules was about 1/6 and 1/5, respectively, that of roots and stems. This study indicated that roots have a low capacity to synthesize ureides, which is in agreement with inferences from sap N composition studies. In another report Matsumoto et al. (17) demonstrated that, when nodulated plants were exposed to $^{15}N_2$ for 24 hr, the atom percent excess of the ureide-N fraction of nodules was eight-fold that of the roots. These results collectively support the view that the primary site of synthesis of ureides is within the nodules.

Activities of xanthine oxidase, the enzyme that oxidizes xanthine to uric acid, and of uricase, the enzyme that oxidizes uric acid to allantoin, were measured in nodules of four-week old plants while neither enzyme activity was detected in the roots, stems and leaves (32). Significant activity of allantoinase, the enzyme that oxidizes allantoin to allantoic acid, was measured in all plant tissues (32). The apparent preferential localization of xanthine oxidase and uricase in nodules provides corraborative support for the inference from ^{15}N labeling and sap N composition studies that nodules are the primary sites of ureide synthesis.

Perhaps the most convincing support for nodules as the primary site of ureide synthesis is provided by the observations of (a) very high ureide concentrations (2 to 5 mg ureide-N/ml) in sap collected from detached nodules (31), and (b) increased xanthine and decreased ureide levels in nodules of plants exposed to allopurinol, an inhibitor of the conversion of xanthine to uric acid by xanthine oxidase (xanthine dehydrogenase) (4). The question of whether the enzyme responsible for conversion of xanthine to uric acid is an oxidase or dehydrogenase has not been fully resolved. Allopurinol would inhibit either activity.

COUPLING OF UREIDE SYNTHESIS TO CURRENT FIXATION OF N_2

The question of how closely ureide synthesis in the nodules and subsequent export in the xylem are coupled to reduction of N_2 has received little experimental attention. In the only $^{15}N_2$ labeling study for soybean plants (17) the labeling period was too long to permit inferences about the relationship between ureide synthesis and transport and N_2 reduction. A more definitive examination of this question has been made with the ureide-forming plant *Vigna unguiculata* (7). Two hr after exposure of nodulated plants to $^{15}N_2$, 91% of the total ^{15}N recovered in the xylem sap was in allantoin and allantoic acid. Unfortunately, the atom percent ^{15}N excess of N in allantoin and allantoic acid of the sap was not presented. An atom percent ^{15}N excess of allantoin and allantoic acid in xylem sap approaching the atom percent ^{15}N excess of the N_2 gas supplied to the nodules would provide definitive evidence for a close connection between N_2 reduction and ureide synthesis and transport. Obviously, short term (30 min - 4 hr) $^{15}N_2$ labeling studies with N_2-fixing soybean plants are needed.

In preliminary experiments (unpublished) we have measured 60% decreases in total N and ureide-N concentrations of sap within 6 hr of inhibiting N_2 reduction by nodules with saturating levels of acetylene (0.1 to 0.2 atm). If acetylene has no detrimental affect on xylem loading processes, this decline in total N and ureide-N concentration of the sap also suggests a relatively close connection between N_2 reduction and transport of fixed N in the form of ureides.

USE OF UREIDES DELIVERED TO THE SHOOT IN THE XYLEM

If relative N composition of sap collected from cut stems is a reasonable reflection of N composition in the xylem stream of intact plants, it becomes apparent that shoot tissues must have the capacity to utilize massive amounts of ureide-N for growth and development. Molar allantoin: allantoic acid ratios in xylem sap of nodulated plants range from 30:70 to 5:95 (Table 4) (31) while the ratio in the soluble N fraction of stems and leaves has been reported to be 60:40 (31). This change from allantoic acid as the predominate ureide in the xylem sap to allantoin as the predominate ureide in the soluble N fraction of stems and leaves implies a much greater use of allantoic acid to support vegetative growth processes than allantoin.

A rapid decline in ureide-N content of stems and pod walls during pod and seed development, low ureide-N concentrations in mature seeds relative to other tissues, and negative correlations between ureide-N concentration and relative growth rates of vegetative tissues have been observed (3,8,16). Since total ureide-N concentrations rather than specific concentrations of allantoin and allantoic acid were presented in these reports, the interpretation is at best tenuous. However, if the molar allantoin: allantoic acid ratio in the soluble N fraction of tissues in these experiments were similar to the 60:40 value reported by Streeter (31), the observations are consistent with the hypothesis that allantoin has a relatively limited role in providing N for vegetative growth, but upon remobilization from vegetative tissues to developing fruits, it has an important role in providing N for seed protein synthesis (8).

BIOCHEMICAL AND PHYSIOLOGICAL IMPLICATIONS OF UREIDES AS PREDOMINATE XYLARY NITROGEN TRANSPORT FORMS

The predominance of ureide-N in bleeding sap from nodulated, N_2-fixing soybean plants suggests a route of N assimilation that differs from that of soybean plants assimilating NO_3 or from that of some other nodulated legumes. Asparagine has been shown to be the major N transport form in nodulated, N_2-fixing plants of *Lupinus albus* (24), *Pisum arvense* (23) and *Pisum sativum* (20). The scheme in Figure 1 is presented as a means of considering the possible route of N assimilation in soybean plants dependent on N_2-fixation. The nodule is considered to consist of two compartments, the bacteriods and the "cytosol." The cytosol refers to the plant cell cytoplasm of the nodule including that which immediately surrounds the bacteroids.

It has been established clearly that bacteroids contain nitrogenase which catalyzes the ATP and reductant dependent reduction of N_2 to NH_4^+ (9,11). The many fold higher activity of glutamine synthetase in nodule "cytosol" fractions than in bacteroids of soybean plants (19) and other legumes (2) suggests that most of the NH_4^+ formed during N_2 reduction is assimilated into glutamine by glutamine synthetase of the "cytosol" fraction. Sap composition studies suggest that glutamine has at least three fates:

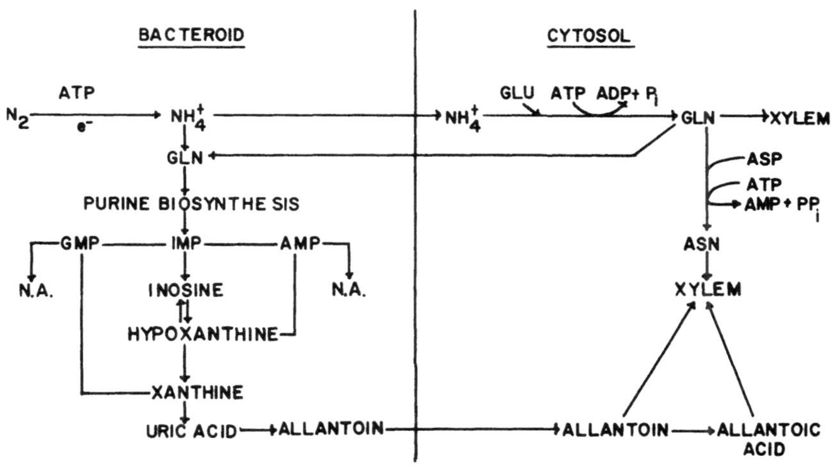

Figure 1. **Proposed nitrogen assimilation pathway for nodulated N_2-fixing soybean plants. N.A. = nucleic acids.**

(a) direct transport to the xylem, (b) support of synthesis, asparagine and other amino acids and (c) support of ureide synthesis as it donates N at positions 3 and 9 in the purine ring of compounds presumed to be precursors of ureide synthesis. The observation that 80% of the N in xylem sap is ureide-N (18) suggests that support of ureide synthesis is by far the most important fate of glutamine. Asparagine synthesis is proposed to occur via asparagine synthetase which catalyzes ATP dependent transfer of the amide group of glutamine to aspartic acid. This enzyme has been observed in the "cytosol" of lupin nodules (27), but to our knowledge it has not been reported to occur in soybean nodules. Regardless of the enzymatic reaction employed, it seems reasonable to propose that asparagine is synthesized in the "cytosol" fraction because the bacteroid fractions of soybean nodules have very high activities of asparaginase, an enzyme which degrades asparagine to aspartic acid (30).

The proposal that ureide synthesis occurs via a purine synthesis and degradation pathway is based on: (a) the occurrence of purine catabolic enzymes (xanthine oxidase-xanthine dehydrogenase, uricase and allantoinase) in nodule tissues (32), (b) accumulation of high levels of xanthine and decreases in the concentration of allantoin and allantoic acid in nodules of plants exposed to allopurinol, an xanthine oxidase-xanthine dehydrogenase inhibitor (4), and (c) lack of any evidence for direct condensation of urea and glyoxylate to form allantoic acid in any plant known to transport ureides in the xylem (25).

The proposals that (a) glutamine is transported into the bacteroid from the "cytosol" and (b) purine biosynthesis and the initial steps of purine degradation occur in the bacteroid are primarily hypothetical (Figure 1). The apparent restriction of xanthine oxidase-xanthine dehydrogenase activity to the bacteroid fraction (32) does suggest that xanthine or one of its precursors (such as glutamine as in Figure 1) is transported into the bacteroid from the "cytosol." The observation that DNA content of bacteroids from soybean nodules that developed over a 12-week period was similar to that of free-living *Rhizobium japonicum* cells (21) implies that bacteroids maintain the capability to synthesize nucleic acids and their precursors. Thus there is no reason to suspect that bacteroids lack the capability to synthesize purines. Examination of the distribution of purine biosynthetic enzymes among the bacteroid and "cytosol" fractions of nodules would permit more definitive inferences about where purines that support ureide production are synthesized.

The nucleotide, inosine monophosphate (IMP), represents a branch point in purine biosynthesis (Figure 1). The apparent rapid flux of N through purine pools, implied by the high ureide-N content of xylem sap, suggests that the primary fate of IMP is the direct production of hypoxanthine and

and xanthine rather than in production of GMP and AMP for use in synthesis of nucleic acid macromolecules. Non-specific alkaline and acid phosphatases which seem to be ubiquitous in living tissues (6) could catalyze dephosphorylation of IMP to produce inosine. Inosine could be converted to hypoxanthine by purine nucleoside phosphorylase, an enzyme which catalyzes the reversible formation of hypoxanthine and ribose -1- phosphate from inosine and inorganic phosphate (28). Kinetic studies with the enzyme isolated from animal tissues indicate that inosine formation is the preferred direction of the reaction (28). However, if the hypoxanthine is metabolized rapidly to xanthine and uric acid by xanthine oxidase or xanthine dehydrogenase, net catabolism of inosine would occur. The localization of uricase activity in the bacteroids (33) and the occurrence of a major portion of allantoinase activity in the "cytosol" (32) suggest that allantoin is transported from the bacteroids to the "cytosol" of the nodules where it is either metabolized to allantoic acid or loaded in the xylem for transport.

Admittedly the proposed assimilatory route is based on limited evidence, it is presented as a framework around which new experiments can be designed. Purine nucleoside phosphorylase would be a key enzyme activity to assay in nodule "cytosol" and bacteroids because demonstration of activity would permit inferences about the flow of N at the IMP branch point of purine metabolism and about the location of purine biosynthetic pathway(s). Short term exposure of nodulated plants to $^{15}N_2$ and subsequent measurement of ^{15}N enrichment in N compounds of the soluble and insoluble fractions of nodules and in allantoic acid and allantoin of the stem sap would also permit inferences about the flow of fixed N through purine metabolism to the ureides. For example, an atom percent ^{15}N excess in allantoic acid and allantoin of xylem sap approaching that of the $^{15}N_2$ gas supplied to the nodules within a short time following initial exposure would suggest rapid movement of newly fixed N through purine metabolism to a relatively small ureide pool accessible for xylem loading. On the other hand, slow increases in atom percent ^{15}N excess in ureides of the sap would indicate mixing of newly fixed N with unlabeled N in large pools of purines or nucleic acids before reaching pools accessible for xylem loading.

The use of allantoic acid and allantoin as predominate N transport forms may have important consequences in terms of the carbon economy of nodulated root systems. Allantoic acid and allantoin have a C:N ratio of 1:1 while asparagine, the major N transport form in sap of several other nodulated legumes (20,23,24), has a C:N ratio of 2:1. Thus, twice as much carbon is associated with transport of N as asparagine than with the ureides. In nodulated, N_2-fixing soybean plants, a substantial excess of inorganic cations relative to inorganic anions has been observed (10). This suggested the requirement for transport of organic acid anions in the sap to balance excess cationic charge. Recently, we have found that the organic acid anion, malate, balances about 75% of the excess cationic charge and contains about 50% of the total carbon in the xylem sap (unpublished results). In view of this large requirement for carbon to balance excess cationic charge in sap of N_2-fixing plants, the use of ureides as major N transport forms represents an efficient means of conserving carbon for support of other functions in nodulated root systems.

Because biosynthesis of ureides (Figure 1) and xylem loading and unloading processes are not well characterized, an assessment of the energetic requirements for synthesis and transport of ureide relative to other N transport forms such as asparagine is difficult. If it is assumed that (a) the energetic requirement for loading and unloading of xylem is similar for the various N transport forms and (b) that carbon associated with N transport forms would otherwise be metabolized to CO_2 and H_2O to produce 6 ATP per carbon or 36 ATP per glucose equivalent, an interesting inference can be made. The energetic equivalent of carbon associated with the transport of allantoic acid, allantoin and asparagine would be 24 ATP equivalents per molecule. However, since a molecule of ureide has twice as much N per unit carbon, four units of N as ureide would be associated with 24 ATP equivalents, whereas four units of N as asparagine would be associated with 48 ATP equivalents. Thus, although the amount of energy used in biosynthesis of ureide is not known, as much as 24 ATP equivalents per molecule could be used in its biosynthesis without the total energetic requirement for its synthesis and use in N transport exceeding that of asparagine. Although quite complex, synthesis and transport of ureide may not represent any greater energetic burden to the nodulated, N_2-fixing soybean plants than would synthesis and transport of asparagine.

PRACTICAL IMPLICATION OF CLOSE ASSOCIATION BETWEEN SAP NITROGEN COMPOSITION AND SOURCE OF NITROGEN NUTRITION

Several lines of evidence suggest that relative sap ureide content would be a reliable index of the relative contribution that N_2-fixation makes to total N input by nodulated soybean plants growing in the presence of combined N. These include (a) definitive effects of source of N assimilated on N composition of xylem sap, (b) the relative constancy of predominate N transport forms throughout development under constant conditions of N nutrition, (c) the predominance of allantoic acid and allantoin in sap from nodulated N_2-fixing plants, and (d) positive correlations ($r^2 > 0.90$) of parameters associated with nodule development and function with percent of total sap N as ureide. It might be feasible to make reasonably accurate calculations of seasonal N_2 fixation by soybean plants growing in the presence of soil N from measurements of seasonal patterns of relative sap ureide content and of total N accumulation. It is emphasized that this proposal is strictly for the soybean plant and could not be extended to other leguminous plants without detailed understanding of how sap N composition is influenced by the source of N nutrition.

NOTES

D. W. Israel, USDA, SEA, North Carolina State University, and P. R. McClure, North Carolina State University, Raleigh, North Carolina 27650.

The authors wish to thank Drs. W. A. Jackson, S. C. Huber and R. F. Wilson and Mr. Tom Rufty for helpful suggestions during preparation of the manuscript; Ms. Virginia Flynt for expert technical assistance; and Ms. Agnes Luciani for an excellent job of typing the manuscript.

REFERENCES

1. Bollard, E. G.1957. Translocation of organic nitrogen in the xylem. Aust. J. Biol. Sci. 10:292-301.
2. Brown, C. M. and M. J. Dilworth. 1975. Ammonia assimilation by *Rhizobium* cultures and bacteroids. J. Gen. Microbiol. 86:39-48.
3. Fujihara, S., K. Yamamoto and M. Yamaguchi. 1977. A possible role of allantoin and the influence of nodulation on its production in soybean plants. Plant Soil. 48:233-242.
4. Fujihara, S. and M. Yamaguchi. 1978. Effects of allopurinol [4-hydroxypyrazolo (3,4-d) pyrimidine] on metabolism of allantoin in soybean plants. Plant Physiol. 62:134-138.
5. Fujihara, S. and M. Yamaguchi. 1978. Probable site of allantoin formation in nodulating soybean plants. Phytochem. 17:1239-1243.
6. Henderson, J. F. and A. R. P. Paterson. 1973. Catabolism of purine nucleotides, pp. 152-170. In Nucleotide Metabolism—An Introduction, Academic Press, New York, NY.
7. Herridge, D. F., C. A. Atkins, J. S. Pate and R. M. Rainbird. 1978. Allantoin and allantoic acid in the nitrogen economy of the cowpea [*Vigna unguiculata* (L.) Walp.]. Plant Physiol. 62:495-498.
8. Ishizuka, J. 1977. Function of symbiotically fixed nitrogen for grain production in soybean, pp. 617-624. In Proceedings International Seminar on Soil Environment and Fertility Management in Intensive Agriculture, Tokyo, Japan.
9. Israel, D. W., R. L. Howard, H. J. Evans, and S. A. Russell. 1974. Purification and characterization of the molybdenum-iron protein component of nitrogenase from soybean nodule bacteroids. J. Biol. Chem. 249:500-508.
10. Israel, D. W. and W. A. Jackson. 1978. The influence of nitrogen nutrition on ion uptake and translocation by leguminous plants, pp. 113-129. In C. S. Andrew and E. J. Kamprath (Eds.) Mineral Nutrition of Legumes in Tropical and Subtropical soils, CSIRO, Melbourne.
11. Koch, B., H. J. Evans, and S. Russell. 1967. Reduction of acetylene and nitrogen gas by breis and cell-free extracts of soybean root nodules. Plant Physiol. 42:466-468.
12. Kushizaki, M., J. Ishizuka and F. Akamatsu. 1964. Physiological studies on the nutrition of soybean plants. I. Effects of nodulation on growth, yield and nitrogen content. J. Sci. Soil Manure Japan 35:319-322.

13. Kushizaki, M., J. Ishizuka, and F. Akamatsu. 1964. Physiological studies on the nutrition of soybean plants. II. Effect of nodulation on the nitrogen constituent. J. Sci. Soil Manure Japan 35:323-327.
14. Matsumoto, T., Y. Yamamoto and M. Yatazawa. 1975. Role of root nodules in the nitrogen nutrition of soybeans. I. Fluctuation of allantoin and some other plant constituents in the growing period. J. Sci. Soil Manure Japan 46:471-477.
15. Matsumoto, T. Y., M. Yamamoto, M. Yatazawa. 1976. Role of root nodules in the nitrogen nutrition of soybeans. II. Fluctuation in allantoin concentration of the bleeding sap. J. Sci. Soil Manure Japan 47:463-469.
16. Matsumoto, T., M. Yatazawa and Y. Yamamoto. 1977. Distribution and change in the contents of allantoin and allantoic acid in developing nodulating and non-nodulating soybean plants. Plant Cell Physiol. 18:353-359.
17. Matsumoto, T., M. Yatazawa and Y. Yamamoto. 1977. Incorporation of $^{15}N_2$. Plant Cell Physiol. 18:459-462.
18. McClure, P. R. and D. W. Israel. 1979. Transport of nitrogen in the xylem of soybean plants. Plant Physiol. (In press).
19. McParland, R. H., J. G. Guevara, R. R. Becker and H. J. Evans. 1976. The purification and properties of glutamine synthetase from cytosol of soya-bean root nodules. Biochem. J. 153:597-606.
20. Minchin, F. R. and J. S. Pate. 1973. The carbon balance of a legume and the functional economy of its root nodules. J. Exp. Bot. 24:259-271.
21. Paau, A. S., J. Oro and J. R. Cowles. 1979. DNA content of free-living rhizobia and bacteroids of various *Rhizobium*-legume associations. Plant Physiol. 63:402-405.
22. Pate, J. S. 1973. Uptake, assimilation and transport of nitrogen compounds by plants. Soil Biol. Biochem. 5:109-119.
23. Pate, J. S. and W. Wallace. 1964. Movement of assimilated nitrogen from the root system of the field bean (*Pisum arvense* L.). Ann. Bot (N.S.) 28:83-99.
24. Pate, J. S., P. J. Sharkey and O. A. M. Lewis. 1975. Xylem to phloem transfer of solutes in fruiting shoots of legumes, studied by a phloem bleeding technique. Planta 122:11-26.
25. Reinbothe, H. and K. Mothes. 1962. Urea, ureides and guanidines in plants. Ann. Rev. Plant Physiol. 13:129-150.
26. Schubert, K. R., N. T. Jennings and H. J. Evans. 1978. Hydrogen reactions of nodulated leguminous plants. II. Effects on dry matter accumulation and nitrogen fixation. Plant Physiol. 61:398-401.
27. Scott, D. B., K. J. F. Farnden and J. G. Robertson. 1976. Ammonia assimilation in lupin nodules. Nature 263:703-705.
28. Stoeckler, J. D., R. P. Agarwal, K. C. Agarwal and R. W. Parks. 1978. Purine nucleoside phosphorylase from human erythrocytes, pp. 530-538. In S. P. Colowick and N. O. Kaplan (Eds.) Methods in Enzymology. Vol. 41. Purine and Pyrimidine Nucleotide Metabolism, Academic Press, New York.
29. Streeter, J. G. 1972. Nitrogen nutrition of field-grown soybean plants. I. Seasonal variations in soil nitrogen and nitrogen composition of stem exudate. Agron. J. 64:311-314.
30. Streeter, J. G. 1977. Asparagine and asparagine transaminase in soybean leaves and root nodules. Plant Physiol. 60:235-239.
31. Streeter, J. G. 1979. Allantoin and allantoic acid in tissues and stem exudate from field-grown soybean plants. Plant Physiol. 63:478-480.
32. Tajima, S. and Y. Yamamoto. 1975. Enzymes in purine catabolism in soybean plants. Plant Cell Physiol. 16:271-282.

33. Tajima, S., M. Yatazawa and Y. Yamamoto. 1977. Allantoin production and utilization in relation of nodule formation in soybeans. Enzymatic studies. Soil Sci. Plant Nutr. 23:225-235.
34. Thomas, R. J., U. Feller and K. H. Erismann. 1979. The effect of different inorganic nitrogen sources and plant age on the composition of bleeding sap of *Phaseolus vulgaris*. New Phytol. 82:657-669.
35. Young, E. G. and C. F. Conway. 1942. On the estimation of allantoin by the Rimini-Schryver reaction. J. Biol. Chem. 142-839-853.

COUPLING BETWEEN PHOTOSYNTHESIS AND NITROGEN FIXATION

J. G. Streeter, H. J. Mederski, and R. A. Ahmad

The first experimental demonstration of a coupling between photosynthesis (PS) and nitrogen fixation (NF) appeared in the literature 45 yr ago and involved a demonstration of increased N content of clover and alfalfa plants grown in N-free media with CO_2 enrichment (24). Early CO_2 enrichment studies with soybeans (2) were done with soil-grown plants and nodulation and nitrogen fixation were not accurately quantified. However, several studies published within the past 5 yr demonstrate clearly that these two vital processes are coupled in soybean plants. We now find ourselves in a position where this coupling can and should be more thoroughly explored at the cellular and chemical levels.

RECENT EVIDENCE FOR THE COUPLING OF PHOTOSYNTHESIS AND NITROGEN FIXATION

Carbon dioxide enrichment of field-grown soybean plants results in multifold increases in shoot dry wt, seed yield/unit area, nodule wt, and specific NF activity of nodules (7). Providing more substrate for PS led to an estimated 4-fold increase in the quantity of N fixed during the growing season, suggesting that the principal limitation for symbiotic N fixation in soybeans is the availability of photosynthate (carbohydrate). Specific activity of soybean nodules can also be doubled for periods of a few days after doubling the shoot/root ratio (21).

One of the clearest and most comprehensive demonstrations of the coupling between PS and NF was a comparison of several methods for alteration of the photosynthetic source/sink ratio (12). These alterations generated greater than a 4-fold range of nodule activities and, because of concomitant changes in nodule wt, a greater than 6-fold range in acetylene reduced/plant/hr (Table 1). Except for the pod removal treatment, the change in NF activity could be largely explained by the effect of the treatment on photosynthetic rate. In the case of pod removal, the effect on nodule activity was probably due to reduced consumption of carbohydrate for fruit growth. The effect of shade on nodule activity of field-grown soybeans has recently been confirmed in a study involving 5 different levels of shading (23).

Several studies have documented the translocation of photosynthate to nodules following $^{14}CO_2$ assimilation (10,11,18). Most $^{14}CO_2$ incorporated is accumulated in actively growing regions of the plant (10,11) and relatively little (<5%) is recovered in nodules 3 to 18 hr after exposure to $^{14}CO_2$ (11). Recovery of radioactivity in nodules is reduced by application of combined nitrogen and during the period of rapid fruit growth (Table 2). Since availability of combined N and development of reproductive sinks are commonly associated with lowered nodule activity (8,11,12), results with $^{14}CO_2$ support the coupling of PS and NF.

Considerable effort has recently been devoted to similar studies of *Pisum sativum*. The influence of stage of development, supply of combined N, variable source-sink ratios, and light intensity on nodule activity and photosynthesis have been documented. These studies clearly demonstrate a coupling of PS and NF and only a few key references are cited for the convenience of the reader (4,14,15).

Table 1. **Effect of photosynthetic source-sink manipulation with soybeans grown in the field. Treatments were imposed at the end of flowering. Values are means of two varieties and three replications [from Lawn and Brun (12)].**

Treatment	Photosynthesis[1]	Seed Yield[2]	Nodule Wt[3]	Specific Nodule Activity[3]
	mg CO_2/dm$_2$ of ground area	*g/plant*	*g/plant*	*µmoles C_2H_2/g fresh wt · hr*
Supplemental light	47.8	16.0	5.37	23.6
Partial depodding	26.3	11.7	4.13	17.6
Control	40.1	12.9	3.90	12.5
Partial shade	22.0	10.1	2.87	8.6
Partial defoliation	24.7	9.9	2.33	5.6

[1]Nine days after treatment application.

[2]Mature plants.

[3]Mean of 2 sampling times, 10 and 19 days after treatment application.

Table 2. Acetylene reduction and incorporation of $^{14}CO_2$ by soybean nodules as influenced by growth stage and application of combined N. Plants grown outdoors in sand culture. [Data from Expt. 2 of Latimore, et al. (11)].

Growth Stage	N Supplied[1]	^{14}C in Nodules[2]	Acetylene Reduction
		%	*μmole/plant · hr*
Vegetative	none	4.7	4.2
	$(NH_4)_2SO_4$	1.0	1.8
	$NaNO_3$	0.2	0.6
Bloom	none	5.0	4.9
	$(NH_4)_2SO_4$	1.7	1.3
	$NaNO_3$	1.4	0.2
Pod development	none	2.2	3.8
	$(NH_4)_2SO_4$	2.0	0.7
	$NaNO_3$	0.5	0.2

[1]450 mg N supplied 10 days before sampling.

[2]18 hr after exposure to $^{14}CO_2$; expressed as percent of total ^{14}C in plant.

MEASUREMENTS OVER SHORT TIME PERIODS AND DISAGREEMENTS ON THE COUPLING

Studies reviewed above seem to show conclusively that the expected coupling between PS and NF does exist in the soybean plant. One would, therefore, predict that when repeated measurements of NF and PS are made over periods of several days, coupling of the two processes could be more precisely quantified. There are complications, however, as illustrated by several recent experiments.

When CO_2 concentration or water stress was varied to alter photosynthetic rate, daily measurements indicated remarkably similar increases or declines in the rates of NF and PS, leading to a conclusion that the two processes are very closely coupled (9). Using a continuous assay for PS and NF, we have found that NF during a 10-hr dark period declines to only 50% or less of the rate observed in the light (16). The conclusion here was that NF depends not only on concomitant PS, but also on carbohydrate from storage pools. Most recently, reports on field grown or hydroponically grown soybeans indicate no diurnal fluctuation in NF (3,20).

It appears that growth conditions, pre-assay conditions, and conditions during the experimental period can have a pronounced influence on the results obtained. Results of an experiment which will help to illustrate some potential complications are shown in Figure 1. The period of darkness was varied in an attempt to allow all acetylene reduction rates to reach a similar

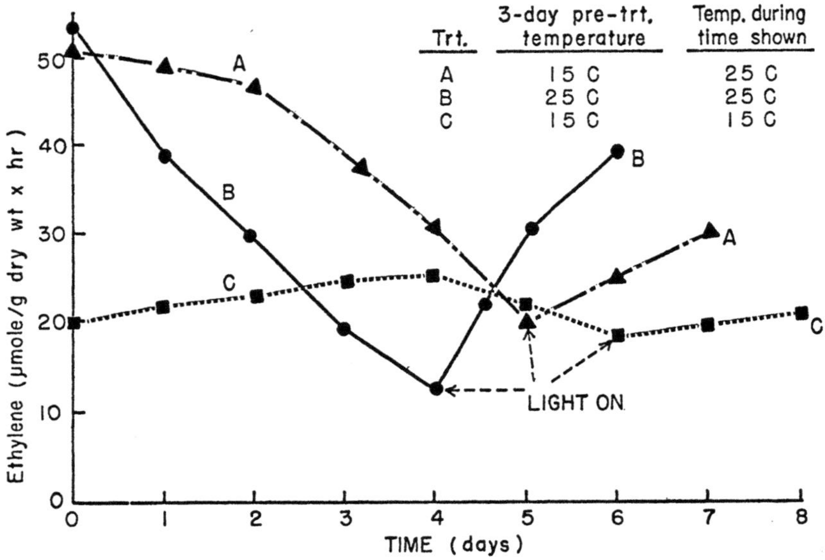

Figure 1. **Acetylene reduction by nodulated soybean roots during periods of darkness and illumination as influenced by root temperature. Fifty-day-old plants were grown in a greenhouse in sand culture with N-free nutrient solution. Roots were washed free of sand and transferred to a spray chamber (16) at the beginning of the pretreatment period during which the photoperiod was 16 hr. Average of two replicates; results from Ahmad (1).**

level before turning the light on. The decline in NF in darkness was relatively rapid when pre-treatment and treatment temperatures were 25 C. When root temperature during the 3-day pre-assay period was 15 C, the dark decline of NF was slowed markedly. When pre-treatment and treatment temperatures were both 15 C, there was no decline in NF during 4 days of darkness and only a small decline during 2 additional days of darkness.

The response of these differently treated plants to a resumption of photosynthesis is even more instructive (Fig. 1). The increase in the rate of NF was directly proportional to the previous rate and magnitude of decline. One might conclude from curve B that the PS-NF coupling is close, from curve A that the coupling is only moderate, and from curve C that the coupling is almost nonexistent.

Nodules were harvested throughout the treatment period shown in Figure 1 for analysis of individual carbohydrate compounds. The results (discussed in more detail in the next section) indicate that much of the divergence in the response of these groups of plants to darkness and re-illumination treatments can be explained by carbohydrate composition. All growth and experimental conditions should be carefully and completely described in experiments of this type and some carbohydrate analysis of nodules is highly desirable.

CARBOHYDRATE COMPOUNDS IN NODULES AND SOME SPECIFIC RELATIONSHIPS TO NITROGEN FIXATION

In spite of the somehwat confusing results in experiments over short time periods, we can conclude that PS and NF are coupled. It is only logical to assume that the chemical basis for the coupling is carbohydrate, and we might predict that carbohydrate concentration in nodules should be positively related to NF activity. This turns out to be true or not true, depending on what sort of carbohydrate analysis is performed.

Initially, we found that standard chemical analyses for reducing sugars or for total "available" carbohydrate (sugar + starch) in nodules yielded values which were not related to NF activity. Similar results had been obtained with pea nodules (13). A recent study has shown that higher total carbohydrate in nodulated roots of male-sterile soybean plants is not associated with higher rates of NF (25). Our results seemed especially puzzling because we were comparing two treatments, one of which had nodules with over twice the NF activity of the control treatment (21,22). This result suggested that nodules rely not only on recently imported photosynthate but also on stored carbohydrate. Another interpretation was that our chemical analyses were not recording compounds which play a role in NF.

A program was begun to identify carbohydrates in nodules and study their individual roles. A current list of compounds which have been identified and an estimate of their seasonal mean concentration in mg/g fresh wt of field grown nodules include: sucrose, 2.84; (+) - pinitol (5-0-methyl-D-inositol), 1.14; D-*Chiro*-inositol, 1.27; glucose, 1.40; α,α - trehalose, 1.34; *myo*-inositol, 0.65; maltose, 0.31; fructose, 0.21; sequoyitol (5-0-methyl-*myo*-inositol), 0.11; and starch (may include some glycogen), 8.30. The principal carbohydrates are sucrose, the major carbohydrate translocated in the phloem of soybean plants, and three cyclitols: pinitol, *chiro*-inositol, and *myo*-inositol. Only glucose, maltose, and fructose respond to tests for reducing sugars, and the cyclitols do not respond to any of the commonly used methods of analysis for total carbohydrate. Pinitol is a major constituent of other parts of the soybean plant (17), but distribution of the other cyclitols is restricted largely to nodules.

In studies where root/shoot ratio was manipulated, shoots were decapittated, or plants were exposed to prolonged darkness, a positive relationship between sucrose or pinitol concentration and NF has been consistently observed (5,22). However, a significant correlation may not be obtained if some other factor (water stress, temperature) is limiting NF rate. For example, analysis of nodules from treatment C in Figure 1 revealed no relationship between pinitol or sucrose concentration and NF. But in situations where carbohydrate is limiting NF rate, we have usually found statistically significant correlations between NF rate and sucrose or pinitol concentration. This is illustrated by analyses of nodules relating to curves A and B (Fig. 1); correlation co-efficients are given in Table 3. In this experiment, correlations between glucose or *myo*-inositol concentrations and NF rate were also highly significant. However, in most of the experiments in which NF rate has been altered, we have found that the glucose and *myo*-inositol correlations were not significant.

We still know little about the localization and metabolism of carbohydrates in soybean nodules. Even with correlations like those in Table 3, we can only speculate that certain carbohydrates are used more directly than others for generating energy and reducing power for NF. While chemical details of the PS-NF coupling mechanism in nodules remain to be elucidated, we can conclude that a basis for coupling of PS and NF has at least been demonstrated at the level of nodule carbohydrate compounds.

THEORETICAL AND "ACTUAL" CONSUMPTION OF CARBOHYDRATE IN NITROGEN FIXATION

Is there justification for continuing to attempt to describe the details of the carbohydrate coupling mechanism in nodules? Perhaps the best way to try to answer this question is to compare estimates of carbohydrate required for NF based on properties of the N fixing enzyme with estimates of carbohydrate actually consumed by legume nodules.

Beginning with the estimate from in vitro studies of nitrogenase that 12 moles of ATP are required/mole of N fixed (5,6) and assuming that one mole of glucose will yield 38 moles of ATP, we can calculate that .32 moles of glucose are required/mole of N fixed. This value is convertable to 57 g glucose/mole N or about 4.1 g glucose/g of N fixed. Estimates of actual

Table 3. Correlations between concentration of carbohydrate compounds and nitrogen fixation (acetylene reduction) rate of soybean nodules. Nodules were obtained from plants treated as described in Figure 1, curves A and B. All correlations are significant at the 1% level of confidence.

Carbohydrate	Range of Concentrations	Correlation Coefficient (Carbohydrate Conc. vs. N Fixation Rate)
	mg/g fresh wt	
Pinitol	.30 - .99	.81
Sucrose	1.30 - 4.16	.90
Glucose	.20 - .40	.94
Myo-inositol	.40 - .60	.86

carbohydrate consumption for NF by nodules are obtained by careful simultaneous measurements of respiration and NF. A recent study of pea nodules led to an estimate of 17 g glucose/g N fixed (15). In another recent study, carbohydrate consumed/gram N fixed by nodulated soybean roots varied according to growth stage; an average carbohydrate requirement of about 20 g/g N fixed was observed during the period of rapid NF (19).

A comparison of the values for carbohydrate consumption based on respiration measurements with values based on the theoretical energy requirement of nitrogenase suggests that there is wasteful consumption of carbohydrate in nodules. Some of the difference can probably be accounted for by the presumably wasteful reduction of protons and evolution of hydrogen by nodules (6). In addition, some carbohydrate is required for nodule growth and for maintenance respiration. But even after allowances are made for these processes, there appears to be at least twice as much carbohydrate consumed as is theoretically required for N fixation, per se.

If these approximations are correct, what might be the reason for this inefficient utilization of carbohydrate and what are the possibilities for increasing the efficiency of N fixation? Our present level of understanding of carbohydrate utilization in nodules does not justify speculation on these questions. We need to know how and where various compounds are metabolized in nodules and especially, how the import of carbohydrate is controlled.

SUMMARY

Half-century-old suggestions for the coupling of PS and NF have been confirmed by studies involving CO_2 enrichment, source/sink manipulations and variations in light level. Studies involving PS and NF measurements over short time periods (1 to 10 days) have been conducted to evaluate the closeness of the coupling, and results have been inconsistent. This divergence of results in probably largely due to differences in carbohydrate "status" of the plants (especially the nodules) which, in turn, is due to the way plants were treated before and during the experiments.

Identification of carbohydrates in nodules has allowed us to trace the coupling of PS and NF to the level of specific compounds. This knowledge has raised new questions about the role of individual compounds in nodules. The consumption of carbohydrate by nodules appears to be two or more times as great as the theoretical energy requirement for N fixation. More information on carbohydrate chemistry and its control in nodules is required before we can decide if the amount of N fixed/gram of carbohydrate can be increased.

NOTES

John G. Streeter, Henry J. Mederski, and R. A. Ahmad, Ohio Agricultural Research and Development Center, Wooster, Ohio 44691.

Approved for publication as Journal Article No. 16-79 of the Ohio Agricultural Research and Development Center, Wooster, Ohio 44691.

REFERENCES

1. Ahmad, R. A. 1978. The effects of water stress, root temperature, and carbohydrate supply to the nodules on nitrogen fixation in soybeans [*Glycine max* (L.) Merr.] plants. Ph.D. Thesis, The Ohio State University.
2. Arthur, J. M., J. D. Guthrie, and J. M. Newell. 1930. Some effects of artificial environments on the growth and chemical composition of plants. J. Bot. 17:416-482.
3. Ayanabe, A., and T. L. Lawson. 1977. Diurnal changes in acetylene reduction in field-grown cowpeas and soybeans. Soil. Biol. Biochem. 9:125-129.
4. Bethlenfalvay, G. J., and D. A. Phillips. 1978. Interactions between symbiotic nitrogen fixation, combined N application and photosynthesis in *Pisum sativum*. Physiol. Plant. 42:119-123.
5. Ching, T. M., S. Hedtke, S. A. Russell, and H. J. Evans. 1975. Energy state and dinitrogen fixation in soybean nodules of dark-grown plants. Plant Physiol. 55:796-798.
6. Evans, H. J., and L. E. Barber. 1977. Biological nitrogen fixation for food and fiber production. Science 197:332-339.
7. Hardy, F. W. F., and U. D. Havelka. 1975. Nitrogen fixation research: a key to world foods? Science 188:633-643.
8. Harper, J. E. 1974. Soil and symbiotic nitrogen requirements for optimum soybean production. Crop Sci. 14:255-260.
9. Huang, C. Y., J. S. Boyer, and L. N. Vanderhoff. 1975. Limitations of acetylene reduction (nitrogen fixation) by photosynthesis in soybeans having low water potentials. Plant Physiol. 56:228-232.
10. Hume, D. J., and J. G. Criswell. 1973. Distribution and utilization of ^{14}C-labelled assimilates in soybeans. Crop Sci. 13:519-524.
11. Latimore, M., Jr., J. Giddens, and D. A. Ashley. 1977. Effect of ammonium and nitrate nitrogen upon photosynthate supply and nitrogen fixation by soybeans. Crop Sci. 17:399-404.
12. Lawn, R. J., and W. A. Brun. 1974. Symbiotic nitrogen fixation in soybeans. I. Effect of photosynthetic source-sink manipulations. Crop Sci. 14:11-16.
13. Lawrie, A. C., and Wheeler. 1973. The supply of photosynthetic assimilates to nodules of *Pisum sativum* L. in relation to the fixation of nitrogen. New Phytol. 72:1341-1347.
14. Lawrie, A. C., and Wheeler. 1974. The effects of flowering and fruit formation on the supply of photosynthetic assimilates to the nodules of *Pisum sativum* L. in relation to the fixation of nitrogen. New Phytol. 73:1119-1127.
15. Mahon, J. D. 1977. Respiration and the energy requirement for nitrogen fixation in nodulated pea roots. Plant Physiol. 60:817-821.
16. Mederski, H. J., and J. G. Streeter. 1977. Continuous automated acetylene reduction assays using intact plants. Plant Physiol. 59:1076-1081.
17. Phillips, D. V., and A. E. Smith. 1974. Soluble carbohydrates in soybean. Can J. Bot. 52:2447-2452.
18. Russell, W. J., and D. R. Johnson. 1975. Translocation patterns in soybeans exposed to $^{14}CO_2$ at four different time periods of the day. Crop Sci. 15:75-77.
19. Ryle, G. J. A., C. E. Powell, and A. J. Gordon. 1978. Effect of source of nitrogen on the growth of Fiskeby Soya bean: the carbon economy of whole plants. Ann. Bot. 42:637-648.

20. Schweitzer, L. E., and J. E. Harper. 1977. The effect of light, dark, and temperature on the root nodule activity (C_2H_2 reduction) of soybean (*Glycine max* L. Merr.). ASA Meeting Abstr. p. 91.
21. Streeter, J. G. 1974. Growth of two soybean shoots on a single root. Effect on nitrogen and dry matter accumulation by shoots and on the rate of nitrogen fixation by nodulated roots. J. Exptl. Bot. 25:189-198.
22. Streeter, J. G., and M. E. Bosler. 1976. Carbohydrates in soybean nodules: identification of compounds and possible relationships to nitrogen fixation. Plant Sci. Lett. 7:321-329.
23. Wahua, T. A. T., and D. A. Miller. 1978. Effects of shading on the N_2-fixation, yield and plant composition of field-grown soybeans. Agron. J. 70:387-392.
24. Wilson, P. W., E. B. Fred, and M. R. Salmon. 1933. Relation between carbon dioxide and elemental nitrogen assimilation in leguminous plants. Soil Sci. 35:145-165.
25. Wilson, R. F., J. W. Burton, J. A. Buck, and C. A. Brim. 1978. Studies on genetic male-sterile soybeans. I. Distribution of plant carbohydrate and nitrogen during development. Plant Physiol. 61:838-841.

REGULATION OF SENESCENCE

L. D. Noodén

As plants (or individual organs) age, they undergo degenerative changes which impair their vital functions, limit growth, and eventually cause death (12,18,23,25,38). In many crop plants, this degeneration seems too sudden to represent a simple time-dependent accrual of lesions induced by the environment. Instead the physiological deterioration seems to be controlled by internal factors, a process which has been termed senescence (12,25). Many crop plants, particularly soybeans, die as their fruit mature and are therefore limited to one reproductive phase, a life cycle pattern called monocarpy (9,25,26,38). The senescence phase of a monocarpic life cycle has been termed monocarpic senescence (26) to distinguish it from other patterns of senescence (12). The fact that monocarpic senescence may begin while the fruit are still developing suggests that it may limit the productive capacity of these plants, and that problem will be the primary thrust of this paper. Although soybeans will be emphasized here, it should be noted that monocarpic senescence in some other species may differ a bit from soybeans (23,25,26).

CORRELATIVE CONTROLS: SOURCE AND TARGET

Long ago, horticulturists discovered that removal of the reproductive structures could prolong life in many monocarpic plants and cause them to initiate more flowers or fruit (9,18,38). The very rapid degeneration of soybean plant during pod ripening can be prevented by removing the flowers or young pods (13,14). In soybeans, early deseeding prevents monocarpic senescence as effectively as removal of the young pods (16). Neither age nor size are important determinants of longevity in soybeans, for soybeans (Biloxi variety) can grow vigorously for more than 15 months reaching a height of at least 23 ft if prevented from flowering with a long-day photoperiod (22). Thus monocarpic senescence is controlled by an internal developmental program, and the control center (the source?) is the developing seeds. For convenience, this yet unidentified influence which the seeds exert will be referred to as the senescence signal (16).

The most conspicuous manifestation of monocarpic senescence is the yellowing of the leaves, and this has formed the basis of a rapid, nondestructive, visual measure of senescence (percent of leaves $\leq$ 1/2 yellow [14]). We have also developed a visual measure of pod development in order to facilitate correlation of senescence-related changes with pod development. The pods are classified into five stages as follows and a numerical average of stage numbers (fruit maturity index, FMI) is calculated. Stage 1: pod length greater than 1 cm but pod not full width and the characteristic bulges of the seed cavity not yet developed. Stage 2: pod full width but the largest seed less than 1 cm long (can be estimated by examining the fruit in front of a bright light). Stage 3: largest seed in the pod near maximum length ($>$ 1 cm for the variety Anoka) but pod 1/4 yellow or less. Stage 4: pod greater than 1/4 yellow but less than or equal to 1/4 brown. Stage 5: pod greater than 1/4 brown. These visual procedures are highly sensitive and reproducible and have been checked against other parameters (14,29,30). As shown in Fig. 1, the percent of green leaves on a plant starts to decrease rapidly at about FMI 2.7. Depodding prevents this yellowing and abscission of the leaves.

Although the visible changes in soybean foliage reflect monocarpic senescence well, the question remains: what is the primary target of the senescence signal from the seeds: the leaves, the stem, the roots? Some evidence already indicates that the roots are not the primary targets through which the seeds induce the death of the plant (16). Among the changes occurring during (or prior to) monocarpic senescence are: cessation of stem elongation and root growth, inhibition of flower formation, inhibition of pod initiation and induction of pod abortion (8,14,23,33,35). Figs. 1 and 2 show that stem elongation stops at about FMI 1, when the pods are still elongating and long before the rapid phase of leaf yellowing starts. While depodding prevents foliar senescence, it does not reinstate stem elongation and leaf production. Other experiments show that apex removal at various times starting about 2 wk before flower opening does not accelerate, but it may delay senescence slightly through retardation of pod development (22). Thus cessation of vegetative growth, e.g., production of new leaves, does not seem to cause monocarpic senescence in soybeans; depodded plants do not senesce even though they do not produce new leaves. The target of the senescence signal is therefor not the shoot apex; moreover, these and other data indicate that the senescence signal as defined here does not cause cessation of growth in soybeans.

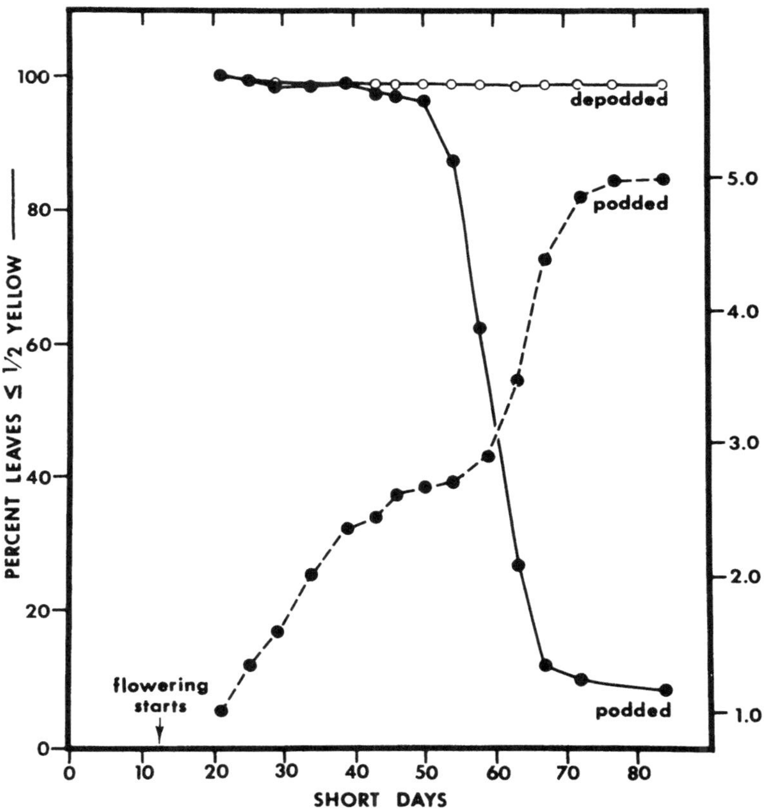

Figure 1. Foliar senescence (yellowing) and pod development kinetics in Anoka soybeans. Grown in pots of soil in environmental control chambers and depodded as described elsewhere (14,27). The procedures for scoring foliar senescence and pod development are discussed in the text. Pods were removed as they reached stage 2.

Two fairly direct lines of evidence now suggest that the leaves are the primary target of the senescence signal (22). The first involves grafted plants (Fig. 3) where the bottom part is of the variety Anoka (will flower and fruit under long days) and the top part is Biloxi (will not flower under long days). Under long days, pods develop on the Anoka (but not the Biloxi) portion, and the Anoka leaves show normal senescence, but the Biloxi continues to grow and carry green leaves. The Anoka stem and root system remain alive, because the Biloxi leaves supply them with photosynthate. Similar results were obtained when Biloxi soybeans were given short days (8 hr light) for

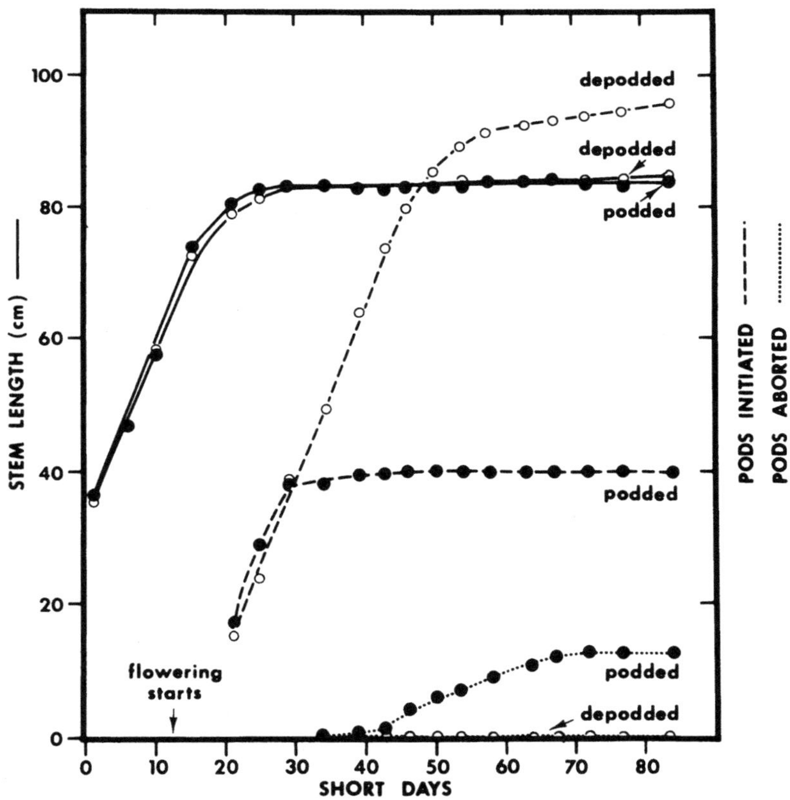

Figure 2. **Stem elongation, pod initiation and pod abortion kinetics in Anoka soybeans. Same plants as in Fig. 1. Pod initiation is scored as pods reaching a length of 1 cm, and pod abortion represents abscission of pods 1 cm or longer.**

3 wk and then transferred back into long days (16 hr light). The lower portion produced fruit, and the leaves senesced normally (but the stem did not turn yellow and die). The upper portion, which did not flower or fruit, continued to produce green leaves and to grow. These findings are consistent with our earlier observations that the senescence signal shows limited mobility and moves mainly downward if at all (16,27). The conclusion is that the plant as a whole can survive if enough leaves are maintained and the senescence signal causes death of the whole plant through its effect on the leaves.

Figure 3. Foliar senescence and pod formation in grafted soybean plants with Biloxi (maturity group VIII) over Anoka (maturity group I). The plants were joined with a tongued approach graft and cut before flowering started. The plants were grown as described earlier (14) except that they were given 8 hr per day fluorescent and incandescent light and then 8 hr of incandescent light only.

REPRESSION OF REPRODUCTIVE DEVELOPMENT

Although repression of flower initiation, inhibition of pod initiation, and induction of pod abortion do not cause monocarpic senescence (and may or may not be mediated by the senescence signal), this aspect of soybean maturation is of considerable agronomic interest, which justifies its inclusion here. Fig. 2 shows depodding allows more pods to be initiated (reach 1 cm in length); however, the pods exert this influence at an earlier stage in their development (before stage 2) (22) compared with stage 3 (end of pod fill and start of pod yellowing) for induction of monocarpic senescence (16,27).

By allowing only a designated number of pods to develop on a plant, it can be shown that only a small proportion (less than 40%) of the maximal pod load is sufficient to give the full senescence response (16,27). The pod dose effect on pod abortion is very similar (22), which again indicates that the most advanced pods can exert an inhibitory influence on less developed pods and as well as inducing the death of the entire plant. The plant is therefore caught in a sort of a vicious circle where the very structures we seek to increase destroy the capacity to support additional pods and repress reproductive development.

IS THE DEATH OF THE PLANT NECESSARY TO SUPPORT SEED DEVELOPMENT IN SOYBEANS?

Many studies indicate that the developing fruit of monocarpic plants including soybeans may withdraw or divert needed resources from the vegetative parts (see 18,20,23,25,28,35,36,38). Of particular interest is the transfer of nitrogen from the leaves (the major vegetative N depot) to the seeds in normal developing soybeans (3). On the surface, these deficiencies seem fully capable of causing the death of the plants. In fact, early botanists recognized the growth of seeds at the expense of the vegetative parts and called monocarpic senescence "exhaustion death" (18).

Through surgical modifications, we have shown that seed development can be uncoupled from monocarpic senescence in soybeans (27). The plants in Fig. 4 illustrate one type of surgical modification used. The senescence curves (Fig. 5) show that the single leaf responds very differently depending on whether it is above or below the pod cluster. About 50% of the leaves senesce (and drop) when they are below the pod cluster but do not senesce when they are above. The fruit maturity index curves show that the altered senescence is not due to interference with pod development. Furthermore, the seed yield (dry weight or nitrogen) per cluster is the same in both arrangements, and the yield of these single pod clusters is actually greater than a comparable node of an unmodified plant (27). Not only do these experiments

Figure 4. **Surgical modifications of single axis soybean plants. Depodded and defoliated as described elsewhere (27).**

argue against the nutrient deficiency theories and favor the idea of a senescence hormone(s) produced by the seeds (26,27), but more important, they show that the seeds can develop without killing the leaves and the rest of the plant. These and the earlier defoliation experiments of Weber (39) also mean that pod development does not necessarily depend on mobilization of reserves accumulated in the leaves over the life of the plant, but assimilation during pod development can be sufficient to sustain pod development given the right circumstances.

HORMONAL CONTROL

Although rather little information has been published on hormonal control of monocarpic senescence, it is tempting to speculate that hormones

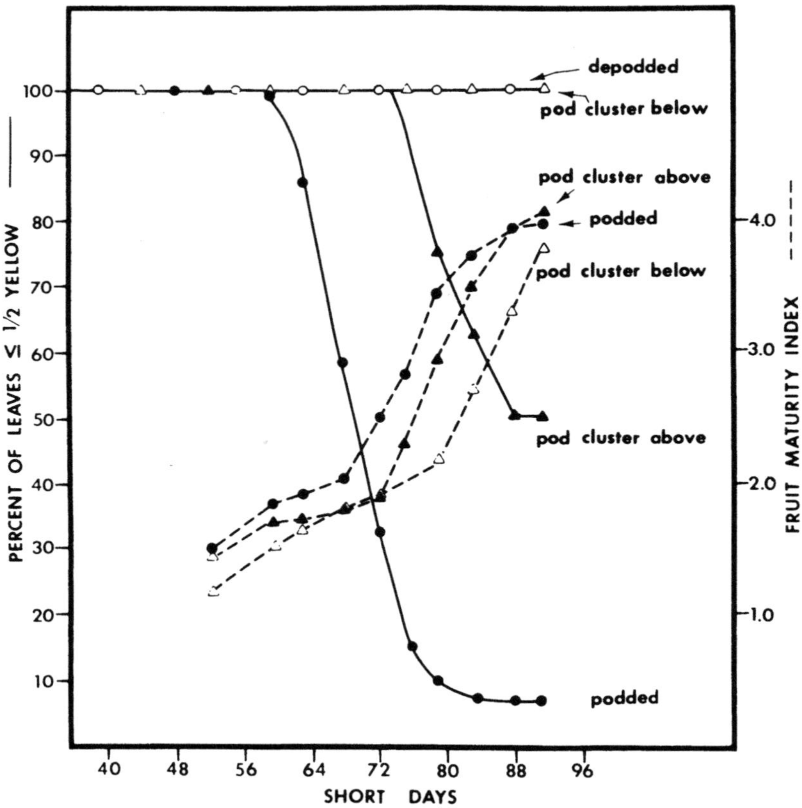

Figure 5. **Foliar senescence and pod development in controls (podded and depodded) and surgically-modified (as in Fig. 4) soybeans.**

mediate (or are at least secondarily involved) in the correlative influence of the seeds in soybeans. We have attempted to correlate changes in hormone levels in the leaves with foliar senescence (Fig. 6). Of course, other hormones may be involved, but zeatin and abscisic acid (ABA) are emphasized A) because of their importance in plant senescence processes (25) and B) because focusing on only two hormones provides a manageable starting point. While these data are based on bioassays of partially-purified materials, the patterns of change appear to be very similar to those reported by Oritani and Yoshida (31), who used gas-liquid chromatography (flame ionization detector) and/or a different group of bioassays. Foliar zeatin-like activity drops quite early during pod development, long before visible yellowing. The argument that

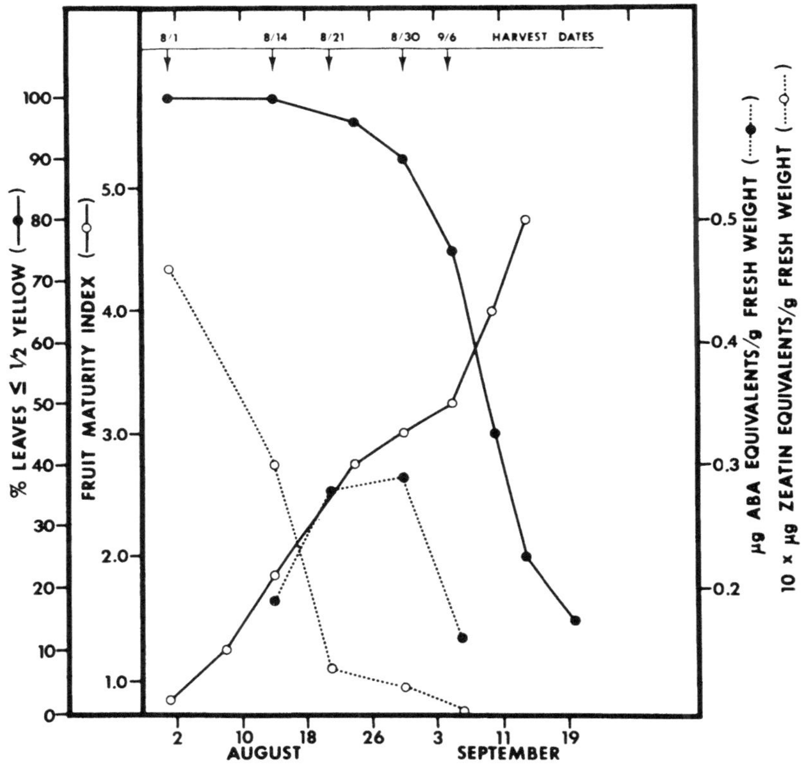

Figure 6. **Correlation of foliar ABA- and zeatin-like activity with pod development and foliar senescence. Anoka soybeans grown in the field (3). Drawn from the data of Lindoo and Noodén (17).**

the developing seeds cause foliar senescence primarily by diverting the supply of cytokinins from the roots (see 23,25,26) does not seem compelling here, for A) the foliar cytokinin activity is substantially decreased before the seeds begin to exert their most critical effects (Aug. 14-10) (17), B) the cytokinin levels in the seeds decline rather than increase from August 14 to August 21, C) developing fruit themselves seem to be capable cytokinin synthesizers (7,21), D) the plants with a single pod cluster below the single leaf (Fig. 4) should senesce faster than the reverse arrangement if the pods were depriving the leaves of cytokinins from the roots, and E) foliar applications of cytokinin cannot prevent monocarpic senescence (15,17). Some evidence also indicates that foliar IAA (indoleacetic acid) also declines during monocarpic senescence (19), but its role in monocarpic senescence is uncertain.

The seeds may not induce monocarpic senescence by functioning as passive sinks (in fact, it is hard to imagine that the seeds can exert such dramatic effects through a purely passive mechanism); ample evidence indicates that fruit can export solutes including hormones (1,2,4,6,26). The prime candidate for the senescence signal is abscisic acid, though even if it is involved, it may not act alone. Of particular interest is the rise in foliar ABA levels preceding leaf yellowing (Fig. 6) when the pods seem to exert their most important inductive effect. Furthermore, during the growth of soybean seeds, they have been shown to develop high ABA levels, which then drop before pod maturation, possibly preceding the increase in foliar ABA (32). Whether or not this ABA is translocated to the leaves remains to be determined. One weak point in this hypothesis is the inability of foliar applications of ABA to induce whole plant senescence in depodded plants, i.e., to substitute for pods, although it does accelerate monocarpic senescence in podded plants (17). ABA may act in conjunction with other factors, e.g. nutrient deficiency.

A HORMONAL ANTIDOTE FOR MONOCARPIC SENESCENCE IN SOYBEANS

Even though cytokinins usually have relatively little senescence-retarding effect on attached leaves and auxins are generally inactive (23,25,38), both cytokinins (especially benzyladenine, BA) and the auxin α-naphthalene acetic acid (NAA) have been shown to delay foliar senescence in soybean (10,11,17,22). Detailed time course studies on foliar yellowing and abscission suggest that BA and NAA act differently, BA having a greater effect on yellowing and NAA a greater effect on abscission (11,24). Thus we tried combinations of BA and NAA and found that these could completely prevent both the yellowing and abscission aspects of monocarpic senescence (11,24) (Fig. 7). Not only does the NAA-BA combination prevent the loss of starch and nitrogen from the leaves (redistribution to the seeds), but it sustains the accumulation of starch in the leaves (Table 1), which suggests that photosynthesis is maintained.

In as much as any treatment which alters pod development may secondarily affect monocarpic senescence, it is important to determine if pod development is affected. Fig. 7 shows that the NAA-BA treatment does not alter pod maturation (FMI curves). Moreover, Table 2 shows that the NAA-BA treatment does not decrease seed yield in terms of dry weight or nitrogen (in some experiments, there is an increase, but this is inconsistent). In any case, pod development is clearly able to proceed in the NAA-BA-treated plant without killing the rest of the plant. Implicit from the data on foliar starch and nitrogen is the probability that this treatment maintains not only photosynthesis but nitrogen accumulation (presumably N_2 fixation) by the roots, and current production from these processes (not withdrawal of life-sustaining resources in the leaves) can supply the needs of the growing seeds.

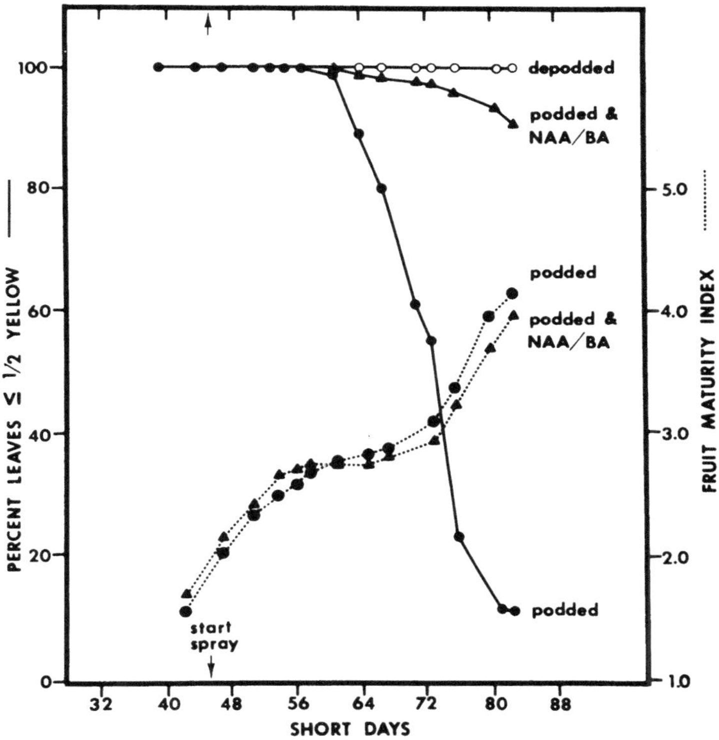

Figure 7. Effect of NAA and BA on pod development and foliar senescence. Anoka soybeans grown in environmental control chambers (14). Hormones at 50 μM in 0.05% (v/v) Tween 80 were sprayed on the foliage every other day starting when flowering began. After Noodén et al. (24).

Table 1. Effect on a combined auxin (NAA)-cytokinin (BA) treatment on starch and nitrogen content in soybean leaves.

	Starch Content		Nitrogen Content	
Short Day	Control	NAA-BA	Control	NAA-BA
	mg glucose/g fr wt		*mg N/g fr wt*	
46	23	--	13	--
57	20	40	12	12
73	2	89	4	16
88	--	100	--	17

Same plants as in Figure 7. Starch and nitrogen were determined as described elsewhere (3,11, 24,37) except that BA and its glycosides (41) were extracted with ethylacetate before the nitrogen determination. After Noodén et al. (24).

Table 2. **Effect of a combined auxin (NAA)-cytokinin (BA) treatment on soybean pod and seed growth.**

Treatment	Pods/Plant	Seeds/Plant	Seeds·Dry Wt/ Plant (g)	Seeds·N/ Plant (g)
Control	17	33	6.6	0.53
NAA-BA	14	28	6.0	0.49

Same plants as in Figure 7. Seed nitrogen was measured as described by Derman et al. (3).

Whether the senescence signal is an auxin-cytokinin deficiency which is corrected by NAA-BA treatment or a senescence hormone which is overridden remains to be determined; however, the evidence that the senescence signal is not a nutrient deficiency applies equally well to a hormone deficiency. Therefore, the override hypothesis seems more likely.

Why then doesn't the prevention of monocarpic senescence increase yield? Probably, the answer lies in earlier suppression of pod initiation and induction of pod abortion which prevents the prolonged productive capacity from being utilized. The NAA-BA treatment does not increase pod initiation or reduce pod abortion even though it is started when flowering begins (10,22,34). Therefore any strategy to increase soybean yields by preventing (or delaying) monocarpic senescence must also provide a capacity to utilize the sustained photosynthesis and nitrogen fixation. On the other hand, setting more pods with no capacity to support them also seems unlikely to increase yields, and this may account for some of the problems with TIBA (2,3,5-triiodobenzoic acid) which seemed effective in initiating more pods (see e.g. 5).

In any case, much remains to be learned about monocarpic senescence in soybeans and other crops, but enough information exists to indicate that more study of monocarpic senescence may open new strategies to crop management and yield improvement.

NOTES

Larry D. Noodén, Dept. of Botany, University of Michigan, Ann Arbor, MI 48109.
Studies reported here were supported in part by research grant 416-15-79 from the USDA Cooperative State Research Service under P.L. 89-106.

REFERENCES

1. Antoszewski, R. and E. Lis. 1968. Translocation of some radioactive compounds from the strawberry receptacle to the mother plant. Bull. Acad. Pol. Sci. 16: 444-446.
2. Bourbouloux, A., and J. L. Bonnemain. 1972. Transport de l'auxine-^{14}C en provenance de jeunes gousses de *Vicia faba* L. Planta (Berlin) 115:161-172.

3. Derman, B. C., D. C. Rupp, and L. D. Noodén. 1978. Mineral distribution in relation to fruit development and monocarpic senescence in Anoka soybeans. Am. J. Bot. 65:205-213.
4. Gianfagna, T. J., and P. J. Davies. 1978. Fruit-induced apical senescence in a genetic lines of peas. Plant Physiol. 61 (suppl.):26.
5. Greer, H. A. L., and I. C. Anderson. 1965. Response of soybeans to triiodobenzoic acid under field conditions. Crop Sci. 5:229-232.
6. Grochowska, M. J. 1968. The influence of growth regulators inserted into apple fruitlets on flower bud initiation. Bull. Acad. Pol. Sci. 16:581-586.
7. Hahn, H., R. DeZacks, and H. Kende. 1974. Cytokinin formation in pea seeds. Naturwissensch. 61:170.
8. Hicks, D. R., and J. W. Pendleton. 1969. Effect of floral bud removal on performance of soybeans. Crop Sci. 9:435-437.
9. Hildebrand, F. 1882. Die Lebensdauer und Vegetationsweise der Pflanzen, ihre Ursachen und ihre Entwicklung. Bot. Jahrb. 2:51-135.
10. James, A. L., I. C. Anderson, and H. A. L. Greer. 1965. Effects of naphthaleneacetic acid on field-grown soybeans. Crop Sci. 5:472-474.
11. Kahanak, G. M., Y. Okatan, D. C. Rupp, and L. D. Noodén. 1978. Hormonal and genetic alteration of monocarpic senescence in soybeans. Plant Physiol. 61 (suppl.): 26.
12. Leopold, A. C. 1961. Senescence in plant development. Science 134:1727-1732.
13. Leopold, A. C., E. Niedergang-Kamien, and J. Janick. 1959. Experimental modification of plant senescence. Plant Physiol. 34:570-573.
14. Lindoo, S. J., and L. D. Noodén. 1976. The interrelation of fruit development and leaf senescence in 'Anoka' soybeans. Bot. Gaz. (Chicago) 137:218-223 (1976).
15. Lindoo, S. J., and L. D. Noodén. 1976. Studies on the role of cytokinin- and ABA-like substances in monocarpic senescence of soybeans. Plant Physiol. 57 (suppl.): 27.
16. Lindoo, S. J., and L. D. Noodén. 1977. Behavior of the soybean senescence signal. Plant Physiol. 59:1136-1140.
17. Lindoo, S. J., and L. D. Noodén. 1978. Correlations of cytokinins and abscisic acid with monocarpic senescence in soybean. Plant Cell Physiol. 19:997-1006.
18. Molisch, H. 1938. The longevity of plants. H. Fulling (transl.). Science Press, Lancaster, Pa., 226 p.
19. Mondal, M. H., W. A. Brun, and M. L. Brenner. 1978. IAA levels and photosynthesis in leaves of control and depodded soybean plants. Plant Physiol. 61 (suppl.):8.
20. de Mooy, C. J., J. Pesek, and E. Spaldon. 1973. Mineral nutrition. p. 267-352. In B. E. Caldwell (ed.) Soybeans: Improvement, Production and Uses. Am. Soc. Agron., Madison, Wis.
21. Nitsch, J. P. 1970. Hormonal factors in growth and development. p. 427-472. In A. C. Hulme (ed.) The Biochemistry of Fruits and Their Products. Academic Press, London.
22. Noodén, L. D. Unpublished data.
23. Noodén, L. D. 1979. Senescence in the whole plant. In press, K. V. Thimann (ed.) Plant Models for Aging Research. CRC Press, West Palm Beach, Fla.
24. Noodén, L. D., G. M. Kahanak, and Y. Okatan. 1979. Prevention of monocarpic senescence in soybeans with auxin and cytokinin. In preparation.
25. Noodén, L. D., and A. C. Leopold. 1978. Hormonal control of senescence and abscission. Phytohormones and Related Compounds, p. 329-369. In D. C. Letham, T. J. Higgins and P. B. Goodwin (eds.), Elsevier, Amsterdam.

26. Noodén, L. D., and S. J. Lindoo. 1978. Monocarpic senescence. What's New in Plant Physiol. 9:25-28.
27. Noodén, L. D., D. C. Rupp, and B. D. Derman. 1978. Separation of seed development from monocarpic senescence in soybeans. Nature (London) 271:354-357.
28. Ogren, W. L., and R. W. Rinne. 1973. Photosynthesis and seed metabolism. p. 391-416. In B. E. Caldwell (ed.) Soybeans: Improvement, Production and Uses. Am. Soc. Agron., Madison, Wis.
29. Okatan, Y., G. M. Kahanak, and L. D. Noodén. 1977. Interrelation of seed development and foliar senescence in Anoka soybeans. Plant Physiol. 59 (suppl.):112.
30. Okatan, Y., G. M. Kahanak, and L. D. Noodén. 1979. Characterization and interrelation of foliar senescence and pod development in soybean. In preparation.
31. Oritani, T., and R. Yoshida. 1973. Studies on nitrogen metabolism in crop plants: XII. Cytokinins and abscisic acid-like substance levels in rice and soybean leaves during their growth and senescence. Proc. Crop Sci. Soc. Japan 42:280-287.
32. Quebedeaux, B., P. B. Sweetser, and J. C. Rowell. 1976. Abscisic acid levels in soybean reproductive structures during development. Plant Physiol. 58:363-366.
33. van Schaik, P. H., and A. H. Probst. 1958. Effects of some environmental factors on flower production and reproductive efficiency in soybeans. Agron. J. 50:192-197.
34. van Schaik, P. H., and A. H. Probst. 1959. Effect of six growth regulators on pod set and seed development in Midwest soybeans. Agron. J. 51:510-511.
35. Shibles, R., I. C. Anderson, and A. H. Gibson. 1975. Soybean. p. 151-189. In L. T. Evans (ed.) Crop Physiology, Some Case Histories. Cambridge University Press, London.
36. Sutcliffe, J. F. 1976. Regulation in the whole plant. p. 394-417. In U. Lüttge and M. G. Pitman (eds.) Encycl. Plant Physiol., New Series, Vol. IIB. Springer-Verlag, Berlin.
37. Tetley, R. M. 1974. Studies on hormonal control of growth and metabolism in cultured tobacco pith explants. Ph.D. Dissertation. University of Michigan, Ann Arbor. 301 p.
38. Wangermann, E. 1965. Longevity and ageing in plants and plant organs. p. 1037-1057. In W. Ruhland (ed.) Handbuch der Pflanzenphysiologie, Vol. 15(2). Springer-Verlag, Berlin.
39. Weber, C. R. 1955. Effects of defoliation and topping simulating hail injury to soybeans. Agron. J. 47:262-266.
40. Williams, R. F. 1955. Redistribution of mineral elements during development. Annu. Rev. Plant Physiol. 6:25-42.
41. Wilson, M. M., M. E. Gordon, D. S. Letham, and C. W. Parker. 1974. Regulators of cell division in plant tissues XIX. The metabolism of 6-benzylaminopurine in radish cotyledons and seedlings. J. Exp. Bot. 25:725-732.

EFFECT OF POD FILLING ON LEAF PHOTOSYNTHESIS IN SOYBEANS

W. A. Brun and T. L. Setter

Soybean yield may be considered as a product of photosynthetic rate, integrated over time, and the partitioning of the resulting photosynthate between physiological or morphological yield components, often referred to as assimilate sinks.

Published reports indicate strong interactions between activities in assimilate sources and sinks (7,11,12). For example, clear interactions have been documented to occur between the processes of pod filling and leaf photosynthesis (14)., as well as between reproductive growth and photosynthetic duration (i.e., the onset of senescence) (14), and also between pod filling and N_2 fixation activity in root nodules (10).

The mechanisms commonly assigned to such interactions are either nutritional or hormonal, or both. Nutritional mechanisms may involve either a competitive situation between alternative sinks for a limited supply of substrates, such as has been postulated to govern the interaction between pod filling and N_2 fixation in nodules (10), or the sink may modulate the source activity by withdrawing more or less nutrients from it and thereby regulating its activity either through mass action or other means (17).

Hormonal mechanisms are also often invoked, particularly when the data fail to support nutritional mechanisms. Hormonal mechanisms of source/sink interactions may be envisioned as acting either on the source, on the sink, or on the transport system between the two. For example, a hormone may stimulate metabolic activity in the sink thereby increasing its nutrient utilization and steepening the nutrient gradient to the source, thus stimulating source activity [Nitsch (16) considers the evidence for such a mechanism to be poor] ; or a hormonal signal may go to the source and there act either directly on source activity (1), or indirectly as we will propose later. Finally, a hormonal mechanism migh involve the transport system between the source and the sink and thereby influence either one through the resulting perturberance of nutrient gradients (18).

The interaction which we wish to discuss is that occurring between the filling soybean pod and the photosynthetic activity in the leaves. It has been observed by several investigators working with several plant species, including soybean (11,14,17), that leaf photosynthetic rates are often stimulated by the presence of a strong assimilate sink on the plant (either vegetative or reproductive), and that photosynthesis is often partially inhibited when such sinks are removed.

Specifically, Thorne and Koller (17) have shown that when a strong sink was created by shading all but one leaf on a soybean plant, the photosynthetic rate of that one illuminated source leaf increased nearly 50% over an 8-day period. This was accompanied by a 3-fold increase in source leaf sucrose concentration and a 10-fold decrease in source leaf starch concentration. RuDP carboxylase activity was found to increase significantly while stomatal conductivity was unaffected.

Conversely, Mondal et al. (14) have reported that when sink strength was decreased by removing the filling pods of field grown 'Hodgson' soybeans, leaf photosynthesis was decreased by an amount which remained fairly constant over the pod filling period. Figure 1 shows the phyotosynthetic rate of soybean leaves of control and depodded plants. One of the two depodding treatments was performed just 32 hr prior to making the photosynthetic measurement, while in the other treatment the plants were kept pod-free from mid bloom.

Starch, soluble carbohydrate, protein, RuDPCase, inorganic phosphate, specific leaf wt, and IAA concentration (15) were also determined on these leaves, but none of these parameters correlated well with the observed photosynthetic changes in response to the two depodding treatments. It thus seemed that the photosynthetic responses to depodding were not mediated by changes in nutritional status of the leaves.

In order to investigate further the mechanism of the responses reported by Mondal et al. we decided to examine two alternative mechanisms. If, as Mondal's work seemed to show, the effect of depodding was not a question of end product accumulation in leaves of depodded plants, then an interruption of phloem transport in the petiole, which surely would cause an accumulation of photosynthate in the leaf blade, might not have any effect on leaf photosynthesis. Secondly, since it had been reported that following fruit removal stomatal conductances decreased in both grape (13) and *Capsicum* (9) leaves, we decided to measure stomatal conductances in soybean leaves on control and depodded plants as well as in leaves of control and petiole girdled plants.

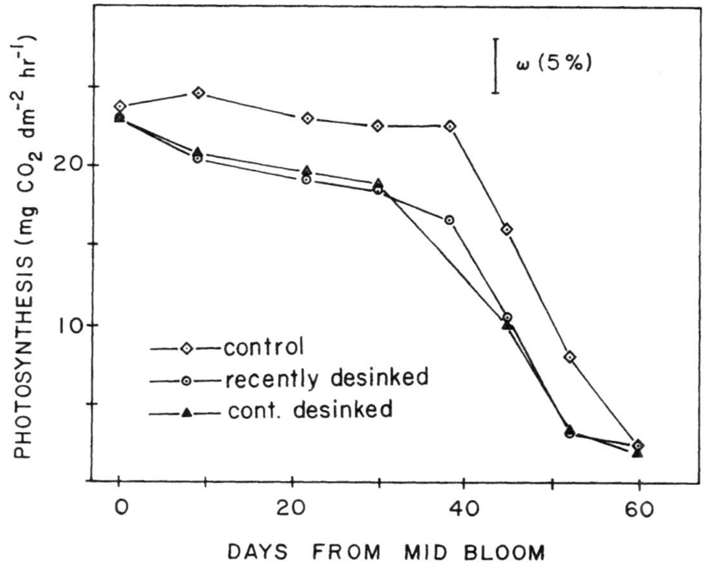

Figure 1. Leaf photosynthetic rate of control, recently desinked, and continuously desinked soybean plants. All means differing by a volume greater than w are significantly different at the 5% level (Tukey's test).

Depodding consisted of removing all of the pods from the plants, and petiole girdling involved exposing a 1 cm segment of the leaf petiole to live steam for a period of 15 seconds. This treatment killed all of the living cells in the petiole segment as evidenced by a complete collapse of the petiole at this point. A small wooden splint was placed on the girdled petiole to prevent it from collapsing. Pressure bomb measurements on the leaf blade indicated that the petiole girdle had no effect on the water potential of the blade.

The plants used were 8 wk old 'Clay' soybeans, raised in a growth chamber with a 12 hr photoperiod consisting of 500 μ Einsteins m^{-2} s^{-1} of mixed fluorescent and incandescent light. Pod development was well under way when the plants were used.

Carbon exchange rates (CER) and stomatal conductivities were measured on the fifth trifoliolate leaf of 9 plants per treatment. CER was measured with a clamp-on leaf cuvette and differential infrared CO_2 analysis. Stomatal conductivity was measured with a diffusion porometer.

The results (Table 1) showed that pod removal caused a significant 71% decrease in CER within 48 hr; this was accompanied by a 63 and 54% decrease in stomatal conductivity on the upper and lower leaf surface, respectively.

Table 1. The effect of soybean pod removal and petiole girdling on CO_2 exchange rates (CER) and stomatal diffusive conductivities. Figures are means of 9 replications.

	CER	Stomatal conductivity	
		Upper surface	Lower surface
	(mg CO_2 dm^{-2} h^{-1})	*(cm s^{-1})*	
Control	22.2	0.38	1.28
Pods removed	6.5	0.14	0.59
LSD (.05)	3.2	0.12	0.73
Control	24.7	0.39	0.70
Steam girdled	2.1	0.05	0.18
LSD (.05)	4.2	0.12	0.17

These results fully confirm the recent report by Koller and Thorne (8) of similar responses in 'Amsoy' and 'Wells' soybeans, although we did not see the leaflet reorientation they observed in 'Amsoy'.

Steam girdling the petiole caused an even more pronounced effect. CER rate was reduced 91% within 24 hr, accompanied by an 87 and 74% decline in stomatal conductivity of the two leaf surfaces, respectively.

A time course study was then conducted using similar plant material and growing conditions. Carbon exchange rate, transpiration rate, and leaf temperature were measure at 30 min intervals for the first 12 hr after treatment. Leaf conductivity to water vapor diffusion was then calculated from the transpiration and leaf temperature data.

Figure 2 shows the mean values obtained from 6 replicate plants. The graphs are plotted in percent of initial value as a function of time after treatment. The initial values for CER were 28, 31, and 27 mg CO_2 dm^{-2} hr^{-1} for the control, depodded and steam girdled treatments, respectively. Initial values for leaf conductivity were 0.69, 0.72, and 0.73 cm s^{-1} for the 3 treatments, respectively.

It can be seen that both treatments caused marked responses which in the steam girdled plants were evident within the first 30 min. For the depodded plants, there was an increase in both parameters in the first 30 min, followed by a decrease. The cause or interpretation of the initial increase is not understood, but perhaps it is a manifestation of the so-called Iwanoff effect (6) in which release of tension by severing the vascular system causes a temporary increase in water potential throughout the plant which leads to a transient stomatal opening response.

It appears from the data presented in Table 1 and Figure 2 that at least a large part of the CER response of both treatments may be attributable to decreased stomatal opening. Calculations of so-called mesophyll resistance to

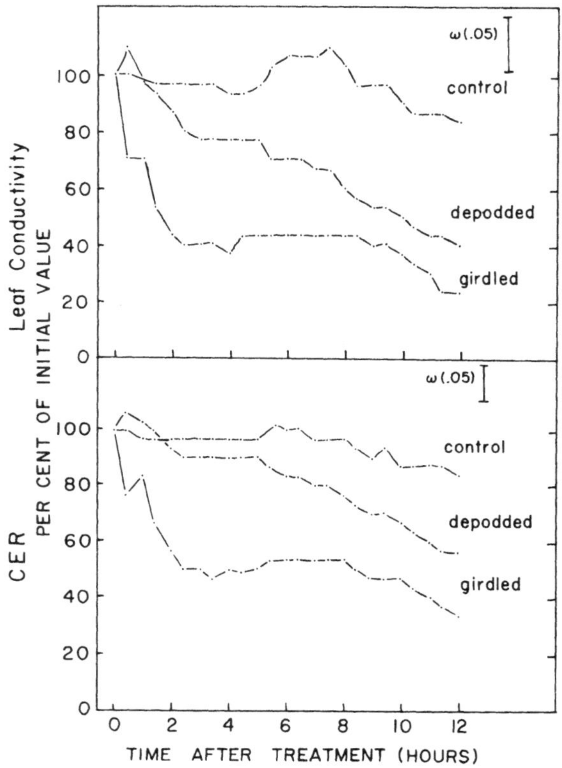

Figure 2. **Carbon exchange rates (CER) and leaf conductivities to water vapor of control, depodded and steam girdled soybean leaves as a function of time after treatment. Each data point represents 6 replicate leaves.**

CO_2 diffusion (3) indicated that there was little if any change in this parameter with the treatments. Such calculations, however, are open to interpretation so we sought a more rigorous examination of the question.

In order to assess the magnitude of treatment-induced changes in non-stomatal, partial reactions of CO_2 fixation, the rate of ^{14}C bicarbonate uptake was determined in 250-micron thick leaf slices. In this leaf slice technique, CO_2 diffusion into the tissue would presumably be through the cut edges of the slice, and thus not be under stomatal control.

The leaf slices were cut from the same 9 leaves per treatment as were used to collect the data shown in Table 1. They were cut with a sliding microtome, suspended in sealed vials with assay buffer at various bicarbonate concentrations, and irradiated with saturating light. After 10 min, the reaction was stopped with acid, the excess CO_2 exhausted, and the amount of ^{14}C fixed into acid-stable products determined by liquid scintillation spectrometry.

Figure 3 shows the results. The ^{14}C uptake data have been transformed to carbon assimilation rates (CAR) expressed as mg CO_2 dm^{-2} of leaf surface hr^{-1} and have been plotted as a function of HCO_3^- concentration. It can be seen that carbon assimilation rates by leaf slices from control and depodded plants were identical, and nearly identical in leaf slices from control and steam girdled leaves. We remind you that the CER from these same leaves before they were sliced were depressed by 71 and 91% in the depodded and steam girdled plants, respectively.

We conclude from this experiment that the observed changes in CER by the intact leaves were entirely due to changes in stomatal opening and were not a function of changes in photosynthetic parameters of the mesophyll tissue.

We are currently considering what the mechanism of these responses might be. Both of the treatments involve a blockage of translocation from the photosynthetic source leaves to the sinks of the plant, but it is not clear how translocation can have such a rapid and drastic effect on stomatal aperture. Further, it is not clear whether it is the translocation of photosynthate which is important or perhaps the translocation of other substances, such as endogenous hormones.

We see no obvious way in which an accumulation of photosynthate in the mesophyll could have such a rapid effect on stomatal behavior.

The hormone abscisic acid (ABA), however, is known to be able to cause rapid stomatal responses and to be produced primarily in leaves (5,19) from where it is translocated to both reproductive (5) and vegetative sinks (19).

We therefore propose, as a working model, that the observed effects of pod removal and petiole girdling are caused by an accumulation of ABA in the treatment leaves which leads to partial stomatal closure. The fact that petiole girdling is more effective than is pod removal we interpret to mean that petiole girdling completely inhibits translocation, whereas pod removal only inhibits translocation to the pods and not to other sink activities in the plant.

In a previous report from our laboratory (2) it was shown that young developing soybean seed contain as much as 13 μg ABA/g fresh weight. This we believe is the highest reported value from any plant tissue. In the previous report we were unable to see any effect of depodding on leaf ABA content, but the plants were grown in the variable environment of the field and the data showed considerable variability.

We have now conducted a preliminary experiment using 'Clay' soybeans in the growth chamber. In this experiment, free and bound ABA were determined in 5 replications of individual leaves from control, depodded, and

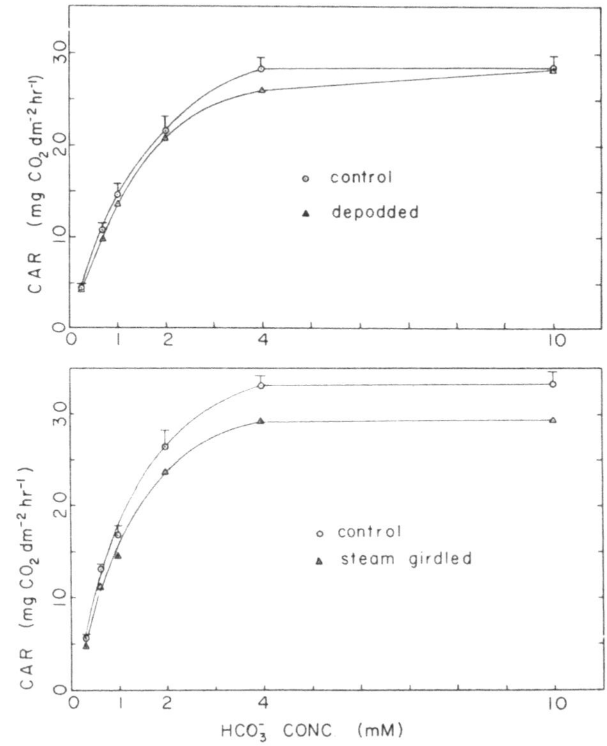

Figure 3. **Carbon assimilation rates at saturating light intensity by leaf slices from control, depodded and steam girdled soybean leaves, as a function of HCO_3^- concentration. Each data point represents 9 replicate leaves.**

steam girdled treatments. The ABA analyses were performed using the HPLC facilities of Dr. M. L. Brenner in our Department of Horticultural Science and Landscape Architecture, essentially as described by Ciha et al. (2).

The results are shown in Table 2. You can see that there was substantially more free ABA in the depodded and steam girdled leaves than in the control. Bound ABA also increased in steam girdled leaves. This we take to support our working hypothesis.

Table 2. **The free and bound ABA contents of soybean leaves on control, depodded and steam girdled leaves. Each figure represents 5 replicate leaves.**

	Control	Depodded	Girdled
		ng ABA/g fresh weight	
Free ABA	2.1 ± .5	19.4 ± 5.5	51.2 ± 11.9
Bound ABA	21.0 ± 4.2	23.3 ± 3.5	45.8 ± 13.3

NOTES

W. A. Brun and T. L. Setter, Department of Agronomy and Plant Genetics, University of Minnesota, St. Paul, MN 55108.

REFERENCES

1. Bidwell, R. G. S., and Chin-Kwong Quong. 1975. Indoleacetic acid effect on the distribution of photosynthetically fixed carbon in the bean plant. Biochem. Physiol. Pflanzen 168(s):361-370.
2. Ciha, A. J., M. L. Brenner, and W. A. Brun. 1978. Effect of pod removal on abscisic acid levels in soybean tissue. Crop Sci. 18:776-779.
3. Gastra, P. 1962. Photosynthesis of crop plants as influenced by light, carbon dioxide, temperature, and stomatal diffusion resistance. Mededel. Landbouwhogesch. Wageningen 59:1-68.
4. Hoad, G. V. 1973. Effect of moisture stress on abscisic acid levels in *Ricinus communis* L. with particular reference to phloem exudate. Planta 113:367-372.
5. Hoad, G. V. 1978. Effect of water stress on abscisic acid levels in white lupin (*Lupinus albus* L.) fruit, leaves and phloem exudate. Planta 142:287-290.
6. Iwanoff, L. 1928. Zur Methodik der Transpirationbestimmung am Standort. Berl. deut. botan. Ges. 46:306-310.
7. King, R. W., I. F. Wardlaw, and L. T. Evans. 1967. Effects of assimilate utilization on photosynthetic rate in wheat. Planta 77:261-276.
8. Koller, H. R., and J. H. Thorne. 1978. Soybean pod removal alters leaf diffusion resistance and leaflet orientation. Crop Sci. 18:305-307.
9. Kriedemann, P. E., B. R. Loveys, J. V. Possingham, and M. Satoh. 1976. Sink effects on stomatal physiology and photosynthesis. In J. F. Wardlaw and J. B. Passioura (eds.) Transport and Transfer Processes in Plants, Acad. Press, N. Y.
10. Lawn, R. J., and W. A. Brun. 1974. Symbiotic nitrogen fixation in soybeans. I. Effect of photosynthetic source-sink manipulations. Crop Sci. 14:11-16.
11. Lenz, F. 1974. Fruit effects on formation and distribution of photosynthetic assimilates. XIX Intern. Hort. Cong., Warsaw, p. 155-166.
12. Loveys, B. R., and P. E. Kriedemann. 1974. Internal control of stomatal physiology and photosynthesis. I. Stomatal regulation and associated changes in endogenous levels of abscisic and phaseic acids. Aust. J. Plant Physiol. 1:407-415.
13. Loveys, B. R., and P. E. Kriedemann. 1974. Hormonal regulation of gas exchange. In R. L. Bieleshi, A. R. Ferguson, and M. J. Cresswell (eds.) Mechanism of regulation of plant growth. The Royal Society of New Zealand, Wellington.
14. Mondal, M. H., W. A. Brun, and M. L. Brenner. 1978. Effects of sink removal on photosynthesis and senescence in leaves of soybean (*Glycine max* L.) plants. Plant Physiol. 61:394-397.
15. Mondal, M. H., W. A. Brun, and M. L. Brenner. 1978. IAA levels and photosynthesis in leaves of control and depodded soybean plants. Plant Physiol. 61(s):39.
16. Nitsch, J. P. 1970. Hormonal factors in growth and development. In A. C. Hulme (ed.) The biochemistry of fruits and their products. Academic Press, N. Y.
17. Thorne, J. H., and H. R. Koller. 1974. Influence of assimilate demand on photosynthesis, diffusive resistances, translocation, and carbohydrate levels of soybean leaves. Plant Physiol. 54:201-207.
18. Williams, A. M., and R. R. Williams. 1978. Regulation of movement of assimilate into ovules of *Pisum sativum* cv. Greenfeast: A 'remote' effect of the pod. Aust. J. Plant Physiol. 5:295-300.
19. Zeevaart, J. A. D. 1977. Sites of abscisic acid synthesis and metabolism in *Ricinus communis* L. Plant Physiol. 59:788-791.

POSTPONEMENT OF SEVERE WATER STRESS IN SOYBEANS BY ROOTING MODIFICATIONS: A PROGRESS REPORT

H. M. Taylor

Distribution of vegetation over the earth's surface is controlled more by water availability than by any other single factor (21). Consequently, much research is being conducted on methods to increase crop water supplies under dryland conditions.

Considerable research is being conducted to increase precipitation through weather modification (16), to increase the total quantity of water that infiltrates the soil surface (3), to control deep percolation through the soil profile (24), to reduce evaporation of water from soils (17), and to alleviate soil conditions that are unfavorable for root exploration (27). Considerable research also is being conducted to adapt the plant so it can grow where chemical insufficiency or toxicity occurs (49). Objective of this latter research often is to increase the quantity of available water by allowing the root system to grow into previously unsuitable soil volumes.

A program was started at Ames, Iowa, in 1973 to test the hypothesis that an additional quantity of water can be made available to soybean *[Glycine max* (L.) Merr.] plants by modifying their root systems even when the plants are grown in the deep fertile soils of Iowa. This research project complements those programs designed to increase plant available water through other mechanisms.

BACKGROUND ON CLIMATE AND SOILS

I will first provide some background information on climate and soils of Iowa. Similar precipitation patterns and deep soils exist at many other locations throughout the world. Annual precipitation at Omaha, Nebraska (on the western border of Iowa) averages about 69 cm, but one year in ten it is as much as 89 cm or as little as 47 cm (Fig. 1). Evapotranspiration at Omaha annually averages about 61 cm; therefore, annual streamflow averages about 8 cm. At Moline, Illinois (on the eastern boarder of Iowa), annual precipitation averages about 84 cm, but one year in ten it is as much as 105 cm or as little as 62 cm (Wadleigh et al., 1966). Annual evapotranspiration averages about 65 cm and streamflow averages about 19 cm. Annual precipitation and streamflow both increase as one travels eastward in Iowa (Fig. 2).

Most of Iowa's soils are deep, fertile, and productive, but some of them require tile drainage for water table control during the early growing season. After July 1, however, the water tables (when within 100 cm of the ground surface) usually decline because transpiration exceeds precipitation. Evapotranspiration from a fully developed soybean canopy at Ames, Iowa, often will exceed 0.70 cm/day (38), but average daily precipitation during July and August at Ames is only 0.25 cm/day. Substantial water is withdrawn from soil profile storage during normal or dry years to satisfy the plant's demand during July, August, and early September. The quantity of water available to satisfy that demand depends upon soil profile characteristics and upon the

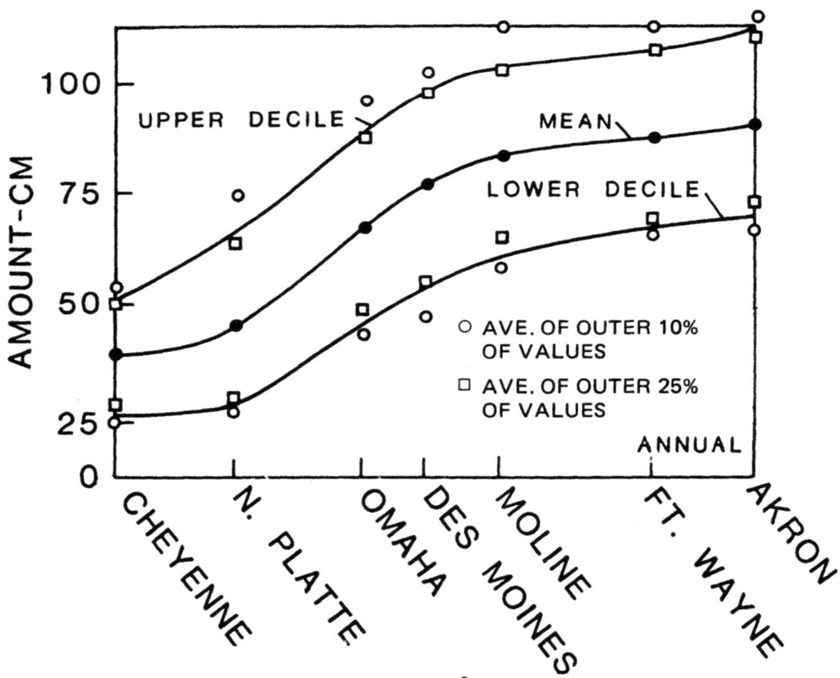

Figure 1. **Annual water supply across the 11° parallel. Redrawn from Wadleigh et al. (1966).**

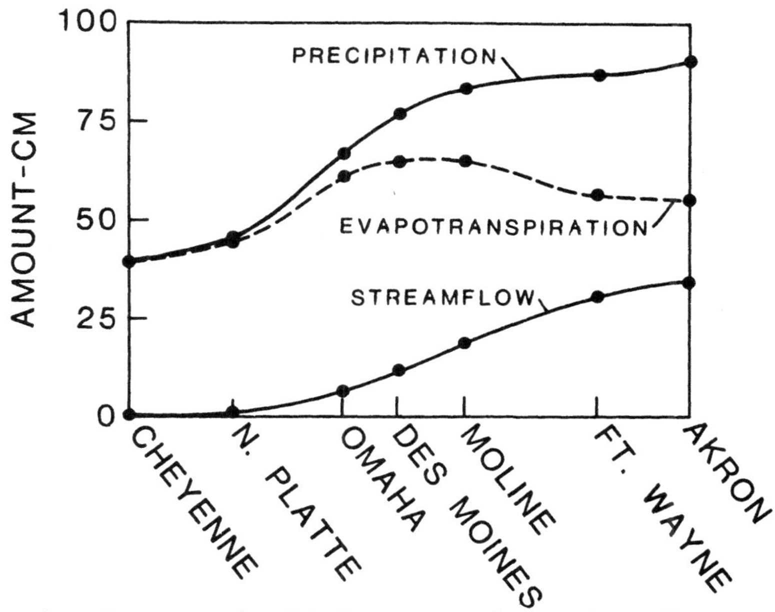

Figure 2. **Average annual precipitation, evapotranspiration and streamflow across the 41° parallel. Redrawn from Wadleigh et al. (1966).**

volume of soil where rooting occurs. Iowa soils often contain about 0.10 to 0.15 cm^3/cm^3 available water at field capacity.

DEEP PERCOLATION

One of the earliest questions encountered in our program was, "From the individual farmer's perspective, is all of the annual precipitation used effectively in crop production?" Obviously, the answer is "no," because considerable runoff (37) and soil water evaporation occur. Reducing these losses increases precipitation effectiveness (17). Precipitation effectiveness also can be increased by reducing annual deep percolation. A recent model (40) showed that about 2.5 cm of water annually percolates through a 150-cm rooting depth for corn (*Zea mays* L.) grown near the western border of Iowa (Fig. 3). This deep percolation increases toward the east and becomes about 8 to 10 cm in the southeastern part of Iowa. If more precipitation is transpired during soybean growth, less percolates through the rooting zone.

To further illustrate that point, I will discuss water retention and extraction data for an actual experiment at Castana, Iowa, during 1976, a dry year. The experiment was located on Ida silt loam soil [fine, silty, mixed (calcareous) mesic family of Typic Udorthents]. Soybeans were grown in rows

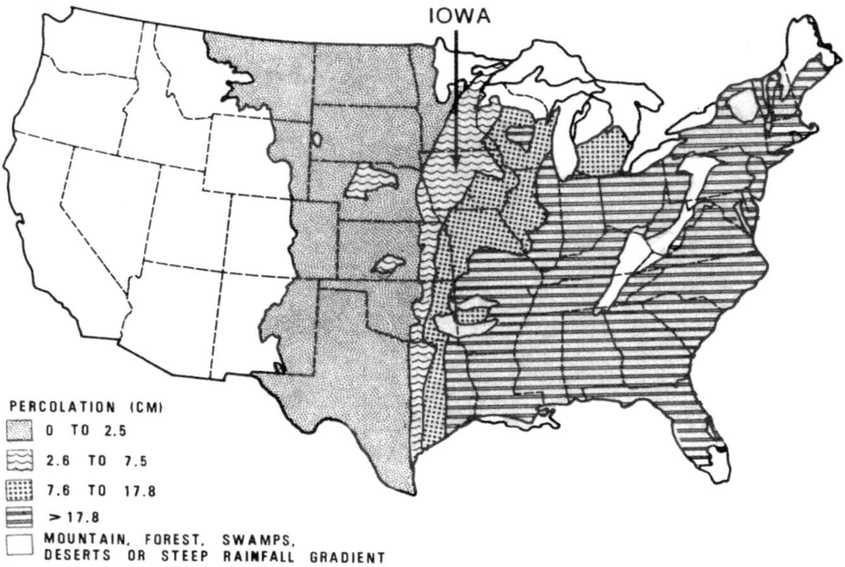

Figure 3. Average annual potential percolation as estimated by Stewart et al. (40).

spaced 100 cm apart. The soybeans were planted on May 12, but on June 23 the profile remained at field capacity except in the surface 15 cm (Curve A, Fig. 4). By August 23, the soil was at or below the wilting point to the 150-cm depth, and some water had been extracted from the 210- to 240-cm depth (Curve B, Fig. 4). Soil water contents were not determined at the 240- to 270-cm depth, but we assumed that no water was extracted from there. If, by some technique, we could extend the depth of water extraction by 30 cm (Curve C, Fig. 4), 2.85 cm more water would be extracted from a profile initially at field capacity (Curve A, Fig. 4). This extraction, in turn, would reduce deep percolation losses between harvest and the next growing season.

A second early question was, "Is the water ordinarily lost to deep percolation a reliable source of extra water for plant growth?" Another question pertinent to the same basic point was, "If we develop techniques to increase transpiration by 5 cm, what is the likelihood that the soil profile will be fully recharged before the next growing season?" The large quantities of water removed by tile drainage in the eastern half of Iowa almost guarantee that these profiles will be completely recharged in about 19 out of 20 years even if an additional 5 cm is used. However, no reliable estimates exist for recharge probabilities in western Iowa. We guess that most of those profiles will be fully recharged in about half of the years. Additional studies are in progress at the Western Iowa Research Farm, Castana, Iowa, to estimate how frequently profiles are fully recharged.

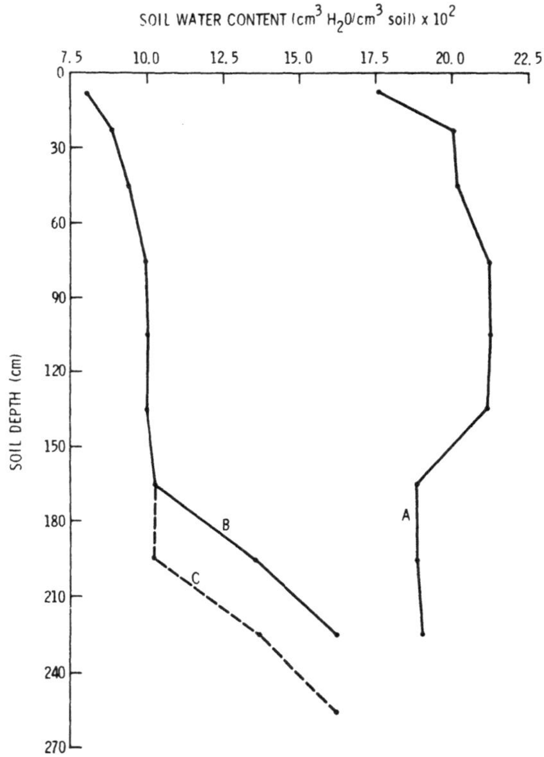

Figure 4. **Soil water content as a function of soil depth for 3 situations. Curve A was the actual soil water content profile on June 23, 1976, for soybeans grown in Ida silt loam soil at Castana, Iowa. Curve B was the water content profile on August 23, 1976, and Curve C is the postulated profile if rooting depth could have been increased by 30 cm.**

One major complicating factor in obtaining estimates of recharge frequency is the presence in loess soils of worm holes, root channels, cracks, and other stable continuous voids. These voids allow some water (an unknown quantity) to bypass relatively dry layers and increase soil water content at some lower depth (10,14). Changes in rooting patterns thus might be of benefit in increasing the total annual plant water supply even though upper soil layers are not completely recharged to field capacity.

ROOTING PARAMETERS AND WATER UPTAKE

Another key question was "What root system parameters determine the rate and duration of water extraction from a soil profile?" The pioneering works of Philip (29) and Gardner (13) provided the framework for many conceptual models that describe water uptake by root systems. One of these partially tested models (43) provides a framework for discussing the importance of the various inputs.

The model incorporates three basic equations. First, transpiration rate (T_a, $cm^3 H_2O/cm^2$ land surface/min) equals rate of decrease of water stored in plants occupying the specified land surface ($\Delta\theta_p$, cm^3/cm^2/min) plus total water uptake from soil by roots of plants occupying the specified land surface (U_a, cm^3/cm^2/min).

$$T_a = \Delta\theta_p + U_a. \qquad [1]$$

Second, total uptake is the sum of uptakes from individual soil volumes within that land surface (U_i, cm^3/min).

$$U_a = \sum_{i=1}^{n} U_i. \qquad [2]$$

Third, water uptake rate by roots located within a single soil volume is described by the equation

$$U_i = (V_i)(L_{vi})(q_i)(\psi_{si} - \psi_p + \psi_{zi} + \Sigma\psi_{fi}). \qquad [3]$$

Where V_i is a rooted volume of soil (cm^3) with uniform properties including uniform rooting density and water potential. L_{vi} is root length density in V_i (cm roots/cm^3 soil). q_i is average root water uptake rate in V_i ($cm^3 H_2O$/cm root/bar/min). ψ_{si} is total soil water potential in V_i (bars). ψ_p is total water potential in plant xylem at the ground surface (bars). ψ_{zi} is water potential loss due to depth below the ground surface (bars), and $\Sigma\psi_{fi}$ is the sum of water potential decreases due to frictional forces from midpoint of V_i to the ground surface.

This model assumes that the important rooting parameters are (a) rooting density, (b) hydraulic conductivity of water as it flows from bulk soil to the xylem lumen (q_i is an expression related to this conductivity), (c) frictional loss of energy as water flows upward through the xylem, and (d) rooting depth (which controls the number or size of volume increments to be considered). Use of this model, and others, is discussed in greater detail elsewhere (19,43).

Several people have been involved in a program to determine, for Iowa conditions, the range of values, relative significance, and methods for modifying each parameter. Field, rhizotron, greenhouse, laboratory, and growth chamber facilities have been used in the various experiments.

ROOTING DENSITY

The model predicts that rooting density plays an important role in determining rate of water uptake from a specific soil volume. I question, however, the importance of rooting density in determining the total annual water supply available to soybeans grown in Iowa except for two circumstances to be discussed later. Let me explain.

Our research group has conducted various experiments on Ida silt loam soil (a loess) at Castana, Iowa, during the past 4 yr, 1975 through 1978. We measured, among other variables, soil water contents and rooting densities every few days for extended periods to a 200- to 250-cm depth. Some of these data are published (4,5,35,47), some are contained in dissertations (23,34), and some are not yet published.

Maximum rooting densities (the greatest root length density that occurred at any sampling time) varied widely (up to four- or five-fold) from one year to the next at a specific depth and site. Regardless of these variations in rooting densities, soil water contents to a 150-cm depth were reduced to the same level as in succeeding years. What is the practical significance of increasing the rooting density above the lowest value found in any of those years? None, I believe.

The first exception to that conclusion arises in the surface 15 cm of soil. Sometimes we have several small rainfalls before a complete canopy develops. High rooting densities (or high conductivities radially in the root) in the surface soil are important so that the soybeans can use this water for transpiration before it evaporates directly from the soil. Use of that water for transpiration reduces use of water stored more deeply.

Rooting density also is important when the soil layers are so deep that roots are present for only a short time before pod-fill is complete. As an example, consider curve B of Fig. 4. Presumably, the root system also could have depleted the water content in the 180- to 240-cm depth to about 10% by volume if rooting density had been greater (or if the roots had arrived there earlier).

The conclusion about the relative ineffectiveness of increased rooting density (in situations other than these two examples) was reinforced by data (previously unpublished) collected during an investigation into the effect of plant spacing on soybean root development (4). Dr. Böhm counted the number of soybean roots that projected from smoothed soil profile walls at various depths. On June 25, he found 234 roots between the soil surface and a depth of 15 cm in a 100-cm wide transect (Fig. 5). On July 25, he found 665 roots in the same area. However, on August 4, the rooting intensity in that area had decreased to 183 roots, only 28% of those found on July 25. This major decrease in rooting density occurred even though rooting densities were increasing at lower depths. Data obtained by Sivakumar et al. (35) also showed major decreases in rooting densities over short periods of time. These decreases in rooting density presumably are caused by decreases in water content of that soil volume (20). A philosophical question occurs,

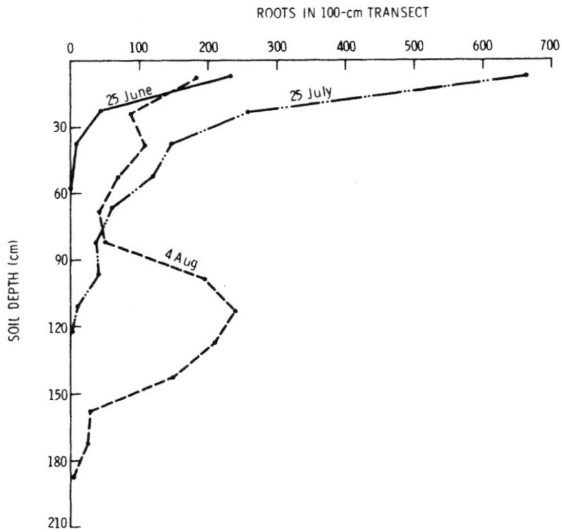

Figure 5. Soybean rooting intensities as functions of soil depth for 3 dates in 1975 at Castana, Iowa. See text for further details.

"Why would such a major loss of roots occur in this layer if rooting density were of major ecological significance?"

SOIL TO XYLEM HYDRAULIC CONDUCTIVITY

The model used in this article states that water uptake rate from a specific soil volume is proportional to the root length in that volume ($L_{vi}V_i$), to the total water potential difference between bulk soil (ψ_s) and root xylem ($\psi_{xy\ell}$) at the same depth, and to the soil-to-xylem hydraulic conductivity (q_i).

These q_i values are available for two isolines of Harosoy soybeans grown in southwestern Minnesota (2) and can be estimated with a reasonable degree of accuracy for the controlled water table experiment of Reicosky et al. (31) and for the field experiment with the 'Wayne' cultivar grown at Castana, Iowa (47). The q_i values decreased with soil water content (or soil water potential) in all 3 experiments. Earlier models of water uptake by root systems (13,29) suggested that this decrease in conductivity was associated with the development of a relatively dry soil shell, with an attendant low soil hydraulic conductivity, immediately outside the root epidermis. Recent experimental evidence, however, suggests that the decrease in q_i is associated with incomplete contact at the soil-root interface (11,15) or with the root tissue itself (32,42).

The q_i term not only varies with soil water content but also seems to vary inversely with total root length of the plants on which it is measured. One clue to the magnitude of that variation is found in a comparison of the Allmaras et al. (2) data with that of Willatt and Taylor (47). Although the water uptake rates (cm^3 H_2O)/(cm^2 land surface)/day were about the same, total root length was about 10 times greater on the soybeans of Willatt and Taylor (47) as on those of Allmaras et al. (2). Water uptake rate for each unit root length averaged about 10 times less for the plants of Willatt and Taylor (47).

This comparison tentatively suggested that plant properties are major determinants of q_i. Eavis and Taylor (9) tested the hypothesis that total length of roots and water uptake for each unit length are inversely related. We hoped to answer the question "How many roots does a soybean plant need to supply an adequate quantity of water for transpiration?" Because transpiration rate of the plant (with incomplete ground cover) increases with leaf area and with evaporative demand, a corollary question in the experiment was "What will be the effect on plant water relations if we alter, at equal evaporative demands, the ratio of root length to leaf area by varying the growth conditions?"

We grew soybeans for 6 wk in soil-filled containers, either of 100 liters or 162 liters, that were located in a grass sod plot. Tops of the containers were 69 cm above the soil surface; container size, level of fertility and previous history of water were the duplicated treatments.

After the 6-wk initial growth period, the plants had different leaf areas and presumably different root lengths. We watered the soil to an average of -1/3 bar soil matric potential and sealed each container to prevent soil water evaporation. We weighed the containers every 2 or 3 days, measured leaf areas on the same days, and periodically determined leaf water potential (ψ_1) (Scholander pressure chamber) in mid-afternoon. When all of the plants in each container stopped increasing in leaf area, we harvested the total plant tops, washed the roots from the soil, and determined root length with a root counting machine (35). At harvest, we observed that roots had thoroughly permeated each container. Our results showed that, at any one soil water content, leaf water potential in midafternoon did not depend on the ratio of root length to leaf area. In addition, transpiration for the period was not significantly influenced at any soil water content by root length in the container. We concluded that transpiration rate (cm^3 H_2O/day/container) was controlled entirely by evaporative demand (not much of a variable in our experiment), by leaf area, and by soil water content (or more likely by soil matric potential) (Fig. 6). In any specific container, specific root water uptake rate (cm^3 H_2O/cm root/day) progressively declined with soil water content. At any particular water content, specific root uptake rates declined with increases in the root length-leaf area ratios (Fig. 7).

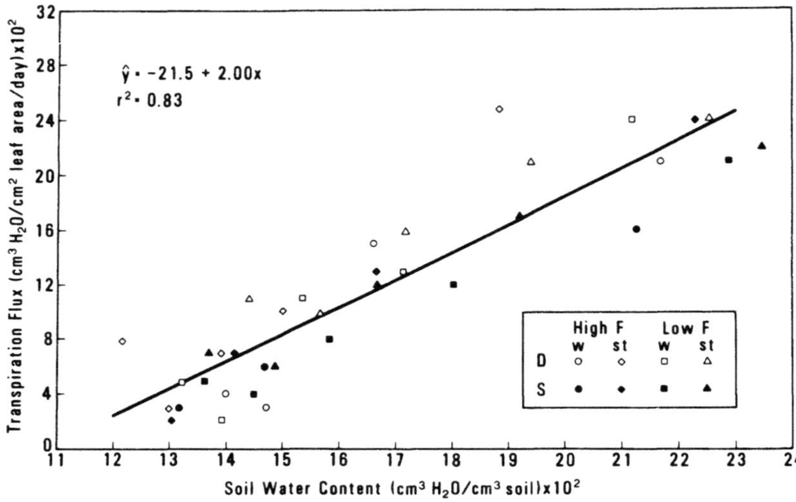

Figure 6. **Effect of soil water content on transpiration flux as affected by fertilizer (F), by deep (D) versus shallow (S) containers and by frequency watered (w) versus water stressed (st) plants. From Eavis and Taylor (9).**

A researcher must always be careful not to extrapolate results beyond the data base. The root length to leaf area ratios ranged from 39 to 140 cm roots/cm^2 leaf area in the Eavis-Taylor experiment (Fig. 7). In addition, all roots were surrounded with soil at about the same water content. The same cultivar of soybeans (Wayne) grown in fertile Ida silt loam soil in a field experiment had root length to leaf area ratios that ranged from 3 to 8 cm roots/cm^2 leaf area (35). In addition, the deep roots in the field experiment usually were located in wetter soil than the shallow roots. It is possible that the root length-leaf area ratios are important when there are 3 to 8 cm roots/cm^2 leaf area but are unimportant when there are 39 to 140 cm roots/cm^2 leaf area. Further experiments should be conducted to help resolve the question "Will soybeans suffer less water stress during a drought if they have a large root length to leaf area ratio?"

FRICTIONAL LOSS OF ENERGY IN THE XYLEM

The possibility that substantial losses of energy can occur during upward flow of water through the xylem of soybean plants needs to be investigated in detail. Taylor and Klepper (43) suggested, on the basis of the Willatt and Taylor (47) experiment, that frictional losses might be on the order of

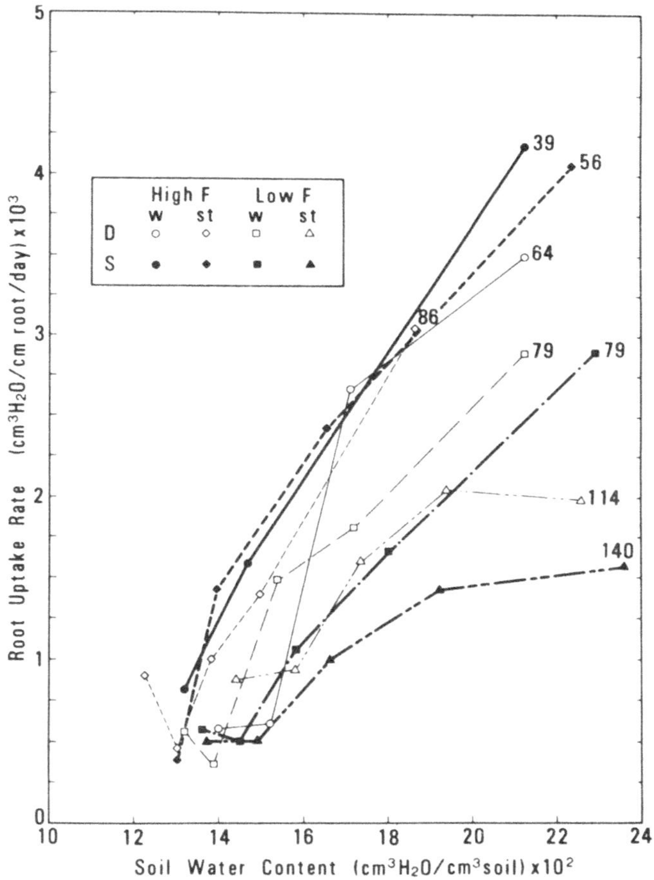

Figure 7. **Effect of soil water content on root water uptake rate as affected by fertilizer (F), by deep (D) versus shallow (S) containers and by frequently water (w) versus stressed (st) plants. The numbers near data trend lines are centimeters of root length per square centimeter of leaf area. From Eavis and Taylor (9).**

0.07 bars/cm. This magnitude of energy loss will drastically lower the water potential difference between soil and xylem of roots located deep within the profile. So (36) used thermocouple psychrometers to estimate root xylem water potentials at various depths from the soil surface. He concluded that substantial frictional losses occur during periods of high transpiration. If high frictional losses in soybeans are real, roots located deep within the profile will not be as effective, at the same soil water potentials, as those located nearer to the soil surface. For our Iowa conditions, these deep roots are still an efficient use of photosynthate because they are located in wetter soil than any shallower roots produced at the same stage of growth.

ROOTING DEPTH

Increases in rooting depth increase the total number or increase the thickness of the soil layers considered in the model of water uptake (43). These increases in rooting depth, at equivalent rooting densities and evaporative demands, increase seasonal water supplies for soybeans grown in Iowa.

Several articles that described soybean rooting patterns and function were available when this project started in 1973 (6,25,30,31,48). Despite this literature, there were not enough data to evaluate the hypothesis that onset of severe water stress in Iowa soybeans can be delayed by increasing rooting depth. We conducted several experiments to answer specific questions involved in testing that hypothesis.

One question was "Are there substantial differences in downward root extension rates among various cultivars of soybeans?" Taylor et al. (41) determined growth rates of soybean taproots in 7-cm diameter, 120-cm long clear acrylic tubes, inclined at 11 degrees from the vertical. There were significant differences ($p < 0.95$) among the 29 cultivars. After a 27-day growth period, the taproots of Blackhawk cultivar averaged 83.5 cm deep while those of Sciota cultivar averaged 117.3 cm deep, an increase of 40% over the Blackhawk cultivar depth. Correlation coefficients indicated that seed size and dry weight of tops at harvest were associated with less than half of this variation. Unpublished experiments conducted in the Ames rhizotron by T. C. Kaspar and C. D. Stanley showed that downward penetration rates for nine cultivars were in the same ranking as they had been in the acrylic tube study. Substantial chances exist that a breeding and selection program can increase the rooting depth for soybeans. We are examining the root extension characteristics of several hundred soybean genotypes each year.

A commonly accepted dogma in crop physiology is that root growth essentially ceases when seeds are developing. If this principle were true for Iowa soybeans, the roots would be required to reach their maximum depth within 60 to 70 days after planting. The cultivars of soybeans grown here, however, are indeterminate in growth habit so we felt that root growth might continue during reproduction. Kaspar et al. (18) determined the root extension rates for seven soybean cultivars during reproductive development. They found that roots extended deeper at average rates of 1.1 to 1.8 cm/day during vegetative development (when the soil was cooler) to 3.3 to 8.1 cm/day as the plants progressed from developmental stage R1 to R2, 2.6 to 3.0 cm/day from R3 to R4, and 2.1 to 3.6 cm/day from R4 to R5. These development stages, as defined by Fehr et al. (12), covered the period from when a flower was present at any node (R1) to the stage where beans could be felt when the pods were squeezed at any of the four uppermost nodes (R5). Some new roots developed almost until physiological maturity. These results were obtained in a rhizotron where plant populations were lower than those usually found in the field. However, Willatt and Taylor (47) found that 'Wayne' soybean roots extended downward 25 cm in 12 days during the period when the tops developed from stage R4 to R5 (35). These results were obtained during 1975, a relatively dry year at Castana, Iowa. During 1978, a relatively wet year, root extension during this period was about half of that during 1975 (data not shown). We therefore concluded that substantial downward root extension occurs during pod-filling, especially during dry years. This fact is extremely important in any selection program to increase rooting depth because it shows that we should examine plants for both rate and duration of taproot extension.

Several soil conditions slow the rates of increase in rooting depth. As one example, fertilization of a soil deficient in one or more essential elements usually increases growth of plants, both tops and roots (45). Sometimes, but not always, the greater root growth results in deeper water extraction (45). Pesek et al. (20) found unfertilized corn (*Zea mays* L.) extracted water from less than 150 cm of soil in Iowa, whereas well-fertilized corn extracted water to 210 cm. In an experiment at Castana, Iowa, D. G. Woolley investigated the effects of two fertility levels on the water extraction patterns of 'Wayne' soybeans grown on a loess soil newly plowed from bromegrass (*Bromus inermis* Leyss.). The two fertility levels were (a) no fertilizer applied for at least 10 yr and (b) an application of 168 kg/ha of both nitrogen and phosphorus (both expressed as the elemental form). The soybean roots extracted significantly more water below the 120-cm depth in the heavily fertilized than in the unfertilized plots, but considerable water was extracted from the 180- to 210-cm soil depth in both treatments (Fig. 8).

Another soil factor that might control rates of rooting depth increases is waterlogging. Anaerobiosis, brought about by waterlogging the soil profile, kills roots, reduces top growth and often reduces crop yield (7). The evidence is not clear, however, on effects of temporary waterlogging at various stages of development on soybean root and shoot responses. Stanley (39) imposed water tables at 45- or 90-cm depths on 'Wayne' soybean roots for seven days during vegetative growth (V8 to V10), during post-flowering, prepod set (R2) and during post-pod set (R4).

During vegetative growth, roots located below the water table ceased growing downward when they were inundated but again grew downward at about the same rate when the water table was drained. When the water table was imposed at post-flowering (R2), inundated roots looked normal as long as the water table was present, but these roots decomposed rapidly when the

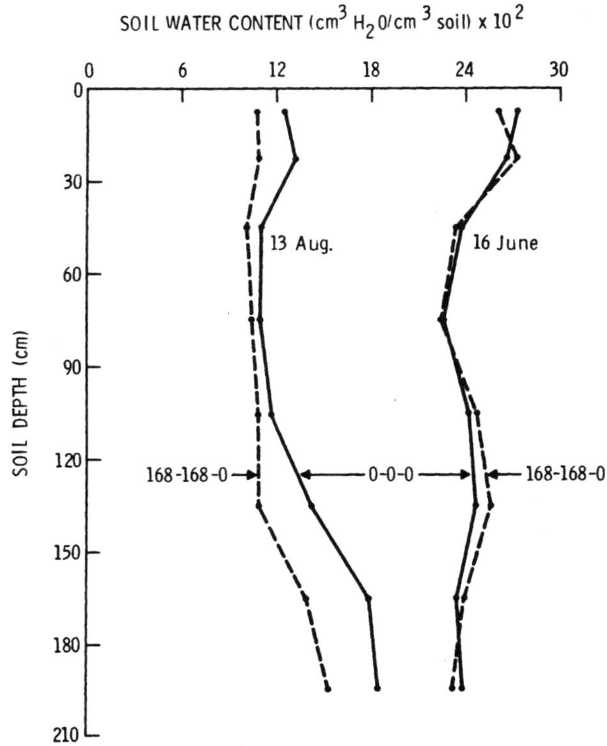

Figure 8. Soil water content as a function of soil depth for 2 fertilizer treatments and 2 sampling dates on a loess soil at Castana, Iowa, during 1975.

soil was drained. Compensatory root growth occurred in a 10- to 15-cm soil layer immediately above the water tables imposed at this time. Vegetative growth had almost stopped before the water tables were imposed at R4. Roots located below the water table soon looked dead. No new root growth occurred when the soil was drained.

Root losses associated with the water tables during the two reproductive periods did not reduce soybean yield. Indeed, the extra water added by the water table treatment actually increased the soybean yield. The 45-cm water tables caused significantly ($p < .95$) greater yields than the 90-cm ones at both the vegetative period (V8) and post-flowering period (R2). There was no difference in yield between the two depths at R4.

Root pruning associated with temporary increases in water table level probably is not nearly as damaging to yield as is often feared.

Many experiments have determined the effects of soil temperature on plant root development (26,33,44). Some of these experiments have studied effects of soil temperature on soybean root growth and function (1,8,22). The available data indicate that the rates at which soybean rooting depths increase (for a specific genotype) in Iowa often are controlled by soil temperature. However, Maduakor (23) investigated the effects of various corn (*Zea mays* L.) mulch rates on soybean top and root development. He concluded that effects of the mulch on soil water relationships were at least as important as those on soil temperature in root growth and soybean yields. We do not have enough data to predict the effects of soil temperature on soybean root extension at soil depths of 150 to 250 cm. Soil temperatures vary with soil depth, with time during the growing season, and among seasons. These variations will affect root extension rates, but further experiments are needed to obtain quantitative relationships in the field.

CONCLUDING REMARKS

I have presented a progress report on our efforts to test the hypothesis that onset of severe water stress for soybeans grown in Iowa can be postponed by rooting modifications. The paper has presented my conclusions that:

(A) Some deep percolation occurs each year in eastern Iowa and in at least half of the years in western Iowa.

(B) In eastern Iowa, this deep percolation represents a dependable source of extra water for soybeans, but in western Iowa an extra 2.5 cm of water might be available only in about half of the years.

(C) Increasing the rooting depth is more likely to successfully postpone water stress than increasing the rooting density or the water uptake rate per unit root length.

(D) There are substantial differences among genotypes in the rate that rooting depth increases during the growing season.

(E) Some soybean cultivars currently grown in Iowa have a rooting depth of more than 200 cm.

(F) Rooting depths continue to increase during reproductive stages of development in the indeterminate soybeans usually grown in Iowa.

(G) Adding fertilizer to a deficient soil will increase the depth of water extraction.

(H) Pruning of soybean roots by temporary high water tables sometimes may have relatively minor consequences on soybeans, and

(I) Soil temperature probably controls rates of rooting extension in May and June, but more quantitative data are needed to prove this hypothesis.

This type of project is a high-risk one. Many unknowns can interfere with attaining the goal. The research described in this paper has increased the probability that soybean genotypes and production practices can be combined to delay the onset of severe plant water stress for a few days. If this delay occurs during pod-filling, higher yields should result.

NOTES

Howard M. Taylor, Agricultural Research, Science and Education Administration, U.S.D.A. and Dept. of Agronomy, Iowa State University, Ames, Iowa 50011.

REFERENCES

1. Allmaras, R. R., A. L. Black, and R. W. Rickman. 1973. Tillage, soil environment and root growth. p. 62-86. In Conservation Tillage: The proceedings of a national conference. Soil Consev. Soc. of Am., Ankeny, IA.
2. Allmaras, R. R., W. W. Nelson, and W. B. Voorhees. 1975. Soybean and corn rootin in southwestern Minnesota. II. Root distributions and related water inflow. Soil Sci. Soc. Am. Proc. 39:771-777.
3. Bertrand, A. R. 1966. Water conservation through improved practices, p. 207-235. In W. H. Pierre et al. (ed.) Plant environment and efficient water use. Am. Soc. Agron., Madison, WI.
4. Böhm, W. 1977. Development of soybean root systems as affected by plant spacing. Z. Acken-und Pflanzenbau (J. Agron. and Crop Sci.) 144:103-112.
5. Böhm, W., H. Maduakor, and H. M. Taylor. 1977. Comparison of five methods for characterizing soybean rooting density and development. Agron. J. 69:415-419.
6. Borst, A. L., and L. E. Thatcher. 1931. Life history and composition of the soybean plant. Ohio Agr. Expt. Sta. Bull. 494.
7. Cannell, R. Q. 1977. Soil aeration and compaction in relation to root growth and soil management. Appl. Biol. 2:1-86.
8. Earley, E. B., and J. L. Cartter. 1945. Effect of the temperature of the root environment on growth of soybean plants. J. Am. Soc. Agron. 37:727-735.
9. Eavis, B. W., and H. M. Taylor. 1979. Transpiration of soybeans as affected by leaf area, root length, and soil water content. Agron. J. 71:441-445.
10. Ehlers, W. 1975. Observations on earthworm channels and infiltration on tilled and untilled loess soil. Soil Sci. 119:242-249.
11. Faiz, S. M. A., and P. E. Weatherley. 1978. Further investigations into the location and magnitude of the hydraulic resistances in the soil:plant system. New Phytol. 81:19-28.
12. Fehr, W. R., C. E. Caviness, D. T. Burmood, and J. S. Pennington. 1971. Stage of development descriptions for soybeans, *Glycine max* (L.) Merrill. Crop Sci. 11: 929-931.
13. Gardner, W. R. 1960. Dynamic aspects of water availability to plants. Soil Sci. 89:63-73.
14. Goss, M. J., J. T. Douglas, K. R. Howse, J. M. Vaughn-Williams, and M. A. Ward. 1978. Effect of cultivation on soil physical conditions, p. 37-41. In ARC Letcombe Lab. Ann. Rep. 1977. Wantage, England.
15. Herkelrath, W. N., E. E. Miller, and W. R. Gardner. 1977. Water uptake by plants. II. The root contact model. Soil Sci. Soc. Am. J. 41:1039-1043.
16. Hess, W. N. 1974. Weather and climate modification. John Wiley & Sons, New York.
17. Hillel, D. I. 1971. Soil and water physical principles and processes. Academic Press, Inc., New York, NY.
18. Kaspar, T. C., C. D. Stanley, and H. M. Taylor. 1978. Soybean root growth during the reproductive states of development. Agron. J. 70:1105-1107.
19. Klepper, B. and H. M. Taylor. 1979. Limitations to current models describing water uptake by plant root systems, p. 51-65. In The soil-root interface. Academic Press, Inc., London, England.

20. Klepper, B., H. M. Taylor, M. G. Huck, and E. L. Fiscus. 1973. Water relations and growth of cotton in drying soil. Agron. J. 65:307-310.
21. Kramer, P. J. 1969. Plant and soil water relationship' A modern symthesis. McGraw-Hill Book Co., New York, NY.
22. Mack, A. R., and K. C. Ivarson. 1972. Yield of soybeans and oil quality in relation to soil temperature and moisture in a field environment. Can. J. Soil Sci. 52:225-235.
23. Maduakor, H. O. 1978. The effect of mulch on the top and root growth of soybeans. Ph.D. Dissertation, Iowa State Univ., Ames, IA.
24. Miller, D. E. 1973. Water retention and flow in layered soil profiles, p. 107-117. In R. R. Bruce (ed.) Field soil water regime, Soil Sci. Soc. Am. Publ. 5, Madison, WI.
25. Mitchell, R. L., and W. J. Russell. 1971. Root development and rooting patterns of soybean [*Glycine max* (L.) Merr.] evaluated under field conditions. Agron. J. 63: 313-316.
26. Nielsen, K. F., and E. C. Humphries. 1966. Effects of root temperature on plant growth. Soils Fert. 29:1-7.
27. Pearson, R. W. 1974. Significance of rooting pattern to crop production and some problems of root research, p. 247-270. In E. W. Carson (ed.) The plant root and its environment. Univ. of VA Press, Charlottesville, VA.
28. Pesek, J. T., Nicholson, R. P., and Spies, C. 1955. What about fertilizers in dry years? Iowa Farm Sci. 9(10):3-6.
29. Philip, J. R. 1957. The physical principles of soil water movement during the irrigation cycle. 3rd Congr. Int. Comm. on Irrig. and Drain. Quest. 8 p. 8.125-8.154.
30. Raper, C. D., and S. A. Barber. 1970. Rooting systems of soybeans. I. Differences in root morphology among varieties. Agron. J. 62:581-584.
31. Reicosky, D. C., R. J. Millington, A. Klute, and D. B. Peters. 1972. Patterns of water uptake and root distribution of soybeans (*Glycine max*) in the presence of water table. Agron. J. 64:292-296.
32. Reicosky, D. C., and J. T. Ritchie. 1976. Relative importance of soil resistance and plant resistance in root water absorption. Soil Sci. Soc. Am. J. 40:293-297.
33. Richards, S. J., R. M. Hagan, and T. M. McCalla. 1952. Soil temperature and plant growth. In B. T. Shaw (ed.) Soil physical conditions and plant growth. Agron. Monog. II, p. 303-480.
34. Sivakumar, M. V. K. 1977. Soil-plant-water relations, growth and nutrient uptake patterns of field-grown soybeans under moisture stress. Ph.D. Dissertation, Iowa State Univ., Ames, IA.
35. Sivakumar, M. V. K., H. M. Taylor, and R. H. Shaw. 1977. Top and root relations of field-grown soybeans. Agron. J. 69:470-473.
36. So, H. B. 1979. Water potential gradients of resistances of a soil-root system measured with the root and soil psychrometer, p. 99-113. In J. L. Harley and R. S. Russell (eds.) The soil-root interface, Academic Press, London, England.
37. Spomer, R. G., W. D. Shrader, P. E. Rosenberry, and E. L. Miller. 1973. Level terraces with stabilized backslopes on loessial cropland in the Missouri Valley: A cost-effectiveness study. J. Soil and Water Conserv. 28:127-130.
38. Stanley, C. D. 1975. The relationship of evapotranspiration to open pan evaporation throughout the growth cycle of soybeans [*Glycine max* (L.) Merr.] M.S. Thesis, Iowa State Univ., Ames, IA.
39. Stanley, C. D. 1978. Soybean top and root response to static and fluctuating water table situations. Ph.D. Dissertation, Iowa State Univ., Ames, IA.

40. Stewart, B. A., D. A. Woolhiser, W. H. Wischmeier, J. H. Caro, and M. H. Frere. 1975. Control of water pollution from cropland. Vol. I. A manual for guideline dedevlopment. Agr. Res. Ser. Rept. ARS-H-5-1, Washington, DC.
41. Taylor, H. M., E. Burnett, and G. D. Booth. 1978. Taproot elongation rates of soybeans. Z. Acker-und-Pflanzenbau 146:33-39.
42. Taylor, H. M., and B. Klepper. 1975. Water uptake by cotton root systems: An examination of assumptions in the single root model. Soil Sci. 120:57-67.
43. Taylor, H. M., and B. Klepper. 1978. The role of rooting characteristics in the supply of water to plants. Adv. Agron. 30:99-128. Academic Press, Inc., New York.
44. van Bavel, C. H. M. 1972. Soil temperature and crop growth. In D. I. Hillel (ed.) Optimizing the soil physical environment toward greater crop yields. p. 23-33. Academic Press, Inc., New York.
45. Viets, F. G. 1962. Fertilizers and the efficient use of water. Adv. Agron. 14:223-264. Academic Press, Inc., New York.
46. Wadleigh, C. H., W. A. Raney, and D. M. Herschfield. 1966. The moisture problem. In W. H. Pierre et al. (ed). Plant environment and efficient water use. p. 1-19. Am. Soc. Agron., Madison, WI.
47. Willatt, S. T., and H. M. Taylor. 1978. Water uptake by soyabean roots as affected by their depth and by soil water content. J. Agr. Sci. (Camb.) 90:205-213.
48. Winter, S. R., and J. W. Pendelton. 1968. Soybean roots: How deep do they go? Soybean Digest 28(7)14.
49. Wright, M. J. 1976. Plant adaptation to mineral stress in problem soils. Spec. Publ., Cornell Univ. Agr. Exp. Sta., Ithaca, NY.

ROLE OF CLASSICAL BREEDING PROCEDURE IN IMPROVEMENT OF SELF-POLLINATED CROPS

V. A. Johnson and J. W. Schmidt

Breeders rely heavily on classical procedures for improvement of self-pollinating crops. Classical breeding has numerous variations, all of which presumably are designed to improve the breeder's chances for success in identifying productive agronomically acceptable phenotypes. The particular procedure followed by the breeder usually is a compromise between what he perceives to be the most sound from a theoretical point of view and what he recognizes as practically feasible. The philosophy of the breeder, the crop species, breeding objectives and priorities, and economic constraints all are involved.

The cooperative Nebraska-USDA program of hard winter wheat improvement will serve as an example of the classical breeding approach to improvement of a self-pollinated crop in which procedures that are significant departures from normal are practiced. We do not suggest that the system we follow would be the best for wheat improvement in all circumstances, nor do we suggest that it could be applied with equal success to other self-pollinated crops. We offer it as one approach for consideration and discussion.

PROGRAM OBJECTIVES AND PRIORITIES

The Nebraska-USDA wheat improvement program has three main objectives: (a) Maximize grower return per unit of investment by developing varieties with improved yield potential and improved responsiveness to production inputs; (b) Stabilize production by reducing varietal vulnerability to diseases, insects, and environmental hazards; and (c) Provide to the processing industry wheat that possesses good milling and baking characteristics and improved nutritive value.

The program also encompasses relevant genetic and cytogenetic studies, production studies and breeding methodology studies. Breeding priorities are largely dictated by the problems of wheat production in Nebraska. The following are emphasized in our program:

(a). *Winterhardiness* — The climate of Nebraska is continental. It is characterized by highly variable and frequently very low winter temperatures that may occur without snow cover and may be accompanied by severe soil moisture stress. To be acceptable to producers, Nebraska wheat varieties must tolerate these conditions.

(b) *Stem rust resistance* — Stem rust *(Puccinia graminis tritici* L.) occurs annually in Nebraska. It has enormous destructive potential and can devastate wheat in the plains area from mid-Kansas northward into the prairie provinces of Canada. Stem rust-susceptible varieties cannot be grown safely in Nebraska.

(c) *Yield and performance stability* — Identification of varieties that perform well relative to other varieties under the conditions of wheat production in Nebraska is a key component of the breeding effort. Spring and summer temperatures fluctuate widely; precipitation is unpredictable and can range from less than 10 in. to more than 40 in. annually; heavy nitrogen fertilization of wheat usually is uneconomical and can be counter-productive; low relative humidity may be accompanied by high-velocity winds in spring. Varieties able to tolerate such conditions as well as respond to favorable conditions with high grain yields are sought.

(d) *Processing and nutritional quality* — Milled flour from hard red winter wheat is utilized mainly for bread-making which in the U.S. is highly automated. Automation imposes relatively rigid and complex milling and baking requirements. Nutritional quality of wheat is influenced strongly by grain protein content which is a heritable trait strongly influenced by production practices and soil fertility.

(d) *Lodging resistance* — Because of frequent high winds during grain development, useful resistance to lodging in hard winter wheat requires a high degree of straw resiliency to avoid straw breakage. Short plant height contributes to lodging resistance but, of itself, does not assure adequate resistance.

Numerous additional traits are monitored. Each contributes to the performance and acceptability of new varieties but has somewhat lower priority in the Nebraska program than the five traits already identified. Included in the latter group are maturity, plant height, leaf rust resistance, wheat streak and soil-borne mosaic resistance, tillering, seed size, test weight and spike size.

BREEDING ASSUMPTIONS

The following assumptions provide the rationale for the Nebraska-USDA wheat improvement program and strongly influence the breeding system that is followed: (a) Useful genetic variability in common wheat *(Triticum aestivum* L.) has not been exhausted; (b) Genetic variance for yield in common wheat is mostly additive and can be fixed in true-breeding lines; (c) Significant increases in yield can result from the control or removal of such yield constraints as diseases and insects; (d) Progeny yield performance cannot be predicted with any degree of certainty from known attributes of the parental varieties; (e) Experimental wheats that remain in the breeder's nursery contribute little to agriculture; (f) Performance stability is as important as high yield potential of varieties grown in the hard winter wheat region of the U.S.; (g) Varietal heterogeneity can serve as a buffer against variations in production environment and can effectively contribute to performance stability of wheat; (h) Genetic vulnerability constitutes a continuing potential threat to production in the U.S. high plains where wheat dominates the crop acreage. Availability and use by growers of many varieties, relatively rapid turnover of varieties, and intra-varietal heterogeneity can reduce genetic vulnerability; (i) New improved varieties should be percentived as transitory. They should be released with expectation of early replacement with better varieties; (j) Wide-scale regional testing provides in a few years reliable information on the area and breadth of adaptation and performance stability of experimental varieties that would require many years if the evaluation were confined to a single state; and (k) Nebraska wheat producers are well-informed, innovative farm operators who are capable of making sound judgements about selection of varieties and other production inputs. Hence, the historical system of strong recommendation by the state experiment station of a few varieties is less desirable than the availability of several well-described varieties from which the producer can choose.

BREEDING PROCEDURE

Three to four hundred crosses are made annually. The relatively large number reflects our belief that we cannot predict reliably the worth of progeny on the basis of known parental attributes. Experience has shown that as many as 400 hybrid populations can be managed adequately with our available land, facilities, personnel, and other resources.

We identify two categories of crosses. For variety development we rely on crosses of adapted by adapted parental lines, whereas crosses of adapted by exotic lines are intended primarily for development of germplasm. The likelihood of finding lines with variety potential in the latter type of hybrid populations is extremely low. The high level of winterhardiness required of Nebraska winter wheat varieties is a major obstacle. Many genes condition winterhardiness. Most exotic wheats are non-winterhardy by Nebraska standards and are poorly adapted to the Nebraska production environment. Odds for finding winterhardy adapted lines in populations involving the exotic wheats are poor.

The Nebraska breeding and selection procedure can be summarized as follows: (a) Make 300 to 400 crosses annually; (b) Grow F_1s in the field and greenhouse; (c) Grow F_2s as bulk hybrids; evaluate for agronomic, disease and quality traits. No discard; (d) Grow F_3s as bulk hybrids; continue agronomic and disease evaluation. Head select approximately 50% of the populations; 100 to 300 heads per population; (e) F_4-head-row nursey comprised of 50- to 60-thousand entries. Screen for stem rust reaction; advance without reselection best 5% on basis of stem rust resistance and plant type; (f) F_5—Preliminary observation plots; evaluate for agronomic value, disease resistance and quality. Retain approximately 25%; (g) F_6—Duplicate plots at three Nebraska sites; initial performance evaluation; continue quality evaluation; (h) F_7—Triplicated yield trials at five Nebraska test sites; continue performance and quality evaluation; (i) F_8—Four-replication yield trials at five Nebraska test sites; continue performance and quality evaluation; (j) F_9—Continue station performance trials; initiate on-farm tests, regional evaluation and seed increase; and (k) F_{10}—Continue state and regional testing; foundation seed increase; large-scale milling and baking evaluation; name and release to certified producers.

Note that hybrid populations are managed as bulks through the F_3 generation. The bulk hybrids are seeded at the normal rate to allow their evaluation under conditions that approximate commercial production. Seed from the F_2 bulks is evaluated for quality by means of milling evaluation, mixing curves, and protein determination. No discard of populations is practiced in the F_2 generation to allow a second year of evaluation before the decision to select or discard is made. Based upon 2 yr of visual field evaluation and F_2 quality tests, we make head selections in the F_3 from the best bulk populations, usually about half of the total number grown. In a limited number of crosses in which we have special interest, we have initiated head selection as early as the F_2. In other crosses of less interest, selection has been delayed until the F_4. However, F_2 and F_4 selection is not as common in our system as F_3 selection.

Head selection is essentially random, although, as the situation may permit, attention is paid to such attributes as maturity, height, disease reaction and plant type. Our use of head selection instead of plant selection is in part dictated by facilities and financial resources. Plant selection would require space planting of populations with attendant requirement for more land, special seeding equipment, problems of weed control and stand loss due to winter killing. Plant selection, because it requires heavy use of hand labor compared with head selection, is substantially more expensive than head selection.

The relatively small number of heads selected per population (100 to 300) is, in part, compensated by the selection of a large number of populations. Head-rows usually number 50 - to 60-thousand, although in some years as many as 100 thousand have been grown. The large number exceeds our capability to evaluate adequately. Head-rows are subjected to artifically created stem rust epidemics. Since reselection is not practiced, head-rows that exhibit reasonable uniformity for maturity, plant height, plant type, and disease reaction are sought.

Lines remain in observation plots for only 1 yr in which their apparent agronomic value and disease resistance are further assessed. Approximately 25% of the lines are advanced to duplicated plots at three Nebraska test sites from which initial performance determinations are made. The number of test sites is increased to 5 and the number of replications to 3 and 4, respectively, in the F_7 and F_8 generations. Ten to 20% of the lines are advanced in each of these generations. Station testing normally is continued into the F_{10} generation and on-farm tests are initiated in the F_9.

Regional evaluation for winterhardiness at northern plains test sites may be initiated as early as the F_8 generation. Lines also will be entered in special disease trials. Those with the best state performance records are nominated for evaluation in one or both of two regional performance nurseries, each involving 20 to 30 test sites in the hard red winter wheat region. Seed increase and large-scale milling and baking evaluation usually occur simultaneously with regional testing.

RESULTS AND DISCUSSION

Twenty improved varieties of hard red winter wheat developed in the cooperative Nebraska-USDA program have been released for commercial production since 1963 (Table 1). Ten of the varieties were selected in the F_3, three in the F_2, and four in the F_4 generation. Three varieties (Scout 66, Scoutland, and Centurk 78) are reselections from Scout and Centurk. Two varieties were released in foreign countries and one in another state.

Development time for the varieties ranged from 10 to 16 yr. This is longer than normal development time for spring wheat and other spring-sown small grains. Because of the requirement of winter wheat for vernalization, two generations per year are not feasible except on a limited scale. Use of greenhouses during the winter does not accelerate the advance of generations unless two generations can be grown. We attach no particular significance to somewhat slow time for development of 10 or more yr for winter wheat compared with as little as 6 ot 7 yr for the spring-sown small grains. Once a breeding program is established, the amount of new material on stream at a given time is the same whether development time is 10 or 6 yr.

Table 1. **Commercially-grown hard winter wheat varieties from the cooperative Nebraska-USDA breeding program, 1963-1978.**

Name	Generation of Selection	Year of Release	Development Time (Yr)
	From Early Generation Selection		
Gage	F_3	1963	16
Scout	F_3 .	1963	14
Lancer	F_3	1963	14
Guide	F_2	1967	10
Trader	F_3	1967	10
Trapper	F_3	1967	10
Centurk	F_4	1971	12
Sentinel	F_3	1973	11
Homestead	F_3	1973	11
HiPlains	F_3	1973	11
Buckskin	F_3	1973	11
Lancota	F_2	1975	10
Agate	F_4	1976	14
Bennett	F_2	1978	10
	By Re-selection		
Scout 66	85 heads from Scout	1967	--
Scoutland	Single head from Scout	1970	--
Centurk 78	5 heads from Centurk	1978	--
	Released Elsewhere		
Bolal (Turkey)	F_3	1970	--
Belinda (Rep. of So. Africa)	F_4	1971	--
Capitan (New Mexico)	F_4	1978	--

Progress in improving the yield of Nebraska wheat since 1920 is shown in Figure 1. Three periods can be identified in terms of the varieties grown. For the first 20 yr (1921 to 1940) Turkey wheat and Nebraska 60 (a selection from Turkey) dominated the acreage. During the next 25 yr (1941 to 1965) two improved selections from Turkey (Cheyenne and Nebred) and Pawnee produced an average yield increase of 7.6 bu/acre. Since 1965, varieties selected in early generations from our program have comprised most of the Nebraska acreage. During the period, the state average yield was 34.2 bu/acre—a 13 bu increase over the previous period or an increase of 60%. The latter group of varieties, with exception of Warrior, possess effective field resistance to stem rust. We believe that these demonstrable yield advances lend strong support to the validity of our breeding assumptions that there is useful genetic variability for yield in wheat which can be fixed in true-breeding lines, and that yields can be increased significantly by removal of production constraints.

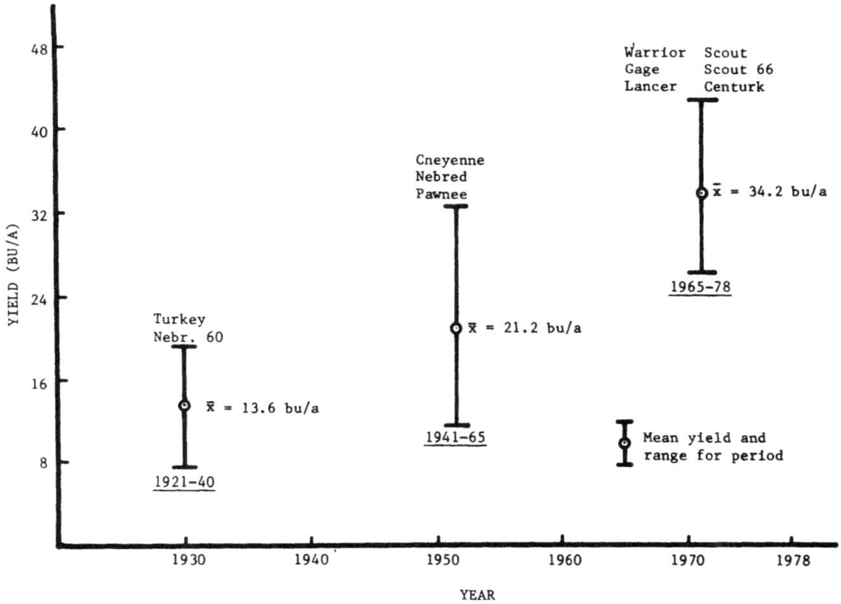

Figure 1. Trend in Nebraska winter wheat yields, 1921 to 1978.

New improved varieties from our program have been widely accepted by producers throughout the hard winter wheat region of the U.S. As shown in Table 2, seven early generation Nebraska varieties and selections from them have constituted nearly one-fourth of the total U.S. wheat acreage for the last decade. Within 6 yr of its release, Scout had become the most extensively grown single wheat variety in the country, occupying an estimated 14.2% of the U.S. wheat acreage in 1969. Although the heaviest concentration of Scout production has been in Nebraska, Kansas, and Colorado, there also has been significant production of Scout in New Mexico, Texas, Oklahoma, South Dakota, and Wyoming. Because of the superior performance and stability of Centurk in regional performance trials, seven other states joined Nebraska and the USDA in the release of Centurk in 1971. By 1974, its use had grown to an estimated 3.1 million acres.

The regional performance of Scout and Centurk in relation to the mean performance of experimental varieties in the Southern Regional Performance Nursery (SRPN) is plotted in Figure 2 (2). Experimental varieties since 1963 together have shown a continuing significant increase in productivity over the 'Kharkof' check variety. Since 1974, their yield superiority over Kharkof ranged from 30 to more than 40%. During the 5 yr that Scout was tested in the SRPN it was the most productive variety in each of the years. Centurk was the most productive in 4 out of 7 yr of testing. Our assumption that

Table 2. **Acceptance by U.S. wheat producers of some early-generation Nebraska varieties and selections from them.**

			1964		1969		1974	
Variety	Generation of Selection	Year of Release	Est. Acres	% U.S. Total	Est. Acres	% U.S. Total	Est. Acres	% U.S. Total
			millions		*millions*		*millions*	
Warrior	F_5	1960	1.5	2.7	1.5	2.8	0.7	1.0
Scout	F_3	1963	--	--	7.1	14.2	7.1	9.9
Lancer	F_3	1963	--	--	2.0	3.7	0.9	1.2
Gage	F_3	1963	--	--	1.3	2.5	0.7	1.0
Scout 66	Sel.	1967	--	--	0.1	0.2	1.9	2.7
Eagle[a]	Sel.	1970	--	--	--	--	2.3	3.2
Centurk	F_4	1971	--	--	--	--	3.1	4.4
Total	--	--	1.5	2.7	12.0	23.4	16.7	23.4

[a]Selected in Kansas.

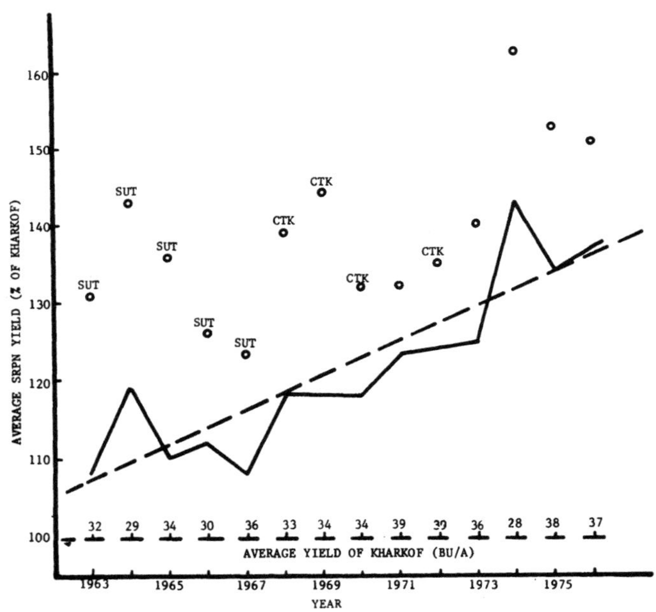

Figure 2. **Hard red winter wheat breeding progress as measured by nursery mean yields and yields of the most productive varieties in the Southern Regional Performance Nursery, 1963 to 1976. (SUT = Scout, CKT = Centurk).**

regional testing can reliably predict productivity and performance stability of new varieties is supported by these data and the acceptance of Scout and Centurk by producers in several states.

Varieties selected in an early generation are likely to possess a substantial amount of heterogeneity—particularly for complex traits. Simply inherited observable or measurable qualitative traits can be readily monitored in the selections to assure reasonable uniformity and non-segregation. This is particularly so in crosses of adapted x adapted parents that are similar in type, plant height, maturity and other readily observed traits. In such hybrid combinations the identification of early generation lines that appear to be uniform is not difficult.

Scout, selected in the F_3 generation, and Centurk, an F_4 selection, are examples of varieties that look uniform but undoubtedly possess substantial heterogeneity. Five varieties were selected from Scout and two from Centurk (Table 3). All the selections from Scout are similar in appearance to Scout but each has been demonstrated to be measurably different from Scout in at least on attribute. Scoutland and Eagle are distinctly superior to Scout in processing quality; Scout 66, Baca, and Eagle are somewhat more productive than Scout in their particular areas of primary adaptation; and Rall exhibits greater tolerance to the wheat streak mosaic virus than Scout. None of the selections, however, is known to be as widely adapted as Scout and none has achieved a production acreage as large as that achieved by Scout. To date, two varieties have been selected from Centurk. As with Scout, each is similar in appearance to Centurk but differs in at least one trait from Centurk. Neither is expected to achieve the widespread popularity of Centurk among producers.

Table 3. **Commercially-grown hard winter wheat varieties selected from 'Scout' and 'Centurk'.**

Name	Year of Release	Selected By
	From Scout	
Scout 66	1967	Nebraska
Scoutland	1970	Nebraska
Eagle	1970	Kansas
Baca	1972	Colorado
Rall	1977	Oklahoma
	From Centurk	
Centurk 78	1978	Nebraska
Rocky	1978	No. Am. Plant Breeders

Why the wide acceptance and popularity of the Nebraska varieties among wheat producers in the hard winter wheat region of the U.S.? To what extent are early selection and associated heterogeneity involved? We believe that they are. As a group, the varieties from Nebraska selected in the F_2, F_3, or F_4 generations exhibit broad adaptation to the environments of the hard winter wheat region and can be characterized as having outstanding stability of performance. We postulate that their heterogeneity does buffer them against changes in production environment as suggested by Allard and Bradshaw (1).

The productiveness and performance stability of Scout and Gage are shown graphically in Figure 3. Their performance in relation to that of Triumph and Kharkof, which is predicted from the linear regressions of individual variety yields on nursery mean yields in the SRPN, is plotted (3). Based on 3 yr of data, the predicted performance of Scout is much superior to that of Kharkof and Triumph in low-yielding to very high-yielding environments. Gage, on the other hand, while not so productive as Scout in any environment, is clearly more productive than Kharkof in all environments. Triumph, an old early-maturing variety particularly adapted to southern Kanasas, Oklahoma, and north-central Texas, is much superior in predicted performance to Kharkof in the low-yielding environments but rapidly loses its yield advantage over Kharkof in the better production environments. The long-time past popularity of Triumph among southern plains producers may be attributable to its good relative performance during a period in which yields seldom exceeded 30 bu/acre. The rapid commercial acceptance of Scout suggests that its performance in the hands of growers was similar to that in Nebraska and regional trials.

Shebeski and Evans postulate many yield genes in wheat—perhaps hundreds (4). Such a large number of genes imposes severe limitations on selection for yield, particularly if selection is delayed until an advanced generation. According to Shebeski and Evans, if parents differ by 25 genes for yield, only one plant in 1330 F_2 plants will carry all the favorable genes in the homozygous or heterozygous condition. If selection is delayed until the F_4 generation, only one plant in 1.8 million can be expected to carry all the favorable alleles. Thus the frequency of superior genotypes in a population decreases rapidly and predictably with the advance of generations. Likewise, the odds for recovering superior genotypes decrease rapidly in a pedigree system in which a relatively small number of reselections is made within lines in each succeeding generation.

The Nebraska-USDA procedure of selecting in the F_2, F_3, and F_4 generations with no reselection may provide the best odds for maximizing and maintaining the frequency of yield genes in each population (F_2-, F_3-, or F_4-derived line). This would not be so on an individual plant basis but only in

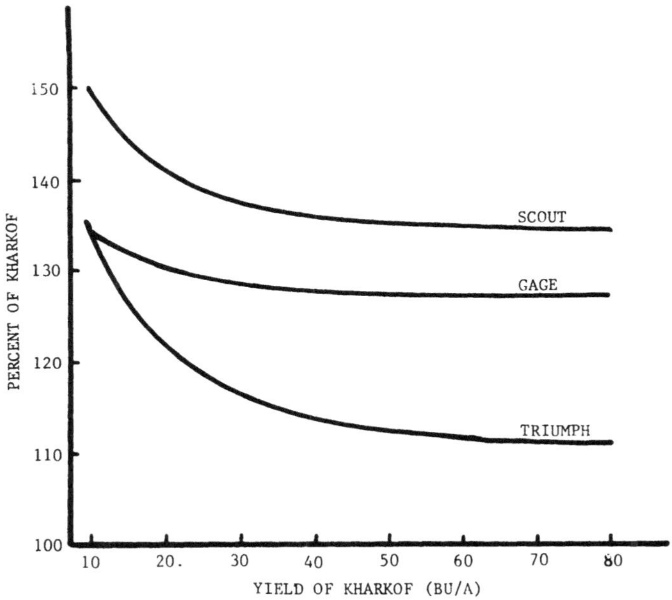

Figure 3. **Yield relationships of three winter wheat varieties to 'Kharkof' predicted from the linear regressions of the variety yields on nursery mean yields in the Southern Regional Performance Nursery, 1961 to 1963.**

the population. If heterogeneity buffers against environmental change and contributes to performance as well as to stability of performance, it may then provide a key to the wide acceptance and popularity of Nebraska-developed varieties like Scout and Centurk. Further, if our assumption that new winter wheat varieties should be perceived as transitory is a valid one, then the segregation and predictable emergence of non-productive individual plants within the varietal population is less significant than would be the case if the varieties were perceived as permanent.

Early generation selection, as practiced in the Nebraska system, has been most successful in crosses of productive adapted x adapted parents. In such hybrid combinations the number of yield genes by which the parents differ could be expected to be substantially smaller on the average than in crosses of widely different adapted x exotic parents. If so, the frequency of plants carrying all of the favorable alleles would be higher in each generation.

The Nebraska selection procedure has been challenged by some breeders and seed certification officials on the basis that varietal identity cannot be maintained. They suggest that natural selection is likely to occur each year to shift the population significantly from what it was at the time of release of the variety. We have not disclaimed this possibility, but we agree with Allard that, should such a shift occur, it would likely improve the population rather than make it poorer. If natural selection pressure is different in different production areas, it could be detrimental to move seed wheat long distances. However, most seed wheat remains in the locality in which it is produced. We have no evidence that Scout has changed measurably during the 15 yr since its release.

Consistent with the belief that genetic vulnerability in hard winter wheat can be reduced by grower use of many varieties and relatively rapid turnover of varieties, the Nebraska Agricultural Experiment Station follows a liberal wheat release policy. Since 1963, an average of one new variety per year from our program has been made available to Nebraska certified producers. The new varieties are released without strong Station recommendations for their use over already available varieties in Nebraska. They are described thoroughly to permit growers to make sound judgements regarding their use. The State Experiment Station publishes annually a listing of available certified varieties as well as the results of statewide on-farm performance trails to guide growers in making wise choices. Fifteen winter wheat varieties appear on the list for 1979. As many as 10 varieties are listed for individual cropping districts.

NOTES

V. A. Johnson, Wheat Research, U.S. Department of Agriculture, SEA/AR, and J. W. Schmidt, Department of Agronomy, University of Nebraska, Lincoln, Nebraska 68583.

REFERENCES

1. Allard. R. W. and A. D. Bradshaw. 1964. Implications of genotype-environmental interactions in applied plant breeding. Crop Sci. 4:503-508.
2. Johnson, V. A. 1978. Breeding for yield and protein content in hard winter wheat. Cereal Foods World 23(2):84-86.
3. Johnson, V. A., S. L. Shafer, and J. W. Schmidt. 1968. Regression analysis of general adaptation in hard red winter wheat *(Triticum aestivum* L.). Crop Sci. 8:187-191.
4. Shebeski, L. H. and L. E. Evans. 1973. Early-generation selection for wide-range adaptability in the breeding program, pp. 587-593. In Proc. 4th Int. Wheat Genetics Symposium, Columbia, MO.

POPULATION IMPROVEMENT IN SELF-POLLINATED CROPS

D. F. Matzinger and E. A. Wernsman

Traditional plant breeding programs in the naturally self-fertilizing species differ considerably from those utilized with the naturally cross-fertilizing species. In both cases, necessary crosses are made by bringing together parents which diverge in gene content for desirable traits. A common procedure with the cross-pollinated species has been to initiate recurrent selection to develop improved populations. Usually one or more generations of random mating of selected genotypes is practiced between cycles. In contrast, with the self-fertilizing species, breeding methods are usually directed toward selfing or backcrossing a series of lines to homozygosity with little, if any, intercrossing beyond the initial source hybrid or population.

Recurrent selection programs are designed to improve the average performance of the population for selected traits by increasing the gene frequency of the desired alleles in the population. Most recurrent selection procedures utilize the additive genetic variance in the source population. It is not the purpose of this chapter to review estimates of genotypic variances in all species. However, in studies where genotypic variances have been estimated in naturally self-fertilizing species, a major proportion of the variance has been additive genetic variance.

The purpose of this presentation is to review some recurrent selection studies that we have conducted in the naturally self-fertilizing species, *Nicotiana tabacum* L. Because of its flower structure and reproductive capacity, this species allows flexibility in mating design and population development not possible with many autogamous species. The results should have application to other naturally self-fertilizing species.

Population improvement studies usually are designed to combine the estimation of genotypic and environmental variances and covariances with selection applied in successive cycles. Estimates of variances and covariances in the base population allow one to predict expected responses from selection procedures. Long-term selection studies allow tests of the adequacy of the genetic model as well as provide improved genetic material for initiating varietal development.

When planning selection studies one must choose a procedure which is compatible with mating limitations of the organism and one which allows comparisons of selection procedures of interest. Once the scheme has been chosen, the procedures are: (a) form desired families necessary in mating design chosen; (b) evaluate progeny in yield test; (c) estimate genetic and environmental variances; (d) predict gain for alternative selection procedures; (e) choose selection procedures to be verified experimentally; (f) identify superior families to be selected; (g) regrow these families from remnant seed; (h) intercross selected families at random; (i) form mating design in this selected population ; (j) evaluate progeny in yield test; and (k) compare observed with predicted gain. This completes one cycle of selection. The procedure may be repeated through additional cycles to determine eventual selection limits or until the population has achieved the performance desired.

RECURRENT FAMILY SELECTION

Experimental studies were established in the initial phases of our tobacco program to allow jointly the estimation of genotypic and environmental variances and the conduct of recurrent selection. Since tobacco is a naturally self-fertilizing species, initial interest was devoted to evaluating the magnitude of additive x additive epistatic variance (σ^2_{AA}) as well as additive genetic (σ^2_A) and dominance (σ^2_D) variances.

The mating design incorporated simultaneous selfing and partial diallel test crossing (6). Beginning with the F_2 generation of a cross between two pure line cultivars, eight F_2 plants were assigned at random to a mating set; four were designated as male parents and four as female parents. The 16 possible crosses were produced to form 16 full-sib families and the eight parent plants were also self-pollinated to produce eight self families. Additional samples of eight F_2 plants were mated by this design to provide 13 sets for field evaluation. Each set was maintained separately in the field and evaluated in two replications at a single location. Also included in each set were check plots of the two initial cultivars and the F_1 hybrid. Joint analyses of self families, full-sib families, and the covariance between the self progeny and the half-sib progeny of each parent provided estimates of σ^2_A, σ^2_D, and σ^2_{AA} (6).

Superior parent plants were identified from joint information on their self and half-sib progenies by utilizing an index procedure (2). Family and check plot data were incorporated in the index to allow identification of superior parents across the entire experiment.

From progeny data of the 104 F_2 plants, 15 superior parents were identified for producing the next cycle. The self families of these selected parents were regrown from remnant seeds and intercrossed. A partial diallel design was used in intercrossing with 14 plants of each family used in pollinations, seven as male parents and seven as female parents. This insured equal representation of each family in crosses and balanced the crosses for maternal contribution. A single plant of each of these crosses served as a parent in constructing the simultaneous selfing and partial diallel design for the next cycle of selection.

This procedure allows for continued cycles of recurrent selection. In addition, from each cycle of selection one may initiate homozygous line isolation from the superior parents which have already been self-pollinated once, and for which progeny evaluation has been obtained.

An example of two cycles of selection with this design was reported by Matzinger et al. (9). The source population was the F_2 generation of the cross between two pure line cultivars, Dixie Bright 244 and Coker 139. The genetic correlation between percent total alkaloids in the leaf and leaf yield in the base population was −0.33. Tobacco alkaloids are nitrogenous constituents and their behavior during selection for increased leaf yield in tobacco has similarities to protein constituents in grain crops. Selection was practiced for increased percent total alkaloids. Predicted and observed responses for alkaloids and yield are presented in Table 1. The cumulative predicted increase in percent total alkaloids for two cycles of selection was 0.466 and the observed response was 0.445. In the base population, the weighted self and half-sib families were distributed between the two parental cultivars for total alkaloids. Following two cycles of selection, C_2, most families were equal to or higher in alkaloids than the higher parental cultivar, Dixie Bright 244. The observed correlated decrease in yield of 10.93 g/plant was only slightly in excess of the predicted decrease of 8.57 g/plant.

Estimation of σ^2_A, σ^2_D, and σ^2_{AA} for three cycles in each of three populations with this design indicated that σ^2_A was significant for most

Table 1. Two cycles of recurrent family selection for increased total alkaloids in Dixie Bright 244 x Coker 139.

	$C_2 - C_0$	
Trait	**Predicted**	**Observed**
Total alkaloids, %	0.466	0.445**
Yield, g/plant	- 8.57	10.93**

**Observed change from C_0 to C_2 significant at 0.01 level.

characters evaluated, whereas only an occasional estimate of σ^2_D or σ^2_{AA} was significant and they were negligible compared to σ^2_A. Since σ^2_A predominated in all populations, the selection method was changed to full-sib family selection. This placed less emphasis on genotypic variance estimation but reduced the time to complete a selection cycle from three to two generations. Full-sib families were evaluated, selected families grown from remnant seeds and intermated in a partial diallel to evaluate in the next cycle. Following four additional cycles of selection in the Dixie Bright 244 x Coker 139 population, the selected population, C_6, was compared with the two original cultivars of the population (Table 2). The alkaloid level in C_6 was 41% above the high parent. The expected correlated decrease in yield reduced C_6 to slightly above the lower yielding parental cultivar.

RECURRENT MASS SELECTION

The predominance of σ^2_A, low genotypic x environmental interaction, and high heritabilities prompted interest in recurrent mass selection. To initiate the studies, eight cultivars were chosen from different breeding programs and crossed into an 8-line hybrid. This hybrid was random mated and designated Black Shank Synthetic. Legg et al. (5) estimated variances in Black Shank Synthetic and compared expected selection response for recurrent mass selection with a number of recurrent family selection procedures. Mass selection gave the least expected progress per cycle; however, the family selection methods each required two or three generations to complete a cycle. If plants to be selected for crossing in mass selection could be identified prior to pollination, a cycle could be completed each generation. An additional benefit could be obtained from intermating only selected plants since the expected progress would be

$$G = \frac{k\sigma^2_A}{\sigma_p}$$

in which k is the selection differential in standard units, σ^2_A is the additive genetic variance in the base population, and σ_p is the phenotypic standard

Table 2. Six cycles of recurrent family selection for increased total alkaloids in Dixie Bright 244 x Coker 139.

Entry	Alkaloids	Yield
	– % –	*– g/plant –*
Dixie Bright 244	2.65	165.6
Coker 139	2.31	212.1
C_6	3.75	174.6
LSD .01	0.56	4.7

deviation of individual plants. This expected progress is twice that obtained when selection is only on female parents in mass selection, a common procedure in corn studies (3).

Legg et al. (5) predicted a gain in yield of 3.75% from one cycle of recurrent mass selection of both male and female parents assuming a 10% selection intensity. Selection for increased yield would be expected to decrease the total alkaloids in the population, reflecting the estimated genetic correlation between yield and total alkaloids of −0.50 in Black Shank Synthetic.

Results from four cycles of recurrent mass selection for increased yield in Black Shank Synthetic were presented by Matzinger and Wernsman (8). Individual fresh wts of leaves were obtained from 1000 plants in each cycle. The three highest yielding plants in each row of 28 plants were selected for the next cycle, providing a theoretical selection differential (k) of 1.64. Random crosses were made among pairs of selected plants on flowers which developed late in the season on leaf axil suckers. Following four cycles of selection, each population was regrown from remnant seeds and intermated to prepare equal age seeds for all cycles in evaluating selection responses. The material was evaluated at three locations in experiments consisting of 10 replications.

The average response for yield and total alkaloids is presented in Table 3. There was a linear increase in yield of 4.29%/cycle, slightly in excess of the predicted value. The observed decrease in total alkaloids of 2.74% was in close agreement with the prediction. Since the selection response was linear, there was no evidence that there had been a significant reduction in genetic variance following selection.

Additional cycles of selection have been conducted, but all cycles have not been evaluated together in a common experiment since the fourth cycle was completed. A single comparison of the initial population and C_{10} is shown in Table 4. A continued increase in yield has been obtained through 10 cycles of selection. The average response per cycle of 4.23% is very similar to the average of 4.29% after four cycles. This continued yield increase suggests that after a selection response of 42% through 10 cycles of selection, there is still no evidence of the exhaustion of genetic variance. It appears that

Table 3. Predicted and observed responses per cycle following 4 cycles of recurrent mass selection for increased yield in Black Shank Synthetic.

Trait	% Response	
	Predicted	Observed
Yield	3.52	4.29**
Total alkaloids	- 2.76	- 2.74*

*,** Linear regression significant at 0.05 and 0.01 levels, respectively.

Table 4. Response to 10 cycles of recurrent mass selection for increased yield in Black Shank Synthetic.

Trait	C_0	C_{10}	%/Cycle
Yield, g/plant	159.4	226.9**	4.23
Total alkaloids, %	3.80	1.79**	5.29

**Significant change from C_0 at 0.01 level.

the correlated decrease in total alkaloids has accelerated between cycles 4 and 10.

RECURRENT INDEX SELECTION

The ability to achieve a selection response for both primary and correlated characters in reasonably good agreement with predictions in our tobacco program suggests that procedures for estimating population parameters are sufficient to enable evaluation of more complex selection procedures. In all of our tobacco populations, significant estimates of genetic correlations exist among a number of characters. Such correlations are particularly troublesome in population improvement if the sign of the correlation is opposite to the selection goals. We have extended the recurrent selection studies to selection indexes where an attempt is made to break antagonistic genetic correlations.

A number of the genetic correlations range from 0.5 to 0.7. Primary interest is in evaluating methods of selection in opposition to correlations of this magnitude. In the Black Shank Synthetic population the genetic correlation between plant ht and number of leaves was 0.67. Three separate recurrent mass selection programs were initiated for (a) shorter plants, (b) increased number of leaves, and (c) an index to combine more leaves on shorter plants (7). Assignment of economic wts was accomplished by equating a 1% of the mean decrease in plant ht to a 1% of the mean increase in number of leaves. The population size of each experiment was 400 plants, arranged in 20 rows of 20 plants each. Three plants were selected in each row. Paired crosses were made among selected plants to initiate the next cycle. Following five cycles of selection, fresh seeds were obtained from each cycle and they were evaluated in four replications at two locations in two yr.

Selection response was linear over the five cycles in all three selection studies for both the selected and the correlated characters. Predicted and observed responses are summarized in Table 5. Selection for shorter plants decreased plant ht 6.52 cm/cycle for a 4.9% reduction of the mean per cycle. The correlated decrease in leaf number was 2.5%/cycle. Selection for increased leaf number increased the number of leaves 7.0%/cycle with a

Table 5. Five cycles of selection for decreased plant ht, increased number of leaves, and a selection index.

	Ht Selection		Leaf Selection		Index Selection	
	Ht	Leaves	Ht	Leaves	Ht	Leaves
	– cm –	– no. –	– cm –	– no. –	– cm –	– no. –
Predicted	- 10.19	- 0.95	6.63	1.35	- 4.26	0.21
Observed	- 6.52**	- 0.46*	4.74**	1.28**	- 2.00*	0.54*
% Gain/Cycle	- 4.9	- 2.5	3.6	7.0	- 1.5	2.9

*,** Response significantly different from zero at 0.05 and 0.01 levels respectively.

correlated increase in plant ht of 3.6%. In the index population there was an increase of 0.54 (2.9%) leaves per plant and a decrease in plant ht of 2.0 cm (1.5%) per cycle. The index was effective in selecting in opposition to the genetic correlation; however, response for number of leaves was greater than predicted and the decrease in plant ht was less than predicted.

RECURRENT RESTRICTED INDEX SELECTION

Restricted selection indexes were proposed by Kempthorne and Nordskog (4) for the situation in which one wishes to change one character and hold one or more correlated characters at the population mean. This is a common breeding objective but very little experimental data have been accumulated in plant populations to verify the theory. We have initiated studies to evaluate the efficacy of this procedure for two negatively correlated traits, yield and percent total alkaloids. The objective of the study was to increase yield while maintaining total alkaloids at the population mean.

The source population was Black Shank Synthetic. Estimates of genetic variance were obtained from replicated full-sib families, assuming all genetic variance was additive. The genetic correlation between yield and total alkaloids in the base population was -0.50. Construction of a restricted index is simple, since no economic wts are required. The index is of the form

$$I = Y + \frac{\sigma A_{yt}}{\sigma^2_{A_t}} T$$

in which Y = mean yield of a full-sib family; T = % total alkaloids of the full-sib family; σA_{yt} = additive genetic covariance between the characters, and $\sigma^2_{A_t}$ = additive genetic variance for total alkaloids.

Following five cycles of selection, each population was regrown from remnant seeds and sib-mated at random. Evaluation of all cycles was obtained from 10 replications at two locations in 2 yr. There was a linear increase in yield from 186.1 to 212.6 g/plant through 5 cycles of selection (Table 6). No significant differences existed for percent total alkaloids among

Table 6. **Restricted selection index for maximizing yield and holding total alkaloids at the original population mean.**

Trait	C_0	C_5
Yield, g/plant	186.1	212.6**
Total alkaloids, %	3.22	3.23

**Significant change from C_0 at 0.01 level.

cycles. The C_5 population contained 3.23% total alkaloids compared with 3.22% in the base population. Thus, the restricted selection index was successful in achieving approximately 3%/cycle gain in yield while restricting any change in total alkaloids.

DISCUSSION

These studies represent examples of the use of several different types of recurrent selection in tobacco. Selection responses have been essentially linear through the cycles completed. In all studies, including the one study which had progressed through 10 cycles of recurrent mass selection, there was no evidence that selection response was becoming limited because of the depletion of genetic variability. It would appear that the forced random mating of selected parent plants or families is releasing much usable genetic variation for improvement of the populations. A serious problem may arise in advanced cycles of selection if other traits are correlated in an antagonistic manner since the population may lack usability for those characters. The use of selection indexes in recurrent selection appears to be effective in combining desirable traits in the improved population. These populations then offer an abundant source of genetic material to initiate pure line breeding programs to isolate superior homozygous lines.

The use of recurrent selection programs in most naturally self-fertilizing species, including soybeans is more difficult than in tobacco because of limitations in obtaining sufficient seeds from hybrid matings. However, if the variability released by these programs and the rapid response to selection obtained in tobacco is applicable to other self-fertilizing species, additional efforts directed toward increased crossing may be justified economically. A promising approach in soybeans was presented by Brim and Stuber (1) which utilizes genetic male-sterility as an aid in intermating populations. Even with this assistance in crossing, there may not be sufficient hybrid seeds for yield tests and an intervening generation of selfing may be necessary.

NOTES

D. F. Matzinger, Department of Genetics, and E. A. Wernsman, Department of Crop Science, N. C. State University, Raleigh, NC 27650.

Paper Number 5934 of the Journal Series of the North Carolina Agricultural Research Service. This research was supported in part by a grant from the North Carolina Tobacco Foundation, Inc.

REFERENCES

1. Brim, C. A. and C. W. Stuber. 1973. Application of genetic male sterility to recurrent selection schemes in soybeans. Crop Sci. 13:528-530.
2. Cockerham, C. C. and D. F. Matzinger. 1966. Simultaneous selfing and partial diallel test crossing. III. Optimum selection procedures. Australian J. Biol. Sci. 19:797-805.
3. Gardner, C. O. 1961. An evaluation of effects of mass selection and seed irradiation with thermal neutrons on yield of corn. Crop Sci. 1:241-245.
4. Kempthorne, O. and A. W. Nordskog. 1959. Restricted selection indices. Biometrics 15:10-19.
5. Legg, P. D., D. F. Matzinger and T. J. Mann. 1965. Genetic variation and covariation in a *Nicotiana tabacum* L. synthetic two generations after synthesis. Crop Sci. 5:30-33.
6. Matzinger, D. F. and C. C. Cockerham. 1963. Simultaneous selfing and partial diallel test crossing. I. Estimation of genetic and environmental parameters. Crop Sci. 3:309-314.
7. Matzinger, D. F., C. C. Cockerham and E. A. Wernsman. 1977. Single character and index mass selection with random mating in a naturally self-fertilizing species, pp. 503-518. In E. Pollak, O. Kempthorne and T. B. Bailey, Jr. (Eds.) Proc. Int. Conf. Quant. Genet., Iowa State Univ. Press, Ames, Iowa.
8. Matzinger, D. F. and E. A. Wernsman. 1968. Four cycles of mass selection in a synthetic variety of an autogamous species *Nicotiana tabacum* L. Crop Sci. 8:239-243.
9. Matzinger, D. F., E. A. Wernsman and C. C. Cockerham. 1972. Recurrent family selection and correlated response in *Nicotiana tabacum* L. I. 'Dixie Bright 244' x Coker 139'. Crop Sci. 12:40-43.

LONG- AND SHORT-TERM RECURRENT SELECTION IN FINITE POPULATIONS—CHOICE OF POPULATION SIZE

J. O. Rawlings

One of the primary concerns in designing a recurrent selection program is the choice of effective population size. The problem is to strike some balance between keeping the effective population size large enough to avoid undue loss of genetic variance by random drift and maximizing the short-term genetic progress within the constraints of resources available.

When one speaks of long-term selection response, the paper most often cited is that of Robertson (18) dealing with the limits to selection in which he gives the frequently cited expressions for expected total advance and half life of the recurrent selection process. The fact that both are directly proportional to the effective population size has led to a general impression that in recurrent selection programs the effective population size must be large. On the other hand, maximizing the expected genetic gain in the short-term requires high intensities of selection which because of limited resources suggests that only a few individuals be selected as parents of the next generation.

This conflict between the short-term and long-term objectives is more apparent than real insofar as most breeding programs are concerned. The purpose of this chapter is to demonstrate that very reasonably sized recurrent selection programs very nearly satisfy the objectives of maximizing both short-term and long-term gains from selection.

The approach used in this chapter is first, to present a discussion on the effective population size needed for retaining in the population the major part of the genetic potential. This discussion utilizes for the most part Kimura's (8) formulation for the probability of fixation of an allele in a population with an additive model. In order to relate Kimura's formulation to truncation selection, the selective advantage of an allele, s, is related to the parameters of truncation selection and some discussion is devoted to what might be assumed to be reasonable values for these parameters. Then it is shown that very reasonably sized selection programs with reasonably high intensities of selection retain most of the long-term potential of the population. All of this discussion assumes the simplest of genetic models. The final comments will relate to some of the findings on the impact of linkage on these effective population sizes.

EFFECT OF FINITENESS ON SHORT-TERM SELECTION

The usual predictor for the change in the mean of a population from one cycle of truncation selection is the product of the selection differential and the linear regression coefficient for the regression of the genetic values of the offspring on the phenotypic values of the parental selection units. The regression coefficient can be altered within limits by the choice of selection units but I do not propose to dwell on this aspect.

The selection differential is often expressed as the product of the standardized selection differential, $\bar{i}$, and the phenotypic standard deviation among the selection units (9). For the case being discussed, the completely additive model, finiteness has no effect on the *mean* progress other than that reflected in the adjustment of $\bar{i}$ as the mean of the appropriate order statistics in a finite sample.

The standardized selection differential, $\bar{i}$, increases as the intensity of selection increases; that is, as the proportion saved becomes smaller. Thus, the formula suggests that the optimum strategy in short-term selection is to use as high an intensity of selection as possible. If the selection program is limited in resources, increasing the intensity of selection would entail a decrease in the number of individuals retained since the total number that can be tested has some practical limit.

The major effect of finiteness on short-term recurrent selection is on the variance of the genetic change, which is inversely related to effective population size. Consequently, the prediction of genetic change is less reliable in small populations; that is, the variation between replicate selection programs is larger.

EFFECT OF FINITENESS ON LONG-TERM SELECTION

The individuals selected in one stage of the program constitute the genetic base for the next cycle and it is well known that an undue restriction in the genetic base is undesirable. The net effect is that many loci may become fixed for the unfavorable alleles simply due to the vagaries of genetic sample, random drift.

One of the effects of drift is to *reduce* the genetic variance within replicates. For gene neutral to the selective forces, the rate of loss of variance due to drift is 1/(2N), where N designates throughout this chapter the effective population size. Since the progress expected from selection is proportional to the genetic variance in the population at the time, it is of interest in cyclic selection to conserve the within replicate genetic variance, at least insofar as that part of the variance being lost by drift is concerned.

The classical reference relating to long-term selection is Robertson (18) in which it is stated that, under certain conditions, the total selection advance in a particular population is 2N times the advance in the first cycle and the half life of the selection process, the time required for the population mean to be moved half way to the limit, is 1.4N. The derivation of this is straight forward. If s is the selective advantage of a particular allele assuming an additive model then the change in gene frequency in one generation of selection is given by

$$\Delta q_1 = ½\, sq(1-q). \qquad \text{(I)}$$

Drift theory says that the product q(1-q) will decline by the factor of 1/2N each generation and, if Δq is small enough that the product q(1-q) is essentially unchanged, the change in the second cycle of selection is given by

$$\Delta q_2 = ½\, sq(1-q)(1-\frac{1}{2N}). \qquad \text{(II)}$$

Continuing this argument Robertson obtains his result:

$$\text{Total change} = \sum_{t=0}^{\infty} \frac{s}{2} q(1-q)\,[1-\frac{1}{2N}]^{t} = Nsq(1-q) = (2N)(\Delta q_1). \qquad \text{(III)}$$

This is 2N times the change in the first generation. The half life of the selection process is determined from the same formulation except summation is only to time t' and then the quantity is solved for the t' which gives half the total change. This turns out to be 1.4N.

The same results can be obtained using the probability of fixation of an allele with initial frequency q which for the additive model takes the following form (8):

$$\mu(q) = \frac{1-e^{-2Nsq}}{1-e^{-2Ns}} \qquad \text{(IV)}$$

Again, s is the selective advantage of the allele and N is the effective population size. A Taylor's series expansion of equation IV with the assumption that the product Ns is small gives the same limit to selection and half life. The relevant point is that the limit to selection is proportional to effective population size if Ns is small. The fact that Robertson's result holds only if Ns is small is widely recognized and is of no particular concern here.

Kimura's formulation for the probability of fixation is appropriate if Ns is finite, not necessarily small, and can be studied directly to determine a reasonable effective population size. Several studies have investigated the usefulness of Kimura's result. Latter (11) reports results using Ns = 4.4, with N as small as 5 in one case, which gave probabilities of fixation varying from 0 to .03 units below those predicted by $\mu(q)$. Ewens (2) also found excellent agreement, maximum discrepancies of .0112, using N = 12 and values of s between 0.0 and 0.1; generally much smaller than the values used by Latter. Hill (5) also gives some comparisons with the diffusion equation over a wide range of conditions for N = 2, 4, and 10. The discrepancy is sizable, say .05, for situations with N = 2 and 4 but agreement is generally good, discrepance less than .02, for N = 10. Since N will be much greater than 10 in most breeding programs, we can accept equation VII as providing sufficiently reliable, although slightly biased, estimates of probabilities of fixation.

Using Kimura's formulation, Robertson (18) states that at least 93% of the total possible gain will be realized if the product Nsq>2. This formulation would seem to be subject to misinterpretation since the lower limit, 93% in this case, is reached as initial gene frequencies approach 1.0, not zero as one might expect on first encounter. For high initial gene frequencies, it is relatively easy to keep Nsq>2. It is at the other extreme where effective population size becomes more critical. For this reason it would seem more informative to use expected gene frequency at fixation directly, Kimura's $\mu(q)$, rather than $\mu(q)$ expressed as a ratio to the total possible advance.

Treating Nsq as a fixed quantity, the probability of fixation has a lower limit of $1-e^{-2(Nsq)}$ as q approaches zero. Of course, $\mu(q)$ approaches one as q approaches 1.0. Figure 1 gives the probability of fixation curves for three values of Nsq. The lower limit for Nsq = 2.0 is .9817, i.e., if Nsq = 2.0, the expected gene frequency at fixation is greater than .98 regardless of the starting gene frequency (assuming of course that $q_0>0$). Similarly, $\mu(q)$ is greater than or equal to .9503 if Nsq = 1.5, and greater than or equal to .8647 if Nsq = 1.0. Values as high as the first two constitute a rather high degree of attainment of the population potential.

The curves in Figure 1 are virtually flat for gene frequencies below ½ so that little error is introduced by taking the minimum point on the curve as the point of reference. Doing this, the minimum value of the product Nsq for attaining a given probability of fixation is given by

$$Nsq > -\tfrac{1}{2} \ln [1-\mu(q)] \qquad \text{(V)}$$

Recall that s is the selective advantage of a particular allele. Falconer (3) has shown that in truncation selections is approximately rated to $\bar{i}$ as follows:

$$s \doteq \bar{i}\, a^* \qquad \text{(VI)}$$

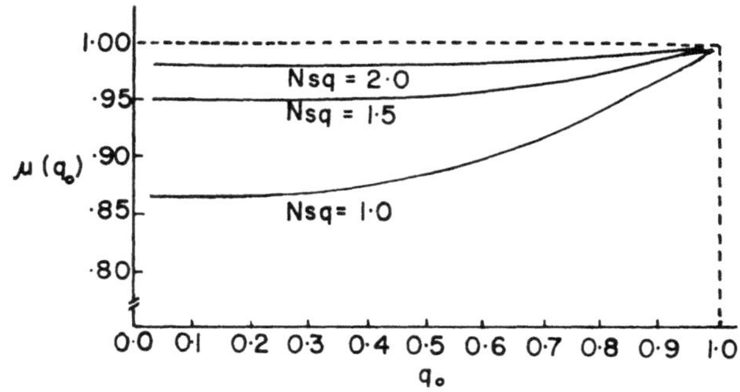

Figure 1. Probabilities of fixation for fixed values of Nsq.

where a^* is the proportional effect of a locus or it is the difference between the genotypic values of the homozygous genotypes for the locus divided by the phenotypic standard deviation. Latter (10) has given some adjustments to this formulation which for our purposes we will not use. They become important as a^* gets large. If one assumes a symmetric genetic model, that is, a model in which all loci have equal effect and gene frequency, a^* can be related to heritability, number of loci, m, and gene frequency as follows:

$$a^* \doteq \frac{2(h^2/m)}{q(1-q)}^{1/2}, \qquad \text{(VII)}$$

It is convenient to think of the ratio (h^2/m) as the per locus contribution to heritability.

Substituting the parameters for truncation selection into equation V gives

$$N\bar{i} > -\frac{1}{2a^*q} \ln [1 - \mu(q)]. \qquad \text{(VIII)}$$

N and $\bar{i}$ are the parameters under the breeders control. If we choose $\mu(q)$ to be 0.95,

$$N\bar{i} > 1.5/a^*q = 1.5 \left(\frac{(1-q)/q}{2(h^2/m)} \right)^{1/2}. \qquad \text{(IX)}$$

Making practical use of equation IX requires some judgment by the individual of the magnitude of s and q or, alternatively, (h^2/m) and q in truncation selection. To aid in developing this judgment, Table 1 gives some indication of the magnitude of a^* for various combinations of gene frequency and (h^2/m). The values chosen for (h^2/m) would seem to cover most genetic models. For example, 1/40 could mean a heritability of a half with 20 loci or a heritability of a tenth with 4 loci. At the other extreme, 1/10,000 could mean, for example, a heritability of a tenth with 1000 loci. The middle

Table 1. Proportionate effect of a locus, a*, expressed in terms of h^2, m, and q. (Assuming m loci with equal gene frequency and effects.)

q	(h^2/m) = Contribution Per Locus to Heritability				
	1/40	1/200	1/1000	1/2000	1/10,000
.5	.448	.200	.090	.064	.029
.25, .75	.514	.232	.104	.072	.032
.10, .90	.746	.336	.150	.108	.048

column, 1/1000 which could be taken to be a heritability of one tenth with 100 loci or a heritability of one half with 500 loci, would seem to be a reasonably conservative model in terms of the per locus contribution to heritability. Thus, a* of the order of 1/10 would seem to be a reasonable lower value to use in developing some idea of minimum population sizes for selection systems.

Table 2 gives the minimum values of $N\bar{i}$ necessary to give $\mu(q)$ greater than or equal to .95 for various genetic situations. There are two points to notice: (a) The values are rather small, say 50 or less, over a wide range of genetic situations; and (b) the very rapid increase in $N\bar{i}$ occurs for very low gene frequencies. To give a feeling for the effect of choosing $\mu(q) = .95$, increasing $\mu(q)$ to $\mu(q) = .98$ increases all of these values by 1/3 and decreasing $\mu(q)$ to $\mu(q) = .865$ decreases all values by 1/3.

It must be emphasized that $\mu(q)$ is an expected value over many replications for a particular locus or, alternatively, it is the average gene frequency within a replication for many loci of similar nature. The variance of the ultimate limit is inversely related to effective population size so that with small populations any one particular replicate may deviate considerably from the specified $\mu(q)$.

Again, using 1/1000 as a reasonable per locus contribution to heritability the third column in Table 2 gives the size $N\bar{i}$ must be to attain the objective. It remains now to remind you of the usual values that $\bar{i}$ takes in truncation selection systems. Table 3 shows for various intensities of selection the

Table 2. Minimum values of $N\bar{i}$ to give $\mu(q) \geq 0.95$.

q_0	(h^2/m)				
	1/40	1/200	1/1000	1/2000	1/10,000
3/4	4	9	20	28	63
1/2	7	15	33	42	104
1/4	12	26	58	84	188
1/10	21	45	100	139	313

Table 3. **Minimum effective population sizes for $h^2/m = 1/1000$.**

Proportion Selected	$\bar{i}$	Initial Frequency of Favorable Allele (q_0) .75	.50	.25	.10
.5	0.80	25	42	73	125
.25	1.27	16	26	46	79
.10	1.76	12	19	33	57
.01	2.66	8	13	22	37
.001	3.50	6	10	17	29

minimum effective population sizes necessary to attain $\mu(q) \geq .95$ assuming $(h^2/m) = 1/1000$. The second column gives the standardized selection differentials corresponding to the stated intensities of selection. First, it will be noted that, as expected, the required effective population size decreases as intensity of selection increases. It is clear from Table 3 that, for reasonably intense selection, maintaining a sufficiently large population size even when the initial gene frequency is as low as .25, or even .10, should be possible in most plant selection programs where hundreds of plants can be observed. For example, an effective population size of only 22 where the intensity of selection is 1/100 is sufficient to give a probability of at least .95 of eventually fixing the favorable allele even if the initial gene frequency is as low as .25. For gene frequencies of .10, the required effective population size increases to 37.

A second point evident in Table 3 is the high price paid by the plant breeder for increasing the intensity of selection, and consequently the rate of progress, above an intensity of say 1/10. To illustrate this point note that two cycles of selecting approximately 19 out of 190 gives as much progress as one cycle of selecting approximately 10 out of 10,000 if q_0 is one half. Or, two cycles of selecting 33 out of 330 gives as much progress as one cycle of 17 out of 17,000 if q_0 is 1/4. In both cases this is more than a 50-fold increase in the total population size in order to double the rate of progress and at the same time hold effective population sizes at a level so as to retain most of the long-term potential.

As plant breeders, the primary objective must be one of obtaining an appreciable advance in a reasonable period of time. The idea of conserving the genetic potential in the population for the long-term response is a secondary objective and one which will be considered seriously only if it does not jeopardize the primary one. If the parameters we have assumed for this illustration are reasonable, it appears that it would be possible, in fact, usual, to attain both objectives. It would appear that an effective population size of the order of 30 would be reasonably adequate for many genetic systems.

There are several aspects of this discussion that must be emphasized. First, the decision as to what magnitudes of a^* or (h^2/m) are reasonable is purely subjective. There is virtually no information providing guidelines as to the number of loci involved or the magnitude of individual effects for genes controlling quantitative traits. It is hoped that expressing a^* in terms of (h^2/m) and q will aid in developing a "feeling" for reasonable values of a^* For this expression, assumptions of all effects being equal and initial gene frequencies being equal are most certainly invalid. So, at best, the suggested effective population sizes are presented only as approximate guidelines. One cannot be explicit about effective population sizes required for various breeding systems unless one can be more explicit about the joint distribution of gene frequencies and locus effects in the population. For our purposes, however, we need be concerned only with finding a reasonable lower bound for the proportionate effect of a locus a^*, which, in turn, defines the *lower limit* on the probability of ultimate fixation; higher values of a^* leading to even higher probabilities of fixation. From the figures given in Table 1, $a^* \geq 0.1$ would seem to me to be a reasonable working hypothesis. We have looked at the effective population sizes required to give $\mu(q) \geq 0.95$ if $(h^2/m) = 1/1000$. This is the column in Table 1 most nearly corresponding to $a^* \doteq 0.1$.

To summarize this aspect, it is informative to list the probabilities of fixation for an array of initial gene frequencies using values of N and a^* in the range that has been suggested. Table 4 gives $\mu(q)$ for $N\bar{i}a^* = 4.0$ and 6.0. While these values apply to any combination of $\bar{i}$, a^*, and N giving the stated product excepting very small values of N, they were arrived at by letting $\bar{i} = 2.0$ (approximately 6% selection), $a^* = 0.1$ and N = 20 and 30, respectively. With an effective population size of 20, $N\bar{i}a^* = 4.0$, we see that virtually all of the loci with gene frequencies above 1/2 will be fixed at the limit. Over 1/2 of those loci with gene frequencies as low as 1/10 will be fixed for the positive alleles, but almost none of the loci with frequencies below 1/100 would be fixed for the positive allele. If the effective population size is increased to 30, $N\bar{i}a^* = 6.0$, 95% of all loci with frequencies above .25 will be fixed while about 1/10 of those with gene frequencies at .01 will be fixed.

Table 4. Probabilities of ultimate fixation for two values of $N\bar{i}a^*$.

	$N\bar{i}a^*$	
Gene Frequency	**4.0**	**6.0**
.90	1.000	1.000
.75	.998	1.000
.50	.982	.998
.25	.865	.950
.10	.551	.699
.01	.077	.113

While a relatively small proportion of the loci with gene frequencies below 0.10 are being fixed in the limit for effective population sizes of 20 and 30, these same loci contribute little to the progress from selection in the early stages. That is, it will take many cycles of selection to move their frequencies up to the point where their Δq per cycle will be appreciable.

A second point that must be kept in mind is that the results given are for loci being subjected to the stated selection pressure. Loci neutral to the selective forces are still being fixed at the rate 1/2N with the probability of fixation of any particular allele being proportional to its initial gene frequency. Thus, if a population is being "carried" through time with the idea of applying at a later date selection pressure for certain traits, it probably would be desirable to increase the effective population size so as to increase the proability of retaining the favorable alleles at the loci controlling the currently neutral traits. If the effective population size is 25, for example, the rate of fixation of 2% per generation during the time the trait is effectively neutral would be more costly in the long term than the loss during the time the trait is being subjected to selection. If simultaneous multiple trait selection is used rather than the tandem selection mentioned above, there is bound to be some form of "conservation of total pressure" unless resources are unlimited. That is, as selection pressure is applied simultaneously to more and more traits in the form of either an index or multiple truncation, the selection pressure per locus must decrease. (This concept applies to the tandem selection as well if one averages selection pressures over time.) It is for this reason that I chose in the preceding discussion, values of (h^2/m) which would allow the number of loci to be rather large. Such factors must be taken into account in developing your ideas of what constitutes a reasonable effective population size consistent with your own selection methods and goals.

A third point to be made is that effective population size in selection systems depends on many factors in addition to the number of individuals selected, e.g., the mating system, the method of selection, the intensity of selection, and the heritability of the selected trait. The simple textbook formulae apply only in populations not subject to any selective forces, artificial or natural. For example in a simple mass selection program on a monoecious organism where n individuals are selected and pair mated to give n/2 full sib families of equal size, the effective population size can be shown to be

$$N_e = \frac{2(n-1)}{1+\sigma^2_f/2} + \tfrac{1}{2} \qquad \text{(X)}$$

where σ^2_f is the variance of family size after selection. If there is no selection, i.e., if the n individuals are chosen at random, the expected variance of family size is known from probability theory. However if selection is operative the variance of family size becomes a rather complicated function of the selection system. The net effect of truncation selection is to increase the variance of family size and, consequently, reduce the effective population size compared to what it would have been if the individuals had been chosen at random (19).

Figure 2 illustrates the effect of selection on effective population size for the selection system I just described. You can see that increasing the intensity of selection, moving from right to left on the horizontal axis, decreases the effective population size. Part of this decrease or part of the upward tilt in this graph is due to the fact that all families are of equal size before selection in the system we have described. Therefore, with no selection the effective population size is approximately 2 n or to be more explicit it is 2(n – 1) + 1/2 for this system. If the family size before selection is random with all families having equal chance of contributing to the offspring, then effective population size of such systems approaches the number saved. Thus equalizing the number of offspring per parent in the testing program will tend to increase the effective population size. This increase becomes small, however, as the intensity of selection increases.

The long-term effect of selection on the effective population size is seen by noting the vertical distance between the lines in the graph. The long-term effect of selection *on effective population size* for a particular system is to increase the effective population size toward what would be obtained if selection were completely at random, $h^2 = 0$, since selection in the long run is depleting the genetic variance. For the most part this long-term change is trivial so that the error introduced by assuming a constant effective population size over time for a selection system is not great.

EFFECTS OF LINKAGE

Finally, I would like to mention some results from several papers which have studied the effects of linkage on progress from selection and the limits

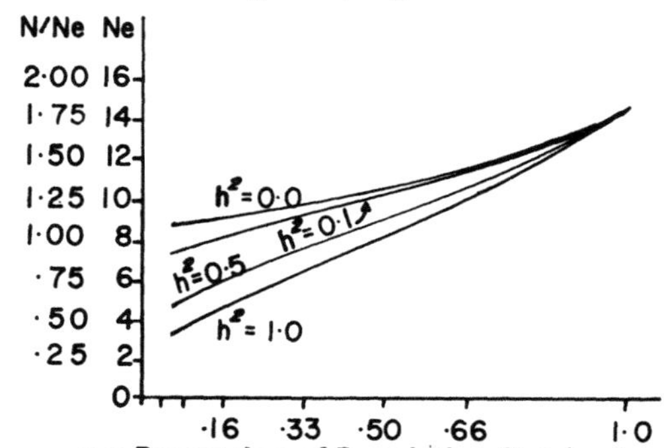

Figure 2. **Effective population size for selecting N = 8 best individuals out of N/α individuals at varying levels of heritability.**

to selection; e.g., Latter (11-13), Hill and Robertson (7), Qureshi (15), Qureshi and Kempthorne (16), Qureshi, Kempthorne, and Hazel (17), and Rebertson (20). Inclusion of linkage in the genetic model greatly complicates the problem and it is difficult to be very general and, at the same time, precise as to the impact of linkage. It is clear that truncation selection introduces a negative linkage disequilibrium among loci and this occurs, only to a lesser degree, even if the loci are segregating independently (14). The net result is that the probability of fixation of the favorable allele is reduced; i.e., the expected long-term gain is less than it would be if the loci were, somehow, always in equilibrium. This reduction in the ultimate gain is directly related to the degree of linkage but in a very nonlinear manner; the major impact occurring only for small frequencies of recombination. Latter (11) looked at the selection situation where two loci start in linkage equilibrium. Both loci have the same proportional effect, that is, the a^* we have talked about is the same for both loci, either 0.1 or 0.5. Recall the .1 is the same order of magnitude we were considering earlier so that $a^* = 0.5$ would represent, comparatively speaking, a major gene. Latter found that this reduction in total response is greatest when the limit that would be expected without linkage is between 60 and 70% of the maximum possible advance. Therefore in his studies he concentrated on the choices of initial gene frequencies which would give limits of this order of magnitude if the loci were segregated independently. Latter's conclusion was that the reduction in total response was very minor unless the recombination probability was less than 0.1. If the recombination probability is 0.1 the reduction in response due to finiteness alone of 20% becomes 22.6% with linkage, 30% becomes 32.7%, and 40% becomes 42.5%, or approximately an additional reduction of 2.5% in each case. If the recombination probability is 0.05 the additional reductions due to linkage become approximately 6% in each case.

Hill and Robertson (7) also studied two linked loci and found that linkage had an important effect on the limits to selection only when the two loci had roughly equal effects. Using the situation where they considered the effects of linkage to be most easily detected (two loci with equal effects at initial frequencies where the effect of linkage is maximum and with linkage distance chosen so that the advance will be most sensitive to changes in Nc) they found that the maximum decrease in chance of fixation was 15% [from $\mu(q) = .78$ to $\mu(q) = .66$] when the linkage distance was changed from $c = ½$ to $c = 0.025$. This was equivalent to a decrease in effective population size of about ¼. As the selective advantage, s, approached the magnitudes we have been considering in this chapter, the decrease in chance of fixation from linkage, $c = 0.025$, reduced to about 5% or an equivalent decrease in effective population size of less than 5%. Recalling that it was expected to be the worst situation (and that the linkage is very tight, $c = 0.025$)

it would appear that the long-term effects of linkage on limits to selection could be offset by a relatively modest increase in effective population size of about 20%.

With respect to the effects of linkage on the rate of response in early generations, the conclusion seems to be rather general that for populations which start in approximate linkage equilibrium, linkage will have a trivial effect on the short-term rate of progress (7,12,20).

REPLICATION OF SELECTION PROGRAMS

I would like to mention the work of Baker and Curnow (1), not so much for the basic idea they present but because in their investigations they relaxed slightly the assumption of equal gene frequencies at all loci and provided at least a Monte Carlo demonstration of some of the points we have mentioned. Their procedure was aimed at maximizing the gain in a reasonable number of generations in such a way so as not to restrict too severely the long term potential genetic improvement. The basic idea was to subdivide the available resources and use several replicate selection programs simultaneously each with relatively small effective population size. The small effective population size will lead to considerable drift or divergence among replicates. At the end of 5 or 10 generations of selection it was assumed that the best replicate could be identified by adequate testing. Thus, by taking advantage of the replicate variation at the end of some reasonably short period of time Baker and Curnow postulated that more progress could be realized in the limited time than by maintaining a single large replicate. To test this hypothesis they used a genetic model which assumed 150 *unlinked* loci all having equal proportional effects with 50 loci having an initial gene frequency of 0.1, 50 with initial gene frequency of 0.2 and 50 with initial gene frequency of 0.3. The phenotypic variance was fixed so that the heritability among individual plants within plots was 0.2. This was approximately equivalent to $a^* = 0.13$ and, with their selection program, $\bar{i} = 1.08$.

Table 5 shows the expected mean progress from selection for 5, 10 and an infinite number of generations for various population sizes (N) and the expected progress in the best of the r replicates. There is little advantage in terms of the mean progress at 5 or 10 generations of increasing the effective population size beyond 16. These results illustrate the point made previously that the main impact of the finiteness is on the limit to selection and not on the early gains.

The reason for the impact of finiteness coming more at the limits of the selection program than at the early gains can be seen by considering each of their three classes of loci separately. Taking for example the population size of 32, there is a total loss in the limit of 63 units compared to what would have been obtained with an infinite population. Of this loss, 64% is due to the

Table 5. **Expected mean progress and progress of best replicate for Baker-Curnow Model.**

		Number of Generations					
		5		10		∞	
N	r	Mean	Best Rep.	Mean	Best Rep.	Mean	Best Rep.
4	64	13.4	30.5	20.7	42.3	28.4	53.9
16	16	16.4	23.6	32.3	42.8	114.5	134.7
32	8	17.0	21.1	35.1	41.3	177.1	190.4
64	4	17.3	19.4	36.7	39.9	220.9	225.0
256	1	17.6	17.6	37.9	37.9	240.0	240.0

loss of the positive allele in the set of loci having initial gene frequencies of 0.10. Twenty-five percent is due to loss of the favorable allele in the group having initial gene frequencies of 0.20 and less than 10% due to loss in the group with initial gene frequencies of 0.30. Thus, the main impact at the limit comes from the loss of alleles in the group of loci with the lowest gene frequency. At the same time the group with the lowest gene frequencies contribute very little to the short-term gains. In this particular example, at the tenth generation approximately 1/6 of the total progress can be attributed to gene frequency changes in this group. One-third of the total progress is attributable to gene frequency changes in the second group and 1/2 of the progress is attributable to the set of loci with starting gene frequencies of 0.3. Thus, most of the short-term gain is realized from loci with the highest initial gene frequencies, in this particular problem, and virtually all such loci will be fixed for the favorable allele with relatively small effective population sizes.

Turning to their primary objective, the expected value of the best individual replicate seems to indicate that there is some intermediate time around 5 to 10 generations when choice of the best of several small replicates will give greater gain than straight mass selection within a single large replicate. There is one aspect of their comparison which should be pointed out. Baker and Curnow assume that the best line through sufficient replication and testing can be identified without error. Up to this point the selection intensity and the amount of effort involved in each of the regimes has been equal. However, testing 16 replicates versus testing 4 replicates introduces some imbalance in both selection intensity and amount of effcrt if the best of the 16 is to be identified with the same degree of certainty as the best of the 4. There is, of course, the other advantage in multiple replicates of some insurance against losing the entire selection program through some mishap.

SUMMARY

In summary, the point I wish to emphasize is that the effective population size needed to retain the long-term genetic potential in a population is well within the reach of any reasonable recurrent selection program. A combination of high selection intensity and a reasonable effective population size can be used so that there need be no major conflict between the objectives associated with short-term and long-term selection programs. The relevant information from studies involving linkage would seem to suggest that the effective numbers we discussed earlier should perhaps be adjusted upward slightly to allow for the linkage depression. I want to emphasize that the models that have been investigated to date, and particularly so for the results I have presented, are extremely simple in relation to the biological systems of interest. The results and suggestions presented should be viewed with that in mind.

NOTES

J. O. Rawlings, Department of Statistics, North Carolina State University, Raleigh, North Carolina 27650.

REFERENCES

1. Baker, L. H., and R. N. Curnow. 1969. Choice of population size and use of variation between replicate populations in plant breeding selection programs. Crop Sci. 9:555-560.
2. Ewens, W. J. 1963. Numerical results and diffusion approximations in a genetic process. Biometrika 50:241-249.
3. Falconer, D. S. 1960. Introduction to Quantitative Genetics. The Ronald Press Company, New York.
4. Finney, D. J. 1958. Plant selection for yield improvement. Euphytica 7:83-106.
5. Hill, W. G. 1969. On the theory of artificial selection in finite population. Genet. Res., Camb. 13:143-163.
6. Hill, W. G. 1974. Variability of response to selection in genetic experiments. Biometrics 30:363-366.
7. Hill, W. G., and A. Robertson. 1966. The effect of linkage on limits to artificial selection. Genet. Res., Camb. 8:269-294.
8. Kimura, M. 1957. Some problems of stochastic processes in genetics. Ann. Math. Stat. 28:882-901.
9. Kojima, K. 1961. Effects of dominance and size of population on response to mass selection. Genet. Res., Camb. 2:177-188.
10. Latter, B. D. H. 1965 a. The response to artificial selection due to autosomal genes of large effect. I. Changes in gene frequency at an additive locus. Aust. J. Biol. Sci. 18:585-598.
11. Latter, B. D. H. 1965 b. The response to artificial selection to autosomal genes of large effect. II. The effects of linkage on limits to selection in finite populations. Aust. J. Biol. Sci. 18:1009-1023.
12. Latter, B. D. H. 1966 a. The response to artificial selection due to autosomal genes of large effect. III. The effects of linkage on the rate of advance and approach to fixation in finite populations. Aust. J. Biol. Sci. 19:131-146.

13. Latter, B. D. H. 1966 b. The interaction between effective population size and linkage intensity under artificial selection. Genet. Res., Camb. 7:313-323.
14. Neeley, D. L. R., and J. O. Rawlings. 1971. Disequilibria and genotypic variance in a recurrent truncation selection system for an additive genetic model. Institute of Statistics Mimeo Series No. 729, Raleigh, NC.
15. Qureshi, A. W. 1968. The role of finite population size and linkage in response to continued truncation selection. II. Dominance and overdominance. Theor. and Appl. Genet. 38:264-270.
16. Qureshi, A. W., and O. Kempthorne. 1968. On the fixation of genes of large effects due to continued truncation selection in small populations of polygenic systems with linkage. Theor. and Appl. Genet. 38:249-255.
17. Quershi, A. W., O. Kempthorne, and L. N. Hazel. 1968. The role of finite population size and linkage in response to continued truncation selection. I. Additive gene action. Theor. and Appl. Genetics 38:256-263.
18. Robertson, A. 1960. A theory of limits in artificial selection. Proc. Royal Soc. B. 153:234-249.
19. Robertson, A. 1961. Inbreeding in artificial selection programs. Genet. Res., Camb. 2:189-194.
20. Robertson, A. 1970. A theory of limits in artificial selection with many linked loci, pp. 246-288. in K. Kojima (Ed.) Mathematical Topics in Population Genetics, Springer-Verlag, New York.

STRATEGIES FOR INTROGRESSING EXOTIC GERMPLASM IN BREEDING PROGRAMS

W. J. Kenworthy

The economic losses suffered by the U.S. corn crop in 1970 due to the southern corn leaf blight graphically demonstrate the hazards of genetic uniformity. For soybean breeders these losses serve as a reminder of the narrow genetic base present in today's cultivars. The lack of genetic diversity in the parentage of the most widely grown U.S. cultivars has been discussed by Johnson and Bernard (7) and Hartwig (6). Both report that relatively few plant introductions have contributed genetic material to the parents of today's U.S. cultivars. Strains of Manchurian origin form the genetic base of many of the U.S. cultivars, especially those grown in the North Central region. Not only does the hazard of pest susceptibility increase with genetic uniformity, but also a potential constraint to further genetic improvement can occur with a narrow genetic base. Our dependence on a few lines as sources of resistance to major soybean pests predisposes today's cultivars to a narrow genetic base. Furthermore, conventional breeding methods such as backcrossing and pedigree selection within selfing generations of crosses between closely related elite lines can contribute little to improving the diversity of new cultivars. The objective of this chapter is to review recent efforts to incorporate exotic soybean germplasm into breeding programs designed to improve seed yield in soybeans, and to suggest an alternative breeding procedure for the utilization of productive exotic germplasm. For this discussion, exotic germplasm will refer to plant introductions of diverse origin.

The utilization of agronomically desirable exotic germplasm in breeding programs can increase the genetic diversity of new cultivars. Furthermore, its utilization can assist the breeder in meeting short- and long-term selection objectives for increasing seed yield. Increased genetic variation in breeding populations is usually recognized as the greatest benefit that can be realized from the introgression of genes from plant introductions. It is also possible, however, to develop productive breeding lines from populations which contain some exotic germplasm even though such lines occur infrequently. Exotic germplasm has been utilized in the following types of breeding schemes for yield improvement in soybeans: two-way crosses, three-way crosses, intermating populations, and intermating populations with selection.

TWO-WAY CROSSES

Early soybean improvement programs made direct use of desirable plant introductions by using them as parents in crosses with adapted parents. As the productivity of newly-released cultivars increased, the utilization of plant introductions as parents in crosses for additional cycles of selection decreased. Today breeders rarely make use of exotic parents as direct sources of yield genes. None of the ten most widely grown U.S. soybean cultivars in 1971 (6) were from crosses which had a plant introduction as one parent. Some cultivars in production, however, are from two-way crosses which contain a plant introduction as one parent. For instance, PI 80837 was one parent in three different two-way crosses from which the cultivars 'Shore', 'Celest', and 'Ware' were selected. Even though PI 80837 is a fairly-productive accession of Maturity Group IV, its outstanding seed quality was the criterion which identified it as a desirable parent. Although productive cultivars have been developed from adapted X exotic crosses, the greatest utilization of exotic germplasm is as sources of genes for pest resistance (6).

THREE-WAY CROSSES

Three-way (adapted X exotic) X adapted crosses increase the amount of adapted germplasm in breeding populations, compared to two-way crosses. Thorne and Fehr (15) compared yields of randomly selected homozygous lines from two-way and three-way populations. Three plant introductions with high protein were used as parents in crosses with two high-yielding adapted cultivars to form six two-way populations. The three-way populations were formed by crossing each of the six two-way F_1's with an adapted cultivar. The average yield of the three-way populations was significantly greater than the two-way population mean. Moreover, the three-way populations had more of the highest yielding lines in the test than the two-way populations. Larger estimates of genetic variance for yield were also observed in the three-way populations than in the two-way populations. They concluded that three-way populations were better sources than two-way populations for the selection of high-yielding lines.

INTERMATING POPULATIONS

Intermating in breeding populations containing exotic and adapted germplasm maximizes the potential for genetic recombinations and enhances the break-up of linkage blocks. Fehr and Clark (3) described five intermated populations containing different levels of exotic germplasm. The populations were formed by crossing four adapted lines and four high-yielding plant introductions in combinations to obtain populations with 100, 75, 50, 25, and 0% exotic germplasm. Schoener (12) studied the effect of the presence of exotic germplasm in these intermated populations by evaluating randomly selected lines of similar maturity from each of the five populations. The population means showed a small, but statistically significant decrease in yield with increasing percentages of exotic germplasm. Most of the superior lines were identified from the populations with 0 and 25% exotic germplasm; however, the highest yielding line observed was from the population with 75% exotic germplasm. She concluded that superior progeny can be selected from a population with a large percentage of exotic germplasm, but the frequency of such lines is low. Genetic variance for yield was reported to be greatest in the populations that had both adapted and exotic germplasm.

INTERMATING POPULATIONS WITH SELECTION

The addition of an intermating generation in cyclic selection is termed recurrent selection. A basic recurrent selection cycle consists of selecting and intermating a group of superior individuals from a base population. Progeny from the intermating generation become the base population for the next cycle of selection. In work at North Carolina (8,9) nine plant introductions of Maturity Groups VI and VII with high protein content and diverse geographical origin were crossed to a highly productive experimental line, D49-2491. The F_1 plants were backcrossed to D49-2491 to form a base population containing 25% exotic germplasm. A recurrent selection procedure was conducted with this population. Each cycle of selection consisted of an intermating generation, a selfing generation, and a testing generation. Selection was based on a S_1 progeny test in which the progeny of a F_1 plant was evaluated for yield.

Evaluations of composites of the lines selected as parents for each of the three cycles of selection indicated a significant increase in yield occurred from the base cycle to cycle three. An average increase in yield of 134 kg/ha/cycle of selection was observed. The yield of the cycle three composite was 20% greater than the adapted parent. Additional information on the performance of some of the highest yielding individual lines from each cycle of selection was obtained on a yield test grown at one location in 1977. The 10 highest yielding lines from cycle three averaged 3568 kg/ha compared to

2855 kg/ha for D49-2491 in this environment. The favorable yield response observed in this recurrent selection program suggests that this breeding procedure may be used to develop populations of both greater diversity and higher productivity.

PROPOSED BREEDING SCHEME

Breeding lines with greater genetic diversity than is present in currently grown cultivars could be derived from populations developed by any of the breeding procedures which utilize diverse exotic germplasm. No information is available on the potential for identifying high-yielding lines from segregating populations synthesized by using highly productive exotic germplasm. The information available for soybeans gives only fragmentary evidence for the potential usefulness of breeding populations containing exotic germplasm. Although productive exotic germplasm was used in the populations developed by Fehr and Clark (3), Schoener (12) evaluated only randomly chosen lines of 'Corsoy' maturity from the populations. It is quite possible that more productive lines exist in these populations. The plant introductions utilized in the studies by Thorne and Fehr (15) and Kenworthy (8) were selected for their diversity of origin and high protein content, not for their productivity. Even with these limitations, the information from these studies suggests that productive lines can be identified from breeding populations containing exotic germplasm incorporated either by two- and three-way crosses or by intermating schemes.

To increase the likelihood of productive recombinants in the breeding populations, breeding procedures which maximize genetic recombination between the exotic and adapted germplasm are the most desirable methods to use. Recurrent selection procedures offer the best technique for the stepwise increase in frequency of desirable genes from the genetic recombination of adapted and exotic germplasm. Recurrent selection in a population containing 25% exotic germplasm was found to meet the short-term selection objective of developing productive lines and the long-term objective of maintaining genetic variability for continued cycles of selection. Therefore, I suggest a two-step procedure to follow when incorporating exotic germplasm into populations to improve yield. The first step is the evaluation and selection of productive plant introductions to use in forming the base population for selection. The second step is the utilization of this population in a recurrent selection procedure to develop productive breeding lines and an improved population for further selection.

EVALUATION AND SELECTION OF EXOTIC GERMPLASM

The large number of unadapted plant introductions in the U.S. germplasm collection has generally discouraged extensive yield evaluations of this material. Cooperative regional yield evaluations of a portion of the collection have been initiated and coordinated by Dr. Clark Jennings of Pioneer Seed Company, Waterloo, Iowa. These yield evaluations offer promise in identifying productive plant introductions; however, plant introductions to be used in recurrent selection programs should also have high general combining ability for yield. Whether these two criteria identify the same plant introductions remains to be seen. Therefore, evaluation of the productive plant introductions for general combining ability is also a logical objective of cooperative regional programs.

Rather than depending upon a series of diallel crosses to quantify combining ability estimates, a top cross screening procedure would be more manageable for the evaluation of a large number of accessions. Limited information is available on the type of tester parent to use for this situation in self-pollinated crops (4,13). Most likely, a related low-yielding tester with a broad genetic base should be used as the tester parent in a cross-pollinated species (2,10,11). For soybeans, a reasonable compromise for estimating combining ability might be to use the mean top cross performance of a plant introduction crossed to a group of tester parents. Cultivars or breeding lines which have some exotic germplasm in their background might be good tester parents to use. Cultivars such as 'Delmar', 'Shore', and 'Ware' might represent a group to use to evaluate the combining ability of plant introductions in Maturity Group IV. If the breeding objective is to form populations from two- or three-way crosses between adapted cultivars and exotic germplasm, the adapted cultivars can be used as testers to identify good combining plant introductions to use in making these crosses.

A series of top crosses between a group of plant introductions and the tester parents occurs in the first generation of the plant introduction evaluation procedure as shown in Table 1. The cooperative efforts of several breeders can permit the evaluation of a number of plant introductions. The F_1 seeds from each top cross are advanced in bulk to obtain enough seeds for planting a yield test. The F_2 seeds of each top cross are bulked. The top crosses developed by several breeders can be included as entries in a test which is grown by cooperating breeders at several locations. Plant introductions which have the highest mean top cross performance are used as parents in forming the base population for the recurrent selection procedure.

Table 1. Summary of operations in the plant introduction evaluation procedure.

Generation	Operation
1	PI's X Tester(s)
2	Advance F_1 seeds for each top cross in bulk
3	Yield test bulks of each top cross

RECURRENT SELECTION PROCEDURE

Parents selected for inclusion in the base population are intercrossed in a fashion that results in the desired proportion of exotic germplasm in the population. The F_1 seeds from the intercrossing are advanced. Ideally, at least one random mating generation should be completed before selection is initiated in the base population (5) as shown in Table 2. Crossing 20 to 30 pairs of randomly chosen F_2 plants has been suggested by Baker (1) to approximate random mating and constitute a population large enough to minimize the effects of genetic drift. Seeds obtained from the final intermating are advanced and individual S_0 plants are harvested. The S_1 lines from each S_0 plant are grown in a yield evaluation. We used a 3 x 3 hill plot (14) at one location for our yield evaluations. Other plot arrangements can be used, but the testing procedure is limited by the amount of seeds produced on one S_0 plant unless additional selfing generations are used to increase seed supplies. We selected 20 high-yielding S_1 lines from a test population of approximately 200 in each cycle. Seeds from the high-yielding entries in the yield test (S_2) were grown and the S_2 plants randomly intermated to form the base population for the next cycle of selection. The selected S_1 lines can be advanced by single seed descent or the pedigree method to identify homozygous breeding lines in each selection cycle.

SUMMARY

Reducing the genetic vulnerability of commercially available soybean cultivars by increasing genetic diversity has heightened interest in developing cultivars from broad-based genetic material. The incorporation of exotic

Table 2. **Summary of operations in the formation of the base population and in the recurrent selection procedure.**

Generation	Operation
1	Intercross selected parents.
2	Advance F_1 seeds.
3	Randomly intermate F_2 plants. (Intermating generation can be repeated.)
4	Advance F_1 seeds from final intermating. Harvest individual F_1 plants (S_1 lines). Begin recurrent selection cycle 1.
5	Yield test S_1 lines.
6	Randomly intermate high-yielding S_1 lines (S_2 plants). Continue selection within S_1 lines.
7	Begin recurrent selection cycle 2. Advance F_1 seeds from intermating.
8	Continue as in generation 5.

germplasm into breeding programs for improving yield per se raises many unanswered questions concerning best procedures to follow. Furthermore, information on agronomic performance of many plant introductions is incomplete. Cooperative regional evaluations offer the first step in gaining more information about the performance of several plant introductions in the U.S. collection. Continued cooperative evaluations can provide additional information concerning combining ability that will permit the development of broad-based populations containing exotic germplasm. Recurrent selection procedures can be used to develop populations and cultivars not only of greater diversity but also of greater productivity.

NOTES

W. J. Kenworthy, Department of Agronomy, University of Maryland, College Park, Maryland 20742.

Scientific Article No. A2582, Contribution No. 5620, of the Maryland Agric. Exp. Sta., Dept. of Agronomy, College Park, MD 20742.

REFERENCES

1. Baker, R. J. 1968. Extent of intermating in self-pollinated species necessary to counteract the effects of genetic drift. Crop Sci. 8:547-550.
2. Eberhart, S. A., M. N. Harrison, and F. Ogada. 1967. A comprehensive breeding system. Zuchter/Genet. Breed. Res. 37:169-174.
3. Fehr, W. R., and R. C. Clark. 1973. Registration of five soybean germplasm populations. Crop Sci. 13:778.
4. Gebrekidan, B., and D. C. Rasmusson. 1970. Evaluating parental cultivars for use in hybrids and heterosis in barley. Crop Sci. 10:500-502.
5. Hanson, W. D. 1959. The breakup of initial linkage blocks under selected mating systems. Genetics 44:857-868.
6. Hartwig, E.E. 1973. Varietal development, pp. 187-210. In B. E. Caldwell (Ed.) Soybeans: Improvement, Production and Uses. Amer. Soc. of Agron., Madison, Wisc.
7. Johnson, H. W., and R. L. Bernard. 1963. Soybean genetics and breeding, pp. 1-73. In A. G. Norman (Ed.) The Soybean. Academic Press, New York.
8. Kenworthy, W. J. 1976. A recurrent selection program for increasing seed yield in soybeans. Ph.D. Thesis. N. C. State Univ., Raleigh, N.C. Univ. Microfilms, Ann Arbor, Mich. (Diss. Abstr. 37:661-B).
9. Kenworthy, W. J., and C. A. Brim. 1979. Recurrent selection in soybeans. I. Seed Yield. Crop Sci. 19:315-318.
10. Lonnquist, J. H., and M. F. Lindsey. 1970. Tester performance level for the evaluation of lines for hybrid performance. Crop Sci. 10:602-604.
11. Rawlings, J. O., and D. L. Thompson. 1962. Performance level as criterion for the choice of maize testers. Crop Sci. 2:217-220.
12. Schoener, C. S. 1978. Utilization of unadapted germplasm in soybean breeding populations. Ph.D. Thesis. Iowa State Univ. Univ. Microfilms, Ann Arbor, Mich. (Diss. Abstr. 39:487-B).
13. Shebeski, L. H. 1966. Quality and yield studies in hybrid wheat. Can. J. Genet. Cytol. 8:375-386.
14. Schutz, W. M., and C. A. Brim. 1967. Inter-genotypic competition in soybeans. I. Evaluation of effects and proposed field plot design. Crop Sci. 7:371-376.
15. Thorne, J. C., and W. R. Fehr. 1970. Exotic germplasm for yield improvement in 2-way and 3-way soybean crosses. Crop Sci. 10:677-678.

ROLE OF PHYSIOLOGY IN SOYBEAN BREEDING

D. N. Moss

When asked to address this conference on the topic of the role of physiology in soybean breeding, I hesitated long before replying; and today I question the wisdom of the reply I finally gave, for I have never worked with soybeans; certainly, therefore, my credentials for addressing this body can, at best, be questioned. The topic as it applies to plant breeding in general is of major interest to me, however, and so, somewhat tentatively, I review some thoughts with you.

May I begin by addressing the question, 'What is the problem?' To do that I will use an example once used by a reputable soybean breeder, Dr. H. W. Johnson, in a speech I heard him give. The example happens to be for corn but, obviously the principle applies universally.

Imagine that a plant breeder wishes to incorporate resistance to two diseases into a potential new variety which will also have the necessary genetic components for high yield. What is the magnitude of the problem he faces?

In Dr. Johnson's illustration he used fairly well-known probabilities for gene recombination to estimate that the desired gene combination should occur in one plant in about two acres of corn.

It is not news to any plant breeder that the chance is remote that, in two acres of corn plants where every plant is different, the breeder will be able to identify the highest yielding plant. It is for that very reason that plant breeders have jobs and that, in breeding programs, large populations must be evaluated. The following facts should be considered: (a) Nature does not select for high yield; (b) The number of crosses a breeder makes is always small relative to the combinations that, ideally, should be considered; (c) Yield is a complex *process*. Selection criteria available to the breeder to evaluate potential on an individual plant basis are unprecise; and (d) Generally, limiting factors are unknown. Therefore, it is only by chance if genetic variability for a limiting factor exists among the progeny of a particular cross.

As one considers the consequences of these facts, it quickly becomes obvious why a plant breeder's work is never done and why the plant breeding process is a "numbers game." In general, the successful breeder will have a large program because the more genetic combinations a breeder sees the greater will be the chance that he will have a superior genotype in that population.

The questions before us today are simply, "Can physiological information have any effect on the efficiency of this breeding process?" If so, "What is its role?"

For 25 years I have been hearing those questions. Over that time period, I have come to believe that it is essential that we understand how plants grow; it is essential, too, that new varieties arise as a result of informed decisions both in the selection of parents and in choosing appropriate individuals among the offspring of those parents. I visualize this happening only when breeders and physiologists are working directly together and I would suggest that the process has at least four steps. May I refer to this process as "plant design."

PLANT DESIGN RESEARCH

Determining Yield Limiting Factors

The goal of physiologists who have dealt with crop plants has always been to identify yield limiting factors. If we don't know them now, what evidence is there that we will ever know them?

A careful consideration of that question leads inevitably to the conclusion that we will never know what factors limit yield, for it is obvious that, as we have learned about a specific yield limiting factor in the past, we have devised methods to minimize the limitation. At what level would our yields be today if we had not come to understand the role of phosphorus, potassium, nitrogen, soil pH, etc., and devised ways to minimize their limitation. Things we have learned are always the easy answers but think for a moment about the years of research that went into study of factors such as copper or molybdenum before an understanding of the role of these elements in plant nutrition led to relatively simple methods to correct problems caused by them.

Several of these problems I have cited have been solved by use of soil amendments. Can we think of an example, though, where physiological information has been utilized to advance the science of plant breeding. One example I have cited before is the example of the high-yielding semi-dwarf wheat and rice varieties which have given rise to the term, "the green revolution." I suppose that wheat and rice plants have been known for as long as man has grown these species; but, historically, they had little impact on man's food supply. When Orville Vogel and, later on, Norm Borlaug sought high yielding wheats and Peter Jennings and others were seeking the same goal for rice, it was only natural that they devoted considerable attention to dwarf types, because lodging has always been a problem in high yielding cereal fields. Work continued for more than 20 years to develop high yielding, short-strawed wheat. The key to developing the new high yields turned out to be selection for a specific physiological trait—nitrogen responsiveness. The reasoning can be summarized as follows: High quality wheat and rice contain significant quantities of protein; protein contains nitrogen; since one can't make something out of nothing, nitrogen must be supplied to the growing crop. However, supplying nitrogen is of no value if the plants can't use it. Thus, the remarkable progress in plant breeding came when the breeders started selecting among the semi-dwarf germplasm for high nitrogen responsiveness.

An example cited by Chandler (1) at a crop physiology symposium in Nebraska a few years ago which shows the yield of 'IR 8' and its tall parent 'Peta' as a function of nitrogen fertilization rate is given in Table 1. When no nitrogen was added, 'Peta' outyielded 'IR 8.' As increasing amounts of nitrogen were added, however, the yield of 'Peta' decreased while the yield of 'IR 8' increased sharply. Again, the key to the selection program was to select for nitrogen responsiveness. Had selection not been done at high N fertility, it is unlikely that the "miracle" rise genotypes ever would have been identified.

We could cite many other examples where physiological knowledge of limiting factors has been put to use in plant breeding. In many areas of the world crop yields are limited by toxic concentration of various elements in the soil. One of the opportunities for breeders is to use the known genetic variability for tolerance to these toxins to develop varieties that can do well in what are now hostile environments.

Table 1. Yields of 'IR 8' and 'Peta' rice as a function of nitrogen applied. Data were collected during the dry season (1).

	Nitrogen Applied (kg/ha)				
Yield	**0**	**30**	**60**	**90**	**120**
			— t/ha —		
'IR 8'	4.8	6.3	7.2	8.0	8.9
'Peta'	5.2	5.6	5.0	4.5	4.1

Other limiting factors are more difficult to deal with, but solutions to problems we don't understand are always difficult. Scientists are studying nitrogen fixation in soybeans and its relationship to carbon metabolism; they seek to understand the photosynthesis process in this crop and its role in yield. Hypothesize with me that some bright biochemist identifies a rate limiting enzyme in one of these processes. What marvelous advances in yield could come then if a physiologist and breeder teamed up and found a simple way to screen for the enzyme, were able to identify genetic variability for the rate at which the enzyme would operate, and developed germplasm which not longer had the limitation.

Would we then continue to hear a question, as I was asked a couple of months ago, "Can you name a single variety of any crop that has ever been developed because a physiologist told the breeder which plant to choose?" Probably we would, because once a limiting factor has been identified and appropriate screening procedures have been devised, a plant breeder still must carry out the variety development phase of the work. It may never be very obvious that physiology has any part of that phase of a process.

Identifying Genetic Variability

For some limiting factors, knowledge of the specific limitation is really all the plant breeder need have from the physiologist, because simple screening procedures are obvious. You can all think of numerous examples and we need not belabor that point. For some limitations, however, the key to progress in improving a crop's capacity with regard to a specific limitation will be whether or not an appropriate screening procedure can be devised. Little progress will be made if someone discovers that a certain poorly adapted genotype carries a capacity for superior rates of nitrogen fixation, as far as its biochemical reactions are concerned, if the only test for that capacity involves extraction and purification of the enzyme, and subsequent complex laboratory determinations of reaction rates. The knowledge remains an academic curiosity until someone discovers a simple way to identify individual plants which carry the desired trait.

Yield is a complex process. The rate of an enzymatic process or the effect of some other trait may be a yield limiting factor only in a genetic background of high yielding genes and when environmental limitations to yield are insignificant. If the desired genetic aberration for reaction rate in the case of our hypothetic enzyme were present in the species in low frequency, then the chances are slight that it would ever be identified or utilized in a breeding effort. It is highly probable that the desired genes would be in the parental stock only if they were identified specifically and transferred into appropriate breeding stock by an informed selection procedure.

Evaluation of Traits

As we direct our attention to more illusive limiting factors such as an enzyme, it may require a major plant breeding program to identify yield limitations. Basic research may suggest specific factors that could limit yield, but to adequately assess the degree to which a specific trait may be limiting yield, may require extracting desirable genes from grossly unadapted genotypes or even from related species and transferring them to a genetic background under which the limitation is perceived to function. What does that mean? It means that we may have to transfer genes from poor backgrounds into "near varieties" before it is even valid to run the experiments necessary to determine the degree to which the trait in question may be a limiting yield. And, testing a trait in one genetic background may not be sufficient to generate a valid evaluation of a trait. It may require moving the gene or genes into several different backgrounds. Obviously, each transfer could require as much effort as developing a disease resistant variety. Thus, the amount of work this concept implies is awesome.

Developing New Varieties

If a research team does identify important yield limiting factors, and ways to screen for the traits in a segregating population, then the problem still remains of developing new varieties. It is entirely possible that yield limiting factors may be identified for which genetic variability cannot be found. Alternatively, it is possible that heritability of a trait may be so low that it is difficult to accumulate a desirable level of the trait in an appropriate genetic background. Identifying the problem may not lead automatically to a solution. And, some traits which may be identified as potentially yield limiting may not prove amenable to genetic manipulation.

May I review with you two examples from work with barley at the University of Minnesota which illustrate some of these problems. When I joined the staff of the Department of Agronomy and Plant Genetics at the University of Minnesota in 1967, Dr. Donald C. Rasmusson and I began a collaborative effort to follow a plant design program as I have outlined here today. One of the characteristics of barley leaves which we believed could be important in yield was the physical process of gas exchange by leaves. We hypothesized that one could possibly affect the rate of gas exchange processes by varying the stomatal frequency on leaves. Therefore, we began to look for genetic differences in stomata frequency. We were able to identify genetic lines which differed in frequency of stomata by a factor of two (2). We then took the genes for controlling stomatal frequency and developed near isogenic lines of barley in a genetic background of the commercial cultivar Dickson, a late maturing cultivar, and Primus, an early barley.

Under test conditions favoring open stomata (high humidity and irradiance), we found that lines with more stomata lost more water and absorbed more carbon dioxide (2,3). However, the yield of lines differing in stomatal frequency was not different; furthermore, subsequent tests suggested that lines which differed in stomatal frequency compensated for these differences in field environments by different degrees of stomatal opening. Frequency was not the controlling factor which regulated gas exchange. In the case of these particular genetic lines, a lot of work over a period of several years did not result in any lines that showed promise as parental material for future cultivars.

Was the work devoted to stomatal frequency done in vain? The same question could have been asked of us about work on a dozen or more traits in barley. The interesting answer that Dr. Rasmusson gives is that being involved in a plant design program has caused him to evaluate, to a relatively thorough degree, a wide spectrum of genetic material that never would have been in his program otherwise. Where reason has gone into selection of traits and where many traits are being tested, logic would say the gene pool from which cultivars are being assembled should lead to superior cultivars. Dr. Rasmusson believes firmly that his progress toward higher yielding cultivars has been more rapid because of "spin-off' from the plant design work. He cites the example of his recent high yielding cultivar release called 'Manker.' The name is a contraction "many kernels;" the cultivar arose from material he brought into his program to evaluate the role of kernel number in yield. From the genetic lines he was developing which differed in size of heads he selected lines which were yielding well to go into his cultivar development program. Thus, the parent material for the cultivar 'Manker' came out of his plant design work. Dr. Rasmusson's enthusiasm for plant design research, where his pay off must come in a cultivar development program, is more valid testimony than anything I can say here today.

Is spin-off the only benefit of a plant design program? Obviously if traits are chosen carefully some will certainly be important in determining yield. One example that comes to mind is height in barley. Lodging is often a problem in barley fields. Short straw cultivars should be more resistant to lodging than tall cultivars. In 1955, Dr. Jean Lambert began a breeding program to develop short-strawed barley cultivars. This work was taken over by Dr. Rasmusson in 1959 and has continued up to the present time. Table 2 shows the results of more than two decades of intensive work on semi-dwarf barley at Minnesota. It is only in the past 2 or 3 yr that yields of semi-dwarf genotypes have equaled the yield of tall genotypes and there is no evidence that superior yields will result from the work.

How can anyone justify nearly 25 yr of sustained research on a subject that leads to few publications and has not shown any economic advantage

Table 2. The yield of the 3-highest yielding semi-dwarf barley genotypes as a percentage of the 3-highest yielding tall entries in Minnesota nurseries from 1970 to 1977. About 1970 was when semi-dwarf lines which had reasonable agronomic characteristics were first identified (2,3).

	1970	1971	1972	1973	1974	1975	1976	1977
% Yield	79	90	86	97	98	98	109	100

during that period of time? We can probably all agree that the researcher and his administrators must have solid reasons if such effort is to be justified in accounting for use of resources; and, how many such efforts (i.e., how many traits) can be carried simultaneously by a research group? What is the personal risk involved to the researchers and those responsible for the programs? As one analyzes these questions, it would be easy to say that it is highly improbable that effective physiology-breeding programs will ever be involved. My personal belief is that few will be sustained by industry. Perhaps it is only in the environment of graduate education programs that any really major sustained program of physiology-breeding research is even possible. Such topics as the heritability of a particular trait or its physiology do present viable thesis research projects of limited scope where one thesis can provide the base for another. In that atmosphere where the researchers may have teaching responsibilities and graduate training functions, it is possible and, I believe, highly productive to have team research on a sustained theme. For the breeder involved, an ongoing cultivar development program may be part of the responsibility which justifies the continued existence of the plant research. If high quality scientific publications are coming out of the program and if a cultivar development program is benefiting from the spin-off then I can see physiology-breeding programs being a reality.

I sincerely hope that the university environment in our land-grant settings does indeed lead to such programs being established and that they are effective. From what I have seen of this world it is essential that somehow we learn to make progress more rapidly in crop breeding programs and that we do things that have not been possible up to now. I see only a bleak future for the world if we don't learn to minimize limitations to crop yield.

NOTES

Dale N. Moss, Department of Crop Science, Oregon State University, Corvallis, OR 97331.

REFERENCES

1. Chandler, R. F., Jr. 1969. Plant morphology and stand geometry in relation to nitrogen. In J. D. Eastin, F. A. Haskins, C. Y. Sullivan, and C. H. M. van Bavel, (eds.) Physiological Aspects of Crop Yield, Amer. Soc. Agron., Madison, WI, pp. 265-289.

2. Miskin, K. E., D. C. Rasmusson, and D. N. Moss. 1972. Inheritance and physiological effects of stomatal frequency in barley. Crop Sci. 12:780-783.
3. Yoshida, T., D. N. Moss, and D. C. Rasmusson. 1975. Effect of stomatal frequency in barley on photosynthesis and transpiration. Bull. Kyushu (Japan) Agr. Exp. Sta. 18:71-80.

ACCOMPLISHMENTS AND PRIORITIES IN PLANT BREEDING

G. F. Sprague

There is little need for any extended presentation of accomplishments of plant breeding. Adequate reviews of the pertinent literature are available. However, some special aspects will be mentioned as a point of departure for speculation on research priorities.

Plant breeding is often arbitrarily considered under two separate headings: the self-pollinating and the cross-pollinating species. In both groups of organisms plant breeding began with simple phenotypic selection. The recognition that cross-pollinating species might best be improved by other procedures possibly had its inception with the reports of Shull in 1908 and 1909 (10,11) on the effects of inbreeding and hybridization in corn. An extension of this work culminated eventually in the commercial use of hybrids. The distinction between the two classes of organisms is becoming less meaningful through the use of special techniques to permit hybrid development in self-pollinating species, e.g. sorghum and wheat.

Possibly of necessity and possibly by inclination, workers interested in the cross-pollinating species have developed a wide range of special interests. This has been evidenced by participation in an early adoption of improved experimental designs and field plot techniques followed by interest in quantitative genetics as a basis for understanding and interpretation of past developments and as a guide for further research. Differences in operating procedures, however, have little relevance to research priorities in the two groups.

The effectiveness of plant breeding is conditioned by two attributes; the presence of adequate genetic variability and the selection scheme employed to effect changes in gene frequency.

GENETIC VARIABILITY

Genetic variability will likely always be less than the breeder desires. Disregarding this craving, readily available genetic resources do vary widely from crop to crop. Possibly one needs to make a distinction between potential genetic variability and the more restricted sample actually used by breeders. For example, it has been reported that commercial hybrids grown in the Corn Belt involve a very restricted genetic base; heavy reliance being placed on two sources, Stiff Stalk Synthetic and Lancaster derivatives. In sorghums, prior to the U.S.D.A.-Texas conversion program genetic diversited used by breeders represented a very small sample of the Milo, Kafir and Feterita germplasm. Current varieties of soybeans trace back to a very limited number of importations.

If progress is limited by lack of genetic variability several options are available. Choice of parental material, regardless of the breeding system to be employed, is commonly based on perceived differences in genotype or gene frequency. Such evaluations are highly conditioned by a subset of genes which influence adaptation. Except for simply inherited traits (e.g. disease resistances) it is often quite difficult to evaluate the potential of unadapted material. This limitation accounts, in large part, for the past heavy reliance on adapted materials. Such concentration may lead to genetic vulnerability, hence the desirability of broadening the genetic base. Some of the alternatives, roughly in order of descending attractiveness, would include, (a) alternative adapted material from similar ecological zones, (b) unadapted exotic material, (c) mutation breeding, and (d) the currently widely advocated techniques of protoplast fusion. Each of these alternatives will result in reduced performance at the F_2 or population levels. This fact partially accounts for the limited use made of these procedures.

Mutation breeding deserves special mention. Here again there is a reduction in mean performance level after treatment but many of the genes responsible for this reduced performance are simply inherited and therefore eliminated easily. However, increased variability of this type plays little role in varietal improvement. The mutational approach to increasing variability has in general, found less favor in the U.S. than in many other areas of the world. This is true in spite of a few notable successes.

The popular press has often stressed the new technique of protoplast fusion as offering promise of removing barriers posed by sexual incompatability and thus, potentially, making all of the diversity within the plant kingdom available for the plant breeders use. These optimistic projections overlook the many difficulties still to be resolved. First, protoplast fusion has been accomplished with only a very limited array of species. The most successful case, to date, involves species of tabacum, but these combinations are available through the sexual route. When techniques are further developed to permit repeatable fusions of protoplasts of choice the next hurdle will involve procedures to foster cell divisions with subsequent development of a callus and regeneration of a functional sporophyte. Chromosome integrity must be maintained during these successive stages. When each of these successive steps can be performed regularly and on a large scale the problems of developing an efficient and competitive cultivar remain.

The options for choosing protoplasts for fusion are tremendous and each investigator, depending on his interests, may make different choices. One combination of theoretical interest might be soybeans and corn; soybeans to provide nitrogen fixing capabilities and a better amino acid balance in the grain and corn to supply increased yield potential. Even if a fertile sporophyte can be produced from fused protoplasts one can only speculate on floral morphology of the sporophyte and the net interaction of the two differing nitrogen metabolism systems. One can say with certainty that problems of developing a useful type will not be simple.

One wonders if advocates of the protoplast fusion route have any real appreciation of the difficulties in developing a successful cultivar by the normal sexual route. The case of Triticale may serve as a useful reminder. Wheat x rye hybrids are made easily and have been known for 100 yr and studied intensively for at least the last 20 yr by a large number of people with better than average financial support. In spite of this sizeable effort Triticale has yet to find a substantial place in world agriculture. If the ability to cross is a measure of relationship and compatibility of interacting genomes, then wider crosses impossible by the sexual route will likely pose even greater problems in the eventual isolation of a fully fertile, useful and acceptable cultivar. Plant breeders are confirmed optimists but few have evidenced any strong interest in protoplast fusion as a solution to the problem of expanding genetic diversity.

SELECTION TECHNIQUES

With adequate genetic variation, progress in developing improved cultivars is dependent upon the efficiency of selection procedures used and the scale of operations. I shall consider first the techniques used in line development and intra-population improvement. The additional procedures appropriate for estimating combining ability will be considered later.

In both the self and cross-pollinating species the obvious first step was an evaluation of existing cultivars. This process led to the identification of strains which became widely used, e.g. Richland soybeans, the several selections of Turkey winter wheat, Reid dent corn, etc. This source of variability (land race varieties) was soon depleted and emphasis was shifted to selection within and among advanced progenies from planned crosses. This pedigree system is common to both the self and cross-pollinating species although in corn it is more commonly called second-cycle selfing. In the self-pollinating species bulk populations can substitute for the pedigree system thus by-passing the opportunity for selection at intermediate levels of heterozygosity and concentrating the selection process among highly homozygous lines.

The pedigree system has been effective in isolating improved cultivars in all crops. It is the standard method of handling self-pollinating material and until recent years has been used widely for corn. Duvick (4) has published a resume of Pioneer's experience with this breeding system. In spite of the success achieved the method has obvious limitations.

Bailey and Comstock (1) have published results from a computer simulation study involving the effects of both linkage and selection on developing superior true-breeding cultivars. They conclude "when one parent line is substantially better than the other, there will be more coupling linkage and the average probability of fixation will be increased. Nevertheless the average genetic value of selected lines will be less than or, at best, little above the level of the superior parent line." Outcome probabilities are inproved somewhat by increasing population size and by progeny testing but the problem of identifying the best genotype available remains a limiting problem.

They suggest two alternatives as improvements over the standard system. One involves a more rapid cycling, i.e. intercrossing selected S_1's to develop a new population. A further refinement would be to use double-cross equivalents to increase the probability of retention of a larger sample of the desired alleles. The second alternative involves an increase in effective population size through random intercrossing among individuals representing a series of selected S_1 families. The increased probability of fixation of desirable alleles by this procedure was judged to be substantial.

This second alternative is essentially similar to some of the intra-population recurrent selection schemes. In species where either genetic or cytoplasmic sterilities are available an artificial random-mating system is established easily. Where such systems are not yet available, the production of crossed seeds by hand pollinations may be feasible. Several investigations have suggested how breeding systems may be established with a minimum number of cross-pollinations (2,3,5,7). These may not satisfy all of the requirements for quantitative genetics analyses but would serve adequately the need of the plant breeder.

Whether one uses the pedigree system or some type of recurrent selection one of the major limitations to progress is the selection intensity which may be applied. An extensive literature is available on the effectiveness of single trait selection. Population size and selection intensity can be adjusted to the facilities available. However, it is seldom that the breeder can confine his interest to a single trait; he must be interested in a whole assemblage of traits which make for commercial acceptability. To satisfy these requirements the breeder has three main alternatives: independent culling levels, tandem selection, or the use of some type of index. In practice the breeder may use all three, though the index used may represent a value judgment rather than a formalized model.

Theoretical considerations have indicated that an index is the more efficient of the three if appropriate weights can be established. The original proposal of a selection index by Smith (12) utilized economic weights. This system found little favor with plant breeders due partly to the problem of establishing economic weights. The procedure became much more attractive when Pesek and Baker (9) substituted the concept of desired gain for economic weights. A number of studies have been reported and others are in progress designed to evaluate the correspondence between expected and realized gain. Thus far this correspondence appears to be roughly comparable to that obtained in other short term recurrent selection studies.

Progress with a selection index is dependent upon selection intensity and the magnitude of the individual genetic variances and the magnitude and sign of the covariances among the traits of interest. When correlations with yield are low, selection intensity for a single phenotypic trait or group of traits, e.g. leaf diseases, insect resistance, etc., can be increased materially by growing the population (or F_2's) under conditions of controlled inoculation or infestation. If the base populations are large the numbers of individuals selected for recombination can be sufficiently great so as to minimize genetic drift effects associated with inbreeding. If phenotypic traits can be fixed in two or three cycles the efficiency of subsequent selection using an index involving yield should be increased.

Two techniques sometimes included under the term genetic engineering have been suggested as possible replacements for conventional plant breeding. These are the routine production of monoploids and cell culture. Naturally occurring monoploids have been reported for a number of genera and species. The advantage of selection among a gametic rather than a zygotic array has long been recognized. The problem has been one of producing monoploids in quantity and subsequent doubling to restore sexual fertility.

New techniques have partially resolved the numbers problem. The ease with which microspores can be induced to produce plantlets appears to have some genetic basis; some varieties in wheat and rice have thus far been completely intractable. This limitation must be resolved as the breeder must have complete freedom of choice in forming hybrids to effect the recombinations of interest. A further limitation affecting potential population size has been the high incidence of chromosome instability and of albino and other chlorophyll deficient types among the monoploid progeny. The rates of spontaneous doubling and the effectiveness of colchicine in inducing doubling varies widely among species and genotypes within species. When each of these several limitations have been resolved and adequate numbers can be produced at will, the problem of selection among individuals will be comparable to that of alternative breeding systems. The problems associated with evaluation and final selection of genotypes are several orders of magnitude greater than the problems involved in their production.

Cell culture has also been suggested as a potentially useful plant breeding adjunct. It was established several years ago that cells of the carrot root were totipotent; under proper conditions they could be induced to regenerate a sexually competent plant. This approach is being used extensively in ornamentals and extensive research is underway with a number of field crop species. The potential impact of this approach is still uncertain and depends on the resolution of some of the technical problems.

The ability to regenerate a plant from vegetative cells does not lead to plant breeding progress. Some provision must be made for the induction and selective propagation of desired genetic variants. This approach may eventually be useful in isolating cells and finally plants that are resistant to heat, salinity, amino acid composition and pathogens which produce phytotoxins. It is not clear at this time how selection schemes could be devised for maturity, dry matter distribution, grain yield and other important criteria required of a successful cultivar.

One should be aware that the molecular biologist and the plant breeder view the developing field of "genetic engineering" from different points of view. The molecular biologist is further encouraged and increasingly optimistic with each success in protoplast fusion or demonstration of successful selection in suspension cell culture. The plant breeder recognizes the many criteria which must be satisfied to obtain a new commercially accaptable cultivar. He may be intensely interested in future possibilities arising from genetic engineering approaches but feels there is little in the current state of "genetic engineering art" that is immediately useful to him. The molecular biologist is enthused over progress made and the plant breeder is awed by the problems remaining to be solved before genetic engineering can become a routinely useful tool.

EVALUATION TECHNIQUES

In several instances we have alluded to selection as the primary factor in modifying gene frequency. Effective selection is dependent upon reliable evaluation techniques. Plant breeders, as a group, have always been interested in new techniques and are quick to adopt these as their effectiveness is demonstrated. A few notable instances may be mentioned. The first involved an increasing emphasis on experimental design and statistical analysis which permitted a greater precision in the estimation of differences among means. The concepts and utilization of quantitative genetics followed attention to improved designs. Artificial inoculation techniques for many important diseases have now become routine procedures. Where adequate techniques for the evaluation of differences in insect resistance have been available significant progress has been achieved, e.g. first brood resistance to the European corn borer and spotted aphid in alfalfa. The amino acid analyzer, micro-milling and baking procedures in wheat, NMR for estimation of oil, near infra-red analysis for estimation of oil and protein are other examples of new procedures which are used widely where appropriate.

If the claims for new develops in genetic engineering (protoplast fusion, anther culture, plasmid transfer, improvements in photosynthetic efficiency) become useful to the plant breeder it will be through developments of new and simple evaluation techniques suited to plant breeding operations. It is only through plant breeding manipulations that any of these new developments can contribute to the development of improved cultivars.

INTER-POPULATIONS AND HYBRID EVALUATION

Where the possibility exists for the production of commercial hybrids, some form of intra-population improvement may be desirable as a prelude to inter-population evaluation programs. This approach appears reasonable as gene frequency changes affecting such things as disease resistance may be effected more quickly and easily than for the more complex trait yield.

In any event at least two base populations are desirable for most rapid progress. If these are chosen or developed to ensure a reasonable degree of inter-population heterosis, subsequent testing operations may be simplified. Current data suggest that there may be little difference in effectiveness among the several inter-population recurrent selection schemes. If this is true then the choice of system to be used becomes one of convenience. It should be emphasized however, that current data are possibly not very definitive and that a longer period of use may be necessary for the true differences among systems to become apparent.

NOTES

G. F. Sprague, University of Illinois, Urbana, Illinois 61801.

REFERENCES

1. Bailey, T. B., Jr. and R. E. Comstock. 1976. Linkage and synthesis of better genotypes in self-fertilizing species. Crop Sci. 16:363-370.
2. Brim, C. A. and C. W. Stuber. 1973. Application of genic male sterility to recurrent selection schemes in soybeans. Crop Sci. 13:528-530.

3. Compton, W. A. 1968. Recurrent selection in self-pollinated crops without extensive crossing. Crop Sci. 8:773-774.
4. Duvick, D. N. 1977. Genetic rates of gain in hybrid maize yields during the past 40 years. Maydica 22:187-196.
5. Fehr, W. R. and L. B. Ortez. 1975. Recurrent selection for yield in soybeans. Jour. Agr. Univ. of Puerto Rico 59:222-232.
6. Hanson, C. H., T. H. Busbice, R. R. Hill, Jr., O. J. Hunt, and A. J. Oakes. 1972. Directed mass selection for developing multiple pest resistance and conserving germplasm in alfalfa. J. Environ. Qual. 1:106-111.
7. Hanson, W. D., A. H. Probst, and B. E. Caldwell. 1967. Evaluation of a population of soybean genotypes with implications for improving self-pollinated crops. Crop Sci. 7:99-103.
8. Matzinger, D. F. and E. A. Wernsman. 1968. Four cycles of mass selection in a synthetic variety of an autogamous species of *Nicotiana tabacum* L. Crop Sci. 8:239-243.
9. Pesek, J. and R. J. Baker. 1969. Desired improvement in relation to selection indices. Can. J. Plant Sci. 49:803-804.
10. Shull, G. H. 1908. The composition of a field of maize. Am. Breed. Assoc. Rep. 4:296-301.
11. Shull, G. M. 1909. A pure line method of corn breeding. Am. Breed. Assoc. Rep. 5:51-59.
12. Smith, H. F. 1936. A discriminant function for plant selection. Ann. Eugen. Lond. 7:240-250.

MOBILIZATION, CONSERVATION AND UTILIZATION OF SOYBEAN GERMPLASM IN THE USSR

N. I. Korsakov

The plant kingdom of our planet is one of the essential biosphere elements and it is the source of our natural resources and the prosperous existence of mankind. The future of our agriculture and that of scientific-technical progress largely depends on the extent of utilization and conservation of specific, intraspecific and varietal diversity of plants existing on Earth. All this was brilliantly foreseen by N. I. Vavilov, who organized an extensive introduction activity in the Soviet Union. For this purpose in 1920, shortly after the Great October Socialist Revolution, a special Institute of Applied Botany and New Crops was established, which later was renamed to the All-Union Institute of Plant Industry. At present the Institute bears the name of its founder and first direct, N. I. Vavilov.

From 1920 to 1940 over 180 expeditions were organized by the Institute, including 40 expeditions to 64 foreign countries. The expeditions were planned according to Vavilov's theories on the origin of crops, their evolution, and their variability in space and time, which he had put forward in "The Centres of Origin of Cultivated Crops," "The Law of Homologous Series in Hereditary Variability" and in "Paths of Soviet Plant Breeding." The Vavilov Institute of Plant Industry (VIR) activities on mobilization of plant resources of the world have become more active in recent years. From 1971 to 1975 26 countries located in centers of origin of cultivated plants were explored, including many countries in Africa, Central and South America, Anterior and Central Asia, Scandinavia, Australia, Great Britain, and Belgium. In addition, many workers of the Institute made short business trips to DDR, Italy, Poland, Mexico, and the USA. As a result, new samples were added to the collection. Periodically, the Institute carries out the exchange of seed samples and planting material with more than 750 scientific institutions and seed breeding firms in most countries of the world. The exploration of different regions of the Soviet Union has become more active. All this made possible the replenishing of the Institute's collection with more than 63,650 samples of various plants during 1971 to 1975. Presently, the Institute's collection numbers 267,000 accessions, including 3,087 soybean samples.

The soybean collection involves all species of the genus *Glycine* L. from all loci of the crop formation. However, samples up to the maturity group IV predominate. Maturity groups V, VI, and later ones are represented merely by separate samples. This material is conserved carefully in special seed storages and is rejuvinated from time to time to maintain seed viability. Before being included into the collection, all introduced material is checked at quarantine nurseries.

On the basis of the study of the collection, scientists of the Institute work on a number of theoretical questions pertaining to the problems of origin, evolution, ecology, systematics and classification of the soybean, its intraspecific differentiation, regularities of geographical variability and reaction to different factors of environment. In addition, studies are conducted on genetic potential, methods of breeding, and distribution of soybeans in the USSR. Special attention is being paid to the study of samples of the collection in relation to the main breeding problems, such as resistance to diseases, pests and unfavorable environmentcal conditions, quality of seeds, response to high fertilizer application, and irrigation.

All the world diversity of soybeans arises from South-East Asia, Australia and East Africa. The most ancient locus of the soybean is in East Africa, the mean areal of the subgenus *Bracteata* Verd. However, the specific diversity of soybeans in this locus is not great. Only one species, *G. wightii* (R. Grah. ex Wight and Arn.) Verdc., has been identified. Nevertheless, this species is superior to many of its relatives from other formation loci by the wide range of forms and diversity of types represented in it. They are of interest as potential sources of a multi-florous effect (50 and more flowers) and resistance to diseases.

The Australian locus where species of the *Glycine* Willd. subgenus were formed is ancient and the richest in number of species. Six of the eight species of the genus *Glycine* L. which are distributed widely here are *G. clandestina* Wendl., *G. falcata* Benth., *G. latrobeana* (Meisn.) Benth., *G. canescens* F. J. Herman, *G. tabacina* (Labill.) Benth., and *G. tomentella* hayata. At present, representatives of this geographic soybean formation locus are also spread on the islands of the Pacific Ocean in the Southern hemisphere, in the Philippines and South China. Breeders of the world have not yet involved species from this locus in their breeding programs, although some of them manifest such interesting traits as resistance to diseases, root rots, drought resistance and polyspermous characteristics of beans (8 to 9 seed).

The South-East Asian locus, where a great diversity of the subgenus *Soja* (Moench) F. J. Herm. is concentrated, is the primary world source from which the progenitors of all cultivated soybean varieties originated. In the southern part of this locus, representatives of three Australian species occur, *G. tabacina, G. tomentella,* and *G. falcata.*

We share F. J. Hermann's and B. Verdcourt's views on the volume of the genus *Glycine* L., but we consider only one species within the range of the *Soja* subgenus—*G. soja* (L.) Sieb. et Zucc, rather than two. This is due to both species *G. max* and *G. ussuriensis* having a common primary area of origin. Wild, semicultivated and cultivated forms are crossed easily and produce fertile progeny, constantly inheriting characters of the parents. They have identical chromosome numbers (2n = 40) and morphology, and the same spectrum of protein immunoelectrophoregrams. All the diversity of soybeans in the subgenus *Soja* is subdivided at the VIR according to phylogenetic principles into five subspecies, *korajensis* Enk., *indochinensis* (Enk.) Kors., *manshurica* Enk., *gracilis* (Scv.) Enk. (semicultivated forms), and *soja* (a wild form).

The plants of spp. *koraensis* Enk. are characterized by the expression of many economically valuable traits. They most fully display signs of the impact of century-old breeding and cultivation of plants under optimum ecological conditions (fertile soil, sufficient moisture, heat and light supply). Plants of this subspecies are distinguished by thick stems, large seeds, and often many flowers. The negative points include pod shattering after ripening, increased hydrophilic capacity, and seed coat splitting. On the basis of utilizing samples of this subspecies, a number of varieties have been developed in the USA, USSR, and other countries, and with the use of extremely early ripening samples, some Swedish varieties of the Fiskeby type.

The plants of ssp. *manshurica* Enk. are characterized by an average expression of most morphological traits. Old land races of this subspecies proved to be the main initial material for soybean breeding in many countries of the world. It is from this center of conjugated evolution of soybean as a host plant and agents of a number of its diseases that valuable donors with genes for resistance to the following diseases were discovered: Bacterial pustule—*Xanthomonas phaseoli* var. *sojensis* (Hedges) Starr et Burk., gene rxp, donor k-6347; Bacterial blight—*Pseudomonas glycine* Coerper, gene Rpg, donors of Manchurian origin, samples k-4365, k-5495; Frogeye leaf spot—*Cercospora sojina* Hara, gene Rcs_1 to race I, donor k-4360; gene Rcs_2 to race 2, donor k-5689; Downy mildew—*Peronospora manshurica* (Naum.) Sydn., gene Rpm, donor k-5764 resistant to all 14 races; and Phytophthora root rot—*Phytophthora sojae* Kauf. et Gerdem., gene Rps, donors k-3977, k-4365.

The above mentioned donors of resistance, as well as other Manchurian varieties are widely used by plant breeders of the USSR, Canada and USA in particular. Thus 98% of the soybean planting area in the USSR is covered by varieties with the parentage from spp. *manshurica*—Amurskaya 41, Amurskaya 42, Primorskaya 529, and Khabarovskaya 4. The genetic base of varieties covering over 80% of the USA soybean acreage is represented by six varieties of Manchurian origin: Dunfield, Manchu, Mandarin, Mucden and Richland (Hartwig, 1973).

The plants of spp. *indochinensis* (Enk.) Kors. are medium- to very late-ripening, medium sized to very tall statured, very leafy and branchy, and the stems and branches are thin and inclined to entanglement and twisting. Among the cultivated varieties, they are distinguished by the majority of dominant expressions of their characteristics. From the breeding standpoint, they might be of value in producing high protein and oil content, as well as sources of resistance to virus diseases.

The plants of spp. *gracilis* (Scv.) Enk. are thin-stemmed, extremely bushy, leafy, slightly pubescent, and have lower temperature requirements at the germination stage. Forms with increased protein content in the seeds (47 to 50%) are found. All of this makes them very valuable components for hybridization in order to create varieties suitable for northern soybean growing regions and varieties for fodder purposes, as well as a source of high protein content. However, their utilization as breeding material is limited due to a high degree of pod shattering, plant lodging and other negative dominant characteristics.

The plants of spp. *soja* belong to wild forms. One species is distinguished by extreme thin-stemness, lodging and pod shattering, and exceptional small seededness (the mass of 1000 seeds is as small as 30 g). There are other undesirable characteristics for cultivated varieties, which as a rule are conditioned by the dominant allels. At the same time, lower temperature requirements at the germination stage, resistance to root rots, large numbers of pods on one plant, and relative resistance to bacterial diseases are of interest to breeders.

As a result of studying the geographical variability of the soybean-plant characteristics and the influence of environmental factors on variability, we have subdivided them into six groups. The first group of characteristics is concerned primarily with the color of flowers, pods, seeds, and the degree of pubescence. The influence of meteorological factors and agrotechnical methods of cultivation on the variability of these characteristics is not higher than 10%. The size of the seeds and pods, and the width of the leaflet comprise the second group of characteristics. Their variability of 11 to 22%

depends on the growing conditions. Characteristics of this particular group correlate rather closely (r = 0.7 to 0.8) with many other plant characteristics. The third group of characteristics such as stem thickness, leaflet length and number of days to maturity changes by 23 to 30% under the influence of environmental conditions.

The number of seeds per pod, bush shape and angle of branch declination from the stem make the fourth group of characteristics. Such characteristics as the length of internodes, branches per plant, height of plant and pods per plant make up the fifth group. The influence of growing conditions on the variability of these characteristics is 47 to 55% and 59 to 69% correspondingly.

The length of the flower cluster, pod shattering, the level of attachment of lower pods, leaves per plant and seed mass from one plant form the sixth group of characteristics. The growing conditions influence their manifestation by 70 to 80%. The coefficient of their variation ranges from 18 to 28%, while that in the first group does not surpass 0.7%; in the second group, 4 to 8%; in the third group, 9%; and in the fourth and fifth this coefficient ranges from 18 to 29%.

The interrelation of characteristics indicated that for the level of correlation coefficients equal to 0.5 all the plant characteristics represent one pleiad. At the r = 0.7 level the size of leaflet, size of seeds, thickness of the stem, length and width of the pod are closely interrelated. In this case the pairing coefficient of the characteristic "seed size" and in some years "thickness of stem" is highest. Therefore they are treated in pleiad as units reflecting peculiarities of other characteristics. Each of them can be used as a characteristic indicator.

The plants of the species of the genus *Glycine* L., possessing high polymorphism, display clear homologous series in hereditary variability of morphological, physiological, and biochemical characteristics. Within each of the species, there have been found typical short day forms, as well as forms with a relatively poor reaction to photoperiod changes, small- and large-seeded forms with dark and light seed scars, types with shortened and normal stems, thick and thin stems, and plants with narrow and comparatively broad leaflets. Homology is also observed in the variability of the components composition of the protein complex that has been revealed in homologous and heterologous reactions of immune serum in different species of soybean. The number of components for each species is rather specific.

From the evolutionary point of view the most ancient soybean species (African) possess a rather small number (6) of protein components, while the evolutionary younger species (South-East Asian) have a greater number of protein components (22). Australian species are intermediate (11 to 14). This proves that the evolution of the genus *Glycine* L. took place in the direction of an increase in protein components and a complexity of protein.

The variability that cannot be kept within the usual conception of the mutagenesis nature and within the regularities of mutation manifestation can be explained partially by the theory of treptions, proposed by J. A. Serra. Taking into consideration the additive importance of treptions (heterochromatization of chromosomes, ontogenetic polyploidization, the chromosome set reduction, genotrophes, and depression of gene functions) it is possible to speculate that they are of some importance in plant evolution. But the combination of the phenomena of a different nature under one term cannot reveal the true nature of these phenomena. The study of the homologous series in the hereditary variability of characteristics, taking into account the evolutionary age (if it is possible to say so) of different species groups and within species and subspecies, revealed that, as a rule, a definite evolutary age corresponds to a definite manifestation of a characteristic. For example, small seeds, relatively small leaves, thin stems, dark color of seed coat and scar, pod shattering, and marked day length reaction are typical for evolutionary old forms of the characteristic. The manifestations of the characteristic, appearing at later stages of evolution, especially during the crop evolution is marked only within cultivated forms. Thus, it is possible to speak not only on the phylogeny of species, but also on the phylogeny of characteristics and on the evolution of hereditary factors. During the process of evolution, as a result of mutations, gene combinations, and the formation of new genetic material, a continuous formation of new alleles takes place that stipulates finally the new manifestations of the characteristics. The complex of inheritance factors is a dynamic system. Evolutionary old forms always dominate over evolutionary young ones.

On the basis of the soybean collection and according to theoretical methods worked out at VIR, new varieties are being developed by breeding institutions of our country. For this purpose VIR distributes annually, catalogs, methods, and over a thousand samples from the seed collection. All soybean varieties cultivated in the USSR were developed with the assistance of the VIR collection. Among them are such varieties as Amurskaya 262, Saljut 216, Amurskaya 310, Jubileinaya and others that converted the region of Priamurje, previously considered to be of little use for soybean cultivation, into the primary region of soybean production in our country. The varieties Komsomolka, Peremoga, Lanka, and Terezinskaya 2 have a potential yield capacity of 30 to 35 centers/ha without irrigation and over 40 centers/ha with irrigation.

This short presentative gives a description of the investigations carried out in the USSR on mobilization, study, and utilization of soybean germplasm. However, the Institute is in need of new material and therefore new expeditions are being planned. Contacts with scientific institutions and scientists of the whole world are being established to increase the exchange of seed samples and data on collection material.

NOTES

N. I. Korsakov, N. I. Vavilov All-Union Scientific Research Institute of Plant Industry, Leningrad, USSR.

SYSTEMS APPROACH TO PEST MANAGEMENT IN SOYBEANS

J. L. Stimac and C. S. Barfield

The systems approach to problem solving has a strong basis in general systems theory and provides a sound framework for developing methods to solve pest problems (2,4,6,7). The terms pest control and pest management are often used as if they were synonymous but control is actually a subset of management. Management strategies may involve the use of many control tactics. The systems approach is useful in evaluating the compatibility of the tactics. However, before discussing the systems approach to pest management in soybeans, we would like to point out some differences between pest control and pest management.

In pest control, actions are taken to suppress pests when population levels are unacceptably high (suppression tactics) or are anticipated to become so (prevention tactics). Rabb (3) has discussed suppression and prevention tactics as remedial and preventive procedures used to reduce pest population levels. Suppression tactics are referred to as remedial because the objective is to kill some proportion of individuals in the pest population. Prevention could be viewed as a suppression action taken when pest populations are below damage levels but anticipated to become unacceptably high. In this case suppression and prevention tactics are distinguishable only in terms of the time at which the pest control action is taken. However, prevention tactics may be aimed at making the habitat less favorable for the pest populations. A characteristic of pest control involving suppression tactics is that control is short term so suppression action must be taken frequently.

Pest management is much broader in scope than pest control. In pest management, actions are taken to dominate or direct the system toward achievement of a particular state of behavior by incorporation or preservation of homeostatic regulatory mechanisms. Ruesink (4) described this action as "design," involving restructuring of the system by adding new objects or modifying existing ones. The objects include biotic components that use feedback to adjust intensity of action as the system changes states. Watt (8) made an analogy between feedback loops in agricultural systems and a person adjusting hot and cold water valves to achieve desired temperature of bathwater. In pest management, the manager becomes the designer and operator of the control valves (tactics).

Pest management is characterized by the use of multiple tactics in an integrated and compatible manner. Some tactics suppress pest levels (chemical, cultural, and some biological control procedures), while other tactics (biological) maintain pests at low levels by naturally adjusting their intensity when pest populations begin to resurge. As a result, pest management can be long lasting in the absence of severe natural or man-induced perturbations of the crop or agroecosystem.

A POPULATION DYNAMICS VIEW OF PEST CONTROL AND MANAGEMENT

The results of pest control and pest management are reflected by dynamic changes in populations of the pest species. Suppressive and preventive pest control are hypothetically contrasted with pest management in Figure 1. In suppressive pest control, numbers of pest organisms exceed an action threshold, action is taken to "knock down" pest levels but the population resurges and must be suppressed again before harvest date. Note that there can be time lag between recognition of the threshold being exceeded and taking of suppresive action. During the time lag substantial crop damage may occur. Also note that successive suppressive actions may disrupt natural suppressive factors such that resurgence of the pest population intensifies through time. This phenomenon is reflected by the increasing amplitude of consecutive pest population peaks.

Preventive pest control is represented by an initial decrease in pest density, followed by a gradual increase in pest numbers as the season progresses. Such a pattern may be indicative of what would occur in an area where cultivation is used to reduce an overwintering pest population and the host crop is a resistant variety which prevents rapid increase in the pest level. If a significant amount of immigration into such an area were observed, the increase shown in Figure 1 would probably be more pronounced such that the pest level might exceed the action threshold prior to crop harvest date.

Under pest management, the trajectory of pest density is shown to remain during the entire season in an equilibrium region below the action threshold (Figure 1). Pest level fluctuates mildly due to variations in weather conditions but the pest population is maintained below the action level.

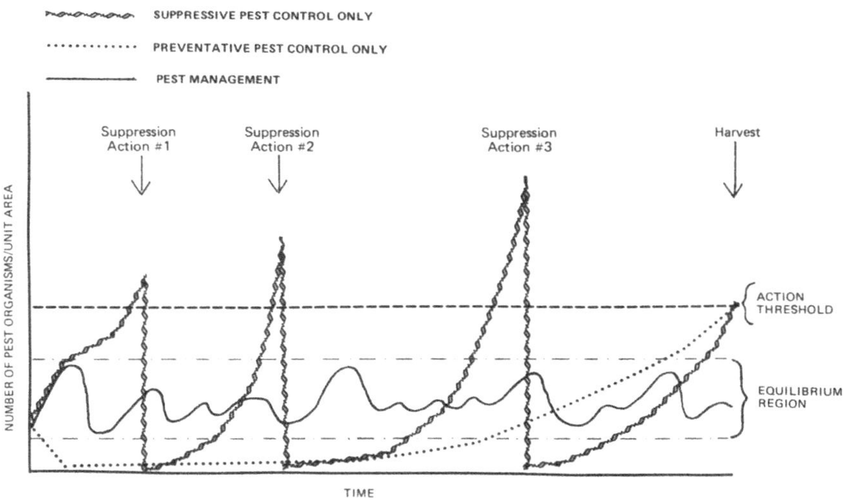

Figure 1. Hypothetical changes in pest population levels under programs of only preventive pest control, only suppressive pest control and pest management, which may include both preventive and suppresive control tactics.

Suppressive control tactics may be used in a remedial manner to supplement regulatory mechanisms if these mechanisms are "swamped" by large influxes of pests into the field. The critical assumption here is that the system possesses homeostatic regulatory mechanisms capable of intensifying action when conditions are suitable for increase in the pest population density. The obvious question arising is, how does one identify and incorporate such mechanisms into a crop system? The answer is not so obvious and is the primary reason why scientists are looking to the systems approach for development of pest management.

The primary distinction between pest control and pest management is that control is directed at suppression of pests, while management is also concerned with maintaining pests at specified levels. In pest control programs we often treat the symptoms of the problems rather than attempting to identify the causes. In pest management the causes of pest problems must be understood before appropriate actions can be taken to direct the system to a desired state and maintain this state through time. Consequently, development of a pest management program in any agricultural production system requires a vast amount of knowledge of the system inputs, components, internal processes, and couplings to other systems. The systems approach provides a means of structuring the knowledge base and specifying gaps in understanding of important causal and regulatory mechanisms.

SYSTEMS APPROACH TO MANAGEMENT OF PESTS IN THE SOYBEAN SYSTEM

Management of pests in the soybean system requires that pests be considered as part of the system or as inputs into the soybean system. The systems approach to pest management in soybeans must begin by defining the boundaries and identifying pest and other components of the soybean system, specifying inputs into this system and specifying how the soybean system is coupled to other crop and non-crop systems. Couplings are outputs from one system that become inputs into another system (2). Soybean pests can be considered as components of the soybean system because some pests of soybeans reside in fields where soybeans are grown. Examples of these pests are insects, nematodes, and pathogens that overwinter in soybean fields. However, some pests of soybeans reside only temporarily in soybeans and are appropriately defined as inputs into the soybean system.

Consider insect pest species that might overwinter or breed continuously in warmer regions of the southeastern United States and migrate to the northern temperate regions when warm, favorable conditions are present there (1). These species specify a coupling between soybean systems in the temperate regions and crop and non-crop systems in subtropical regions. That is, outputs of the subtropical regions provide inputs of pest inocula into the soybean system. Also, pests common to soybeans and other crop and non-crop hosts in the region may provide pest inocula to soybeans indirectly. Pests can migrate to alternate hosts such as corn, peanuts, tobacco and non-crop plants (weeds) and then disperse into soybeans at various times during the soybean season (1). Rabb (3) pointed out that the ebb and flow of a population in a single field requires an understanding of the dynamics of that population defined large enough geographically to account for movement into and out of the field. The point is that understanding pest population dynamics in a soybean field must come about by viewing the soybean field as a subunit of the agroecosystem.

A conceptual view of soybean fields as subsystems within the agroecosystem is represented in Figure 2. This conceptual model can be used to hypothesize how pest problems in soybeans are influenced by other components in the agroecosystem. Consider flows of pest inocula into a particular soybean field. Migratory pests enter the target ecosystem from an unspecified source. Think of this source as an input into a regional pool of pest inocula. The regional pool provides inputs of pest inocula into different areas, each of which can be viewed as an agroecosystem, containing various crop and non-crop habitats for the pest species. Pest inocula in a given area are distributed into crop and non-crop habitats, each habitat consisting of numerous fields or comparable spatial units for non-crop habitats. Inputs of pest inocula can be from the area pool, local dispersal of pests between fields

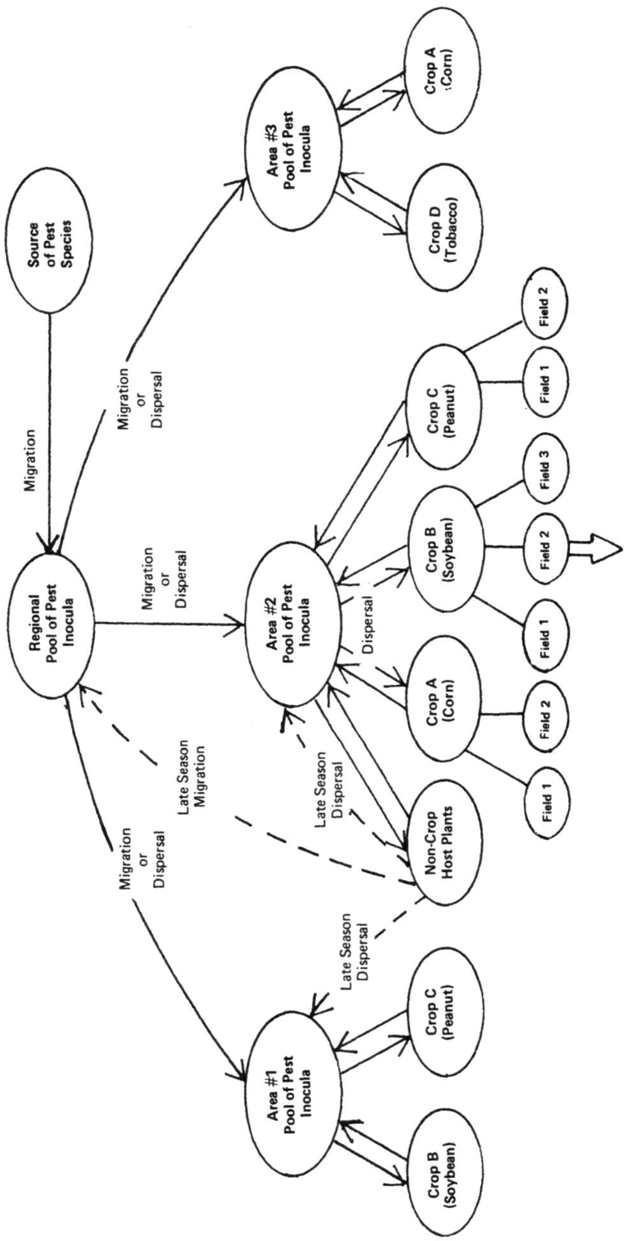

Figure 2. Spatial hierarchy model of inputs of pest inocula into fields.

and resident pest populations. These inputs specify the potential for pest problems at any given soybean field site. Therefore, the systems approach to pest management in soybeans implies that management strategies consist of tactics that are implemented in both soybean fields as well as other fields and habitats providing inputs of pest inocula into soybean fields. To simply exercise pest control tactics only in soybean fields would be treating the symptoms of pest problems not management of soybean pests over wide production areas. This is an interesting statement because the vast majority of control tactics used against insect pests of soybeans are directed at field level problems.

SYSTEMS APPROACH AT THE FIELD LEVEL

Although a wide area view of pests in soybeans is necessary, developing regional management programs can be hindered by political and socioeconomic constraints. Such constraints will slow development of interstate regional pest management programs. In the foreseeable future, most management programs probably will be directed at the area or field level. Therefore, it is important to consider how the systems approach to pest management might be useful at the field level. A conceptual model of pests in a hypothetical soybean field is represented in Figure 3. The conceptual model is a representation of a hierarchy of pests which could affect soybean roots, leaves, stems, and pods. Each pest population is defined as a module that potentially can alter soybean production by imposing stresses on one or more classes of plant parts. Natural enemies of the pests are represented similarly as modules that potentially can alter growth of pest populations. The hierarchy of pest and natural enemy population modules is intentionally incomplete and is used here to demonstrate the systems approach to identifying the magnitude of potential pest problems in a soybean field. If additional pests are identified they can be added to the species hierarchy by specifying which class or classes of plant parts they affect.

The soybean crop in a particular field is viewed as a collection of plant parts, each of which can be the object of numerous pest complexes. Roots of soybeans potentially are subject to weed, nematode, pathogen, and insect pests. Likewise, leaves, stems, and pods can be subject to stresses imposed by these four types of pests. Each type of pest problem can be subdivided further into multiple target species, each of which may be influenced by complexes of natural enemies. The systems approach provides a method for viewing pests of soybeans as a hierarchy of interacting target populations. Understanding the interactions among target and non-target species is the key to evaluating combinations of control tactics that could be used in various pest management programs. It is important to note that control tactics exercised against one type of pest may influence directly or indirectly (or

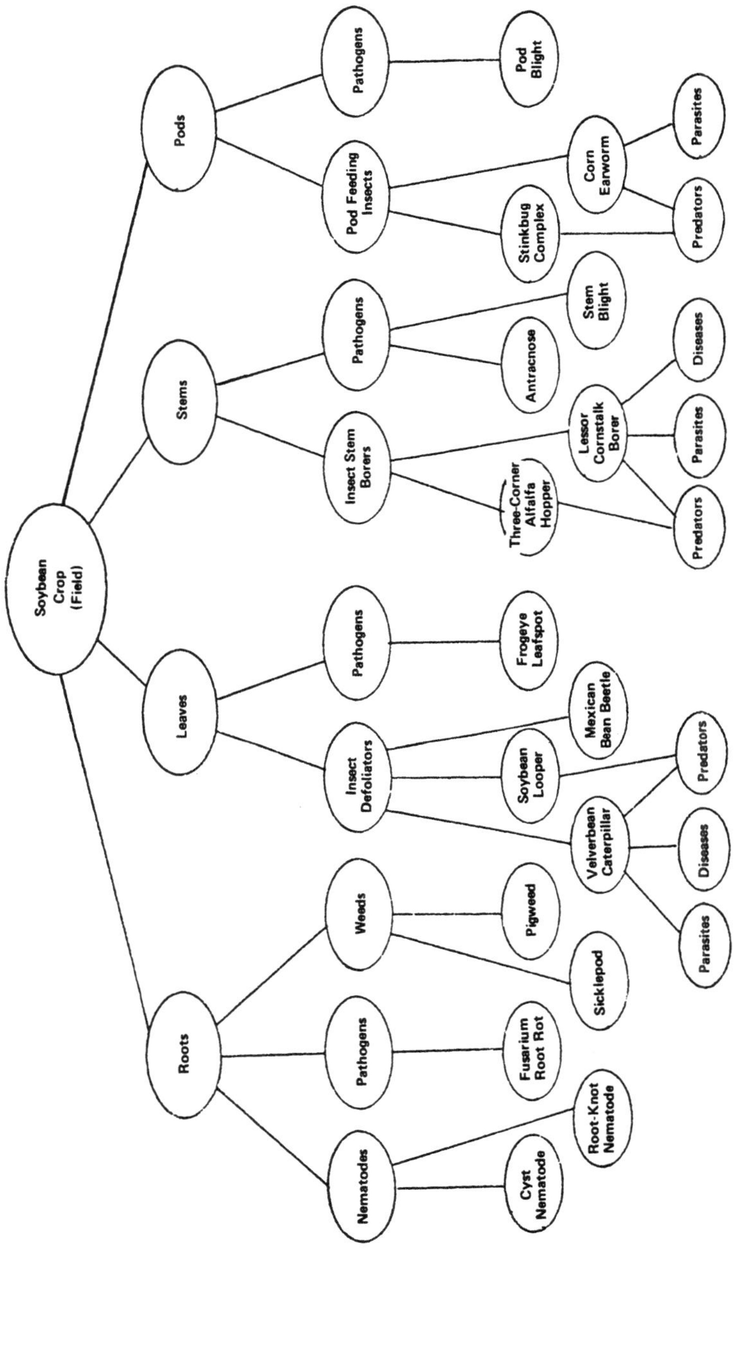

Figure 3. A conceptual model of a species hierarchy of pest populations influencing growth of the soybean crop in a field.

interact with) other classes of pest organisms. This suggests that the systems approach to pest management in soybeans requires that insect pest management be evaluated in the context of other types of soybean pests and the control tactics used against them. The ability to develop comprehensive pest management programs is therefore constrained by one's ability to evaluate interactions among various types of pest control tactics that could be implemented at the area or field level.

EVALUATING CONTROL TACTICS USED IN SOYBEAN PEST MANAGEMENT

Since other chapters in the Proceedings will discuss specific examples of how control tactics interact, we will confine our comments to a general discussion of a method for evaluating combinations of control tacts and some problems associated with using this method.

Assume we are interested in developing a management program for the hypothetical soybean field discussed earlier (Fig. 3). To evaluate specific pest management strategies, the conceptual model must be represented in a more explicit form, a mathematical model of the soybean crop system which includes submodels of each of the target pests. The mathematical system model can then be translated into a computer simulation model, which will serve as a tool for evaluating reponses of pest populations to different combinations of control tactics. The difficulty of developing the field level computer simulation model of the soybean crop-pest system should be emphasized. To the best of our knowledge, there are currently no simulation models for any crop system which contain submodels of all important target pest species. Ruesink (4) provided a review of mathematical and simulation models in pest management but even the most extensively developed models were confined to the crop and several target pests, mostly insect species. However, the fact that no complete crop-pest system models have been developed yet does not preclude discussing how such models can be developed and used in pest management.

Each of the pest modules shown in Figure 3 could represent a field population model of a particular pest species. The soybean crop module could represent a model of the soybean crop with submodels for production of each of the plant parts (roots, leaves, stems, and pods). The submodels of plant parts provide information on the status of those parts to each of the appropriate pest models. In turn the pest models provide the soybean crop model with information on pest activity (damage) in the crop. The system model could be used to evaluate tactics if models of pest populations could respond to the direct and indirect effects of the tactics. Building models to accomplish this task requires much knowledge of how control tactics influence growth and behavior of pests and natural enemies, not simply knowing the levels of mortality inflicted on each target pest species.

As an example, consider a model of one of the insect populations that defoliates soybean leaves (Fig. 3). Each insect module is a model of one of the target pest populations. The structure of one of the insect population models is represented conceptually in Figure 4. The population is divided into age (or stage) classes represented by boxes. Arrows connecting boxes provide a pathway for moving from one stage to another. Mathematical representations of population processes (emergence, mating, oviposition, hatching, feeding, moulting, and mortality) are used to alter the numbers of individuals in each of the stages. Feeding (leaf damage) is a function of consumption rates and numbers of individuals in each larval stage.

Control tactics used against the defoliator are entered into the model by specifying how each tactic affects direct changes in the numbers of individuals in each stage and model parameters representing rates of insect oviposition, development, feeding, and mortality (7). For example, if we wanted to evaluate compatibility of an insecticide application with parasites and predators of the defoliator, the following information must be provided: (a) What proportion of individuals in each life stage of the target pest are killed directly? Are there any residue effects? (b) Will the insecticide application significantly change any important behavioral attributes of survivors? (Example: alter rates or pattern of oviposition and feeding of the pest.) (c) Will the insecticide indirectly alter rates of stage-specific mortalities by reducing numbers of parasites and predators? If so, specify the proportion of parasites and predators that would be killed as a result of the insecticide application. (d) Are there any phytotoxic effects on the soybean plant? If so, explicitly specify how soybean growth will be altered. (e) Will the insecticide kill other target and non-target organisms? If so, explicitly specify the proportions of individuals of each species that will be killed. This list of questions is most certainly incomplete but gives an idea of the level of knowledge needed to evaluate control tactics using computer simulation models.

There are three major problems in using crop-pest system models to evaluate combinations of control tactics and develop management strategies for pests of soybeans at the field level: (a) a tremendous amount of information (knowledge) is needed to develop the soybean crop and pest population models; (b) if such models were available, specific information on the effects of control tactics is not available nor is there any significant amount of research effort being devoted to this area; and (c) the inputs of between field and area dispersal of insect pests and beneficial organisms are not known and methods for measuring such inputs quantitatively are poorly developed. The first two problems apply to field level models and the third problem relates to the need to develop wide are pest population models.

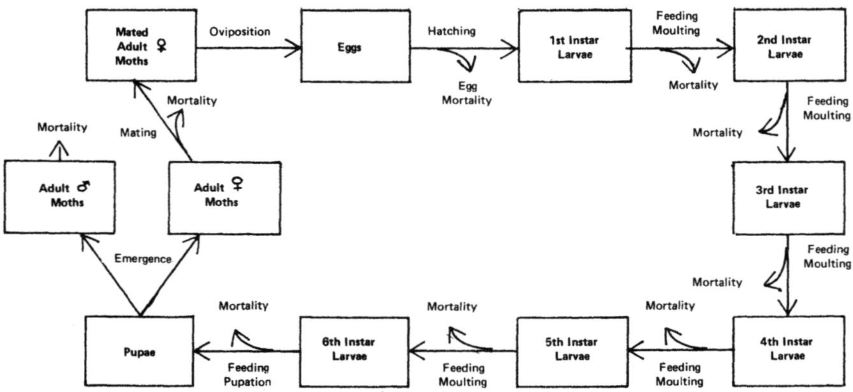

Figure 4. A conceptual population model of an insect defoliator in a soybean crop.

CONCLUSIONS

Development of management strategies and programs for pests of soybeans is a large step from classical pest control applied at the field level. Consequently, pest management requires a tremendously increased level of understanding. The systems approach provides a framework for taking a more holistic view of soybean pest problems in the context of the agroecosystem. This approach can be used to structure our understanding of ecological processes that regulate pest populations and to identify deficiencies in our understanding of how different pest control tactics influence the soybean crop-pest system.

At present the systems approach to pest management in soybeans is primarily at the conceptual stage of development. As more knowledge is accumulated on pest and beneficial species in soybeans, the systems approach will move from the conceptual model stage to the mathematical and computer simulation model stages. At this point of development, the systems approach to pest management in soybeans will begin to show greater value because the simulation models will serve as experimental tools for the pest manager. In this chapter we have used the systems approach to structure current knowledge about pest problems in soybeans and in the process we have identified many deficiencies in the knowledge base needed to comprehensively manage pests of soybeans. We should not be discouraged by our level of understanding of soybean crop-pest systems but rather appreciate the complexity of the tasks before us. Transforming field-oriented pest control practices into wide area pest management programs will be a large leap forward. Hopefully, the systems approach will allow us to take one step at a time.

NOTES

J. L. Stimac and C. S. Barfield, Department of Entomology and Nematology, University of Florida, Gainesville, Florida 32611.

REFERENCES

1. Barfield, C. S. and J. W. Jones. 1979. Research needs for modeling pest management systems involving defoliators in agronomic crop systems. Proc. Symposium on Fall Armyworm. Fla. Entomol. 62:98-114.
2. Klir, G. V. 1969. An approach to general systems theory. Van Nostrand-Reinhold, New York, N.Y. 323 p.
3. Rabb, R. L. 1978. A sharp focus on insect populations and pest management from a wide-area view. Bull. Entomol. Soc. Amer. 24:55-61.
4. Ruesink, W. G. 1976. Status of the systems approach to pest management. Ann. Rev. Entomol. 21:27-44.
5. Stimac, J. L. 1977. A model study of a plant herbivore system. Ph.D. Dissertation, Oregon State University, Corvallis, Oregon. 240 p.
6. Watt, K. E. F. 1961. Mathematical models for use in insect pest control. Can. Ent. 93:1-62.
7. Watt, K. E. F. (Ed.) Systems analysis in ecology. Academic Press, New York, N.Y. 276 p.
8. Watt, K. E. F. 1970. The systems point of view in pest management. In R. L. Rabb and F. E. Guthrie (eds.) Concepts of Pest Management, p.71-83.

INTERACTIONS OF CONTROL TACTICS IN SOYBEANS

L. D. Newsom

The use of pesticidal chemicals plays an important role in soybean production in the U.S. The crop is attacked by a wide variety of pest species for which the only satisfactory method of control currently available is the intelligent use of conventional chemical pesticides. In Louisiana, for example, where pest problems may be considered representative of areas bordering the Gulf of Mexico, soybeans are affected by a large complex of pest species. There are 25 species that are considered to be of major importance (Table 1). Many more are important locally or are sporadic in occurrence.

There are at least as many chemical compounds used for control of the pests as there are species of pests. However, about ten are most commonly used in Louisiana (Table 2). The total amount of pesticides applied per season to soybeans treated at maximum levels for control of the pest complex ranges from about 7.90 to 12.50 kg/ha. Herbicides comprise a large percentage of the total amount. About 90% of the total acreage receives herbicide application each season. An average of probably less than one-fourth receives an application of insecticide. Even less is treated with fungicide. Areas treated with insecticides vary from year to year depending upon variations in pest populations, but there is little variation in areas treated with herbicides and fungicides.

All of these chemicals—herbicides, insecticides, fungicides, and nematicides—are biologically active compounds. Many are highly toxic and possess a broad spectrum of activity. It appears reasonable to believe that such chemicals may have direct effects on physiological and biochemical processes in soybeans. Indeed, it is a well known fact that the margin of safety for many pesticides is relatively narrow between effectiveness for control of pest species and phytotoxic effects on plants.

Table 1. Some major pests of soybeans in Louisiana.

INSECTS

Bean leaf beetle, *Cerotoma trifurcata* (Forster)
Corn earworm, *Heliothis zea* (Boddie)
Soybean looper, *Pseudoplusia includens* (Walker)
Velvetbean caterpillar, *Anticarsia gemmatalis* Hubner
Southern green stink bug, *Nezara viridula* (Linnaeus)
Green stink bug, *Acrosternum hilare* (Say)
Brown stink bug, *Euschistus servus* (Say)

NEMATODES

Soybean cyst nematode, *Herodera glycines* Ichinohe
Rootknot nematode, *Meloidogyne* spp.

FUNGI

Aerial blight, *Rhizoctonia solani* Kuehn
Frogeye spot, *Cercospora sojina* Hara
Purple seed stain, *Cercospora kikuchii* (T. Matsu and Tomoyasu) Chupp
Pod and stem blight, *Diaporthe phaseolorum* (Cke. and Ell.) var. *sojae* Wehm.
Phytophthora rot, *Phytophthora megasperma* (Drechs.) var. *sojae* A. A. Hildebrand
Brown spot, *Septoria glycines* Hemma

VIRUSES

Bean pod mottle virus (BPMV)
Soybean mosaic virus (SMV)

WEEDS

Cocklebur, *Xanthium strumarium* Linnaeus
Johnson grass, *Sorghum halepense* (Linnaeus) Persoon
Morning glory, *Ipomoea* spp.
Hemp sesbania, *Sesbania exaltata* (Rafinesque) Rydberg
Barnyard grass, *Echinochloa* spp.
Broadleaf signal grass, *Brachiaria platyphylla* (Grisebach) Nash
Wild poinsettia, *Poinsettia heterophylla* (Linnaeus) Small
Prickly sida, *Sida spinosa* Linnaeus

When wide varieties of pesticides are applied for control of an equally varied pest complex, bewildering numbers of interactions are made possible. The following will be considered in this chapter: (a) effects of pesticides and mixtures of pesticides on growth and development of soybeans; and (b) effects of different components of pest management systems on other components of the systems.

Table 2. **Some examples of commonly recommended pesticide usage patterns for control of soybean pest complexes in Louisiana[a].**

Common Name	Amount Active Ingredient, kg/ha	Time of Application	Pests Controlled
WEEDS		WEEDS	
Trifluralin	0.56 to 1.68	Preplant or immediately after planting	Annual grasses, johnson grass from seed, some broadleaf weeds
Alachlor	2.24 to 3.36	At planting	Annual grasses, johnson grass from seed, some broadleaf weeds
Metribuzin	0.43 to 0.84	At planting	Annual grasses, most broadleaf weeds
Dinoseb	1.68	When seedlings begin to emerge	Small annual grasses and broadleaf weeds
Bentazon	0.84 to 1.68	When first trifoliate leaves have developed	Cocklebur and other broadleaf weeds
		INSECTS	
Methyl parathion	0.28 to 0.56	When populations reach economic injury thresholds	Bean leaf beetle, stink bugs, and velvetbean caterpillar
Carbaryl	0.56 to 1.12	When populations reach economic injury thresholds	Corn earworm and bean leaf beetle
Methomyl	0.50	When populations reach economic injury thresholds	Soybean looper and corn earworm
Bacillus thuriengiensis	0.28 to 0.56	When populations reach economic injury thresholds	Velvetbean caterpillar and soybean looper
		DISEASES	
Benomyl	0.56	Two applications at 0.28 kg/ha, the first at full bloom and the second two weeks later	Pod and stem blight, anthracnose, and frog-eye spot

[a]Source: Louisiana Cooperative Extension Service Pest Control Guides, 1978.

EFFECTS OF PESTICIDES AND MIXTURES OF PESTICIDES ON GROWTH AND DEVELOPMENT OF SOYBEANS

T. T. Lee (15) has called attention to the fact that direct effects of pesticides on the biochemistry and physiology of crop plants have been studied very little. The effects of mixtures of chemicals such as those occurring in soybean fields have been studied even less.

The evaluation of agricultural chemicals for their effectiveness as pesticides traditionally involves testing single compounds in a variety of systems for efficacy against pest species and safety for use on plants. Toxicity to plants is usually determined by occurrence of gross symptoms such as stunting, necrosis, and discoloration of foliage. Relatively little attention is paid to more subtle effects that may stress the plant in a variety of ways, making it more susceptible to injury by pests, for example. Possible adverse effects of mixtures that do not occur when components of the mixtures are tested singly receive little, if any, consideration.

This continues to be the case despite the obvious fact that pesticides rarely occur as single compounds in agricultural ecosystems. Both chemicals and their metabolites occur as complex mixtures in most cases, and they interact (6,7,16,17). Chang et al. (6) pointed out that metabolism of insecticides may be inhibited in the presence of herbicides. Chio and Sanborn (7) suggested that interactions of herbicides and insecticides in plants are most often synergistic because the insecticides inhibit herbicide degradation. Lichtenstein et al. (16) in an important paper reported that the herbicide atrazine, among several others, enhanced the activity of several insecticides, including carbofuran, to *Drosophila melanogaster* Meigen, *Musca domestica* Linnaeus, and larvae of *Aedes aegypti* (Linnaeus).

A striking example of a herbicide-insecticide interaction is that of propanil and organophosphorus or carbamate insecticides on rice. The discovery that propanil was sufficiently selective to be applied to rice for control of barnyard grass, *Echinochloa* spp., without adverse effects on rice was responsible for substantial increases in yields. Matsunaka (17) reported that tolerance of rice to propanil was based on the presence of a propanil hydrolyzing enzyme. Tolerance was reduced drastically in the presence of organophosphorus or carbamate insecticides. The insecticides enhance herbicidal activity of propanil and drastically reduce, or eliminate, its selectivity to rice. This reaction prevents propanil and the insecticides from being applied in mixtures or at the same time. In order to avoid severe phytoxicity, application of the insecticide has to be delayed for a week, or longer, are propanil is applied.

The fact that very little is known about the effect of individual pesticides on the physiology and biochemistry of soybeans indicates a need for expanded research in this area. Chemicals tested singly may be found to be without obvious adverse effects on the growth and development of soybeans but at the same time may be responsible for more subtle, but important, effects. There is great need for research on chemicals that are safe when applied singly to soybeans but may be hazardous when introduced into biological systems in the presence of other pesticides.

Effects of Pesticides on Nodulation and N_2 Fixation in Soybeans

The possible adverse effects of pesticides on nitrogen fixing organisms has been a major concern in the case of leguminous crops. Herbicides have especially been suspect. Most of the research on this important potential problem has been done in the laboratory at much higher rates of the pesticides than recommended for field use. Results reported have been highly variable. Vincent (25) found that inhibiting doses reported in the literature ranged from 50 to 4000 ppm for some species of *Rhizobium.* He suggested that differences between strains within species and variation in experimental conditions may mask possible adverse effects of pesticides on *Rhizobium.*

Gibson (9), on the basis of an intensive review of the literature, concluded that the newer herbicides such as trifluralin, nitralin, and chlorpropham, for example, when used at normal rates of application have little effect on free living rhizobia but may affect nodulation of soybeans. Dunnigan et al. (8) reported results of extensive greenhouse and field experiments to test effects of herbicides on nodulation. Seven commonly used herbicides were tested on five soils in the greenhouse and one soil in the field. Dosages employed were 1/5, 1, and 5 times normal field rates. Several of the chemicals had detrimental, but transient, effects on nodulation of soybeans in the greenhouse experiments. Severity of effects varied among the different types of soil. Their greenhouse studies were supplemented by a three-year field study in which no significantly detrimental effects on nodulation of soybean plants were observed at dosages as much as twice normal field rates. They concluded that most herbicides registered for use on soybeans would not have adverse effects on nodulation if used at rates recommended by the manufacturer.

Development of the C_2H_2-reduction technique as an indirect method for estimating N_2 fixation for field as well as laboratory studies (10) has provided an efficient method for studying possible adverse effects of pesticides on leguminous plants. Smith et al. (23) used the method to study the effects of seven OP and carbamate insecticides on three species. They studied alfalfa, red clover, and sweet clover in the laboratory on both vermiculite and soil substrates. Effects of the insecticides were strongly influenced by the substrate. Inhibition of C_2H_2-reducing ability by the carbamates carbofuran and aldicarb was limited to plants grown in soil. A decrease in C_2-H_2-reducing ability was correlated with a reduction in nodule formation The insecticides had no adverse effects on *Rhizobium* spp. in broth culture.

In a greenhouse experiment (L. D. Newsom, unpublished), carbofuran, fensulfothion, and chlorpyrifos significantly reduced nodule formation and C_2H_2-reducing ability of Bragg soybeans. Plants were grown in 6-inch clay pots on Mhoon loam soil treated at planting with two dosage rates, one slightly in excess of the field rate and the second double the first. Plants were harvested for analyses at first blossom (Growth Stage R-1) and were about 50 cm tall. Only carbofuran produced overt phytotoxic symptoms on the plants. Both dosage rates produced severe necrosis, puckering at the apex, and downward cupping of leaves. Plants were stunted severely at the high dosage rate but not the lower rate. Plant biomass measured by oven dry wt of both shoots and roots did not differ significantly from the control except for the high rate of carbofuran. The most significant result of this experiment was a severe reduction in nodulation and C_2H_2-reducing ability caused by both fensulfothion andchloropyrifos without any observable symptoms of phytotoxicity (Table 3).

Data available on the effects of single pesticides on nodulation and N_2-fixation consistently show that many commonly used herbicides and insecticides produced adverse effects under some of the conditions existing in the experiments. Thus, it appears reasonable to expect that such effects may be substantially greater where mixtures are involved. Unfortunately, almost no research has been reported on the effects of mixtures of pesticides.

Effects of Pesticides on Endomycorrhizae

The association of endomycorrhizae and soybeans has attracted little attention until recently. This is surprising because it had been reported more than 30 yr ago by Asai (2). He compared the growth of soybeans in sterilized soil to which both micorrhiza and *Rhizobium* had been added, to that inoculated with *Rhizobium* only, and a control. The addition of mycrorrhizae to the soil increased dry wt by more than 70% and nodulation by more than 30% compared to both *Rhizobium* alone and the control.

Nesheim and Linn (19) demonstrated adverse effects of the fungicides arasan and botran at concentrations of 50 ppm, and greater, to *Endogone fasciculata* in corn. Root volume was significantly decreased in treated compared to control plots in greenhouse experiments. Ross and Harper (22) reported large increases in yields of soybean in soil previously fumigated with mixtures of methyl bromide and chlorpicrin or with methyl bromide and mixture of D-D, methylisocyanate, and chlorpicrin. Increases in yield were 34 to 40% in small isolated plots and 29% in field plots. They suggested that results of experiments with chemicals that affect endomycorrhizae adversely should be re-evaluated. Bird et al. (4) found that treatment with the nematicide DBCP (2, 4-dibromo 3-chloropropane) significantly increased development of endomycorrhizae in cotton. They suggested that the chemical may

Table 3. Effects of some insecticide-nematicide compounds on growth and development of soybeans in pot experiments in the greenhouse. Baton Rouge, Louisiana, 1979.

	Carbofuran		Fensulfothion		Chlorpyrifos		Control
Rate of application (kg a.i./ha)	4.50	2.25	4.50	2.25	4.50	2.25	--
Dry wt shoots, g	4.95*	6.84*	7.87	9.56	8.55	7.96	10.07
Dry wt roots, g	0.98*	1.61*	2.46	3.07	2.87	3.04	3.00
No. nodules/ 2 plants	60.4*	136.4*	147.2*	125.0*	121.8*	142.0*	294.4
C_2H_4, μM/ 2 plants	3.62*	19.03*	22.9*	21.0*	21.76*	19.4*	77.16
No. Helicotylenchus/pt. soil	1256*	2440*	880*	1760*	5816*	4776*	9940

*Significantly different from control $P < 0.05$.

have controlled nematodes that affected mycorrhyzae adversely, or that the physiology of the roots was changed in a way that favored mycorrhizae.

Backman and Clark (3) tested the effects of several insecticide-nematicide compounds, three biocides, and one fungicide on an endomycorrhiza of peanuts tentatively identified as *Glomus* sp. The pesticides were tested at field rates in greenhouse and field experiments. Most of the pesticides tested affected the fungus little, or not at all. However, carbofuran and sodium azide reduced growth significantly at 30 days after planting. Levels of mycorrhizae in feeder roots were "normal" 120 days after planting. Data from greenhouse tests supported the field data.

Effects of carbofuran on peanuts are particuarly interesting. It has been used widely on peanuts for control of nematodes for several years with substantial benefits in yields. Backman and Clark (3) suggested that the response in yields may be affected by elimination of endomychorrhizae in two ways: (a) reduced in soils low in fertility, and (b) increased in high fertility soils by reducing "parasitic effects" of the fungus.

Endomycorrhizae, generally believed to benefit soybeans by increasing the phosphorus uptake which benefits nodulation and N_2-fixation, complement and are complemented by the stimulatory effects of *Rhizobium* on plant growth, and lower resistance of roots to water transport (5,9,18). Therefore, adverse effects on endomycorrhizae of soybeans are cause for serious concern. Research on the effects of pesticides and pesticide mixtures on soybeans under field conditions should be accelerated. More research is needed especially now that the insecticide-nematicide aldicarb has been registered for use on soybeans. Also, it appears likely that carbofuran will be registered for similar use soon. Both compounds are likely to be used extensively for control of nematode and insect pests of soybeans in the future.

Stimulatory Effect on Plant Growth of Organophosphorous and Carbamate Insecticide-Nematicide Compounds

Treatment of crops with systemic insecticide-nematicide compounds such as aldicarb, carbofuran, and disulfoton often enhance growth of many species to some extent. Some of the reasons offered to explain this apparent stimulatory effect are the possible control of species that have not been recognized as pests, or direct effects on physiological and biochemical processes in plants.

Effects of carbofuran and disulfoton on maturity and yield of burley tobacco reported by Pless et al. (21) are typical of enhanced growth that often is exhibited by treated plants. In field experiments they observed that tobacco treated with carbofuran and disulfoton reached maturity earlier and outyielded the control by 35 to 38%, respectively. Pests that may have been controlled were not considered to be significant. Similar results with tobacco were obtained in greenhouse tests with carbofuran in sterilized soil and in field plots fumigated with methyl bromide (T. E. Reagan, personal communication).

A particuarly interesting finding reported by Pless et al. (21) was that the "physiological response" observed in their experiments occurred only in the high-phosphate soils of the middle Tennessee basin. This tends to support the intriguing suggestion by Backman and Clark (3) that increased yields might occur on high fertility soils by eliminating effects of parasitism by endomycorrhizae with pesticide treatments. Carling et al. (5) found that when phosphorus was substituted for mycorrhizae infection in soybeans, similar increases in growth and enzyme activity were obtained.

Most of the information on stimulation of plant growth by pesticides has been observational and speculative. Recently, however, Lee (15) reported his findings on the effects of carbofuran on IAA (indole-3-acetic acid) metabolism and plant growth. He found no stimulatory effect of carbofuran on growth of plant callus tissue in the absence of IAA but obtained variable results when it was present. He demonstrated that two metabolites, carbofuran phenol and 3-hydroxy-carbofuran phenol, were stimulatory in pea stem assays in the presence of low concentrations of IAA but not when it was absent. He interpreted these data to suggest that carbofuran can promote plant growth through the effect of its metabolites on IAA. His work offers some interesting avenues that may lead to an understanding of this important phenomenon.

Effects of Different Components of Pest Management Systems on Other Components of the System

Ideally, pest management tactics and strategies for control of the total pest complex of soybeans should be integrated into one overall system. Substantial progress has been made toward achieving that goal, but it is still a painfully slow process. Too often, tactics are developed by one discipline for control of pests in ignorance of, or with indifference to, effects they may have on other components of the overall system.

Adverse Effects of Pesticides

Entomologists responsible for research on soybean insects are dedicated to the objective of avoiding overuse and misuse of insecticides on the crop. Their research on developing insect pest management systems for the crop is based on observation of the following principles: (a) effective monitoring of populations, (b) establishing economic injury thresholds and using them as the basis for applying insecticides, (c) using the most environmentally safe and non-polluting chemicals available in ways that are as ecologically selective as possible, (d) making minimum use of insecticides, both in dosage rates and number of treatments, (e) making maximum use of tactics other than repetitive applications of insecticides, and (f) relying heavily upon the regulatory effects of natural enemies on pest population.

Pest management systems based on these principles cannot be effective if populations of natural enemies are disrupted. Obviously, employment of any tactic that has disruptive effects on these agents should be avoided whenever possible. Unfortunately, members of all classes of agricultural pesticides may have disruptive effects on some natural enemies.

Insecticides. Insecticide use on soybeans in the U.S. has been held to low levels. Entomologists agreed more than 10 yr ago to discontinue recommendations for all uses of organochlorine insecticides on soybeans. This agreement has been implemented with good success on all except for toxaphene. Unfortunately, some states found that it was impossible to eliminate it from their recommendations. Insecticide use thus far has had minimal adverse effects on the general predator and parasite fauna. These agents continue to play a major role in control of soybean pests.

Fungicides. Entomopathogenic organisms play an important role in the control of many soybean pests, particuarly lepidopterous species. Fungal pathogens, *Nomuraea rileyi* (Farlow) and *Entomophthora gammae* (Weiser), for example, are especially valuable agents. Therefore, entomologists were greatly concerned when the fungicide benomyl was registered and recommended for prophyllactic use on soybeans (1). Two applications at the rate of 0.28 kg a.i./ha were recommended, the first at full bloom and a second two weeks later. It was feared that routine applications of this broad spectrum, highly active chemical might have devastating effects on entomopathogens such as *N. rileyi* and *E. gammae.* Adverse effects of various pesticides on several species of entomopathogens have already been well documented (13,20,24,26).

Initial evaluation of the effects of benomyl on *N. rileyi* suggested that concern for possible adverse effects might be well grounded. Johnson et al. (14) reported a delay of about three weeks in development of an epizootic of *N. rileyi* on velvetbean caterpillar larvae in plots treated with benomyl when compared to the control. They also reported that the control outyielded plots treated with benomyl by about 15%. Subsequent research, however, has not confirmed such drastic effects. Herzog (11) and Husin (12) reported results of extensive field and laboratory experiments that showed little impact of benomyl on populations of three species of lepidoptera. However, mortality of larvae from infection with *N. rileyi* in the controls was about three times that in treated plots.

Commercial experience thus far after hundreds of thousands of hectares of soybeans in the U.S. have been treated with benomyl indicates that its impact on entomopathogens has been of little importance. The fact that it is consistently and demonstrably toxic to such organisms as *N. rileyi*, however, should not be ignored. Even if it was much more toxic to such organisms than it has proved to be, benomyl would still be used where fungal diseases such as *Diaporthe* are a problem. Growers would not be willing to sacrifice the substantial increases in yields realized from its use.

Insecticide-Nematicide Mixtures. Beginning with the 1979 season, a broad spectrum, highly toxic insecticide-nematicide, aldicarb, will be registered for use on soybeans. It will be used prinicipally for control of nematodes and subterranean insect pests. Application will be at planting time prinicipally, at about the same time herbicides such as trifluralin, alachlor, and metrabuzin are applied. Use of a compound such as aldicarb in this prophyllactic manner is ample cause for concern by entomologists, weed control specialists, and members of all disciplines interested in the *Rhizobium*-endomycorrhizae-soybean association.

Previous experience with use of aldicarb on cotton has shown that it often has had devastating effects on populations of insect natural enemies. Adverse effects of aldicarb treatments on populations of predators and parasites of *Heliothis* spp. on cotton have been especially troublesome. This experience dictates that growers should proceed with prophyllactic use of aldicarb on large areas with extreme caution.

Weed control specialists may have cause for as much concern, or more, than entomologists over prospective wide scale use of aldicarb on soybeans. Introducing another highly active chemical into the soil in the presence of one or more herbicides, some of which are being used at rates approaching unacceptable levels of phytotoxicity, could result in serious damage. Such effects are especially likely in periods of unfavorable weather conditions during early growth and development of plants.

Although equally speculative, the effects of mixtures of aldicarb and various herbicides on nodulation, N_2-fixation, and endomycorrhizae could be severe. Obviously, much more information on these possible interactions than has been published is needed before growers use it for treating large areas.

The soybean industry cannot tolerate problems that will develop from excessive use of pesticides. Prophyllactic use of any pesticide is likely to create new problems with pests or to exacerbate old ones. Therefore, it is critically important that caution be observed in introducing a broad spectrum chemical such as aldicarb into soybean ecosystems before its overall effects on pest management systems have been evaluated adequately. Such an evaluation has not been made yet.

Insecticide Used as a Herbicide. One of the most interesting examples of the effects of one tactic on other components of the pest control system on soybeans is that of the use of the insecticide toxaphene as a herbicide. Events leading to this use are obscure. Apparently it developed as a result of growers observing that toxaphene applied for insect control to cotton and soybeans gave control of sicklepod, *Cassia obtusifolia,* equal to that provided by the most effective herbicides for control of this important pest. At any rate, no weed control specialist has been willing to claim credit for developing the herbicidal use of toxaphene. It has been used illegally for many years on large acreages of soybeans in the southeastern U.S. for sicklepod control.

The manufacturer has recently initiated an effort to obtain registration for toxaphene as a herbicide. Use of this persistent organochlorine insecticide as a herbicide is highly detrimental to predatory and parasitic insects, especially the ground-dwelling predators. Its use on soybean should be discontinued for that reason alone. However, its use as a herbicide will continue to add to the burden of toxaphene residues in the environment that are already at intolerable levels. Lake Providence in Northeast Louisiana has already been closed to fishing because of unacceptably high residues of toxaphene in fish. Consideration is presently being given to closing parts of Pearl River in Southeast Louisiana for the same reason. Either the problem of adverse effects on insect natural enemies or its further pollution of the environment provides ample reason for denying registration for its use as a herbicide in soybeans.

CONCLUSIONS

Data available on the subjects discussed, though fragmentary and inadequte, suggest the following questions: (a) Does injury to soybeans result from current uses of pesticides to the extent that growth and yield are affected but the adverse effects are masked by benefits from pest control; (b) Do pesticides introduced into soybean ecosystems, singly or in mixtures, cause direct physiological and biochemical injury and stresses that are unrecognized; (c) Do mixtures of pesticides and their metabolites in soybean ecosystems potentiate toxic effects; act synergistically, additively, or antogonistically on nodulation, N_2-fixation, endomycorrhizae, plant growth and development, pest species, non-target species; and (d) How can problems resulting from development of tactics in one discipline, but that have adverse effects on tactics developed for other components of pest management systems, best be resolved?

A greatly expanded research effort will be required to provide answers to these questions. However, the need is so great that soybean researchers can no longer remain unresponsive. The known and possible interactions within and between complexes of pests and tactics used for their control overwhelm the imagination. Research of the sort required to deal with them effectively will require the highest level of interdisciplinary cooperation and coordination. Traditional approaches along disciplinary lines are inadequate to solve such complex problems.

NOTES

L. D. Newsom, Department of Entomology, Louisiana Agricultural Experiment Station, Louisiana State University, Baton Rouge, LA 70803.

REFERENCES

1. Anonymous. 1975. Agricultural Bulletin. Supplemental Labeling EPA Reg. No. 352-354. Benlate-Benomyl fungicide for control of certain diseases of soybeans. E. I. DuPont de Nemours and Co., Inc., 1 p.
2. Asai, T. 1944. Uber die mykorrhizenbildung leguminosen Pflanzen. Jap. J. Bot. 13:463-485.
3. Backman, P. A. and E. M. Clark. 1977. Effect of carbofuran and other pesticides on vesicular-arbuscular mycorrhizae in peanuts. Nematropica 7:14-17.
4. Bird, G. W., J. R. Rich, and S. V. Gover. 1974. Increased mycorrhizae of cotton roots in soils treated with nematicides. Phytopathology 64:48-51.
5. Carling, D. E., W. G. Riehle, M. F. Brown, and D. R. Johnson. 1978. Effects of a vesicular-arbuscular mycorrhizae fungus on nitrate reductase and nitrogenase activities in nodulating and non-nodulating soybeans. Phytopathology 68:1590-1596.
6. Chang, F. Y., G. R. Stephenson, and L. W. Smith. 1971. Influence of herbicides on insecticide metabolism in leaf tissues. J. Agric. Food Chem. 19:1183-1186.
7. Chio, H. and J. R. Sanborn. 1977. Atrazine inhibition of carbofuran metabolism in the house cricket. J. Econ. Entomol. 70:544-546.
8. Dunigan, E. P., J. P. Frey, L. D. Allen, Jr., and A. McMahon. 1972. Herbicidal effects on the nodulation of *Glycine Max* (L.) Merrill. Agron. J. 64:806-808.
9. Gibson, A. H. 1977. The influence of the environment and managerial practices on the legume—*Rhizobium* symbiosis, pp. 393-450. In R. W. F. Hardy and A. H. Gibson (eds.) A treatise on dinitrogen fixation, Section IV, John Wiley & Sons, New York, 527 pp.

10. Hardy, R. W. F. and R. D. Holsten. 1977. Methods for measurement of dinitrogen fixation, pp. 451-486. In R. W. F. Hardy and A. H. Gibson (eds.) A treatise on dinitrogen fixation, Section IV, John Wiley & Sons, New York, 527 pp.
11. Herzog, D. C. Impact of foliar fungicide applications on the incidence of entomopathogenic fungi in defoliating caterpillar populations on soybean. Paper presented at Third Annual Meeting of the Southern Soybean Disease Workers Council, Baton Rouge, Louisiana, March 18, 1976.
12. Husin, A. H. 1978. Effect of foliar fungicides on fungi infecting the soybean looper, *Pseudoplusia includens* (Walker), and the velvetbean caterpillar, *Anticarsis gemmatalis* Hubner, in the field. M. S. Thesis, Dept. of Entomology, Louisiana State Univ., Baton Rouge, Louisiana, 149 pp.
13. Ignoffo, C. M., D. L. Hostetter, C. Garcia, and R. E. Pinnell. 1975. Sensitivity of the entomopathogenic fungus *Nomuraea rileyi* to chemical pesticides used on soybeans. Environ. Entomol. 4:765-768.
14. Johnson, D. W., L. D. Kish, and G. E. Allen. 1976. Field evaluation of selected pesticides on the natural development of the entomopathogen, *Nomuraea rileyi*, on the velvetbean caterpillar in soybeans. Environ. Entomol. 5:964-966.
15. Lee, T. T. 1976. Insecticide-plant interaction: Carbofuran effect on indole-3-acetic acid metabolism and plant growth. Life Sciences 18:205-210.
16. Lichtenstein, E. P., T. T. Liang, and B. N. Anderegg. 1973. Synergism of insecticides by herbicides. Science 181:847-849.
17. Matsunaka, S. 1968. Propanil hydrolysis: Inhibition in rice plants by insecticides. Science 160:1360-1361.
18. Mosse, B., C. L. Powell, and D. S. Hayman. 1976. Plant growth responses to vesicular-arbuscular mycorrhiza, rock phosphate, and symbiotic nitrogen fixation. New. Phytol. 76:331-342.
19. Nesheim, O. N. and M. B. Linn. 1969. Deleterious effect of certain fungitoxicants on the formation of mycorrhizae on corn by *Endogone fasciculata* and corn root development. Phytopathology 59:297-300.
20. Olmert, I. and R. G. Kenneth. 1974. Sensitivity of the entomopathogenic fungi, *Beauvaria bassiana, Verticillium lecanii,* and *Verticillium* sp. to fungicides and insecticides. Environ. Entomol. 3:33-38.
21. Pless, C. D., E. T. Cherry, and H. Morgan, Jr. 1971. Growth and yield of burley tobacco as affected by two systemic insecticides. J. Econ. Entomol. 64:172-175.
22. Ross, J. P. and J. A. Harper. 1970. Effect of *Endogone* mycorrhiza on soybean yields. Phytopathology 60:1552-1556.
23. Smith, C. R., B. R. Funke, and J. T. Schulz. 1978. Effects of insecticides on acetylene reduction by alfalfa, red clover and sweet clover. Soil Biol. Biochem. 10: 463-466.
24. Soper, R. S., F. R. Holbrook, and C. C. Gordon. 1974. Comparative pesticide effects on *Entomophthora* and the phytopathogen *Alternaria solani.* Environ. Entomol. 3:560-563.
25. Vincent, J. M. 1977. Rhizobium: General microbiology, pp. 277-366. In R. W. F. Hardy and W. S. Silver (eds.) A treatise on dinitrogen fixation, Section III, John Wiley & Sons, New York, 675 pp.
26. Wilding, N. 1972. The effect of systemic fungicides on the aphid pathogen, *Cephalosporium aphidicola.* Plant Pathol. 21:137-139.

PEST PROBLEMS OF SOYBEANS AND CONTROL IN NIGERIA

M. I. Ezueh and S. O. Dina

The soybean, *Glycine max* (L.) Merrill, was first introduced into Nigeria in 1910 (3). Apart from Benue State where it is grown on a commercial scale, limited quantities are produced in Southern Zaria, Kwara and Niger States (6) (Fig. 1). About 170,000 ha of land are planted to soybeans with an annual production of about 70,000 metric tons. Cultivation is usually in mixtures with sorghum, millet and citrus. The popular variety grown is known as the Malayan, which is a tall vigorous indeterminate type. In the traditional management of soybeans, no crop protection measures are usually applied and the average yield was reported to be about 300 kg/ha (1). However, a more recent survey indicates that yields of about 1,000 kg/ha are obtainable probably due to the adoption of better technology.

The National Cereals Research Institute (NCRI) began an active program on soybean improvement in 1974 with cultivars obtained from IITA and some local sources. As part of this effort, a survey of the insect fauna of the crop was carried out between 1974 and 1977 with an objective to identify insects of economic importance in the production of soybeans. Preliminary control trials were carried out at Moor Plantation, Ibadan, with six insecticides during two seasons of 1978 in an attempt to develop a chemical control schedule for soybean monocrops.

MATERIALS AND METHODS

Insect Survey

An insect collection was made weekly at Ibadan headquarters of NCRI. At Uyo and Mokwa outstations a collection was made every two weeks.

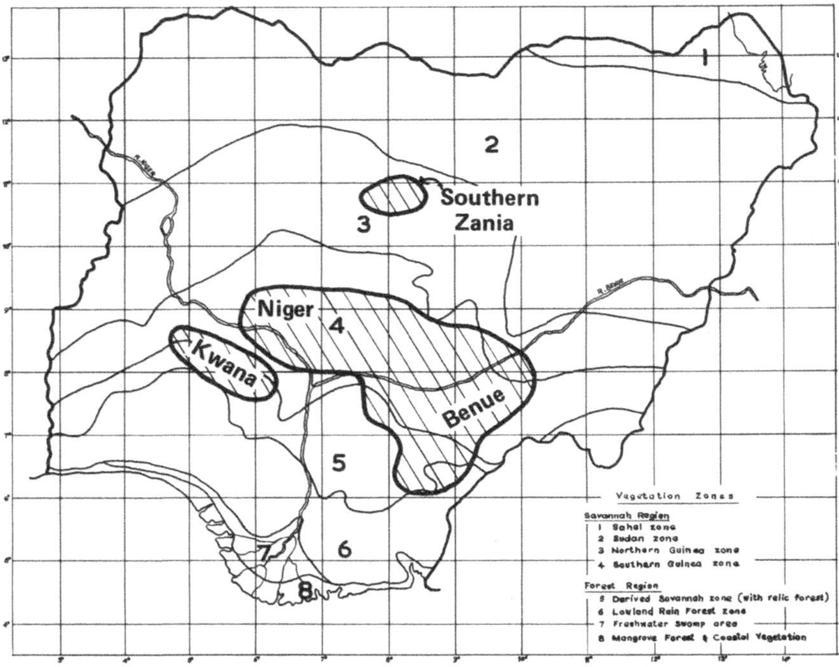

Figure 1. Soybean producing areas in Nigeria.

Most of the collections were made on breeders' plots which usually contained different cultivars with widely different maturity characteristics. Collection of adult specimens was made with a sweep net. Larval specimens were collected from the field in glass vials and reared to adults in the laboratory on soybean leaves, stem portions, and pods. The laboratory temperature was 26 ± 1.5 C. Thrips, leafhoppers, and other small insects were collected with the D-Vac suction sampler.

Insecticide Trials

Monocrotophos, lindane, carbaryl (Vetox), curacron, permethrin, and decamethrin (Decis) were screened in both early and late seasons of 1978 using the soybean variety Bossier. An untreated plot served as a check. Plot size was 15.36 m^2 and the test was replicated three times in the early season and four times in the late season. The layout used was a randomized complete block design. Planting distances were 5 cm on ridges 60 cm apart. Dose rates tested were between 50 and 1500 kg a.i./ha. Applications were made three times at weekly intervals commencing at flower initiation (about 35 days after planting). Assessments for insect damage were made on the stems and

200 randomly picked dry pods. Comparisons were made on insect damage and yield of dry grains among the treatments after analysis of variance.

RESULTS

Insect Survey

A total of 135 insect species were collected in the survey. The 120 phytophagous species were from 44 families and 7 orders. Two of these attacked roots, another 5 were endophagous, and the remaining 115 were ectophagous feeders, damaging foliage and reproductive parts. Eleven species of insect predators from 5 families and 3 orders were also encountered during the collection. A list of the more regular and potential pests and predators is given in Table 1. The following insects are highlighted because of their importance as pests of cowpea which is more widely grown in Nigeria.

Maruca testulalis Geyer attacks soybean stem, leaves and pods. The larvae bores into the stem and causes die-back of the affected branch on the plant. Although older larvae were observed feeding on flowers, damage seems to be concentrated mainly on foliage. *Cydia ptychora* Meyrick is also a serious pest of soybean. This moth caused about 32% pod damage and 20% stem damage in a crop planted on June 22 which matured on October 20, 1977. Bored stems however showed no signs of die-back of the affected parts. *Lagria villosa* F. and *Chrysolagria cuprina* were found during the late vegetative growth phase. They attained a peak population at about crop maturation and caused extensive defoliation. Since most of the pods have matured at this time, it is unlikely that their attack may cause any depression in yield. These species also attacked flowers and pods. Feeding on flowers was restricted to the non-essential parts while damage to the pods consisted of mere abrasions of the epidermal layer. Both the larvae and adults of *L. villosa* had the same feeding characteristics.

Several pentatomidae were encountered frequently. These were: *Nezara viridula* L., both *viridula torquata* F. and var *smaragdula* F; *Acrosternum acuta* Dallas; and *Piezodorus* spp. The *Nezara* spp. occurred in large numbers at about 11 weeks after planting and continued to build up until harvest. *N. viridula* var *smaragdula* was the most abundant type found. Adults and nymphs of this species feed extensively on the leaves and pods causing shrivelling and incomplete pod-fill. *Acrosternum acuta* Dallas had the same feeding habits as *Nezara* spp. *Piezodorus* spp. attacked soybean plots at about 5 weeks after planting and infestation continued until harvest. They fed extensively on young flowers, leafbuds and shoots, and later transferred to the pods. *Riptortus dentipes* F. (Alydidae) was also abundant. Adults and nymphs fed on pods and their attack seemed to have been seasonal, occurring after 6 weeks on the early season crop planted on April 21, 1977. They were hardly found on soybeans planted on June 22, 1977.

Table 1. Insect pests associated with soybeans in Nigeria.

Names	Plant Part Attacked	Pest Status
COLEOPTERA		
Galerucidae		
Aulacophora africana Weise	Leaves	Minor
**Barombia humeralis* Laboiss	Leaves	Minor
**Paraluperodes quaternus* Fairm	Leaves	Minor
**Ootheca mutabilis* Sahlb	Leaves	Minor
Curculionidae		
Alcides dentipes F.	Leaves	Minor
Nematocerus acerbus Forster	Leaves	Minor
Omotrachelus sp.	Leaves	Minor
Lagriidae		
**Chrysolagria cuprina* Thomson	Flower, leaves and pods	Minor
**Lagria villosa* F.	Flower, leaves and pods	Minor
Rutelidae		
Anomala denuda Arrow	Leaves	Minor
Meloidae		
Coryna hemanniae F.	Flowers	Minor
Cocinellidae		
Epilachna similis Thunberg	Flowers	Minor
LEPIDOPTERA		
Arctiidae		
**Diacrisia aurantiaca* Hollande	Leaves	Minor
**Spilosoma maculosa* Stoll	Flowers, leaves and pods	Minor
Noctuidae		
**Heliothis armigera* Hubner	Pods and leaves	Minor
Mocis mayeri Boisduval	Leaves	Minor
**Spodoptera littoralis* Boisduval	Leaves	Minor
Olethrentidae		
Cydia ptychora Meyrick	Pods and stem	Major
Cydia sp.	Shoot	Minor
Pyralidae		
Hedylepta indicata F.	Leaves	Major
**Maruca testulalis* Geyer	Flowers, pods and leaves	Minor
Lycaenidae		
**Virachola antalus* Hopkins	Pods	Minor
HEMIPTERA		
Pentatomidae		
Acrosternum acuta Dallas	Pods and leaves	Major
Aspavia armigera F.	Pods and leaves	Major
Aspavia brunae F.	Pods and leaves	Minor
Aspavia hastator F. var	Pods and leaves	Minor
Nezara viridula L. var	Pods and leaves	Minor
Smaragdula F.	Pods and leaves	Major
Nezara viridula torquata F. var	Pods and leaves	Major

Names	Plant Part Attacked	Pest Status
HEMIPTERA (Continued)		
Piezodorus sp.		
punctiventris Dallas	Pods and leaves	Major
Piezodorus sp.	Pods and leaves	Major
Coreidae		
**Acanthomia horrida* Germar	Pods and leaves	Minor
**Anoplocnemis curvipes* F.	Pods and leaves	Minor
Leptocorisa elegans Blote	Leaves	Minor
Leptoglossus membranaceous F.	Leaves	Minor
Stenocoris southwoodi Ahmad	Leaves	Minor
Alydidae		
Mirperus sp.	Pods and leaves	Minor
Riptortus dantipes F.	Pods and leaves	Major
Cicadellidae		
Empoasca sp.	Leaves	Minor
Miridae		
Halticus tibialis Reuter	Leaves	Minor
Pyrhocoridae		
Dysdercus superstitiosus F.	Leaves	Minor
Lygaeidae		
Lygaeus sp.		
Oxycarenus sp.		
multiformis Samy	Leaves	Minor
Aphididae		
Aphis craccivora Koch	Pods and leaves	Minor
Aphis gossypii Glover	Leaves	Minor
Aleyrodidae		
Bemisia sp.	Leaves	Minor
THYSANOPTERA		
Thripidae		
Caliothrips impurus Priesner	Flowers and shoots	Minor
Sericothrips occipitalis Hood	Flowers and shoots	Minor

PREDATORS

COLEOPTERA

Coccinellidae

Cydonia lunata F.

Cydonia vicina Mulsant

Exochomus flavipes Thunberg

Hyperaspis pumila Mulsant

HEMIPTERA

Pentatomidae

Macrorhaphis infuscata Walker

HEMIPTERA (Continued)

Lygaeidae

Geocoris ambillis Stål

Reduviidae

Rhinocoris bicolor F.

Rhinocoris sp.

Nagusta sp.

HYMENOPTERA

Vespidae

Belanogaster junceus F.

The relative damage by any single species of the above-mentioned hemipterans was not quantified but collectively they appeared to be the most important pests of this crop. In addition to occasional other hemipterans such as *Aspavia armigera* F. and *A. brunae* F., they accounted for up to 50.5% partially filled or shrivelled pods in soybean plots. A list of parasites and predators occurring in the collections is given in Table 1. These included 4 coccinellids, 1 pentatomid, 1 lygaeid, 3 reduvids, and 1 vespid.

Insecticide Trials

In both early and late season trials, none of the insecticidal treatments resulted in a significant increase in grain yield over the control as shown in Table 2. Insect control was also not significantly different at both seasons but carbaryl appeared to have controlled hemipteran damage better than the other compounds in the late season ($p<0.01$) (Table 3). Insect infestation levels were much lower on the late season crop.

DISCUSSION AND CONCLUSIONS

There is very little published information on the pests of soybeans in Nigeria probably because the crop is not yet widely grown. However, a number of beetles and hemipteran insects have been identified as major pests of soybean at the International Institute of Tropical Agriculture, Ibadan (Ogunlana, unpublished data, 1974). Raheja (4) recorded 47 insect species on 10 minor legume crops in Northern Nigeria including soybean. In this study the major insect problems are highlighted with the pentatomids constituting the most serious group. In general the insect problems are similar to those of cowpea.

In traditional agriculture, Nigerian farmers obtain grain yields of 300 to 800 kg/ha without crop protection, probably because the commonly grown variety, Malayan, is an indeterminate, long duration variety that appears to have become tolerant to the major pests of this crop. With the prospects of

Table 2. **Grain yield of soybeans at different insecticidal treatments.**

Insecticide	Rate of a.i./ha	Yield in g/plant	
		Early Crop	Late Crop
Monocrotophos	750	24.5	3.9
Decamethrin	50	25.3	3.8
Permethrin	50	25.5	3.6
Curacron (CGA 15324)	1,000	26.3	4.2
Lindane	1,000	--	4.4
Carbaryl	1,000	26.5	3.9
Control	--	24.3	3.8

Table 3. Effects of different insecticidal treatments on insect damage on soybean.

	Early Season		Late Season	
Insecticide	% Cydia Damaged Seeds	% Pods Damaged By Insects	% Cydia Damaged Seeds	% Pods Damaged By Insects[1]
Monocrotophos (Nuvacron)	1.5	43.0	0.25	12.9
Decamethrin (Decis)	1.1	47.7	0.13	12.9
Permethrin	0.5	50.0	0.0	12.4
Curacron (CGA 15324)	3.1	54.8	0.0	12.2
Carbaryl (Vetox)	1.4	46.2	0.0	8.9
Control	1.9	43.7	0.5	13.0

[1] $P<0.01$; LSD = 0.59.

an active improvement program there is no doubt that many improved soybean varieties will be introduced which may not withstand the challenge by insect pests to the same degree. Furthermore, when commercial production becomes established insect problems are bound to escalate and some of the seemingly unimportant insects may eventually become economically important.

Cydia ptychora, which usually attacks mature and drying pods of cowpea in Nigeria, behaved differently on soybeans by assuming a stem-boring habit. This may be an off-season adaptation for survival in the absence or reduced abundance of its natural host, cowpea. This is similar, however, to the behavior of *Laspeyresia fabivora* which bores petioles and stems of soybean in Brazil and Peru (5) and of *Matsumuraeses phaseoli* Mats. on stems on pods in Japan (2).

A number of predators were found among the insects collected, indicating a good potential for the control of these pests by natural enemies. This observation should be borne in mind when chemical control measures are envisaged in order to prevent a serious disturbance of the agro-ecosystem which might result from indiscretional use of pesticides. This is particularly significant when managing the cereal-soybean intercrop systems practiced in Nigeria because of the potential value of such crop mixtures in promoting biological control.

Raheja (4) obtained higher yields in soybean sprayed with insecticides than in control plots and noted some damage by coreids. In the present investigation insecticide treatments did not result in higher grain yields over the untreated control. There are possible explanations for this result. For example, (a) the spray regime imposed might have been inadequate for a season long protection (see Fig. 2); (b) insect populations might have been too low to make the spray effective; and (c) the plants might have compensated for

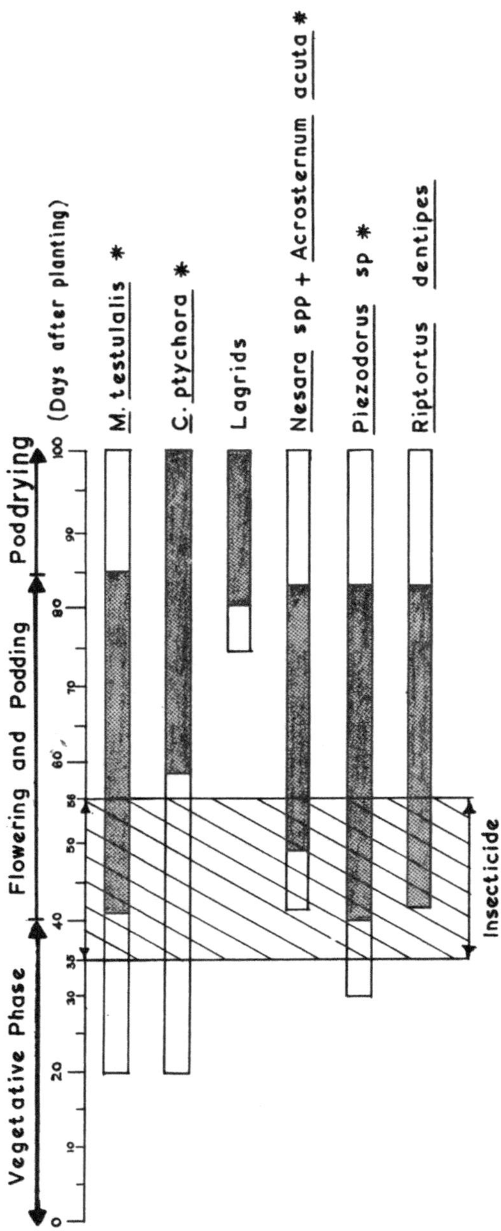

Figure 2. Schematic diagram showing pest succession on the soybean crop in relation to insecticide schedule. Shaded areas correspond to peak activity. * = major pests.

insect injury particularly in relation to the pod-sucking insects by exhibiting some prolificacy in pod-formation. These aspects will be further investigated and it is expected that a more effective insecticide scheduling will emerge after a better understanding of the pest succession and build-up in relation to the phenology of this crop.

NOTES

M. I. Ezueh and S. O. Dina, National Cereals Research Institute, Moor Plantation, Ibadan, Nigeria.

REFERENCES

1. Edem, E. V. 1975. Soybean in Nigeria (a review article). Memo. Federal Dept. Agric. Res. No. 116.
2. Kobayashi, Y. 1976. Insect pests of soybean in Japan. PANS 22:336-349.
3. Mayo, J. K. 1945. Soybeans in Nigeria. Trop. Agric. 22:226-229.
4. Raheja, A. K. 1976. Insect pests of grain legumes other than cowpea grown at Samaru. Samaru Agric. Newsletter. 18:98-101.
5. Turnipseed, S. G. and M. Kogan. 1976. Soybean entomology, Ann. Rev. Ent. 21:247-282.
6. Van Rheenen, H. A. 1975. Soybeans in the Northern States of Nigeria. Proc. IITA. 158-159.

SOYBEAN INSECTS AND THEIR NATURAL ENEMIES IN BRAZIL – A COMPARISON WITH THE SOUTHERN UNITED STATES

S. G. Turnipseed

The production of soybeans in Brazil has increased dramatically during the late 1960's and throughout the 1970's. Initial hectarage centered in the southernmost state of Rio Grande do Sul, but has expanded northward through Parana into southern Goias and westward into the state of Mato Grosso. Although drought has restricted production levels during the past two seasons, approximately 14 million mt were estimated for 1979, making Brazil the second largest soybean producing nation.

Interesting production comparisons exist between Brazil and the southern U.S. Latitudes involved in major production areas of Brazil are similar to the southern U.S. to somewhat more subtropical in Brazil; determinate cultivars of maturity groups V-IX are grown in both areas; and pest problems, particularly with respect to insects, have numerous similarities. In Brazil the crop is planted in November and December, and harvest occurs from late March throughout April, depending on the latitudes and varieties involved. In the southern U.S. soybeans are grown during Brazilian winters, being planted in May and June and harvested in October and November.

INSECT PESTS

Soybean insects in Brazil were discussed by Panizzi et al. (4). They presented photographs and described important pests, their natural enemies and recommended insect management practices. Feeding niches of insects in soybeans were compared for North America, South America and the Orient by Turnipseed and Kogan (8). A listing of important foliage, pod and stem feeding insects of soybean in Brazil and the southern U.S. is shown in Table 1.

Table 1. Soybean insect pests in Brazil and the southern U.S.[a]

Area of Plant Fed Upon	Brazil	Southern U.S.
Foliage	*Anticarsia gemmatalis* (M)	*Anticarsia gemmatalis* (M)
	Pseudoplusia includens (M)	*Pseudoplusia includens* (M)
		Plathypena scabra
	Diabrotica speciosa	*Diabrotica* spp.
	Cerotoma sp.	*Cerotoma trifurcata*
	Colaspis sp.	
		Epilachna varivestis
Pods	*Nezara viridula* (M)	*Nezara viridula* (M)
	Piezodorus guildinii (M)	*Piezodorus guildinii*
	Euschistus heros	*Euschistus* spp.
	Acrosternum spp.	*Acrosternum hilare*
		Heliothis zea (M)
	Etiella zinckenella	
Stems	*Epinotia aporema* (M)	
	Elasmopalpus lignosellus	*Elasmopalpus lignosellus*
		Spissistilus festinus
		Dectes texanus texanus

[a](M) = major pest for which substantial hectarage is treated.

Major pests in both countries are denoted by (M) which indicates that substantial hectarage requires treatment.

The velvetbean caterpillar, *Anticarsia gemmatalis,* is the most important foliage feeding pest in both countries. The soybean looper, *Pseudoplusia includens,* although a major pest, is much less important than *A. gemmatalis* in Brazil. *P. includens* is reported to cause significant damage in the state of Sao Paulo where soybeans are grown in association with cotton. This pest is much more important in the southern U.S. where soybeans are frequently produced near cotton. Foliage feeders of less importance in Brazil include the chrysomelid beetles, *Diabrotica speciosa, Cerotoma* sp. and *Colaspis* sp. A similar group of chrysomelids found in the southern U.S. includes *Diabrotica* spp. and *Cerotoma trifurcata.* In the upper South and in northern areas of the U.S., *Colaspis brunnea* has been described as a pest species. It should be noted that in addition to feeding on foliage, pod-feeding is frequently observed by *Cerotoma* in both countries, and that larvae of all these beetles feed on roots and/or nodules, causing undetermined damage to the crop. The green cloverworm, *Plathypena scabra,* is an occasional pest on foliage in the upper South and Midwest. Also, the Mexican bean beetle, *Epilachna varivestis,* is an occasional pest in the Atlantic coastal areas and in southern Indiana in the U.S.

Of the insects that attack pods, the southern green stink bug, *Nezara viridula,* is of major importance both in Brazil and the southern U.S. These stink bugs damage seed within pods by direct feeding or by transmitting a yeast spot disease. Another stink bug, *Piezodorus guildinii,* is a major pest in Brazil, but is not important in the U.S., being found only occasionally in the southernmost latitudes. Two other genera, *Euschistus* and *Acrosternum,* often occur on soybeans in Brazil as members of the stink bug complex. Of these two genera, *Euschistus servus* may contribute to economic damage throughout the southern U.S. and *Acrosternum hilare* in the upper South. *Heliothis zea* is a major pest and often consumes whole pods or parts of pods in the southern U.S., but is not found on soybeans in Brazil. Also, *Etiella zinckenella* may occasionally damage pods in the southernmost part of Brazil by feeding through the pod, consuming the seed and webbing pods together. It is not a pest in the U.S.

The most important stem-feeding insect in Brazil is the shoot and axil borer, *Epinotia aporema.* This insect bores into the upper stem and feeds on the growing tips of the plant, often webbing the foliage of the bud together. Its major damage occurs in late vegetative growth, but feeding on flowers has been observed. It does not damage soybeans in the U.S. Another pest in Brazil that is less important is the lesser cornstalk borer, *Elasmopalpus lignosellus,* which attacks seedlings at the soil surface, leaving a characteristic webbing. It is widely distributed and is reported to cause considerable damage during the first year of cultivation on newly cleared land in the Cerrado region of central Brazil. The same species is widespread in the southern U.S. and occasionally causes extensive damage in sandy soils on late planted beans when conditions are hot and dry. Two species which do not occur in Brazil often cause damage in the southern U.S. These are a stem borer, *Dectes texanus texanus,* that is more of a problem in areas of the upper South, and the more widely distributed three-cornered alfalfa hopper, *Spissistilus festinus,* which damages stems by girdling with its piercing mouthparts.

NATURAL ENEMIES OF INSECT PESTS

Entomopathogens, predators and parasites play a major role in the natural regulation of pest species in both Brazil and the southern U.S. (4,8). The most spectacular of these natural enemies are the pathogens, which cause disease outbreaks in lepidopterous larvae, in many instances inflicting virtually 100% mortality in high populations. Complexes of polyphagous predators are probably equally important in providing natural control since they consume a high percentage of eggs and small larvae of pest species. An abundance of hymenopteran and dipteran parasites are probably least spectacular, but they often destroy eggs, larvae, pupae or adults of pests, exerting effective control of certain species.

The most important pathogen of *A. gemmatalis* in all areas of Brazil is the fungus, *Nomuraea rileyi.* It is also an important natural enemy of *P. includens.* Both of these lepidopterous pests are also attacked by *Entomophthora* spp. Several pest species, including *D. speciosa, Cerotoma* sp. and *N. viridula*, are killed by *Beauvaria* spp. A nuclear polyhedrosis virus was observed in populations of *A. gemmatalis* in Santa Catarina and Rio Grande do Sul. The virus has been imported into the U.S. where it has excellent potential as a biological insecticide against *A. gemmatalis* (1). As in Brazil, the most important pathogen in the southern U.S. is *Nomuraea rileyi.* It attacks *P. scabra, H. zea, P. includens* and *A. gemmatalis.* Also, *Entomophthora gammae* is an important pathogen in *P. includens* in the U.S. *Beauvaria* is also present, but is much less important than *Nomuraea* and *Entomophthora.*

The major predators of soybean insect pests in Brazil include *Nabis* spp., *Geocoris* spp., various spiders, *Callida* spp., *Lebia* spp. and ants. A somewhat similar situation exists in the southern U.S. in that nabids, geocorids and spiders (especially *Oxyopes salticus* and *Pardosa saxatilis)* are the most important predators (3), and, although *Callida* spp., *Lebia* spp. and ants are important, they do not seem to be as prevalent as in Brazil. The above predators attack eggs and small larvae, often to the extent that pest outbreaks are avoided. Larger larvae are also attacked in both countries by various predacious stink bugs (particularly in Brazil), earwigs, wasps, and larger carabids. *Calosoma* sp. were observed by the writer in very high numbers preying upon *A. gemmatalis* on soybeans in southern Goias. Adults were observed searching foliage as well as the soil surface during the day and larvae of the beetle were common on the soil surface, attacking large *A. gemmatalis* larvae and pupae.

The regulating effect of parasites on certain pest insects is evident both in Brazil and the southern U.S. Some species are common to both countries, whereas others are not. In Brazil two of the most important parasites are *Microcharops bimaculata* (Ichneumonidae) and *Copidosoma truncatellum* (Encyrtidae) parasitizing *A. gemmatalis* and *P. includens,* respectively. The scelionids *Telenomus morideae* and *Microphanurus scuticarinatus* are important egg parasites of *P. guildinii,* parasitism by the former being as high as 27% in areas of Parana. The tachinid *Eutrichopodopis nitens* is a common parasite of several species of stink bugs. In the southern U.S., stink bug eggs (primarily *N. viridula*) are commonly parasitized by a scelionid, *Trissolcus basalis,* and adults by a tachinid, *Trichopoda pennipes.* Parasites of lepidopterous larvae in South Carolina (southern U.S.) have been described by Sanders (6) and include *Copidosoma truncatellum* in *P. includens, Apanteles marginiventris* in several species, and several tachinids.

CURRENT INSECT MANAGEMENT STRATEGIES

The many similarities in production practices, insect pests and their natural enemies would suggest that similar insect management programs might be effective for Brazil and the southern U.S. Also, research has demonstrated that application of high dosages of broad spectrum insecticides will cause resurgences of pest species in both countries through interference with natural enemies (5,7).

Kogan et al. (2) developed a management strategy for insect pests in southern Brazil which was generally the same as strategies employed in the southern U.S. This consisted of scouting fields once a week to monitor populations of defoliating lepidopterous caterpillars and pod sucking stink bugs. Damage thresholds and treatment decisions were based on assessment of defoliation levels, pest populations, and presence of pathogens, mainly the fungus *N. rileyi.* Minimal dosages of nonpersistent pesticides were prescribed in fields reaching damage thresholds. Insecticide applications were reduced 78% in the experimental pest management fields compared with conventional grower treated fields.

Success of the above-described insect management strategy has been recognized and it was officially adopted for the state of Parana (11) and later by the federal agricultural research organization, EMBRAPA (4). Recently, considerable improvements have been made in the program in Brazil, including information delivery through state-wide television networks. Currently, insect management is practiced on a substantial amount of soybean hectarage in Brazil.

OPPORTUNITIES FOR COOPERATIVE BRAZILIAN—U.S. RESEARCH

Even though associations between Brazilian and U.S. soybean entomologists have been brief and intermittent, substantial benefits have accrued. The applicability of the above-described insect management strategies developed in the southern U.S. to largely similar insect problems in southern Brazil is an excellent example (2). Early work with Dimilin® was conducted jointly in Brazil and the southern U.S. and this insecticide now offers considerable potential for control of *A. gemmatalis* in both countries (10). Cooperative work has speeded the development of genotypes with resistance to insect pests (9). A virus of *A. gemmatalis* collected from Chapeco, Santa Catarina in southern Brazil (1) shows promise for use against the same pest in the southern U.S. where the virus is not native. It is effective as a microbial insecticide and may be introduced into certain soybean agroecosystems as a permanent natural control agent.

Provided with these examples, it is obvious that increased, well-planned, and uninterrupted cooperation among entomolgists in both countries would produce much greater benefits in managing insect pests. Joint efforts in the development of pest resistant soybean genotypes and in on-site research-development-exchange programs with natural enemies (entomopathogens, predators and parasites) of pest species should be explored. More virulent strains or more effective types of pathogens and more effective species of predators and parasites could be developed for both countries.

Finally, this cooperation need not be limited to insect pests, although it would probably be most beneficial in this pest class. There are sufficient similarities in weeds, plant pathogens, and nematode pests to warrant consideration of cooperative work in the whole area of soybean crop protection.

NOTES

Sam G. Turnipseed, Department of Entomology, Clemson University, former Visiting Professor, University of Wisconsin, on assignment in Brazil (Aug. 1975 to July 1976).

REFERENCES

1. Carner, G. R. and S. G. Turnipseed. 1977. Potential of a nuclear polyhedrosis virus for control of the velvetbean caterpillar in soybean. J. Econ. Entomol. 70:608-610.
2. Kogan, M., S. G. Turnipseed, M. Shepard, E. B. de Oliveira, and A. Borgo. 1977. Pilot insect pest management program for soybean in southern Brazil. J. Econ. Entomol. 70:659-663.
3. McCarty, M. T. 1979. Identification of predaceous arthropods in soybeans using autoradiography. M.S. Thesis, Clemson Univ., 30 pp.
4. Panizzi, A. R., B. S. Correa, D. L. Gazzoni, E. B. de Oliveira, G. G. Newman, and S. G. Turnipseed. 1977. Insetos da Soja no Brazil. EMBRAPA Boletim Tecnico no. 1. 20 pp.
5. Panizzi, A. R., B. S. Correa, G. G. Newman, and S. G. Turnipseed. 1977. Efeito d insecticidas na populacao das principais pragas da soja. Anais da Sociedade Entomologica do Brazil no. 2. pp. 264-275.
6. Sanders, G. J. 1978. Parasites of lepidopterous larvae in insect resistant and susceptible soybeans in South Carolina. M.S. Thesis, Clemson Univ., 72 pp.
7. Shepard, M., G. R. Carner, and S. G. Turnipseed. 1977. Colonization and resurgence of insect pests of soybean in response to insecticides and field isolation. Environ. Entomol. 6:501-506.
8. Turnipseed, S. G. and M. Kogan. 1976. Soybean entomology. Ann. Rev. of Entomol. 21:247-282.
9. Turnipseed, S. G. and M. J. Sullivan. 1976. Plant resistance in soybean insect management, pp. 549-560. In L. D. Hill (Ed.) World Soybean Research, Interstate Printers and Publishers, Inc., Danville, Ill., 1073 pp.
10. Turnipseed, S. G., E. A. Heinrichs, R. F. P. da Silva, and J. W. Todd. 1974. Response of soybean insects to foliar applications of a chitin synthesis inhibitor TH 6040. J. Econ. Entomol. 67:760-762.
11. Villacorta, A., E. B. de Oliveira, and M. Kogan. 1976. Pragas e seu controle, pp. 213-227. In IAPAR, Manual agropecuario para o Parana, Fundacao Instituto Agronomico do Parana, Londrina, Parana, Brazil, 387 pp.

SOYBEAN INSECT PROBLEMS IN INDIA

A. K. Bhattacharya and Y. S. Rathore

According to the experience of Indian researchers and farmers about 60 to 100 species of insects attack soybeans at various stages of development. However, not all insects found in soybean fields are necessarily injurious to the crop. Under our prevalent field conditions few species may be considered key pests; some are potential pests and others are sporadic pests which seldom have a major economic impact. It is essential to define what is an injurious population and the economic loss that it can cause to avoid unnecessary expenditure on pesticides and the detrimental effects resulting from their misuse. The objective of this chapter is to summarize our current knowledge of the major soybean insect pests in India and to define the criteria that we use in deciding on the judicious application of insecticide to prevent the economic loss due to these pests.

FIELD PESTS

Seed and Seedling Stage

Hylemya platura (Meigen) – Seedcorn maggot: These small, yellowish-gray flies lay elongated, whitish eggs in the soil. Eggs hatch into small, creamy maggots. Under laboratory conditions maggots fed moist swollen seeds reach the pupal stage in 7 to 10 days. Pupation occurs in the soil. Puparia are elongated and brown. The pupal period lasts 6 to 10 days. Maggots feed on the germinating seeds in the ground and also on cotyledons of emerging seedlings. Each seed can be infested by up to 10 maggots. Infested seeds can be found in the furrows a few days after sowing. Maggots may also attack the plumule causing germination failure and irregular or patchy stands. Sometimes only partially eaten cotyledons emerge above the soil surface and are easily identified by the presence of brownish patches. Severe infestations by this species occur in the spring season (February through May) at Pantnagar.

Agrotis ipsilon (Hufnagel) — Black cutworm: The adult is a medium-sized moth (about 2.5 cm long) with dirty greyish wings marked by black or grey spots. Moths are attracted to light. Female moths lay about 200 to 400 spherical, creamy white eggs singly on the lower surface of leaves (18). Eggs hatch in 2 to 13 days. Fully grown caterpillars are about 45 cm long and dirty black in color. The larval stage takes approximately 30 days and the life cycle is completed in 35 to 65 days (15). The caterpillars live in cracks and crevices of the soil near the plant and feed during the night. They cut stems of seedlings and feed on leaves. Sometimes they drag the cut plant portion into cracks near the base of the stem. Young seedlings may also show stem cuts 7.5 to 10 cm above ground level. Close examination of cracks near cut seedlings reveals the presence of smooth, grey larvae which exhibit a characteristic coiling behavior when they are touched. At Pantnagar severe infestation due to this insect was observed in spring 1972.

Stem Borers

Melanagromyza sojae (Zehntner) — Stem fly: These are about 2 mm long flies, light brown soon after emergence from the puparium and turning dark metallic black after tanning and hardening of the integument. The host range of the stem fly includes *Glycine max* (L.) Merrill, *G. sojae* Sieb and Zucc, *Cajanus indicus* Spreng., *Indigofera suffruticosa* Mill., *I sumatrana* Gaertn., *Phaseolus calcaratus* Roxb., *P. radiatus* L., *Swainsonia galegifolia* (And.) R. Br., *Melilotus* sp., and *Medicago sativa* L. (21). It is a serious pest of soybeans in northern regions of India. Adults feed by making multiple punctures which appear as white spots on leaves. Eggs are laid in the soft tissues of the leaf and hatch in 2 to 7 days. Larvae start feeding on the leaf and move towards the center of the stem, penetrating through the petiole. In 2 to 3 days maggots reach the stem and undergo 3 to 4 moults. Duration of the larval stage is approximately 10 to 15 days and pupation occurs in the stems. When an infected stem is split open a distinct tunnel can be seen corresponding to the area eaten by the maggot. Before pupation the maggots prepare a hole at the base of the stem for exit of the future adult fly. The puparia are barrel-shaped and dark yellowish brown. Duration of the pupal stage is approximately 7 to 10 days.

This insect is a major pest of the rainy (July through October) and spring (February through May) season crop. In the early stages of crop growth only 20 to 30% of the plants are affected. As plants grow the infestation level increases gradually and at harvest 100% of the plants may be infested. Maggots cause severe damage by tunneling the main stem, but may also tunnel petioles and side branches. In general 20 to 30 day old plants do not withstand this type of damage which results in drying of leaves and withering death of plants after a while. Sometimes plants produce secondary branches which compensate losses to some extent. Early damage can be compensated by replanting. However, fully grown plants when infested show no morphological symptoms or abnormalities and it is almost impossible to identify such plants from uninfested ones. Goot (13) reported that about 70 to 100% of *G. sojae* plants were attacked by this pest. It was also observed that common native plants such as *Aeschynomene indica, Flemingia* sp., and *Phaseolus sublobatus* act as a permanent reservoir for this species. In Japan soybean plants are also attacked by this pest during the summer and early autumn. In Kyushu (Japan) up to 100% infestation was observed by Kato (14). Recently, at Pantnagar, an attempt was made to study the correlation between tunnel length and various other characters including plant ht, number of pods/plant, grain wt, and wt/grain. It was concluded that tunnel length fails to show direct loss in the yield of soybean grains. In Delhi and Jabalpur, the stem fly *M. phaseoli* Tryon, is the most destructive pest of soybeans (Table 1). This insect damages 95 to 100% of plants in the rainy season and about 60% of plants in the spring season. Screening of 1700 cultivars of soybeans for stem fly resistance indicated that all these cultivars are highly susceptible to this pest (11).

Oberea brevis Swed. — Girdle beetle: This is a serious pest in Madhya Pradesh Provence where it damages up to 30% of the soybean plants (10). Gangrade (11) found that when this pest infests a soybean plant it reduces pods and grains by 50%. The mean wt of 100 seeds from the healthy plants, and the bottom and top cut portions of the infested plant showed significant differences. Gangrade also estimated that a net loss of 9.42 kg of pods and 5.43 kg of grains/ha occurred for every 1.0% of infestation. Our recent surveys reveal that crops grown in the hills of Uttar Pradesh, Tarai and Delhi also suffer some losses due to this insect. The adult is a medium-size cerambycid beetle with the anterior half of the elytra deep brown and the posterior half

Table 1. Susceptibility of 'Bragg' to *Melanagromyza phaseoli* at different intervals of time (January through April, 1972).

Period of Closing	Infested Plants	Tunnel Length	Yield
— days —	*— % —*	*— cm —*	*— g —*
5	0.0	–	524.60
10	5.6	10.8	328.48
15	0.0	–	426.24
20	8.3	83.3	187.57
25	5.0	46.6	222.10
30	13.2	87.1	207.19
35	10.1	71.2	110.15
40	36.6	138.4	145.51
Closed till maturity	0.0	0.0	432.39

deep black. This is a polyphagous species which infests cowpea, bittergourd and chilies (12,20).

After mating, females make two parallel girdles, usually on the petiole or at times on the main stem or side branches. Females may bore several trial holes between these girdles before inserting a single yellow egg inside the stem. Each female can lay 8 to 72 eggs, each measuring 1.25 to 2.25 mm. Hatching occurs in 3 to 8 days. The newly emerged larvae are pale whitish, while the full grown grub is about 18 to 20 mm long and deep yellow. The larvae feed inside the stem for approximately 40 to 60 days during the months of June and July (12). The total life cycle is completed in 40 to 70 days.

It appears that the girdles serve to adjust the moisture content of the oviposition site. This observation is based on the fact that in many cases an egg or the larva was not found inside the stem between the freshly-made girdles. In these instances it is possible that the female did not lay an egg because of the lack of a suitable environment. Field studies revealed that one plant may have as many as three or four pairs of girdles. Early stage infestations are characterized by fresh, green-colored girdles which later turn brown.

If a petiole is girdled, the trifoliate leaves begin drying around the edges, which results in the curling of the leaf margin. Finally the entire leaflet dries up, and can be easily spotted in the field. This type of drying is restricted only to the girdled petiole. On the other hand, if a branch of the main stem is girdled, it may result in the drying of all leaves above the girdled area. If an infested area of the plant is split open, a wide, deep brown tunnel can be observed.

Defoliators

Diacrisia obliqua (Walker) – Bihar Hairy Caterpillar: Adult moths have light yellow wings with black dots on the anterior margin of the fore wings. Their abdomens are pink with black dots on the dorsal and lateral surfaces. The male moths measure approximately 40 to 50 mm while females measure approximately 50 to 60 mm. A single female may lay 1000 to 1500 eggs, arranged in closely-packed batches of 400 to 750 on the leaf surfaces. These pale greenish eggs hatch in 3 to 7 days from August to September and 11 to 15 days from November to December. Newly emerged caterpillars are gregarious feeders on the leaf epidermis, skeletonizing entire leaves. At the end of the third instar, larvae migrate to adjacent plants and become solitary feeders consuming large amounts of leaf area. There are 5 to 6 distinct instars which last approximately 20 to 30 days. The full-grown caterpillar measures approximately 40 to 50 mm. Duration of the pupal state is approximately 10 to 15 days. Under normal conditions, the life cycle is completed in 5 to 6 weeks, but during cold periods of the year a single generation may take up to 10 weeks. This is a major pest of soybeans, feeding on a wide range of host plants including soybeans, green gram, black gram, jute and maize (9).

Spodoptera litura (Fabricius) – Tobacco caterpillar: Adults are medium-sized moths (35 to 40 mm wing span) whose fore wings are greenish brown with scattered wavy markings. The hind wings are shiny white with light ciliated boarders. Females lay 1000 to 2000 eggs in batches covered with buff-colored hairs. Eggs hatch in 3 to 7 days. The newly emerged larvae have green bodies and black heads with dorsolateral black dots on the abdominal segments. Fully grown caterpillars are 20 to 30 mm in length with rows of yellow-green dorsolateral stripes and lateral white bands. Young larvae feed gregariously up to the third instar when they migrate to adjacent plants. At temperatures below 27 to 30 C, larvae moult 6 to 7 times, while within these temperatures, only 5 to 6 instars have been recorded (2). On soybeans, the larval stage is approximately 14 to 20 days at 27 C. Pupation occurs in the soil and adults emerge in 7 to 10 days. Under normal conditions, the life cycle is completed in 30 to 37 days. This caterpillar has a wide variety of host plants and is considered to be one of the most important insect pests of soybeans (1,17). Early larval instars feed on the epidermis, leaving the main veins and thus skeletonizing the leaves. Later instars feed voraciously on young and old leaves and can completely defoliate plants.

Spodoptera exigua (Hubner) – Lucerne caterpillar: These greyish brown moths measure about 25 to 35 mm long. The fore wings are usually greyish brown with a light spot in the center. The hind wings are pale brown, darkening near their ciliated margins. Eggs are laid on either the upper or lower surfaces of leaves in irregular clusters of 20 to 2000 eggs which are usually covered with whitish-yellow scales. A single female may lay 1300 to 1500 eggs which hatch in 1 to 3 days. Newly emerged larvae are light green, but darken with age and develop white longitudinal stripes. Fully grown, they measure approximately 25 mm. Larvae moult 5 to 7 times on soybeans during a 17 to 22 day period. Pupation occurs among dried, fallen leaves or in the soil and lasts approximately 6 to 8 days. The entire life cycle is completed in 4 to 5 weeks at 27 C. This is a polyphagous insect which feeds on alfalfa, potato, maize, cabbage, indigo, tomato, egg plant, cotton, and linseed (10,18), and is also an important pest of soybeans. Early instars skeletonize leaves by feeding on the epidermis. Mature larvae chew small holes in older leaves and near the growing tip. High population levels may completely defoliate the plant.

Plusia orichalcea Fabricius – Semilooper: This insect appears from August into the spring season. Adults are medium sized moths with a metallic-yellow patch in the fore wing. Following a 3 to 5 day preoviposition period, eggs are laid singly on both sides of leaves. In 3 to 4 days, the pale greenish-white larvae emerge and feed singly on the soft tissues, leaving the veins. Fully developed larvae are green with distinct black and deep green longitudinal lines along the body. The larval period lasts approximately 13 to 24 days and pupation occurs inside folded leaves. Adults emerge in 8 to 9 days and live 2 to 5 days. The life cycle is completed in approximately 27 days (7). A severe infestation will leave the plant with only its main branches.

Helicoverpa (Heliothis) armigera Hubner — Gram pod borer: This is a serious pest of soybeans in Madhya Pradesh (10). Adults are medium sized moths, light yellow to brown, with distinct black dots in the fore wings and lighter colored hind wings. Females lay 1200 to 1550 eggs singly on leaf surfaces. Eggs hatch in 3 to 4 days. The newly emerged larvae are light green with dark longitudinal stripes. Fully grown larvae measure approximately 30 mm and larval development is completed in 20 to 25 days. Pupation occurs in the soil. Adults emerge in 9 to 13 days and the life cycle is completed in 31 to 35 days (10). In addition to soybeans, this insect feeds on gram, peas, cotton, and lucerne. Caterpillars may feed on tender leaves, but are most damaging to developing pods.

Mocis undata Fabricius — Brown striped semilooper: This insect is establishing itself as a soybean pest in Madhya Pradesh. Adults are stout moths with a wing span of 20 to 25 mm. The fore wings have three black stripes and the hind wings have definite smoky black stripes along their margins. Female moths lay 50 to 200 round, yellowish-green eggs which hatch in 3 to 5 days. Newly hatched larvae are pale green with pinkish brown heads. Larvae moult 6 times before pupation and are fully developed in 17 to 22 days, reaching a length of 41 to 53 mm. Pupation takes place in a cocoon fashioned of leaves stuck together with silken threads. The pupal period lasts for approximately 5 to 9 days in September and 38 to 43 days during October through December (10). This pest can also attack such plants as *Pueraria,* derris, and velvetbean, and at times causes serious damage to soybeans during the rainy season.

Lamprosema indicata Fabricius — Leaf roller: This insect occurs during the rainy season and has established itself as a major problem in Madhya Pradesh. Damage by this pest was also noticed in Tarai area during the rainy season of 1975. The adult moths have yellowish wings marked with distinct wavy brown lines. Following a preoviposition period of 1 to 4 days, females lay eggs either singly or in groups of 5 to 20 on the leaf surface. Eggs hatch in 7 to 8 days. The small white first instar larvae fold leaves around themselves as protection from natural enemies. They feed on the mesophyll resulting in an intact papery skeleton of folded leaves. There are 6 larval molts, and fully grown larvae measure 16 to 20 mm. The pupal period lasts 5 to 16 days.

Stomopteryx subsecivella Zeller — Leaf miner: This is a specific pest of groundnut which has begun attacking soybeans and is an important pest in Madhya Pradesh. Adults are small, dark brown moths with two white specks near the costal margin of the fore wings. Female moths lay 5 to 260 white rectangular eggs on the lower leaf surface which hatch in 2 to 4 days (10). Newly hatched larvae mine into leaves and feed below the epidermis. Larvae molt 3 times over an 8 to 12 day period of growth. They then spin cup-shaped cocoons by sticking adjacent leaves together. The pupal stage lasts 4 to 6 days, and the total life cycle is completed in 16 to 22 days. Larval damage is characterized by narrow tunnels between the layers of epidermis. Heavily mined areas within a field appear crinkled and distorted. Severe infestations may cause yield loss.

Sucking Insects

Bemisia tabaci Gennadius – Sweetpotato whitefly: This is a serious pest of soybeans in the Tarai and plains of the U.P. and Delhi regions. Adults are small (approximately 1.0 mm) yellow-bodied insects with whitish-grey wings and are densely covered with a waxy powder. Both nymphs and pupae are black and oval. The purpae have marginal bristles. Besides soybeans, this insect also feeds on tomato, okra, cotton, black gram, and green gram (16). Females lay 38 to 106 yellow eggs on the lower surface of leaves. These change to dark brown before hatching in 1 to 2 days. Nymphs feed by sucking sap from the leaves. The nymphs feed for approximately 7 to 14 days before molting into a quiescent "pupal" stage. Adults emerge in 8 to 14 days. The life cycle is completed in 13 to 72 days (16). Adults are known to transmit the yellow mosaic virus disease of soybeans. Field surveys did not reveal this disease in the plains of M.P. or the hills of U.P., but high levels of disease were found in the Tarai and the plains of the U.P. and Delhi regions. However, in Tarai regions, in spite of the presence of a large amount of inoculum and white fly, the disease does not appear during the spring season; it is most prevalent in July and August. Plants suffering from yellow mosaic virus disease have reduced pod formation and yield (Table 2). The early infested plants showed considerable reduction in the yield. Field studies reveal that 30 day infested plants produced significantly less grains as compared to plants infested on 60, 70, 80 and 90 days after sowing. Recently at Pantnagar it was observed

Table 2. **Effects of different levels of yellow mosaic infestation on the yield of soybeans.**

Exposure Period (Days)	Plant Ht (cm)	No. Pods/Plant	No. Grains/Plant	Wt Grains/Plant	Wt (g/grain)
30	57.24	12.3	25.5	2.27	0.1009
40	61.44	31.2	60.7	7.53	0.1149
50	58.48	30.4	60.4	7.21	0.1160
60	61.88	36.4	68.9	9.29	0.1335
70	61.44	40.4	79.8	10.70	0.1324
80	59.52	45.1	88.3	12.13	0.1384
90	49.10	42.4	82.2	11.98	0.1418
Healthy	60.68	40.9	83.9	11.71	0.1422
C. D. at 5%	4.90	11.0	23.3	3.09	0.0129

that PK-71-21 and PK-71-24 were less susceptible to yellow mosaic disease. Therefore, these two cultivars can be effectively used for developing yellow mosaic resistant varieties.

Nezara viridula Linnaeus – Southern green stink bug: Although adults are green, there are three distinct color variations in Madhya Pradesh. Males are generally smaller than females. Eggs are laid in masses of 42 to 113 eggs. Hatching occurs in 4 to 6 days. There are 5 nymphal instars. Early instars remain in clusters but disperse to neighboring plants by the third instar. Both nymphs and adults feed on green seeds, by piercing the developing pods and seeds within. Field trials conducted at Madhya Pradesh revealed that one insect per plant may cause 80% seed injury (19).

STORAGE PESTS

Raw Soybeans

Researchers have recently recorded the successful development of the almond moth, *Ephestia cautella* (Walker); Khapra beetle, *Trogoderma granarium* Everts; Pulse beetle, *Callasobruchus chinensis* (L.); and cigarette beetle, *Lasioderma serricorne* (F.) on stored soybean seeds. Of these, *E. cautella* is the most potentially damaging (8). It appears that larvae are unable to penetrate undamaged seeds, but can complete development if seed is offered as ground flour. Tests using flours of some of the most common varities (i.e. Bragg, Hood, Lee, Harosoy-63 and Hardee) resulted in 46 to 91% adult development within 28 to 32 days. It is possible that these insects may become serious problems in the future (4). Bhattacharya et al. (6) also reported that larvae of *L. serricorne* failed to develop on whole undamaged grains. But this insect develops quickly when seeds were offered as flours. Addition of yeast further hastened the development, resulting in 99% adult emergence. On the other hand larvae of *T. granarium* which attack a wide variety of substances of animal and plant origin showed poor development on soybean flour. Prolongation in the developmental period and poor adult emergence were also recorded on 179 germplasms (5). Addition of yeast and steaming significantly improved the quality of the diet (3). Therefore, apart from nutritional deficiencies, heat labile inhibitors are also responsible for the suppression of development of *T. granarium* on raw soybeans.

Soy Products

At present a number of soy products are being marketed in India by the Soya Production and Research Association, Bareilly. The major products are nutrinugget (a textured product made from 100% defatted soyflour), protesnack (flavored and spiced snack), nutriahar (extruded product made from cooked soybeans), pausticahar (ready to eat food made by extruding a mixture containing 44% corn, 40% full fat soy flour, 15% sugar and 1% vitamin-mineral premix), and protein plus (ready to eat product made by mixing 67.5% corn, 29.8% defatted soy flour, 1% salt and 1.7% vitamin-mineral premix).

Growth response of 4 stored grain insects on these soy products revealed that larvae of *T. granarium* failed to complete development on nutrinugget chunks, granules and powder while supplementation of yeast improved the quantity of the diet. Similar improvement due to the addition of yeast was also observed on protesnack and nutriahar powder. Larvae of *Tribolium castaneum* also did not show good response on nutrinugget chunks, granules and powder. Addition of yeast also failed to influence the quality of the nutrinugget powder. Significant improvement in growth and development of this insect on protesnack and nutriahar powder was observed when 5% yeast was supplimented in the diet. Larvae of *E. cautella* behaved differently on these products. On nutrinugget chunks 11% of adults emerged in 62 days while on nutrinugget granules or powder, or powder plus 5% yeast, larvae indicated poor development. Protesnack chunk was also infested by this insect but powder form of this diet was better utilized by the larvae. Results also revealed that pausticahar, nutriahar and protein plus powder were also severely infested by this insect. Larvae of *L. serricorne* indicated better development on all soy products as compared to the rest of the 3 species of insects. In all cases adult emergence varied from 76 to 100%. However, it appears that pausticahar was efficiently utilized by the larvae as compared to the rest of the test diets.

CONTROL

Proper pest management decisions are essential for obtaining the highest yield with minimum economic loss. Our goal in pest management is to prevent an increasing pest population from exceeding the economic injury level by intervening with appropriate control.

Fields should be scouted weekly to determine if potential pests are present in sufficient numbers to warrant control. Specific insecticidal control options are summarized in Table 3. A basic knowledge of the behavior of a pest species is important for a successful management program. For example, *S. litura, D. obligua,* and *S. exigua* lay eggs in clusters. Freshly hatched larvae feed gregariously through the third instar. During this stage of infestation the recommended insecticides may be applied sparingly in these infested areas. It may also be possible to use a hand-picking and killing technique in these locally infested areas. By taking advantage of the insects' behavior and spot-treating the early instar clustered population, one could avoid spraying the entire field later to control the widely dispersed older instars.

Table 3. **Summary of recommended insecticides and suggested rates for control of the major soybean insect pests in India.**

Insect	Insecticide	Rate	Application and Remarks
Stem fly	Thimet 10G	10 kg/ha	Soil insecticides
	Disyston 5G	20 kg/ha	
Girdle beetle	Thiodan (endosulfan) 35% EC	0.05% EC spray	Dilute with water to required concentration
	Dimethoate 30% EC	0.03% EC spray	
	Methyl demeton 30% EC	0.03% EC spray	
White flies	Methyl demeton 25% EC	1.0 liter/ha	Apply, mixed with sufficient water carrier, 20, 30, 40 and 50 days after planting to prevent yellow mosaic virus disease transmitted by white flies
Lepidopterous spp.	Thiodan (endosulfan) 35% EC Ekalux (quinalphos) 25% EC	1.0 to 1.25 liter/ha	Apply, mixed with sufficient water carrier, 4, 6, 8, and 10 weeks following planting
Defoliators and sucking insects	Methyl demeton 25% EC + Thiodan 35% EC	1.0 liter/ha 1.0 liter/ha	Apply, mixed with sufficient water

NOTES

A. K. Bhattacharya and Y. S. Rathore, Department of Entomology, G. B. Pant University of Agriculture and Technology, Pantnagar (Nainital), India.

Some research findings reported in this article are based on work carried out under a grant provided by the USDA under the PL-480 (FG-In-461) research project. The authors are thankful to Dr. M. Kogan, Entomologist, and C. G. Helm, Assistant Supportive Scientist, Section of Economic Entomology, Illinois Natural History Survey, Urbana, Illinois, for offering valuable comments and critical examination of this manuscript.

REFERENCES

1. Babu, M. H., A. K. Bhattacharya, and Y. S. Rathore. 1978. Developmental behavior of three lepidopterous pests on soybean and green gram. Zeit. ang. Entomol. 85: 108-112.
2. Bhatt, N. S. and A. K. Bhattacharya. 1976. Development of *Spodoptera littoralis* (Boisd.) (Lepidoptera: Noctuidae) at constant temperatures on two host plants. Zeit. ang. Entomol. 70:201-206.
3. Bhattacharya, A. K. and S. R. Yadav. 1978. Effect of heat treatment on the nutritive value of soybeans for *Trogoderma granarium* Everts. Bull. Grain Tech. 16: 169-173.
4. Bhattacharya, A. K., R. R. P. Chaudhary, and R. R. S. Rathore. 1976. Susceptibility of several varieties of soybean to *Ephestia cautella* (Walker) (Lepidoptera: Phycitidae). J. Stored Prod. Res. 12:143-148.

5. Bhattacharya, A. K., R. R. P. Chaudhary, S. R. Yadav, and Y. S. Rathore. 1978. Susceptibility of soybean germplasms to *Trogoderma granarium* Everts. Bull. Grain Tech. 16:87-93.
6. Bhattacharya, A. K., H. J. Vyas, and Y. S. Rathore. 1977. Damage of soybean by *Lasioderma serricorne* (Fabricius). Bull. Grain Tech. 15:151-152.
7. Bhattacharya, A. K., S. R. Yadav, and R. R. P. Chaudhary. 1977. Biology of *Plusia orichalcea* Fabricius (Lepidoptera: Noctuidae). Sci. and Cult. 43:173-174.
8. Chaudhary, R. R. P. and A. K. Bhattacharya. 1976. Larval developmental behavior of *Ephestia cautella* (Walker) on several food commodities. Bull. Grain Tech. 16:3-8.
9. Deshmukh, P. D., Y. S. Rathore, and A. K. Bhattacharya. 1977. Studies on growth and development of *Diacrisia obliqua* Walker (Lepid.:Arctiidae) on sixteen plant species. Zeit. ang. Entomol. 84:431-435.
10. Gangrade, G. A. 1974. Insects of soybean. Directorate of Research Service, JNKVV, Jabalpore, Madhya Pradesh, India, 88 pp.
11. Gangrade, G. A. 1976. Assessment of effects on yield and quality of soybeans caused by major arthropod pests. Dept. of Entomology, Jawaharlal Nehru Agricultural University, Jabalpur, Madhya Pradesh, India, 142 pp.
12. Gangrade, G. A., K. N. Kapoor, and J. P. Gujrati. 1971. Biology, behavior, diapause and control of *Oberia brevis* Swed. (Coleopotera: Cerambycidae-Lamiidae) on soybeans in Madhya Pradesh. Entomologist 104:260-264.
13. Goot, P. van der. 1930. De *Agromyza yliegjes* de inlandsche Katjanggewassen op Java. Meded. Inst. Plziekt., Buitenz. 78:1-97.
14. Kato, S. 1961. Taxonomic studies on soybean leaf and stem mining flies (Diptera: Agromyzidae) of economic importance in Japan with description of three new species. Bull. Natn. Inst. Agric. Sci. (c). 13:171-206.
15. Narayanan, E. S. 1954. The greasy cutworm, *Agrotis ypsiolon* Rott., a serious pest of *rabi* crops. Indian Fmg. 3:8-10.
16. Nene, Y. L. 1972. Viral diseases of pulse crops in Uttar Pradesh, GBPUAT, Pantnagar, Distt. Nainital, India, 190 pp.
17. Prasad, J. and A. K. Bhattacharya. 1975. Growth and development of *Spodoptera littoralis* (Boisd.) (Lepidoptera: Noctuidae) on several plants. Zeit. ang. Entomol. 79:34-38.
18. Rai, B. K. 1976. Pests of oilseed crops in India and their control. I.C.A.R., New Delhi, India, 121 pp.
19. Singh, Z. 1973. Southern green stink bug and its relationship to soybeans. Bionomics of the southern green stink bug, *Nezara viridula* Linn. (Hemiptera: Pentatomidae) in central India. Metropolitan Book Co (Pvt.) Ltd.Delhi, India, 105 pp.
20. Singh, D. P. and K. M. Singh. 1966. Certain aspects of bionomics and control of *Oberea brevis* Swed. Labdev. J. Sci. Tech. 4:174-177.
21. Spencer, K. A. 1973. Agromyzidae (Diptera) of economic importance. E. S. Gottingen (ed.), Dr. W. Junk, B. V., The Hague, The Netherlands, 418 pp.

INSECT PROBLEMS OF SOYBEANS IN THE UNITED STATES

M. Kogan

Nearly 25 million ha were planted to soybean in the U.S. in 1978 and the tendency is for more hectareage increases in coming years. This huge area encompasses very diverse agroecological conditions. Production is highly concentrated in two major natural regions known as the Coastal Plains and the Interior Plains, with smaller areas in soybean production also found in the Interior and the Appalachian Highlands (Figs. 1 and 2). These regions extend from the Atlantic and Gulf coasts westward and northward between latitudes 28 and 46° north. Little is planted west of the 98° meridian (1). These natural regions have a variety of soil types, temperature and precipitation regimes (3); and soybean is grown in association with several other crops forming characteristic agroecosystems. Some of the most typical agroecosystems are soybean-corn-forage in the Midwest and soybean-corn-cotton-small grain, soybean-grassland, soybean-rice, and soybean-sugarcane in the South. Other crops that often appear in association with soybean are tobacco and peanut in the Atlantic coast and vegetables, conifers and hardwood forests over the entire production area. Fallow and abandoned fields often contribute to the diversity of these agroecosystems which are further enriched by weeds that grow along hedgerows and ditch-banks and within fields when the weedy plants escape control (36).

A perception of the heterogeneity of the agroecological conditions in the U.S. is essential to explain the qualitative and quantitative diversity of arthropod communities associated with soybean in this part of the world. Furthermore, one cannot cover insect pests of soybean in the U.S. as a whole, because each growing region has unique pest problems and faunal characteristics. Broad generalizations may lead to a distorted picture. It is, therefore, of interest to classify the major soybean production regions in terms of those

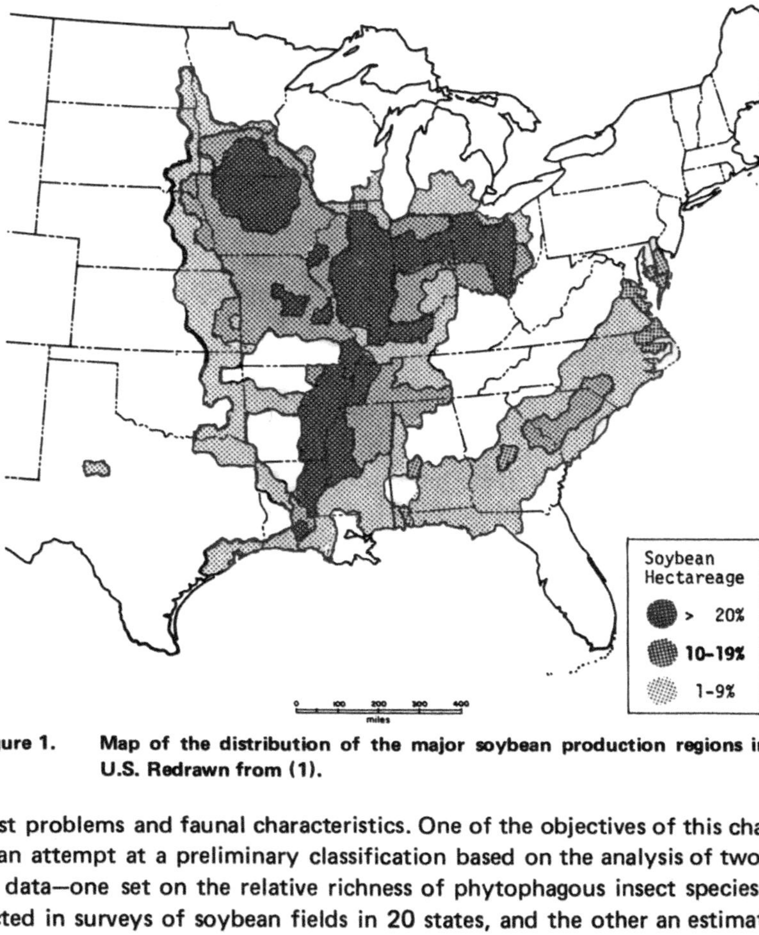

Figure 1. **Map of the distribution of the major soybean production regions in the U.S. Redrawn from (1).**

pest problems and faunal characteristics. One of the objectives of this chapter is an attempt at a preliminary classification based on the analysis of two sets of data—one set on the relative richness of phytophagous insect species collected in surveys of soybean fields in 20 states, and the other an estimate of the pest status of some 24 species in 18 states. The purpose of this exercise is to establish a pattern upon which some generalizations can be made on the current status and possible trends in the evolution of soybean/arthropod associations in the U.S. Identification of these natural regions may help develop intraregional programs to advance our understanding of the dynamics of pest populations and it may help develop interregional programs for investigating tendencies of arthropod community structure by comparing species packing and niche occupancy in the various regions.

This study has been limited to the above-ground phytophagous insect fauna on which survey data and economic thresholds are available more readily. Although ground dwelling nodule and root feeding species have been included in some of the analyses, their economic impact and the actual role of many of

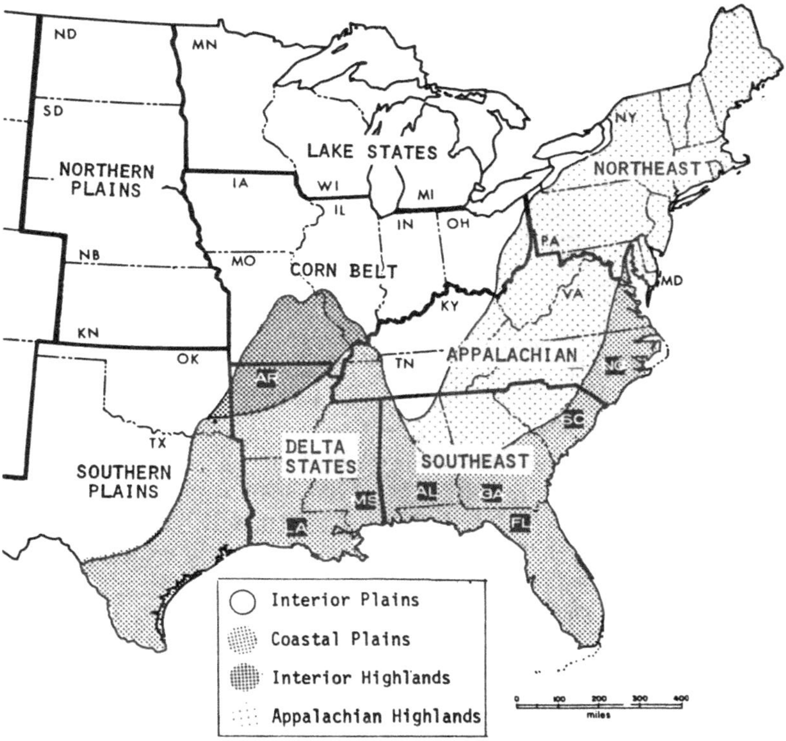

Figure 2. **Map of the major natural zones of the U.S. Much of the soybean production is concentrated in areas of the Coastal Plains and the Interior Plains, mainly along the Mississippi River Valley (see Fig. 1).**

them are not known. For example, the platystomatid fly, *Rivellia quadrifasciata* (Macquart) has been collected in surveys since the early 1970's from various states but larval feeding on soybean roots and nodules has been ascertained only recently (12). Similarly the effect of insects as vectors of diseases has not been considered because little is known about their economic role, except perhaps for the yeast spot disease transmitted by stink bugs and the bean pod mottle and cowpea mosaic viruses transmitted by the bean leaf beetle.

INSECTS AND INSECT PESTS OF SOYBEAN

Several extensive surveys of the soybean-associated arthropod fauna have been conducted in various states (4,5,10,34). Since 1969 we at Illinois have compiled data on soybean arthropods throughout the U.S. (and also abroad) and a reference collection has been assembled (21). The collection is known by the acronym IRCSA—International Reference Collection of Soybean-Associated Arthropods. These various sources have identified over 700 species of phytophagous insects collected in soybean fields, although not all of these species are soybean feeders and the actual role of many of the species is not known.

Despite the relative richness of the insect fauna found in association with the crop, a survey conducted among entomologists in 18 states has shown that 83.2% of the total amount of insect damage to soybean in the U.S. is produced by no more than 8 species. Seven of these species belong to 3 major guilds or complexes—(a) lepidopterous defoliators including: the velvetbean caterpillar, *Anticarsia gemmatalis* (Hubner), the soybean looper, *Pseudoplusia includens* (Walker), and the green cloverworm, *Plathypena scabra* (F.); (b) coleopterous defoliators including: the Mexican bean beetle, *Epilachna varivestis* (Mulsant), and the bean leaf beetle, *Cerotoma trifurcata* (Forster); and (c) pod feeding Pentatomidae including the southern green stinkbug, *Nezara viridula* (L.), and the green stinkbug, *Acrosternum hilare* (Say). The eighth species, the corn earworm, *Heliothis zea* (Boddie), is locally important as a foliage and pod feeder. Six other species or group of species account for 15.4% of the potential damage of insects to soybeans. These species are: the tobacco budworm, *Heliothis virescens* (F.), the beet armyworm, *Spodoptera exigua* (Hubner), the three-cornered alfalfa hopper, *Spissistilus festinus* (Say), the spider mite, *Tetranichus* spp., the lesser corn stalk borer, *Elasmopalpus lignosellus* (Zeller), and various species of grasshoppers mainly in the genus *Melanoplus*. In addition to these 14, another 10 species or species complexes may cause sporadic damage in very limited areas (see list of species in Table 1).

The biology and economic impact of the 8 major species have been studied with increasing interest since the surge of soybean in American and world economies in the past 20 yr. Several review articles have been written on these species (16,33,35,36,39) and comprehensive bibliographies are now available for all except the soybean looper and the green stink bug (11,15,19, 25,26,32).

An attempt has been made to summarize some biological information on the 8 major species with references to the more basic literature (Table 2). This summary is supplemented by Figure 3 diagramming the most common patterns of seasonal abundance of 12 of the species regarded as important soybean insect pests in the U.S.

The second part of this paper will present an analytical approach to quantify the relative impact of the various species and to assess regional differences (and similarities) in terms of the relative economic impact of the pest species and general characteristics of the phytophagous fauna. Simple

Table 1. Pest impact index (PI) based on estimates of the frequency at which arthropods reach pest status over a 10 yr period, and the area affected by the pests. Estimates based on replies to a questionnaire.

Species	AL	AR	FL	GA	IA	IL	IN	KY	LA	MD	MO	MS	NC	OK	SC	TN	TX	VA
Anticarsia gemmatalis	7.5	1.0	95.0	24.0	0	0	0	0	7.0	0	0.1	26.0	1.0	2.0	5.0	0	6.0	0
Pseudoplusia includens	7.5	1.0	21.0	20.0	0	0	0	0	6.0	0	0.2	20.0	1.0	2.0	10.0	0	3.0	8.0
Plathypena scabra	0	2.0	0	0	0.8	2.8	2.6	6.0	0.2	2.0	1.5	18.0	0.6	10.0	0	0	30.0	40.0
Heliothis zea	15.0	3.0	21.0	24.0	0	0	0	0	0.7	8.0	0.7	23.0	37.0	10.0	17.5	1.0	5.0	40.0
H. virescens	6.2	0	0	0	0	0	0	0	0	0	0.7	11.2	0.1	0	0	0	0	0
Epilachna varivestis	1.0	0	0	0	0	0	6.4	2.5	0	14.0	0	0	5.0	0	2.5	7.5	0	24.0
Cerotoma trifurcata	1.0	0	0	0	0	0.7	0.2	0.1	10.5	0	0	5.0	5.5	4.8	0	0	0	5.0
Nezara viridula	1.0	2.0	24.0	25.0	0	0	0	0	27.0	0	0	10.0	0	0	16.0	0	25.0	0
Acrosternum hilare	0	0	0.6	0.1	0	0	0	1.5	0.1	0	0	7.2	0.8	12.0	5.6	0	2.5	4.0
Spodoptera exigua	2.5	1.0	6.0	2.0	0	0	0	0	0.5	0	0	8.7	3.6	0	0.3	0	0	1.5
Vanessa cardui	0	0	0	0	0.1	0	0	0	0	0	0	0	0	0	0	0	0	0
Trichoplusia ni	0	0	0	0	0	0	0	0	0	0	0	0	0	0	0	0	0	2.0
Epargyreus clarus	0	0	0	0	0	0	0	0	0	2.5	0	0	0	0	0	0	0	0
Arctiidae	0	0	0	0	0	0	0	0	0	2.5	0	0.1	0	0	0	0	0	0
Spodoptera eridania	0	0	0	0	0	0	0	0	0	0	0	0	0.1	0	0	0	0	0
Delia platura	0	0	0	0	0	0	0	0.1	0	0	0	0	0	0	0	0	0	0
Agrotis ipsilon	0	1.0	0	0	0	0.1	0.2	0.2	0	0.3	0.2	0.1	0	0	0	0	0	0
Elasmopalpus lignosellus	0	0	20.0	7.5	0	0	0	0	0	0	0	0.1	0	1.5	1.0	0	0	0.3
Colaspis spp.	0	0	0	0	0	0	0	0	0	0	0	0	0.3	0	0	0	0	0
Sericothrips variabilis	0	0	0	1.0	0	0.3	0.1	0.1	0	0	0	0.1	0.4	0	0	0	0	2.0
Tetranichus spp.	0	0	4.0	0.1	0.1	0.4	0.2	0.1	0	6.0	0	0.1	0	2.0	0	0	0	0.6
Spissistilus festinus	0	0	0	9.0	0	0	0	0	1.5	0	0	2.2	0	0	9.0	0	0	0
Dectes texanus	0	0	0	0	0	0	0	0	0	0	0	0	0.5	0	0	0	0	0
Melanoplus spp.	0	0	0	0	0.1	2.1	9.0	1.5	0	0	0.1	0.7	0.2	2.5	0	0	0	0

Table 2. **General biological features of and nature of damage caused by eight insect species responsible for about 83% of the damage caused by insects to soybeans in the U.S.**

Guild and Species	Common Name	Biological Features	Nature of Damage	Key References
LEPIDOPTEROUS DEFOLIATORS				
Anticarsia gemmatalis	Velvetbean caterpillar	Overwinter southern Florida, Caribbean Islands; migrate north in spring. 3-4 generations in south; 2 generations north. Mostly legume hosts, but feed also on plants of 3 other families.	Larvae feeding on foliage.	(7,9,10,15,16,35 36)
Pseudoplusia includens	Soybean looper	Overwinter as pupae. 3-4 generations/yr. Rather polyphagous including hosts in various plant families but preferring legumes.	Larvae feeding on foliage.	(6,8,9,10,18,35, 36)
Plathypena scabra	Green cloverworm	Overwinter as adults in southern latitudes; migrate north. 2-4 generations/yr. Mostly legume hosts.	Larvae feeding on foliage.	(9,10,24,28,32, 35,36,41)

COLEOPTEROUS DEFOLIATORS				
Epilachna varivestis	Mexican bean beetle	Overwinter as adults. 2-4 generations/yr. Oligophagous restricted to Faboideae.	Adults and larvae feeding on foliage.	(9,10,25,35,36)
Cerotoma trifurcata	Bean leaf beetle	Overwinter as adults. 2-3 generations/yr. Oligophagous restricted to legume hosts.	Adults: leaf feeders; occasionally pod feeders; may transmit bean pod mottle and other viruses. Larvae: root and nodule feeders.	(10,17,22,26,35, 36,38,39)
POD FEEDING PENTATOMIDAE				
Nezara viridula	Southern green stinkbug	Overwinter as adults. 1 generation on soybeans. Polyphagous.	2-5th instar nymphs and adults mostly on pods and seeds. May transmit yeast spot disease.	(10,11,33,35,36)
Acrosternum hilare	Green stinkbug	Same as *N. viridula*	Same as *N. viridula*	(10,23,35,36)
POD FEEDING LEPIDOPTERA				
Heliothis zea	Corn earworm	Overwinter as pupae in soil. 1-2 generations on soybeans. Polyphagous.	Larvae mostly pod feeders but may feed on foliage and grow-tips in early infestation.	(10,19,35,36)

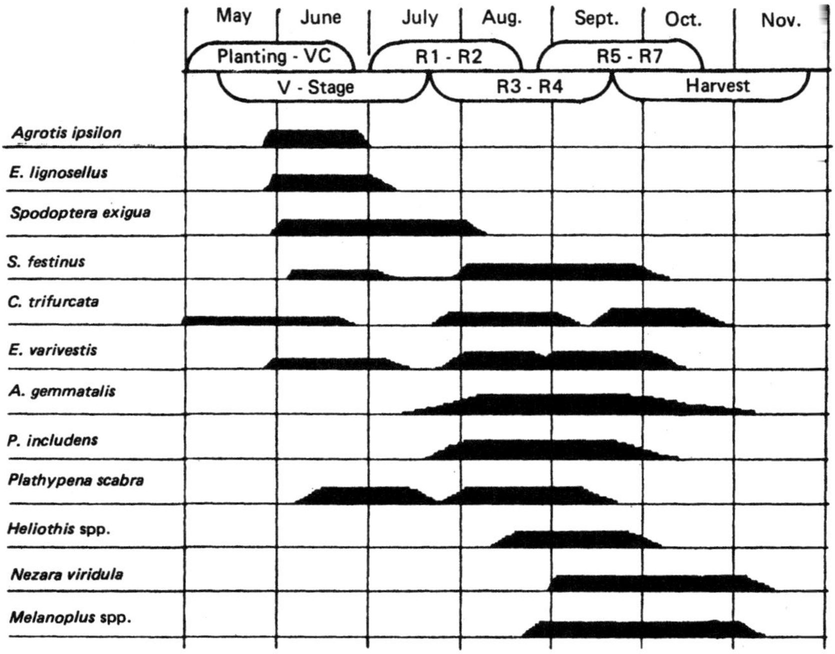

Figure 3. **Phenological relationship of some of the most important pests of soybean in the U.S. in relation to the growing cycle of the crop. Marked differences exist in crop cycle and insect phenology among regions.**

correlations of pest impact with mean latitude and mean annual temperature are included as a preliminary step in interpretation of results of these analyses.

MATERIALS AND METHODS

Data used in the analyses that follow come from two main sources: (a) insect surveys conducted in individual states or through the efforts of the Illinois soybean entomology program (data stored with the IRCSA unit); and (b) a questionnaire sent to 25 research and extension entomologists in each of 18 major soybean-producing states in the U.S.

Faunal Composition Analysis

IRCSA's computerized files were searched and listings of the species and number of individuals/species collected in 20 states were obtained. Most collections were made using a set of 100 sweeps/field as the sampling unit. The number of sampling units/state was variable. Parasitoids, predators and obviously incidental non-phytophagous species were excluded. Aphids were excluded because the general sampling procedures used in the surveys usually failed to detect the rich aphid fauna in the aerial plankton that lands continually on soybeans (M. E. Irwin and S. T. Halbert unpublished). The data that remained and were used in the analyses consisted of 42,968 specimens representing 301 phytophagous species.

The IRCSA data were supplemented with published records from extensive surveys in 4 states—Arkansas (34), Missouri (5), North Carolina (10), and Ohio (4). Again only phytophagous species were used. The total survey data included 453 species; no species abundance data were used, and only the presence or absence of the species in the state records was analyzed. These expanded survey data were used to compute a matrix of Sorensen's similarity quotients (31) for all 20 states compared two by two. (Similarity quotient $SQ = 2j/(a+b)$, where a = number of species in state A, b = number of species in state B, and j = number of species common to states A and B). A cluster analysis on the untransformed data was performed using the method of unweighed pair groups using simple averages (30).

Pest Impact Analysis

A questionnaire was sent to 25 entomologists in 18 states which accounted for 80% of the total hectareage planted to soybeans and were responsible for 78.5% of the total production in the U.S. in 1977. The questionnaire presented a list of 18 species and it requested for each species (a) an estimate of the probability of occurrence of outbreaks over a 10-yr-period (outbreaks were defined as population levels that had triggered a treatment response from farmers, but not necessarily populations that exceeded currently accepted economic thresholds—Table 3 (20); and (b) an estimate of the mean percentage of the area of the states affected by these outbreaks. An estimate of the relative abundance (on a 1 to 3 scale) during 3 growth periods attempted to provide data for comparative seasonal damage risk. The 3 growth periods were defined as early (up to R1), mid (R2 to R4), and late (R5 to harvest) (14).

These data were used to compute a pest impact index (PI) by multiplying the probability of outbreaks (P_o) by the percent of the area affected by these outbreaks (P_a), thus $PI = (P_o \times P_a) \times 100$. Pest impact indices for the 18 states and 24 insect pests (18 listed in the questionnaire and 6 added by various respondents) were tabulated (Table 2). A matrix of paired comparisons among all states was computed using the PI's for each species such that

$$C = [\sum_{i=1}^{n} (2A_i)/(A_i+B_i)]\,/n,$$

where A_i = the smaller, and B_i = the larger of the PI's for species i in a pair of states, and n = number of species in a pair of states with at least one PI value > 0.

Table 3. Percent of economic impact of 24 species based on rank index.

Species in Rank of % RPI	States Reporting	% Economic Impact
Heliothis zea	AL, AR, FL, GA, LA, MD, MO, MS, NC, OK, SC, TN, TX, VA	17.9
Anticarsia gemmatalis	AL, AR, FL, GA, LA, MO, MS, NC, OK, SC, TX, VA	14.0
Nezara viridula	AL, AR, FL, GA, LA, MS, SC, TX	13.4
Plathypena scabra	AR, IA, IL, IN, KY, LA, MD, MO, MS, NC, OK, TX, VA	12.2
Pseudoplusia includens	AL, AR, FL, GA, LA, MO, MS, NC, OK, SC, TX, VA	10.7
Epilachna varivestis	AL, IN, KY, MD, NC, SC, TN, VA	7.6
Cerotoma trifurcata	AL, IL, IN, KY, LA, MS, NC, OK, VA	4.5
Acrosternum hilare	FL, GA, KY, LA, MS, NC, OK, SC, TX, VA	2.9
	Sub total	83.2
Melanoplus spp.	IA, IL, IN, KY, MO, MS, NC, OK	4.2
Spodoptera exigua	AL, AR, FL, GA, LA, MS, NC, SC, VA	3.6
Heliothis virescens	AL, MO, MS, NC	3.5
Spissistilus festinus	GA, LA, MS, SC	2.3
Elasmopalpus lignosellus	FL, GA, MS, OK, SC, VA	1.2
Tetranichus spp.	FL, GA, IA, IL, IN, KY, MD, MS, OK, VA	0.6
	Sub total	15.4
Other 11 spp. (see Table 1)	Sub total	1.4

PI values were used to rank the species in a manner that took into account the area of the region affected by a given pest. This was done by computing a rank index RPI (or rank of pest impact) by adding the product of the PI of a species for a given state with the area of that state planted to soybean as a percentage (in decimal form) of the total area planted in the U.S. Thus the rank index for species i was

$$RPI_i = \sum_{j=l}^{n} PI_{i,j} x (A_j / 2.32 x 10^7),$$

where $PI_{i,j}$ = pest impact index for species i in state j; A_j = area in state j planted to soybeans; $2.32 x 10^7$ = U.S. soybean hectareage in 1977; n = number of states reporting on species i. The statistics for area planted were extracted from (2) and the base year was 1977.

RESULTS

Results of the various types of analyses are presented in tables and graphs; the textual material is limited to indicating those aspects that will be most relevant in the discussion. Because of space limitations only higher categories (families and orders) of the materials used in the faunal analyses are tabulated.

Faunal Composition Analysis

Table 4 presents the number of species of phytophagous insects by higher categories/state based on published surveys and the IRCSA data. The cluster analysis using similarity quotients (SQ's) with number of species/state is presented as a dendrogram with the levels of similarity indicated on the scale on the top of the graph (Fig. 4). The dendrogram is interrupted by three vertical lines at similarity levels .35, .42 and .45. OH and MD did not cluster above .30 level with any other units. The most meaningful aspect of the analysis at the .35 similarity level is the fact that the 4 states with the richest representation (NC, AR, MO, and IL) clustered together but were separated logically at the .46 level where IL and MO remained clustered but NC and AR had split. The .42 level was then used in partitioning the states in regions of common faunal characteristics. Obvious discrepancies from expected groupings occurred in the separation of GA from the cluster that included FL, LA, MS and SC. However, it is noticeable that even with less than adequate representation from some states, 4 major groupings were formed on the bases of faunal relationships that correlate reasonably well with what would be expected from geographic proximity alone.

Pest Impact Analysis

The PI's (Table 1) computed from the questionnaire data were used as input to generate a matrix of "C" paired comparisons for all possible pairs of states (144). This matrix was the input for a cluster analysis that generated the dendrogram shown in Figure 5. A vertical line indicates the .23 similarity level to the right of which there are 5 clusters and 4 individual states separated in a fashion that corresponds to a scale of mean pest impact values ranging from 7.98 (for FL) to .05 (for IA). The mean PI's per cluster being: cluster I (FL, GA, MS) = 6.09; cluster II (AL, LA, SC, TX) = 2.28; cluster III (OK, VA) = 4.71; cluster IV (NC) = 2.34; cluster V (MD) = 1.47; cluster VI (AR, MO) = .30; cluster VII (KY, IN, IL) = .52; cluster VIII (IA) = .05; and cluster IX (TN) = .35. This clustering formed the basis for the identification of regions of common pest impact values (PI>4.0; 4.0>PI>1.0; 1.0>PI>.1; PI<.1).

Ranking of PI by RPI index is summarized in Table 5 for the 8 top species.

Table 4. Number of species of phytophagous insects collected in soybean fields in 20 states. Totals grouped by order and family.

Orders and Family	AL	AR	FL	GA	IA	IL	IN	LA	MD	MN	MO	MS	NB	NC	OH	SC	SD	TN	VA	WI
Coleoptera	6	31	12	6	18	31	10	15	11	13	47	19	5	30	29	12	7	10	10	1
Cerambycidae	--	--	1	--	--	--	--	1	--	--	1	1	--	1	--	--	--	--	--	--
Chrysomelidae	5	22	6	3	15	25	9	13	2	10	32	15	5	18	21	8	7	7	7	1
Coccinellidae	--	1	1	1	--	1	1	--	1	--	1	--	--	1	1	1	--	--	1	--
Elateridae	1	6	3	2	1	2	--	1	6	2	2	2	--	5	3	2	--	2	2	--
Meloidae	--	1	--	--	2	3	--	--	1	1	11	1	--	3	2	1	--	1	--	--
Scarabaeidae	--	--	1	--	--	--	--	--	1	--	--	--	--	2	2	--	--	--	--	--
Diptera	1	2	--	1	2	11	2	1	--	2	7	1	--	2	1	--	2	1	1	--
Agromyzidae	--	1	--	--	--	2	--	--	--	--	4	--	--	--	--	--	--	--	--	--
Anthomyidae	--	--	--	--	2	8	2	--	--	1	2	--	--	1	1	--	2	--	--	--
Plathystomatidae	1	1	--	1	--	1	--	1	--	1	1	1	--	1	--	--	--	1	1	--
Hemiptera	4	38	5	5	13	22	6	8	4	8	39	20	3	23	14	6	8	6	7	4
Coreidae	--	2	--	--	--	--	--	1	--	--	6	4	--	2	--	1	--	--	--	--
Cydnidae	--	5	--	--	1	--	--	--	--	--	1	1	--	1	--	--	--	--	--	--
Miridae	2	20	1	4	9	13	4	2	2	7	15	5	2	8	8	2	6	3	4	3
Pentatomidae	2	11	4	1	3	7	1	5	2	1	15	8	1	11	6	3	1	1	3	1
Piesmidae	--	--	--	--	--	1	--	--	--	--	1	1	--	1	--	--	1	--	--	--
Tingidae	--	--	--	--	--	1	1	--	--	--	1	1	--	--	--	--	--	2	--	--
Homoptera	10	60	21	13	24	53	12	17	3	13	56	41	8	25	15	24	15	16	12	5
Aleyrodidae	--	1	1	--	--	1	--	1	--	1	1	1	--	--	--	1	1	1	--	--
Cercopidae	--	3	1	--	1	1	1	--	1	1	--	1	--	2	1	--	--	--	--	1

Cicadellidae	10	44	15	9	14	38	10	13	2	8	38	32	6	18	6	16	10	12	9	4
Coccidae	--	--	--	1	--	--	--	--	--	--	--	--	--	--	--	--	--	--	--	--
Delphacidae	--	5	3	1	3	3	1	2	--	1	3	4	1	4	1	5	1	1	1	--
Dictyopharidae	--	1	--	--	--	2	--	--	--	--	2	--	1	--	--	--	--	--	--	--
Flatidae	--	1	--	--	--	2	--	--	--	1	2	1	--	--	--	1	--	1	1	--
Issidae	--	--	--	--	1	--	--	--	--	--	1	--	--	--	--	--	--	--	--	--
Membracidae	1	5	1	1	5	6	--	1	--	1	7	2	--	1	7	1	2	1	--	--
Psyllidae	--	--	--	1	--	--	--	--	--	--	2	--	--	--	--	--	1	--	1	--
Lepidoptera	2	23	7	2	10	32	1	7	--	5	8	15	1	49	1	7	2	4	5	1
Arctiidae	--	2	--	--	2	1	--	1	--	--	1	2	--	1	--	1	--	--	1	--
Geometridae	--	--	--	--	--	3	--	--	--	--	1	--	--	5	--	--	--	--	--	--
Hesperiidae	--	2	1	--	--	1	--	--	--	--	1	--	--	2	--	1	--	--	--	--
Lymantriidae	--	--	--	--	1	1	--	--	--	--	--	--	--	1	--	--	--	--	1	--
Noctuidae	2	14	5	2	7	14	1	6	--	4	3	11	1	25	1	5	1	2	2	1
Nymphalidae	--	--	--	--	--	1	--	--	--	--	--	1	--	1	--	--	--	--	--	--
Pieridae	--	1	--	--	--	3	--	--	--	1	1	--	--	1	--	--	--	--	--	--
Pyralidae	--	2	1	--	--	5	--	--	--	--	1	--	--	6	--	--	1	2	1	--
Tortricidae	--	2	--	--	--	3	--	--	--	--	--	1	--	7	--	--	--	--	--	--
Orthoptera	--	6	1	--	3	3	--	3	2	1	8	7	2	3	6	3	2	--	2	2
Acrididae	--	6	1	--	3	3	--	2	2	1	7	7	2	3	6	3	2	--	2	2
Tetrigidae	--	--	--	--	--	--	--	1	--	--	1	--	--	--	--	--	--	--	--	--
Thysanoptera	2	3	2	1	1	6	1	--	--	3	12	2	1	4	1	4	3	1	2	2
Thripidae	2	3	2	1	1	6	1	--	--	3	12	2	1	4	1	4	3	1	2	2
Total no. spp/state	25	163	48	28	71	157	32	43	20	43	177	105	20	136	68	56	39	38	39	13

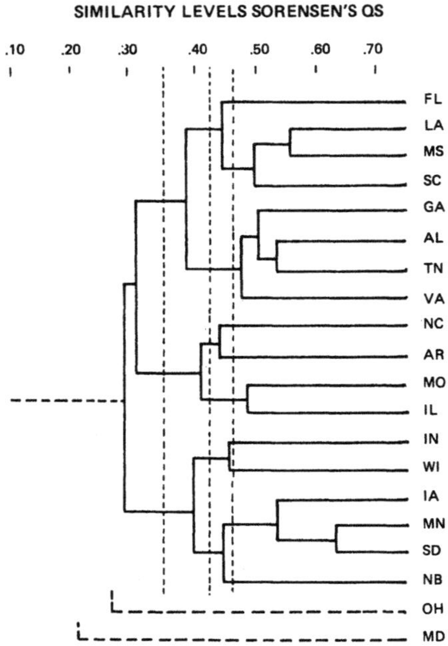

Figure 4. **Dendrogram of similarity quotients computed by cluster analysis using data on the occurrence of 453 species from 20 states.**

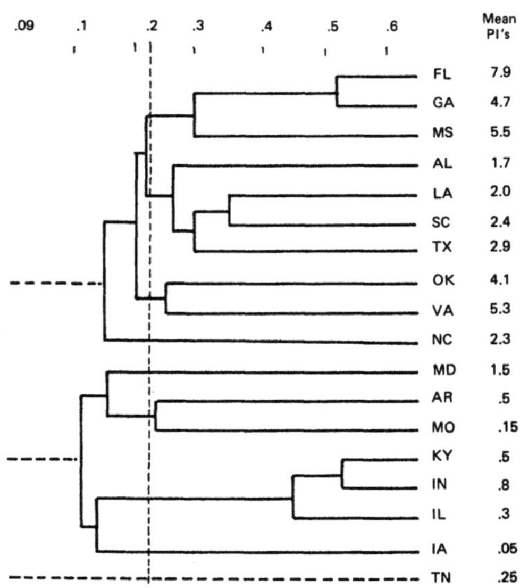

Figure 5. **Dendrogram of distances of mean PI values for 24 species in 18 states, computed by cluster analysis.**

Table 5. **Rank of economic impact of eight major soybean pests in the U.S. Ranking based on an index computed by multiplying the "PI" value for each pest in each state by the area of the state as a percentage of the U.S. area in soybean production in 1977 (total U.S. area in soybean production ca. 23.2 million ha).**

Pest Species	Mean "PI"[1]	No. States[2]	% of Total U.S. Area	Rank Index
Anticarsia gemmatalis	9.7	11	38.6	3.49
Pseudoplusia includens	5.5	12	39.3	2.69
Plathypena scabra	6.4	13	69.2	3.01
Epilachna varivestis	3.5	8	21.4	1.90
Cerotoma trifurcata	1.3	9	41.4	1.13
Nezara viridula	7.2	8	27.5	3.35
Acrosternum hilare	1.9	10	22.7	0.72
Heliothis zea	11.4	14	43.6	4.47

[1]Mean "PI" = ΣPI for each state/18; PI = pest impact index (see text for explanation).
[2]Number of states reporting economic infestations out of 18 included in the analysis.

Heliothis zea ranked first (rank index 4.47) as it had a mean PI (pest impact) of 11.4 in 14 states accounting for 43.6% of the soybean hectarage in 1977. In order of impact the species following *H. zea* were: *A. gemmatalis, N. viridula, P. includens, P. scabra, E. varivestis, C. trifurcata,* and *A. hilare.*

The ranking index for each species expressed as a percentage of the total RPI ($[RPI_i/\Sigma RPI] \times 100$) is shown in Table 3.

DISCUSSION

The 453 phytophagous species used in the faunal analysis included representatives of 4 major categories: (a) oligophagous species native to the American continent and adapted to feeding on herbaceous legumes, (b) polyphagous species more or less adapted to feeding on soybean, (c) oligophagous species associated with nonlegume plants that might have shifted food-habits to incorporate soybean, and (d) species feeding on plants other than legumes (weeds, surrounding crops, wild plants) that visit soybean fields but seldom feed and never establish colonies on this plant. The capacity of a species to colonize is an important criterion to determine its relationship to soybean. However, insects such as grasshoppers and other polyphagous species may be quite damaging to the crop and use it as a normal resource without establishing colonies. A special category of potentially damaging colonists is that of exotic species adapted to feeding on soybean in other parts of the world. So far no invaders from abroad have been detected on soybean in the U.S.

The number of true colonists of soybean fields at any given time and place does not seem to be large. Price (29) recorded no more than ca. 32

species in an Illinois field, near the border of a field at peak development. The center of the field had only about 20 species with the difference being accounted for by sporadic visitors that did not colonize the crop.

As host records were not readily available for all the species used in the analysis the following summary is just an approximation. Among the 453 species in this study there were 40 species of oligophagous habit more or less restricted to feeding on legumes; 101 polyphagous species, at least in part, accepting soybean as host; 7 species of uncertain oligophagous habit some of which may have switched to soybean feeding in the absence of more preferred hosts; 85 incidental species, not likely to feed on soybean; and the rest of uncertain host association. It seems, therefore, that only 9% of the fauna represents oligophagous legume feeders. However, among the polyphagous species are included some of the most serious pests of soybean, such as *Heliothis zea* and *Nezara viridula.*

Because of the uneven number of samples included in the faunal analysis the dendrogram in Figure 4 is most accurate in reflecting the relationships among the states that had the best survey data (AR, IL, MO, NC). For the needs of our discussion it will suffice to indicate that common faunal composition partially revealed in the dendrogram in Figure 4 roughly matches the clustering obtained through the pest impact analysis.

Replies to the questionnaire were used to rank the 24 species by order of potential economic impact. This was done with a rank index that combined PI values for each species and the area planted to soybean in each state as a percentage of the total soybean area in the U.S. (Table 3). Based on this analysis it is evident that 83.2% of the potential economic impact is due to 8 pest species and that 14 species account for 98.6% of the total damage potential to soybean by insect pests.

The pest impact analysis provided a quantitative criterion for grouping the various soybean-producing states in regions of similar risk of damage by a group of pests occurring at greater frequency within the region. States themselves are very heterogeneous because they are political units and not necessarily ecological units (see Fig. 2). Therefore, groupings of interstate regions should better reflect natural ecological uniformity than groups of states. The available data, however, do not permit refinement of the analysis to the level necessary to reveal these natural regions.

The heterogeneity of the agroecosystems is responsible to a large extent for the populational fluctuations and hence the relative economic impact of an insect pest. Two elements responsible for this heterogeneity were singled out to illustrate this relationship. Figure 6 is a tridimensional plot with mean latitude of the major producing areas in each of the 18 states on the X axis, mean annual temperatures over a period of 30 yr (37) on the Z axis, and mean pest impact on the Y axis (represented on a log-scale).

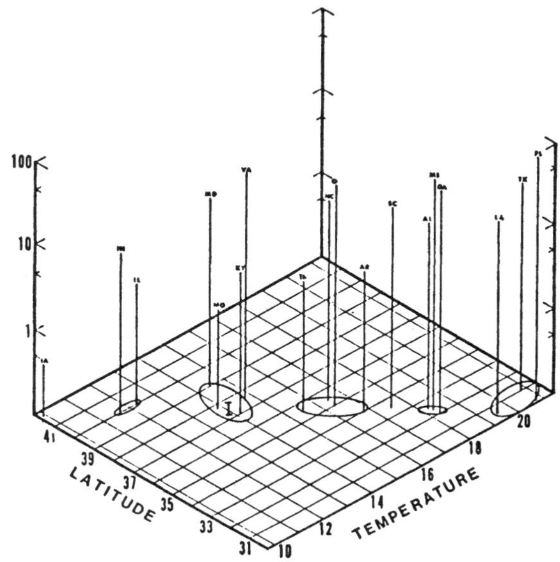

Figure 6. Tridimensional plot showing correlations of mean latitude and mean annual temperature (in C) to PI.

It is evident that mean annual temperature is highly inversely correlated with latitude (r = -.988). However, the moderating effect of the sea on the climate (if we abstract other environmental factors that may be more directly important to insects and that are also correlated with temperature) is just enough to change the pest impact by almost one order of magnitude. For example, the elypsoid marked *I* in Figure 6 includes 4 states (MO, KY, MD, and VA). The mean latitude of MO (ca. 38.5°N) is lower than that of MD (ca. 39.5°N) but the coastal position of MD causes a mean annual temperature difference of 1 C. The mean pest impact in MO is estimated at .15; that in MD is 1.47. The same relationship is evident when comparing KY and VA in the same elypsoid *I*, and it is repeated in the comparisons of NC and TN, and to some extent SC and AR.

Pest population fluctuations are a consequence of environmental conditions but they reach pest status as a result of man's decisions (27). It is possible that the pest impact data reported by different states reflect differences in their interpretation of the economic effect of fluctuating populations of phytophagous species on soybean. Much of these differences are due to discrepancies in the adopted economic thresholds for the various species. Table 6 presents a summary of economic thresholds adopted by various states in 1970 to 1972, and those adopted after 1976. It is interesting to note that the total hectareage sprayed with insecticides in 1971 was the same as

Table 6. **Economic threshold information found in extension control recommendations for four soybean pest guilds in several states (prior to 1974). A summary of current economic thresholds used by most states is presented for comparative purposes. Thresholds for soybean at pod fill stage.[1]**

States and Year	Pest Guilds or Complexes: Lep. Defol.	Coleop. Defol.	Stink Bugs	*Heliothis* spp.
AL (72)	20% defol.	--	3/m	1/m
AR (70)	40% defol.[2]	--	1/m	2/m
GA (71)	20% defol.	--	3/m	1/m
IA (72)	20% defol.	--	--	--
IL (72)	No E.T.	No E.T.	No E.T.	--
MD (74)	6-15 larvae/m	No E.T.	--	--
MO (72)	--	No E.T.	2/m	1/m
VA (72)	20% defol.	15/m[4]	1/m	6/m
1976[3]	24 worms/m + 15% defol.	15-20 beetles/m + 15-20% defol.	3/m	9/m[5]

[1]No E.T. means no established economic thresholds although control recommendations are presented. E.T. extracted from extension publications from each state.
[2]No growth stage specified.
[3]Based on (20).
[4]For Mexican bean beetle.
[5]6 in AR.

the hectareage treated in 1976 [1.4 million ha according to Eichers et al. (13)] although the total hectareage planted to soybean in 1976 was 15% greater than the hectareage in 1971 (2).

The adoption of higher economic thresholds has in all likelihood reduced the pest impact index of certain species in some regions. For example, in 1973 in Illinois there was an outbreak of green cloverworm, and ca. 300,000 ha were sprayed. Economic thresholds for that species were then set at a very low level of 9 larvae/m of row. In 1977 there was an outbreak nearly as extensive but only ca. 136,000 ha were sprayed (40), reflecting the 1974 change in economic threshold for the green cloverworm to 36 larvae/m of row and 15% defoliation at the pod set stage. The pest impact of this species, therefore, was reduced drastically although absolute populations still fluctuated within limits regulated by the environmental factors.

Finally, based on combined data on the faunal analysis and on pest impact, a tentative map of pest impact zones is shown in Figure 7. The map extrapolates from data in the faunal and PI analyses particularly in subdividing the Corn-Belt region into two PI zones. The zone of highest pest impact corresponds to a large extent to the Coastal Plain regions and the eastern reaches of the Appalachian region (see also Fig. 2). This major zone could be

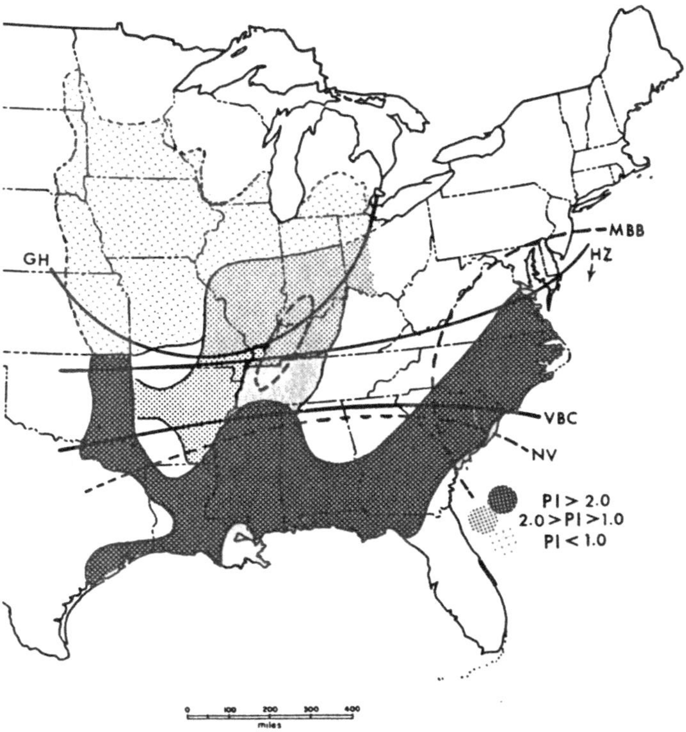

Figure 7. Zones of similar mean pest impact showing the broad ranges of distribution of some of the major soybean pests. (HZ = *Heliothis* spp.; VBC = *Anticarsia gemmatalis;* MBB = *Epilachna varivestis;* NV = *Nezara viridula;* GH = *Melanoplus* spp. The elypsoid along southern Indiana, Illinois, Kentucky and into Tennessee shows a pocket of incidences of *E. varivestis* in the Interior Plains.

subdivided into sub-zones in which the velvetbean caterpillar and the southern green stink bug represent the most dominant pests with soybean loopers, bean leaf beetles, and Mexican bean beetles being more or less serious in some particular area. Much of this zone is affected by *Heliothis zea* and *H. virescens.* The Corn-Belt has the green cloverworm and the grasshoppers as the species with the highest PI's with other pests either being very localized or occurring very infrequently.

Reports on insecticide use (13) in the various regions further substantiate results of the pest impact analysis. In 1976, 78.5% of the insecticides (active ingredients) used to control soybean insect pests in the U.S. were used in the southeast (AL, GA, FL, SC); the mean PI for these states was 4.17, and they clustered very closely together in Figure 5. The Appalachian and Northeastern regions (including NC, VA, TN, KY, and MD) used 15.5% of the insecticides; in 3 of these states (VA, MD and NC) the mean PI is 3.3 (KY is .5 and probably accounts for only a fraction of the total insecticide use reported for this region although no state records are available to support this statement). The Delta and Southern Plains (AR, LA, MS, TX, OK) used 4% of the total amount of insecticides used in 1976, and the mean PI in these states was 3.00; finally the Corn-Belt (including MO, IL, IN and IA) reported use of 1.46% of the insecticides used in soybeans, and the mean PI value was 0.32. As expected there is clear tendency for increased insecticide use in regions of greater pest impact. The correlation, however, was not significant for the 1976 insecticide use data with use in the southeast disproportionate to the estimated pest impact. One explanation for this discrepancy may be the need for multiple sprays to control the velvetbean caterpillar in Florida.

CONCLUSIONS

Quantitative evidence was presented for the identification of soybean growing zones in the U.S. characterized by a communality of insect faunal composition and pest impact values. Among the various soybean-growing regions in the U.S. there are vast differences in the frequency and amplitude of yearly fluctuations of populations of phytophagous insect species. Some of them often reach pest status.

The pest impact index (PI) may be a convenient criterion to compare the relative importance of various species and allow state and federal agencies to better allocate research priorities.

The following are a few conclusions from the set of analyses discussed above: (a) About 9% of the phytophagous insect fauna collected in soybean fields are oligophagous species closely associated with herbaceous legumes. Although these represent a rather rich fauna several feeding niches remain unoccupied or are only partially occupied by species not adapted perfectly to the soybean. (b) There are no records of exotic pests of soybean in the U.S. (c) None of the soybean insect pests has reached a dominant or key position in any of the regions (a PI value of 50.0 seems to be a good criterion for considering a species as a key pest). Only the velvetbean caterpillar in Florida would qualify as a key pest of soybean in that state. (d) Pest impact analysis has shown that 14 species account for 98.6% of the damage potential to soybean in the U.S. (e) Based on the RPI ranking index *Heliothis zea* was established as the number one pest of soybean in the U.S. and corresponds to 17.9% of the total pest impact per area. The stinkbug complex is second with 16.3% of the total RPI, and (f) as a guild the lepidopterous defoliators are the most important pests with a combined RPI of 40.5%.

This overview of the insect pest situation in the U.S. should be interpreted as a description of one steady state in a very dynamic situation. It will be interesting to reassess this situation 5 or 10 yr hence.

NOTES

Marcos Kogan, Office of Agricultural Entomology, University of Illinois and Illinois Natural History Survey, Urbana, Illinois 61801.

This work could not have been done without the cooperation of soybean entomologists who have established a remarkable working relationship through the USDA regional project S-74 and the NSF- and EPA-sponsored Huffaker project. Many entomologists have contributed samples for the IRCSA collection, and the work of George Godfrey (from 1970 to 1976) and John Bouseman (since 1976) as curators of the collection has been invaluable. Those who provided state estimates for the pest impact analysis were: M. Bass (AL); A. Mueller and W. Yearian (AR); D. Herzog (FL); J. Todd (GA); L. Pedigo (IA); D. Kuhlman and M. Irwin (IL); R. Edwards (IN); H. Raney, P. Sloderbeck, and K. Yeargan (KY); R. McPherson and D. Newsom (LA); G. Dively (MD); C. Ignoffo (MO); G. Beland, L. Lambert, and H. Pitre (MS); B. Hill (OK); J. R. Bradley, Jr., W. Campbell and J. Van Duyn (NC); S. Turnipseed (SC); G. Lentz (TN); C. C. Bowling (TX); and J. Smith (VA). Ellen Brewer and Howard Ojalvo helped with computer programming; Jenny Kogan, information specialist (SIRIC), assisted with searches for agricultural maps, and Sandy McGary typed the manuscript.

This work was supported in part by the USDA regional project S-74; the NSF and US-EPA, through a grant (NSF GB-34718) to the University of California; and the International Soybean Program (INTSOY) of the University of Illinois College of Agriculture and the University of Puerto Rico. Opinions expressed herein are those of the author and not necessarily those of the supporting agencies.

REFERENCES

1. American Soybean Association. 1976. Soybean production maps. Soybean Digest, Hudson, IA.
2. American Soybean Association. 1978. Soybean digest blue book. A.S.A., Hudson, IA. 177 pp.
3. Anonymous. 1970. World atlas of agriculture. Vol. 3. Americas, p. 409-465. Instit. Geografico de Agostini, Novarra, Italy, 4 vols.
4. Balduf, W. V. 1923. The insects of the soybean in Ohio. Ohio Agric. Exp. Sta. Bull. 366:145-181.
5. Blickenstaff, C.C. and J. L. Huggans. 1962. Soybean insects and related arthropods in Missouri. Mo. Agric. Exp. Sta. Res. Bull. 803. 51 pp.
6. Burleigh, J. G. 1972. Population dynamics and biotic controls of the soybean looper in Louisiana. Environ. Entomol. 1:290-294.
7. Buschman, L. L., W. H. Whitcomb, T. M. Neal, and D. L. Mays. 1977. Winter survival and hosts of the velvetbean caterpillar in Florida. Fla. Entomol. 60:267-273.
8. Canerday, T. D. and F. S. Arant. 1967. Biology of *Pseudoplusia includens* and notes on biology of *Trichoplusia ni, Rachiplusia ou* and *Autographa biloba.* J. Econ. Entomol. 60:870-871.
9. Carner, G. R., M. Shepard, and S. G. Turnipseed. 1974. Seasonal abundance of insect pests of soybeans. J. Econ. Entomol. 67:487-493.
10. Deitz, L. L., J. W. Van Duyn, J. R. Bradley, Jr., R. L. Rabb, W. M. Brooks, and R. E. Stinner. 1976. A guide to the identification and biology of soybean arthropods in North Carolina. N. C. Agric. Exp. Sta., Techn. Bull. 238. 264 pp.
11. DeWitt, N. and G. Godfrey. 1972. The literature of arthropods associated with soybeans. II. A bibliography of the southern green stink bug, *Nezara viridula* (L.). Ill. Nat. Hist. Survey Biol. Notes 78. 23 pp.
12. Eastman, C. E. and A. L. Wuensche. 1977. A new insect damaging nodules of soybean: *Rivellia quadrifasciata* (Macquart). J. Ga. Entomol. Soc. 12:190-199.

13. Eichers, T. R., P. A. Audrilenas, and T. W. Anderson. 1978. Farmers' use of pesticides in 1976. U.S. Dept. Agric., Economics, Statistics and Cooperative Services, Ag. Econ. Dept. Rept., 418. 58 pp.
14. Fehr, W. R. and C. E. Caviness. 1977. Stages of soybean development. Coop. Ext. Serv., Ag. & Home Ec. Exp. Sta., Iowa State Univ., Ames, IA 80. 12 pp.
15. Ford, B. J., J. R. Strayer, J. Reid, and G. L. Godfrey. 1975. The literature of arthropods associated with soybeans. IV. A bibliography of the velvetbean caterpillar, *Anticarsia gemmatalis* (Hubner) (Lepidoptera: Noctuidae). Ill. Nat. Hist. Survey, Biol. Notes 92. 15 pp.
16. Greene, G. L. 1976. Pest management of the velvetbean caterpillar in a soybean ecosystem. p. 602-610. In L. D. Hill (ed.) World Soybean Research, Proc. World Soybean Res. Conf., Interstate Printers, Danville, IL. 1073 pp.
17. Isely, D. 1930. The biology of the bean leaf beetle. Arkansas Agric. Exp. Sta. Bull. 248. 20 pp.
18. Jensen, R. L., L. D. Newsom, and J. Gibbens. 1974. The soybean looper: Effects of adult nutrition on oviposition, mating frequency and longevity. J. Econ. Entomol. 67:467-470.
19. Kogan, J., D. K. Sell, R. E. Stinner, J. R. Bradley, Jr., and M. Kogan. 1978. The literature of arthropods associated with soybean. V. A bibliography of *Heliothis zea* (Boddie) and *H. virescens* (L.) (Lepidoptera: Noctuidae). Univ. of Illinois, College of Agric., INTSOY Ser. 17. 242 pp.
20. Kogan, M. 1976. Soybean disease and insect pest management. p. 114-121. In R. Goodman (ed.) Expanding the use of soybean—A conference for Asia and Oceania. Chiangmai, Thailand. Univ. of Illinois, INTSOY Ser. 10. 261 pp.
21. Kogan, M. 1978. The role of systematics in integrated pest management. Paper presented in symposium. "Systematics Resources and the User Community."—Section A. Annual Meetings—Entomol. Soc. Amer. 16 pp.
22. Kogan, M., W. G. Ruesink, and K. McDowell. 1974. Spatial and temporal distribution pattern of the bean leaf beetle, *Cerotoma trifurcata* (Forster), on soybeans in Illinois. Environ. Entomol. 3:607-617.
23. Miner, F. D. 1966. Biology and control of stink bug on soybeans. Arkansas Agr. Exp. Sta. Bull. 708. 40 pp.
24. Myers, F. V. and L. P. Pedigo. 1978. Winter survival of the green cloverworm, *Plathypena scabra* (F.) in central Iowa (Lepidoptera: Noctuidae). J. Kan. Entomol. Soc. 51:288-293.
25. Nichols, M. P. and M. Kogan. 1972. The literature of arthropods associated with soybeans. I. A bibliography of the Mexican bean beetle, *Epilachna varivestis* (Mulsant) (Coleoptera: Coccinellidae). Ill. Nat. Hist. Survey, Biol. Notes 77. 20 pp.
26. Nichols, M. P., M. Kogan, and G. P. Waldbauer. 1974. The literature of arthropods associated with soybeans. III. A bibliography of the bean leaf beetles *Cerotoma trifurcata* (Forster) and *C. ruficornis* (Olivier). Ill. Nat. Hist. Surv. Biol. Notes. 85. 16 pp.
27. Pedigo, L. P. 1975. Insect threats and challenges to Iowa agroecosystems. Iowa State J. Res. 49:457-466.
28. Pedigo, L. P., J. D. Stone, and G. L. Lentz. 1973. Biological synopsis of the green cloverworm in central Iowa. J. Econ. Entomol. 66:665-673.
29. Price, P. W. 1976. Colonization of crop by arthropods: None equilibrium communities in soybean fields. Environ. Entomol. 5:605-611.
30. Sokal, R. E. and P. H. A. Sneath. 1967. Principles of numerical taxonomy. W. H. Freeman, San Francisco, CA. 359 pp.

31. Southwood, T. R. E. 1978. Ecological methods. Halsted Press, John Wiley & Sons, New York. 524 pp.
32. Stone, J. D. and L. P. Pedigo. 1972. Selected bibliography of the green cloverworm, *Plathypena scabra*. Bull. Entomol. Soc. Amer. 18:24-26.
33. Todd, J. W. 1976. Effect of stinkbug feeding on soybean seed quality. p. 611-618. In L. D. Hill (ed.) World Soybean Research, Proc. World Soybean Res. Conf., Interstate Printers, Danville, IL. 1073 pp.
34. Tugwell, P., E. P. Rouse, and R. G. Thompson. 1973. Insects in soybeans and a weed host (*Desmodium* sp.). Arkansas Agric. Exp. Sta., Rept. Ser. 214. pp. 1-18.
35. Turnipseed, S. G. 1973. Insects. p. 545-572. In B. E. Caldwell (ed.) Soybeans: Improvement, Production, and Uses. Amer. Soc. Agron., Madison, WI. 681 pp.
36. Turnipseed, S. G. and M. Kogan. 1976. Soybean entomology. Annual. Rev. of Entomol. 21:25-60.
37. United States Department of Agriculture. 1972. Agricultural Statistics. U.S. Gov. Print. Off., Washington, D.C. 759 pp.
38. Waldbauer, G. P. and M. Kogan. 1976. Bean leaf beetles: Phenological relationship with soybean in Illinois. Environ. Entomol. 5:35-44.
39. Waldbauer, G. P. and M. Kogan. 1976. Bean leaf beetles: Bionomics and economic role in soybean agroecosystems. p. 619-628. In L. D. Hill (ed.) World Soybean Research, Interstate Printers, Danville, IL. 1073 pp.
40. Wedberg, J. L. and K. D. Black. 1978. Insect situation and outlook and insecticide usage. 30th Illinois Custom Spray Operators Training School. Summaries of presentations. Univ. of Ill., College of Agriculture. pp. 119-134.
41. Wellik, J. M. and L. P. Pedigo. 1978. Innate capacity of increase and ovipositional pattern of the green cloverworm. Environ. Entomol. 7:171-177.

ENTOMOPATHOGENS FOR CONTROL OF INSECT PESTS OF SOYBEANS

C. M. Ignoffo

Many major soybean insect pests are attacked by pathogenic microorganisms (14). Included among these microorganisms are 4 bacterial species, at least 5 fungi, about 5 species of protozoa and close to 2 dozen viruses (Table 1). Many of the serious pests of soybeans are caterpillars that can be controlled selectively, with minimal environmental impact, by use of 3 of these entomopathogens, i.e. a fungus, *Nomuraea rileyi;* a bacterium, *Bacillus thuringiensis* and a virus, *Baculovirus heliothis.* Two widely different control concepts might be used on soybeans, a preventative approach in which a program is initiated before caterpillar pests reach economic threshold levels and an insecticidal approach in which a program is initiated when pest insects reach threshhold levels. Thus a season-long program as envisioned would begin with an early preventative application of the fungus, followed, if needed, by applications of the bacterium for defoliating caterpillars, and then by the virus, for podworms. This preventative-type approach would have its maximum usefulness only for those areas in which caterpillar pests are known to be a chronic problem. You might ask about the feasibility of this kind of program on soybeans. To answer this I will look at each phase of the program and relate how the pathogens might be produced; whether they are safe to use; and how effective they are in suppressing damaging soybean insects.

PHASE I: PREVENTION OF CATERPILLAR DAMAGE

Nomuraea rileyi is an imperfect fungus that primarily infects caterpillars although at least two species of beetles are also susceptible (Table 2). Natural epizootics have been observed in all insects listed in Table 2 except *Leptinotarsa*

Table 1. Species of pathogenic microorganisms isolated from arthropod pests of soybeans.[a]

BACTERIA	
Bacillus thuringiensis	*Coccobacillus acridiorum*
Bacillus popilliae	*Serratia marcescens*
FUNGI	
Beauveria bassiana	*Metarhizium anisopliae*
Entomophthora gammae	*Nomuraea rileyi*
Entomophthora spp.	
PROTOZOA	
Pleistophora spp.	*Malameba locustae*
Leptomonas serpens	*Nosema* spp.
	Vairimorpha necatrix
VIRUSES	
Cytoplasmic polyhedroses *Heliothis zea, Plathypena scabra, Spodoptera exigua, Spodoptera frugiperda, Trichoplusia ni*	Granuloses *Heliothis zea, Plathypena scabra, Spodoptera exigua, Spodoptera frugiperda, Trichoplusia ni*
Entomopox viruses *Agriotes* spp., *Melanoplus* spp.	Nucleopolyhedroses *Anticarsia gemmatalis, Autographa californica, Heliothis zea, Psuedoplusia includens, Spodoptera exigua, Spodoptera frugiperda, Trichoplusia ni.*

[a]Abstracted from 4,7,10,20-25, 29.

decemlineata, Bombyx mori, Peridroma saucia, and *Spodoptera exigua.* These natural epizootics occur in insect pests of pastures (clover and alfalfa) and many row and grain crops (corn, rice, cotton, soybeans, and tobacco) (17,19). Epizootics occur every year in spite of the fact that there is a wide difference in susceptibility to *N. rileyi.* For example, there is a 25-fold range in relative susceptibility to *N. rileyi* from the most susceptible host, the beet armyworm, *Spodoptera exigua*, to the least susceptible host, the velvetbean caterpillar *Anticarsia gemmatalis* (11,14). *N. rileyi* is the only pathogen of the three under consideration that is not available commercially.

Production of *Nomuraea rileyi*

Only small quantities (<5 kg) of conidia of *N. rileyi* have been produced. However, since *N. rileyi* can be produced on natural ingredients commercial production most likely would utilize a submerged process or a combination of surmerged and surface production processes. Conidia produced using these

Table 2. **Species of insect pests that are reported to be susceptible to the fungus *Nomuraea rileyi*.**

	COLEOPTERA
Hypera punctata	*Leptinotarsa decemlineata*
	LEPIDOPTERA
Agrotis gladiaria	*Hyphantria cunea*
Agrotis ipsilon[a]	*Lymantria dispar*
Amathes badinodis	*Ostrinia nubilalis*
Anticarsia gemmatalis[a]	*Peridroma saucia*[a]
Bombyx mori	*Pieris rapae*
Chrysodeixis eriosoma	*Plathypena scabra*[a]
Cosmia spp.	*Pseudaletia unipuncta*
Diaphania pyloalis	*Pseudoplusia includens*[a]
Evergestis forficalis	*Spodoptera exigua*[a]
Feltia ducens	*Spodoptera frugiperda*
Heliothis armigera	*Spodoptera litoralis*
Heliothis virescens[a]	*Spodoptera ornithogalli*
Heliothis zea[a]	*Trichoplusia ni*[a]

[a]Reported pests of soybeans.

techniques can be formulated with various diluents and adjuvants to provide a relatively stable, viable product. Production costs based on preliminary estimates of laboratory production of conidia are within the realm of commercial feasibility.

Specificity of *Nomuraea rileyi*

Eighteen species, mostly pests representing six insect orders (Coleoptera, Lepidoptera, Diptera, Hymenoptera, Neuroptera, Hemiptera) were not susceptible to high concentrations of conidia. More importantly, however, three beneficial predators of pest caterpillars (*Hippodamia convergens, Chrysopa carnea, Podisus maculiventris*) and three parasites of pest caterpillars (*Voria ruralis, Apanteles marginiventris, Campoletis sonorensis*) which are natural residents of soybean fields were not susceptible when exposed to levels about 25x higher than those used in field experiments (15). Thus, most beneficial insects found in soybeans should not be affected adversely by artificial applications of *N. rileyi.*

Indirect evidence (temperature profiles for germination, growth and sporulation) indicates that *N. rileyi* will not infect warm-blooded vertebrates (11). More recently, this indirect evidence has been corroborated by direct challenge to mammals (Table 3). As, for example, conidia of *N. rileyi* were inactivated rapidly (half-life ca. 20 min) in human gastric juices (10^9 conidia/person); mice were not affected by stomach intubation of 2.4 billion conidia/

Table 3. Summary of tests to evaluate possible risks of use of *Nomuraea rileyi* to mammals (18).

Test System	Route of Administration	Million Conidia/Animal	Hectare Equivalents (Man)[a]
Human gastric juice	*In vitro*	10[b]	0.001
Mice	Stomach intubation	2440	0.4
Rats	Inhalation	10	0.05
Rabbits	Eye instillation	122	0.001
	Dermal application	3000[c]	0.02

[a]Based upon 10 trillion conidia/0.4 ha.
[b]Per ml of gastric juice.
[c]Per cm^2 of skin.

mouse (the 2.4 billions were the equivalent of a 70-kg man receiving a 0.4-ha application of conidia); and greater than 99.9% of the infectivity was lost during passage through the alimentary tract (18). In other tests, abnormalities (clinical, pathological, or histological) were not observed in white rats exposed to 10 million viable conidia/liter of air for 1 hr (equal to a 70-kg man inhaling conidia applied to ca. 1/500th ha) or in rabbits treated with 120 million conidia/eye or 3 billion conidia/cm^2 of skin.

Effectiveness of *Nomuraea rileyi*

The application of *N. rileyi* as an insecticide probably will not control heavy populations of caterpillars unless directed against young larvae and, even then, high concentrations would probably be needed (2,5,8,9,17). An early, heavy application of conidia or release of infected larvae, however, will induce an earlier than normal epizootic and thus could suppress caterpillar pests when soybeans are most sensitive to insect feeding (13,15,26).

Thus in one experiment to test this induced-epizootic concept, a heavy application of conidia (ca. 10 trillion conidia/0.4 ha) was sprayed on soybeans when half of the plants had at least one flower (15). In a second experiment (15), soybeans were sprayed when pods were 6 mm long at one of the four uppermost nodes. In a third experiment (26), fungus-infected cadavers of larvae were cut into 3-mm sections, mixed with dry vermiculite and distributed over soybeans by a manually operated grass-seed spreader at a rate of 3360 cadavers/ha (ca. 2 to 3 pieces of cadaver/1.5 m row). In all three experiments the artificial introduction of the fungus, timed to a specific phenological stage of soybean growth, significantly altered the epizootic pattern of *N. rileyi*. The initial detection and peak incidence of infected caterpillar pests in treated plots were advanced at least 14 days compared with untreated plots. The peak in treated plots, therefore, occurred prior to and during the stages of soybean growth that were most sensitive to defoliation, i.e. the stages we wanted to protect from insect damage.

Since it has been proven that *N. rileyi* can be used to induce an epizootic (15,26) and major caterpillar pests of soybeans are susceptible to *N. rileyi* (11), it could be used as the first step in a three-pronged program to suppress caterpillar pests of soybeans below the economic threshhold level. In the midwest on indeterminate soybeans we would first apply conidia of *N. rileyi* at initial flowering or when about one-half of the plants had flowers which is about 2 to 4 weeks before the stage of soybeans that we are trying to protect from damaging caterpillars. If environmental conditions are adequate (presence of dew for 2 to 4 hours) (19), then susceptible caterpillars will be infected and will provide additional conidia to suppress a build-up of caterpillar pests. This fact was demonstrated vividly in central Missouri in 1974 (15); however, we have not proved whether this preventative concept is feasible throughout the U.S. soybean belt. Thus, the proper timing of an application of *N. rileyi* relative to a specific phenological event would have to be determined for each of the major classes of soybeans and the region in which they are grown.

PHASE II: INSECTICIDAL CONTROL OF DEFOLIATING CATERPILLARS

The second phase of our three-pronged program (use of the bacterium *B. thuringiensis)* would be initiated only if *N. rileyi* fails to induce an epizootic or enzootic that results in season-long suppression of soybean caterpillars. In that case, *B. thuringiensis* will be used as a microbial insecticide to selectively suppress incipient, damaging populations of defoliating soybean caterpillars. Products based upon *B. thuringiensis* are available commercially and in use against soybean caterpillars. Use rates of these products on soybeans range from 114 to 454 g/0.4 ha (1.8 to 7 billion IU/0.4 ha) for control of the green cloverworm *(P. scabra),* velvetbean caterpillar *(A. gemmatalis),* soybean looper (*Pseudoplusia includens*), and cabbage looper (*Trichoplusia ni*) (1,30).

Production of *Bacillus thuringiensis*

Two types of industrial processes (surface and submerged) are used to produce *B. thuringiensis* products (6). In the surface process, the organism is cultured on a semi-solid medium composed of wheat bran, expanded perlite, soybean meal, glucose, and inorganic salts. Harvesting of the bran after 36 to 48 hr and subsequent formulation provide an insecticidally active product. In the submerged process, test-tube slants of *B. thuringiensis* are used as the initial inoculum for a stepwise increase in volume through a series of shake flasks and a seed fermentor. Cultivation to produce the product is done in large fermentation tanks in a liquid medium that provides various sources of carbon (molasses, dextrose, grain mash, grain by-products), and nitrogen (cottonseed meal, fish meal, soybean meal, whey, yeast, or casein hydrolysates). Essential minerals and growth factors are also included when necessary. The total volume of solids is generally limited to about 5%.

Insecticidally active ingredients of *B. thuringiensis* are recovered by centrifugation, filtration, precipitation, spray-drying, or a combination of these methods. Safety of each production batch is confirmed by intraperitoneal administration to white mice. Production costs of the surface or submerged process are about $0.50/454 g and $1.69/454 g, respectively (6). Formulated, wettable powder products retail for about $20.00/kg and use rates on all commodities range from about 57 to 454 g/0.4 ha.

Specificity of *Bacillus thuringiensis*

Although *B. thuringiensis* first became commercially available ca. 1960, it is not a new organism. *B. thuringiensis* is a naturally occurring bacterium that exists nearly everywhere and probably was around when man first made his appearance. Products based on *B. thuringiensis* are among the safest insecticides in use today and have a complete exemption from the requirements of a tolerance. Repeated tests with *B. thuringiensis* have been conducted against plants, invertebrates, and vertebrates including man with no signs of toxicity or pathogenicity (Table 4). In some instances, slight dermal and inhalation irritation was observed in test animals but the irritation was attributed to the diluents and adjuvants and not to *B. thuringiensis* per se. Since products of *B. thuringiensis* are not toxic or pathogenic, it is not necessary to wear protective clothing when handling them and workers may re-enter treated fields or forests immediately after spraying.

Effectiveness of *Bacillus thuringiensis*

Products of *B. thuringiensis* are used throughout the world to control about 100 caterpillar pests of vegetables, field crops, pastures, trees and ornamentals. Although most of the major caterpillar pests of soybeans are susceptible to *B. thuringiensis* (Table 5), they are not all equally susceptible (16). The relative order of susceptibility varies from 0.01 for the most susceptible species, the green cloverworm to 3.29 for the most resistant species, the podworm *(Heliothis zea)* and, as expected, young caterpillars are more susceptible than older caterpillars (Table 6). Surprisingly, velvetbean caterpillars become more resistant (relative to the other species) as they mature.

Laboratory studies (16) indicated that *P. scabra* should be controlled easily on soybeans at levels normally used to control *T. ni* (227 to 454 g/0.4 ha, the equivalent of 3.6 to 7.3 billion IU/0.4 ha). Also, lower rates

Table 4. **Summary of the kinds of tests used to evaluate the possible risks of *Bacillus thuringiensis* and *Baculovirus heliothis* to plants, invertebrates and vertebrates.**

Type of Test	*Bacillus thuringiensis*	*Baculovirus heliothis*
	Acute Toxicity-Pathogenicity	
Per os	Fish, birds, mouse, rat, guinea pig, pig, man	Rat, mouse, birds, fish, oyster, shrimp
Intraperitoneal	Birds, mouse, rat, guinea pig, rabbit	Rat, mouse
Subcutaneous	Mouse, guinea pig	Rat
Inhalation	Mouse, man	Rat, guinea pig
Intravenous	— —	Rat, mouse
Intracerebral	— —	Rat, mouse
	Sensitivity-Irritation	
Eye	Rabbit	Rabbit
Skin	Guinea pig, rabbit	Rat, guinea pig, rabbit, man
	Subacute Toxicity-Pathogenicity	
Diet	Chicken, rat, man	Monkey, dog, rat, mouse
Inhalation	Rat, man	Monkey, dog, rat, mouse
Subcutaneous	— —	Monkey, dog, rat, mouse
Allergenicity-sensitivity	Guinea pig, man	Man, guinea pig
	Teratogenicity/Carcinogenicity	
	— —	Rat, mouse
	Serial Passage/Replication Potential	
	Mouse	Man, primate tissue culture
	Invertebrate Specificity	
	Beneficial insects, arthropods	Beneficial insects, arthropods
	Phytotoxicity	
	Vegetables, field crops, ornamentals, trees	Vegetables, field crops

Table 5. Species of some of the major caterpillar pests of soybeans that are susceptible to the bacterium *Bacillus thuringiensis*.

Achyra rantalis	*Loxostege commixtalis*
Agrotis ipsilon	*Plathypena scabra*[a]
Anticarsia gemmatalis[a]	*Pseudoplusia includens*[a]
Colias eurytheme	*Rachiplusia ou*
Diacrisia virginica	*Spodoptera exigua*
Estigmene acrea	*Spodoptera ornithogalli*
Heliothis zea	*Trichoplusia ni*[a]

[a]Registered for use against these pests on soybeans.

Table 6. Estimated LC-50, and relative susceptibility of 1-day-old and 4-day-old larvae of soybean pests exposed continuously to diets surface-treated with *Bacillus thuringiensis*.

	1-Day-Old		4-Day-Old	
Species	LC-50[a]	RS[b]	LC-50	RS
Trichoplusia ni	85.7	1.00	1792.2	1.00
Heliothis zea	272.7	3.18	6103.9	3.40
Spodoptera exigua	233.8	2.73	5194.8	2.90
Pseudoplusia includens	93.5	1.09	1649.4	0.92
Anticarsia gemmatalis	7.8	0.09	1428.6	0.80
Plathypena scabra	0.9	0.01	1.1	0.01

[a]ng/cm^2.
[b]Relative susceptibility based on LC-50 of *T. ni*.

(¼ to ¾ per 0.4 ha; 1.8 to 5.4 billion IU/0.4 ha) may be equally as effective, especially if the treatment is directed at young *P. scabra*. Caterpillars of *A. gemmatalis* and *P. includens* also could be controlled at levels normally used to control *T. ni*, and, as with *P. scabra*, effectiveness would be enhanced by treating when young caterpillars are first found. These laboratory studies corroborated field studies conducted on soybeans in Arkansas (30). In Arkansas, larvae of *P. scabra* were reduced 76% (range 58 to 93%) by using *B. thuringiensis* at 1 to 4 billion IU/0.4 ha and *A. gemmatalis* and *P. includens* were reduced 65 to 69%, respectively. Thus, populations of defoliating caterpillars as high as 21 larvae/30-cm row occurring during the most sensitive stages of soybeans can be suppressed below damaging levels by use of *B. thuringiensis*. On the other hand, larvae, of *S. exigua* or *H. zea* cannot be controlled on soybeans with *B. thuringiensis* at rates normally used to control *T. ni* (16). Laboratory results indicate that rates of 1.4 to 1.8 kg/0.4 ha

(22 to 30 billion IU/0.4 ha) at a cost of $27.00 to $36.00/0.4 ha would be needed. In 1977, *B. thuringiensis* products were registered for use on soybeans to control four of the most susceptible species, i.e., *P. scabra, A. gemmatalis, P. includens* and *T. ni.* Recommended rates are from 113 to 227 g/ 0.4 ha (eq. to 1.82 to 3.63 billion IU/0.4 ha) for the more susceptible *A. gemmatalis* and *P. scabra* to 227 to 454 g/0.4 (3.63 to 7.26 billion IU/0.4 ha) for the most resistant *P. includens* and *T. ni* larvae. Thus costs for one application of *B. thuringiensis* would range from ca. $2.00 to $9.00/0.4 ha.

PHASE III: INSECTICIDAL CONTROL OF POD-FEEDING *HELIOTHIS* SPP.

The last phase of our three-pronged program would be the use of a specific nucleopolyhedrosis virus, *Baculovirus heliothis.* This virus is available commercially for control of *Heliothis* species on cotton but has not been registered for use against *Heliothis* on soybeans or other crops. Species of *Heliothis* are the most serious pod-feeding pests of soybeans. As with *B. thuringiensis* this virus would be used as a microbial insecticide only if the early preventative application of *N. rileyi* fails to provide control.

Production of *Baculovirus heliothis*

Insect viruses are more difficult to produce than are fungi or bacteria. They are obligate parasites and thus must be produced in living hosts. In practically all cases, this means mass producing the virus in the target pest insect, harvesting the virus from the insect, and then formulating this preparation into a product.

Production of *B. heliothis* is a typical example of an *in vivo* process. A semi-synthetic diet is used to rear bollworms to be used to produce the virus. The diet is a water-based mixture of casein, sucrose, wheat germ, yeast, Wesson salts, growth factors, and a filler with antimicrobial substances often added to inhibit growth of secondary bacteria, yeast, or fungi (12). Industrial food machinery is used to automate the virus production process. Hot liquid diet, prepared in a large mixing tank is piped to a filling machine that automatically dispenses a uniform volume of diet into individual cells of plastic trays. Neonatal caterpillars obtained from the insectary are place individually in the cell of each tray. Hundreds of these trays are thus processed semi-automatically and stacked on mobile racks for incubation (5 to 7 days at 26 C). During incubation the virus replicates within the caterpillars and produces 5,000 to 10,000 times more virus than that originally used. One caterpillar can produce as much as 36 billion polyhedral inclusion bodies (PIB), ca. 30% of its dry wt (12). Virus-killed larvae collected from individual cells by use of a suction tube are slurried with water, and the suspension is filtered through a screen to remove large particles. Further concentration of the virus is done by centrifugation, precipitation, filtration, or spray-drying.

As with other microbial insecticides, the safety of each batch of virus is confirmed by a mouse injection test and by evaluation of the types and levels of secondary microbial contaminants. Also, the insecticidal activity of each batch is determined by a bioassay and the activity is related to a standard preparation. Insecticidal activity is reported as an LD-50 or LC-50 with 95% fiducial limits and recorded as the number of insecticidal units and PIB/wt of virus product.

Costs of producing diseased caterpillars is about 0.1 to 4 cents per larva; the product from 1 to 100 larvae is generally used to treat 0.4 ha. The current commercial product of *B. heliothis* (ELCAR™) is a wettable powder formulated with several adjuvants to increase stability and efficacy. Cost of ELCAR to the grower is about $1.55/31 g and recommended rates for control of *Heliothis* on cotton are from 62 to 124 g/0.4 ha per application.

Specificity of *Baculovirus heliothis*

Once produced, viruses are evaluated to determine if they pose any risk to man, other animals, and plants. Their ubiquitous presence on fruits and vegetables and the lack of toxicity or pathogenicity to persons handling them is considered indirect evidence of safety. However, this indirect evidence has been corroborated by direct experimentation. All such evidence to date demonstrates that *B. heliothis* is probably the most specific and safest insecticide in use today (12). It only attacks species of *Heliothis* (7 are reported susceptible) and could not be transmitted to 36 other insects, spiders or mites at rates equivalent to 10 to 100x the recommended field rate (Table 7). Results of an extensive series of studies against 18 vertebrates and 24 plants demonstrated that *B. heliothis* would pose no harm to man and the environment (12).

Effectiveness of *Baculovirus heliothis*

During the last two decades between 150 and 200 field tests were conducted with *B. heliothis* (12). About 60% of these tests were on cotton, 30% on corn and the balance on soybean, sorghum, tobacco or tomato. Although seven studies have been conducted on soybeans, only three of these were field tests. These field studies demonstrated that *B. heliothis* is probably the most effective of the available entomopathogens in suppressing populations of *H. zea* larvae feeding on soybeans.

In a test conducted in 1975 (17), the number of surviving larvae decreased steadily as rates were increased from 6 to 600 x 10^9 PIB/0.4 ha (Table 8). Larval populations were reduced from 92 to 100%. In another test conducted in 1976 (17), moth populations from all virus treatments were statistically lower than populations in the larval-infested check (Table 8). Reductions in moth populations ranged from about 70 to 90%. Although increasing the number of applications of virus did not produce statistically

Table 7. Insect species that were not susceptible when exposed to *Baculovirus heliothis* at doses 10 to 100 times the recommended field rate.

HYMENOPTERA	NEUROPTERA
Apis mellifera[a]	*Chrysopa carnea*[a]
Brachymeria intermedia[a]	
Campoletis sonorensis[a]	LEPIDOPTERA
Chelonus blackburni[a]	
Meterorous leviventris[a]	*Agrotis segetum*
	Anthela varia
DIPTERA	*Cactoblastis cactorum*
	Chrysodeixis chalcites[a]
Musca domestica	*Euproctis subflava*
Voria ruralis[a]	*Galleria mellonella*
	Manduca quinquemaculata
COLEOPTERA	*Manduca sexta*
	Nomophila neartica
Hippodamia convergens[a]	*Orgyia anartoides*
Phaedon brassicae	*Papilio xuthus*
	Pieries brassicae
HETEROPTERA	*Pieris rapae*
	Plusia nigrisigna[a]
Geocrois pallens[a]	*Plutella surinamensis*
Laodelphax striatella	*Spodoptera exigua*[a]
Orius tristicolor[a]	*Spodoptera frugiperda*[a]
Nephotettix cinticeps[a]	*Spodoptera litura*
Nilaparvata lugens	*Spodoptera mauritia*
	Spodoptera ornithogalli[a]
	Trichoplusia ni[a]

[a]Species found in soybeans.

significant differences, one application of 20 LE/0.4 ha (ILE = 6×10^9 PIB) resulted in fewer moths than either two applications of 10 LE/0.4 ha or four applications of 5 LE/0.4 ha. Apparently one application provided sufficient residual virus to control subsequent larval reinfestations. All viral treatments also significantly reduced larval populations of *H. zea*; reductions ranged from ca. 84 (5 LE x 1) to 98% (5 LE x 4).

SUMMARY

How effective must microbial insecticides be to be used on soybeans? They need not completely eliminate the pest but only suppress it so that the pest will not reduce soybean yield or quality. But what is this level of suppression? To answer this, soybean entomologists have obtained data on the interrelationships between soybean plants and caterpillar pests, the damage these pests cause, and the subsequent effects of insect damage on yield

Table 8. **Effectiveness of *Baculovirus heliothis* in controlling *Heliothis zea* feeding on soybeans (17)[a].**

Treatments		Average Percent Reduction		
		1975	1976	
LE/0.4 ha	Number	Larvae	Moths	Larvae
1	4	92.0	--	--
10	4	96.7	--	--
100	4	100.0	--	--
5	1	--	69.2	83.8
5	4	--	76.8	97.8
10	2	--	80.0	94.9
20	1	--	82.2	93.5
40	1	--	88.9	96.6

[a]Plots artifically infested with 23x and 67x the estimated economic threshold (3.5 larvae/30 cm-row) for 1975 and 1976, respectively.

and quality of soybeans. To summarize, we now can estimate what each caterpillar pest is capable of doing to soybeans and how this relates to population density, phenological stage of soybean growth and its effect on yield and quality (27,28).

These data, plus the known effectiveness of entomopathogens, indicate that entomopathogens can be used successfully to suppress caterpillar pests of soybeans below levels established as economic thresholds. More specifically, I believe that major caterpillar pests of soybeans can be controlled by a preventative application of *N. rileyi* followed by use of *B. thuringiensis* to control defoliators and *B. heliothis* to control pod-feeding bollworms (Table 9). The fungus *N. rileyi* would be the first phase of the pest management strategy. As an example, under midwest conditions (indeterminate soybeans) one preventative and heavy application of conidia would be made during flowering which is before economic levels of insect defoliators are anticipated. If environmental conditions are adequate, this initial inoculum should increase to control defoliators during stages when soybeans are most sensitive to defoliation and also should provide control of damaging *H. zea* later in the season (15). Indeed, no more than this one application might be needed throughout the entire season. However, if environmental conditions are not optimum for development of an epizootic of *N. rileyi* this preventative application of *N. rileyi* would be useless and we would have to rely on monitoring for timely applications of *B. thuringiensis* (from when pods were just visible to when full-sized green beans were present) to control or suppress incipient, damaging populations of defoliating caterpillars (16). Timing and rate of applications of *B. thuringiensis* would be based upon estimates of anticipated damage predicted from data on larval consumption and yield

Table 9. A proposed program for use of entomopathogens to control caterpillar pests of soybeans.

Species of Entomopathogen	Soybean Phenology[a]	Pests to be Controlled
Nomuraea rileyi	Initial flowering to ½ of plants with flowers (R1)	Defoliating and pod-feeding caterpillars
Bacillus thuringiensis	Pods just visible (R3) to pods containing full-size green beans (R6)[b]	Defoliating caterpillars
Baculovirus heliothis	Pods ¾ inch long (R4) to pods beginning to yellow (R7)[b]	Pod-feeding caterpillars

[a]Stage at which entomopathogen will be used; classification system of Fehr et al., 1971.
[b]At 1 of 4 uppermost nodes.

losses (3,27,28). Later in the season (when pods were developing to when they turn yellow), we would rely on *B. heliothis* to control *Heliothis* larvae feeding on pods. With this program we would only use quick-killing chemical insecticides as a last resort and then only at reduced rates to suppress populations of defoliating or depodding caterpillars that were not controlled by *N. rileyi, B. thuringiensis,* or *B. heliothis.*

NOTES

Carlo M. Ignoffo, U.S. Department of Agriculture, SEA/AR, Biological Control of Insects Research, Columbia, MO 65205.

REFERENCES

1. Beegle, C.C., L. P. Pedigo, F. L. Poston, and J. D. Stone. 1973. Field effectiveness of the granulosis virus of the green cloverworm as compared with *Bacillus thuringiensis* and selected chemical insecticides on soybeans. J. Econ. Entomol. 66:1137-1138.
2. Bell, J. V. 1978. Development and mortality in bollworms fed resistant and susceptible soybean cultivars treated with *Nomuraea rileyi* or *Bacillus thuringiensis.* J. Georgia Entomol. Soc. 13:50-55.
3. Boldt, P. E., K. D. Biever, and C. M. Ignoffo. 1975. Lepidopteran pests of soybeans: Consumption of soybean foliage and pods and development time. J. Econ. Entomol. 68:480-482.
4. Carner, G. R., M. Shepard, and S. G. Turnipseed. 1975. Disease incidence in lepidopterous pests of soybeans. J. Georgia Entomol. Soc. 10:99-105.
5. Chamberlin, F. S., and S. R. Dutky. 1958. Tests of pathogens for control of tobacco insects. J. Econ. Entomol. 51:506.
6. Dulmage, H. T. 1971. Economics of microbial control, pp. 581-590. In H. D. Burges and N. W. Hussey (ed.) Microbial control of insects and mites. Academic Press, New York, NY.
7. Fuxa, J. R., and W. M. Brooks. 1978. Persistence of spores of *Vairimorpha necatrix* on tobacco, cotton, and soybean foliage. J. Econ. Entomol. 71:169-172.
8. Getzin, L. W. 1961. *Spicaria rileyi* an entomogenous fungus of *Trichoplusia ni.* J. Insect Pathol. 3:2-10.

9. Hostetter, D. L., and C. M. Ignoffo. 1978. Utilization of entomopathogens as control agents against cabbage looper larvae in various agroecosystems. Florida Monogr. Series (In press).
10. Hughes, K. M. 1957. An annotated list and bibliography of insects reported to have virus diseases. Hilgardia 26:597-629.
11. Ignoffo, C. M. 1979. Fungi: Pest control by *Nomuraea rileyi.* In H. D. Burges (ed.) Microbial control of insects, Vol. II, Academic Press, London (In press).
12. Ignoffo, C. M. and T. L. Couch. 1979. *Baculovirus heliothis:* A nucleopolyhedrosis virus of *Heliothis* species. In H. D. Burges (ed.) Microbial control of insects, Vol. II, Academic Press, London (In press).
13. Ignoffo, C. M., B. Puttler, N. L. Marston, D. L. Hostetter, and W. A. Dickerson. 1975. Seasonal incidence of the entomopathogenic fungus *Spicaria rileyi* associated with noctuid pests of soybeans. J. Invertebr. Pathol. 25:135-137.
14. Ignoffo, C. M., N. L. Marston, B. Puttler, D. L. Hostetter, G. D. Thomas, K. D. Biever, and W. A. Dickerson. 1976. Natural biotic agents controlling insect pests of Missouri soybeans, pp. 561-578. In L. D. Hill (ed.) World Soybean Research, Proc. World Soybean Res. Conf., Interstate Printers and Publ., Danville, IL.
15. Ignoffo, C. M., N. L. Marston, D. L. Hostetter, B. Puttler, and J. V. Bell. 1976. Natural and induced epizootics of *Nomuraea rileyi* in soybean caterpillars. J. Invertebr. Pathol. 27:191-198.
16. Ignoffo, C. M., D. L. Hostetter, R. E. Pinnell, and C. Garcia. 1977. Relative susceptibility of six soybean caterpillars to a standard preparation of *Bacillus thuringiensis* var. *kurstaki.* J. Econ. Entomol. 70:60-63.
17. Ignoffo, C. M., D. L. Hostetter, K. D. Biever, C. Garcia, G. D. Thomas, W. A. Dickerson, and R. Pinnell. 1978. Evaluation of an entomopathogenic bacterium, fungus, and virus for control of *Heliothis zea* on soybeans. J. Econ. Entomol. 71:165-168.
18. Ignoffo, C. M., C. Garcia, R. W. Kapp, and W. B. Coate. 1979. An evaluation of the risks to mammals of the use of an entomopathogenic fungus, *Nomuraea rileyi* (Farlow) Samson, as a microbial insecticide. Environ. Entomol. 8:354-359.
19. Kish, L. P., and G. E. Allen. 1978. The biology and ecology of *Nomuraea rileyi* and a program for predicting its incidence on *Anticarsia gemmatalis* in soybean. Florida Agric. Exp. Stn. Bull. 795. 48 pp.
20. Martignoni, M. E., and R. L. Langston. 1960. Supplement to an annotated list and bibliography of insects reported to have a virus disease. Hilgardia 30:1-40.
21. McLeod, P. J., S. Y. Young III, and W. C. Yearian. 1978. Effectiveness of microbial and chemical insecticides on beet armyworm larvae on soybeans. J. Georgia Entomol. Soc. 13:266-269.
22. Muller-Kogler, E. 1965. Pilzkrankheiten bei Insekten, Vol. 1. Paul Parey, Hamburg and Berlin. 460 pp.
23. Newman, G. G., and G. R. Carner. 1975. Factors affecting the spore form of *Entomophthora gammae.* J. Invertebr. Pathol. 26:29-34.
24. Steinhaus, E. A. 1951. Report on diagnoses of diseased insects, 1944-1950. Hilgardia 20:629-678.
25. Steinhaus, E. A., and G. A. Marsh. 1962. Report of diagnoses of diseased insects, 1951-1961. Hilgardia 33:349-490.
26. Sprenkel, R. K., and W. M. Brooks. 1975. Artificial dissemination and epizootic initiation of *Nomuraea rileyi,* an entomogenous fungus of lepidopterous pests of soybeans. J. Econ. Entomol. 68:847-851.
27. Thomas, G. D., C. M. Ignoffo, K. D. Biever, and D. B. Smith. 1974. Influence of defoliation and depodding on yield of soybeans. J. Econ. Entomol. 67:683-685.
28. Thomas, G. D., C. M. Ignoffo, and D. B. Smith. 1976. Influence of defoliation and depodding on quality of soybeans. J. Econ. Entomol. 69:737-740.

29. Weiser, J. 1961. Die microsporidien als parasiten der insekten monogra. Z. Angew, Entomol. No. 17, Paul Parey, Hamburg and Berlin. 149 pp.
30. Yearian, W. C., J. M. Livingston, and S. Y. Young. 1973. Field evaluation of *Bacillus thuringiensis* for control of selected lepidopterous foliage feeders on soybeans. Arkansas Agric. Exp. Stn. Rep. Ser. 212. 8 pp.

INSECT PEST MANAGEMENT IN NORTH CAROLINA SOYBEANS

J. R. Bradley, Jr. and J. W. Van Duyn

The soybean industry in North Carolina, as we know it today, is a recent development. Within the last 2 decades, land planted to soybeans has risen from 0.5 to 1.5 million acres in response to the increased economic value of the crop. This three-fold increase in soybean acreage came about as more cropland was planted to soybeans in the coastal plain and piedmont regions and large tracts of coastal swampland were placed into a soybean-corn rotation.

Soybean insect pest problems were handled through the 1960's on an ad hoc, individual field basis (if at all), largely because entomological research and extension efforts were too limited to allow for the conceptualization, development, and implementation of a more comprehensive pest management program. The entomology program consisted of little more than evaluating insecticides for efficacy and recommending those which gave desired levels of population reduction. Furthermore, most farmers concerned themselves with losses to insects only on crops of higher value (i.e., tobacco, cotton and peanut), so the entomologists' attitude of "let the sleeping dog lie" was probably the most appropriate strategy for that period. As the price of soybeans rose from \$2 - 3 a bushel (where it had been for decades) to \$7 or more in the early 1970's, farmers became much less tolerant of their insect competitors and looked to entomologists to provide solutions to their problems.

Entomologists at North Carolina State University responded to the increased value of soybean and changing farmer philosophy toward insect pests on the crop by greatly expanding research. It was apparent that dynamic insect pest management systems, undergirded by a base of sound biological knowledge of the soybean agroecosystem, must be developed as an alternative to continued dependency on the pesticide tactic as the first and only line of defense against insect pests. An entomological team largely supported by the National Science Foundation through a grant (GB-28855) to North Carolina State University (NCSU) and by the National Science Foundation and the Environmental Protection Agency through a grant (NSF GB-34718) to the University of California (which subcontracted research at NCSU) began to develop the necessary biological base for insect management on soybean.

The corn earworm, *Heliothis zea* (Boddie), had historically been the major insect pest of soybean in North Carolina and over most of the state it was the only insect pest that had to be routinely controlled with insecticides. Its pod feeding often causes almost complete crop loss in many soybean fields throughout the coastal plain of the southeastern U.S. when insecticides are not applied to reduce larval populations that have exceeded economic thresh-holds (7). Therefore, we initiated a research program designed to provide an understanding of the relationship between soybean and *H. zea.* We also sought to identify the major factors (abiotic and biotic) that affect that association. This paper concerns the progress of that research program and the develeopment of a management system for *H. zea* on soybean in North Carolina. We propose an approach to insect pest management (as well as our system) applicable not only to *H. zea* in North Carolina but also to other insect pests here and in different agroecosystems throughout the world.

The major objectives of our research were (1) to develop a description of the spatial and temporal patterns of *H. zea* on soybean and other crops which contribute to its general population and to identify factors affecting these patterns, (2) to identify key arthropod natural enemies of *H. zea* in soybean fields and to determine their effect on *H. zea* populations, and (3) to develop a dynamic management system for *H. zea* on soybean utilizing a multi-tactical approach based on the strategy of preventing the problem, where possible, through manipulation of crop production techniques and utilizing natural controls to the fullest possible extent.

DEVELOPMENT OF THE BIOLOGICAL BASE FOR MANAGEMENT

The research target for objective 1 was the general *H. zea* population over a wide area as well as the specific component of that population that occurred on soybeans. This holistic approach was essential if the objective was to be accomplished, because population potentials at each local focus (i.e., a particular field of a host) are largely determined by patterns of invading moths originating from other, and at times distant, sources. Since the *H. zea* life system extends into, and is a part of, many divergent communities, both managed (cultivated) and unmanaged (uncultivated), its complexity can be analyzed and understood only through a modeling effort (6).

Our intention was to use modeling procedures as an organizational framework and as a system through which we could manipulate vast quantities of data toward the accomplishment of our objective of understanding *H. zea* population patterns. We sought to partition the *H. zea* life system into its component parts and to understand the contribution each part (i.e., specific host crop or phenological type of a specific host crop) made to any other part or to the whole.

HELSIM-1 (5) and HELSIM-2 (6) presented the production flow of *H. zea* life stages through its four principal host plants--corn, cotton, soybean, and tobacco. These models were generated from a vast data base which we assembled from 1970 through 1976 by intensive larval sampling in hundreds of fields of the four principal hosts as well as adult monitoring with blacklight traps and wide-area monitoring of the phenological development of the host plants. The generalized production of *H. zea* large larvae on the major hosts in North Carolina revealed that corn is the major host of *H. zea* in North Carolina through the F2 generation. As corn matures toward mid-summer, soybean becomes the dominant host and occupies the bulk of the *H. zea* population during August and September.

When we concentrated our attention on that portion of the *H. zea* population associated with soybean, we found larval distribution among fields to be highly clumped. The most obvious factors affecting distribution were planting date and phenological stage and extent of canopy development during the flight of second generation moths which had emerged from corn fields.

Early planted soybean fields rarely sustained damaging larval populations of *H. zea*. The planting date phenomenon is clearly illustrated from data collected at a 1974 study site in Halifax County (Table 1). None of the soybean fields planted by May 20 required insecticide applications for control of *H. zea* larvae, whereas 79% of the late planted fields reached *H. zea* larval threshhold levels and had to be sprayed.

The relationship between phenological stage of a soybean field and *H. zea* infestation is illustrated through data collected during 1972 and 1973 which showed that larval populations (in fields where larvae were found) consistently peaked at the medium pod stage (Fig. 1). These soybean fields were

Table 1: Relationship between soybean planting dates and fields requiring insecticide application for control of *H. zea* (Halifax Co., N.C. 1974).

Planting Date Class	Number of Fields in Each Class	% Fields Requiring Insecticide Application
Early (By May 20)	15	0
Mid (May 21 - June 10)	13	33
Late (After June 10)	14	79

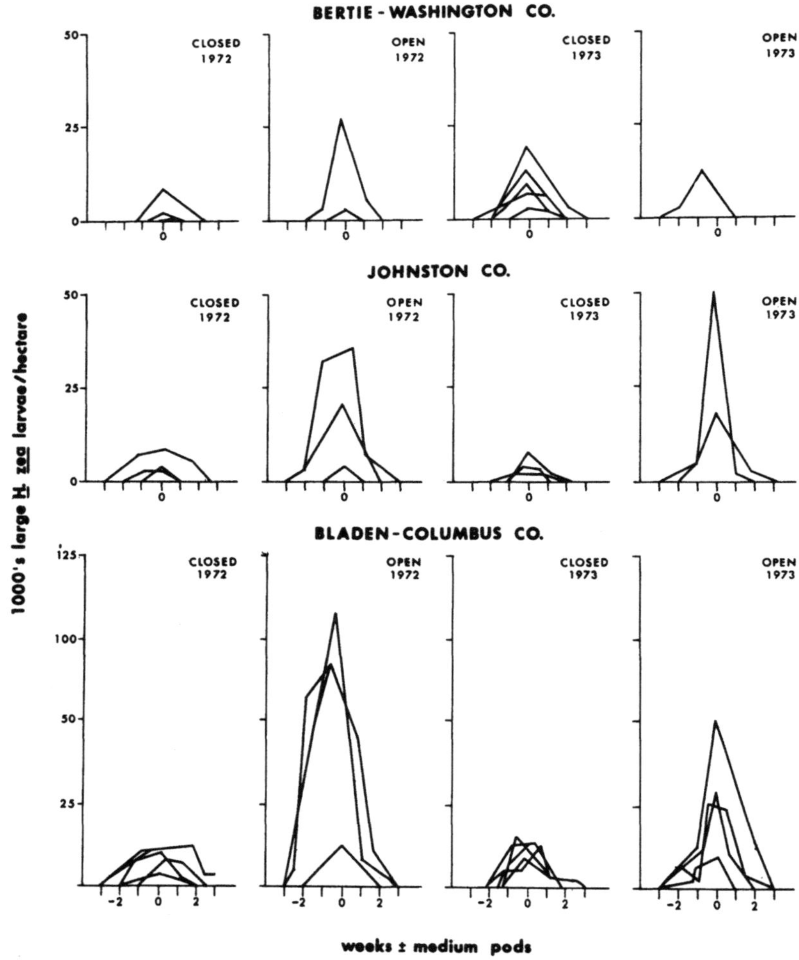

Figure 1. ***Heliothis zea*** **larval populations on open and closed canopy soybean fields in 3 regions of North Carolina (1972 and 1973).**

at peak bloom 2 weeks previously when egg deposition occurred (based on average developmental time from egg to large larva). Therefore, fields at peak flower during the period when second generation moths were ovipositing were more likely to be subject to heavy egg deposition than fields in which flowering was completed or extremely late planted fields which had not begun to flower.

The third factor affecting *H. zea* population levels on soybean was the extent of canopy development during second generation moth flight. Fields that had closed canopies were much less likely to have *H. zea* populations that reached damage threshold levels than fields with open canopies (Fig. 1).

We categorized 1500 soybean fields in a 1973 survey (2) into closed and open canopy types during the peak of second generation *H. zea* moth flight. In that survey, damage threshold levels of *H. zea* larvae developed in only 2.8% of the closed canopy fields whereas the threshold was exceeded in 22% of open canopy fields.

The relationship between canopy development in soybean and *H. zea* larval populations may be explained by the following factors or any combination of these factors.

1. Canopy development is a function of planting date and the extent of canopy development is indicative of soybean maturity, i.e., more mature plants are less attractive to ovipositing moths and less acceptable as a substrate for larval establishment.
2. The closed canopy may be mechanically interfering with *H. zea* flight and ovipositional behavior.
3. The closed canopy may be obscuring a visual clue originating from flowers (i.e., infrafred radiation) that is a necessary ovipositional stimulus.
4. Natural enemy populations (arthropod parasites and predators and pathogens) may be greater in closed canopy soybean fields (4).

With any dynamic biological system, there are likely to be exceptions to generalized population processes during years characterized by atypical weather patterns. The significance of modeling technology to *H. zea* management on soybean in North Carolina was demonstrated during 1977 when rainfall was less than normal and temperature was higher than normal during the first half of the growing season.

Computer simulations (using HELSIM) had predicted that under hot and dry conditions, *H. zea* populations on early-planted soybeans might exceed those normally encountered on late-planted soybeans. These simulations included direct effects of temperature on *H. zea* development as well as indirect effects on crop growth rate and canopy development. During the summer of 1977, much of North Carolina was subjected to prolonged hot, dry weather. In those areas, there were numerous observations of high infestation levels of *H. zea* in early-planted soybean fields, but not in late-planted soybean fields. The explanation for this occurrence is threefold.

1. Corn matured much more rapidly than normal. Also, in many areas over 40% of the total corn acreage (particularly late-planted fields) dried up without producing ears.
2. Since blooming in soybeans is largely photoperiodically controlled, early planted soybean fields bloomed only 2-3 days earlier than normal (as opposed to corn which matured 2-3 weeks earlier). In addition, the prolonged drought prevented normal soybean canopy

development and many early planted fields had greatly reduced foliage at time of bloom (the peak attractive state for *H. zea* oviposition).

3. Overwintered *H. zea* adults emerged about 2 weeks earlier than normal and development was about 20% faster than at usual temperatures, resulting in the peak second generation moth flight coincident with early blooming soybeans, rather than 2 weeks later than peak-bloom in early planted soybean fields (the usual situation).

The lack of attractive late corn, the availability of highly attractive, open canopy, blooming soybean fields, and the earlier second generation moth flight coincident with peak bloom in these early-planted soybeans resulted in major infestations of *H. zea* in early planted soybean fields in the drought-striken regions. In those areas where there was adequate rainfall, *H. zea* was found at high larval densities only in late-planted fields, as had been observed and simulated for normal weather patterns. The unusual weather pattern experienced in North Carolina in 1977 may be only a very rare occurrence; however, it did illustrate the dynamics of the agroecosystems we seek to manage and demonstrated the necessity of modeling technology as an integral part of insect pest management.

Another major factor contributing to the regulation of *H. zea* populations on soybean is the arthropod natural enemy complex. This potentially valuable management tool is too often ignored or improperly utilized because of lack of knowledge of the key species involved and their impact upon pest populations. We sought to overcome this problem, in part, through an extensive, 3-year survey of the arthropod fauna found in North Carolina soybean fields (2). The most commonly encountered entomophagous arthropods identified in that survey were: *Orius insidiosus* (Say), *Geocoris punctipes* (Say), *Nabis* spp., *Coleomegilla maculata* (DeGeer), and miscellaneous Araneida.

The value of these predaceous arthropod species toward limiting development of *H. zea* populations on soybean was then established in a large-plot field experiment during 1975. In that experiment the arthropod natural enemy complex in one treatment was decimated by a broad spectrum insecticide, methyl parathion, applied just prior to soybean flowering and *H. zea* egg deposition. Treatment of other plots consisted of the application of a recommended insecticide, carbaryl, when the *H. zea* larval population reached the damage threshold (2 large larvae/row ft) or no insecticide application. The significance of the deliberate destruction of natural enemies to subsequent *H. zea* populations and soybean yields in that experimental field is shown in Table 2. A reduction in total number of certain predator species by approximately 90% resulted in 2 times more *H. zea* larvae and only one-third the soybean yield when compared to the untreated check.

Table 2: Effects of properly and improperly timed insecticide applications on corn earworm, its predators and yields in soybean (Halifax Co., N.C., 1975).

Insecticide Program	No. Insect Predators/Row Ft	Corn Earworm Larvae/Row Ft.	Soybean Yield Bu/Acre
Single application properly timed (at damage threshold 2 larvae/row ft.)	1.07	0.94	30
Single application improperly timed (just prior to *H. zea* egg deposition)	0.15	9.88	6
Untreated check	1.13	4.05	18

Soybean yield in the treatment where insecticide application was based on the *H. zea* larval threshold was 5 times greater than yield in the treatment where predators were eliminated before they exerted their maximum effect on the *H. zea* population. However, natural enemies alone did not prevent the corn earworm population from reducing soybean yield, since the control yielded only 60% as much as the treatment where natural enemies were supplemented by a properly timed insecticide application. This experiment specifically documented the important contribution arthropod predators make toward regulation of phytophagous caterpillar species in soybean. Also, this experiment substantiated previous observations that predator removal by untimely insecticide applications (mistimed for *H. zea* or made to control other pests) could have a profound effect on *H. zea* numbers and soybean yield.

We investigated the effects of soil-applied systemic pesticides (i.e., aldicarb, carbofuran, fensulfothion, phorate) on insect predators of *H. zea* in soybean during 1976 and 1977. Aldicarb applied in the seed-furrow at planting for Mexican bean beetle control or in a band for nematode control reduced hemipterous predator populations as severely as did methyl parathion applied as a foliage spray (3). Corn earworms were as much as 7 times more numerous in aldicarb treatments as in the untreated check where predator populations were undisturbed.

THE MANAGEMENT SYSTEM

Now that we had developed a spatial and temporal description of that portion of the *H. zea* life system associated with soybean and had identified and understood some of the most important factors affecting that distribution, a sound management system was designed.

The insect management system we developed for *H. zea* on soybean is structured by the four principal strategies of prediction, prevention, detection, and suppression. This structure can accommodate all tactics and provides an organized and flexible procedure for the development of compatible management systems for individual insect species or complexes. The organization also is conveniently used for discussion purposes.

Prediction

Predictions concerning the species of insect which may occur, probabilities of occurrence (frequency and intensity), spatial and temporal distributions, and the consequences (potential for damage) are based upon historical occurrences, knowledge of biology and behavior, monitoring data, and any other relevant information (i.e., weather). Most initial predictions are made prior to the growing season (i.e., corn earworm will occur in soybeans in 1979) and are updated through the season as new information becomes available (i.e., because of the hot, dry weather the corn earworm will occur earlier than typical and at higher densities). Although predictions may vary greatly in accuracy and significance they are valuable tools for guiding subsequent strategies and tactics. The fundamental prediction made for most of the soybean growing area in North Carolina is that corn earworm will occur at damaging levels and that late planted (after June 1) soybean cultivars of maturity group VII (i.e., Bragg and Ransom) with open canopies at flowering will be most severely attacked. This prediction can be revised as information gathered from corn-field surveys, light traps, and from the *H. zea* development model (HELSIM) becomes available. Revised predictions reflect current population development trends and may range from severe outbreaks with atypical temporal and spatial distributions (i.e., severe infestation will occur in early and late maturing soybeans) to low populations with atypical temporal and spatial distributions (i.e., only the most attractive late soybeans will develop threshold populations).

Prevention

The ultimate objective is to prevent yield loss which is greater than control costs. Yield loss is directly influenced by (1) the population density of pest insects and by (2) the ability of the host crop to tolerate insect pests. Therefore, preventative techniques are designed to maintain pest insect populations at low levels and to maintain crop tolerance at high levels.

Populations of pest insects are mainly regulated by reducing the immigration and establishment of colonizers and by increasing biological control within the crop. Crop attractiveness to corn earworm moths is greatly affected by time of flowering and canopy development and both of these factors can be regulated through cultural techniques (i.e., cultivar selection, planting date, row spacing) and insuring adequate crop health. Maturity group V soybean cultivars typically bloom prior to the corn earworm immigration from field corn and are not highly attractive to ovipositing moths. Consequently, these cultivars (i.e., Forrest, York, Essex) usually are not colonized by corn earworms at a level beyond the capability of the natural control agents to keep larval populations below the damage threshold. Early planting (by May 20) of any cultivar, including those which often bloom in synchrony with the *H. zea* moth flight, reduces crop attractiveness and colonization. Presumably, early planting has this effect because of a more complete canopy development, a slight desynchronization of flowering with the moth flight, and a greater proportion of more mature foliage on which larval establishment is reduced. Row spacing influences canopy development in later planted soybeans and thus is also used to reduce crop attractiveness (later planted soybeans only). Additionally, plant growth greatly affects canopy development and for that reason fertility, pH, nematodes and other pests, plow-pans, and other growth-limiting factors are managed in order to stimulate vegetative growth. Through the use of these various cultural tactics *H. zea* often can be relegated to a low frequency pest through reduced egg deposition and subsequent larval establishment.

Attempts to achieve high mortality of *H. zea* eggs and larvae within the soybean crop mainly involve promotion of indigenous biological control agents. The significance of this biological control complex upon the corn earworm population is influenced by the presence of hosts over an extended time period, by the length of time from planting until corn earworm invasion (development time), by the presence or lack of toxic chemicals, by temperature and humidity conditions, and many other factors. Consequently, recommendations are made to (1) use no soil insecticides, (2) plant early, (3) promote vigorous growth resulting in rapid canopy development, (4) use foliar insecticides only when needed, and (5) use selective insecticides (i.e., low rates of carbaryl) for pre-bloom spraying whenever possible. These tactics help insure a complex of biological control agents which usually exerts substantial control pressure on the immatures of invading lepidopterous pests.

Detection

Detection procedures are used to describe the *H. zea* population in the general environment as well as in individual fields. Monitoring relies upon light trapping of moths, wide-area surveys, and scouting of specific fields. Light trap and wide-area surveys provide information on corn earworm abundance, development, and distribution among soybean fields differing in phenological stage. Information is mainly used to confirm and revise prediction and to inform Agricultural Extension Agents, farmers, and others concerning the corn earworm problem. Methods of relaying monitoring information and revised predictions (derived from the *Heliothis* computer model HELSIM) include special "Insect Alert" bulletins, a weekly "Insect

Survey Note," mass media, and a special toll-free "Tele-Tip" telephone recording (updated three times weekly). These communications are mainly directed to Agricultural Extension Agents, who in turn notify their respective clientele.

Although corn earworm problems could usually be prevented through the use of the previously mentioned techniques, many farmers cannot take full advantage of the practices due to conflicts within their overall production program. Therefore insect population and damage monitoring (scouting) is necessary to identify the need to treat individual fields. Early season scouting is done for various foliage feeders (Mexican bean beetle, bean leaf beetle, green cloverworm, beet armyworm, etc.) and utilizes a foliage loss estimating technique; the scout samples by estimating leaf damage at random sites (one sample per five acres with a minimum of three samples—thresholds; prebloom 35%, bloom-pod fill 15%).

Scouting for corn earworm is initiated when soybean blooming is approximately 90% completed in individual fields or when corn earworm moth populations peak. For scouting purposes, a field can be defined as connected or adjacent soybean plantings of approximately equal potential host characteristics (i.e., same cultivar and planting date, with similar vegetative growth). *H. zea* moths are strong fliers and distribute their eggs in a random fashion within fields, except along field margins and in areas of poor growth where oviposition is likely to be greater. Scouting is thus done in a random fashion, but avoiding field margins and weak spots within fields. Samples are taken using the beat cloth technique (1) and the same number of samples are required as when making foliage loss estimates. Intensive sampling is begun in fields which are highly attractive to ovipositing moths (blooming—open canopy) and is deemphasized as fields of lesser degrees of attractiveness are encountered. Sequential sampling is often useful and sampling can be terminated after 3 samples if corn earworm numbers average less than 0.50 per row foot or more than 8 per row foot and samples are not highly variable (no one sample comprising 50% or more of the total number) (Van Duyn, unpublished data). Thresholds for corn earworm in North Carolina are 2 large larvae per row foot and are considered to be damage thresholds, as opposed to economic injury thresholds.

Suppression

Suppression refers to the quick reduction of a population and, in this instance, relies solely upon insecticides. Since few soybean insect problems occur later in the season than corn earworm, insecticide selectivity to favor arthropod predators and parasites is not a major factor governing choice of insecticide. Efficacy against corn earworm, cost, application ease, mammalian toxicity, and availability are the main criteria influencing insecticide selection. Rate adjustments tailored to individual needs can be accomplished if scouting information is reliable. Since the desired effect is to lower the population below the threshold, the degree of control needed is dependent upon initial population density. Therefore higher rates (e.g., methomyl, 0.45 lb/ai/A) are used for severe infestation (greater than 8 larvae per row foot) but lower rates can be used for less severe infestations (e.g., methomyl 0.22 or 0.16 lb/ai/A for 5 and 3 earworms per row foot, respectively).

Application is most frequently accomplished with an airplane although high clearance sprayers and tractor mounted sprayers are often used. The corn earworm is easily controlled in soybeans with recommended insecticides and spray volumes do not appear to be highly important as long as adequate coverage is achieved.

This insect management system was developed and tested within the context of a pilot pest management project in eastern North Carolina. The results of using the system were (1) lowered frequency of soybean fields with above-damage threshold populations of corn earworm larvae, (2) more efficient overall insecticide use, (3) reduced number of insecticide applications in individual fields (through preventing mistimed sprays), and (4) reduced grower costs without associated yield losses.

The organizational framework of prediction, prevention, detection, and suppression accommodates all useful entomological information and, more importantly, is suitable for the management approaches of plant pathology, nematology, weed science, and crop production. The structure is now being used to compatibly organize pest management procedures (for all pests) into overall crop management systems in North Carolina.

NOTES

J. R. Bradley, Jr. and J. W. Van Duyn, Dept. of Entomology, North Carolina State Univ., Raleigh, NC 27650.

REFERENCES

1. Boyer, W. P. and W. A. Dumas. 1963. Soybean insect survey as used in Arkansas. U.S.D.A. Coop. Econ. Insect. Rept. 13:91-2.
2. Deitz, L. L., J. W. Van Duyn, J. R. Bradley, Jr., R. L. Rabb, W. M. Brooks, and R. E. Stinner. 1976. A guide to the identification and biology of soybean arthropods in North Carolina. N.C. Agr. Exp. Sta. Tech. Bull. 238.
3. Morrison, D. E., J. R. Bradley, Jr., and J. W. Van Duyn. 1979. Populations of corn earworm and associated predators after applications of certain soil-applied pesticides to soybeans. J. Econ. Entomol. 72:97-100.
4. Sprenkel, R. K., W. M. Brooks, J. W. Van Duyn, and L. L. Deitz. 1979. The effects of three cultural variables on the incidence of *Nomuraea rileyi*, phytophagous Lepidoptera and their predators on soybeans. Environ. Entomol. In press.
5. Stinner, R. E., R. L. Rabb, and J. R. Bradley, Jr. 1974. Population dynamics of *Heliothis zea* (Boddie) and *H. virescens* (F.) in North Carolina: A simulation model. Environ. Entomol. 3:163-8.

6. Stinner, R. E., R. L. Rabb, and J. R. Bradley, Jr. 1976. Natural factors operating in the dynamics of *Heliothis zea* in North Carolina. Proc. XV Int. Cong. Entomol., Washington, D.C. p. 622-42.
7. Turnipseed, S. G. 1973. Insects. In B. E. Caldwell (ed). Soybeans: Improvement, production, and uses. Amer. Soc. Agron., Madison, Wis. p. 545-72.

ROLE OF *PHOMOPSIS* SP. IN THE SOYBEAN SEED ROT PROBLEM

A. F. Schmitthenner and K. T. Kmetz

Soybean seed decay and poor seed quality have commonly been attributed to the pod and stem blight pathogen, *Diaporthe phaseolorum* var. *sojae (Dps)* (5, 6, 12, 16, 19, 20, 21, 22, 24, 27), and the soybean stem canker pathogen, *D. phaseolorum* var. *caulivora (Dpc)* (16, 24). A third inciting agent, an undescribed *Phomopsis* sp., generally has been overlooked or considered to be a *Diaporthe* sp. based on cultural and asexual characteristics. Lehman (20), Lutrell (21), and Hildebrand (12) have described non-perthecial isolates of *Phomopsis*. The importance of an imperfect *Phomopsis* as the most prevalent component of the soybean pod and stem blight and seed decay complex was emphasized by Kmetz et al. (16). The importance of *Phomposis* in the soybean seed decay complex has not been evaluated. Until recently, attention has been focused on distinguishing between *Dps* and *Dpc* (4, 12, 24) and the identity and role of the non-perithecial Phomopsis isolates from soybean was neglected.

DISTINGUISHING CHARACTERISTICS OF *PHOMOPSIS* SP.

Imperfect soybean *Phomopsis* isolates can be distinguished readily from *Dps* and *Dpc* on acidified potato dextrose agar (APDA) (17). Single, alpha spore isolates produce dense, fluffy, white mycelium, turning yellowish green to brown with age. Erumpent, multi-chambered pycnidia develop in dense, carbonaceous, pulvinate stromata. Pycnidia with prominent beaks are produced in old cultures. Alpha spores are extruded in a gelatinous matrix from the ostiole but beta spores are rarely found. Perithecia never develop. Stromata are very well developed, frequently extending under the entire culture.

Isolates of *Dps* produce colonies of loose-growing mycelium at first white, then turning dull orange to pink and loosely uniting into strands. Mycelium and agar become dark in the underside with age. Single pycnidia with slightly protruding beaks form in echinulate stromata. Multi-beaked pycnidia can be found in old cultures, but the beaks never elongate. Perithecia are produced after 1 month either singly or intermingled with pycnidia. Incubation in light is important for perithecial production.

Isolates of *Dpc* produce colonies similar to *Dps* but mycelium remains white and frequently forms dense strands or tufts with age. Pycnidia form infrequently and are usually sterile. Perithecia form in clusters associated with pycnidia or in mycelial tufts in old cultures exposed to light. Alpha spores of the *Phomopsis* sp. and *Dps* are similar as are ascospores of *Dps* and *Dpc*. *Phomopsis* produces predominantly alpha spores after seven successive transfers, whereas *Dps* produces increasing numbers of beta spores with successive subculturing.

PREVALENCE OF *PHOMOPSIS* IN SOYBEAN STRAW

Phomopsis sp., *Dps,* and *Dpc* all can be found associated with mature soybean straw (16, 19). *Phomopsis* was isolated more frequently than *Dps* and *Dpc*. It was isolated from 92% of 1,200 pycnidia from mature soybean tissue and *Dps* from 8%. *Phomopsis* was recovered from the gelatinous spore matrix or the entire sporocarp of all erumpent, beaked and sometimes multichambered pycnidia sampled, while *Dps* was obtained from sublobose pycnidia with short, rounded to conical beaks. Occasionally, beta spores were found in *Dps* pycnidia but never in those of the *Phomopsis* sp. Perithecia were induced to form on mature soybean straw by placing stem sections in moist chambers for 30 or more days at the end of the growing season. Of 200 cultures obtained from these perithecia, 97% were *Dps* and 3% were *Dpc*. Perithecia also were observed on soybean stems that had overwintered in the field. Out of 72 single ascospores obtained from six perithecia from each of four stems, 69 were *Dpc* and three were *Dps*.

PREVALENCE OF *PHOMOPSIS* IN SOYBEAN SEED

In Ohio Kmetz reported that *Phomopsis* was the most prevalent soybean seed fungus (15). It constituted 77% of seed isolates compared to 18% and 5% for *Dps* and *Dpc,* respectively. Similar results were obtained from seed isolations in 1977 (Table 1). Percentage seed infected with fungi was quite high. *Phomopsis* was the most prevalent fungus isolated; *Diaporthe (Dps + Dpc), Alternaria* and *Cercospora* were isolated less frequently except in one test where *Cercospora* was unusually high. Other fungi constituted a very small percentage of the isolates obtained (Table 2). *Fusarium (F. acuminatum* and *F. equiseti)* was the predominent minor genus obtained.

Table 1. Percentage fungus infected soybean seed and major fungi (percentage of total fungi) isolated from soybean seed in Ohio in 1977.

	Test 1	Test 2	Test 3	Test 4	Test 5	Test 6
Number of seed plated	9,600	5,300	6,900	8,600	6,000	11,900
Infected seed	78.9	37.0	69.4	43.9	43.4	61.9
Phomopsis	75.4	70.8	42.8	28.7	60.7	65.8
Diaporthe	8.6	4.6	10.1	13.0	11.5	8.6
Cercospora	1.2	3.8	26.7	47.4	7.8	10.5
Alternaria	13.4	14.6	18.4	7.3	15.1	13.2
Others	1.4	6.2	2.0	3.6	3.9	1.9

Table 2. Minor fungi (percentage of total fungi) and other microorganisms (%) isolated from soybean seed in Ohio in 1977.

	Test 1	Test 2	Test 3	Test 4	Test 5	Test 6
Percentage minor fungi	1.4	6.2	2.0	3.6	3.9	1.9
Fusarium	17	30	50	25	47	48
Colletotrichum	0	11	13	5	3	6
Chaetomium	1	22	5	3	22	7
Epicoccum	1	0	4	10	5	4
Rhizoctonia	1	2	4	9	3	7
Nigrospora	0	0	1	4	3	4
Bacteria	0	9	5	2	1	2
Others[a]	80	26	18	42	16	14

[a]Includes *Aspergillus* sp., *Botryodiplodia* sp., *Cladosporium* sp., *Macrophomina* sp., *Penicillium* sp., and *Trichoderma* sp., as less than 1 % of minor fungi and unidentified fungi.

The imperfect *Phomopsis* sp. also may be the most prevalent seed decay fungus in other soybean growing areas. The literature is difficult to evaluate on this point since most reports refer to *Dps* as the soybean seed decay pathogen and do not indicate if isolates obtained were identified on the basis of perithecia produced or the imperfect Phomopsis stage. Lehman (20), Lutrell (21) and Hildebrand (12) reported *Diaporthe* and imperfect Phomopsis types from soybean. Kilpatrick (14) found *Phomposis sojae* to be a major soybean seed fungus, but it is not clear whether this fungus was different from the *Phomopsis sojae* described by Lehman that is now called *Dps.* Ross (25) reported on a *Phomopsis* similar to Kmetz et al. (16) in North Carolina. Ellis et al. (10) reported on the occurrence of similar *Phomopsis* sp. on seed of several legumes in the tropics. Comparisons of *Phomopsis* isolates from soybeans from different sections of the USA and different parts of the world need to be made before the relative prevalence of *Phomopsis, Dps,* and *Dpc* on soybean seed can be determined.

VIRULENCE OF *PHOMOPSIS*

Neither the imperfect *Phomopsis* sp. nor *Dps* produce symptoms on vigorously growing plants although occasional isolates were found by Kmetz (15) that produced stem canker-like symptoms. Generally, only *Dpc* isolates consistently produce typical stem canker in 7-wk-old inoculated soybeans. Both *Phomopsis* sp. and *Dps* form rows of pycnidia on mature soybean stems (pod and stem blight symptoms), whereas *Dpc* only produced perithecia and only on overwintered stems. According to Kmetz (15) who used the toothpick method of inoculation, *Phomopsis* rotted more green pods than *Dpc* or *Dps,* and *Phomopsis* and *Dpc* rotted more green and mature seed than *Dps.* Also, *Phomopsis* sp. and *Dpc* were capable of rotting mature soybean seed before they germinated, while *Dps* colonized germinated seed after the radicle had formed. Germination was arrested by all three fungi if seed were placed on APDA cultures for 48 hr prior to planting in soil. Percentage survival of *Dps* infected seedlings was higher than for *Phomopsis* sp. or *Dpc* infected seedlings. We have confirmed these results by wound inoculating mature soybean seed with blocks of agar cultures after the seed had incubated overnight on wet filter paper. *Phomopsis, Dpc, Dps, Fusarium* and *Nigrospora* are virulent seed pathogens. Germination progresses further before seedlings are rotted when they are infected by *Dps* than by *Phomopsis* and *Dpc. Phomopsis* sp. and *Dps* are virulent only on developing or germinating seed. *Phomopsis* may be more virulent than *Dps.*

DISEASE CYCLE OF THE *PHOMOPSIS* SP.

The disease cycle of the imperfect *Phomopsis* sp. may differ from both *Dps* and *Dpc* (15, 19). The *Phomopsis* sp. overwinters on pycnidia on soybean straw having the characteristic symptoms of pod and stem blight. The following spring alpha spores ooze in a dense matrix from mature pycnidia and are splashed onto young plants where latent infection occurs. Later in the season, alpha spore inoculum from overwintered pycnidia is supplemented with secondary incoculum from pycnidia developing on fallen cotyledons and petioles. Latent infections of young pods occur. As the pods mature, *Phomopsis* spreads from the pod wall to the seed, especially under wet conditions (25, 26). Colonization of the seed coat and seed decay occurs.

Pycnidia of *Dps* are much less prevalent in soybean straw than those of the *Phomopsis* sp. Both *Dps* and *Dpc* may produce perithecia on overwintered straw, but they are much less evident than pycnidia. It is not yet completely clear how much of seed infection from *Dps* originates from ascospores or alpha spores. The distribution of *Phomopsis* and *Diaporthe* on soybean plants may be different (18). *Phomopsis* was more prevalent in seed on the lower parts of the plant, while *Diaporthe* was more prevalent in seed on the upper parts. Thus there is evidence that *Phomopsis* and *Diaporthe* inoculum may not be equally prevalent throughout the growing season. Typical *Phomopsis* distribution data obtained from an aerial spray test in Ohio in 1978 are presented in Table 3.

Percentage seed infection with *Diaporthe* as well as *Phomopsis* becomes progressively higher as harvest is delayed (8, 15, 16, 28). Sources of the late infections are not yet established. They could come from colonization of mature pods by mycelium from mature stems after latent infections become active when the plants die, from activation of latent infections in the pods (19), or from splashing of spores from plant debris onto seed through cracks in deteriorating pods' walls (5).

CONTROL OF *PHOMOPSIS*

Crop Rotation

Crop rotation and residue management can influence the severity of *Phomopsis* seed infection and have been recommended for control of pod and stem blight (1). Incidence of *Phomopsis* in seed was less in soybeans following corn than in soybeans following soybeans in four locations in Ohio (15). Crop history had no effect on seed infection with *Dps* except when harvest was delayed 2 months. Then, *Dps* was higher in seed from soybeans following corn than from soybeans following soybeans. Levels of *Dps* did not exceed 30% in this test while *Phomopsis* percentage reached 70%. Thus, there is evidence that the effects of rotation on levels of *Phomopsis* and *Dps* are different. *Phomopsis* could be isolated more frequently from soybean straw recovered from unplowed plots of soybean following soybean than from plowed plots but that more perithecia of *Diaporthe* could be induced to form on soybean straw recovered from fall-plowed than from spring-plowed or unplowed soil (15).

Table 3. **Distribution of *Phomopsis* infected seed in soybean plants sprayed aerially with Benlate in 1978.**

Treatment	Top 1/3 of Plant	Middle 1/3 of Plant	Bottom 1/3 of Plant
None	2[a]	4	20
Benlate (two 0.45 kg/ha applications)	3	7	21
Benlate (one late 0.9 kg/ha application)	5	7	23

[a]Mean of five samples of 200 seed each.

Ripening Under Dry Conditions

The disease caused by *Phomopsis,* Phomopsis seed infection, can be reduced or eliminated if soybeans mature under dry conditions. The effect of environment during seed maturation was first noted by Lehman (20). High temperature and humidity in September and October in Indiana favor fungus development in seed (28). Severe *Phomopsis* infection of seed in North Carolina was obtained only in irrigated plots (25). Spilker (26) studied the effects of moisture and temperature independently in growth chambers. The environmental variables were imposed at the green bean stage (R6). Significant seed infection was obtained at both high (33 day - 24 night) and low (26 day - 16 night) temperatures if the relative humidity was high (90%), but at neither temperature if the relative humidity was low (40%). Moisture during ripening apparently is more important than temperature. Growing pidgeon pea in a dry region of Puerto Rico significantly reduced Phomopsis seed infection (11). This approach to *Phomopsis* control in soybean might be useful in areas with distinct wet and dry seasons, but would not be applicable to soybean seed production areas in the U.S.

Late Planting and Late Cultivars

In humid, temperate areas some success in reducing Phomopsis seed infection has been obtained by planting late cultivars or early cultivars late so that seed matures late in the season (14, 23). Seed infection in an early maturing cultivar, Amsoy 71, decreased from 70% to 10% when the planting date was changed from 2 May to 14 June in Ohio (15). Planting date had no effect on seed infection in the late maturing cultivar Williams and percentage seed infection never exceeded 12%. No planting date effect was noted with *Dps,* but percentage infection was uniformly low. A typical maturity response in soybean to Phomopsis seed infection is summarized in Table 4 (Jeffers and Schmitthenner, unpublished). Wilcox et al. (28) have suggested that cooler temperatures late in the fall inhibit colonization of seed by *Dps* but Spilker (26) was not able to show a large low temperature response in controlled environment tests. The mechanism by which soybean seed maturing late in the season escape Phomopsis seed infection still needs to be determined.

Table 4. **Effect of soybean cultivar maturity on percentage moldy seed, Phomopsis seed infection and seed germination.**

Cultivar	Days to Maturity	Moldy Seed (%)	Phomopsis (%)	Germination (%)
Amsoy 71	124	59	83	6
Beeson	127	35	71	14
Wayne	134	14	46	44
Williams	140	9	30	30

Delayed Harvest

The effect of delayed harvest on seed quality and seed infection with *Dps* has been well established (8, 28). Kmetz et al. (16) have reported a similar relationship with *Phomopsis* sp. The biggest increase in seed infection occurred between the yellow pod and mature pod stages in the early maturing cultivar Amsoy 71. Thus, significant seed infections may occur before soybean seed are mature enough to harvest and prompt harvest may not completely control *Phomopsis.*

Resistance

Several sources of resistance to *Dps* are known (3, 13). Resistance to *Phomopsis* has been found in maturity groups VIII, IX, and X. It is possible that resistance to Phomopsis seed infections occurs in all soybean groups, but care must be taken to separate out environmental and maturity effects before such resistance can be evaluated.

Fungicide Sprays

Control of *Dps* and *Phomopsis* in soybean seed by foliar application of Benlate has been reported (7, 8, 25). *Phomopsis* sp. in soybean seed also can be reduced by foliar sprays of other fungicides, and seed germination can be significantly improved. Results of 4 yr of spraying for control of *Phomopsis* in seed and improved germination in Ohio are summarized in Table 5. Severity of *Phomopsis* varied each year. Fungicides were applied at stages R3 (young pod) and R5 (green bean). Benlate was most effective. Other fungicides decreased Phomopsis seed infection and improved seed germination some years.

Fungicide Seed Treatments

Improvement of soybean stands and yields from fungicide seed treatment of poor quality seed has been reported. In Volume 33 of Fungicide and Nematicide Tests (2) there are seven tests summarized. *Alternaria, Fusarium, Dps* and *Phomopsis* are variously listed as seed-borne pathogens in the samples tested. In our tests in Ohio, the effectiveness of the seed treatment has been related to the severity of infection. Results for 1977 and 1978 for two different seed qualities are summarized in Table 6. It is concluded from these data that common seed treatments improve emergence of *Phomopsis*-infected seed and that plants from *Phomopsis*-infected seed can yield as much as plants from uninfected seed.

SUMMARY

An undescribed *Phomopsis* sp. is the most prevalent soybean seed decay fungus in Ohio and probably elsewhere. *Diaporthe phaseolorum* var. *sojae*

Table 5. Effect of foliar application of fungicide on percentage Phomopsis seed infection and seed germination in soybean.

Fungicide	Rate (kg/ha)	1975		1976		1977		1978	
		Phomopsis (%)	Germ. (%)	*Phomopsis* (%)	Germ. (%)	*Phomopsis* (%)	Germ. (%)	*Phomopsis* (%)	Germ. (%)
None	-	25[a]	76	33.0	49	74	44	55	38
Benlate 50W	.45	6	80	0.5	71	39	70	12	85
Bravo 6F	1.8	10	69	5.8	78	60	54	12	79
Difolitan	1.8	11	62	13.7	76	57	54	61	54
Duter	.45	15	81	8.8	74	73	45	52	42
Mertect 340F	.45	12	83	0.7	71	67	54	56	48
LSD ($p = .05$)		6.8	8.9	4.4	6.9	8.5	13.3	16.1	17.4

[a]Mean of four replications of 100 seed each.

Table 6. Effect of fungicide seed treatment on stands and yields of soybean from good and poor quality seed.

	1977				1978			
	Good Quality Seed[a]		Poor Quality Seed[b]		Good Quality Seed[c]		Poor Quality Seed[d]	
	Stand (%)	Yield (kg/ha)	Stand (%)	Yield (kg/ha)	Stand (%)	Yield (kg/ha)	Stand (%)	Yield (kg/ha)
None	55	3,200	43	2,700	86	2,700	43	1,900
Captan	91	3,600	66	3,100	83	2,800	74	2,600
Thiram	84	3,600	67	3,300	85	2,700	68	2,500
Vitavax 200	83	3,700	66	3,300	86	2,700	74	2,800
	LSD Stand (.05) = 9.5% LSD Yield (.05) = 450 kg				LSD Stand (.05) = 8.5% LSD Yield (.05) = 335 kg			

[a]Seed germination 90 %, Phomopsis seed infection 8%.
[b]Seed germination 62%, Phomopsis seed infection 37%.
[c]Seed germination 95%, Phomopsis seed infection not tested.
[d]Seed germination 55%, Phomopsis seed infection 63%.
[e]Mean of four rows 6 m long, spaced .75 m, planted with 200 seed/row.

(Dps), the pod and stem blight pathogen and *D. phaseolorum* var. *caulivora (Dpc),* the stem and canker pathogen are less prevalent. The *Phomopsis* sp. differs from the latter in the absence of the perfect stage (perithecia), presence of distinctive beaked pycnidia in old cultures and soybean stems, extensive stroma development and distinctive mycelium on acid PDA. *Phomopsis* can rot developing and germinating seed but is latent in growing plants and, frequently, in mature seed. It persists as pycnidia in soybean straw and may form secondary inoculum on fallen cotyledons and petioles. *Phomopsis* forms latent infections in young pods, primarily on the lower half of the plant, from where it colonizes seed as they mature under wet conditions. Both *Dpc* and *Dps* form perithecia on overwintered straw and are found more frequently on seed from the top third of plants. Phomopsis seed infection can be reduced by controlling inoculum through rotation and residue management, ripening soybeans under dry conditions, planting late cultivars or planting early cultivars late so that the crop matures late in the fall, and application of fungicides such as Benlate to pods when they are 1 cm long and again at the green bean stage. Delaying harvest increases Phomopsis seed infection. Some sources of cultivars that are resistant to pod and stem blight have been reported that may be resistant to *Phomopsis.* Fungicide seed treatment will improve stands of *Phomopsis*-infected seed and prevent yield loss. More work is needed on *identification* of *Phomopsis* sp. from soybeans and other seed legumes from U.S. and other countries to determine what species are involved and if all can be controlled in the same manner.

NOTES

A. F. Schmitthenner, Department of Plant Pathology, Ohio Agricultural Research and Development Center, Wooster, Ohio; and K. T. Kmetz, E. I. Du Pont de Nemours & Co., Columbus, Ohio.

Journal Article No. 49-79 of the Ohio Agricultural Research and Development Center, Wooster, Research funded in part by the Ohio Seed Improvement Association Research Foundation, Dublin, Ohio.

REFERENCES

1. Anonymous. 1975. p. 18-20. In J. B. Sinclair and M. C. Shurtleff (Eds.) Compendium of soybean diseases. Am. Phytopath. Soc., Inc., St. Paul MN.
2. Anonymous. 1978. p. 173-178. In C. W. Averre (Ed.) Fungicide and nematicide tests. Volume 33. Am. Phytopath. Soc., Inc., St. Paul, MN.
3. Athlow, K. L. 1973. Fungus diseases, p. 459-490. In B. E. Caldwell (Ed.) Soybeans: improvement, production, and uses. Am. Soc. of Agronomy, Inc. Madison, WI.
4. Athow, K. L. and R. M. Caldwell. 1954. A comparative study of Diaporthe stem canker and pod and stem blight of soybeans. Phytopathology 44:319-325.
5. Athow, K. L. and F. A. Laviolette. 1973. Pod protection effects on soybean seed germination and infection with *Diaporthe phaseolorum* var. *sojae* and other microorganisms. Phytopathology 53:1021-1025.

6. Ellis, M. A., C. C. Machado, C. Prasartsee, and J. B. Sinclair. 1974. Occurrence of *Diaporthe phaseolorum* var. *sojae (Phomopsis* sp.) in various soybean seedlots. Plant Dis. Reptr. 58:173-176.
7. Ellis, M. A., M. B. Ilyas, F. D. Tenne, J. B. Sinclair, and H. L. Palm. 1974. Effect of foliar spplications of benomyl on internally seed-borne fungi and pod and stem blight in soybean. Plant Dis. Reptr. 58:760-763.
8. Ellis, M. A. and J. B. Sinclair. 1976. Effect of benomyl field sprays on internally-borne fungi, germination, and emergence of late-harvested soybean seeds. Phytopathology 66:680-692.
9. Ellis, M. A., E. H. Paschall, II, and S. R. Foor. 1977. Varietal differences in seed quality of soybean *(Glycine max)* and dry bean *(Phaseolus vulgaris).* Proc. Am. Phytpath. Soc. 3:295 (Abstr).
10. Ellis, M. A., E. H. Paschal, II, and E. Rosario. 1978. Similarities in the internally seed-borne fungi of four leguminous crops. Proc. Am. Phytopath. Soc. 4:176.
11. Ellis, M. A. and E. H. Paschal, II. 1978. Methods of controlling internally seed-borne fungi of pigeon pea *(Cajanus cajan).* Proc. Am. Phytopath Soc. 4:176.
12. Hildebrand, A. A. 1954. Observations on the occurrence of stem canker and pod and stem blight fungi on mature stems of soybeans. Plant Dis. Reptr. 38:640-646.
13. Hymowitz, T., S. G. Carmer, and C. A. Newell. 1976. Soybean cultivars released in the United States and Canada. Morphological descriptions and responses to selected foliar, stem, and root diseases. INTSOY Series No. 9. Univ. of Illinois, Urbana, IL.
14. Kilpatrick, R. A. and E. E. Hartig. 1955. Effect of planting date on incidence of fungus infection of Ogden soybean seeds grown at Walnut Hills, Florida. Plant Dis. Reptr. 39:174-176.
15. Kmetz, K. T. 1975. Soybean seed decay: studies on disease cycles, effects of cultural practices on disease severity and differentiation of the pathogens *Phomopsis* sp., *Diaporthe phaseolorum* var. *sojae* and *Diaporthe phaseolorum* var. *caulivora.* Ph.D. Thesis. The Ohio State University, Colubmus, OH, 120 p.
16. Kmetz, K., C. W. Ellett, and A. F. Schmitthenner. 1974. Isolation of seed-borne *Diaporthe phaseolorum* and *Phomopsis* from immature soybean plants. Plant Dis. Reptr. 58:978-982.
17. Kmetz, K, A. F. Schmitthenner, and C. W. Ellett. 1975. Identification of *Phomopsis* and *Diaporthe* isolates associated with soybean seed decay by colony morphology, symptom development and pathogenicity. 67th Ann. Meetirg of the Am. Phytpath. Soc., Houston, TX. (Abstr).
18. Kmetz, K., C. W. Ellett, and A. F. Schmitthenner. 1979. Soybean seed decay: prevalence of infection and symptom expression of *Phomopsis* sp., *Diaporthe phaseolorum* var. *sojae* and *D. phaseolorum* var. *caulivora.* Phytopathology 68: 836-839.
19. Kmetz, K., C. W. Ellett, and A. F. Schmitthenner. 1979. Soybean seed decay: sources of inoculum and nature of infection. Phytopathology (In press).
20. Lehman, Samual G. 1923. Pod and stem blight of soybean. Annu. Missouri Botan. Garden 10:111-169.
21. Luttrell, E. S. 1947. *Diaporthe phaseolorum* var. *sojae* on crop plants. Phytopathology 37:445-465.
22. Nicholson, J. F., C. D. Dhingra, and J. B. Sinclair. 1972. Internal seed-borne nature of *Sclerotinia sclerotiorum* and *Phomopsis* sp. and their effects on soybean seed quality. Phytopathology 62:1261-1263.
23. Nicholson, J. F. and J. B. Sinclair. 1973. Effect of planting date, storage conditions and seed-borne fungi on soybean seed quality. Plant Dis. Reptr. 57:770-774.

24. Peterson, J. L. and R. F. Strelecki. 1965. The effect of variants of *Diaporthe phaseolorum* on soybean germination and growth in New Jersey. Plant Dis. Reptr. 49:228-229.
25. Ross, J. P. 1975. Effect of overhead irrigation and Benomyl sprays on late-season foliar diseases, seed infection and yields of soybeans. Plant Dis. Reptr. 59:809-813.
26. Spilker, D. A. 1977. The effect of foliar potassium and phosphorus, and temperature and humidity during maturation on the reduction of soybean seed quality by *Phomopsis* sp. M.Sc. Thesis, The Ohio State University, Columbus, OH, 35 p.
27. Wallen, V. R. and T. F. Cuddy. 1960. Relation of seed-borne *Diaporthe phaseolorum* to the germination of soybeans. Proc. Assoc. Official Seed Analysts 50:137-140.
28. Wilcox, J. R., F. A. Laviolette, and K. L. Athow. 1974. Deterioration of soybean seed quality associated with harvest delay. Plant Dis. Reptr. 58:130-133.

RESEARCH ON PHYTOPHTHORA ROOT AND STEM ROT: ISOLATION, TESTING PROCEDURES, AND SEVEN NEW PHYSIOLOGIC RACES

B. L. Keeling

Root and stem rot of soybeans [*Glycine max* (L.) Merr.] caused by *Phytophthora megaspera* Drechs. var. *sojae* Hildeb., can be one of the crops' most destructive diseases. It is most severe in low, poorly-drained, clay soils. In the lower Mississippi River Valley alone, there are over two million ha of alluvial soils where phytophthora rot can cause a severe reduction in soybean yields if susceptible cultivars are grown. Recently, it has become evident that natural populations of this pathogen are made up of numerous pathogenic or physiologic races. My comments will be limited to experimental techniques that worked well for us in the study of this physiological specialization and the testing of breeding material for resistance. I will also present evidence of seven new races of this pathogen.

ISOLATION OF THE PATHOGEN

The pathogen may be isolated from infected soybean plants at any stage of growth. Small pieces of infected tissue taken from the edge of a diseased area are surface disinfected for one minute in an aqueous solution containing 0.5% sodium hypochlorite and 10% ethyl alcohol and rinsed in sterile water. A layer of selective medium is then inverted over the pieces of infected tissue in a petri plate (3,8). After 3 to 4 days incubation at 22 to 24 C, uncontaminated isolates of the fungus may be transferred as it grows through the selective medium. The selective medium consists of 40 ml V-8 juice (Campbell Soup Company), 0.6 g $CaCO_3$, 0.2 g yeast extract, 1.0 g sucrose, 10 mg cholesterol, 20 mg 50% benomyl, 27 mg PCNB 75 wp, 100 mg neomycin sulfate, 30 mg chloramphenicol, 20 g agar, and 1000 ml water (7). All ingredients are mixed and autoclaved for 20 minutes.

TESTING PROCEDURES

The two methods used at our research station to test soybean cultivars for resistance to phytophthora rot are hypocotyl puncture and hydroponic culture. Inoculum for both methods is produced by growing *Phytophthora megasperma* var. *sojae* in a semisolid corn meal medium (2.5 g Difco corn meal agar in 100 ml water) for 10 days at 20 C.

Hypocotyl Puncture

This inoculation method is a variation of the technique described by Kaufman and Gerdemann (4). It consists of dipping a spear-shaped needle through a culture of the fungus in semisolid medium to pick up strands of mycelia. The needle is then inserted through the hypocotyl of 10-day-old plants approximately 1 cm below the cotyledons. The strands of mycelia are deposited on and within the wound when the needle is withdrawn. Inoculated plants are placed in a moist chamber for 16 to 18 hr and then transferred to a greenhouse bench. Greenhouse temperatures are maintained at 22 to 24 C. Susceptible plants are killed within 4 to 5 days after inoculation.

Hydroponic Culture.

Seeds of cultivars to be tested by this method are germinated in vermiculite. When the seedlings reach a total length of approximately 10 cm (4 to 5 days after planting), they are transferred to holes 5 mm in diamether punched 50 mm apart in sheets of styrofoam 25 mm thick. The styrofoam sheets are floated on a 25%Hoagland's nutrient solution approximately 15 cm deep. Two days after the seedlings are placed in the styrofoam sheets, a 10-day-old fungus culture growing in semisolid corn meal agar is added to the nutrient solution at a rate of 100 ml/10 liters of nutrient solution. The test is then shaded with brown Kraft wrapping paper until disease symptoms begin to appear (3 to 5 days). The plants are classified as dead or alive 7 days after adding the pathogen. These tests are done in a greenhouse with a night temperature of 21 C and a day temperature of 21 to 30 C.

PHYSIOLOGICAL RACES

Physiologic Race Determination

Physiologic races of *P. megasperma* var. *sojae* are identified on the basis of their virulence or nonvirulence on individual cultivars included in a set of differentially resistant or susceptible soybean lines. These differential host cultivars are Harosoy, Sanga, Harosoy 63, Mack, Altona, PI 103091, PI 171442, and Tracy. They were selected by a group of pathologists and breeders at a meeting held at Harrow, Ontario, in 1976.

Nine physiologic races of *P. megasperma* var. *sojae* have been reported. The first one recognized (1) was designated race 1. Subsequently, race 2 was recognized in 1956 (6), race 3 in 1972 (9), race 4 in 1974 (10), races 5 and 6 in 1976 (2), and races 7, 8, and 9 in 1977 (5). Of the 9 races described previously, only races 1, 2, and 4 have been found in the Mississippi River delta area of Mississippi, Arkansas, and Louisiana. However, several new virulent strains have been recovered from soybeans in Mississippi and are proposed as races 10, 11, 12, 13, 14, 15, and 16. Their reaction and those of the 9 races described previously are listed in Table 1. Races 10 through 14 were isolated from soybeans grown on the Delta Branch Experiment Station at Stoneville, MS, race 15 from soybeans growing near Swiftwater, MS, and race 16 from soybeans in experimental plots at the Delta and Pine Land Plantation at Scott, MS.

These new races were identified using the hypocotyl inoculation technique.

Table 1: Physiologic races of *Phytophthora megaspera* var. *sojae* from soybeans.

Differential Cultivar	Physiologic Race															
	1	2	3	4	5	6	7	8	9	10	11	12	13	14	15	16
Harosoy	S[a]	S	S	S	S	S	S	S	S	S	S	R	S	S	S	R
Sanga	R[a]	S	R	R	R	R	R	R	R	S	S	S	R	R	R	S
Harosoy 63	R	R	S	S	S	S	S	S	S	R	R	R	R	R	R	R
Mack	R	R	R	S	S	R	R	R	R	R	R	S	R	S	R	S
Altona	R	R	R	R	S	S	S	S	S	R	S	R	S	R	R	R
PI 103091	R	R	R	R	R	S	R	S	R	R	R	R	R	R	R	R
PI 171442	R	R	R	R	R	S	S	R	R	S	R	S	R	R	S	R
Tracy	R	R	R	R	R	R	R	R	R	S	R	S	R	R	R	R

[a]Symbols: S = susceptible, and R = resistant.

NOTES

B. L. Keeling, Soybean Production Research, U. S. Delta States Agricultural Research Center, Stoneville, Mississippi 38776.

REFERENCES

1. Barnard, R. L., P. E. Smith, N. J. Kaufmann, and A. F. Schmitthenner. 1957. Inheritance of resistance to phytophthora root and stem rot in soybeans. Agron. J. 49:391.
2. Haas, J. H. and R. I. Buzzell. 1976. New races 5 and 6 of *Phytophthora megasperma* var. *sojae* and differential reactions of soybean cultivars for races 1 to 6. Phytopathology 66:1361-1362.

3. Hoitink, H. A. J. and A. F. Schmitthenner. 1969. Rhodendron wilt caused by phytophthora citricola. Phytopathology 59:708-709.
4. Kaufmann, M. J. and J. W. Gerdemann. 1958. Root and stem rot of soybean caused by phytophthora sojae N. Sp. Phytopathology 48:201-208.
5. Laviolette, F. A. and K. L. Athow. 1977. Three new physiologic races of *Phytophthora megasperma* var. *sojae.* Phytopathology 67:267-268.
6. Morgan, F. L. and E. E. Hartwig. 1965. Physiologic specialization in *Phytophthora megasperma* var. *sojae.* Phytopathology 55:1277-1279.
7. Schmitthenner, A. F. Personal communication.
8. Schmitthenner, A. F. and J. W. Hilty. 1962. A modified dilution technique for obtaining single-spore isolates from contaminated material. Phytopathology 52: 582-583.
9. Schmitthenner, A. F. 1972. Evidence for a new race of *Phytophthora megasperma* var. *sojae* pathogenic to soybeans. Plant Disease Reporter 56:536-539.
10. Schwenk, F. W. and T. Sims. 1974. Race 4 of *Phytophthora megasperma* var. *sojae* from soybeans proposed. Plant Disease Reporter 58:352-354.

ECOLOGICAL FACTORS AFFECTING WEED COMPETITION IN SOYBEANS

C. G. McWhorter and D. T. Patterson

In 1976 the losses caused by weeds in soybeans and the cost of their control amounted to about $1.3 billion annually (58). Weeds are estimated to cause a 10 to 15% reduction in yield and quality of soybeans, which accounts for a loss of more than $350 million annually. Farmers treat about 80% of the planted acreage with herbicides, at a cost of more than $400 million. Farmers also spend more than $550 million annually on cultural weed control practices including seedbed preparation and cultivation. With inflation conservatively estimated at only 10% since 1976, the losses caused by weeds in soybeans in 1979 will exceed $1.5 billion annually in the United States.

Weeds cost soybean producers more than do all other pests combined (58). The excessive cost caused by weeds is largely due to the complexity of the weed control problem and the lack of selective cost-effective techniques that permit farmers to control weeds efficiently. This chapter summarizes the present weed problems experienced by American soybean producers and the major factors affecting present and future weed problems in soybeans.

PRESENT WEED PROBLEMS

More than 60 individual weed species infest soybean fields in the U.S. Major losses are caused in all regions by annual broadleaf and grass weeds. Perennial weeds including grasses, broadleaves, and sedges are increasingly troublesome. In this section we provide common characteristics of existing weeds, summarize the level of competition normally expected from some of the major weeds, discuss the extent to which American producers have been able to control weeds in soybeans, and provide examples of recent ecological shifts that have occurred with weeds in soybeans.

Characterization of the Present Weed Problem

The major weeds infesting soybean fields in the U.S. that were described by Wax (85) are listed in Table 1. Sixty-two % of the major weeds are dicotyledons while 38% are monocotyledons. Sixty-five % of the major weeds are annuals while 35% are perennials. Fifty-five % of the major weeds are exotic in origin while only 45% are native to the U.S. Allelopathy has been demonstrated in 54% of the major weeds in soybeans (Table 1). As will be discussed later, plants with the C_4 dicarboxylic acid pathway of photosynthesis are much more efficient in carbon-gaining capacity than are the more common C_3 plants (8). A surprisingly high number, 93%, of the grasses listed as major weeds in soybeans are C_4 plants; only quackgrass is a C_3 plant. Conversely, all of the dicotyledons listed are C_3 plants except pigweed.

A summary of some of the characteristics mentioned above for the major weeds in soybeans versus similar characteristics for the world's worst weeds, as listed by Holm et al. (30) is presented in Table 2. Seventy-two % of the world's worst 18 weeds are monocots as compared to only 38% of the major weeds in soybeans. Also a lower percentage of the major weeds in soybeans are perennial, have vegetative reproduction, have rhizome production, and have C_4 photosynthesis than do the world's worst 18 weeds. Comparisons of this type suggest that there are many other highly competitive "weedy" plants throughout the world that have not been introduced in the soybean fields in the U.S.

Holm et al. (30) list 34 weeds that are considered "serious," "principal," or "common" weeds in soybeans in various regions of the world. Of these, the following 18 are not included by the list of Wax (85): *Ageratum conyzoides* L.; *Commelina benghalensis* L.; *Cyperus iria* L.; *Echinochloa colonum* (L.) Link; *Eclipta prostrata* (L.) L.; *Equisetum arvense* L.; *Euphorbia hirta* L.; *Galium aparine* L.; *Heliotropium indicum* L.; *Imperata cylindrica* (L.) Beauv.; *Leptochloa chinensis* (L.) Nees; *Leptochloa panicea* (Retz.) Ohwi; *Mimosa invisa* Mart.; *Mimosa pudica* L.; *Oxalis corniculata* L.; *Physalis angulata* L.; *Rottboellia exaltata* L. f.; *Solanum nigrum* L. Many of these species are already present in the U.S. and represent a potential threat to soybean production. The five weeds that Holm et al. (30) singled out as "serious or principal" soybean weeds through the world [*Cyperus rotundus* L., *Echinochloa colonum* (L.) Link, *Echinochloa crusgalli* (L.) Beauv., *Eleusine indica* (L.) Gaertn., and *Rottboellia exaltata* L. f.] are monocots with the C_4-dicarboxylic acid pathway of photosynthesis.

SOYBEAN LOSSES DUE TO WEEDS

In 1965 it was estimated that weeds cause average annual losses of about 17% of the potential value of the crop (31). A number of improved and more efficient weed control practices have been made available since 1965, but the average annual loss due to weeds has probably not been reduced appreciably, due to increased infestations of weeds resistant to herbicides, local introductions of new weeds, and ecological shifts in weed populations. This review of losses due to weeds will not be extensive because other reviews are available (11,85,86).

Weeds compete actively for light, nutrients, and moisture. They reduce both the quantity and quality of the harvested product. The presence of weeds increases the incidence of diseases and insects and frequently hampers the efficient operation of equipment. Unfortunately, the specific competitive effect of individual weeds in soybeans is largely unknown, but reduction in soybean yields from weeds is well documented (85).

The level of weed competition in soybeans is generally acknowledged to be greater in the southern U.S. than in the northern producing areas. Barrentine and Oliver (11) reported that common cocklebur reduced soybean yields up to 76% with seasonal competition. McWhorter and Hartwig (46) also reported reduced soybean yields by 63 to 75% but found that some soybean cultivars competed more effectively with cocklebur than others. In Georgia, Hauser et al. (27) reported that mixtures of common cocklebur and yellow nutsedge reduced soybean yields by 75%. Hemp sesbania reduced soybean yields by 60 to 80% (45) and sicklepod reduced soybean yields by 35% (76). Yield reductions of the type reported may be higher than those experienced by many farmers; but, even so, these research findings indicate the yield losses that occur frequently in heavily infested spots within fields.

In field research with natural weed infestations in Iowa, yield losses were 6 to 27%, even with good cultural practices (67,70). Several annual weeds including pigweed, smartweed, velvetleaf, and foxtail were also found to reduce soybean yields by 10%, even when acceptable cultural practices were used (70). In Iowa, Staniforth (69) showed that giant foxtail reduced soybean yields by 25%. In Illinois, Knake and Slife (33) reported that giant foxtail reduced soybean yields by 30%.

The level of competition provided by individual weeds may depend on soil moisture levels. Relatively little research has been reported on the interaction of soil moisture and weed competition, but Staniforth (66) found greater reductions in soybean yields when limited moisture was present than when adequate moisture was available. The level of competition provided by weeds is also related to the level of weed infestation and to the duration

Table 1. Characteristics of the major weeds in soybeans in the U.S.

		Plant Characteristic[2]							
		Photosynthetic Pathway		Class		Growth Characteristics			
Common name[1]	Scientific Name	C_3	C_4	Monocot	Dicot	Annual	Perennial	Exotic	Allelo-pathic
Barnyardgrass	*Echinochloa crusgalli* (L.) Beauv.		X	X		X		Yes	No
Bindweed, field	*Convolvulus arvensis* L.	X			X		X	Yes	Yes
Cocklebur	*Xanthium pensylvanicum* Wallr.	X			X	X		No	No
Crabgrass	*Digitaria* spp.		X	X		X		Yes	Yes
Crotalaria	*Crotalaria* spp.	X		X		X	X	Yes	No
Foxtail, giant	*Setaria faberi* Herrm.		X	X		X		Yes	Yes
Foxtail, green	*Setaria viridis* (L.) Beauv.		X	X		X		Yes	Yes
Foxtail, Yellow	*Setaria lutescens* (Weigel) Hubb.		X	X		X		Yes	Yes
Goosegrass	*Eleusine indica* (L.) Gaertn.		X	X		X		Yes	No
Jimsonweed	*Datura stramonium* L.	X			X	X		Yes	Yes
Johnsongrass	*Sorghum halepense* (L.) Pers.		X	X			X	Yes	Yes
Lambsquarters	*Chenopodium album* L.	X			X	X		Yes	Yes
Milkweed, common	*Asclepias syriaca* L.	X			X		X	No	Yes
Milkweed, honeyvine	*Ampelamus albidus* (Nutt.) Britt.	X			X		X	No	No
Morningglory, annual	*Ipomoea* spp.	X			X	X		Yes	No
Morningglory, bigroot	*Ipomoea pandurata* (L.) G.F.W. Mey	X			X		X	No	No
Mustard, wild	*Brassica kaber* (DC.) L. C. Wheeler var. *pinnatifida* (Stokes) L. C. Wheeler	X			X	X		Yes	Yes

Nutsedge, purple	*Cyperus rotundus* L.		X	X			X	Yes	Yes
Nutsedge, yellow	*Cyperus esculentus* L.		X	X			X	Yes	Yes
Panicum, Texas	*Panicum texanum* Buckl.		X	X		X		No	No
Pigweed	*Amaranthus* spp.		X		X	X		Yes	Yes
Pusley, Florida	*Richardia scabra* L.	X			X	X		Yes	No
Quackgrass	*Agropyron repens* (L.) Beauv.	X		X			X	No	Yes
Ragweed, common	*Ambrosia artemisiifolia* L.	X			X	X		No	Yes
Ragweed, giant	*Ambrosia trifida* L.	X			X	X		No	Yes
Redvine	*Brunnichia cirrhosa* Gaertn.	X			X		X	No	No
Sandbur	*Cenchrus* spp.		X			X	X	No	No
Sesbania, hemp	*Sesbania exaltata* (Raf.) Cory	X			X	X		No	No
Shattercane	*Sorghum bicolor* (L.) Moench		X	X		X		Yes	No
Sicklepod	*Cassia obtusifolia* L.	X			X	X		Yes	No
Sida, Prickly	*Sida spinosa* L.	X			X	X		Yes	No
Signalgrass, broadleaf	*Brachiaria platyphylla* (Griseb.) Nash		X	X		X		No	No
Smartweed	*Polygonum* spp.	X			X	X	X	Yes	Yes
Sunflower, common	*Helianthus annuus* L.	X			X	X		No	Yes
Thistle, Canada	*Cirsium arvense* (L.) Scop.	X			X		X	Yes	Yes
Trumpetcreeper	*Campsis radicans* (L.) Seem.	X			X		X	No	No
Velvetleaf	*Abutilon theophrasti* Medic.	X			X	X		Yes	Yes
TOTAL PLANTS		23	14	14	23	26	14	22 Yes	20 Yes
PERCENT OF TOTAL		62	38	38	62	65	35	55	54

[1] List of major weeds in soybeans adapted from Wax (86).
[2] Plant characteristics determined from a literature survey by D. T. Patterson.

Table 2. A comparison of plant characteristics of the world's worst weeds as listed by Holm et al. (30) and the major weeds in U.S. soybean production (86).

Plant Characteristic[1]	World's Worst Weeds from Holm et al. (30) Worst 18	World's Worst Weeds from Holm et al. (30) Worst 76	Major Weeds in Soybeans from Wax (86) Worst 37
		– % –	
Monocot	72	53	38
Dicot	28	47	62
Perennials	44	55	32
Vegetative reproduction	61	51	35
Rhizome production	33	25	19
C_4 photosynthesis	78	42	62
Exotic	--	--	55
Allelopathic	--	--	54

[1] Plant characteristics were determined from a literature survey by D. T. Patterson.

of competition as indicated in research on giant foxtail (34,35), cocklebur (11), hemp sesbania (45), sicklepod (76), wild common sunflower (6), morningglory (93), pigweed (48), and wild mustard (12). These and other studies generally show that major competition occurs during the first 4 to 6 wk after soybean emergence. Soybeans become much more competitive to weeds at 4 to 8 wk after emergence, and soybean yields usually are not reduced if adequate control is provided for the first 4 to 6 wk even when weeds emerge later.

Relatively few studies have been conducted to define the effect of soil nutrients as a limiting factor in soybean production in conjunction with competition by weeds. However, Vengris et al. (83) showed that many weeds have a higher mineral content than crops and that weeds may be able to use minerals at the expense of crop plants. Increasing the fertility level may not overcome the competitive effects of weeds, because the increased fertility often causes the weeds to grow more rapidly than the crop (68,84).

Heavy infestations of weeds in soybean fields at harvest interfere greatly with timeliness and efficiency of harvest (17,52,85). Weeds in the field at harvest also result in reduced soybean grades (4,5). In Mississippi (43,44), common cocklebur resulted in foreign matter content of up to 5.1% in harvested soybeans. At least 70% control of common cocklebur was required to keep seed moisture levels from exceeding 13%. Small discounts for damaged kernels occurred when common cocklebur control was less than 40%. With adequate control of common cocklebur, the estimated U.S. soybean grade was 1.3; whereas, in the absence of control, common cocklebur resulted in a grade of 3.9 (4,5). In these studies soybean yields were increased by about 6% for each 10% increase in common cocklebur control, and net returns to land management and general farm overhead nearly doubled as a result of adequate cocklebur control (43). Also, hemp sesbania was found to reduce soybean grades at weed populations that did not reduce soybean yields (45). Seeds of jimsonweed, showy crotalaria, and morningglory are often poisonous to both livestock and humans. The presence of these weed seeds in soybean seed not only lowers grades but may prevent sale until the soybean seed are cleaned. This is often a very expensive process.

Relationship of Weed Control to Soybean Production

Since 1968 there have been no national studies to describe the extent of the weed control problem in soybeans and to establish the cost of weed control in soybeans (82). The only current information available that provides insight into the farmers' ability to control weeds in soybeans is in soybean growers' surveys conducted in 1967, 1971, and 1977 by the National Soybean Crop Improvement Council (NSCIC) (49,50,51). These surveys indicated that weeds caused greater difficulties in soybean production than all of the other pests combined. Farmers acknowledged that improved weed control techniques and herbicides have contributed more to increasing soybean yields during the past few years than have any other practice. Farmers place weed control highest among their many priorities in soybean production.

The surveys by NSCIC show that 60 to 70% of the soybean producers do not feel that they obtain excellent weed control with the herbicides available (51). The percentage of farmers who obtain excellent control of weeds with herbicides has increased only slightly during the last 10 yr (49,50) (Table 3). The difficulty in obtaining excellent weed control is probably due to the large number of broadleaf weed species present in farmers' fields. The farmers in all 4 soybean production regions listed cocklebur, morningglory, and pigweed (Table 4) as most difficult to control in the 1968 NSCIC survey (51). Many other weed species occur in all soybean producing regions but apparently they were not sufficiently troublesome in each of the 4 regions to be listed as difficult to control. The percentage of farmers in the regions who list cocklebur as difficult to control was higher in 1976 than in 1970. In the western and midwestern regions, velvetleaf was also listed as a troublesome weed about 10% more frequently in 1976 than in 1970.

The large number of different weed species present in soybean fields contributed to the overall difficulty that farmers have in providing adequate weed control programs (Table 4). Also farmers were unable to control individual weed species, resulting in an increased level of severity in another weed that was not controlled. These so called "ecological shifts" contribute to the complexity of the weed control problem confronted by the farmer.

Table 3. Levels of weed control in soybeans obtained by farmers in surveys conducted by the National Soybean Crop Improvement Council (49,50,51).

Level of Control Obtained with Herbicides	Farmers in Each Region Indicating Various Levels of Weed Control		
	1966	1970	1976
		– % –	
SOUTHEAST[1]			
Excellent	--	31	30
Good	--	58	50
Fair	--	10	18
Unsatisfactory	--	2	1
DELTA[2]			
Excellent	38	26	42
Good	52	48	44
Fair	5	21	9
Unsatisfactory	5	5	5
WESTERN[3]			
Excellent	32	40	40
Good	45	44	38
Fair	18	12	19
Unsatisfactory	5	4	4
MIDWEST[3]			
Excellent	26	42	35
Good	49	48	45
Fair	19	8	9
Unsatisfactory	6	3	1

[1]NJ, DE, MD, VA, NC, SC, GA, AL, FL.
[2]KY, TN, AR, MS, LA.
[3]ND, SD, NB, OK, TX.
[4]MO, IA, MN, WI, IL, IN, OH, MI.

Recent Examples of Ecological Shifts in Weed Populations in Soybeans

Plants previously unreported either as weeds in soybeans or in rotational crops appear to be reported at an increased rate. Weeds that are incidental in rotational crops are thus introduced into soybean fields where they become highly competitive. Newly introduced weeds include wild poinsettia (*Euphorbia heterophylla* L.) (9,10), purple moonflower (*Ipomoea turbinata* Lagasca y Segura) (18), Texas gourd (*Cucurbita texana* A. Gray) (26), and hophornbeam copperleaf (*Acalypha ostryaefolia* Riddell) (63). Plants that are becoming increasingly troublesome as weeds in soybean fields that were originally present in other crops include fall panicum *(Panicum dichotoniflorum* Michx.) (55), jimsonweed (62), shattercane (14), wild cucumber (*Echinocystis lobata* (Michx.) Torr. & Gray), artichoke (*Helianthus tuberosus* L.),

Table 4. Weeds listed as most difficult to control in soybeans by farmers in surveys conducted by the National Soybean Crop Improvement Council (49,50,51).

Weeds Listed as Difficult to Control	Farmers Listing Individual Weeds as Difficult to Control							
	Delta States[1]		Southeastern States[2]		Western States[3]		Midwestern States[4]	
	1970	1976	1970	1976	1970	1976	1970	1976
				– % –				
Cocklebur	42	62	29	39	22	25	14	16
Morningglory	20	6	9	14	4	3	5	4
Prickly sida	14	3	--	<1	--	--	--	--
Johnsongrass	11	21	--	<1	0	7	--	--
Pigweed	9	3	14	9	8	6	8	2
Hemp sesbania	5	3	12	13	--	--	--	--
Nutsedge	0	3	--	--	--	--	1	3
Jimsonweed	--	--	4	8	0	1	7	7
Ragweed	--	--	0	5	0	1	2	2
Sicklepod	--	--	0	5	--	--	--	--
Velvetleaf	--	--	--	--	28	33	22	34
Sunflower	--	--	--	--	12	6	4	4
Smartweed	--	--	--	--	10	6	17	6
Foxtail	--	--	--	--	8	1	7	3
Canada thistle	--	--	--	--	2	0	2	2
Burrnettle	--	--	--	--	0	1	--	--
Devils claw	--	--	--	--	0	1	--	--
Other grasses	--	--	--	--	--	--	6	2
Lambsquarters	--	--	--	--	--	--	3	5
Milkweed	--	--	--	--	--	--	1	5
Wild mustard	--	--	--	--	--	--	1	1

[1]KY, TN, AR, MS, LA.
[2]NJ, DE, MD, VA, NC, SC, GA, AL, FL.
[3]ND, SD, NB, OK, TX.
[4]MO, IA, MN, WI, IL, IN, OH, MI.

beggarweed (*Bidens* spp.), and firebrush (*Myrica faya* Ait.) (51). Shifts in weed populations also occur as a result of the continuous use of herbicides and from the use of certain crop rotations (1,13,16,28,38,89).

Minimum- and no-tillage techniques for corn and soybean production have resulted in many ecological shifts. Even with the best herbicide practices, weeds increase in severity with continuous no-tillage production (90); and weed populations develop that are resistant to the herbicides being used (22,29,36,79). In no-tillage corn research in Ohio, weeds that increased included common dandelion (*Taraxacum officinale* Weber), Canada thistle, common milkweed, horsenettle (*Solanum carolinense* L.) groundcherry (*Physalis* spp.), and tall ironweed (*Vernonia altissima* Nutt.) (79). A farmer's survey in Kentucky indicated that many perennial weeds became much more troublesome in no-tillage corn (29). Perennials such as johnsongrass and bermudagrass [*Cynodon dactylon* (L.) Pers.] spread so rapidly in no-tillage production that it generally has been conceded that no-tillage should not be used in fields infested with these weeds (37,57). Worsham (94) listed several pitfalls with no-tillage farming and emphasized that no-tillage production should not be attempted on land on which perennial weeds are established and ecological shifts in weed populations can appear quickly. It is primarily the inability of farmers to control perennial weeds and annuals that are resistant to herbicides that prevents more extensive use of no-tillage production.

Johnsongrass has been a major troublesome weed in cotton and soybeans for many decades and serves as an example of how a single weed species may undergo a series of ecological shifts. Until about 1960, johnsongrass was considered to be the worst weed in soybeans in the southeastern U.S. It was ranked among the top 5 worst weeds in cotton, although johnsongrass was less troublesome in cotton than in soybeans because of the intensive level of hand labor used to control weeds in cotton. As farmers discontinued use of hand labor for weed control in cotton, the level of johnsongrass infestation found in cotton fields increased. Highly selective herbicides for the control of johnsongrass were not available in the late 1950's, when it appeared that johnsongrass would be a major limiting factor in the production of both cotton and soybeans. The arsenical and dinitroaniline herbicides were introduced in the early 1960's. These were so effective in controlling johnsongrass that its relative importance as a weed generally diminished. A survey in 1968 indicated that not a single southern state listed johnsongrass as the most important weed in soybean production (82). In 1977, a subsequent survey listed johnsongrass as the worst weed in soybeans in Louisiana, the second worst weed in Mississippi and Tennessee, the third worst weed in Arkansas, the fourth worst in Virginia and Alabama, the fifth worst in Oklahoma and Georgia, and the seventh worst weed in North Carolina (39). Johnsongrass is one of the most costly weeds to control in many of the southeastern states (42), irrespective of crop.

In 1968 a survey indicated that johnsongrass was troublesome in cotton, but of the 12 states reporting, 4 states reported that the level of johnsongrass infestation was stationary, 4 states reported that johnsongrass infestations were on a downward trend, 2 states reported that johnsongrass was no significant problem, and only 1 state reported that johnsongrass infestations were increasing (82). In 1977 another survey showed that johnsongrass was the worst weed in cotton and caused 16.4% of all losses caused by weeds (47). In an additional survey in 1977 (39), johnsongrass was listed as 1 of the 7 most commonly occurring weeds in cotton in 9 of the 10 southern states reporting. Thus, while widespread use of highly effective herbicides for johnsongrass control in the early 1960's temporarily minimized its importance, johnsongrass has again increased as a troublesome weed throughout the southeastern U.S. Unfortunately, no research has been conducted to study whether new strains of johnsongrass may be present now that were not present when the arsenical and dinitroaniline herbicides were first introduced.

Three additional surveys also indicate that johnsongrass is increasing as a weed problem. In 1977, johnsongrass was listed as 1 of the 5 worst perennial weeds in 6 of the 13 states represented in the North Central Weed Control Conference (NCWCC) (23). It was listed as the most severe perennial weed problem in 4 of the NCWCC states. Johnsongrass was also 1 of the 2 worst weeds in crops in West Virginia (91). The severity of the problem caused by johnsongrass in states as far north as Illinois, Indiana, Kansas, Kentucky, Missouri, and Ohio raises the possibility of new strains being evolved that are more tolerant of cool temperatures. A survey conducted by the senior author showed that johnsongrass was in more cotton fields in the delta areas of Mississippi, Arkansas, and Louisiana in 1978 than in 1976 (Table 5). In the north delta of Mississippi, for example, 46% of the cotton fields evaluated were listed as not having johnsongrass in 1976 but only 11% of the fields were free of johnsongrass in 1978. The survey summarized in Table 5 includes evaluation of over 1300 fields representing nearly 200,000 ha of cotton and soybeans. Because of the scope of the survey, it should provide a reliable estimate of the extent of johnsongrass infestation.

Common milkweed is another perennial that has increased rapidly both in the area infested and in the severity of its competition (23). All of the states represented by the North Central Weed Control Conference had more cropland infested with common milkweed in 1977 than in 1969. In Nebraska, for example, common milkweed infested only 28% of the cropland in 1969 but it infested 50% in 1977. Probable causes listed for these increases were reduced usage of mechanical cultivation, widespread use of preemergence herbicides, and removal of annual weeds, providing a more favorable environment for establishment and growth of common milkweed (23). These and other factors affecting ecological shifts are discussed below.

FACTORS AFFECTING FUTURE WEED PROBLEMS

Weeds in soybeans have been the subject of much research, directed mostly toward the discovery of methods of control. The frequency and extent to which individual species of weeds occur in soybeans in the U.S. are largely unknown. Essentially no research has documented the factors that contribute to the "weediness" of unwanted plants nor of the factors that affect the ultimate distribution of "new" weeds in soybeans. McNeill (40) stated that there are many unanswered questions about the past and present evolution of most weeds. These questions include how both weedy and nonweedy races of the same species coexist and how certain species seem suddenly to become serious weeds for no evident crop management reasons.

Table 5. Levels of infestation of johnsongrass fields in August, 1976 to 1978 in cotton and soybean in three states[1] in four areas of the alluvial floodplain of the lower Mississippi River Valley (Delta).

Johnsongrass Infestation	Estimated Yield Loss	Northwest Mississippi: North Delta 1976	Northwest Mississippi: North Delta 1977	Northwest Mississippi: North Delta 1978	Northwest Mississippi: South Delta 1976	Northwest Mississippi: South Delta 1977	Northwest Mississippi: South Delta 1978	Southeast Arkansas 1977	Southeast Arkansas 1978	Northeast Louisiana 1977	Northeast Louisiana 1978
						– % –					
						Cotton fields					
0	0	46	22	11	48	24	14	26	27	23	15
1 to 5	<1	25	42	53	30	36	43	36	32	49	45
6 to 10	1 to 5	15	21	23	11	25	26	25	23	22	20
11 to 40	6 to 20	11	14	10	9	12	16	13	16	6	17
41 to 70	21 to 40	4	1	3	2	4	1	0	3	0	4
71 to 100	41 to 60	0	0	0	0	0	0	0	0	0	0
						Soybean fields					
0	0	15	5	5	20	10	5	17	20	7	12
1 to 5	<1	18	25	29	17	22	27	24	23	28	25
6 to 10	1 to 5	23	27	35	23	27	28	24	21	29	26
11 to 40	6 to 20	29	31	28	29	27	29	27	30	27	25
41 to 70	21 to 40	8	11	4	10	12	11	9	7	7	12
71 to 100	41 to 60	5	1	<1	2	2	<1	0	<1	2	<1

[1]The survey included evaluations of approximately 596 individual cotton fields and 772 soybean fields each year. The same route was followed each year representing a total of nearly 1600 km. Total annual acreage represented by all fields evaluated was cotton–34,000 ha, and soybeans–41,300 ha.

Anderson (3) has stressed the need for more ecological and taxonomical research in agricultural situations to study the ecological shifts.

Three of the most interesting recent reviews on the evolution of weeds and factors relating to the prediction of new weed problems are by Baker (8), McNeill (40), and Parker (54). Baker (8) developed the following list of characteristics that might be expected in "ideal weeds": germination requirements fulfilled in many environments; discontinuous germination (internally controlled) and great longevity of seed; rapid growth through vegetative phase to flowering; continuous seed production for as long as growing conditions permit; self-compatible but not completely autogamous or apomictic; when cross-pollinated, unspecialized visitors or wind utilized; very high seed output in favorable environmental circumstances; produces some seed in wide range of environmental conditions: tolerant and plastic; has adaptations for short- and long-distance dispersal; if a perennial has vigorous vegetative reproduction or regeneration from fragments; if a perennial has brittleness so not easily drawn from ground; and has ability to compete interspecifically by special means (rosette, choking growth) (allelochemics). These characteristics appear to describe the major weeds that occur in soybeans in the U.S. Plants without a few of these characteristics are unlikely to be troublesome as weeds. These characteristics are primarily botanical indicators of "weediness" in plants, but there are many other factors that affect the presence and frequency of weeds in soybeans, as discussed below.

Cultural Selection

Variables under this heading include different rotational patterns, production of soybeans without tillage, production of soybeans with different tillage practices, use of different soybean cultivars, techniques in harvesting and seed cleaning, and a wide variety of other production variables. Probably no other single practice has resulted in such rapid ecological shifts of weeds in soybeans and corn as has the use of no-tillage production (7,15,24,29,64, 88,94). Annual weeds may not be troublesome during the first year of production, but later, crabgrass, fall panicum, and other annuals become serious problems (29). In addition to annuals, several perennials become major problems including some that are normally not troublesome when cultivation is utilized, such as greenbriar (*Smilax* spp.), brambles (*Smilax* spp.), mulberry (*Morus* spp.), and poison ivy (*Rhus radicans* L.) (29).

The use of different crop rotations often has a significant impact on the composition of specific weed populations (13,16,20,28). The weed problems in corn and soybeans are now much more similar than in the past, perhaps because of the extent to which these crops are rotated (72). The extent to which different crop rotations affect populations of troublesome weeds depends on the herbicides used for weed control in the individual crops and the extent to which mechanical cultivation may or may not be used within a given crop. The effect of row spacing can influence dramatically the level of weeds in soybeans, because the percentage of the uncultivated area increases as row width decreases. Usually, soybeans in rows 50 cm wide or less do not receive cultivation, and weeds are more troublesome because they must be controlled with herbicides (87).

Postemergence tillage systems have an important effect on weed populations in soybeans, but the use of different preplanting tillage systems may be equally important. In Indiana, comparisons of no-tillage, conventional moldboard plowing, and chisel plowing showed rapid shifts in populations of fall panicum, lambsquarters, and pigweed (92). Without the use of herbicides for weed control, plots receiving conventional tillage in the plots had twice the populations of fall panicum than plots with either chisel tillage or no-tillage, but no-tillage plots had much greater populations of fall panicum than plots with conventional tillage or chisel plowing when certain herbicide programs were used (92). A much greater research effort is needed to establish optimum rotational patterns, the use of pre- and postemergence herbicides and tillage patterns, and many other associated variables to define those cropping systems producing maximum yields with minimum weed interference.

Distribution of weed seed during the harvesting operation accounts probably for the recent rapid spread of sicklepod throughout Tennessee, Arkansas, and Mississippi. Earlier, this weed was confined primarily to southern Mississippi, Alabama, Georgia, and the Carolinas, but within the last 3 to 4 yr its rate of spread has increased rapidly. This spread occurs probably because combines are cleaned too infrequently and, transport seeds to fields uninfested previously. Inadequate removal of weed seeds from harvested soybeans probably accounts for rapid increases in the areas infested with purple moonflower and balloonvine (*Cardiospermum halicacabum* L.). Seeds of these weeds are very difficult to remove from soybean seeds.

Chemical Selection

As discussed by Parker (54), the entire history of herbicide development since the early 1950's has been one of changing weed floras. The selection of resistant weeds is the direct result of using chemically similar herbicides over a period of years. This may result also in many local weed populations showing variation in susceptibility. These genetic changes are discussed later.

Common cocklebur was frequently the predominant species in many soybean fields but metribuzin [4-amino-6-*tert*-butyl-3-(methylthio)-*as*-triazine-5(4*H*)one] applied preemergence and bentazon [3-isopropyl-1*H*-2,1,3-benzothiadiazin-(4)3*H*-one 2,2-dioxide] applied postemergence has reduced greatly populations of common cocklebur. Other species that were suppressed formerly by common cocklebur, such as johnsongrass, and morningglory, pigweed, and other annuals, are apparently becoming severe weeds as a direct result of farmers' developing more effective control programs for common cocklebur. Before the present extensive use of metribuzin and bentazon for control of cocklebur and other broadleaf weeds, the repeated use of herbicides more specific for control of grasses, such as trifluralin (α,α,α-trifluoro-2,6-dinitro-*N*,*N*-dipropyl-*p*-toluidine) and alachlor [2-chloro-2′,6′-diethyl-*N*-(methoxymethyl)acetanilide] had increased greatly the difficulty caused by broadleaf weeds. In corn, the continuous use of triazine herbicides over a period of many years has caused greatly increased populations of annual grasses including crabgrass, *Setaria* spp., and *Panicum* spp. (54). Continuous use of propanil (3′,4′-dichloropropionanilide) in rice has increased the level of interference caused by red rice and other annual grasses. There are increasing numbers of reports on the effect of herbicides on species diversity (77,80,89), but it is acknowledged generally that detrimental ecological shifts can be prevented in soybean production by alternating the use of appropriate herbicides. Alternating herbicides of different chemical classes may prove more efficient in preventing shifts of annual weeds than perennials. Perennial weeds in soybeans continue to present a great threat to row-crop production as the use of hand labor continues to decline.

Distribution

The accidental introduction and further distribution of weeds from other areas has caused many of the major weed problems in American soybean production as indicated by the number of exotic species listed in Table 1. Many of these were introduced because it was felt earlier that the species might be useful. Many were introduced probably as scientific curiosities. Other introductions were accidental and the exact manner of introduction usually was not documented. Most of the exotic weeds that are troublesome in soybeans were introduced many decades ago, but a few more recent introductions, including itchgrass (*Rottboellia exaltata* L. f.) and cogongrass [*Imperata cylindrica* (L.) Beauv.] pose a threat to soybean production.

Cogongrass is confined presently to portions of Florida, southern Alabama, and Mississippi. Holm et al. (30) lists cogongrass as the seventh worst weed in the world, but it does not appear to be sufficiently competitive in the U.S. to create a major weed control problem in conventionally cultivated soybeans. Even so, genetic changes in cogongrass or the greater use of minimum tillage could result in cogongrass becoming a very troublesome weed for soybean producers. The area infested with cogongrass is increasing in size, but the potential of cogongrass as a major weed in the U.S. is unknown.

Itchgrass could be more troublesome than even johnsongrass in certain portions of the U.S., and the total land area that could be infested eventually with this weed appears to be great (56). Itchgrass is confined presently to southern Florida and southern Louisiana but in all likelihood the area infested with itchgrass will continue to increase, probably through its presence in crop seeds, soils, and farm equipment.

The Federal Noxious Weed Act was passed by the U.S. Congress in 1974 but was not funded until 1978. Continued funding of this Act should be highly beneficial not only in reducing future accidental introductions but also in curtailing the movement of more recently introduced exotic weeds that are presently of limited distribution.

Genetic Changes

The existence of ecotypes or races among various species of weeds has been recognized for many years. The increasing frequency with which different ecotypes are reported as resistant to herbicides is disturbing (19,25,31, 32,41,53,60,61,65,74,75,78,82). The variation within species often has been ascribed to environmental conditions, but recently the formation of ecotypes has been shown to be more complicated than thought previously. King (32) has characterized a number of ecotypes including regional, altitudinal, latitudinal, physiological, and edaphic ecotypes. A large number of local populations among various species of weeds are reported that are resistant to herbicides; this could result in reduced control of weeds that were controlled previously without appreciable difficulty. New ecotypes result not only from the continuous use of certain individual herbicides but also from inadequate doses of herbicides that provide only partial control. As pointed out by Parker (54), the use of inadequate doses of herbicides increases greatly the risk of resistant types surviving and increasing. The literature on resistant species and strains has grown so rapidly in recent years that we have not attempted to provide a comprehensive survey but a number of review articles are available (8,32,54,82).

Photosynthetic Pathways

Only in recent years has the method of carbon fixation used by individual plants been shown to contribute to "weediness." The importance of the method of carbon fixation is shown by the fact that all of the monocotyledonous weeds except quackgrass listed as troublesome in soybeans (Table 1) possess the C_4 (dicarboxylic acid) photosynthetic pathway. The number of grasses with the C_4 pathway seems unusually high in view of the fact that comparatively few of the monocotyledons reported to date possess this pathway (73). Plants with the C_4 pathway differ from C_3 plants in a large number of anatomical, biochemical, and ultrastructural characteristics (71).

Plants with the C_4 pathway are in highest relative abundance in areas with high minimum temperatures during the growing season (73). The percentage of C_4 dicotyledonous species in a geographic region is predicted best by a combination of summer-pan evaporation and dryness ratio (71). Because of their traits, C_4 plants are particularly well represented in hot areas with high solar radiation, whereas C_3 plants may have the greatest advantage in cooler temperatures. Many of the C_4 plants are annuals, have the ability to produce rapid growth, and are often drought resistant, thus increasing their "weediness" (8). Plants with C_4 and C_3 pathways have been reported within the same genus which, according to Baker (8), indicates the evolution of the C_4 system from a C_3 ancestry. Research relating the method of carbon fixation used by plants to weediness and competition is relatively new; more research is needed to relate the different photosynthetic systems and the associated features of water relations and respiration to competition, interference, and weediness. Several recent reviews on this subject are available (8,21,71, 73).

Allelopathy

Plants may release chemicals into the environment that are secondary relative to such primary compounds as carbohydrates and proteins. Some of these secondary chemicals (allelochemics) cause direct and indirect effects, either harmful or beneficial, to other plants (2). Rice (59) suggested that allelopathy could be important within the general field of vegetation control because allelochemics prevent seed decay, prevent infection by pathogens, affect dormancy of seeds and buds, inhibit the growth of adjacent plants, inhibit preharvest seed germination, inhibit nitrification, and generally affect patterns of vegetation and plant succession. Although the general concepts within the field of allelopathy are nearly 300 yr old (59), very little research has been conducted to establish the importance of allelopathy within the framework of weed control in soybeans. A surprisingly high number of the major weeds in soybeans have been associated with the production of allelochemics (Table 1). None of the research reported on allelopathy to date has had a direct influence on weed control in soybeans, but undoubtedly, allelochemics play an important ecological role in soybean production and agriculture in general. This area of research is so new that the first USDA, SEA/AR funding for allelopathy research was in FY79. Recommendations for future allelopathy research and summaries on previous research were published in 1977 (2).

SUMMARY STATEMENT

Much research has been conducted on weed control in soybeans but probably more than 90% of that reported to date concerns the development of technology to provide better means of control. About 8% of the total research conducted to date concerns the competitive effect of individual weed species in soybeans, the remaining 2% is a combination of other variables. Research to determine the effect of weeds on crop yields has primarily been limited to the effect of weeds on yield reductions; only a minimum amount of research has been reported on the effect of weeds on the quality of the crop produced. Very little research has been reported on the specific factors that affect soybean-weed interactions or those factors that may affect future weed problems including cultural selection, chemical selection, distribution, genetic changes, photosynthetic pathways, and allelopathy.

Much greater attention should be given to the procedures needed to prevent problems arising from changes in weed flora that occur from continuous use of individual herbicides and combinations of herbicides. Included in this research should be the effect of rotating crops and using different herbicides with different modes of action.

Little research has been conducted to define the environmental conditions under which specific weeds are most competitive to soybeans. Work of this type needs to be increased, especially with those weeds such as itchgrass and cogongrass that have the greatest potential of spreading into areas uninfested previously. These studies should include work on life cycles and reproductive patterns as affected by environment to provide information that would prevent "minor" weeds from becoming "major" weeds.

Many observations are made by research scientists and extension specialists on changes in weed populations, but too often this information is summarized in abstracts if it is published at all. More effort is needed in monitoring the changes in weed populations on an annual basis and in publishing this information.

Research relating genetic changes, the different photosynthetic pathways, and allelopathy to "weediness" has been very modest and research in these areas should be fruitful and productive. Likewise, research on cultural selection in weed-soybean ecology would be beneficial. As suggested by Parker (54), the ultimate aim is a full study of population dynamics and the response of individual weed species to anticipated changes in chemical and cultural practices. The studies should include the monitoring of all parameters of seed production, seed loss, and buried seed reserves. Computerized modeling programs could be used to make projections on rates of spread and the conditions conducive to spread. A few studies of this type have been initiated with individual weeds, but this research is in its infancy. The total SY and dollar effort needed to complete an undertaking of this type for all of the more common weed species in each of the major crops in the U.S. would be very great, but it may be justified for a few of the major weeds, particularly those which have been shown to pose the greatest future threat. Losses in soybeans due to weeds and their control exceed the combined losses caused by insects and diseases, and this fact appears to provide ample justification to increase the effort to establish the economic role of weeds in soybean production, to establish the necessity for controlling certain specified levels of individual weeds in soybeans, and to provide criteria for prediciting new weed problems that could damage American soybean production.

NOTES

C. G. McWhorter and D. T. Patterson, Southern Weed Science Laboratory, Agricultural Research, SEA/USDA, Stoneville, Mississippi.

REFERENCES

1. Agricultural Research Service, USDA. 1965. A survey of extent and cost of weed control and specific weed problems. Publ. U.S. Dept. Agric., ARS 34-23-1, 80 p.
2. Agricultural Research Service, USDA. 1977. Report of the Research Planning Conference on the role of secondary compounds in plant interactions (allelopathy). 124 p.
3. Anderson, E. 1952. Plants, Man and Life. Little Brown, Boston, 252 p.
4. Anderson, J. M. and C. G. McWhorter. 1976. The economics of common cocklebur control in soybean production. Weed Sci. 24:397-400.
5. Anderson, J. M. and C. G. McWhorter. 1977. Effects of cocklebur control on soybean production in the Delta of Mississippi. Mississippi Agric. and For. Exp. Stn. Tech. Bull. 80, 20 p.
6. Auwarter, G. E. and J. D. Nalewaja. 1976. Volunteer sunflower competition in soybeans. Proc. North Cent. Weed Control Conf. 31:34.
7. Ayers, V. 1975. Weed control and no-till farming now and in the future. Proc. South. Weed Sci. Soc. 28:15-18.
8. Baker, H. G. 1974. The evolution of weeds. Annu. Rev. Ecol. Syst. 5:1-24.
9. Bannon, J. S., J. B. Baker, T. R. Harger, and R. L. Rogers. 1976. Weed watch—wild poinsettia. Weeds Today 8(1):12.
10. Bannon, J. S., R. L. Rogers, J. L. Killmer, and P. R. Vidrine. 1975. Controlling wild poinsettia in soybeans. Proc. South. Weed Sci. Soc. 28:50.
11. Barrentine, W. L. and L. R. Oliver. 1977. Competition, threshold levels, and control of cocklebur in soybeans. Mississippi Agric. and For. Exp. Stn. and Arkansas Agric. Exp. Stn. Techn. Bull. 83, 27 p.
12. Berglund, D. R. and J. D. Nalewaja. 1971. Competition between soybeans and wild mustard. Abstr. Weed Sci. Soc. Am., p. 116-117.
13. Buchanan, G. A., C. S. Hoveland, V. L. Brown, and R. H. Wade. 1975. Weed population shifts influenced by crop rotations and weed control programs. Proc. South. Weed Sci. Soc. 28:60-71.
14. Burnside, O. C. 1973. Shattercane . . . A serious weed throughout the central United States. Weeds Today 4(2):21.
15. Burnside, O. C. 1978. Have you considered close-drilled soybeans? Weeds Today 9(2):9-10.
16. Burnside, O. C. 1978. Mechanical, cultural, and chemical control of weeds in a sorghum-soybean (*Sorghum bicolor*)-(*Glycine max*) rotation. Weed Sci. 26:362-369.
17. Burnside, O. C., G. A. Wicks, D. D. Warnes, B. R. Somerhalder, and S. A. Weeks. 1969. Effect of weeds on harvesting efficiency in corn, sorghum and soybeans. Weed Sci. 18:438-441.
18. Chandler, J. M. 1976. Today's weed—purple moonflower. Weeds Today 7(2):30.
19. Charles, A. H., J. L. Jones, and P. J. Ryan. 1978. Dalapon resistance in *Lolium perenne* populations. J. British Grassland Soc. 33:93-97.
20. Dowler, C. C., E. W. Hauser, and A. W. Johnson. 1974. Crop-herbicide sequences on a southeastern Coastal Plain soil. Weed Sci. 22:500-505.
21. Downton, W. J. S. 1975. The occurrence of C_4 photosynthesis among plants. Photosynthetica 9:96-105.
22. Everson, A. C., D. N. Hyder, H. R. Gardner, and R. E. Bement. 1969. Chemical versus mechanical fallow of abandoned croplands. Weed Sci. 17:548-551.
23. Evetts, L. L. 1977. Common milkweed—the problem. Proc. North Central Weed Control Conf. 32:96-99.
24. Frans, R. 1977. No-till: it's not for everyone. Weeds Today 8(3):17-19.

25. Gupta, U. and P. S. Ramkrishnan. 1977. Inter-ecotypic competitive relationships within *Cynodon dactylon* (L.) Pers. Acta Bot. Indica 5:107-113.
26. Harrison, S. and L. R. Oliver. 1977. Texas gourd, a potentially serious weed problem to Arkansas producers. Arkansas Farm Res. 26(3):7.
27. Hauser, E. W., C. C. Dowler, and W. H. Marchant. 1969. Progress report: systems of weed control for soybeans. Proc. South. Weed Sci. Soc. 22:97.
28. Hauser, E. W., M. D. Jellum, C. C. Dowler, and W. H. Marchant. 1972. Systems of weed control for soybeans in the Coastal Plain. Weed Sci. 20:592-598.
29. Herron, J. W., L. Thompson, Jr., and C. H. Slack. 1971. Weed problems in no-till corn. Proc. South. Weed Sci. Soc. 24:170.
30. Holm, L. G., D. L. Plucknett, J. V. Pancho, and J. P. Herberger. 1977. The world's worst weeds. Distribution and biology. Univ. Press of Hawaii, Honolulu. 609 p.
31. Jensen, K. I. N., J. D. Bandeen, and V. Souza Machado. 1977. Studies on the differential tolerance of two lamb's-quarters selections to triazine herbicides. Can. J. Plant Sci. 57:1169-1177.
32. King, L. J. 1966. Weed ecotypes—A review. Proc. Northeast. Weed Control Conf. 20:604-611.
33. Knake, E. L. and F. W. Slife. 1962. Competition of *Setaria faberii* with corn and soybeans. Weeds 10:26-29.
34. Kanke, E. L. and F. W. Slife. 1965. Giant Foxtail seeded at various times in corn and soybeans. Weeds 13:331-334.
35. Knake, E. L. and R. W. Slife. 1969. Effect of time of giant foxtail removal from corn and soybeans. Weed Sci. 17:281-283.
36. Kosovac, Z. 1968. Application of herbicides as an alternative to ploughing and cultivation in sunflower and wheat production. Proc. Br. Weed Control Conf. 9: 849-854.
37. Lawton, K. and M. B. Tesar. 1953. Preliminary studies on the root distribution pattern of alfalfa and bromegrass with respect to phosphorus placement. Proc. Ann. Phosphorus Conf., North Central Region 5:85-90.
38. Lunsford, J. N., C. Cole, and D. R. Zarecov. 1976. Sesbania control in soybeans with bentazon. Proc. South. Weed Sci. Soc. 29:107.
39. McCormick, L. K. 1977. Weed survey—Southern states, p. 184-215. In Res. Rep. of the Southern Weed Sci. Soc.
40. McNeill, J. 1976. The taxonomy and evolution of weeds. Weed Res. 16:399-413.
41. McWhorter, C. G. 1971. Control of johnsongrass ecotypes. Weed Sci. 19:229-233.
42. McWhorter, C. G. 1976. Johnsongrass and its control., p. 426-434. In L. D. Hill (Ed.) Proc. World Soybean Res. Conf., The Interstate Printers and Publishers, Inc., Danville, IL.
43. McWhorter, C. G. and J. M. Anderson. 1976. Effectiveness of metribuzin applied preemergence for economical control of common cocklebur in soybeans. Weed Sci. 24:385-390.
44. McWhorter, C. G. and J. M. Anderson. 1976. Bantazon applied postemergence for economical control of common cocklebur in soybeans. Weed Sci. 24:391-396.
45. McWhorter, C. G. and J. M. Anderson. 1979. Hemp sesbania (*Sesbania exaltata*) competition in soybeans (*Glycine max*). Weed Sci. (In press).
46. McWhorter, C. G. and E. E. Hartwig. 1972. Competition of johnsongrass and cocklebur with six soybean varieties. Weed Sci. 20:56-59.
47. McWhorter, C. G. and T. N. Jordan. 1979. Limited tillage in cotton production. Weed Sci. Soc. Am. Monograph (In press)
48. Moolani, M. K., E. L. Kanke, and F. W. Slife. 1964. Competition of smooth pigweed with corn and soybeans. Weeds 12:126-128.

49. National Soybean Crop Improvement Council. 1969. A survey of 317 of America's top soybean growers. Conducted by the National Soybean Crop Improvement Council, Urbana, IL, 22 p.
50. National Soybean Crop Improvement Council. 1971. 1971 Top Soybean Grower Survey. Conducted by the National Soybean Crop Improvement Council, Urbana, IL, 25 p.
51. National Soybean Crop Improvement Council. 1977. 1977 Trop Soybean Grower Survey. Conducted by the National Soybean Crop Improvement Council, Urbana, IL, 26 p.
52. Nave, W. R. and L. M. Wax. 1971. Effect of weeds on soybean yield and harvesting efficiency. Weed Sci. 19:533-535.
53. Oliver, L. R. and M. M. Schreiber. 1971. Differential selectivity of herbicides on six *Setaria taxa.* Weed Sci. 19:428-431.
54. Parker, C. 1976. Prediction of new weed problems, especially in the developing world., p. 249-264. In J. M. Cherrett and G. R. Sager (Eds.) Origins of Pest, Parasite, Disease and Weed Problems, The 18th Symp. of the Brit. Ecol. Soc., Bangor,
55. Parochetti, J. V. 1975. Control of fall panicum in agronomic crops in the mid-Atlantic region, p. 15-17. In R. A. Peters (Chrm) Fall Panicum Workshop. Suppl. to Proc. Northeast. Weed Sci. Soc.
56. Patterson, D. T., C. R. Meyer, E. P. Flint, and P. C. Quimby, Jr. 1979. Temperature response and potential distribution of itchgrass (*Rottboellia exaltata* L. f.) in the United States. Weed Sci. 27 (In press)
57. Phillips, R. E. 1968. Minimum seedbed preparation for cotton. Agron. J. 60:437-441.
58. Report of the ARS, USDA Research Planning Conference on Weed Control in Soybeans. 1976. Tifton, GA. 54 p.
59. Rice, E. L. 1977. Allelopathy—An overview, p. 7-16. In Report of the ARS, USDA Research Planning Conference on the Role of Secondary Compounds in Plant Interactions (Allelopathy).
60. Roche, B. E. and T. J. Muzik. 1964. Ecological and physiological study of *Echinochloa crusgalli* and the response of its biotypes to sodium 2,2-dichloropropionate. Agron. J. 56:155-160.
61. Rocheconste, E. 1962. Studies on the biotypes of *Cynodon dactylon* (L.) Pers. II. Growth response to trichloroacetic acid and 2,2-dichloropropionic acid. Weed Res. 2:136-145.
62. Ross, M. A., J. L. Williams, and T. T. Bauman. 1973. The jimsonweed problem in Indiana soybean production. Proc. North Cent. Weed Control Conf. 28:48.
63. Rutz, G. 1978. Species shifts may bring new weeds. Delta Farm Press 35(36):1.
64. Sanford, J. O., D. L. Myhre, and N. C. Merwine. 1973. Double cropping systems involving no-tillage and conventional tillage. Agron. J. 65(6):977-982.
65. Santelmann, P. W. and J. A. Meade. 1961. Variation in morphological characteristics and dalapon susceptibility within the species *Setaria lutescens* and *S. faberii.* Weeds 9:406-410.
66. Staniforth, D. W. 1958. Soybean-foxtail competition under varying soil moisture conditions. Agron. J. 50:13-15.
67. Staniforth, D. W. 1962. Responses of nodulating and non-nodulating soybeans to weed competition. Proc. North Cent. Weed Control Conf. 19:49-52.
68. Staniforth, D.W. 1962. Responses of soybean varieties to weed competition. Agron. J. 54:11-13.
69. Staniforth, D. W. 1965. Competition effects of three foxtail species on soybeans. Weeds 13:191-193.

70. Staniforth, D. W. and C. R. Weber. 1956. Effects of annual weeds on the growth and yield of soybeans. Agron. J. 48:467-471.
71. Stowe, L. G. and J. A. Teeri. 1978. The geographic distribution of C_4 species of the dicotyledonae in relation to climate. Am. Nat. 112:609-623.
72. Strand, O. E. 1977. A survey of weed problems in Minnesota. Proc. North Cent. Weed Control Conf. 32:64-67.
73. Teeri, J. A. and L. C. Stowe. 1976. Climatic patterns and the distribution of C_4 grasses in North America. Oecologia (Berlin) 23:1-12.
74. Thomas, S. M. and B. G. Murray. 1978. Herbicide tolerance and polyploidy in *Cynodon dactylon* (L.) Pers. (Gramineae). Ann. Bot. (London) 42:137-143.
75. Thompson, L., Jr., R. W. Schumacher, and C. E. Rieck. 1974. An atrazine resistant strain of redroot pigweed. Abstr. Weed Sci. Soc. Am., No. 196.
76. Thurlow, D. L. and G. A. Buchanan. 1972. Competition of sicklepod with soybeans. Weed Sci. 20:379-384.
77. Tomkins, D. J. and W. F. Grant. 1977. Effects of herbicides on species diversity of two plant communities. Ecology 58:398-406.
78. Tomkins, D. J. and W. F. Grant. 1978. Morphological and genetic factors influencing the response of weed species to herbicides. Can. J. Bot. 56:1466-1471.
79. Triplett, G. B., Jr. 1966. Herbicide systems for no-tillage corn (*Zea mays* L.) following sod. Agron. J. 58:157-159.
80. Triplett, G. B., Jr. and G. D. Lytle. 1972. Control and ecology of weeds in continuous corn grown without tillage. Weed Sci. 20:453-457.
81. U.S. Department of Agriculture. 1965. Losses in agriculture. U.S. Dept. Agric. Handbook No. 291. 120 p.
82. U.S. Department of Agriculture. 1972. Extent and cost of weed control with herbicides and an evaluation of important weeds, 1968. ARS-H-1, 227 p.
83. Vengris, J., W. G. Colby, and M. Drake. 1955. Plant nutrient competition between weeds and corn. Agron. J. 47:213-216.
84. Vengris, J., M. Drake, W. G. Colby, and J Bart. 1953. Chemical composition of weeds and accompanying crop plants. Agron. J. 45:213-218.
85. Wax, L. M. 1973. Weed control, p. 417-257. In B. E. Caldwell (Ed.) Soybeans: Improvement, Production, and Uses, Am. Soc. Agron., Madison, WI.
86. Wax, L. M. 1976. Difficult-to-control annual weeds, p. 420-425. In L. D. Hill (Ed.) Proc. World Soybean Res. Conf., The Interstate Printers and Publishers, Inc., Danville, IL.
87. Wax, L. M., W. R. Nave, and R. L. Cooper. 1977. Weed control in narrow and wide-row soybeans. Weed Sci. 25:73-78.
88. Wax, L. M. and J. W. Pendleton. 1968. Effects of row spacing on soybeans. Weed Sci. 16:462-465.
89. Weber, J. B., J. A. Best, and W. W. Witt. 1974. Herbicide residues and weed species shifts on modified-soil field plots. Weed Sci. 22:427-433.
90. Wellings, L. W. 1968. Minimum cultivations for cereals on the experimental husbandry farms. Proc. Br. Weed Control Conf. 9:842-848.
91. West Virginia Department of Agriculture. 1978. 1977 West Virginia Farm Weed Survey., Charleston, WV, 4 p.
92. Williams, J. L., M. A. Ross, and T. T. Bauman. 1972. Tillage influences weed control in corn. Proc. Indiana Food and Agric. Chem. Conf., Purdue Univ., West Lafayette, IN.
93. Wilson, H. P. and R. H. Cole. 1966. Morningglory competition in soybeans. Weeds 14:49-51.
94. Worsham, A. D. 1977. No-till: worth trying. Weeds Today 8(3):16.

WEED CONTROL SYSTEMS IN THE CORN BELT STATES

F. W. Slife

Soybean acreage has grown rapidly in the Corn Belt States in the past 10 yr. This area produces about 2/3 of the total U.S. production. Improved weed control technology has been a major factor in allowing soybean acreage to expand. The intensity of both annual and perennial weed species has decreased and yield losses due to weeds have been reduced dramatically.

PERENNIAL WEEDS

Much of the soybean producing area in the Corn Belt is free of serious infestations of perennial weeds (7). In the northern third, scattered infestations of Canada thistle (*Cirsicum arvense*) and quackgrass (*Agropyron repens*) are present. The southern one third of the area has scattered infestations of johnsongrass (*Sorghum halepense*). This weed is more prevalent in the river bottom areas. The central one third of the Corn Belt is relatively free of these serious perennial weeds.

Yellow nutsedge (*Cyperus esculentus*) occurs in all parts of the Corn Belt. It is a problem that needs added attention, but it does not reduce yields to the extent of the previously mentioned species (8).

The following list of perennial weeds are found in soybeans in some parts of the region but under the current crop and weed management systems, these have not spread rapidly. There is considerable evidence that some of these species may become more vigorous under reduced tillage systems (5).

Climbing Milkweed	*Ampelamus albidus*
Common Milkweed	*Asclepias syriaca*
Wirestem Muhly	*Muhlenbergia frondosa*
Field Bindweed	*Convolvulus arvensis*
Hedge Bindweed	*Convolvulus sepium*
Wild Sweet Potato	*Ipomoea pandurata*
Horse Nettle	*Solanum carolinense*
Trumpet Creeper	*Campsis radicans*
Hemp Dogbane	*Apocynum cannabinum*

ANNUAL WEEDS

Summer annual weeds in soybeans are the dominant problem in the Corn Belt (7). They have become prominent because large areas produce only corn and soybeans. Since the life cycles of the summer annual weed coincide with these crops, the weeds have the opportunity to increase unless good control measures are practiced. Traditional rotations which included crops with different life cycles have decreased rapidly.

Some of the more common annual weeds found in soybeans in the Corn Belt are:

Crabgrass	*Digitaria* spp.
Giant Foxtail	*Setaria faberii*
Fall Panicum	*Panicum dicotamiflorum*
Common Cocklebur	*Xanthium pensylvanicum*
Jimsonweed	*Datura stramonium*
Common Lambsquarter	*Chenopodium album*
Annual Morningglory	*Ipomea* spp.
Pigweed	*Amaranthus* spp.
Giant Ragweed	*Ambrosia trifida*
Common Ragweed	*Ambrosia artemisifolia*
Smartweed	*Polygonum* spp.
Velvetleaf	*Abutilon theophrasti*

The most dominant annual weed species is giant foxtail. It is found in all the soybean growing areas. Good control measures have reduced the intensity of this species, but it still remains as the most prominent species.

In certain years, the most prominent weed problem in soybeans in the Midwest is volunteer corn. This is due to poor corn harvest conditions or disease where ears drop from the corn plant. The conditions are aggravated by reduced tillage systems on the corn stubble where the ears are not buried in the soil. Volunteer corn can reduce soybean yields, harvested soybeans can be docked for corn contamination, and volunteer corn serves as a host for the corn root worm.

The use of a postemergence spray applied through a recirculation sprayer is now used for this problem and a postemergence herbicide for volunteer corn in soybeans is expected to receive label clearance in 1979.

Preharvest treatments are occasionally used in soybeans in an attempt to dessicate green weeds in mature soybeans. Weed management programs in soybeans have been so successful in the past 5 yr that this practice is seldom used.

New herbicides will be needed to sustain the present level of good weed control. Of particular importance would be new postemergence chemicals that would control both grass weeds and broadleaf weeds. These chemicals could be superior to the present soil applied herbicides used for reduced tillage systems.

WEED MANAGEMENT SYSTEMS

Weed management systems used for soybeans in the Corn Belt vary widely but are based on (1) crop rotation, (2) tillage, and (3) herbicide treatment.

Rotation with Other Crops

Soybeans are not often planted in the same field each year but are more commonly rotated with other crops. The predominant other crop is corn. This is an excellent system insofar as weed control is concerned, since most of the troublesome annual broadleaf weeds in soybeans can be controlled with herbicides in corn. This prevents seed production and reduces the weed potential when soybeans are planted again. Some rotations include at least 2 yr of corn before the land is again planted to soybeans. In the southern half of the Corn Belt, winter wheat is frequently included in the cropping sequence. Winter wheat discourages the growth of summer annual weeds common to soybeans and corn and allows control measures to be initiated on perennial weeds after wheat harvest.

The acreage devoted to each crop in the rotation may vary each year due to anticipated returns for each component of the rotation. An increasing number of producers vary the cropping sequence to control a particular weed problem in a particular field. Soybeans are planted in johnsongrass infested areas so that a double rate of a dinitroaniline herbicide can be used for 2 yr for control. Corn can be planted in the same area for 2 or 3 yr to reduce the intensity of the annual broadleaf weed problem, before returning the land to soybeans. Fields heavily infested with annual grasses are frequently planted to soybeans because excellent herbicides are available for soybeans to control this problem.

Crop rotations are now receiving much more emphasis than in the previous 5 yr because of the development of pest resistance. The corn rootworm has developed a high degree of resistance to some insecticides that corn monoculture is decreasing. In addition, the soybean cyst nematode is now prominent in the Corn Belt. Both of these important pest problems can be at least partially solved with more diverse cropping sequences.

Tillage

Tillage is still widely used in the Corn Belt for weed control in soybeans. Preplant tillage is used to destroy one or more crops of annual weeds before planting, a rotary hoe or similar implement is used to break the soil crust to improve soybean emergence and to destroy weed seedlings and one or more row cultivations are customary (4). These tillage practices have not changed greatly even though herbicides are used on over 90% of the soybean acreage (2).

Tillage practices without herbicides have been inconsistent for weed control because rainfall frequently interferes with timely operations. Herbicides used without subsequent tillage practices give a high degree of weed control but herbicide treatment combined with tillage has given more consistent weed control.

The present system of herbicide treatment and row cultivation normally gives good weed control for at least 6 wk. By that time the soybean canopy has developed enough to prevent further weed infestation (1,3).

Tillage practices in the Corn Belt are now undergoing substantial change. Planting soybeans in wheat stubble without tillage (double cropping) has been a standard practice in the lower half of the Corn Belt for some 10 yr. More recently, reduced tillage systems that leave part or all of the previous year's crop residue on the soil surface are being accepted. There is also a trend toward planting soybeans in very narrow rows that cannot be row cultivated.

All of these new practices reduce the cost of tillage operations, but this is partially offset by an increased herbicide requirement. For the most part, herbicides are available to sustain the trend toward reduced tillage at the present time. Weed scientists must continue to develop new weed management practices that will allow this valuable practice to continue.

Perennial weeds are expected to increase in reduced tillage systems and new herbicides will be needed as a new weed spectrum adapts to the tillage changes.

Herbicides

Herbicides, in conjunction with tillage and other good management practices, are a key factor to high yields of soybeans in the Corn Belt. Producers are now experiencing the highest degree of weed control since soybeans were introduced. They have become more precise in the use of herbicides and herbicide treatments are now being initiated for individual fields. Many producers have analyzed their soils for organic matter content and degree of acidity in order to be more precise in herbicide rates. Pest management programs that utilize scouts are aiding in the identification of problem weed species in particular fields.

Since herbicide treatments for soybeans are changed frequently, the possibility of developing weed resistance within a weed species to a particular herbicide is somewhat low. Weed resistance to a particular herbicide can develop if an herbicide is used on the same area for many years (6).

Herbicide treatments used in the Corn Belt include preplant, premergence and postemergence applications. Various combinations of the application methods are used but preplant and preemergence are the dominant treatments. Each of these usually consists of two chemicals mixed together, one for annual grasses and the other for annual broadleaf weeds. As much as 30% of these treatments are applied by custom applicators using ground equipment. These soil applied treatments combined with tillage have performed so well that postemergence treatments have not been widely used as compared to the southern regions. Interest in postemergence treatments is increasing, however, and they are now used for double crop and narrow rowed soybeans.

In certain areas, part of the soybean acreage is hand weeded to remove weeds that have escaped herbicide treatment and row cultivation. This practice reduces weed seed production and it removes the top growth of certain perennials but it is probably done more for esthetic reasons than any other purposes.

SUMMARY

Soybean acreage has increased in the Corn Belt States in the last 10 yr and in some areas it is approaching equal status acreage with corn. Annual weeds pose more of a problem than perennial weeds but weed management programs currently are reducing the total weed problem. Increased research effort is needed in the following areas to maintain and improve the present weed control program in soybeans. Some of those areas are: (1) Integration of weed management systems into other pest management systems (2) Improved weed management systems for reduced tillage systems (3) Development of varieties for improved tolerance to herbicides, and (4) Research on biological control of weeds.

NOTES

Fred W. Slife, Department of Agronomy, University of Illinois, Urbana, Illinois 61801.

REFERENCES

1. Bloomberg, J. R. and L. M. Wax. 1979. Competition of common cocklebur (*Xanthium pensylvanicum*) with soybeans. Weed Sci. (In press).

2. Eichers, T. R., P. A. Andrilenas, and T. W. Anderson. 1978. Farmers Use of Pesticides in 1976. USDA Agricultural Economic Report No. 418.
3. Knake, E. L. and F. W. Slife. 1965. Giant foxtail seeded at various times in corn and soybeans. Weeds 13:331-334.
4. Lovely, W. G., C. R. Weber, and D. W. Staniforth. 1958. Effectiveness of the rotary hoe for weed control in soybeans. Agron. J. 50:621-625.
5. Triplett, G. B. and G. D. Lytle. 1972. Control and ecology of weeds in continuous corn grown without tillage. Weed Sci. 20:453-457.
6. Warwick, D. D. 1976. Plant variability in triazine resistance. Res. Review 65:56-63.
7. Wax, L. M. 1973. Weed control, p. 417-456. In B. E. Caldwell (ed). Soybeans: Improvement, Production, and Uses. Am. Soc. Agron., Madison, Wis.
8. Wax, L. M., E. W. Stoller, F. W. Slife, and R. N. Anderson. 1972. Yellow nutsedge control in soybeans. Weed Sci. 20:194-201.

WEED CONTROL SYSTEMS IN SOUTHERN U.S.

R. Frans

Before proceeding to a discussion of the weed control systems in southern United States, it seems appropriate to describe briefly the region in question as well as types of soybeans grown, something of the crop culture, and something of the weeds infesting the crop. The southern United States is an area of relatively high temperature and moisture during the soybean growing season. Although the region is in the temperate zone, rainfall tends to be abundant during this period and, as indicated, temperatures high, all of which is conducive to luxuriant plant growth. This includes, unfortunately, not only the crops in question but invading weeds as well. Therefore, competition between soybean and weeds tends to be high, necessitating a high level of control.

Soybeans varying widely in maturity are grown in the South, ranging from groups IV to VIII. In many areas of the southern region maturity groups will be mixed as far as adaptation is concerned so that harvest will not need to be accomplished all at the same time. Most of the soybean varieties in the South are of the determinate bushy type. Usually this may be considered a plus for weed control purposes. The bushy varieties tend to shade the soil early, forming complete canopy or lapping of the rows reasonably early. Although some variation may now exist in the width of rows, it is now generally agreed that row width should be adjusted so that lapping will occur approximately at the time of first bloom of the crop.

A decade or more ago soybeans were considered secondary to other field crops grown in the South, such as cotton, rice, or corn. Much of the principal effort and management went to these primary crops with the residual being expended for soybeans. That situation is not so in the present time. There have been steady declines in corn and cotton acreages which have given rise to increased soybean acreages in this ten-yr period. Most recently, however, rice acreages in the Mississippi River Delta have expanded more rapidly and perhaps this expansion is responsible for the somewhat stable situation that now exists in the soybean acreages of the South.

Weedings infesting southern soybeans are as variable perhaps as the variability of plant life to be found in that region. Prior to the high use of herbicides a broad range of both annual broadleaves and grass weeds were to be found in most soybean fields. Perennial weeds were also scattered liberally throughout the region. Ecological weed shifts have been covered in an earlier paper, but suffice it to say that these shifts have left many weeds now difficult to control either through cultural or chemical needs. These include both annual broadleaves and several perennial vines. Common cocklebur, for example, was one of the annual broadleaf weeds that increased sharply following increased use of preemergence herbicides. Better control measures in the past few years, however, have given some relief in its control.

Johnsongrass has always been and continues to be a major problem in soybean fields of the South. It is true that specialized measures have been adapted for its control. Oftentimes, however, the measures are costly and this factor presents an obstacle to a broader use of these practices, to the end that the weed continues to be a problem. Other weeds will be discussed particularly as they relate to the systems of weed control practices described below.

SYSTEMS OF CONTROL BY SOYBEAN GROWTH STAGE

The particular system of weed control adopted by a given farmer should logically be selected upon the basis of the weed infestation occurring in his fields. It must be admitted, however, that practices being employed often are not needed, simply because some farmers are not aware of the extent or type of weed problem present in their fields. Nevertheless, we will discuss the optimum systems that should be selected at this point according to the weeds present.

If the particular weed problem is a perennial such as johnsongrass, the severity of its effect usually requires that all practices possible be employed for its control. These may include herbicides applied before planting, at planting, and postemergence, coupled with good cultural measures. If the weed infestation is primarily small-seeded annuals, then these may be controlled adequately by preemergence herbicides, although some postemergence practices may be required as well. If the weed infestation includes vines, then preplant and preemergence measures are not too successful. The weed usually is best attacked by post emergence practices. Let us review these various systems available at these stages of soybean growth.

Preplanting Measures

Preplanting cultural practices are used extensively in soybean fields of the South. These cultural measures are for the purpose of loosening the soil to facilitate planting and to destroy existing weed growth. There are several herbicides, principally of the dinitroaniline group, that may be applied during this period of time and incorporated into the soil. Incorporation, of course, is both to protect against volatility loss of the herbicides and to bring the chemical into contact with weed seed or vegetative plant parts. Grass weeds are most easily controlled by these herbicides. Johnsongrass seed is particularly susceptible to the dinitroanilines and even rhizome johnsongrass may be controlled when these materials are applied at twice the normal rate according to soil type.

Other materials applied preplanting for johnsongrass control may include dalapon (Dowpon or Basanite) or glyphosate (Roundup). Neither material, however, is consistent for control when applied before planting. Too often johnsongrass growth at this time of the year is slow and foliar development too insufficient for adequate uptake and subsequent root penetration by the herbicides. Also, delays in waiting for sufficient johnsongrass growth before using these materials may unduly delay time of soybean planting.

Another material that has been used preplanting is the contact herbicide paraquat. It is often used for weed knock down prior to planting, particularly under the stale seedbed concept. This is an expensive practice, however, and must be questioned as to efficacy when compared to similar results obtained by conventional, cultural practices.

Herbicide mixtures are also used preplanting and may include the dinitroanaline herbicides mixed with certain preemergence materials and then both incorporated. Trifluralin (Treflan) or profluralin (Tolban), for example, are commonly mixed with better preemergence herbicides and incorporated. It has been found that there is little loss of selectivity from the preemergence material with this practice and that it may result in savings of extra trips over the field if the farmer contemplates using both herbicides anyway.

During this time of preplanting there are cultural practices that may be integrated with herbicide applications to achieve maximum control of weeds either in existence or those that emerge prior to planting. Repeated diskings, for example, have been shown to be effective in helping reduce johnsongrass infestations before the planting of the crop.

Preemergence Practices

Soybean planting time has also been the time at which many of the herbicide control practices of the past, at least for many years, have been used. These practices include the wide variety of preemergence herbicides that are applied immediately following the planting of the crop. Although several of these have been used in the past, mention will be made here only of the two or three more important ones. The acetanilide group contains, at present, two herbicides commercially available in the South: alachlor and metolachlor. There are prospects of other acetanilides becoming available very shortly for soybean producers. These herbicides have been quite useful on grass weeds infesting soybean fields, and generally are of high degree of selectivity for the crop.

The triazine family contains one herbicide, metribuzin, commercially known as Sencor or Lexone. This herbicide has also been quite effective for control of a fairly broad range of weeds that germinate shortly after planting the crop. Rate of application with this herbicide, however, has proven to be quite critical and must be adjusted carefully according to soil type and organic matter in the soil. Again, the use of preemergence herbicide practices depends upon the weeds infesting the soybean field. If these weeds tend to be the small-seeded annuals, then preemergence herbicides should be an integral part of the weed control program for that field.

"Cracking" or Soybean Emergence Stage

Often, weeds emerge with soybeans that may not have been controlled by earlier practices or that may not be susceptible to earlier practices, either cultural or chemical. Two examples of these weeds in the South would include common cocklebur and weeds of the morningglory family. Neither of these, for example, are particularly susceptible to either preplant or preemergence herbicides. It has been found, however, that when these weeds are in the cotyledonary stage of growth, they may be susceptible to such contact herbicides as dinoseb (Preemerge) or to a popular combination used in the South, naptalam and dinoseb (Dyanap). Either of these herbicides can be sprayed directly over the row to both emerged soybean and weed with little damage to the soybean, yet give good control of the seedling weeds. Often, if a preemergence herbicide has not been used previously, it is mixed with these contact materials to give soil residual effect. Alachlor, in particular, has been useful in these combinations. Timing of application of these herbicides is critical. Beyond this stage of growth, the weeds may become more resistant to the herbicides and the crop plant more susceptible to damage. Usually the period of time at which cracking or soybean emergence applications can be made is only about 3 to 4 days.

Early Postemergence Practices

For convenience, the early postemergence stage of soybean growth will be referred to as the V2 or V4 vegetative stage. These stages, of course, are from approximately one trifoliate to three, fully developed trifoliate leaves. Certain overtop practices have become popular at these stages, particularly where such weeds as common cocklebur have not been adequately controlled earlier. Dinoseb, for example, can be applied when common cocklebur is as tall as the soybean plants. If the rate is adjusted carefully, damage to the soybeans usually will be minimal. These practices are suggested only when it becomes impossible to employ other means for weed removal. Bentazon (Basagran), which is considerably more selective than dinoseb, can also be applied overtop at this stage of growth for common cocklebur control.

It is desirable that a differential in height between soybean plants and weeds will have been obtained by the time soybeans reach this stage of growth. With this differential, herbicides can be successfully post-directed. Chloroxuron (Tenoran) is an example of this application although at times it is also used overtop. Dinoseb, or a mixture of naptalam and dinoseb, can be directed successfully on young broadleaf weeds under the soybean canopy. In many situations, repeat applications of these herbicides will be needed to continue to extend the weed-free period as soybeans continue to grow and weeds continue to germinate.

Later Postemergence

By the time soybeans attain the V4 to V6 stage of growth other herbicide practices can be used. A combination that is particularly good for morningglory is that of linuron (Lorox) and 2,4-DB (Butyrac or Butoxone) or 2,4-DB alone. Either of these can be post-directed starting at this approximate stage of growth. Again, for most successful control, there must be a differential in height between soybean and the young morningglory seedlings. Sometimes, the herbicide paraquat is successfully mixed with 2,4-DB and post-directed, if both grass and broadleaf weeds are present. Bentazon can also be applied overtop at these stages if common cocklebur continues to be troublesome.

In nearly all these situations later postemergence applications can be made and are simply a continuation of the earlier practices outlined above. These later practices, of course, are done if the weed in questions continues to be a serious problem.

An exception to the above might be those situations in which weeds continue to grow uncontrolled above the canopy of the soybeans. In the South, johnsongrass fits this category and often will emerge through the canopy at mid-season regardless of practices used before. New innovative techniques are being introduced into this area and include such things as recirculating sprayers and new methods of "wiping on" herbicides. Of particular interest to soybean farmers today is the use of glyphosate with these new application techniques. The wipe-on technique is accomplished through what is called the rope wick applicator, and several variations of this are rapidly becoming available in the mid-South. All of these methods are promising since johnsongrass is considerably more susceptible to glyphosate later in the season than it is in its early stage of growth. Nevertheless, it must be realized that considerable competition has already occurred by this stage. Continued emphasis must continue to be placed upon early control measures for control of this weed. Also, by the time johnsongrass has emerged above the soybean canopy, viable seedheads have been formed and there will still be an adequate seed supply for reinfestation in the following year.

Herbicide Innovations

New herbicide products are coming into existence through company and academic research programs. One of the more interesting herbicide families is that of the diphenylethers. Oxyfluorfen (Goal) and acifluorfen (Blazer) are representative of this family. There is the promise of other diphenylether analogs becoming available in the next year or two. Some of these compounds demonstrate considerable selectivity for soybeans when they are applied overtop, yet may give good grass weed control, or in some cases, good broadleaf control as well. If these compounds continue to be promising, their adoption may radically change present postemergence control programs in use. This will include a definite shift away from post-directed practices to broadcast overtop applications. It must be emphasized, however, that costs of these materials are relatively high and must be taken into consideration before adoption is made.

Along with these new herbicides interest has arisen regarding the possibility of their mixture with more established herbicides. Work is underway in the South on mixtures of these compounds with such selective materials as bentazon. As costs become more in line with need for control we will probably see increased use of these mixtures in the future.

CONTROL ACCORDING TO SOYBEAN CULTURE SYSTEMS

Soybeans are grown in the southern region of the United States under several different cultural systems. A discussion of these systems is necessary in order to understand types of weed control practices utilized.

Full-Season, Standard Row-Width Soybeans

This system, of course, is the method of growing soybeans in conventionally-spaced rows. Various of the practices outlined above will be utilized according to the weed problems present. As has already been indicated, many variations of practices are possible and are in use at the present time.

Full-Season, Reduced Tillage Soybeans

Under this system the land is not disturbed prior to planting. The soybean crop is planted in the trash or stubble from the last year's crop. Most often a preemergence herbicide will be used in mixture with a contact herbicide, such as paraquat. These mixtures are applied at planting and then normal cultivation and postemergence practices proceed as necessary. These reduced tillage systems have been of considerable value on rolling lands where water or wind erosion of soil becomes a problem. They are not used extensively on the flat lands of the Delta region of the mid-South because weed infestations in these areas between cropping seasons are too severe to be handled without cultural operations.

Other limitations are present with this system. There is difficulty in obtaining a good soybean stand on certain soils that tend to be crusted or compacted. In addition, if areas are infested with perennial weeds such as johnsongrass, these methods of planting should not be used.

Double-Cropping Soybeans After Small Grains

Because of the relatively long growing season in the southern United States, many farmers plant small grains in the fall and, upon harvest in the spring, immediately follow with a planting of soybeans. In some cases this may involve planting directly into the small grain stubble after the small grain is harvested with no further seedbed preparation. Although it may be difficult to plant and obtain the stand with conventional planters, new planting equipment designed to break through the stubble promises to make the planting operation easier. There may be difficulty experienced also in cultivation, again where soils have a tendency to become compacted.

Farmers may also burn the stubble from the small grain, disk these fields, and then plant soybeans. This is most often done to attempt to conserve soil moisture at a period of time in the year when rainfall may be limiting. Under either of the above methods, conventional or narrow-row beans may be used. Again, if narrow-row beans are planted with this system, the area must be free of perennial weeds. In these situations, of course, the cultivation option has been lost and one can resort only to the use of herbicides for further control. However, the introduction of the more selective and promising materials for weed control may make this method an attractive one in future years.

Full-Season, Narrow-Row "Broadcast" Soybeans

Many farmers in the South are becoming interested in growing soybeans in grain drill width rows planted at the optimum time of the year. This has also been termed the "no-till" system for growing soybeans. Usually, conventional seedbed preparation is practiced which will allow also the use of preplant incorporated herbicides. In addition to the preplant materials, preemergence herbicides may be applied at the time of planting. No cultivation is done, and weed problems after emergence must be attacked by overtop applications. This is a new concept in the South and is of great interest to many, principally to minimize the necessity for extensive cultural operations throughout the year. Although there are no particular yield advantages from full-season, narrow-row soybeans with the varieties commonly used in the South, the system, if it can be successfully adopted, offers promise of reducing production costs because of the elimination of the need for interrow weeding through cultural means.

This sytem also has limitations which are similar to those listed above. Areas infested with perennial weed problems such as johnsongrass should be avoided. Again, there is no cultivation option under this system, and producers must rely on the preplant or preemergence herbicides and whatever may be available to be applied overtop of soybeans during the rest of the season. The advantage of the new selective herbicides becoming available, however, and as noted above, lends considerable promise to this method for Southern soybean farmers.

THE SOYBEAN-RICE ROTATION

Special emphasis must be given to the growing of soybeans in a rotation with rice. This is a common practice in the mid-South. Emphasis is given to this rotation here because of developing weed problems that are unique to the rice crop. Red rice is one of the more serious of these problems and has increased considerably in recent years with an increasing acreage of rice. New practices, aimed specifically at control of this weed, are in the process of being developed at the present time.

The best method of attack for red rice at present seems to be during the years when soybeans are grown on a particular area. Double rates of the dinitroaniline herbicides, for example, and applied preplant incorporated offer considerable promise. The herbicide alachlor can also be incorporated at a rate nearly double the usual preemergence rate. With either of these applications post-directed herbicides must be used to maintain early control attained. Paraquat, in particular, applied post-directed after soybeans are tall enough, has been one of the better practices to maintain this control. Since paraquat can be used only post-directed, this sytem can only be used with standard row-width soybeans. Generally, 2 to 3 yrs of intensive practices such as these are necessary to gain control over red rice.

SUMMARY

Soybeans grown in southern United States are subject to severe weed infestation due to favorable climatic conditions for plant growth. Because weed growth is most competitive early, extensive practices of early control are used until soybeans are large enough to form a foliage canopy capable of giving complete shading to the soil.

Preplanting, soil-incorporated herbicides are commonly applied for control of grass weeds, often followed by preemergence herbicides at planting, or by "cracking" stage applications for weeds emerging with soybeans. Early control is then extended by utilization of various postemergence applications, either directed or overtop, depending on weed infestation. Increased selectivity of postemergence herbicides offers promise of even improved control during these early soybean growth stages.

Several soybean cultural systems are used in the South. Large areas are still planted to conventionally-spaced rows, although interest is increasing in narrow-row, full season soybeans. Double-cropping soybeans, usually following small grains, is also a common practice. In all these systems weed control practices must be adapted to fit existing weed infestations, and certain systems, such as narrow-row soybeans with no cultivation option, must be avoided where perennial weeds are a problem .

NOTES

Robert Frans, Altheimer Laboratory, Department of Agronomy, University of Arkansas, Fayetteville, Arkansas 72701.

FOLIAR FERTILIZATION OF SOYBEANS DURING SEED-FILLING

J. J. Hanway

Although foliar application of micronutrients is practiced extensively, especially for tree crops, foliar applications of the major nutrients for field grain crops generally have not been effective (10). Therefore, when we reported (3) very significant yield increases of soybeans resulting from foliar application of NPKS solutions during seed-filling it created much interest. Good seed yield increases have been obtained from these foliar applications in several field experiments, but in many of the experiments that have been conducted there were no yield increases and in some experiments there were yield decreases. Obviously, we still have things to learn before foliar fertilization of soybeans during seed-filling can be used as a general field production practice.

At the World Soybean Research Conference in Champaign, Illinois, in 1975 I discussed some interrelated development and biochemical processes in the growth of soybean plants (5). These processes provided a basis for foliar fertilization of soybeans during seed-filling, and we had started field experimentation in 1974 (3). These studies were concerned primarily with foliar application of the nutrients N, P, K and S that were being translocated from the plant leaves and other vegetative plant parts to the developing seeds during seed-filling.

In studies and practices involving fertilizer use and mineral nutrition of soybeans and other grain crops, we have been concerned primarily with the nutrition of the "mother" plant rather than the nutrition of the developing "infant" seeds. For example, we have developed relations between the nutrient content of the leaves on the plants and final seed yields or the increases in yield resulting from soil fertilization. The seeds generally have been considered as "sinks" where materials and nutrients that have been synthesized or accumulated in the plants are deposited. I now ask you to start thinking of the developing seeds as an infant plant with very definite nutrient requirements. The number of seeds that develop, their rate of growth and their size at maturity will depend on the mineral nutrients as well as the organic materials made available to the developing seeds by the plant.

During seed-filling, if there has been a good seed-set, the developing seeds rather than regions of vegetative growth become the primary sinks for materials synthesized in the plant. Soluble carbohydrates produced by photosynthesis in the leaves are channeled primarily to the developing seeds and their concentrations in the stems (and undoubtedly in the roots) decreases. Because of this, N-fixation by the rhizobia in the root nodules slows and stops and root growth generally ceases or is very limited. Although nutrient uptake by the plant continues, the rate of uptake of at least some of the nutrients gradually decreases. The uptake of N, P, K and S is not adequate to supply the needs of the developing seeds so these nutrients that are relatively mobile in the plant are translocated from other plant parts to the seeds resulting in depletion of these nutrients in the leaves. This depletion results in leaves turning yellow and then dying.

Therefore, during the seed-filling period we are concerned not only with maintaining adequate nutrients in the leaves so they remain active photosynthetically, but also with supplying adequate nutrients to the developing seeds. Foliar fertilization during seed-filling appears to offer promising possibilities for doing this.

Since xylem connections to the seeds generally are poor and there is very little transpiration of water from the seeds, the supply of nutrients and other materials to the seeds must be through the phloem of the "mother" plant. N, P, K and S are readily transported in the phloem. However, the transport of several other nutrient elements such as Ca, Mg, Mn, Fe and B is limited. As shown in Table 1, the concentrations of N, P. K and S in soybean seeds (which are primarily cotyledons) are similar to or higher than the concentrations in healthy green leaves on the plants. However, in the seeds the concentrations of Ca, Mg and Mn (and Si) are very low. In addition, concentrations of Fe and B are low. Possibly one or more of these nutrient elements that are relatively immobile in the plant phloem is limiting seed development.

THE "MOBILE" NUTRIENTS

Field experiments were conducted in 1975 and 1976 in which foliar applications of N, P, K and S solutions were made during seed-filling. Urea, potassium polyphosphate and potassium sulfate were used as the primary sources of N, P, K and S. Preliminary tests had shown that a foliar application of these solutions resulted in serious "leaf burn" if more than about

Table 1. Typical concentrations (on a dry wt basis) of different elements in soybean leaves during vegetative growth and at plant maturity and of mature soybean seeds for Hark, Amsoy and Wayne varieties (1).

Element	Leaves		Seeds
	Vegetative	Maturity	
N %	5.0	2.3	6.5
P %	0.35	0.20	0.6
S %	0.25	0.14	0.30
K %	2.0	0.8	1.8
Zn ppm	70	44	40
Cu ppm	17	14	17
Ca %	1.4	3.4	0.3
Mg %	0.5	0.7	0.2
Mn ppm	100	160	20
Fe ppm	400	400	100
B ppm	50	65	30
Al ppm	300	300	<8
Si %	1.8	1.8	<0.15
Na %	0.06	0.6	0.03

20 kg N/ha was applied per spraying. Therefore, the rate of application per spraying was held at this level.

The results of these studies may be summarized as follows: (a) Foliar application of 80 + 8 + 24 + 4 kg of N + P + K + S/ha applied in 4 sprayings during the seed-filling period (stage R5 to R6.5), (2) increased soybean yields very significantly. Where N, P or K was omitted from the foliar application there generally was little or no yield increase and where S was omitted the yield increase was less than where S was included with the N, P and K (Table 2); (b) Increasing the amount of any one of the nutrients above the 10:1:3:05 ratio of N:P:K:S (the ratio found in soybean seeds) did not result in an additional yield increase; (c) The yield increase occurred principally in the top half of the plant; (d) The yield increases resulted from increases in numbers of harvested seeds—not increases in average seed size; and (e) The foliar NPKS applications at the rate which resulted in optimum yields generally did not maintain the N, P and K content of the leaves (Table 3). At higher rates of application the percentages of P in the leaves were maintained, the N percentages decreased although the decreases were less than where no N was applied, but the K percentages were not affected.

The results of one 1975 experiment (3) were especially striking and probably resulted in much of the excitement that resulted in 1976 (Table 4). In this experiment soybean yields of two cultivars were increased by about

Table 2. **Effects on soybean yields and seed size of foliar applications of different nutrient solutions applied between growth stages R5 and R6.5.**

Total Nutrient Application (4 Sprayings)				Experiment		
				Ames 1975	Ames 1976	Kanawha 1976
N	P	K	S	Corsoy	Corsoy	Steele
	– *kg/ha* –				– *Soybean yield (kg/ha)* –	
0	0	0	0	2980	3040	2620
					– *Yield increase (kg/ha)* –	
0	8	24	4	- 170	70	80
80	0	24	4	110	350	70
80	8	0	4	- 140	400	250
80	8	24	0	320	620	250
80	8	24	4	570	730	620
					– *Seed size (g/100 seeds)* –	
0	0	0	0	15.8	15.2	18.4
0	8	24	4	15.7	16.4	18.5
80	0	24	4	15.9	14.9	18.5
80	8	0	4	15.7	16.4	18.5
80	8	24	0	16.7	15.1	18.2
80	8	24	4	16.2	15.4	17.5

1500 kg/ha and average seed size was not influenced by the foliar treatment. Yield increases this large have not been obtained in any other experiment. But these results indicate that such increases are possible.

Field experimental data to evaluate the effects of different variables are still very limited, so definite conclusions are not yet feasible. Some tentative observations are: (a) Increasing the total amount of NPKS applied to greater than 100 + 10 + 30 + 5 kg of N + P + K + S/ha or the number of sprayings with the same total nutrient application resulted in lower yield increases; (b) Although yield responses to foliar fertilization have differed among different varieties in any one experiment, no one variety has consistently responded more than others in different experiments; (c) Foliar spraying of NPKS solutions during midday when the sun is shining often results in much greater leaf burn than similar sprayings in the morning or evening. Although we do not know the reason for this, spraying in the evening is generally recommended; (d) By using sulfuric acid to supply varying amounts of the S, the pH of the spray solution can be varied between pH 2.5 and 7. Although none of the solutions resulted in serious leaf burn, some of the data obtained by T. Smith indicate that a solution pH of 5 to 6 may be most desirable.

Table 3. Effects of different rates of nutrient applications during seed-filling on percentages of N, P and K in youngest mature leaf on soybean plants (3).[a]

Growth Stage	kg N applied/ha			kg P applied/ha			kg K applied/ha		
	0	80	160	0	8	16	0	24	48
	– *% leaf N* –			– *% leaf P* –			– *% leaf K* –		
R5.5	3.05	3.22	3.55	.20	.24	.30	.72	.90	.91
R6.0	2.11	2.66	2.82	.16	.25	.28	.70	.84	.69
R6.5	1.52	2.60	2.93	.15	.26	.33	.62	.72	.61

[a]Foliar applications at growth stages 5, 5.5, 6 and 6.5.

Table 4. Effects of foliar applications of NPKS during seed-filling on yields and seed size of two soybean cultivars.

Soybean Variety	Soybean Yield Not Sprayed	Yield Increase Sprayed[a]	Seed Size Not Sprayed	Seed Size Sprayed[a]
	– *kg/ha* –		– *g/100 seeds* –	
Corsoy	3540	1570	15.2	15.0
Amsoy	3850	1490	16.7	16.0

[a]A total application of 96 + 9.6 + 28.8 + 4.8 kg of N + P + K + S/ha applied at developmental stages R5, R5.7, R6.2 and R6.6.

THE "IMMOBILE" NUTRIENTS

Although research data are limited, it is interesting and may be most profitable to consider the effects of the relatively immobile nutrients on seed yields. The results that have been obtained indicate that foliar applications of the mobile nutrients N, P, K and S during seed-filling often resulted in seed yield increases but in many experiments resulted in no yield increases. Were some other nutrient elements, such as B, Fe, Ca, Mg or Mn that are relatively immobile, limiting seed yields?

Nelson (8) found that boron applications on Iowa soils resulted in alfalfa yield increases but not hay yield increases. Iron deficiencies in soybeans generally become evident in new growth of the plants when older plant parts show no deficiency symptoms. The concentrations of these nutrient elements in soybean seeds are low in comparison to the concentrations in the leaves. Preliminary studies of foliar applications of different forms of these nutrient elements that may be more mobile in the plants were started in 1978 by S. Ritchie.

The marked increases in Ca and Mn concentrations in the leaves during seed-filling (5) and the very low concentrations of these elements and Mg in the seeds (Table 1) indicate that these elements are not retranslocated from the leaves to the developing seeds and may be deficient for optimum growth of the seeds. Konno (6) also reported this increase in Ca, a similar marked increase in Mg in soybean leaves and petioles during seed-filling, and very low concentrations of these elements in the seeds. Certainly, studies should be started concerning the possibilities that these nutrient elements are limiting soybean seed development and, if so, methods should be developed for improving their translocation in the plants to the seeds.

The research at the U.S. Plant, Soil and Nutrition Laboratory in Ithaca, N.Y. (8) is most interesting and offers promise of a technique to study the requirements of the developing soybean seeds for the relatively immobile nutrients. These researchers have separated the cotyledons of small, embryonic soybean seeds and grown them in vitro culture to study the nutrient requirements of the developing cotyledons. The growth rate of the cotyledons in the standard culture medium was greater than on the plants. Omitting P, K, S or Fe from the culture solution reduced growth. Omission of either Ca or Mg had little effect, but omission of both reduced growth and protein synthesis. More information is needed concerning the effects of these and other nutrients such as Mn, B, Zn, Cu and possibly Si as well as the effects of varying concentrations of all the different nutrient elements on growth, protein synthesis and final nutrient content of the cotyledons. For example, the available data indicate that the rate of growth and other characteristics of the seeds may be markedly affected by the availability to the seeds of the nutrient elements such as Ca, Mg, Mn, Fe and B that are relatively immobile in the phloem.

Using a similar technique with cucumber cotyledons, Green and Muir (3) have shown that Ca and K levels markedly influenced the growth response to cytokinins.

We have conducted some very preliminary greenhouse tests in which soybean seeds were planted in acid-washed sand to study the effect of different nutrient elements in the seed on growth of the plant immediately after germination. Where the sand was wetted with water or with Hoagland's solution minus Ca and Mn the plants developed only to the stage where the cotyledons had emerged and unfolded. Where the sand was wetted with a solution containing Ca and Mn the plants continued to grow and develop for a time similar to where the sand was wetted with a complete Hoagland's solution. These results indicate that Ca and/or Mn were deficient soon after the seed germinated. They may have been deficient during the development of the seeds. Some results of McAlister and Krober (7) published in 1951 are pertinent to these observations. They observed a rapid loss in dry weight, N, P and K in soybean cotyledons during the first few days after seed germination and seedling emergence, but a rapid increase in Ca in the cotyledons following seedling emergence.

OTHER FACTORS

Much evidence exists to show that urea is more rapidly absorbed and is less phytotoxic than most other common forms of N which can be applied in a foliar spray. However, foliar fertilization during seed-filling requires larger amounts of N than of other nutrients, and it has been observed that the "leaf burn" which occurs with foliar fertilization of soybeans during seed-filling was associated with the urea in the solution. Since cyanates, carbamates, biuret and cyanamid may be formed during the decomposition of urea, one student, M. Meshi, has been studying the toxicity of these materials when present in the NPKS sprays. Initical observations indicate that cyanate and cyanamid are most toxic followed by biuret. Where 20 kg N/ha was applied per spraying, if more than 5% of the N, <1 kg N/ha, was present as these forms, there was serious leaf burn and a decrease in seed yield. Similar amounts of carbamate N, although sometimes resulting in leaf burn, did not result in seed yield decreases. Because decomposition of urea to form these substances would be expected to be associated with the urease in and on the plant leaves and other plant parts, another student, N. El-Hout, has begun to collect data concerning urease in soybean plants during the season.

Varietal differences in response to foliar fertilization applications may be expected and have been observed. To characterize the nutrient status in different plant parts of different cultivars throughout the season, G. Loberg working with R. Shibles has done extensive samplings and analysis for different nutrient elements in different soybean lines.

Plant physiologists have made tremendous contributions to our knowledge about hormones and growth regulators (auxins, gibberellins, cytokinins, ethylene and abscissic acid) in soybean plants. This knowledge may supply the information needed to devise foliar nutrient applications that will be more consistently successful. For example, foliar applications of a cytokinin to delay nutrient depletion and senescence of leaves in conjunction with foliar nutrient applications to supply the nutrients needed by the developing seeds may be successful. Or, the use of inhibitors of ethylene and abscissic acid in leaves together with foliar nutrient applications may improve seed yields.

Foliar fertilization of soybeans during seed-filling shows real promise. But we still have things to learn before it will be consistently successful and accepted as a general practice.

NOTES

John J. Hanway, Agronomy Department, Iowa State University, Ames, Iowa 50011.

REFERENCES

1. Dunphy, E. J. 1972. Soybean responses to phosphorus and potassium fertility differences. Ph.D. Thesis, Iowa State Univ., Ames, IA.
2. Fehr, W. R. and C. E. Caviness. 1977. Stages of soybean development. Sp. Rept. 80, Iowa Agr. and Home Ec. Expt. Sta., Iowa State Univ., Ames, Iowa.
3. Garcia, R. L. and J. J. Hanway. 1976. Foliar fertilization of soybeans during the seed-filling period. Agron. J. 68:653-657.
4. Green, J., and R. M. Muir. 1978. The effect of potassium on cotyledon expansion induced by cytokinins. Physiol. Plant. 43:213-218.
5. Hanway, J. J. 1976. Interrelated development and biochemical process in the growth of soybean plants, pp. 5-15. In L. D. Hill (Ed.) World Soybean Research, Interstate Printers and Publishers, Inc., Danville, IL.
6. Konnon, S. 1977. Growth and ripening of soybeans. Tech. Bull. No. 32, Food and Fertilizer Technology Center, Taipei City, Taiwan.
7. McAlister, D. F. and O. A. Krober. 1951. Translocation of food reserves from soybean cotyledons and their influence on the development of the plant. Plant Physiol. 26:525-538.
8. Nelson, N. M. 1953. Alfalfa seed and forage production in relation to boron fertilization of certain Iowa soils. M.S. Thesis, Iowa State Univ., Ames, Iowa.
9. Thompson, J. F., J. T. Madison, and A. E. Muenster. 1977. In vitro culture of immature cotyledons of soya bean (*Glycine max* L. Merr.). Ann. Bot. 41:29-39.
10. Wittwer, S. W., and M. J. BuKovac. 1969. The uptake of nutrients through leaf surfaces. Handbuch der Pflanzenernahrung und Dungung. Springer-Verlag/Wien-New York.

FACTORS AFFECTING THE RESPONSE OF SOYBEANS TO MOLYBDENUM APPLICATION

F. C. Boswell

Molybdenum (Mo) is one of the seven micronutrients essential for normal plant growth and development. This element is somewhat unique when compared to the other micronutrients because its availability increases as soil pH increases and it is required in relatively small quantities for correcting deficiencies in plants. Molybdenum is involved in at least two essential plant functions (16). It is required in all plants for protein synthesis and for symbiotic nitrogen fixation in legumes. The specific function of Mo in nitrogen fixation is somewhat obscure but is implicated in the constituent of an enzyme involved in the process of fixing atmospheric nitrogen (34). Bergersen (5) indicated that Mo is an essential constituent of nitrogenase in N_2 fixing organisms of soybean nodules (*Rhizobium japonica*). Although Bortels (8) showed the biological importance of Mo in 1930, its essentiality for plants was not reported until 1939 when Arnon and Stout (3) produced Mo-deficient tomato plants in six successive experiments conducted under rigidly controlled conditions.

Bortels in 1937 (9) and Evans in 1956 (16) alluded to legumes, including soybeans [*Glycine max* (L.) Merr.] making better growth and fixing more nitrogen when supplied with Mo. However, it was not until 1962 that specific soybean responses to Mo application were reported. Berger (4) reported Mo deficiencies in the U.S. on soybeans in Wisconsin, Illinois and Indiana. During that year, Parker and Harris (35) also reported significant soybean yield responses to application of Mo under field conditions in Georgia. The following year, Lavy and Barber (29) reported significant soybean yield responses to Mo applications in Indiana. In subsequent years, soybean yield responses to Mo applications have been reported in at least seven additional states in the

U.S. (2,15,31,32,40,41,44) and more than five countries outside the continental U.S. (7,12,23,30,43). These countries include Japan, China, Taiwan, Russia, and France. Generally, responses in the U.S., as well as other countries, involve soils that are acid in nature, usually with a water pH of 6.2 or less.

In addition to increasing soybean yields, Mo additions to the soil, seeds, or soybean foliage at proper rates and methods may contribute to vigorous plant growth, increased protein contents and increased build-up of nitrogen in the plant and soil which accompany increased nodulation and symbiotic microbial activity.

Soybeans are high in protein and oil and further development in the utilization of soybeans in the human diet will lead to a demand for increased yields per acre as well as an expanded acreage in future years. In this context, it would be helpful to better understand factors affecting the response of soybeans to Mo applications since small quantities of this essential micronutrient have a potential to affect desirable influences on this important crop.

SOIL RELATIONSHIPS TO Mo REQUIREMENTS OF SOYBEANS

Many soils do not have sufficient Mo present in a form soybean plants can use. Generally, these soils are acid with pH levels below 6.2, inherently low in total Mo content, coarse textured (often loamy sands or sandy loams), relatively low in organic matter, with appreciable amounts of iron, and severely eroded and/or heavily weathered. In a review, Cheng and Ouellette (11) stated that results from various countries indicate that the average total Mo content of soils is 2.0 ppm and the average available Mo content is 0.2 ppm. Plants often show Mo deficiency when the available soil Mo is less than 0.1 ppm and the soil is acid. Yield responses in soybeans have been noted when the available soil Mo is above 0.3 ppm (10).

Soil pH

The relatively small fraction of the total Mo content of a soil that is available to plants is dependent on soil pH (25). Unlike other essential micronutrients, Mo availability increases as the soil pH is increased. Therefore, soybean responses to Mo applications are greatest on acid soils with a majority of the responses occurring east of the Mississippi River. Molybdenum responses in Japan, Taiwan, and China have also occurred on acid soils while a study conducted in France showed appreciable yield increases at pH near neutral. Smith and Connell (42) showed a significant response to seed applied Mo (0.5 g/kg seed) at pH values of 6.1 to 6.3 in a 5-yr study on a silt loam soil in Tennessee.

Workers in Louisiana (41) showed that 36% of the variation in yield response to application of Mo at 15 sites with 14 different soils was attributed to the influence of soil pH. They found a curvilinear relation between soil pH and yield response to Mo (Figure 1). The regression equation indicates a very rapid increase in the slope of the curvilinear line as soil pH is decreased from 6.5 downward. They indicated an average yield increase of approximately 323, 175 and 67 kg/ha at soil pH values of 5.0, 5.5 and 6.0, respectively, from application of Mo. They did not show a significant yield increase from the application of Mo when the soil pH was 6.2 or higher. Responses to Mo added as fertilizer are greater on acid soils than on alkaline soils. The application of limestone to acid soils is an effective method of increasing available soil Mo as well as the Mo concentration in plants (1).

Soil Molybdenum

Davies (14) presented a review of factors affecting Mo availability in soils which indicated that alkaline soils, some "young" soils, and most, but not all, organic soils are prone to produce plants of high Mo content. Conversely, soils that are highly podzolized, soils of high anion-exchange capacity, low pH, high in iron, and possibly soils that are Mo depleted by exhaustive cropping may be deficient in Mo. Cox and Kamprath (13) reported that crop responses had been obtained with soil levels up to 0.4 ppm but most values were less than 0.2 ppm.

Soil Mo data relative to soybean yield response is somewhat limited. Sedberry et al. (41) reported soil total Mo values of 0.63 to 1.64 ppm while extractable Mo (Tamm's solution, ammonium oxalate adjusted to pH 3.3) ranged from 0.03 to 0.31 ppm on 14 Louisiana soils where significant yield responses were obtained. Boswell and Anderson (10) obtained significant soybean yield responses from soils of Georgia which ranged from 0.06 ppm to

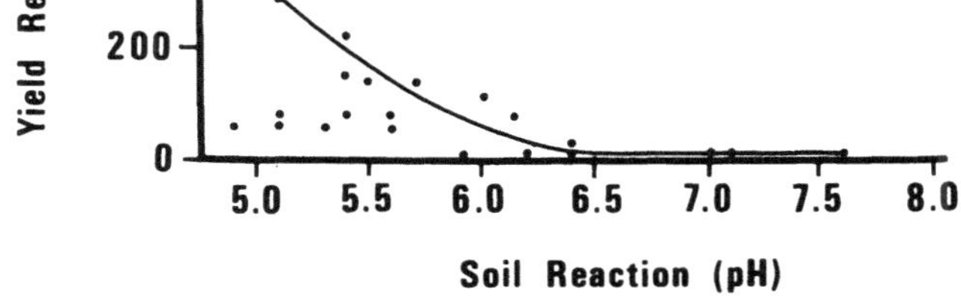

Figure 1. **Influence of soil reaction (pH) on the yield response by soybeans to the application of Mo on 14 soils (41).**

0.32 ppm extractable Mo contents. Hashimoto and Yamasaki (23) reported significant soybean yield increases to Mo applications when applied to Japanese soils containing 0.11 ppm of available Mo.

Soil Texture, Organic Matter and Iron

Several reviewers (11,14,16,20,25) have indicated that soil factors such as texture, organic matter and iron content affect Mo needs by plants. Davies (14) reported that Mo deficiencies have been noted on peat soils in West Australia and New Zealand. He also stated that Mo becomes unavailable in ironstone soils. However, reported data on organic matter or iron levels of soils in relation to Mo response on soybeans are not evident. Although soil texture relations to Mo responses have not been assessed fully, most responses have been on soils of silt loam, sandy loam or loamy sands with most response reported on soils with silt loam textures.

Sedberry et al. (41) related extractable and total soil Mo to several chemical properties of 14 soils of Louisiana by means of correlation coefficients. Significant correlations occurred between extractable Mo and extractable K, organic matter, and cation exchange capacity (Table 1). A significant negative correlation (r-0.502) was found between soil pH and the extractable Mo content of the soils. Significant correlations were not found between the extractable and total Mo contents of the soils. Other soil chemical properties were not significantly related to total Mo contents of the soil.

Table 1. **Relationships as shown by correlation coefficients (r) between extractable and total Mo contents and certain chemical properties of the soils (41).**

Chemical Property	Extractable Mo	Total Mo
	—r—	
Extractable P	-0.364	-0.223
Extractable K	0.523[a]	0.125
Extractable Ca	0.122	0.150
Extractable Mg	0.026	0.128
Organic matter	0.665[b]	0.311
Cation exchange capacity	0.747[b]	0.342
Base saturation	-0.388	0.008
Soil reaction, pH	-0.502[a]	-0.002
Extractable Mo		0.347
Total Mo	0.347	

[a]Significant difference at the 5% probability level.

[b]Significant difference at the 1% probability level.

THE SOYBEAN PLANT AND Mo RELATIONSHIPS

Visual Deficiency Symptoms

Since Mo is an essential element in nitrogen fixation and nitrate reduction, visual deficiency symptoms may be very similar to nitrogen deficiency in nonlegumes. These symptoms include a pale green color in developing leaves. Peterson and Purvis (38) developed Mo deficiency symptoms for soybeans in purified nutrient solution. They indicated that the leaves were light green in color, twisted on the stems with necrotic areas adjacent to the midrib, between the veins, and along the margin. Under field conditions, Mo deficient soybean leaves are light green color, often small in size. The plants are somewhat stunted when compared to plants receiving adequate Mo. The plant often develops sparse, upright canopies that do not allow overlap in normal row widths.

Tissue Levels of Mo

For plants in general, Johnson (28) indicated that a Mo deficiency occurs frequently when the plant concentration is less than 0.10 ppm Mo. Levels of Mo in tissue may exceed several hundred ppm with no evidence of detrimental effects in plant growth. However, Barshad (6) reported that Mo concentrations greater than 15 ppm in pasture plants may be toxic to cattle. James et al. (26) reported the Mo content of alfalfa plants from greenhouse studies was much lower than that of plants grown in the field. Similar comparisons for soybeans have not been reported.

Limited soybean tissue concentrations have been reported. Studies in Taiwan (30) showed the Mo content of soybean leaf blade tissue to be 0.20 ppm on soil with pH levels of 4.4 and was increased to 0.45 ppm when 200 g/ha Mo was applied. Hashimoto and Yamaski (23) increased Mo concentrations in soybean leaf tissue, stem, root, and nodules by the application of 130 mg Mo/liter of seeds (Table 2). They found that Mo concentrations of most plant components decreased as sampling dates were delayed. Boswell and Anderson (10) reported 0.13 ppm Mo concentration of leaf petiole tissue from untreated plants grown on Georgia soils. A seed treatment of 217 g Mo/ha increased Mo concentration of leaf tissue from 0.13 ppm to 0.22 ppm. Foliar spray

Table 2. Molybdenum concentrations of soybean plant components (23).

	Plant Component[a]			
Treatment	Leaves	Stems	Roots	Nodule
		– ppm –		
Control	0.32	0.30	0.32	0.90
Mo Applied[b]	1.13	2.75	3.36	6.10

[a]Sampled 40th day after treatment.
[b]Rate—130 mg Mo/liter of seed.

applications of Mo increased leaf tissue levels to 4.70 ppm Mo by spray application of 217 g Mo/ha at full bloom-early pod growth period [R-3 stage as described by Fehr and Caviness (17)].

Louisiana workers (41) showed a yield response to applied Mo on Calhoun and Crowley sil soils which had tissue contents of 0.14 and 0.19 ppm Mo with added lime or Mo. Seed applied Mo (35 g Mo/ha) increased the tissue to 0.87 and 1.54 ppm Mo, respectively (Table 3). Their data also indicated that the application of limestone to the soil increased both native and applied Mo and increased Mo concentration in the leaf tissue of the soybean plants.

DeMooy (15) obtained significant soybean yield responses to 230 g Mo/67 kg seed on silt loam soils that had leaf tissue levels of 0.87 to 1.65 ppm Mo (spectrographic analyses) when sampled from the control plots at full bloom growth stage.

Soybean Seed Mo Levels

Reported levels of Mo in soybean seeds range from 0.6 to 47.0 ppm. Several workers (19,22,37,38) have shown that the Mo requirements of a soybean crop can be supplied by seed containing a high Mo content. Gurley and Giddens (19) showed that the seed part as well as the location of the seed on the plant influenced the Mo content (Table 4). Parker and Harris (37) reported a significant response to applied foliar Mo (70 g/ha) when the Mo content of the seeds which were planted contained approximately 8 ppm Mo or less. Hawes et al. (24) reported Mo-values for soybean seeds to be 0.68, 1.32, 2.60, 2.49, 2.44 and 3.89 ppm when grown on soils fertilized

Table 3. **Effects of limestone and Mo additions on the Mo content of soybean leaf tissue grown on soils in Louisiana (41).**

Treatment	Mo in Leaf Tissue	
	Calhoun Sil Soil	Crowley Sil Soil
	– ppm –	
0 lime		
0 Mo	0.14	0.19
35 g Mo/ha (seed spplied)	0.87	1.54
560 g Mo/ha (soil applied)	0.60	0.62
4.5 mt/ha lime		
0 Mo	0.18	0.64
35 g Mo/ha (seed applied)	0.93	1.60
560 g Mo/ha (soil applied)	2.24	0.90
9.0 mt/ha lime		
0 Mo	0.28	0.84
35 g Mo/ha (seed applied)	0.84	2.94
560 g Mo/ha (soil applied)	4.22	1.97

Table 4. **Molybdenum content of different parts of soybean seeds and of seeds from different locations on the plant (19).**

Part of Seed[a]	ppm Mo	Location of Seed[b]	ppm Mo
Seed Coat	10	Top 1/3 of plant	0.8
Embryo	28	Middle 1/3 of plant	1.7
Cotyledon	43	Lower 1/3 of plant	2.5

[a]Grown from seeds which contained 44 ppm Mo.
[b]Sampled from plants receiving 70 g Mo/ha as foliar spray.

the previous year with 0, 0.22, 0.44, 0.88, 1.32 and 2.64 kg Mo/ha. Soybean grain Mo from the residual application was considerably higher than barley or wheat grain.

Influence of Mo on Other Soybean Constituents

Additional plant components which have been reported to be affected by Mo applications are seed size, nitrogen, protein, oil, nodule size and number, leaf chlorophyll, inoculants and nucleic acids. Lee et al. (30) reported significantly greater numbers of pods, seeds and seed weight when 200 g/ha of Mo were applied. In addition, total N of tissue at full bloom-early pod stage and crude protein of seeds were increased by Mo applications. Parker and Harris (37) and Sedberry et al. (41) have also shown an increase in seed weight when soybeans respond to Mo application. Numerous other workers (10,21,22,30,35-37,42) have reported increased nitrogen contents of tissues with Mo applications. Parker and Harris (35), Boswell and Anderson (10), and Lee et al. (30) have shown increased bean protein with Mo application but an inverse relationship with oil content of soybean seeds (Figure 2). Generally, the protein increase will be 3 to 4% of each unit decrease in oil content. However, Lee et al. (30) reported a soybean seed protein increase of 8.7% while seed soil was decreased 5% with Mo seed treatments.

Smith and Connell (42) showed a highly significant increase in leaf chlorophyll when 35 g Mo/ha were added as seed treatment. This increase was from approximately 2% for 0 Mo to 4% for plants with added Mo. Hashimoto and Yamasaki (23) reported that nodules on Mo-treated soybean plants were characteristically larger size, had lower water content and higher Mo content than non-treated plants.

Studies reported by Giddens (18) indicate that the addition of 7 or 28 g of molybdenum compounds (sodium molybdate, ammonium molybdate, molybdic oxide or 'Moly-Gro') to 140 g of inoculant applied to 2700 kg of soybean seed significantly reduced nodulation of soybeans grown in the greenhouse. Using the same compounds, Johnson et al. (27) evaluated seed treatment effects on the germination of various soybean varieties. They

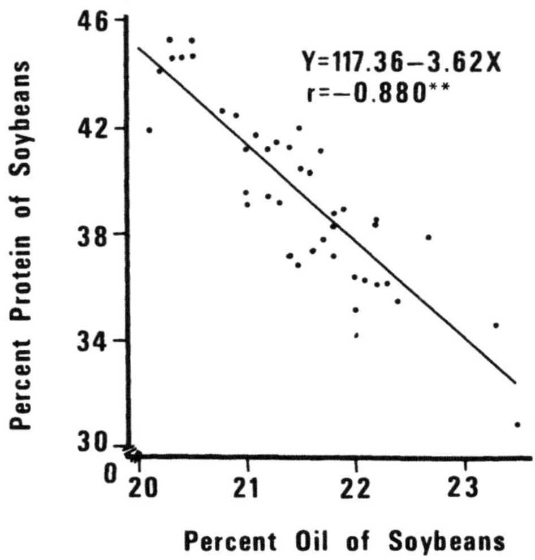

Figure 2. Regression of soybean seed protein on soil when 217 g/ha Mo were applied (10).

found that neither source nor rate of Mo, up to 224 g/2700 kg of seed caused germination losses. Pozsár (39) reported that Mo, as ammonium molybdate in a concentration of 50 ppm, stimulated nucleic acid synthesis in soybean leaves through enhancing the intensity of nitrate reduction.

FACTORS AFFECTING SOYBEAN YIELD RESPONSES

Factors that may influence the soybean yield response to Mo applications are total and available soil Mo, soil pH, rates of application, methods and timing of Mo application. Studies by Parker and Harris (37) and Hashimoto and Yamasaki (23) showed no difference in yield response due to sources of Mo.

Soil Mo and pH

Sedberry et al. (41) obtained significant yield responses to added Mo on Louisiana soils that varied from 0.04 ppm to 0.22 ppm of extractable Mo; however, they were unable to establish a critical concentration level of available Mo in the soil. They found that yield responses were influenced by soil pH and were able to relate Mo concentration of soybean leaves to soil pH. However, concentrations of available Mo in the soil could not be well related to soil pH or Mo content of leaf tissue (Table 5).

The relationship between soybean yields and soil pH with and without applied Mo was depicted by Parker and Harris (35) as shown in Figure 3.

Table 5. Influence of soil pH on the concentration of Mo in the soil and leaves of soybeans grown on a silt loam soil (41).

Soil pH	Mo Concentration	
	In Soil	In Leaves
	– ppm –	
5.1	0.26	0.41
6.0	.25	.40
6.7	.29	.50
7.2	.27	.65
7.4	.29	.86
7.5	.29	1.30
LSD – 5%	ns	0.10

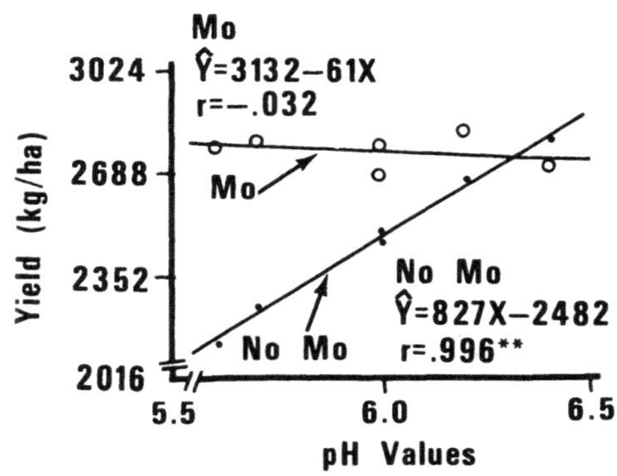

Figure 3. Relationship between soybean yield and soil pH with and without Mo (35).

A high significant correlation (r = 0.996) was found between the mean yields of treatments without added Mo and soil pH. However, this relationship became nonsignificant (r = -0.032) when Mo was added at a rate of 225 g/ha.

Parker and Harris (35) also compared the effects of Mo applied by two methods and dolomitic limestone rates on yield of soybeans in Georgia (Table 6). Molybdenum, without lime, increased yields 46% by foliar application and 55% by seed treatment. However, the yield difference in favor of seed treatment was significant only when averaged for all lime treatments. Lime, without Mo, increased yields 31 and 50% at the 2.24 and 4.48 t/ha rates, respectively. Lime had no significant effect on yield when Mo was applied and yields were not influenced by Mo when more than 2.24 t/ha were applied. They concluded from this study, along with other experiments, that added Mo was equivalent to 4.48 t/ha of lime in promoting yield increases.

Table 6. Effects of molybdenum and lime on the yield of soybeans grown on a loam soil (initial pH of 5.5) in Georgia (35).

Treatment	Dolomitic Limestone (t/ha)					Average
	0	2.24	4.48	6.72	8.96	
– g/ha –			*– yield (kg/ha)* –*			
0	2036 b	2675 b	3050 a	3004 a	3152 a	2782 c
224 (foliar)	2977 a	3125 a	3050 a	3084 a	3158 a	3078 b
224 (seed)	3158 a	3266 a	3266 a	3353 a	3326 a	3273 a
Average	2722	3024	3125	3145	3212	

*Data in the same column followed by the same letter and data in rows underscored by the same line do not differ significantly at the 5% probability level.

Sedberry et al. (41) reported significant soybean yield increases from the application of Mo in 11 of 22 experiments. Where responses occurred, soil pH values ranged from 4.9 to 6.0. Yield increases above nontreated plots ranged from 87 to 786 kg/ha with the highest yield occurring on a Gallion vfsl with a pH of 4.9. They indicated that applications of seeds, foliar and soil Mo treatments proved to be equally effective in correcting Mo deficiency on acid soils, although various Mo rates were used, depending on the treatment method. More Mo is required to attain maximum yields when applied to the soil than when applied to the foliage. Parker and Harris (37) observed that 885 g/ha of row-applied Mo were required to be equivalent to 28 g/ha of foliar-applied Mo or a 32-fold difference.

Utilizing an equation proposed by Mueller et al. (33), Sedberry et al. (41) showed that a response to applied Mo was usually obtained when the value for soil pH + (10x the ppm of ammonium oxalate-oxalic acid extractable Mo content of the soil) was less than 7.5 (Table 7). There were two exceptions to the 7.5 Mueller value; the Calhoun sil with a pH of 4.9 and the Yahola vfsl with a pH of 7.1. The Calhoun sil contained 0.21 ppm of extractable Mo while the Yahola soil was alkaline.

On moderately acid soils (pH 5.1 to 5.6) Hagstrom and Berger (21) and Parker and Harris (35) reported that nitrogen (N) deficiency symptoms were observed on soybean plants grown on plots not treated with lime or Mo. To further assess Mo and N rates on acid soils, Parker and Harris (36) evaluated the influence of N and Mo on yield of non-nodulating and nodulating soybeans for 3 yr. Yields from the final year (1968) of this study are shown in Table 8. Molybdenum did not increase yields with the non-nodulating soybeans. For the nodulating soybeans, significant responses to Mo were obtained up to 67 kg/ha of N. Average yields for the Mo treated and nontreated non-nodulating soybeans were increased for all increments of N

Table 7. Soil pH and extractable Mo content of several Louisiana soils used to predict the soybean yield response to Mo application (41).

Soil Type	Soil pH	ppm Ext. Soil Mo	Mueller Value[a]	Measured Response
Calhoun sil	4.9	0.21	7.0	ns[b]
Gallion vfsl	4.9	0.04	5.3	*[c]
Alligator C	5.1	0.42	9.3	ns
Calhoun sil	5.1	0.17	6.8	*
Gallion vfsl	5.3	0.04	5.7	*
Mantachie fsl	5.4	0.22	7.6	ns
Jeanerette sil	5.4	0.14	6.8	*
Perry C	5.4	0.23	7.7	*
Crowley sil	5.5	0.14	6.9	*
Crowley sil	5.6	0.18	7.4	*
Acadia sil	5.7	0.05	6.2	*
Crowley sil	5.9	0.14	7.3	*
Robinsonville vfsl	6.2	0.18	8.0	ns
Commerce sil	6.4	0.22	8.6	ns
Yahola vfsl	7.1	0.03	7.4	ns
Norwood sil	7.6	0.04	8.0	ns

[a] Mueller value = soil pH + (10 x ppm extractable Mo).
[b] ns denotes a nonsignificant difference in yield.
[c] * denotes a significant difference in yield at the 5% probability level.

Table 8. Influence of N and Mo on yield of non-nodulating and nodulating soybeans (36).

	Yield*							
	Non-nodulating (kg N/ha)				Nodulating (kg N/ha)			
Mo Rate	0	67	134	201	0	67	134	201
– g/ha –	*– kg/ha –*							
0	1714 a	2661 a	2997 a	3145 a	2506 b	2762 b	3084 a	3111 a
34	1620 a	2574 a	2937 a	3158 a	3051 a	3111 a	3226 a	3132 a
Average	1667	2618	2967	3152	2778	2936	3155	3122

*Data in the same column followed by the same letter and data underscored by the same line in a row do not differ significantly at the 5% level by DMRT.

except the 134 kg/ha rate without Mo. Also, Mo without added N increased soybean leaf N. They concluded that N efficiency may be improved by the addition of Mo to nodulating soybeans. They also indicated that the primary function of Mo was to correct a N deficiency and that sufficient soil Mo was available for nitrate reduction but not for symbiosis.

Sources, Rate, and Time of Mo Application

Parker and Harris (37) reported on a 3-yr study with sources and rates of Mo on soybean yield (Table 9). The study was conducted on loam soils with pH values ranging from 5.0 to 5.4, depending on the year and soil. Yields were increased 58% at the highest yield level but 83% of the total increase occurred with the lowest Mo rate (17 g/ha). Average of sources indicated no additional yield increases above the 69 g Mo/ha rate. There were no differences in yield response due to source of Mo.

A comparison of seed treatment versus foliar application of Mo to soybeans by Parker and Harris (35) indicated that seed treatment was superior. However in a subsequent report, these investigators (36) found little or no difference between the two methods. In addition, soil-applied Mo was not different from the seed or foliar applications. Boswell and Anderson (10) showed that time of spray application of Mo was important (Table 10). These data indicate that delayed foliar applications of Mo after the R3 (17) growth stage do not allow maximum yield responses. Little effect is noted on N concentration of leaf petiole or total oil and protein content of seed. When foliar sprays were applied at the proper growth stage, they found no difference in seed and foliar applied Mo relative to yield, protein or oil effects on soybeans. On a Montevallo silt loam Boswell and Anderson (10) reported a 75% yield increase from seed applied Mo.

Table 9. **Effect of sources and rates of molybdenum on soybean yields, 3-yr average (37).**

		Source			
Rate*	Chelate	Ammonium Molybdate	Sodium Molybdate	½ Chelate + ½ Sodium Mo	Average
– g/ha –			*– yield (kg/ha)*† *–*		
00	1532 b	1539 b	1620 c	1499 b	1546 c
17	2285 a	2312 a	2312 b	2251 a	2292 b
69	2339 a	2446 a	2433 ab	2278 a	2372 ab
138	2426 a	2500 a	2446 ab	2379 b	2439 a
277	2426 a	2352 a	2567 a	2439 a	2446 a
Average	2204	2231	2278	2171	

*Rate of each source applied as seed treatment.

†Data in the same column followed by the same letter and data in rows underscored by the same line do not differ significantly at the 5% probability level.

Table 10. Yield total oil and protein of seeds of soybeans as influenced by seed applied and spray application of Mo at various periods (10).

Mo Treatment Period*	Yield†	Protein	Total Oil
	– kg/ha –	*– % –*	
Check	1568 c++	38.6	21.8
Seed treated preplant	1907 a	39.6	21.6
Spray 20-cm ht (V4)	1903 a	39.8	21.2
Spray pre-bloom (R1)	1890 a	39.8	21.5
Spray bloom (R3)	1660 b	40.0	21.3
Spray early pod (R5)	1716 b	39.7	21.5

*Mo source and rate—217 g Mo/ha as sodium molybdate.

†Average from 7 sites with 5 replicates.

++Data in the same column followed by the same letter do not differ significantly at the 5% probability level.

SUMMARY

Soybean yield responses to supplemental Mo application have been reported in the Far East (Japan, China, Taiwan), Europe, and at least 12 states of the U.S. Generally, the most frequently observed responses in the U.S. have occurred east of the Mississippi River where rainfall is moderate to heavy and soils tend to be acid. Yield increases have varied with the largest increases being greater than 75% from the application of Mo to soybean seed or as foliar sprays. Yield responses have been obtained with seed, foliar spray or soil applications of Mo. However, rates must be higher for soil applications than for seed or foliar spray treatment. Responses have been obtained with seed treated Mo rates as low as 17 g/ha, while soil application rates may be greater than 800 g/ha. Molybdenum applications often increase seed protein and decrease seed oil content.

With nodulating soybeans, nitrogen efficiency may be improved by the addition of Mo. The primary function of Mo may be to correct a nitrogen deficiency when sufficient soil Mo is available for nitrate reduction but not for symbiosis.

Generally, soils which may respond to Mo applications are acid with pH levels below 6.2, inherently low in total Mo content, coarse textured to silt loams, relatively low in organic matter, and severely eroded and/or heavily weathered. Applications of limestone to maintain soil pH above 6.2 may effectively correct or prevent molybdenum deficiencies. Molybdenum can be used to reduce the lime requirement of soybeans.

Critical tissue levels of Mo have not been well established although most leaf tissue contents have been less than 0.20 ppm where yield response occurred. Although data are limited, Mo concentrations of soybean plant components, as influenced by Mo application are seed > nodules > roots > stems > leaves.

NOTES

Fred C. Boswell, Department of Agronomy, University of Georgia, Georgia Station, Experiment, Georgia 30212.

REFERENCES

1. Allaway, W. H. 1968. Agronomic controls over the environmental cycling of trace elements. Adv. Agron. 20:235-271.
2. Anthony, J. L. 1967. Fertilizing soybeans in the hill section of Mississippi. Miss. Agri. Exp. Sta. Bull. 743.
3. Arnon, D. I. and P. R. Stout. 1939. Molybdenum as an essential element for higher plants. Plant Physiol. 14:599-602.
4. Berger, K. C. 1962. Micronutrient symposium—Micronutrient deficiencies in the United States. J. Agric. Food Chem. 10(3):178-181.
5. Bergersen, F. J. 1971. Biochemistry of symbiotic nitrogen fixation in legumes. Annv. Rev. Plant Physiol. 22:121-140.
6. Barshad, I. 1948. Molybdenum content of pasture plants in relation to toxicity to cattle. Soil Sci. 66:187-195.
7. Bertrand, D. 1962. De l'emploi pratique du molybdene comme engrais pour la culture du Soja. C. R. Acad. Sci. Paris 255:2814-2816.
8. Bortels, H. 1930. Molybdenum as a catalyst in the biological fixation of nitrogen. Arch. Microbiol. 1:333-342.
9. Bortels, H. 1937. The effect of molybdenum and vanadium fertilization on leguminosae. Arch. Microbiol. 8:13-26.
10. Boswell, F. C. and O. E. Anderson. 1969. Effect of time of molybdenum application on soybean yield and on nitrogen, oil, and molybdenum contents. Agron. J. 61:58-60.
11. Cheng, B. T. and C. J. Ouellette. 1973. Molybdenum as a plant nutrient. Soils and Fert. 36(6):207-215.
12. Chu, C., C. W. Liang and E. F. Chen. 1963. Effect of minor elements on the growth and yield of soybean in some important soil types of Northeastern China. (Chinese, English summary) Acta. Pedol. Sin 11:417-425.
13. Cox, F. R. and E. J. Kamprath. 1972. Micronutrient soil tests, pp. 289-317. In J. J. Mortvedt et al. (Eds.) Micronutrients in agriculture. Soil Sci. Soc. of Am., Madison, Wis.
14. Davies, E. B. 1956. Factors affecting molybdenum availability in soils. Soil Sci. 81:209-221.
15. DeMooy, C. J. 1970. Molybdenum response of soybeans [*Glycine max* (L.) Merrill] in Iowa. Agron. J. 62:195-197.
16. Evans, H. J. 1956. Role of molybdenum in plant nutrition. Soil Sci. 81:199-208.
17. Fehr, W. R. and C. E. Caviness. 1977. Stages of soybean development. Coop. Ext. Ser. Iowa State Univ., Ames, Iowa. Spec. Rept. 80.
18. Giddens, J. 1964. Effect of adding molybdenum compounds to soybean inoculant. Agron. J. 56:362-363.
19. Gurley, W. H. and J. Giddens. 1969. Factors affecting uptake, yield response, and carryover of molybdenum in soybean seeds. Agron. J. 61:7-9.
20. Hagstrom, G. R. 1977. A closer look at molybdenum. Fert. Sol. 21 (July-Aug.).

21. Hagstrom, G. R. and K. C. Berger. 1963. Molybdenum status of three Wisconsin soils and its effects on four legumes. Agron. J. 55:399-401.
22. Harris, H. B., M. B. Parker and B. J. Johnson. 1965. Influence of molybdenum content of soybean seed and other factors associated with seed source on progeny response to molybdenum. Agron. J. 57:397-399.
23. Hashimoto, K. and S. Yamasaki. 1976. Effects of molybdenum application on the yield, nitrogen nutrition and nodule development of soybeans. Soil Sci. Plant Nutr. 22(4):435-443.
24. Hawes, R. L., J. L. Sims, and K. L. Wells. 1976. Molybdenum concentrations of certain crop species as influenced by previous applications of molybdenum fertilizer. Agron. J. 68:217-218.
25. Hodgson, J. F. 1963. Micronutrients in soils. Adv. Agron. 15:119-154.
26. James, D. W., T. L. Jackson and M. E. Harward. 1968. Effect of molybdenum and lime on the growth and molybdenum content of alfalfa grown on acid soils. Soil Sci. 105:397-402.
27. Johnson, B. J., H. B. Harris, M. B. Parker, and E. E. Winstead. 1966. Effects of molybdenum seed treatments on germination of soybeans. Agron. J. 58:517-518.
28. Johnson, C. M. 1966. Molybdenum, pp. 286-301. In H. D. Chapman (Ed.) Diagnostic criteria for plants and soils, Univ. of Calif., Berkeley, CA.
29. Lavy, T. L. and S. A. Barber. 1963. A relationship between the yield response of soybeans to molybdenum applications and the molybdenum content of the seed produced. Agron. J. 55:154-155.
30. Lee, L. T., W. T. Tang, and W. F. Tasi. 1967. Effects of various liming rates and molybdenum seed treatment on soybean yield, chemical composition and uptake of N, P, K, Ca and Mo. (Chinese, English summary). J. Agri. Assn. China Taiwan 59:65-76.
31. Long, O. H., J. R. Overton, T. McCutchen, and L. M. Safley. 1965. Molybdenum on soybeans. Tenn. Farm and Home Sci. Prog. Rept. 54:11-13.
32. Martinez, J. F., W. K. Robertson, and F. G. Martin. 1977. Soil and seed treatments with lime, molybdenum, and boron for soybeans [*Glycine max* (L.) Merr.] production on a Florida spodosol. Soil and Crop Sci. Soc. of Fla. Proc. 36:58-60.
33. Mueller, K. E., E. Wuth, B. Witter, R. Ebeling, and W. Bergmann. 1964. Molybdenum supply of Thuringian soils and the influence of molybdenum fertilization on yield, crude protein, and mineral content of alfalfa. Albrecht-Thaer Arch. 8(4-5):353-373. In L. G. Albrigo, R. C. Scafranek, N. F. Childers (Eds.) The role of molybdenum in plants and soils. A supplemental bibliography with abstract. Hort. Dept., Rutgers—The State Univ. New Brunswick, N.J. 285 pp.
34. Nicholas, D. J. D. 1957. The function of trace metals in the nitrogen metabolism of plants. Ann. Bot. (N.S.) 21(84):587-598.
35. Parker, M. B. and H. B. Harris. 1962. Soybean response to molybdenum and lime and the relationship between yield and chemical composition. Agron. J. 54:480-483.
36. Parker, M. B. and H. B. Harris. 1977. Yield and leaf nitrogen of nodulating and non-nodulating soybeans as affected by nitrogen and molybdenum. Agron. J. 69:551-554.
37. Parker, M. B. and H. B. Harris. 1978. Molybdenum studies on soybeans. Ga. Agri. Exp. Sta. Res. Bul. 215.
38. Peterson, N. K. and E. R. Purvis. 1961. Development of molybdenum deficiency symptoms in certain crop plants. Soil Sci. Soc. Am. Proc. 25:111-117.
39. Pozsár, B. I. 1965. Effect of molybdenum on the synthesis of nucleic acid in soybean leaves. Acta. Agr. Hung. 14:301-308.

40. Rogers, H. T., F. Adams and D. L. Thurlow. 1973. Lime needs of soybeans on Alabama soils. Ala. Agri. Exp. Sta. Bul. 452.
41. Sedberry, J. E., Jr., T. S. Dharma putra, R. H. Brupbacher, S. A. Phillips, J. G. Marshall, L. W. Sloane, D. R. Melville, J. I. Rabb, and J. H. Davis. 1973. Molybdenum investigations with soybeans in Louisiana. La. State Agri. Exp. Sta. Bul. 670.
42. Smith, H. C. and J. Connell. 1973. Lime and molybdenum in soybean production. Tenn. Farm and Home Sci. 85:28-31.
43. Sokolova, M. F. 1966. Results of experiments using molybdenum fertilizers for soya in the Far East. (R) Agrokhimiya 5:88-95.
44. Thompson, L. F. and G. W. Hardy. 1964. Soybean response to applications of molybdenum and limestone. Ark. Farm Res. XIII (1)2.

SOYBEAN HARVESTING EQUIPMENT: RECENT INNOVATIONS AND CURRENT STATUS

W. R. Nave

The rising need for protein in many parts of the world has encouraged an increase in the worldwide production of soybeans during recent years. The estimated production increased from 28.0 to 69.1 million metric tons from 1962 to 1976 (1). Major producers in 1976 were the U.S., contributing 50% of the world's total production; the People's Republic of China, contributing 15%; and Brazil, contributing 14%. In the U.S. alone, over 20 million ha (49 million acres) of soybeans were harvested.

Soybeans have become the leading cash crop in the U.S., but they are still planted and harvested with equipment designed primarily for other crops. Since soybeans are physically unlike other major crops, the use of this equipment causes excessive damage and field losses.

Harvesting is one of the most critical steps in profitable soybean production. Surveys by Lamp et al. (1) and Nave et al. (16) showed that the average soybean producer loses 8 to 10% of his soybean crop in the harvesting operation. More than 80% of this loss is caused by the combine header. Some of the factors that influence header loss are choice of variety, row width, plant population, cultural practices, harvest conditions, combine efficiency, and operator skill.

HISTORY OF SOYBEAN HARVESTING

The earliest harvester designed specifically for soybeans was a two-wheeled, horse-drawn machine that straddled the bean row (19). This special harvester was used in Virginia and North Carolina about 1920, but was never used frequently in the North Central States. Harvesting losses ranged from a low of 20% to a high of 60%. In areas where small grains are grown, the binder or mower and thresher were used for soybean harvesting. When soybeans were cut with the binder or mower and then threshed, harvesting losses ranged from 16 to 35% of the total yield and averaged 25%.

The grain combine harvester was first used for soybeans in the mid-twenties and has been a major factor in the expansion of soybean production. A survey of 12 combines operated in Illinois in 1927 (20) provided the following information: "Losses varied from 5.7 to 22.6%, with an average loss of 11.4%. The greatest loss occurred back of the cutterbar, because of the low pods, thin stands, low branches, down beans, and uneven ground. Where low plant populations occurred, or where soybeans were planted in rows, the pods formed very low on the stems; in many cases, only 2 to 5 cm (1 to 2 in.) above the ground, and often with the pods touching the ground. With very level ground, some of the combines were able to cut as low as 7 to 10 cm (3 to 4 in.), but even at that low cutting height one or two pods may be left on much of the stubble. The header loss averaged 9.8%, while the threshing and separating loss was about 1.8%."

The first major breakthrough in significantly reducing soybean harvesting loss was the introduction of the floating cutterbar attachment developed by Horace D. Hume and J. Edward Love in 1930. In 1933 Hume successfully demonstrated the floating cutterbar and reel in soybeans at Champaign, Illinois, but acceptance in the soybean region was slow. As an example, only 12 flexible floating cutterbars and 12 pickup reels were delivered to farms in Illinois in 1934, and the 12 cutterbars were returned as having no apparent value (24).

Until about 1970, little progress was made in reducing soybean harvesting losses. In recent years, attachments such as floating cutterbars with hydraulic height control and pickup reels with hydraulic height and speed control have become common features on many combines used for soybean harvesting. At least one combine manufacturer has considered the option of automatically controlling reel speed relative to ground speed. Harvesting loss can be reduced 25% or more with proper combine adjustments and the use of these attachments.

HARVESTING LOSS EVALUATION

For several years, agricultural engineers have been evaluating methods of reducing soybean harvesting losses. Initially, the research was conducted primarily at universities in the soybean-producing states. More recently, combine manufacturing companies have undertaken research and equipment development projects. As a result combine headers designed specifically for soybeans are now appearing on the market.

A combine must adapt to specific harvesting needs to achieve high harvesting efficiency. With soybeans the combine must cut close to the ground to reduce stubble and shatter losses. The soybean plants must also be handled gently to reduce losses from shattering. The plant material must flow positively and evenly through the combine's threshing and separating mechanisms. Methods for containing shattered soybeans and picking up lodged stalks are also important.

Studies by researchers at Ohio, Illinois and Iowa proved that major gains in harvesting efficiency must come through the reduction of gathering losses, particularly shatter, lodging and stalk losses. Almost all gathering losses come from knife and reel action. Aggressive handling by a header's cross auger and by the slat conveyors in the feeder housing can thresh a substantial percentage of the soybeans before they enter the combine cylinder. Under such conditions, the slope and tightness of the header's deck and a positive feeding mechanism to the combine cylinder become critical in reducing harvesting losses.

OHIO STUDY

Probably the most extensive study of soybean harvesting was made by Lamp et al. in Ohio (12). In tests conducted from 1956 to 1960, they found that total harvesting losses varied from 9.8 to 19.3%. Gathering losses represented 80% of all losses and consisted of 55% shatter loss, 28% lodging and stalk losses, and 17% stubble loss. They also found that preharvesting loss was negligible when harvesting was completed with kernel moisture above 10%.

The Ohio study proved that operating a standard bat reel 38 to 51 mm (15 to 20 in.) above the ground at a reel speed index of 1.25 resulted in minimal shatter losses. It also showed that gathering losses increased as the width of cut was increased, a problem that was attributed to rigidity of the header. As ground speed increased from 4 to 8 km/h (2.5 to 5 mph) total machine losses increased over 50%. Most of this increase was at speeds greater than 5 km/h (3.2 mph).

ILLINOIS AND USDA STUDY

In 1968 a cooperative research project was initiated between the U.S. Department of Agriculture, SEA/AR, and the Agricultural Engineering Department at the University of Illinois. One of the project goals was to analyze soybean harvesting losses, and subsequent research was direct toward improving harvesting techniques and equipment.

Research at the University of Illinois from 1968 to 1971 helped to identify several factors that cause excessive soybean harvesting losses (16). Weeds cause a reduction in soybean yield; however, they will not cause significant combine losses when care is used at harvest. Ground speed must be reduced when harvesting weedy soybeans unless the weeds are desiccated before harvest begins.

Yield is not affected significantly by population and row spacing in most cases. Higher populations, however, can cause excessive header losses because of lodging. The cultivation practice of throwing soil into the row for weed control may reduce lodging but usually increases harvesting losses since plant material will be cut and lost between the ridged rows. Low population in 20-cm (8-in.) rows may increase harvesting losses from the lower podding.

Field investigations were conducted to isolate and evaluate the percentage of the gathering losses attributable to each component of the header when soybeans were harvested with a combine (8). Of the three major components of the header, the cutterbar caused 81% of the average total header loss in the three varieties tested, the auger caused 13%, and the reel caused the remaining 6%. From this study it was concluded that improving the cutterbar design should be the most effective way to reduce gathering losses. Reel-speed studies showed no significant difference in header losses when reel index was 1.2 to 1.7. Shatter losses were somewhat lower when the reel tines were set at 20 cm (8 in.) above the cutterbar than when they were set at 7.6 cm (3 in.) above the cutterbar.

A 1971 survey of 35 combines operating in four areas of the soybean-producing states showed that harvesting losses ranged from 3.5 to 12.7% (16). Maintaining a low stubble height was the most critical factor in reducing harvesting losses. Careful combine adjustments were observed to be an important factor in obtaining low threshing and separating losses.

More recent studies by Nave and Hoag (13) revealed that conventional cutterbars cause soybean shatter from excessive acceleration during the cutting process. Modification of cutterbars with lipless guards or sickle and guard spacings less than 7.6 cm (3 in.) were effective in reducing shatter losses. Research is continuing to evaluate new methods of cutting to further reduce harvesting loss.

IOWA STUDY

Research at Iowa State University by Quick (21) confirmed the loss values observed at Ohio and Illinois. With the aid of a laboratory test stand and high-speed photography, he determined several factors that influenced header losses. Slow knife speed was found to be the prime limitation on combine ground speed and capacity in soybeans. Reducing guard and cutterbar spacing from 7.6 cm (3 in.) to 3.8 cm (1.5 in.) reduced header losses significantly. Quick concluded from the laboratory and field studies that any attachment that enabled the operator to lower the cutting height and to control the plants during cutting by reducing stalk slippage would be an aid in reducing header losses.

DEVELOPMENT OF IMPROVED EQUIPMENT

Public Institutions

Two basic approaches have been investigated for soybean harvesting. One has been to modify conventional grain platforms; and the other, to develop row-crop harvesting equipment.

In the first approach, research by USDA and state agricultural engineers at the University of Illinois has concentrated on the use of air-jet guards on a header equipped with a floating-cutterbar attachment. The results from 4 yr of testing indicate that the use of air-jet guards is a practical way to keep losses below 4% of the yield, even if the soybean moisture at harvest is less than 12% (17). These tests also showed that the air jets were more effective in reducing harvesting losses in rows spaced at 20 cm (8 in.) than in rows spaced at 76 cm (30 in.).

In work involving the second approach, the development of row-crop harvesting equipment, a student design class at the University of Illinois designed and built a row-crop header in 1969 (15). The row headers floated individually, which theoretically should provide a low cutting height. The design of the header mechanism involved three functional requirements: (a) cutting the plants; (b) conveying the plants, and (c) feeding the plants into the threshing area. A circular cutting mechanism was considered, but a conventional reciprocating cutting mechanism was actually built and tested. Gathering chains equipped with tines and paddles were used for moving material into the cross auger. Considerable difficulty was encountered because material remained on the tines and caused excessive shatter loss.

A puller-header for harvesting soybeans was developed at Purdue University in 1972 (32). Their row crop device used wide-tread, small-diameter rubber tires to pull stalks with or without cutting off the root system. The reduction in harvest losses with the puller unit was about 70% compared to a conventional header, and about 50% compared to a floating cutterbar header (5). Two primary problems with the puller unit were the excessive soil that remained on the plant roots and the high degree of steering accuracy required for satisfactory operation.

Since 1968, agricultural engineers at the University of Minnesota have developed and tested two soybean harvesting devices (4). One device consisted of row-oriented rotary cutters with cylindrical rollers or belts for gathering the plants. Losses were reduced 41 to 67% less than with the conventional header in 1969 tests (28). Their studies with old pull-type, draper combines prompted investigation of the second device, a full-width floating-draper extension. Floating the rigid cutterbar in the lateral plane allowed the cutterbar to follow the ground contour independently of the header frame. Losses were 14 to 59% less with the draper extension than with the conventional floating cutterbar (29).

Industry

An integral, flexible-cutterbar header developed and introduced by John Deere (JD) in 1975 is a modification of the conventional grain platform design (3). It was designed to: (a) Reduce the incline between the knife and auger, (b) Improve the durability of the sickle and cutterbar by floating the wobble drive unit with the cutterbar, and (c) Improve the functional performance of floating-cutterbar headers when harvesting crops other than soybeans. A continuous skid shoe spans the full width of the header, so the combine operator is not restricted to driving on rows and a particular row spacing is not necessary. As a solution to the long-standing problem of crop dividing in soybean harvesting, long floating dividers were designed for the new header.

With the flexible-cutterbar header, total loss was 30% less than with the conventional floating cutterbar in tests conducted by Deere and Company. The reduction was greatest in the amount of shatter loss. During a 2-yr study, the average loss with the flexible cutterbar was 5.5% (3).

A low-profile, row-crop header, was designed and introduced by Deere and Company in 1974 (2). The individual row units were designed to pivot independently about the common drive shaft. Adjustable skid shoes on each side of the row maintained a low cutting height. The adjustable load spring provided with each row unit minimized soil pressure on the shoes. A rotary-knife cutting mechanism was chosen because, during cutting, its impact on the plant was less severe than that of a reciprocating knife and it caused less plant agitation. A pair of meshed, corrugated gathering belts extended ahead of the knife to grip the plant just before it was cut and then extended rearward to continuously convey the cut stalk into the auger. The operator could adjust the speed of the belts and cutting mechanism to match the ground speed of the combine. A closed trough under the gathering belts collected the loose beans, which were then swept up into the auger by the stalks as they were transported by the belts. Lodged and leaning plants were lifted and guided into the gathering belts and knife by low-profile points. Bichel et al. (2) reported that, in a 2-yr test, harvesting losses with this unit were below 4%.

Another approach to reducing soybean harvesting loss was the introduction of a narrow-pitch floating cutterbar by White Farm Equipment Company in 1977. The basic research on the design was conducted at Iowa State

University by Quick and Buchele (23). The sickle and guard spacing was reduced from 7.6 cm (3 in.) to 3.8 cm (1.5 in.). The cutterbar was mounted on a flexible barback with a flat plate knife drive. Sickle sections and guards were manufactured in multiples, and sections were bolted on (instead of rivitted) for easy field servicing (24).

The primary goal in reducing cutterbar pitch by one-half (7.6 to 3.8 cm) was to reduce stem movement during cutting. Reduced stem movement, at the same knife speed, was expected to result in: (a) lower accelerations; (b) less seed shatter; (c) shorter stubble and, subsequently, lower loss; (d) higher forward speed capability (24,25). The results indicated a 50% higher forward speed, lower losses, and a reasonably flat speed-to-loss curve that was superior to that of the conventional floating cutterbar with a sickle and guard spacing of 7.6 cm (3 in.). Losses as low as 1% in narrow-row soybeans were reported (25).

Several other campanies have also responded with improved header designs for soybean harvesting (24). Lynch Manufacturing Company produces a vertical drum reel attachment for row crops such as soybeans. Extended dividers lift lodged plants into the reel drums. Several companies, including Hart-Carter Company, J. E. Love Company and Hiniker Company, produce floating-cutterbar attachments that have conventional knife sections and guards. At least three companies produce headers which contain a flexible floating cutterbar as an integral part. The John Deere integral flexible cutterbar has been discussed previously.

In 1976 Avco New Idea introduced an integral floating cutterbar with a low-profile, continuous-steel plate between the knife and the platform. The cutterbar and deck float on a series of terrain sensors that follow ground contours. Decreased gathering losses and increased ground speeds are claimed (4).

In 1977 International Harvester (IH) introduced a new grain header featuring a 15-cm (6-in.) integral-flex-range floating cutterbar with full-width skid shoes and automatic height control (9). The performance of the IH and Avco New Idea integral floating cutterbars should be similar to that of the JD integral flexible cutterbar; however, data are not available for these comparisons.

The bean pan concept was first manufactured by inventor Frank Baker, and by Hesston Corporation for one season (4). Pans are suspended ahead of the cutterbar between the rows to catch shattered pods, beans, and cut stalks. Collected trash, pods, and beans are swept into the platform by brushes attached to the reel bats. More complete emptying of the pans is achieved by raising the head. Loss reductions of 72% over the conventional header have been measured (30).

HEADER PERFORMANCE COMPARISONS

In 1976 a header loss performance comparison was made between the JD integral flexible-cutterbar header equipped with air-jet guards, the JD row-crop header, and a conventional floating-cutterbar header equipped with air-jet guards (14). The JD integral flexible-cutterbar header and the conventional floating-cutterbar header were operated with and without the air assist engaged. The floating-cutterbar header with the air-jet guards but without the air assist engaged was used as the standard for comparison.

All comparisons were made with the air-jet guards in place since it was not feasible to remove or adjust the guards between the test runs. Therefore, the guards provided some lifting effect on plant material even when the air assist was not engaged. The platform headers were compared in both 76-cm (30-in.) and 18-cm (7-in.) rows in four varieties of soybeans. The row-crop header could be tested only in 76-cm (30-in.) rows.

The losses with all of the new designs were less than with the standard floating cutterbar. When operating in 76-cm (30-in.) rows, the row-crop header reduced total header loss by 84% (Table 1). Total header loss with the flexible-cutterbar header without air was 56% less and with air, 61% less than with the standard floating cutterbar. Operation of the air assist on the floating cutterbar reduced losses by about 45%, which was similar to the value found by Nave and Yoerger (17).

Losses were reduced also when operating in 18-cm (7-in.) rows. Total header loss was 60% less when the flexible cutterbar header was used, with or without air, than when the standard floating-cutterbar header was used (Table 2). Total header loss was 45% less when the floating cutterbar was used with the air assist in operation than when the standard cutterbar was used with the air assist turned off.

The total header loss with both types of platform headers was 30% less for 18-cm (7-in.) rows than for 76-cm (30-in.) rows. The row-crop header was the most efficient header under the test conditions in 1976 (Table 1). However, the row-crop header cannot be used in row spacings less than 76-cm (30-in.). The row-crop header appeared to handle the soybean plants more gently than the platform headers if the speed of the gathering belts and rotary cutters was properly coordinated with the ground speed of the combine.

No significant advantage was found for the use of the air assist with the flexible-cutterbar header; however, a final conclusion cannot be drawn from this experiment since a flexible cutterbar with the air guards removed was not tested. There was also some question about the effectiveness of the air assist on the flexible-cutterbar header since the readings indicated that air velocity from some of the jets was lower than desired.

Table 1. Effect of header design on harvesting loss in 76-cm rows.[a]

	Loss (%)[b]		
	Shatter	Stalk	Total Header[c]
Flexible cutterbar	1.7 a	1.9 a	3.8 a
Flexible cutterbar w/air assist	1.8 a	1.5 ab	3.4 a
Floating cutterbar (standard)	6.2 b	2.3 c	8.7 b
Floating cutterbar w/air assist	2.8 c	2.0 a	4.9 c
Row-crop	0.3 d	1.0 b	1.4 d

[a]Average of test data from eight replications in each of four varieties—Corsoy, Amsoy-71, Beeson, and Williams

[b]All losses are given as a percentage of gross yield. Mean gross yield was 2.85 t/ha (42.4 bu/acre). Within columns, numbers with the same letters do not differ significantly at the 5 percent level, based on the LSD of means.

[c]Total header loss includes shatter, stalk, stubble, and lodged losses.

Table 2. Effect of header design on harvesting loss in 18-cm rows.[a]

	Loss (%)[b]		
	Shatter	Stalk	Total Header[c]
Flexible cutterbar	1.6 a	0.6 ab	2.4 a
Flexible cutterbar w/air assist	1.6 ab	0.5 a	2.4 a
Floating cutterbar (Standard)	4.8 c	1.2 c	6.3 c
Floating cutterbar	2.2 d	0.9 b	3.3 b

[a]Average of test data from eight replications in each of four varieties—Corsoy, Amsoy-71, Beeson, and Williams.

[b]All losses are given as a percentage of gross yield. Mean gross yield was 2.95 t/ha (43.8 bu/acre). Within columns numbers with the same letters do not differ significantly at the 5 percent level, based on the LSD of means.

[c]Total header loss includes shatter, stalk, stubble and lodged losses.

The data for the 1976 season confirmed results from the tests of Nave and Yoerger (17) which indicated that losses from use of a floating-cutterbar header with air assist was 45% less than from use of a conventional floating-cutterbar header. Further tests under different crop conditions and with some modification on the air-jet guards of the integral, flexible-cutterbar header would be required before a final analysis can be made to confirm the effectiveness of air jets on a flexible-cutterbar header.

The level of loss was so low with the flexible-cutterbar header that the air-jet system might not be practical. The flexible-cutterbar header was observed to have the following advantages over a conventional floating-cutterbar header: (a) Long crop-dividing points, which helped in lodged soybeans; (b) An extended platform and low profile, which reduced shatter and stalk loss; and (c) A large-diameter auger, which more rapidly moved the plant material to the center and helped reduce stalk loss.

The advantages of the air-jet system on the integral flexible-cutterbar header and the floating cutterbar header were: (a) The 15-cm (6-in.) long air guards provided more accurate ground sensing than the skid pads provide on a standard floating cutterbar; (b) Plant material was fed more evenly into the combine, an improvement that reduced header loss and could increase combine threshing efficiency; and (c) The air system provided a better view of the cutterbar for the operator, since the plant material is moved away more rapidly.

One disadvantage of the air-jet system was the fan location. The fan should be relocated to give the operator a clear view of the right end of the cutterbar. On the basis of harvesting loss data for 1976, the row-crop header was the most efficient header tested. However, an economic comparison might be useful in assessing the interactions among variety, row spacing, and combine header.

RECENT DEVELOPMENTS IN GRAIN THRESHING

The quality of marketed soybeans is of growing concern to both domestic and foreign grain buyers. Recent increases in the number of complaints about the quality of soybeans exported to foreign countries have led to an evaluation of the magnitude and causes of soybean damage during harvesting and handling. One way to improve the quality of soybeans is to reduce the mechanical damage caused by the combine threshing mechanism during harvesting. Efforts to reduce threshing damage while increasing capacity have resulted in the development of rotary threshing equipment. Rotary combines have one or more longitudinal rotors to replace the conventional cylinder and straw walkers for threshing and separating grain from crop material. The rotor swirls the crop material rearward, passing it over concaves several times. Because the material is apparently subjected to less impact with a rotor than with a conventional cylinder, threshing action is reported to be more gentle with a rotor (7).

In a study by Saiji Paul et al. (27), the seed quality of soybeans harvested with a Sperry New Holland TR-70 rotary combine was compared with that of soybeans harvested with a Sperry New Holland Model 1400 conventional rasp-bar-cylinder combine. In most instances soybeans had a higher germination percentage when harvested with the rotary combine than when harvested with the conventional combine. Also, the ratio of grain loss at the rear of the combine to material other than grain was lower for the rotary combine than for the conventional combine. Tests conducted by IH engineers indicated that soybeans harvested with an IH Axial-Flow combine contained only half as much crackage as those harvested with a conventional rasp-bar-cylinder combine (7).

The main features of the Sperry New Holland TR-70 as described by Quick are: (a) It has two counter-rotating axial rotors for threshing and separation. (b) The two rotors and discharge beater combination provide low separating losses. (c) The compact rotors result in a narrow body width with the capacity of a wider conventional rasp-bar-cylinder combine. (d) It has larger thresher clearances than a conventional rasp-bar-cylinder, since the threshing action depends more on centrifugal force than on rubbing. (e) The crop is spiralled over rotors to pass over concaves several times instead of just once as over the concaves of a conventional cylinder. The crop may pass over the rotor as many as nine times during the approximately 3 seconds it takes to pass through the machine, as compared with the 9-second crop dwell time through a conventional cylinder and walkers; and (f) It has a compact machine envelope, which leaves room for a large grain bin.

COMPARISON OF ROTARY AND CONVENTIONAL THRESHING

A study was conducted at Illinois by Newbery et al. (18) to evaluate the damage to soybeans caused by rotary and conventional threshing mechanisms. In this study an IH 1460 Axial-Flow (single rotor) combine, a Sperry New Holland TR-70 (double rotor) combine, and a JD 7700 (conventional rasp-bar-cylinder) combine were tested under field conditions to evaluate the quality of harvested soybeans and the associated composite threshing and separating losses for each combine. All three combines were equipped with 6.1-m-wide floating cutterbar headers.

The following list summarizes the results of the Illinois study (18) and pertains only to the results of that study with the particular combines used and conditions existing during the 3 days of testing in the Amsoy-71 variety of soybeans: (a) The percentages of splits in the Amsoy-71 soybeans were significantly higher for the conventional cylinder than for the single- or double-rotor threshing mechanisms when results with the three threshing mechanisms at similar peripheral threshing speeds were compared. However, when the mechanisms were operated within the range of cylinder or rotor speeds recommended by the respective manufacturers, the percentages of splits did not exceed the allowable 10% limit for U.S. No. 1 grade soybeans. (b) With all three mechanisms, the percentages of splits increased as the peripheral threshing speed of the cylinder or rotor was increased. The increase in splits was less with the rotary threshing mechanisms than with the conventional cylinder (Figure 1). With all three mechanisms, composite threshing and separating losses decreased as the respective cylinder or rotor speed was increased. Mean threshing and separating losses ranged from 5.3 to 18.2 kg/ha. Threshing and separating losses with the rotary combines were significantly higher at the slowest rotor speed than at the faster speeds. (c) Increasing the concave clearance generally decreased the percentages of splits for all three combines, but the effect was less than that caused by changes in cylinder or rotor speed. The percentages of splits were not significantly affected by concave adjustment until after a minimum clearance was reached for the rotary combines. (e) Soybean susceptibility to breakage, as determined by the Stein breakage test, and seedcoat crack percentages did not differ significantly as a result of the type of threshing mechanism or the cylinder or rotor speed used; neither did other grain inspection grading factors such as test weight, percentage of damaged kernels, and percentage of foreign materials; and (f) Improvements were found to be needed in the design of augers and elevators that convey soybeans from the clean-grain auger to the grain tank. Increases in percentages of splits caused by elevating soybeans from the clean-grain auger to the grain tank averaged 0.96, 0.60, and 1.42% for the conventional cylinder, single-rotor, and double-rotor combines, respectively.

Quick and Buchele (24) concluded that rotary combines have the following advantages and disadvantages as compared with conventional cylinder combines. The advantages are: (a) A more compact design, which allows increased grain bin capacity for a given machine size. (b) A reduced number of working parts, which should result in lower manufacturing costs. For example, IH cited its 1460 Axial-flow design as having 17% higher capacity at only 10% higher cost than its conventional 915 model. (c) Improved accessibility and serviceability, facilitating maintenance and removal of crop residues from the machine. For example, the centrifugal separator of the rotary combine is completely self-emptying and essentially maintenance free. (d) The airflow created by the rotors causes a draft through the machine that reduces the dust nuisance at the feeder in dry harvesting conditions; and (e) Under some conditions, a cleaner soybean sample can be obtained with a considerable reduction in soybean damage. This additional cleanliness and reduction in damage may represent a considerable savings because it should reduce the dockage for unmillable material in loads of grain at the delivery point.

The disadvantages are: (a) Straw breakup is generally much more severe. This excessive breakup may be a drawback when the straw is of value. (b) Manufacturers have acknowledged that the first generation of rotary designs were not yet ready in 1977 for the rice crop in North America or for certain European long straw crops; and (c) When straw is tough or damp, more of it

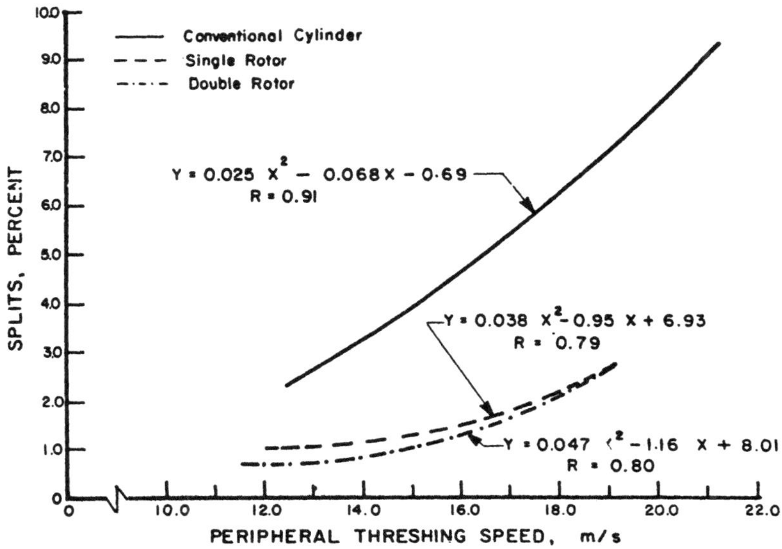

Figure 1. Effect of peripheral threshing speed on percentage of splits in clean grain auger samples in the laboratory averaged over all harvest moistures.

may wrap around the threshing mechanism than when a conventional cylinder is used, and power consumption may be greater.

Results of the Ohio study by Saij Paul et al. (27). and the Illinois study by Newbery et al. (18) indicated that adjustments to rotary combines may be less critical than those to a conventional rasp-bar-cylinder combine. However, results from their research also indicated that a conventional combine will thresh and clean soybeans such that the damage level will be well below that leading to dockage if the combine is properly adjusted. In addition to the rotary combines introduced by Sperry New Holland and International Harvester, one has been introduced by Allis Chalmers.

MONITORING DEVICES FOR COMBINES

The operator of a modern combine must monitor and adjust many parameters affecting the machine's performance. Many of the self-propelled models are equipped with cab enclosures to isolate the operator from the combine noise. Therefore, methods for remote monitoring and controlling of the combine operation must be adopted. Some remote monitoring devices are already available. Powerplant monitors, which consist of warning lights or analog readouts of various engine functions, are quite common on combines. Electronic monitors for sensing shaft speed, ground speed, and grain losses have been developed by several manufacturers (26). A method has been developed by van Loo (31) to measure the moisture content of the straw entering the combine and use it in conjunction with a grain-loss monitor to improve the loss measurement accuracy.

Automatic sensors that are used to monitor the machine and crop status could eventually perform the necessary adjustments now required of the operator and may become an integral part of future harvesting machinery. Monitors for use on process machinery can be divided into two categories: (a) those which monitor mechanical components of the machine, and (b) those which are needed to alert the operator if the intended function is not being carried out efficiently (26). Many of the needed function monitors are not yet available on combines because the necessary parameters have not yet been adequately described. A good example of needed information that is not available is the specific relationship between combine cylinder speed, grain moisture, and grain damage during threshing.

Equipment to monitor grain loss through combines has been developed during the past few years, and at least 10 companies now merchandise grain-loss monitors. Certain limitations of grain loss monitors are discussed in the literature. For example, a certain reading of a given loss rate does not necessarily indicate the level of machine performance because the level of performance depends on the harvesting rate. One manufacturer has partly overcome this monitor limitation by providing various scales on the monitor meter for differing travel speeds. Other monitors use a signal that is proportional to travel speed to indicate the loss rate per unit of field area (26).

Users report that the two most valuable aspects of monitors are: (a) they indicate when grain loss due to overloading or misadjustment begins, and (b) they indicate when corrective action is needed due to sudden changes in loss rate. Without monitors it is virtually impossible for the combine operator to know whether the combine is operating at its maximum separating capacity.

Sensing devices that will indicate unthreshed-grain loss and grain quality are not available and will be difficult to develop. Improvements to the present monitors of separating performance and the development of new sensors may lead to the eventual use of more automatic controls on the combine.

Through the use of integrated circuits and microprocessors complete arithmetic capabilities can be carried on board a combine. Since the basic instructions are permanently locked in the memory of the microprocessor it can automatically adjust various functions such as cylinder speed, concave clearance, and fan speed.

When a combine is equipped with a microprocessor, an interface system is required to connect the microprocessor to the "real world." This interface system consists of drivers for the various machine actuators, a display panel that will allow the operator to monitor and make changes to the microprocessor and data converters to convert the binary/numeric information supplied by the microprocessor to a form intelligible to both the harvester and the operator (10).

Automatic ground speed control is a good example of a typical use for a microprocessor. The required measurements include grain loss, material flow, and powerplant load. The microprocessor processes this information, compares it with a desirable standard, and then makes an appropriate adjustment in ground speed quickly and automatically.

The use of a microprocessor for automatic cylinder-speed control as a function of grain moisture was investigated by Brizgis et al. (6) at the University of Illinois. A continuous-flow moisture sensor was mounted in the grain tank auger. The results indicated that output reading of the continuous-flow moisture-sensor increased sufficiently with grain-moisture increases to accomplish automatic cylinder-speed control with present combine design and moisture-sensing equipment.

Research is underway at Clemson University to further investigate the use of microprocessors to control several functions of the combine under field conditions. The major obstacle in this research may well be the lack of adequate descriptions of combine function parameters.

SUMMARY

The history of soybean harvesting is traced from the mid-twenties through the years of major improvements in the late sixties and seventies. Through a concentrated effort by public researchers, the problem areas were described and suggestions were made for improvements. Due to an intensive research effort by both public researchers and farm equipment manufacturers, major improvements in the design of soybean harvesting equipment have been accomplished. Harvesting losses can be reduced to less than 4% by the use of any of several improved combine headers currently available. The introduction of rotary combines by the farm equipment industry indicates that progress is being made to improve soybean quality and harvesting efficiency. The use of monitoring equipment on combines is progressing, and the automatic control of major combine operating parameters is receiving new emphasis.

NOTES

W. Ralph Nave, U.S. Department of Agriculture, SEA/AR, Urbana, Illinois 61801.

Trade names are used in this chapter solely for the purpose of providing specific information. Mention of the trade name, proprietary product, or specific equipment does not constitute a guarantee or warranty by the U.S. Department of Agriculture or the University of Illinois, and does not imply approval or the exclusion of other products that may be suitable.

REFERENCES

1. Agr. Statistics. 1977. USDA. Washington, D.C.
2. Bichel, D. C., E. J. Hengen, and G. W. Rohweder. 1974. Row-crop head for combine harvesters. ASAE Paper No. 74-1556. ASAE, St. Joseph, MI.
3. Bichel, D. C., E. J. Hengen, and R. E. Mott. 1976. Designing the new concept header. Ag. Eng. 57(9):21-23.
4. Bichel, D. C. and E. J. Hengen. 1977. Development of soybean harvesting equipment in the U.S.A. Proc. Int. Grain and Forage Harvesting Conf., Ames, Iowa. ASAE, St. Joseph, MI.
5. Boddiford, J. K., Jr. and C. B. Richey. 1975. Development of a puller-header for combining soybeans. ASAE Trans. 18(6):1003-1005.
6. Brizgis, L. J., W. R. Nave, and M. R. Paulsen. 1978. Automatic cylinder speed control for combines. ASAE Paper No. 78-1569. ASAE, St. Joseph, MI.
7. DePauw, R. A., R. L. Francis, and H. C. Snyder. 1977. Engineering aspects of axial-flow combine design. ASAE Paper No. 77-1550. ASAE, St. Joseph, MI.
8. Dunn, W. E., W. R. Nave, and B. J. Butler. 1973. Combine header component losses in soybeans. ASAE Trans. 16(6):1032-1035.
9. Kerber, D. R. and O. W. Johnson. 1977. Advances in grain header developments. ASAE Paper No. 77-1547. ASAE, St. Joseph, MI.
10. Kopp, K. A. 1977. Electronic technology—A changing force in harvest machinery monitor and control systems. Proc. Int. Grain and Forage Harvesting Conf., Ames, Iowa. ASAE, St. Joseph, MI.
11. Lamp, B. J., W. H. Johnson, and K. A. Harkness. 1961. Soybean harvesting losses—Approaches to reduction. ASAE Trans. 4(2):203-205.
12. Lamp, B. J., W. H. Johnson, K. A. Harkness, and P. E. Smith. 1962. Soybean harvesting approaches to improved harvesting efficiencies. Ohio Agr. Exp. Sta. Res. Bull. 899.
13. Nave, W. R. and D. L. Hoag. 1975. Relationship of sickle and guard spacing and sickle frequency to soybean shatter loss. ASAE Trans. 18(4):630-632.
14. Nave, W. R., J. W. Hummel, and R. R. Yoerger. 1977. Air-jet and row-crop headers for soybeans. ASAE Trans. 20(6):1037-1041.
15. Nave, W. R., D. E. Tate, and B. J. Butler. 1972. Combine headers for soybeans. ASAE Trans. 15(4):632-635.
16. Nave, W. R., D. E. Tate, J. L. Butler, and R. R. Yoerger. 1973. Soybean harvesting. ARS, USDA. ARS NC-7.
17. Nave, W. R. and R. R. Yoerger. 1975. Use of air-jet guards to reduce soybean harvesting losses. ASAE Trans. 18(4):626-629.
18. Newbery, R. S., M. R. Paulsen, and W. R. Nave. 1978. Soybean quality with rotary and conventional threshing. ASAE Paper No. 78-1560. ASAE, St. Joseph, MI.
19. Norman, A. G. 1967. The soybean. Academic Press, New York, NY. 239 pp.
20. Present status of combine harvesting. 1927. Papers, discussion, and reports presented at the "combine" session of ASAE, Chicago, IL. 38 pp.
21. Quick, G. R. 1973. Laboratory analysis of the combine header. ASAE Trans. 16(1):5-12.
22. Quick, G. R. 1977. Development of rotary and axial thresher/separators. Proc. Int. Grain and Forage Harvesting Conf., Ames, IA. ASAE, St. Joseph, MI.
23. Quick, G. R. and W. F. Buchele. 1974. Reducing combine gathering losses in soybeans. ASAE Trans. 17(6):1123-1129.
24. Quick, G. R. and W. F. Buchele. 1978. The grain harvesters. ASAE, St. Joseph, MI 269 pp.

25. Quick, G. R. and W. M. Mills. 1977. Narrow-pitch combine cutterbar design and appraisal. Proc. Int. Grain and Forage Harvesting Conf., Ames, IA. ASAE, St. Joseph, MI.
26. Reed, W. B. 1977. A review of monitoring devices for combines. Proc. Int. Grain and forage Harvesting Conf., Ames, IA. ASAE, St. Joseph, MI.
27. Saij Paul, K. K., L. O. Drew, and D. M. Byg. 1977. New design combine effects on soybean seed quality. Proc. Int. Grain and Forage Harvesting Conf., Ames, IA. ASAE, St. Joseph, MI.
28. Schertz, C. E. 1975. Soybean harvesting. Deere and Company. Soybean Production Conference, Moline, IL.
29. Schertz, C. E., D. W. Nordquist, D. W. Peterson, T. D. Bebernes, and J. B. Unterzuber. Reduction of soybean gathering losses with floating draper-extension. ASAE Paper No. 76-1553. ASAE, St. Joseph, MI.
30. Schrock, M. D., S. Briggs, and F. Baker. 1974. A pan type harvesting aid. ASAE Paper No. 74-404. ASAE, St. Joseph, MI.
31. van Loo, J. 1977. Measuring the moisture content of straw continuously. Proc. Int. Grain and Forage Harvesting Conf., Ames, IA. ASAE, St. Joseph, MI.
32. Williams, M. M. and C. B. Richey. 1973. A new approach to gathering soybeans. ASAE Trans. 16(6):1017-1019.

CRITICAL FACTORS IN SOYBEAN SEEDLING EMERGENCE

H. D. Bowen and J. W. Hummel

This chapter is presented in two parts. The first part describes laboratory studies at N. C. State Univ. on the emergence by soybean seedlings under several levels of three soil environmental factors that are often critical to emergence due to the vagaries of weather and soil conditions at the time of planting and during the germination and emergence period. The second part reports on field investigations conducted at the Univ. of Illinois on the effect of biological and physical factors on soybean seedling emergence and yield.

ENVIRONMENTAL FACTORS AND THEIR EFFECT ON HYPOCOTYL AND ROOT GROWTH

A truly vigorous soybean seedling can survive almost any weather stress imposed upon it short of freezing, and by contrast a low vigor seedling will succomb to stress of mild intensity and short duration. With high vigor seedlings very few factors are critical except under extremes of weather and soil conditions at planting time, but with low vigor seedlings, a stress of any one or more of the edaphic (soil physical) factors makes all aspects of the planting operation critical. Unfortunately, the average producer buying certified seed assumes high germination percentage to mean high vigor, but in fact low vigor seedlots may show as high an official germination percentage on the seed tag as a high vigor seedlot. This seeming anomaly comes about because the seed producers are required to indicate only the 20 to 30 C germination percentage which does not distinguish between high and low vigor seeds. Under non-stressed conditions low vigor seedlings may perform identically with high vigor seedlings, but under stress conditions the low vigor seedlots may give 10% or less of performance they had under nonstressed conditions, whereas high vigor seedlots may have germination and emergence that is 70 to 80% of the nonstressed performance. Thus it seems that the single most critical factor in ensuring a good stand is to be sure that the seeds have a high vigor and are substantially free from mechanical damage.

Since the seed vigor cannot reliably be estimated from germination percentage reported on the seed tag, the producer should demand vigor information from his seed source and be willing to pay for high performance seeds if seeds are to be planted under stress conditions. A producer should plant his most vigorous seed lots when there is a danger of stress and less vigorous seed lots when conditions are predicted to be less stressful. Most agricultural universities can help the producer determine the vigor of his seedlots or the producer may make his own tests after consulting his county extension agent in the event his seed supplier cannot or will not supply that information.

The critical factors of the soil physical environment during germination and emergence are temperature, moisture, mechanical impedance, and aeration for soybeans. Depending on the geographic area, climate, seasonal and local weather conditions, and soil types and conditions, any one of these factors or any combination of them can be critical. An effort is made here to quantify some of the stresses that result from simulated field conditions and the response of one soybean cultivar (1975 Forest) to these stresses.

Although it is likely that there are quantitative differences in the response of each cultivar to environmental stress the nature of the response will be similar. Six soybean cultivars (Hill, Essex, Bragg, Lee 74, Forest and Ransom) grown in replicated plots for 3 yr at Clayton, N.C., have shown similar responses to wet cold, to hot dry, and to crusting emergence conditions. The response of soybeans to the environment are from the growth chamber study described below. The stress data are from several studies made on the Ap horizon of North Carolina Coastal Plains soils over a 20-yr period. Unfortunately, the soils vary in mechanical analysis, but they are all soils of the Norfolk loamy sand series. They are a Wagram loamy sand with a mechanical analysis of 90.2% sand, 6.6% silt and 3.2% clay; a Ruston loamy sand of mechanical analysis 83.2% sand, 11.0% silt and 5.8% clay; and an Orangeburg loamy sand of mechanical analysis 80% sand, 17.2% silt and 2.8% clay. Even though the mechanical analyses differ considerably, they have in common similar moisture characteristic curves with field capacity water contents between 7 and 10% (d.b.), low organic matter, and severe crusting following high intensity storms (4 + cm/hr) of greater than 1 cm of rainfall.

Growth Chamber Studies

Equipment and Procedure. Seedlings of Forest 1975 soybean *Glycine max* (L.) Merrill were grown at 48 different soil environmental conditions; four levels of temperature (15, 20, 25, and 30 C), three levels of moisture,

(0.3, 1.0, and 10 bars), and four levels of mechanical impedance (0.23, 1.12, 2.24, and 3.36 kg/cm^2) in growth chambers held to ±1.0 C.

Plant growth boxes measuring 62 cm wide by 8 cm deep by 45 cm high were made of plywood on three sides with a window side of transparent material. A plexiglass side piece 10 cm high allowed the observation of the growth of the hypocotyl. The seeds were planted at the juncture of the plexiglass sidepiece with a glass sidepiece 36 cm high for observation of the roots. The boxes were tilted 10° with the vertical to force the roots to the glass sidepiece. The plexiglass sidepiece was angled inward 15° with the glass sidepiece. When the box was in the tilted position, the plexiglass sidepiece was angled 5° inward from the vertical in an attempt to force the hypocotyl against the plexiglass. This was only partially successful and considerable data was lost under some treatments due to the hypocotyl not staying in contact with the plexiglass. Most of the time the roots remained visible from the planting depth to the bottom of the box.

A Ruston loamy sand with a mechanical analysis of 83.2% sand, 11.0% silt, and 5.8% clay was used in the tests. Figure 1 shows the moisture characteristic curve. Water was mixed with air-dry soil to bring the soil to the proper moisture content. The soil was allowed to equilibrate for a minimum of 24 hr. The soil was sieved (4 mm mesh) and poured into the soil boxes in layers and packed with surface compaction and vibration to as uniform mechanical impedance as could be attained. The amount of surface compaction on

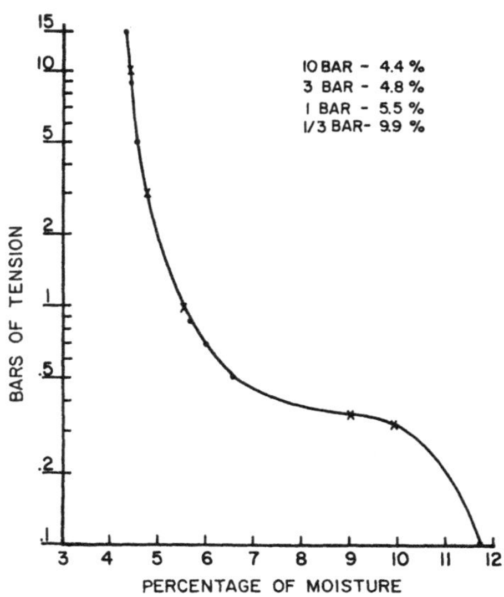

Figure 1. Soil moisture characteristics curve for Ruston loamy sand.

each layer, the magnitude and duration of the vibration, and the thickness of the soil layers used in the filling process depended on both the moisture content and the mechanical impedance level. A soil penetrometer with a tip included angle of 30° and a base cone diameter of 0.46 cm was used to quantify the mechanical impedance and establish compaction procedures for the different moisture levels.

Seeds were inspected visually and those with cracked seedcoats and those which were irregular in size or shape were eliminated. The remaining seeds were pregerminated in vermiculite at 30 C for approximately 18 hr. Seeds with a 3 mm radicle length were chosen for planting.

After the box was filled with packed soil to the top of the glass window the plexiglass strip was fastened to the side of the box and 15 pregerminated soybean seeds were planted against the transparent side of the box at the juncture of the glass and plexiglass parts of the window. The seeds were spaced 4 cm apart with the radicles pointed down and visible through the glass side of the box. The 10-cm space above the seed planting depth was filled with soil and compacted to the same level of mechanical impedance as the soil in the root zone. The hypocotyls and roots were subjected to the same level of impedance from the top soil surface to the bottom of the box. A leaky plastic cover was then fitted and taped to the top of the box to reduce moisture loss and the resulting changes in mechanical impedance due to drying of the soil.

Four replications of 15 seeds each for the 48 combinations of temperature, moisture and mechanical impedance were observed. A factorial design was used, and for each replication the order of preparation of the boxes and the position of the boxes in the growth chamber were randomized completely. Air was supplied to each box at the rate of 0.6 ml/s to insure that no oxygen deficiencies existed in the soil. The air was bubbled through water to reduce the tendency of the soil to dry out from the passage of air through the soil. The boxes were covered with black plastic except when the readings were taken.

There were four runs with all combinations of moisture and impedance at each temperature. Each run extended for 10 days at a constant temperature. Daily readings of hypocotyl and root lengths were taken.

Results and Discussion. Both hypocotyl and radicle length were limited by the size of the boxes. At the lowest impedance, 0.23 kg/cm^2, the hypocotyls had emerged through the 10 cm of soil over the seedlings within 3 to 4 days and presumably would have pushed through several additional centimeters of soil if the seeds had been planted deeper. The roots grew to within 2 to 5 cm of the bottom of the growth box in 4 to 10 days depending on the temperature and stopped short of the bottom. It has been observed in our laboratory that as a penetrometer nears the bottom of a container of soil the resistance to penetration increases because the penetrometer compacts the soil ahead of itself. Thus the data does not show how far the hypocotyls and radicles could have gone against the indicated impedance but only how far they did go before the hypocotyls emerged or the roots were stopped by excessive impedance as they approached the bottom. Even with these limitations in the data a number of things can be observed.

Hypocotyl growth rate (Figure 2B) increased as temperature increased from 15 to 30 C in the same order as reported by Gilman (3) and Hatfield and Egli (4) for seed germinated and grown in paper towels.

The effect of impedance on hypocotyl growth (Figure 2B) is most dramatic in the range of 0 to 1 kg/cm^2. It is much less dramatic in the range of 1 to 3 kg/cm^2. The lowest impedance (0.23 kg/cm^2) is about the minimum impedance level that can be obtained by pouring a sandy soil into a container. The soil is very loose at this impedance and extreme care must be exercised to prevent additional settling and compaction in handling the boxes. However, this very slight resistance is sufficient to reduce the growth rate to about half that of the zero impedance growth rate.

Tables 1 and 2 show the hypocotyl and redicle lengths after 10 days of growth at each temperature, moisture, and impedance level. Temperature had very little effect on the length attained by either the hypocotyl or root except at the lowest impedance and the lowest temperature. A plot of the data (Figure 3) for 0.23 kg/cm^2 for 25 C at the four impedance levels is somewhat scattered but it is easy to see a trend in hypocotyl length as a

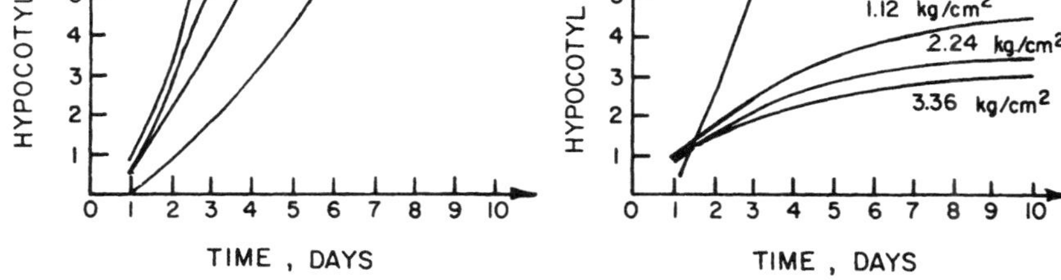

Figure 2. **(A) Hypocotyl growth for Forest soybeans at four levels of temperature as a function of time in days after imbibition. The impedance was 0.23 kg/cm^2 and the moisture was 5.5%; and (B) Hypocotyl growth of Forest soybeans at four levels of impedance at 25 C and 5.5% moisture.**

Table 1. Average hypocotyl length at 10 days after imbibition.

Temp.	Moisture Content	Impedance, kg/cm^2 0.23	1.12	2.24	3.36
– C –	– % d.b. –	– *Hypocotyl length, cm* –			
30	9.0	10.0	3.5	3.8	3.1
25	9.0	9.5[a]	4.0	3.5	2.7
20	9.0	10.0	4.0	3.6	3.2
15	9.0	9.2	4.0	2.7	2.5
30.	5.5	9.5	3.2	2.8	2.6
25	5.5	10.0	4.2	3.0	2.6
20	5.5	8.0	5.0	3.1	2.6
15	5.5	7.0	2.0	2.5	2.5
30	4.4	9.0	3.0	2.0	2.0
25	4.4	7.5	3.0	3.0	2.2
20	4.4	7.0	3.2	3.0	2.5
15	4.4	5.0		2.0	1.7

[a]Only one replication of data. Others are averages of four replications through 10 cm of soil at a given impedance and moisture content. Those which emerged were counted as being 10 cm long.

Table 2. Average radicle length at 10 days after imbibition.[a]

Temp.	Moisture Content	Impedance, kg/cm^2 0.23	1.12	2.24	3.36
– C –	– % d.b. –	– *Radicle length, cm* –			
30	9.0	33	33	33	33
25	9.4	30	32	32	30
20	9.0	30	28	26	27
15	9.0	22	19	15	15
30	5.5	33	33	32	33
25	5.5	30	21	32	30
20	5.5	24	21	13	14
15	5.5	24	21	13	14
30	4.4	33	33	32	32
25	4.4	32	30	24	30
20	4.4	32	32	28	25
15	4.4	23	14	13	15

[a]The boxes were 35 cm high from planting depth to the bottom, but penetrometer readings always increased as the penetrometer approached within 2 to 5 cm of the bottom of the box. Thus radicle growth was retarded after 30 to 33 cm.

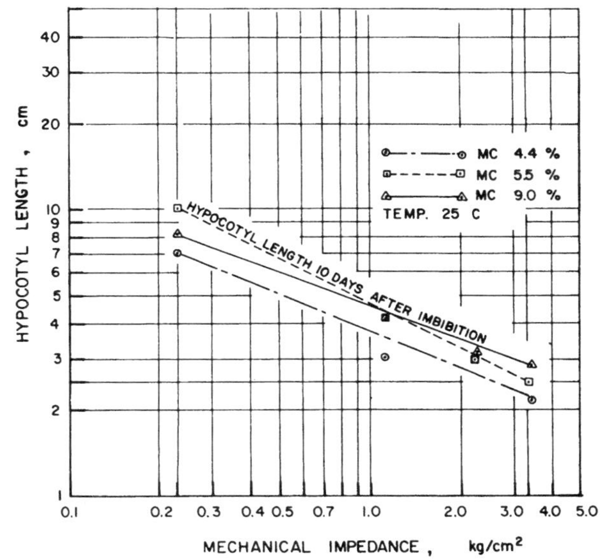

Figure 3. **Hypocotyl length 10 days after imbibition as a function of mechanical impedance for 3 moisture levels at a temperature of 25 C and 0.23 kg/cm^2 mechanical impedance.**

function of impedance. Projecting back to .01 kg/cm^2 would give approximately the 18 to 20-cm hypocotyl length obtained by Gilman (3) for seedlings grown with little or no mechanical impedance. In terms of planting depth it is easily seen (Figure 3) that at 25 C seedlings could not emerge from a depth greater than 2.5 to 3.0 cm in 12 to 13 days after planting if the impedance is 2.0 kg/cm^2, but could emerge from a depth of 7 to 8 cm if the impedance were only 0.23 kg/cm^2.

Moisture affected both hypocotyl and radicle growth, with the 5.5% (d.b.) moisture content generally the most favorable and 4.4% the least favorable, but differences in growth rates were small and would not be statistically significant for this data.

The growth chamber data, though limited by the dimensions of the growth boxes, shows that temperature controls growth rate of both hypocotyl and radicle, but mechanical impedance from 0 to 1 kg/cm^2 reduces hypocotyl growth very dramatically. Further increases in impedance show less reduction of growth rate. Moisture between 4.4 and 9.0% for this soil did not materially influence either hypocotyl or root growth. Within 10 days after imbibition, roots at all conditions had reached or exceeded 25 cm except those at 15 C and these exceeded 12 cm even at the highest impedance. This implies that in tilled soil healthy soybean plants can get their roots down to sufficient moisture and emerge in 12 to 13 days if the moisture was adequate at planting, the average impedance was less than 3 kg/cm^2 and the average temperature was 15 C or greater.

Stress Raising Aspects of Soil, Weather, and Planter Settings

If planting is done at the proper time, temperature per se, will usually not be a problem unless some unusual weather comes in after planting. But cold wet weather will harm weak seeds. Most seeds including soybeans can obtain sufficient water to germinate from soils with moisture tension up to the wilting point if they have sufficient time. Because of the shallow planting depth required to achieve a short emergence time, moisture stress may become a problem.

Seed Imbibition and Visible Moisture Front. To investigate the uptake of moisture by seeds as a function of soil moisture content, cotton (another oil crop) seeds were planted in a Ruston loamy sand with a field capacity of 9.3% and a wilting percentage of 4.4% (d.b.). Seeds were extracted every 2 hr and the moisture content determined gravimetrically. The seeds were planted in soil with moisture contents of 5, 7, and 9% at a depth of 2.2 cm and subjected to the radiation of a clear May 1st day in Raleigh, N.C.

The soil was packed to a uniform bulk density of 1.3 g/cc. One hundred seeds were planted in each 0.09 m^2 soil box and placed on the Edaphotron (2), a 4-m turntable that revolves once/day under a bank of heat lamps that can be adjusted to simulate the radiation for any specific day of the year. The seeds were planted so that the seed top surface was at the planting depth by pressing the seeds into the soil with a finger. Replicated boxes of seeds were alternated with boxes for determining the depth to the visible moisture front. The depth to the front was measured with an ohmmeter with a small wire probe which was pushed into the soil slowly and carefully. One lead of the ohmmeter was thrust deep into the soil to contact moisture. The other lead was attached to the wire probe. When the probe was in the dry soil above the visible moisture front, the ohmmeter showed several megohms resistance but when the visible front was contacted the resistance dropped to a low value. The front measurement was very repeatable and could be meassured to within 0.1 cm.

Each soil box was placed on the Edaphotron at the 6 AM station. A soil box was placed on the Edaphotron every 2 hr once the test was started. In effect this meant that the seeds were planted at 6 AM on a clear May 1st day in Raleigh, N.C.

The visible moisture front (where the soil looks slightly moist) (Figure 4) moves down inversely as the amount of water in the soil and for the 5% moisture content soil it crosses the seed planting depth at exactly noon of the first day. By 6 PM the front has reached its maximum depth for the day and even

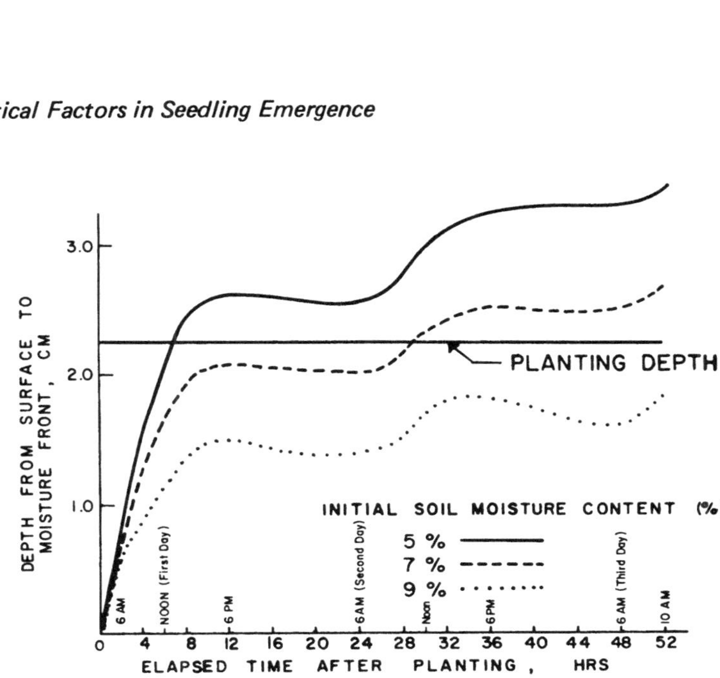

Figure 4. **Depth to moisture front as a function of elapsed time after planting for a Ruston loamy sand.**

moves slightly to the surface overnight by picking up moisture from below. The second day the front moves down again for all moisture levels and by 10 AM of the second day the front in the soil with 7% moisture content crosses the planting depth. The visible front in the soil with the 9% moisture content did not reach planting depth on the second day, but presumably would have crossed it on the third or fourth day.

Seed moisture content (Figure 5) also varied with time. All seeds began picking up moisture at the same rate until noon of the first day at which time the seeds in soil with 5% moisture content began losing moisture. Precisely at the same time, the visible moisture front (Figure 4) reached the top of the seeds. Even though the seeds in the 7% moisture content soil were still in moist soil the first night and continued to pick up moisture until 10 AM of the second day, they began losing moisture rapidly at precisely 10 AM, when the visible moisture front (Figure 4) crossed the planting depth. At 3 PM of the second day the moisture front of the 9% moisture content soil reached its maximum depth for that day at about 0.4 cm above the planting depth. At that depth the seeds lost some 5 or 6% moisture which they picked up again by 11 PM that night. Radicles were coming out the second day so that the 9% moisture content soil germinated all of the seeds. The seeds planted in soil with 5 and 7% moisture content (Figure 5) lost moisture rapidly as the visible moisture front went deeper. When the visible moisture front of the 7% moisture content got within 0.2 cm of the planting

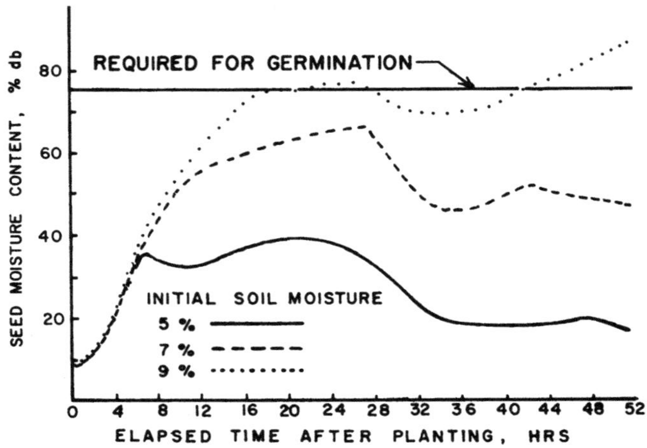

Figure 5. Seed moisture content as a function of time after planting for cotton seed when planted in a Ruston loamy sand.

depth, the seed moisture pickup slowed down indicating the evaporating front is not a sharp line at the beginning of the visible moisture front. This test has not been done on soybeans but their imbibition times are similar to cotton and they must behave in essentially the same way.

From this test we learned that the seeds' rate of pickup of moisture is the same for the soils with 5, 7, and 9% moisture content and that when the visible moisture front crosses the seed planting depth, the seeds begin to lose moisture. We also learned that moisture can be lost from a seed when the visible moisture front is within 0.2 to 0.4 cm of the seed.

Bringing Water to Seeds by Compaction. Most research workers are aware that press wheels generally improve moisture availability under drier-than-optimum soil conditions. They often overlook the fact that it is the volumetric water content rather than gravimetric water content that determines (a) the availability of water to an imbibing seed or growing root and (b) the rates of moisture movement to seeds and roots.

A seed imbibing water from soil is controlled by the volumetric water content of the soil. When soil moisture is reported in percent by wt dry basis, i.e., gravimetric water content, we often overlook that increasing the bulk density of the soil increases the volumetric water content of the soil. The relation of gravimetric water content to bulk density and volumetric water content is illustrated in Figure 6.

The important thing about the moisture-bulk density relation is that an engineer, agronomist, or farmer can change the amount of water within a 1 cm^3 volume of soil simply by changing the bulk density. In a soil with a

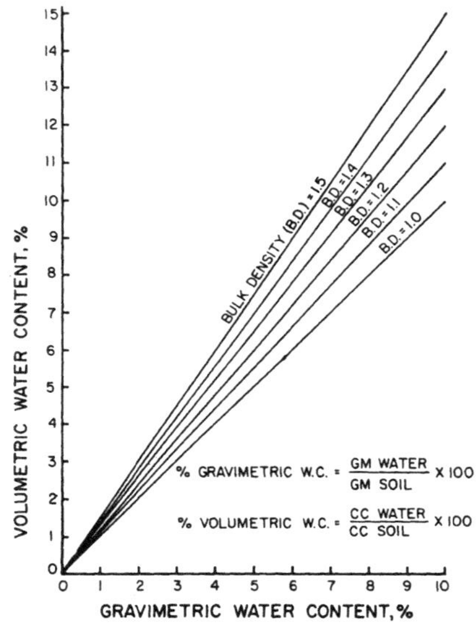

Figure 6. **Chart relating gravimetric water content in % to volumetric water content in % as a function of dry bulk density.**

gravimetric moisture content of 5% packed to a dry bulk density of 1.0 g/c^3, the volume of water is 5% of the volume of soil. That same 5% moisture content soil by wt, when packed to a bulk density of 1.5 g/cm^3 has 7.5% moisture on a volumetric basis.

However, an even more important benefit from an increased moisture content of the soil is the rather dramatic increase of soil moisture diffusivity, D(Θ), that results from an increase in moisture content. In increasing the soil bulk density from 1.0 to 1.5 g/c^3 and the resulting volumetric water content of a 5% by wt (d.b.) soil moisture content from 5% by volume to 7.5% by volume, the increase in bulk density was in the ratio 1.5:1 but the increase in moisture diffusivity (Figure 7) was in the ratio 4:1. Water movement for one-dimensional movement as shown in Coble (2) is given by:

$$J_c = -D(\Theta) \frac{d\Theta}{dx}$$

where J_c is the volume of water moved/sec across a 1 cm^2 area normal to the direction of flow, x; D(Θ) is the moisture diffusivity, cm^2/sec; and dΘ/dx is the volumetric moisture gradient of the soil in the x-direction.

Increasing soil moisture by any means not only increases the water content in the vicinity of the seeds, but more importantly, increases the moisture conductivity in the soil around the seeds and in the root zone.

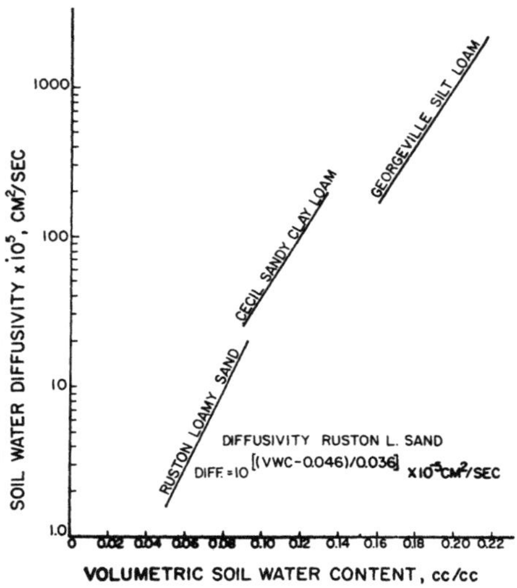

Figure 7. **Soil water diffusivity, D(Θ) as a function of volumetric water content for three soils.**

Summary

In summary, the two most critical environmental factors are temperature and impedance provided we do not plant in extremely low moisture soil. Soybean emergence is adversely effected by even a slight amount of mechanical impedance and deep plantings will require more time for emergence than shallow plantings under any conditions. Planting depth, moisture content in and above the seed zone, and compaction of the soil are very important factors in insuring the availability of moisture for the seeds. Soybean roots are not as sensitive to impedance as are soybean hypocotyls. This suggests that, especially at low moisture contents at planting, higher pressures and bulk densities in the root zone will be an advantage to the seedling. Increased soil compaction over the seed is beneficial from a soil moisture viewpoint, but it is likely to increase the impedance to serious levels.

FIELD EVALUATION OF BIOLOGICAL AND PHYSICAL FACTORS AFFECTING EMERGENCE

Seed Quality and Seed Meters

Field germination levels, while often lower, can be no higher than the percentage of viable seeds placed in the planter. An initial reduction in field populations may occur as the seeds pass through the planter seed meter. Several recently-developed commercial planter meters and grain drill meters were compared by Nave and Paulsen (5) to determine the amount of damage suffered by different varieties of soybean seed in the seed metering process. Planter metering devices evaluated included: (a) JD 7000 Max-Emerge planter single-run feed cup, (b) IH 400 Cyclo-Planter air drum, (c) IH 510 grain drill fluted roll, (d) IH 58 planter horizontal plate meter, and (e) White 5400 Plant/Aire planter air disk.

All seed meters were adjusted to a nominal seeding rate of 400,000 seeds/ha. A slow-speed drive (optional on the IH 510 drill) was used on the fluted roll meter, as recommended for soybean metering. Samples of several varieties of soybeans at three moisture levels were passed through the seed meters. They were evaluated along with a control sample for each moisture level using five germination tests. The percentage of split soybeans was also recorded. None of the germination tests were able to detect differences among the seed meter treatments and the control sample.

A significantly higher percentage of split soybeans occurred with the plate meter and the air drum meter. However, damage levels as measured by percentage of split soybeans never exceeded 1.3% compared to 0.48% for the control sample. Nave and Paulsen (5) concluded that all five commercial meters were satisfactory for planting soybeans. They did note a higher susceptibility of larger seeded varieties to mechanical damage.

Fungicide Seed Treatment and Planting Depth

Once a viable seed has been delivered to the soil, its ability to germinate and produce a seedling is a function of the total seed environment. Assuming the chemical environment is controlled at an adequate level, research was initiated to investigate some aspects of the biological and soil physical environments. The specific objectives of this study were: (a) to evaluate the use of fungicides on soybean seed, (b) to determine how planting depth affects soybean yield, and (c) to assess the effect of seeding rate on yield. A previous study by Athow and Caldwell (1) studied seed treatment and seeding rate effects on yield of soybean varieties grown in the 1950's.

Procedure

This study was conducted during 1974, 1975 and 1976 on Drummer silty clay loam and Flanagan silt loam soils of the Ag. Eng. Farm at the Univ. of Illinois. All plots were planted in corn the previous year and plowed in the fall. Fertilizer was applied in the early spring, herbicides were applied and incorporated twice, and soybeans were planted in 76 cm rows using an IH Cyclo-Planter. Plots were established in a randomized complete bock design with four replications. In 1974, three varieties (Amsoy, Amsoy-71, and Beeson) were planted at depths of 2.5-, 5.0, and 7.5-cm. Seeding rates were adjusted according to germination tests to deliver 348,000 viable seeds/ha.

In 1975, Amsoy-71 was the only soybean variety planted, but two seeding rates were used—348,000 viable seeds/ha and 469,000 viable seeds/ha. Two planting depths, 4.5-cm and 6.4-cm, were used. Thiram (Tetramethylthiuram disulfide) was used at 1 g active ingredient/1 kg of soybean seed as a treatment and became a third variable. In 1976, two soybean varieties (Amsoy-71 and Corsoy) were planted at depths of 2.5-, 5.0-, and 7.5-cm. Two seeding rates, 198,000 viable seeds/ha and 317,000 viable seeds/ha, were used. A fungicide, Captan [N-(tricholormethyl thio)-4-cyclohexene-1, 2, -diacarboximide], was applied as a seed treatment at 1 g active ingredient/1 kg of soybean seed to investigate seed treatment effects.

Results and Discussion

Emergence. The data on the influence of fungicide seed treatment and planting depth on emergence are summarized in Table 3. Significant differences in emergence among planting depths were recorded for all three varieties in 1974. Significantly higher emergence was achieved, based on Duncan's New Multiple Range Test (DMR), in all three varieties at the 2.5 cm planting depth than at the 5.0 cm depth. For the Amsoy and Beeson varieties, a significant decrease in emergence also occurred as planting depth was increased from 5.0- to 7.5-cm. A mean air temperature of 20 C and 12 cm of rainfall in the form of thundershowers during the 10 days following planting resulted in a soil physical environment conductive to germination at more shallow planting depths, but probably limited soil aeration at deeper depths and produced a soil biological environment where disease and other soil micro-organisms could thrive. These factors were along with the greater soil mechanical impedance due to increased planting depth contributed to the significant differences in emergence among planting depths.

The use of Thiram as a seed treatment in 1975 did not increase emergence levels significantly. Emergence percentages were higher for the 4.5 cm depth than for the 6.4 cm depth, and in most cases, the difference was significant according to DMR. Again in 1975, adequate moisture and temperature conditions (4.37 cm of rainfall and mean air temperature of 23 C during the 10 days following planting) resulted in a soil physical environment which favored the more shallow planting depth. The higher seeding rate almost always produced significantly higher emergence percentages according to DMR. The closer spacing of the seedlings in the row probably resulted in greater forces to overcome soil mechanical impedance.

The use of Captan as a seed treatment in 1976 did not have a statistically significant effect on emergence percentage. However, in most comparisons at the same planting depth and seeding rate, higher emergence percentages were attained by the untreated seed lots. Generally, higher emergence percentages were attained with the low seeding rate (198,000 viable seeds/ha).

Table 3. Emergence of fungicide-treated and untreated soybean seed at different planting depths, 1974-76.*

Planting depth, cm	Percent emergence of selected varieties†					
	Amsoy	Amsoy-71		Beeson	Corsoy	
	Untreated	Treated	Untreated	Untreated	Treated	Untreated
1974						
2.5	81 a		85 a	82 a		
5.0	65 b		62 b	61 b		
7.5	34 c		53 b	44 c		
1975§						
4.5		79 bc	77 bc			
		88 a	84 ab			
6.4		67 ce	63 e			
		76 bc	75 cd			
1976§						
2.5		89 ab	86 abc		84 abc	81 abcd
		78 cd	92 a		69 e	87 ab
5.0		81 bcd	92 a		89 ab	91 ab
		77 cd	89 ab		74 de	82 abcd
7.5		72 d	81 bcd		80 bcd	75 cde
		73 d	82 abcd		69 e	73 de

* Thiram was used as the seed treatment in 1975. Captan was used as the seed treatment in 1976.

† Within columns for each year, numbers with the same letters do not differ significantly at the 5 percent level, based on Duncan's New Multiple Range Test.

§Two seeding rates were investigated in 1975 and 1976. The first emergence percentage given within each column for each planting depth in 1975 is for a seeding rate of 348,000 viable seeds/ha, the second emergence percentage is for a seeding rate of 469,000 viable seeds/ha. In 1976, the first emergence percentage given within each column for each planting depth is for a seeding rate of 198,000 viable seeds/ha; the second emergence percentage is for a seeding rate of 317,000 viable seeds/ha.

The lack of measurable precipitation for 12 days following planting resulted in low soil mechanical impedances for all treatments and increased competition for moisture at the high seed rate (317,000 viable seeds/ha). A general trend of lower emergence percentage with deeper planting depth was observed. Within each seeding rate level, no significant difference was evident between the 2.5 and 5.0 cm depths.

Yield. Data on the influence of fungicide seed treatment and planting data on yield are summarized in Table 4. Significant differences in yield due to planting depth were recorded for all three varieties in 1974 using the ANOVA and associated F-Test. The use of DMR identified a significantly lower yield at the 7.5 cm planting depth than at either the 2.5 or 5.0 cm planting depths for both the Amsoy and Beeson varieties. The Amsoy-71 variety, which is resistant to Race 1 of Phytophthora root rot, also produced lower yields as planting depth increased, but these differences in yield were not significant.

In 1975, no significant differences existed among the yields due to planting depth, seed treatment with thiram, or seeding rate. Above normal precipitation and temperature contributed to an ideal soil physical environment during the first six weeks of growth and a difference of only 2 cm between the two planting depth levels resulted in a reduced range of plant populations. The soybean plants were able to compensate in terms of yield for this range of plant populations.

Again in 1976, no significant differences existed among the yields for the Amsoy-71 variety due to planting depth, seed treatment with Captan, or seeding rate. For the Corsoy variety, signficiant differences in yield did exist according to the ANOVA and F-Test for means. However, the DMR did not establish a relationship among the yield differences for any of the independent variables.

While yield is the ultimate measure of the value of adjustment of the total seed environment, its use in evaluating those variables contributing to that environment is limited. Weather variations, cultural practices, equipment malfunctions, and management decisions during the growing and harvesting season influence the yields obtained. In addition, the soybean plants' remarkable ability to compensate in terms of yield for variations in plant population make yield a rather insensitive measure of the adequacy of the total seed environment.

Planter Press Wheels

Planter press wheels may or may not be beneficial to the total seed environment, depending upon the press wheel pressure used, the design of the press wheel tire, the soil type and soil moisture content at planting, and the weather conditions between planting and seedling emergence. A study

Table 4. Yield of fungicide-treated and untreated soybean seed at different planting depths, 1974-76.*

Planting depth, cm	Amsoy	Amsoy-71		Beeson	Corsoy	
	Untreated	Treated	Untreated	Untreated	Treated	Untreated
1974						
2.5	3.22 ab		3.32 a	3.27 ab		
5.0	3.06 b		3.19 ab	3.12 ab		
7.5	2.62 d		3.10 ab	2.83 c		
1975§						
4.5		3.43	3.46			
		3.30	3.28			
6.4		3.42	3.34			
		3.30	3.45			
1976§						
2.5		2.74	2.65		3.03 a	3.08 a
		2.73	2.67		3.11 a	2.95 ab
5.0		2.71	2.74		3.06 a	2.93 ab
		2.68	2.72		3.09 a	3.06 a
7.5		2.59	2.65		2.81 a	2.83 b
		2.73	2.66		2.98 ab	3.02 a

Yield (t/ha) of selected varieties†

* Thiram was used as the seed treatment in 1975. Captan was used as the seed treatment in 1976.

† Within columns for each year, numbers with the same letters do not differ significantly at the 5 percent level, based on Duncan's New Multiple Range Test.

§ Two seeding rates were investigated in 1975 and 1976. The first emergence percentage given within each column for each planting depth in 1975 is for a seeding rate of 348,000 viable seeds/ha, the second emergence percentage is for a seeding rate of 469,000 viable seeds/ha. In 1976, the first emergence percentage given within each column for each planting depth is for a seeding rate of 198,000 viable seeds/ha; the second emergence percentage is for a seeding rate of 317,000 viable seeds/ha.

was conducted in 1978 to compare three different planter press wheel tire designs and three different drill press wheel tire widths at two soybean planting depths. The tests were conducted on Drummer silty clay loam soil with three replications of each treatment. Plots received 5 cm of irrigation water immediately after planting to accentuate differences in soil mechanical impedance resulting from soil compaction differences among press wheel tires.

Data on the effect of planter press wheel tire design and drill press wheel width are summarized in Table 5. Significant differences in emergence percentage and yield were identified using the ANOVA and F-Test for differences among means. Using the DMR, emergence percentage at 16 days after planting was significantly higher for the shallow planting depth than for the deeper planting depth for all press wheel tires except the 25 mm Smooth Crown press wheel tire. This tire had a significantly higher emergence percentage than the other press wheel tires at the deeper planting depth; otherwise no significant differences existed among press wheel tires. Yields generally followed the same pattern as the emergence percentage data, but smaller differences were recorded.

Table 5. Emergence and yield of soybeans planted at different planting depths and with different press wheels, 1978.*

Implement Press Wheel	Planting Depth, cm	Percent† Emergence	Yield† t/ha
Planter			
Mod. Flat Traction	4.2	59 abc	3.17 abcd
	6.4	44 d	2.49 f
Dual Traction	4.0	63 ab	3.09 abcd
	6.4	45 d	2.71 ef
Press & Closure	4.0	66 a	3.07 bcd
	5.9	53 bcd	2.95 de
Drill			
76 mm Dual Crown	3.2	63 ab	3.35 ab
	6.3	47 d	3.00 cde
51 mm Smooth Crown	3.2	65 ab	3.42 a
	6.3	48 cd	3.13 abcd
25 mm Smooth Crown	3.2	63 ab	3.34 abc
	6.3	60 abc	3.43 a

*Soybeans of the Beeson variety were planted in 76-cm rows using an IH Cyclo-Planter and in 18-cm rows using an IH 510 Grain Drill.

†Within columns, numbers with the same letters do not differ significantly at the 5% level, based on Duncan's New Multiple Range Test.

SUMMARY

In summary, a range of planting depths will give nearly equal results in terms of yield when seed quality and seed environmental conditions in the soil are good. When adverse seed environmental conditions prevail due to excess moisture and/or crusting, a more shallow planting depth is favored. Conversely, when poor germinating conditions exist due to less than adequate moisture levels, a deeper planting depth is better. For the soils used in this study, a planting depth of 4 cm appears to be the best compromise. Planters and drills are available from equipment manufacturers that can meter seed accurately, do negligible amounts of damage during metering, and can accurately control planting depth in our present clean seedbed situations. Finally, only when plant populations deviate more than 25 to 30% from recommended optimum populations will yield be seriously affected. Inaccuracies in predicting emergency percentage which result in such large deviations from recommended optimum populations may result from adverse weather conditions during the germination period and/or less than ideal seedbed preparation.

NOTES

H. D. Bowen, Biological and Agricultural Engineering Department, North Carolina State University, Raleigh, N.C. 27650; and J. W. Hummel, U. S. Department of Agriculture, SEA/AR, Urbana, Ill. 61801.

Trade names are used in this chapter solely for the purpose of providing specific information. Mention of the trade name, proprietary product, or specific equipment does not constitute a guarantee or warranty by the USDA or the Univ. of Illinois, and does not imply approval or the exclusion of other products that may be suitable.

REFERENCES

1. Athow, K. L. and R. M. Caldwell. 1956. The influence of seed treatment and planting rate on the emergence and yield of soybeans. Phytopathology 46(2):91-95.
2. Coble, C. G. 1972. A mathematical simulation of water movement in unsaturated soil during drying. Ph.D. Thesis, N. C. State Univ., Raleigh, N.C.
3. Gilman, D. F., W. R. Fehr, and J. S. Burris. 1973. Temperature effects on hypocotyl elongation of soybeans. Crop Sci. 12:246-249.
4. Hatfield, J. L. and D. B. Egli. 1973. Effect of temperature on the rate of soybean hypocotyl elongation and field emergence. Crop Sci. 14:423-426.
5. Nave, W. R. and M. R. Paulsen. 1977. Soybean seed quality as affected by planter meters. ASAE Paper No. 77—1004, ASAE, St. Joseph, MI 49085.

MECHANIZATION ALTERNATIVES FOR SMALL ACREAGES IN LESS-DEVELOPED COUNTRIES

M. L. Esmay and M. Hoki

World soybean production has increased nearly two-thirds in the past 7 to 8 yr, as shown in Table 1 (1). Brazil, which has increased production 7 to 8 fold in that same period of time, is giving special attention to the development of the soybean crop and processing industry (9). Indonesia, even in the tropics, is producing soybeans commercially at locations near the equator. Subtropical and tropical regions are thus being shown to have a great potential for soybean production. That potential will no doubt increase as new adaptable cultivars are made available. The improvement of soybeans is still, however, in an early stage; thus, there is much room for development (12). Technological innovation for higher production and better preservation of soybeans must be implemented to meet diversified regional conditions and requirements.

Soybean characteristics have been studied quite extensively with the possible exception of the engineering properties (2,3,13,14). More extensive and applicable data related to mechanical, hygroscopic and thermal properties of soybeans are necessary for a significant improvement of production and preservation operations (7).

TECHNOLOGY AND LOSSES

Production and post-production methods and techniques for soybeans have evolved in the less developed countries without extensive knowledge or consideration of the effect on the quality of the final harvested product. Some non-mechanized operations which have been used safely for rice can be quite harmful to soybean seeds. For example, threshing by beating or treading provides impacts sufficient to damage many seeds. Physical property data

Table 1. Soybean production data for selected major producer countries (1).

	1969-71			1977		
	Hectares	Production	Avg. Yield	Hectares	Production	Avg. Yield
	– kha –	*– kt –*	*– kg/ha –*	*– kha –*	*– kt –*	*– kg/ha –*
World	35,314	46,747	1,324	49,426	77,502	1,568
U.S.	17,036	31,174	1,830	23,435	46,712	1,993
Brazil	1,314	1,547	1,178	7,059	12,100	1,714
China	13,859	11,398	822	14,236	12,955	910
Indonesia	643	468	728	663	527	795

for soybean seeds show they have less mechanical strength than rice. Table 2 shows that the compressive strength of soybeans was 1/6 to 1/7 of that for rice in the moisture ranges tested. Conventional combines have similar difficulties in minimizing damage to soybean seeds as they were designed originally for small grains with higher resistance to mechanical impact.

Considerable effort has, however, been put into the development of large "modern" soybean harvesting machines. The problem is how to focus attention on the post production quality loss problems in the less developed countries. Engineers like to think of themselves as doing research at the forefront of knowledge. The results of such new and basic research are then quite naturally applied to the largest and most expensive machinery. The cost of research can no doubt be recovered most quickly by including it as a small part of the sale price of the high priced machines. Unfortunately, these research and design procedures do not help the less developed countries that can only afford low cost hand and animal technology. Creating something other than the largest, most complex machine has not been considered very exciting or even worthwhile. Intermediate technology seems only to be interpreted as some routine adoption of a long ago discarded machine from the "advanced" countries. Occasionally such an application might work but generally does not.

APPROPRIATE TECHNOLOGY

Appropriate technology must take on increased meaning even in the "developed" U.S. For the first time ever, engineers must begin to consider energy efficiency. Technology just for the sake of technology is being questioned. Is bigger always better? Do new designs always have to be more expensive and more complex? Is labor efficiency the only index for innovation and mechanization? Can various product quality losses be sacrificed arbitrarily for mechanization? Does mechanization have to be more and more automated? Will technology that makes possible the concentration of the population masses in a few large cities be forever justified?

Table 2. Ultimate compressive strength of soybeans and rice (4,6).

M.C.	Soybeans	Rice
– % w.b. –	– kg/mm^2 –	– kg/mm^2 –
8	2.5	16.0
13	1.1	8.1
17	0.6	4.5

Appropriate technology for a primitive society may mean the design of a better club for beating the soybean vines for threshing. Possibly a wider, softer club would separate more bean seeds easier and without the sharp impact responsible for damage. But, who wants to work on such a mundane problem? It would be argued, I am sure, that engineering skills are not needed for such simple improvements. But why not? Cannot the same laboratory physical property results be applied to solving the threshing by beating problem as for a complex threshing machine? The problems may in fact be more complex than threshing with a machine. For example, uneven maturity at time of harvest is often encountered. This may be due partly to the mixed varieties and partly to the varying conditions of tillage, water management, fertilizer and pesticide application. As these unpredictable conditions may remain for some time, along with the susceptibility to mechanical damage of soybeans, a device for selective harvesting may be justified in some areas. A good example is "Ketap or ani ani" for rice. This is a selective harvesting tool still used widely because of its simplicity and ingenuity incorporated with labor intensity (5). Many people would benefit from improvement of the hand bean threshing operation or the development of a simple selective harvesting device unless the "improved" technology removed itself from their reach by its high cost and/or put them out of a job because of undesirable labor efficiency.

SOLAR DRYING

Pride, recognition and compensation must be promoted for designs that are truly appropriate for given situations under specific conditions. The social and public costs and returns must be considered as well as the private cost and returns. Antoher example of a need for an appropriate technology is for drying of beans and grain crops. Sun drying has been used for centuries. It is a use of free solar energy, but there are problems. Sun drying as generally practiced is an uncontrolled process. Some seeds, due to the random exposure, are overheated and dried too fast and too much, while others are underdried. High temperatures and overdrying cause quality deterioration.

There is a need for a simple, low cost, non-mechanical solar dryer with an essentially self-controlling temperature. The dryer would require no outside mechanical power source, which is always expensive and also burns expensive fossil fuels. The design of such a solar drying requires the careful and skillful application of some basic engineering concepts of heat transfer and fluid flow; however, few engineers are interested. Comparatively, it takes little engineering expertise to assemble mechanical components to force heated air through a fixed bed of beans or grain. The design of a successful low-cost solar grain dryer with clear and black polyethylene plastic and some wood framing materials requires the application of the basic engineering concepts of thermal convection, drying theory, solar radiation and product physical properties. The thermal convection and temperature must be balanced in such a way that during intense solar radiation the dryer does not overheat the product. If designed properly, the air will merely flow through the product faster rather than increasing appreciably in temperature. When clouds shield the sun or during a rain shower air movement and drying would cease because of a lack of thermal convective forces. As compared to open sun drying the solar dryer would have two important advantages; one would be temperature control, which can eliminate overdrying and much seed checking and cracking; and two, protection from rain showers and rewetting.

RESEARCH NEEDS

The hygroscopic and thermal properties of many grains have been studied although less attention has been given to soybeans. The recent rapid increase in production with the associated problems accelerates the need for research on the processing and storage of soybeans (8,10,11).

Soybeans are a high value, high protein crop. Thus, countries around the world are promoting production. Many countries, particularly in Asia, are mainly concerned with supplying their own consumption demands, while others such as Brazil are interested in expanding soybeans into a major export commodity. Bean production in most Asian countries is labor intensive and on small farms with field sizes that average less than one acre in size. In many of the rice producing tropical countries soybean production is an off-season crop following one or two crops of rice. Since labor is adequate and yield levels low, considerable increase in land productivity is necessary. Soybean production may be totally a hand operation in many of these countries. The only tools used would be a hoe-type device for primary tillage and a sickle for cutting the crop. This does not necessarily mean that improved seed varieties are not used. Higher yielding varieties are often planted and plant protection chemicals applied.

The International Crop Research Institute for the Semi-Arid Tropics (ICRISAT) in India is developing a farming system for beans, peas and sorghum utilizing animal power. A permanent ridge and ditch cultivation approach is being tried. The narrow ditches of 1.5 m on center serve as paths for the animals and implement wheels in order to prevent soil compaction in the plant growing ridge area. The ditches may also serve for irrigation.

CONCLUSIONS AND RECOMMENDATIONS

Soybean production in most developing countries is labor intensive and on small farms. Land productivity can be increased through the careful utilization of labor and adoption of selected new varieties and cultural practices. Appropriate mechanization and technological improvements for the post-harvest operations are necessary to preserve high quality and minimize losses. The physical properties of soybeans must be known and considered in the design of improved tools and machines for the post production operations. Greater attention must be given to the design and development of technology that is appropriately simple, low cost and energy efficient for small producers with little capital.

NOTES

M. L. Esmay, Department of Agricultural Engineering, Michigan State University, East Lansing, Michigan 48824; and M. Hoki, Department of Agricultural Engineering, Pennsylvania State University, University Park, Pennsylvania 16802 (on sabbatical leave from Mie University, Tsu, Japan).

REFERENCES

1. FAO Agricultural Statistics. 1978. FAO, Rome, Italy.
2. Goodman, R. M. (Ed.). 1976. Expanding the use of soybeans. Proc. of Conf. for Asia and Ocenia.
3. Hill, L.D. (Ed.). 1975. World soybean research. Proc. World Soybean Research Conf., Interstate Printers and Publ., Danville, IL.
4. Hoki, M. and K. Tomita. 1976. Moisture effects on mechanical strength of soybeans. Bull. Fac. Agric., Mie University 51:103-110.
5. Hoki, M. 1977. Farming operations and labor requirements for paddy cultivation in Sarawak, East Malaysia. Southeast Asian Studies, Kyoto Univ., Japan 15(3): 457-471.
6. Lee, K.-W. and O. R. Kunze. 1972. Temperature and moisture effects on mechanical properties of rice. ASAE paper no. 72-338. ASAE, St. Joseph, MI.
7. Nicholas, C. J. 1978. Enhancing and maintaining soybean quality in the threshing-marketing system: II. Marketing researcher's view. Grain and Forage Harvesting, ASAE, St. Joseph, MI.
8. Norman, A. G. (Ed.). 1978. Soybean physiology, agronomy and utilization. Academic Press, New York.
9. Overhults, D. G., G. M. White, H. E. Hamilton and I. J. Ross. 1973. Drying soybeans with heated air. ASAE Trans. 16(1):112-113.
10. Ramstad, P. E. and W. F. Geddes. 1942. The respiration and storage behavior of soybeans. Minnesota Agr. Expt. Sta. Bull. 156:46.
11. Stephens, L. E. and D. L. Rabe. 1978. Crop properties data for harvesting. Grain and Forage Harvesting, ASAE, St. Joseph, MI.

12. Whigham, D. K. and H. C. Minor. 1978. Agronomic characteristics and environmental stress. In A. G. Norman (Ed.) Soybean physiology, agronomy and utilization. Academic Press, New York.
13. Whigham, D. K. (Ed.). 1975. Soybean production, protection and utilization. Proc. of Conf. for Scientists of Africa, Middle East and South Asia. Univ. of Illinois, Urbana, Illinois.
14. White, G. M., O. J. Loewer, I. J. Ross and D. B. Egli. 1976. Storage characteristics of soybeans dried with heated air. ASAE Trans. 19(12):306-310.

SOIL CONSERVATION PRACTICES IN SOYBEAN PRODUCTION

J. C. Siemens

Several alternative methods can be employed to reduce erosion on land used for the production of soybeans. Three principal alternatives are terracing the land, performing all field operations on the contour, and practicing conservation tillage. Terracing has not been well accepted because it is expensive and not well adapted to mechanization. Performing all operations on the contour is inexpensive, but in many situations it is impractical and not sufficiently effective in controlling erosion. Conservation tillage is an approach aimed at combining efficient row-crop production with adequate control of soil erosion by wind and water. It is the objective of this chapter to discuss that approach.

The effectiveness of conservation tillage in controlling erosion depends upon creating and maintaining a surface that resists erosion. The surface's resistance to erosive forces results from the protective cover of the crop during the growing season and from crop residues or a rough, stable soil surface at other times of the year. Conservation tillage systems employ various tools for primary tillage operations (such as the moldboard plow, chisel plow, subsoiler, or disk) and for secondary tillage (such as the disk, harrow, or field cultivator). Numerous other tools can also be used. The degree of success achieved in controlling erosion depends on when and how the tools are used. Three commonly used conservation tillage systems are the chisel-plow, disk, and no-tillage systems.

Farmer acceptance of conservation tillage depends primarily upon whether such systems provide greater profit than is attained with conventional tillage methods. The profit, of course, is obtained by subtracting the costs from the income, which is directly related to yield.

Although the costs of erosion control practices can be determined quite readily, their benefits are more difficult to quantify. Few data are available to use in evaluating the economic impacts of erosion. Erosion causes damage to the land on which it occurs and, of perhaps even more importance, causes damage downstream. Sediment carried by runoff water from agricultural land is regarded as the major nonpoint source pollutant of streams, rivers, and lakes in the U.S. Legislation in some countries requires that erosion be controlled to some tolerable level.

From the preceding discussion it is apparent that three major factors must be evaluated if one wishes to recommend conservation tillage to farmers. These factors are the effects on (a) income or yield, (b) costs, and (c) erosion.

EROSION CONTROL

Only a few studies have been conducted on the effect of soybean production on soil erosion. Miller (3) showed that a rye cover crop could be used to reduce erosion after growing soybeans. Browning et al. (1) conclude that soybeans leave the soil loose and susceptible to erosion. Van Doren et al. (6) studied erosion on a corn-soybean rotation. The corn plots showed a higher soil loss particularly in the spring following the soybean crop.

Moldenhauer and Wischmeier (4) measured soil erosion caused by natural rainfall. On land where soybeans were grown after corn, the soil loss was similar to that where corn followed corn. Where corn was grown following soybeans the soil loss was 40% greater than with continuous corn. Mannering et al. (2) and Siemens and Oschwald (5) concluded that soil losses in the spring were much greater after soybeans had been grown than after corn. In the study by Siemens and Oschwald (5) a rainfall simulator (rainulator) was used to apply intense "rain storms" to plots on which soybeans had been grown. After the growing season the plots received three different treatments: moldboard plowing, chisel plowing, and no tillage. Rainulator runs were made on the plots the next spring before any spring tillage took place. A similar set of rainulator runs was made for the same tillage treatments after growing corn on the plots.

For all three tillage treatments, a higher percentage of the water applied ran off after soybeans had been grown than after corn (Figures 1 and 2). The soil surface appeared to be more stable after corn that after soybeans. On the fall-moldboard-plowed and chisel-plowed plots the rough surface persisted for a longer period of time following corn than following soybeans. The higher runoff following soybeans also reflects the smaller quantity of residues left after that crop has been harvested.

Soil loss with fall plowing was two to three times greater after soybeans than after corn (Figures 3 and 4). After soybeans, the soil loss caused by 100 mm of rain was over 18 t/ha with the fall-moldboard plow treatment,

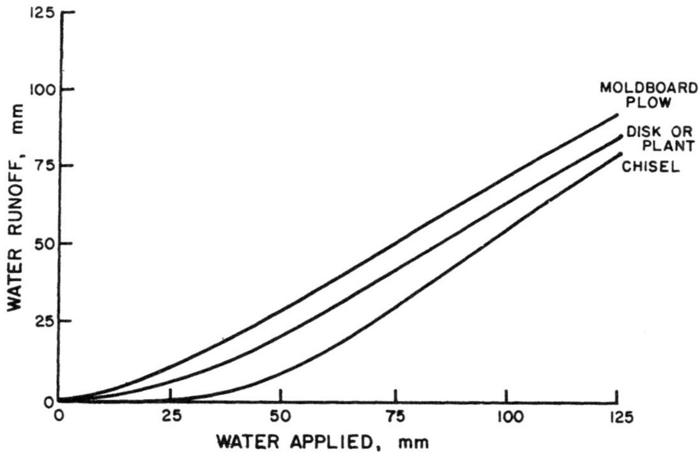

Figure 1. **Water runoff after soybeans.**

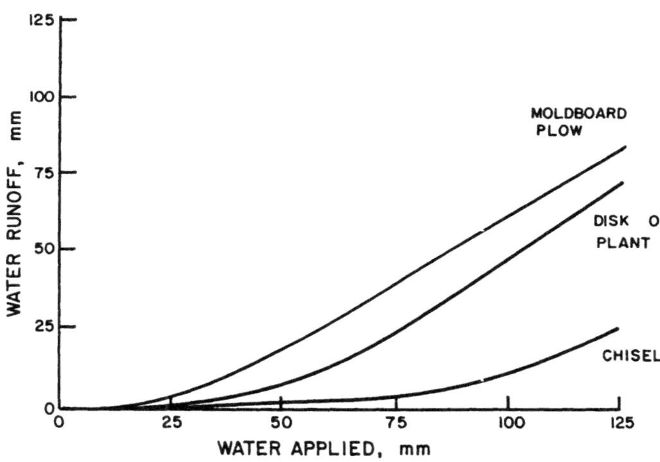

Figure 2. **Water runoff after corn.**

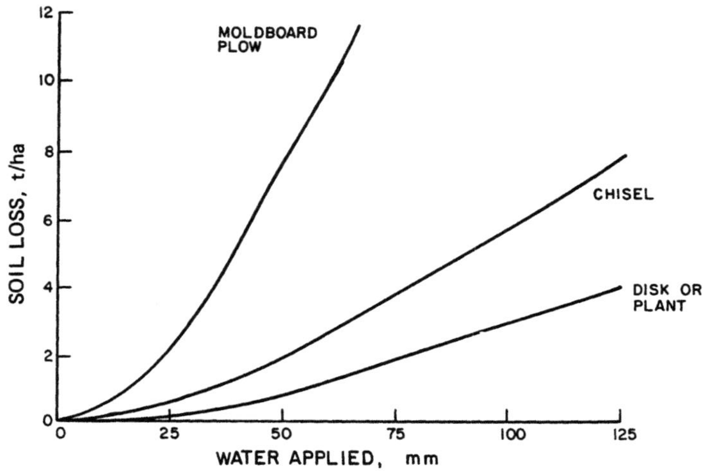

Figure 3. **Soil loss after soybeans.**

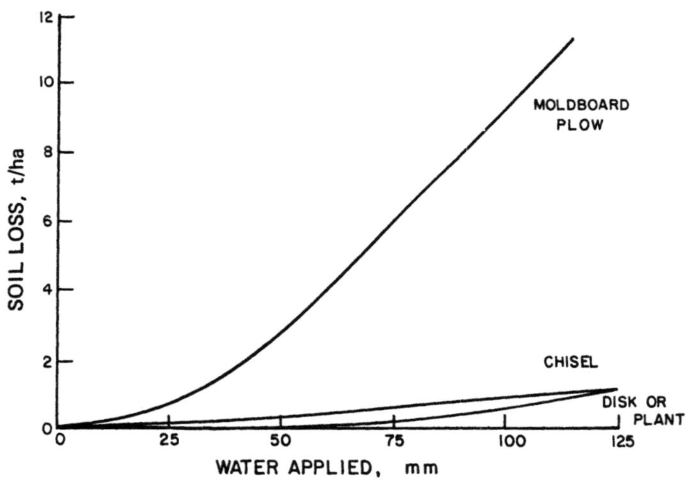

Figure 4. **Soil loss after corn.**

6.2 t/ha with the chisel-plow treatment, and 2.5 t/ha with no tillage. After corn the first 100 mm of rain carried 10 t soil/ha from the surface with the moldboard plow treatment and less than 1.0 t/ha with the conservation tillage treatments. The larger quantities of residue after corn and the relatively loose, easily erodible soil following soybeans account for the greater soil losses after the latter crop. It is concluded that excessive soil erosion is much more likely after soybeans than after corn.

YIELD

Numerous factors affected by tillage have an influence on yield. The main factors are the amount of plant residue on the soil surface, the distribution of nutrients, the weed infestation, the plant population, and possibly the soil density. Plant residues on the soil surface decrease the evaporation rate and the rate at which the soil warms up in the spring. Depending on the conditions, these effects may be detrimental to soybean growth. As far as could be determined, the effects of surface residues on soybean growth have not been assessed.

Fertilizers and lime are mixed uniformly in the tilled layer when tillage systems such as moldboard plowing are used. With chisel plows, disks, and no tillage, less mixing of the soil takes place. As a result, nutrients, especially phosphorus, potassium, and lime are concentrated near the soil surface. It is not clear, however, whether the nonuniformity of nutrient distribution to the depth normally tilled reduces yields.

Weeds must be controlled to obtain maximum yield. As the amount of tillage is reduced below that required for a weed-free seedbed, weed control problems increase. Several factors contribute to weed control problems when using reduced tillage. With conservation tillage systems in which the moldboard plow is not used, weed seeds are left on or near the surface. Also, plant residues on the surface reduce evaporation, making the soil at or near the surface moist enough for weed-seed germination over longer periods of time.

Soybeans are often planted in late spring. With conservation tillage the amount of spring tillage must be reduced. Hence, weed control may not be satisfactory because weeds are allowed to emerge before planting. Realistic herbicide applications may not control weeds that emerge both before and after planting. With most conservation tillage systems, preplant-incorporated herbicides cannot be used because two tillage operations are required for incorporation after applying the herbicide.

Usually, soybean plant populations are nearly the same with conservation and conventional tillage methods. However, the reduced populations sometimes encountered with conservation tillage may be caused by poor seed-soil contact in the rough seedbed and by some seeds being placed in contact with the residue.

Yields obtained with the different tillage systems are of the utmost importance. In our experiments we have obtained slightly lower yields with reduced tillage, especially on poorly drained soils having high organic-matter content. On well-drained soils with little organic matter, the reduction in yield with reduced tillage systems seems to be minimal. Numerous studies have reported yields with conservation tillage systems equivalent to those obtained with the moldboard-plow system.

COSTS

To compare costs for different tillage systems one must determine the types of machinery to be used for each system. The size of the machinery must be selected according to the farm size and the amount of labor available. It is important when selecting the machinery set for a farm that scheduling of the operations be considered so that any decrease in yield caused by untimely operations can be taken into account.

The costs of a tillage system include the fixed and variable machinery costs, labor costs if applicable, and the costs related to untimely field operations. Costs for pesticides, fertilizers, and other inputs affected by the tillage systems being compared must also be determined.

Costs for producing soybeans with different tillage systems have been estimated. In general, as the amount of tillage is reduced, machinery-related costs decrease and pesticide costs increase. Considering only these two variables, the costs for the different tillage systems are about the same. Thus, from the standpoint of cost it makes little difference which tillage system is used.

SUMMARY

Several tillage alternatives are available to soybean producers. The best system for a given producer depends on a number of factors such as location, slope, and soil type. Total cost differences between practical systems are minor, although yields may be slightly lower with conservation tillage systems. However, the greatly reduced soil erosion rate achievable with a conservation tillage system is an important benefit that must be fully considered.

NOTES

J. C. Siemens, Department of Agricultural Engineering, University of Illinois, Urbana, Illinois 61801.

REFERENCES

1. Browning, G. M., M. B. Russell, and J. R. Johnston. 1942. The relation of cultural treatment of corn and soybeans to moisture condition and soil structure. Soil sci. Soc. Proc. 7:108-113.

2. Mannering, J. V., D. B. Griffith, C. B. Johnson, and R. Z. Wheaton. 1976. Conservation tillage-effects on crop production and sediment yield. Paper No. 76-2551, ASAE, St. Joseph, Michigan.
3. Miller, M. F. 1936. Cropping systems in relation to erosion control. Missouri Agr. Exp. Sta. Bull. 16:17-19.
4. Moldenhauer, W. C., and W. H. Wieschmeier. 1969. Soybeans in corn-soybean rotation, permit erosion but put blame on corn. Crops and Soils Mag. 21(6):20.
5. Siemens, J. C., and W. R. Oschwald. 1978. Corn-soybean tillage systems: erosion control, effects on crop production, costs. ASAE Trans. 21(2):293-302.
6. Van Doren, C. A., R. S. Stauffer, and E. H. Kiddler. 1950. Effect of contour farming on soil loss and runoff. Soil Sci. Soc. Proc. 15:413-417.

EQUIPMENT FOR NARROW-ROW AND SOLID PLANT SOYBEANS

R. I. Throckmorton

The American farmer has seen a decade of planter improvement. The changes in seed metering devices have removed the need to match seed and seed plates, have provided monitors for proper operation, as well as changing seed population "on-the-go". However, only one major change in row spacing (to 30 in.) has been general in the industry. Progressive soybean growers desiring to increase yields via narrow rows can well ask, "What has the equipment industry done for me recently?" The purpose of this chapter is to answer that question and present recent developments.

The continued growth in soybean production has placed it in the number two position in hectares for harvest in the U.S. in 1978 (Table 1). This outstanding success record has, in a large part, been due to the U.S. farmer's ability to meet the market needs created for soybeans for domestic and export use. The Soybean Association's market development efforts have been a real catalyst in developing this demand. One would have to question if soybeans could have reached second place in the U.S. without this association's efforts.

Recent data indicate that approximately 50% of the U.S. soybean crop is exported and there is a continued optimism for additional exports to such countries as the Peoples Republic of China where oil use is 7 lb yearly per individual, compared to 55 lb per individual in the U.S.

At the first World Soybean Research Conference at Urbana, Ill., in 1975, Baumheckel (1) presented a paper on soybean row spacing trends. His title describes very well the changes in the last 25 to 30 yr as farmers stopped drilling soybeans due to weed infestation and placed the crop in the same row spacing as other principle crops for mechanical cultivation. With new

Table 1. **Millions of hectares for harvest.**[a]

	1977	1978
Corn for grain	28.3	27.4
Soybeans for beans	23.3	26.5
All hay	24.5	24.8
All wheat	26.8	22.9

[a]USDA Crop Report Board (October 11, 1978).

herbicides there is a recognizable number of farmers completing the circle back to solid seeding. For those interested in the well documented change in soybean row spacing over the years, I strongly recommend that you refer to Baumheckel's paper.

The advantages of 30 in. and narrower rows for soybeans was recognized during the research for narrower corn row spacings. It was known, at that time, that 30 in. was not optimum for soybeans, but the convenience of 30 in. corn and soybean culture for the farmers kept soybeans in 30 in. rows for several years.

ONE STEP FORWARD

In recent years, we have witnessed two predominate milestones in soybean production: (a) herbicides for pre- and post-emergence weed control and (b) the development of semi-dwarf determinant varieties which have the capacity for high plant populations without lodging.

We have also experienced new planting equipment for a variety of row spacings and farming systems. While these developments are not likely to be the final answer, the equipment industry has taken "one step forward" to provide equipment better suited to solid seeded and very narrow row soybean planting. This chapter will deal in detail on new equipment developments for very narrow row and solid seeded soybeans.

I have attempted to group the equipment to be described into four categories. At this stage, many individuals could provide their own equally appropriate set of categories. Before too long, the equipment industry will have to arrive at descriptions acceptable to farmers, to avoid confustion. The categories I have selected are: solid seeding equipment—drills; very narrow row equipment—drills and planters; narrow row equipment—row crop planters; and farmer developed equipment. The equipment and features to be described were largely submitted by the manufacturers.

SOLID SEEDING EQUIPMENT—DRILLS

This equipment is generally represented by equipment which approaches equi-distant planting patterns—which I have chosen to call solid seeding. Rows are from 6 in. to 8 in. wide with approximately 2 to 2½ plants/foot of row.

Crustbuster Drill—The double disk end wheel style is available in 13 ft to 41 ft widths which fold for transport. Several types of gauge or press wheels are available including a 2 in. x 13 in. and 1 in. x 10 in. combination gauge/press. Both of these press wheel units control the opener depth as well as firm the seed from the sides with loose soil over the seed. The single 1 in. x 10 in. spring loader seed firming wheel firms the seed into moist soil and should do a good job when used with a seed bed finishing tool just ahead of the drill. This manufacturer also offers press type drills in 7 ft single to 41 ft three-section folding models.

John Deere (Moline, IL) offers both end wheel and press drills in conventional configuration. In addition, Deere has just announced a new drill in the 10 in. and 12 in. row spacings. This will be reported in the very narrow row category.

The *Great Plains Drill* (Assaria, KS) is a three section, end wheel style which is easily folded for transport. It has been shown at soybean field days, and the manufacturer recommends it for soybean planting.

Haybuster (Jamestown, ND) offers end wheel drills as a single 12 ft or double 24 ft size. Spacing is 6 in. and, as most drills, it can be plugged to plant 12 in. or 18 in. rows. The seed firming wheel presses seeds into moist soil before the seeds are covered. The press wheel also gauges the opener depth. With adjustable opener down pressure to 400 lb/opener the drill can be used for both first crop and double crop.

The *International Harvester* (Chicago, IL) press drill has been well accepted for soybean planting and is preferred as multi-hookups and easy transport which remains on the drill. Spacings from 6 in. to 8 in. (or skips) in widths of 7 ft to 14 ft single units up to multi-hookups of 56 ft are available. Hydraulic track erasers are mounted on the drill hitch frame and lift with the openers for turns or transport. The seed feed shaft speed is standard for soybean seed planting population ranges.

Just introduced for sale in 1979, is the I.H. "Soybean Special" drill with a combination gauge/press wheel with 3/8 in. depth increments. The special gauge press tire has a depressed center to firm the seeds from the sides leaving the center more loose for easier emergence. It is available in 6 in. to 8 in. rows and 8 ft to 13 ft single units. Sowing charts are in seeds per unit length of row rather than lbs per acre.

Melroe recently introduced a multi-unit end wheel type drill in 6 in. or 7 in. spacing and double disk openers. Also from Melroe is the 702-3D no-till drill with 6 7/8 spacing. Sizes range from 10 ft to 13 in. The 11 in. coulter mounter on each pressured opener cuts residues for no-till or double cropping practices. The Melroe press drill is 6 in. or 7 in. spacing in single sections of 5 ft to 14 ft with multi-hookups available. Press tires are 2 in. round steel or 2½ in. semi-pneumatic rubber tires.

The *Tye* drill (Lockney, TX) is a mounted drill in 80 in. to 320 in. widths in 6 2/3 in. rows. The convertible end wheel drill is 160 in. Four types of press wheels are available: 2 in. x 13 in. gauge/press, 1 in. x 10 in. spring loaded seed firming, 4 in. x 12 in. gauge/press or 6½ in. x 20 in. gauge/press. Row spacing variations allow for popular configurations of 14 in. with a 30 in. tire track lane. Track erasers are available and mount on the main frame bar. Coulters are available for no-till or double cropping practices.

EARLY STANDS AND CANOPY CLOSURE

Solid seeded stands appear consistent from the roadside, however, in the earlier stages stands observed in the field may appear irregular. These irregular stands may cause farmers some mental concerns, however, research has generally found that properly drilled seeded stands are adequate.

A 6 in. row can close the plant canopy in 5 weeks, helping to control weeds. At harvest, the lack of cultivator ridges and the higher bottom pod ht reduce harvest losses and increase the ease of harvest.

New varieties being developed for solid seeding are shorter to avoid lodging. Higher plant populations of these new types generally respond with higher yields.

SECONDARY TILLAGE FOR SOLID SEEDING

A secondary tillage operation just ahead of the grain drill has produced some of the best stands in research trials. Such equipment may include packer-mulchers, spring tooth attachments with points or sweeps, and new tools such as the Lilliston "Tillager". Caution must be exercised when using spring tooth attachments with points, the drill opener can easily slide into one of the tooth slots and plant too deeply. Spacing of the spring teeth so the opener runs between the slots is critical.

VERY NARROW ROW EQUIPMENT–DRILLS AND PLANTERS

This equipment category is generally represented by drills and planters with 10 in., up to and including 20 in., spacings. By plugging various feed cups in drills, they can be used for very narrow rows with and without tractor paths in many planting patterns.

Allis Chalmers (Milwaukee, WI) offers the "Air Planter" for 4 through 12 rows in 20 in. spacing with provisions for no-till practices. Just recently introduced is a 15 in. row spacing, "Air Planter" unit for pull or mounted planters in 4 to 12 rows. Opener mounted depth gauge wheels are available for the new 15 in. row unit.

The *Buffalo Slot Planter* is available in 4 to 12 rows in 20 in. spacings for first crop or double crop conditions. A narrow seed firming wheel presses seed into moist soil before covering with loose soil from the disk coverers.

The *Burch "Bean Machine"* (Evansville, IN) has 10 in. row spacings in 14 to 24 rows. The twin row planting units are gravity drop with no knockers, springs or cut offs. Due to the almost solid press wheel feature the Bean Machine plants to a uniform depth even in variable soils.

Cole (Charlotte, NC) offers 20 in. row spacing in tool bar type planters in sizes of 8 through 12 rows.

The *International Harvester* unit planters are available in 14 in. rows up to 12 rows wide.

John Deere (Des Moines, IA) just introduced a new "Tru-Vee" opener for grain drills. The 10 in. row is in zig zag rank in 8 to 13 ft widths. Twelve in. rows can be planted with the straight rank configuration. A variety of spacings can be obtained by plugging the metering cups. Two and three drill hitches to 39 ft are available. The "Tru-Vee" is the same principle for depth control as on row crop planters and utilizes two 2 to 13 in. replaceable rubber gauge wheels. Depths of 3/4 to 3½ in. in 3/8 in. increments are available. Covering is accomplished by the two small spring coverers which knock down the edge-putting soil over the seed. Seed pressing is with gang press wheels.

The *Lilliston* (Albany, GA) planter is available in 20 in. row spacings with 4 through 8 rows. It can be equipped for first crop or no-till double cropping after small grain or in sod.

The *Tye* grain drill can be set up for single very narrow row plantings. The mounting system provides an almost infinite arrangement of row spacings and skip rows for tractor paths. The track remover on the drill assures all units penetrate the soil for planting. The type "Twin Row" system can also be utilized for bed or flat planting.

The *White Company* (Oak Brook, IL) offers 18 in. rows planters in 6 to 16 row versions as well as 20 in. rows in 6 to 16 rows with the "Plant Air" system. Both systems are adaptable to no-till or minimum tillage operations. Also available is a 19 in. row tool bar configuration with skips for tractor paths.

NARROW ROW EQUIPMENT—ROW-CROP PLANTERS

This category includes unit planters and conventional row-crop planters with 21 to 30 in. rows capabilities. These spacings can also be obtained by plugging drill holes as in the previous very narrow group.

While a great deal of activity is apparent in the 6 to 20 in. row spacing range, many farmers have not yet gone from wide rows to 30 in. rows. The Illinois Crop Reporting Service recently published a report on row spacing for soybeans for 1978. The report indicates that 49% of the Illinois acreage is planted in rows over 30 in. The same report indicates that 63% of the farmers use greater than 30 in. rows for soybeans.

Since 30 in. is the most popular single row spacing in the "other than wide row" category, equipment for 30 in. rows should also be included. Almost every planter manufacturer provides a variety of 30 in. row planters. Only a few representative types will be covered here: Allis Chalmers — 28 and 30 in. in 4 through 8 rows; Buffalo — 28 and 30 in. in 2 through 12 rows; Burch — single 30 in. or paired rows in 2 through 8 rows; Cole — 30 in. rows in 6 and 8 rows; John Deere — 30 in. rows in 4 through 16 rows; Ford (Troy, MI) — 24 to 30 in. pull type in plate or air types; International Harvester — 28 and 30 in. pull type 4 through 8 rows, 28 and 30 in. mounted 4 through 16 rows, special tool bars in 6 rows 20, 24, and 26 in.; Kinze bar — 30 in. rows in 16 to 24 rows; Orthman — wide variety of narrow row spacing and sizes; and Tye — single units on tool bar, variety of spacing and number of rows.

FARMER DEVELOPED EQUIPMENT

Those attending the 1975 Conference saw the slide of an experimental precision 7 in. row spacing soybean planter. It was a double drum planter to provide for 16 rows. Cost, weight, and other factors resulted in dropping this development.

Raymond Furrer of White County, Indiana, developed a double drum pull type Cyclo planter to plant eight 38 in. corn rows and fifteen 19 in. soybean rows. His previous experience with a unit planter set up on 19 in. rows out-yielded the 38 in. by about 10 bu/acre. He disliked the in-the-row spacing of the unit planter and also wanted to use the same planter for corn and soybeans. He plants solid rows and uses a post emergence herbicide, although he has no reservations to driving over a row early in the season to cultivate, as the beans generally recover from the early traffic.

On a large scale, Bob Anderson of Warren County, Indiana, converted a 12 row 30 in. Max-Emerge planter to a combination 12 - 30 in. rows for corn and 21 - 15 in. rows for soybeans. There are 4 rows left open for tractor tire paths. The 9 extra rows were mounted in front of the main frame and seed delivered from a soybean hopper and fed through the starter fertilizer tube and into the disk fertilizer opener. The depth was gauged by a pair of cultivator wheels mounted on each side of the starter fertilizer opener. The planter drive was continued over the main frame to a front central drive shaft. Behind each front rank opener the planter carrying tires were changed to dual rib tires similar to those on the end wheel grain drills. This put more of the weight on the sides of the seed row.

In Marshall County, Iowa, J. D. Hunt converted an 8-row Max-Emerge planter to a 12-row 19 in. soybean and 8-row 38 in. corn planter. Four rows were added leaving tractor paths for cultivation. He reported he did not realize any yield increase due to problems of chemical weed control.

Further north in Wright County, Iowa, Bob Kalkwarf rearranged a mounted Cyclo planter to plant eight 30 in. corn rows and fifteen 15 in. soybean rows. This was an experimental machine furnished by International Harvester to plant sixteen 15 in. soybean rows and eight 30 in. corn rows, however, this arrangement required moving the units on the tool bar. Bob wanted an easier system which did not require moving the units. He did this by placing a unit between each 30 in. row and placing the wheel track units on each tip of the tool bar. The soybean units are held up by a pin and clip-guard when planting corn. With the planter lowered the pins can be easily removed to allow all 15 units to plant soybeans. The rear carrying wheels run between rows. The entire soybean planting width matched his combine platform width and he has subsequently purchased the planter.

SEED METERING SYSTEMS

In reviewing the equipment discussed in this chapter, there is a variety of seed metering devices utilized. There has been some question by farmers as to the possibility of seed damage by some types. The Univ. of Illinois conducted seed damage tests with a variety of planters and new grain drills. The results as given in M. R. Paulson's report show no statistical difference in seed damage from the many types of metering devices at normal seed moisture content.

In the future there will be many more steps forward in soybean production as new varieties and practices develop. Increasing yields through the adoption of new practices and equipment options will help the farmer produce more soybeans—at a profit.

NOTES

R. I. Throckmorton, Product Research, International Harvester Company, Chicago, Illinois 60611.

REFERENCE

1. Baumheckel, R. E. 1976. Planting equipment and the importance of depth control, pp. 190-196. In L. D. Hill (Ed.) World Soybean Research, Interstate Printers and Publishers, Inc., Danville, Ill.

SOYBEAN DAMAGE DETECTION

M. R. Paulsen

Recent publicity over complaints about the quality of U.S. soybeans for export has stimulated increased concern for soybean quality research. One of the major problems in measuring soybean quality is the large amount of variability among samples. Such variability may be due to the subjective nature of the damage tests, randomness of the sample, and the relatively small size of the samples collected.

The purpose of this chapter is to discuss some of the soybean damage tests we have conducted in the past few years and to show the application of some of these tests to the evaluation of soybean planter metering mechanisms. Similar tests performed by Newbery et al. (7) were also used in evaluating soybean damage from conventional and rotary combine threshing mechanisms.

Soybean damage may manifest itself in the form of physical breakage, seedcoat crackage, and internal damage. Tests for evaluating these different forms of damage all require a random sample representative of the seedlot being evaluated.

PHYSICAL BREAKAGE

Sieving was a very successful method for determining physical breakage. In our studies a sample of 1000 to 1500 g was weighed and sieved on a Clipper seed cleaner to separate the broken beans from the whole beans. The cleaner was equipped with a 4.76-mm by 19.05-mm (12/64-in. by 3/4-in.) slotted sieve on top and a 3.97-mm by 19.05-mm (10/64-in. by 3/4-in.) slotted sieve below. Material retained on the sieves was handsorted to remove splits and foreign material from the large whole beans on the top sieve and the small whole beans on the bottom sieve. Material passing through both sieves was sieved by 4.76-mm (12/64-in.) round hole and 3.18-mm (8/64-in.) round hole hand dockage sieves. Splits and other material were retained by the two sieves. Material passing through the 3.18-mm sieve was classified as foreign material. Percentages of large whole beans, small whole beans, splits, and foreign material were calculated by wt. Samples for oven moisture tests were also collected at the time of sieving.

Another test, the Stein Breakage test, was developed originally for shelled corn to test for susceptibility to future breakage due to handling. This test was applied to soybeans by placing a 100-g sample of beans (previously sieved on a 3.97-mm by 19.05-mm slotted sieve) into the tested. After the sample was impacted for two min by a rotating impeller, the sample was removed and resieved on the same sieve. Stein breakage was defined as the wt percentage of material that passed through the sieve after resieving. The effectiveness of this test on soybeans has usually been inconclusive in our experiments.

SEEDCOAT CRACKS

Soybeans may be completely whole yet have seedcoat damage in the form of seedcoat cracks. We have used indoxyl acetate, sodium hypochlorite, and tetrazolium tests, although tetrazolium may also be considered an indicator of internal damage.

The indoxyl acetate test consisted of immersing 100 whole soybeans in a 0.1% indoxyl acetate-ethanol solution with a pH of 6.0 for 10 sec (8). The beans were removed from the solution and sprayed with a 20% household ammonia-distilled water solution for 10 sec. Next the beans were dried with unheated air from a small blower. Finally, those soybeans with any blue-green coloring present were hand-separated from the sample. The seedcoat crack percentage was defined as the ratio of the number of blue-green colored beans to the number of beans in the sample, multiplied by 100. The only areas that colored blue-green were seedcoat cracks, scratches, abrasions, or other small imperfections in the seedcoat. French et al. (3) reported that when the seedcoat is intact no enzyme activity occurs to produce indigo. But if the seedcoat is ruptured, the indoxyl acetate solution goes into the parenchyma tissues. (The seedcoat is composed of an outer palisade cell layer, a middle hour-glass cell layer, and an inner parenchyma cell layer.) Enzymes in the parenchyma cells hydrolyze indoxyl acetate and cause indigo to be deposited. Ammonia fumes accelerate the staining reaction and aid in the formation of indigo.

For the sodium hypochlorite test, 100 whole soybeans were immersed in a 0.1% sodium hypochlorite solution for 5 min (9,10). Soybeans with seedcoat cracks readily absorbed the solution and swelled 2 to 3 times their normal size. The enlarged seed size facilitated separation of seedcoat-cracked beans. The percentage of seedcoat cracks was defined as the ratio of the number of swelled soybeans to the number of soybeans in the sample, multiplied by 100.

The tetrazolium test involved premoistening 100 whole soybeans by wrapping them in damp paper towels for 12 hr. Next, the sample was immersed in a 1.0% tetrazolium (2,3,5-triphenyl tetrazolium chloride) solution with a pH of 6.0 for 5 to 7 hr. The seeds were removed from the solution after a bright red stain developed in the healthy tissue. The beans were then rinsed with water, and temporarily immersed in water prior to hand separation for damage. Damaged tissues stained a dark red color, and nonliving tissues remained white (2,5). Moore (5) reported that when living cells absorb a tetrazolium solution, a reaction between the tetrazolium and hydrogen given off during normal respiration causes the outer layer of the seed beneath the seedcoat to stain a bright red color. Bruised or damaged tissues develop a dark red color indicating a rapid respiration rate, while dead tissues remain white. One of the interesting findings about the tetrazolium test was that the most frequent place on the seed for dead tissue to appear was in the area directly beneath a seedcoat crack. For this reason tetrazolium damage, defined as the ratio of number of beans with any white showing to the total number of beans in the sample, multiplied by 100, was also considered a test for seedcoat cracks.

Coefficients of variation based on 10 replications with each of three soybean varieties, Amsoy-71, Corsoy, and Williams, ranged from 1.3 to 22.0, 2.1 to 23.7, and 1.2 to 27.8 for the indoxyl acetate, sodium hypochlorite, and tetrazolium tests, respectively. The relatively high coefficients of variation greatly limited the ability of these tests to show significant differences among various treatments that cause relatively low levels of damage. The indoxyl acetate test was found to be a quick, non-destructive test that did not have harmful effects on warm germination percentages of beans found not to have seedcoat cracks (8).

INTERNAL DAMAGE

Tests which might be considered related to internal damage include warm germination, cold test, accelerated aging, accelerated aging cold test, and the tetrazolium potential germination tests. Other tests for fungi, protein, oil content, and oxygen uptake (1) may also be indicators of internal damage but will not be discussed in this chapter.

The warm germination test involved placing 100 seeds on moist paper tissue that was maintained at conditions of 25 C and 95 to 100% relative humidity for six days.

The cold test consisted of planting 100 seeds in a soil-sand mixture maintained at 10 C for seven days, followed by the conditions of the warm germination test for four days. The results of the cold test generally show a higher correlation with field emergence and final stand than the other types of germination tests (4).

For the accelerated aging test, 100 seeds were subjected to a 40 C heat treatment for two days, followed by warm germination test conditions for six days.

The most severe of all germination tests was the accelerated aging cold test. Here 100 seeds were exposed to 40 C temperatures for 32 hr, 10 C for seven days, and then the conditions of the warm germination test for four days.

The tetrazolium potential germination test consisted of performing a tetrazolium test as previously described, except that those soybeans having some white areas were further classified as germinable or not germinable. If large areas of the seed or small vital areas around the radicle-hypocotyl axis remained white the seed cannot germinate. Seeds were categorized as potentially germinable based on the criteria given by Delouche (2).

APPLICATION OF DAMAGE TESTS

All of the previously described tests for seed damage were used in a study by Nave and Paulsen (6) to determine the effect of planter metering mechanisms on seed damage. Germination tests were performed by the Illinois Crop Improvement Association at Urbana, Illinois.

The planter metering devices evaluated were: (a) a single-run feed cup on a John Deere series 7000 Max-Emerge planter, (b) an air drum on an International Harvester (IH) series 400 Cyclo planter, (c) a fluted roller on an IH series 510 grain drill, (d) a horizontal plate meter on an IH series 58 planter, and (e) an air disk on a White series 5400 Plant/Aire planter.

The meter for each planter was set to provide a normal seeding rate of approximately 400,000 seeds/ha. Meters were operated in a stationary position at normal field planting speeds by placing a drive wheel on an electrically powered set of rollers. Approximately 25 kg of Beeson, Williams, and Wells variety soybeans were passed through the meters and collected in burlap bags. Seed moistures ranged from 10.5 to 13.3% wet basis. In addition to the five planter metering mechanisms, seeds were also collected after being subjected to an 8.2-m drop test, and a control sample was obtained directly from the seed bags. Three replications were performed.

Mean warm germination percentages ranged from 84.7% for the drop test to 89.1% for the air drum meter (Table 1). Statistical analyses indicated that there were no significant differences in warm germination percentages among any of the planter meters, control, or drop test samples at the 5% level

Table 1. Soybean seed damage evaluations by treatments averaged over all varieties and replications.

Treatment	Germination Tests								
	Warm Germ.	Cold Test	Accel. Aging	Accel. Aging Cold Test	Tetra-Zolium Potential	Tetra-Zolium Damage	Indoxyl Acetate	Sodium Hypo-chlorite	Splits
					– % –				
Control	87.1	62.6	30.3	1.1	96.6 b[1]	32.7	28.0	13.7	0.38 a
Feed cup	87.6	60.8	24.7	1.4	96.7 b	30.8	27.7	12.7	0.43 a
Air drum	89.1	63.8	33.3	0.8	96.9 b	29.4	30.6	13.4	1.00 bc
Fluted roller	88.3	60.1	34.1	0.9	96.7 b	32.6	27.6	12.8	0.56 a
Plate	85.4	57.6	26.6	1.4	95.8 a	31.8	24.9	11.7	1.22 bc
Air Disk	87.8	62.3	42.7	1.6	96.4 b	29.9	29.2	12.7	0.52 a
Drop, 8.2 m	84.7	61.3	33.7	1.9	95.8 a	32.7	30.4	13.7	0.85 b
LSD	--	--	--	--	0.8	--	--	--	0.24

[1]Numbers with the same letter within columns do not differ significantly at the 5% level based on the LSD test for differences between means.

Table 2. Soybean seed damage evaluations by varieties averaged over all treatments and replications.

	Germination Tests								
Varieties	Warm Germ.	Cold Test	Accel. Aging	Accel. Aging Cold Test	Tetra-Zolium Potential	Tetra-Zolium Damage	Indoxyl Acetate	Sodium Hypo-chlorite	Splits
					– % –				
Beeson	87.1	44.5 a[1]	29.9	0.6 a	96.2	30.9	29.2	14.3 b	1.02 b
Wells	87.4	76.6 c	39.2	1.2 a	96.1	31.1	28.2	11.0 a	0.59 a
Williams	86.9	62.5 b	27.5	2.1 b	97.0	32.3	27.7	13.6 b	0.52 a
LSD	–	5.4	--	0.9	--	--	--	0.8	0.19

[1]Numbers with the same letter within columns do not differ significantly at the 5% level based on the LSD test for differences between means.

level of probability. Cold test percentages were much lower, ranging from 57.6% for the plate meter to 63.8% for the air drum meter.

The only tests showing significant differences among treatments were those for tetrazolium potential germination and percentages of splits. The plate meter and the drop test had significantly lower tetrazolium potential germination than the other five treatments. The plate meter with 1.2% and the air drum meter with 1.0% were significantly higher in splits than the drop test with 0.85% splits. The control and the feed cup, air disk, and fluted roller meters all had significantly less splits than the other treatments.

Seed damage was found to be affected more by variety than by planter meters (Table 2). While the warm germination percentages did not vary significantly among variables, Beeson with 44.5% cold germination was significantly lower than the Williams and Wells varieties. The level of splits for the Beeson variety (1.02%) was significantly higher than the levels of the other two varieties. Beeson and Williams both had significantly higher percentages of seedcoat cracks than the Wells variety.

SUMMARY

Sieving for percentage of splits is a good test for physical breakage when large samples are used. The indoxyl acetate test is a good non-destructive test for quick indication of the presence of seedcoat cracks; but like the sodium hypochlorite and tetrazolium damage tests, there is considerable variability among replicated experimental units. At the relatively low levels of splits obtained from the planter meters, none of the tests for seedcoat cracks indicated any significant differences among the planter meters tested. Germination tests, with the exception of the tetrazolium potential germination test, also did not vary significantly among planter meters.

NOTES

M. R. Paulsen, Agricultural Engineering Department, University of Illinois, Urbana, IL 61801.

Trade names are used in this chapter solely for the purpose of providing specific information. Mention of a trade name, proprietary product, or specific equipment does not constitute a guarantee or warranty by the University of Illinois or the U.S. Department of Agriculture and does not imply approval of the named product to the exclusion of other products that may be suitable.

REFERENCES

1. Cain, D. F. and R. G. Holmes. 1977. Evaluation of soybean seed impact damage. ASAE Paper No. 77-1552. American Society of Agricultural Engineers, St. Joseph, MI 49085.
2. Delouche, J. C., T. W. Still, M. Raspet, and M. Lienhard. 1962. The tetrazolium test for seed viability. Technical Bulletin 51. Mississippi State Univ., Agric. Exp. Sta., State College, MS.

3. French, R. C., J. A. Thompson, and C. H. Kingsolver. 1962. Indoxyl acetate as an indicator of cracked seedcoats of white beans and other light colored legume seeds. Proc. Amer. Soc. for Horticultural Science 80:377-386.
4. Johnson, R. R. and L. M. Wax. 1978. Relationship of soybean germination and vigor tests to field performance. Agronomy Journal 70:273-278.
5. Moore, R. P. 1960. Tetrazolium testing techniques. Proc. 38th Annual Meeting, Soc. of Commercial Seed Tech., 45-51.
6. Nave, W. R. and M. R. Paulsen. 1977. Soybean seed quality as affected by planter meters. ASAE Paper No. 77-1004. American Society of Agricultural Engineers, St. Joseph, MI 49085.
7. Newbery, R. S., M. R. Paulsen and W. R. Nave. 1978. Soybean quality with rotary and conventional threshing. ASAE Paper No. 78-1560. American Society of Agricultural Engineers, St. Joseph, MI 49085.
8. Paulsen, M. R. and W. R. Nave. 1977. Soybean seedcoat damage detection methods. ASAE Paper No. 77-3503. American Society of Agricultural Engineers, St. Joseph, MI 49085.
9. Rodda, E. D., M. P. Steinberg, and L. S. Wei. 1973. Soybean damage detection and evaluation for food use. Transactions of the ASAE, 16(2):365-366.
10. Young, R. E. 1968. Mechanical damage to soybeans during harvesting, M.S. Thesis, Iowa State University, Ames, IA.

DRYING METHODS AND THE EFFECT ON SOYBEAN QUALITY

G. M. White, I. J. Ross and D. B. Egli

The common practice today is to allow soybeans to field dry to safe storage moisture contents before harvesting. If weather conditions are favorable this generally results in high quality soybeans. If weather conditions are not favorable then field drying will be delayed (or perhaps prevented) and soybean quality will decrease. Under these conditions harvest losses can be expected to increase, and the harvested beans may not be in a storable condition. Even under favorable weather conditions, excessive harvest losses are often incurred due to brittleness of the stems and pods at low harvest moisture contents.

It has been suggested that soybeans be harvested at moisture contents in the 13 to 15% range rather than the usual practice of harvesting at or below 12% moisture. Byg and Johnson (3) found an 8% shatter loss occurred when soybeans were combined at 10% moisture as compared with a 2% shatter loss at 17% moisture. Soybeans harvested above 13% moisture content will require drying before they can be safely stored.

The quality of soybean planting seed is particularly sensitive to weather conditions in the time interval between maturity and harvest. Ambient temperature, relative humidity and rainfall can combine to cause rapid declines in planting seed quality if the seed is left in the field after the date on which it can first be harvested. Soybeans which are harvested as soon as possible after they have reached harvest maturity will have maximum viability, but their moisture content may be too high for safe storage. If so, they need to be dried to or below 13% moisture soon after harvest in order to maintain their viability in storage. Natural air or low temperature drying can be used for this purpose if harvest moisture contents are 16% or below. Drying temperatures, air flow rates and drying time all need to be controlled within certain limits in order to maintain maximum seed quality.

Forced air drying of soybeans is advantageous in that it permits earlier harvest (thereby reducing harvest losses), and it provides a means of obtaining desired moisture contents for long-term safe storage. It may be disadvantageous if improper drying conditions are provided in that the germination of seed beans may be reduced and physical damage to the beans may reduce their market grade and potential storage life. Oil quality may also be affected either directly or indirectly as a result of drying damage.

SCOPE

This paper includes a review of literature related to soybean drying and the factors which influence the drying process. The effect of various drying parameters on seed and grain quality is discussed and the results of research by the authors on the levels of seed coat and cotyledon damage in soybeans dried with heated air is described. Recommended criteria for drying soybeans for both seed and grain are summarized along with a discussion of research needs.

SOYBEAN DRYING RESEARCH REVIEW

Prior to 1970 very little research on soybean drying was reported in literature. In 1945 and 1946 Holman and Carter (7) carried out a limited number of drying studies as part of their work on soybean storage. They found that drying was best accomplished using forced-natural air in mild weather or forced-heated air when ambient temperatures were low and/or relative humidities high. For natural-air drying they recommended that air temperatures should be above 16 C and the relative humidity below 75%. Maximum recommended temperature for heated air drying in a bin was 54 to 60 C with recommended depths of 1.2 m or less. They discussed problems with over drying near the air inlet and with moisture differences in the dried batch, but they did not mention any seed coat cracks which they might have observed as a result of the drying process.

Matthes and Welch (8) dried planting seed (cultivars Dare and Mack) soybeans from 28.2 and 22.7% moisture content, respectively, using 42 and 55% relative humidity air with flow rates from 1.7 to 13.2 m^3/min-t (cubic meters per minute per tonne). They found a definite correlation between drying time and seed germination in the upper levels of their batch dryer. Based on their results they recommended minimum air flow rates of 9.9 to 13.2 m^3/min-t to maintain seed germination for the moisture contents and drying conditions tested.

The hygroscopic and thermal properties of soybeans were measured by Alam and Shove (2), and relationships developed to predict equilibrium moisture content, equilibrium relative humidity, specific heat and latent heat values for soybeans. Using this information with an equilibrium thin-layer drying model they developed a deep-bed drying simulation model for ambient-air and/or low temperature drying of soybeans. A satisfactory agreement was found between experimental and predicted moisture gradients when a 1.8 m column of soybeans was dried with natural air at 1.1 m^3/min-t. Alam's simulation program (1) was further modified to simulate low temperature drying of deep beds of soybeans using hourly weather data.

Overhults, et al. (14) dried soybeans of the Cutler cultivar in thin-layers from 20 to 33% initial moisture content down to 10% moisture with drying air temperatures ranging from 38 to 104 C. A two parameter modified logarithmic drying model was used to fit the data. Both drying parameters were found to be functions of the drying air temperature. Severe physical damage in the form of seed coat and cotyledon cracks was observed. Although the extent of such damage was not quantified it was more pronounced at high temperatures and high initial moisture contents. In a related study, Overhults et al. (15) extracted oil from soybeans dried in the above-described experiments and evaluated its quality by determining free fatty acid content, iodine number, peroxide number, and thiobarbituric acid (TBA) value. Results indicated that although some oxidation of the oil occurred, the overall quality of the oil was not seriously affected by the drying process. However, because some oxidative deterioration was incurred, the potential storage life of the soybeans may have been decreased. Some internal deterioration of the soybeans dried at the higher temperatures may also have occurred as indicated by the fact that it was easier to extract the oil from these samples.

Walker and Barre (20) conducted thin-layer drying tests on two cultivars of soybeans at temperatures ranging from 32 to 66 C and at relative humidities ranging from 20 to 60%. Initial moisture contents were 16 to 20%. They observed considerable cracking of the soybean seed coat in soybeans dried at high temperatures and/or low relative humidities. There were significant differences between cultivars, but no explanation could be given for these differences. They observed little or no cracking of seed coats in either cultivar when the relative humidity of the drying air was 40% or more. Initial moisture contents in the range of conditions tested had no significant effect on the number of seed coat cracks. There was little effect of temperature on germination up to and including 54.4 C; above that, there was a drastic reduction in germination.

Rojanasaroj, et al. (17) impacted soybeans which had been dried with heated air at temperatures ranging from 24 to 74 C. They concluded that heated air drying results in considerable physical damage to soybeans and that subsequent impact damage during handling is related to drying air temperature. Higher drying temperatures can be expected to result in an increase in the number of splits and cracked beans after handling, Moisture content also was found to significantly affect the amount of impact damage, with higher moisture levels resulting in less damage.

The storability of soybeans dried with heated air has been studied by White, et al. (21). Soybeans which had been dried with heated air ranging in temperature from 24 to 74 C were stored at two moisture contents (12 and 17%) and three storage temperatures (10, 21 and 32 C). Molds developed much more rapidly in the high moisture content seed which had been dried at the higher temperatures. This was believed to have been caused by the increased number of seed coat and cotyledon cracks which developed as a result of the higher drying air temperatures. The percentage of seed coat cracks ranged from 30% for the 24 C drying air temperature to 84% for the 74 C air temperature. Such damage can be expected to encourage the growth of fungi and other microorganisms in stored soybeans wherever environmental conditions are favorable for such growth. Higher drying air temperatures not only reduced germination levels at the start of storage but the remaining germination also decreased earlier during storage in comparison to the soybeans dried at lower temperatures. This was thought to be partly due to increased mold growth and a general deterioration in seed quality resulting from the higher percentage of seed coat cracks. Free fatty acid percentages tended to increase with storage moisture, storage temperature and storage time. Wherever seed coat cracks caused by high temperature drying encouraged mold growth there was generally a corresponding increase in free fatty acid.

Pfost (16) studied the effect of environmental and varietal factors on crackage in soybeans dried at temperatures ranging from 32 to 66 C. He included 17 cultivars of soybeans grown at 2 locations in his tests. Cultivar effects were found to be significant at the 0.1% level. For any given cultivar, the relative humidity of the drying air was found to be the most significant factor affecting the level of seed coat cracks. In general, crackage increased with an increase in drying air temperature, initial moisture content, and drying rate and decreased with increased final moisture content and drying air relative humidity. Most cracks were observed to form during the first few minutes of drying.

More recently, Misra and Young (12,13) used finite element analysis to mathematically describe drying and shrinkage of a remoistened soybean seed and to estimate shrinkage stresses in soybeans during drying. They concluded that the prediction of stresses based on elastic material assumptions was not adequate for soybeans and that a consideration of viscoelastic properties would probably improve their stress predictions. Haghighi and Segerlind (4) also use a finite element model to analyze the stresses in soybeans under thermo-hydro loads imposed by the drying process. Their results show the tensile stress on the surface of the soybeans reaching a peak approximately one hour after drying starts. Since ultimate strength properties of the soybean was not known, the time of failure (production of seed coat or cotyledon cracks) could not be predicted. Mensah, et al. (11) in studies of the mechanical properties of soybeans found that the ultimate tensile stress and the relaxation modulus of the soybean seed coat both decreased as seed coat temperature and equilibrium relative humidity increased. A three-term Maxwell model was found to be adequate in characterizing the visco-elastic behavior of the soybean seed coat.

EXPERIMENTAL STUDIES

Introduction

The reivew of literature has indicated that drying methods and drying conditions have considerable bearing on the quality of dried soybeans and their value as seed or grain. It would therefore be advantageous if soybean dryers, particularly heated air dryers, could be designed and operated in such a way as to reduce or minimize adverse effects from the drying process. This would require knowledge of the occurrence of drying damage in soybeans as related to the characteristics of the soybeans being dried, the variation in drying air conditions with time at different locations in the dryer, and drying air flow rates.

The research summarized in this section was undertaken by the authors in an effort to supply some of this data. The main objective was to establish a better understanding of the relationships which exist between various drying parameters, the rate of drying, and the development of seed coat and cotyledon damage in soybeans when dried with heated air. The experimental work was conducted and analyzed as two separate projects: one concerned with drying damage under fully-exposed or thin-layer drying conditions, and the other with the occurrence of drying damage at different locations in a deep-bed soybean dryer. In both studies observed drying damage was classified into two categories: seed coat and cotyledon (cleavage) cracks; the latter being a more severe type of drying damage.

Thin-Layer Drying Damage

Experimental Design and Procedures. In this study thin-layer drying experiments were conducted for various combinations of the following variables: (a) Initial moisture content (16, 20 and 24% wet basis), (b) Drying air temperature (30, 40, 50, 60, and 70 C), and (c) Dewpoint temperature (8, 13, 18, 24, 28, 33, and 38 C). Test conditions which resulted in drying air relative humidities greater than 55% were omitted because of extremely slow or non-existent drying rates.

High quality seed of the Williams cultivar were used in all tests. As received, these seed had a moisture content of 13%. All test samples were remoistened in a high-humidity environmental chamber to their selected initial moisture content and stored in plastic bags at 10 C for at least 2 days prior to their use in a drying experiment. Before drying, the beans were removed from storage and allowed to come to room temperature. For each test a control sample was placed in a wire-mesh tray at room conditions and allowed to dry slowly to the desired moisture content. The levels of damage in these samples were compared to those in the corresponding samples dried under fully-exposed test conditions. Test sample weights were monitored through each test by periodically removing the samples from the dryer to take weight measurements. After drying to 12% moisture content all samples were removed from the dryer and allowed to cool to room temperature. They were then sealed in plastic bags and stored at room temperature until damage determinations could be made.

Damage determinations were made by visually sorting out the damaged seed from 3 sub-samples of approximately 28 g each from each of the controls and from each of the samples dried with heated air. Damaged seed were classified as having only seed coat cracks or both seed coat and cleavage (cotyledon) cracks. For purposes of analysis the sum of both types of damage was considered as seed coat crack damage. The second type of damage was classified as cleavage crack damage. The percentage (by weight) of each of these categories of damage was determined for each sub-sample.

Drying Test Results. In analyzing the drying data obtained in this investigation the drying model employed by Overhults, et al. (14) was used. This model was of the form:

$$MR = \frac{M - M_e}{M_o - M_e} = \exp\left[-(Kt)^n\right] \qquad (I)$$

where MR = moisture ratio; M = moisture content (dry basis) at any time, t; M_o = initial moisture content (dry basis); M_e = equilibrium moisture content, dry basis); t = time, hr; K = empirical drying constant, hr^{-1}; and n = empirical drying exponent. Values for M, M_o and t for each test were available from the experimental drying data. M_e values were calculated using Henderson's (5) equilibrium moisture equation with the constants for soybeans as developed by Alam and Shove (2).

Regression techniques were used to obtain values of K and n. Values of n did not show any relationship to initial moisture content but were related to drying air temperature and relative humidity by the following equation:

$$n = 0.33 + 0.00238\,Rh + 0.00276T \qquad (II)$$

where Rh = drying air relative humidity, %; and T = drying air temperature, C. The R^2 value for this regression was 0.577 and the standard error 0.0268.

Values of K were found to be a function of initial moisture content, drying air temperature and relative humidity, with some interaction between moisture content and temperature. A regression analysis yielded:

$$K = -0.207 + 3.57 \times 10^{-3}T + 2.16 \times 10^{-3}M_o + 2.613 \times 10^{-3}Rh + 3.202 \times 10^{-6}M_oT^2 \qquad (III)$$

where M_o is percent moisture content on a wet basis. The R^2 for this regression equation was 0.964 with a standard error of 0.025.

In general, there was good agreement in all tests between observed moisture contents and those predicted using equations I, II and III. The mean difference between observed and predicted values at the end of drying was 0.34%. The maximum difference was 1.77%, with only 7 out of 87 tests having more than 1% difference between observed and predicted final moisture contents (22).

Drying Damage Results. Observed levels of seed coat and cleavage cracks in all test samples included any damage which existed in the soybeans prior to drying as well as any damage caused by the drying process itself. Therefore, in analyzing drying damage results the observed levels of seed coat and cleavage cracks in the control samples were subtracted from the respective damage categories in each of the corresponding test samples. Drying damage is, therefore, presented in terms of the increase in percent damage resulting from the drying process.

Increases in percent seed coat cracks (SCC) for the various drying tests were analyzed to see if they could be related to the condition of the soybeans, the condition of the drying air or the rate of drying. SCC levels were found to be correlated to the initial moisture content, the drying air relative humidity, the difference between the vapor pressure of the drying air and the saturated vapor pressure at the wet bulb temperature, the dew point depression, $(M_o - M_e)$ and K^n. However, because of interrelationships between variables most of the variation in SCC levels could be explained using only M_o (% initial moisture on a wet basis) and Rh (% relative humidity). Drying air temperature was not found to be significant. A regression analysis resulted in the following equation:

$$SCC = 128.76 - 1.246M_o - 3.06Rh + 2.118 \times 10^{-2}Rh^2 \qquad (IV)$$

R^2 was 0.885 and the standard error was 10.2. Very little improvements in the level of R^2 could be realized by introducing other variables into the regression analysis. All coefficients were significantly different from zero (.1% level).

Figure 1 shows a plot of equation IV for 20% initial moisture along with the experimental data points. As can be seen, the regression predictions are high below 5% relative humidity and slightly low in the 10 to 15% range. Results indicate that little or no increase in SCC should occur when drying 20% Williams soybeans with air at 53% relative humidity or higher. The relationship between observed and predicted SCC was similar for all moisture contents. Figure 2 is a graph of equation IV for each of the tested initial

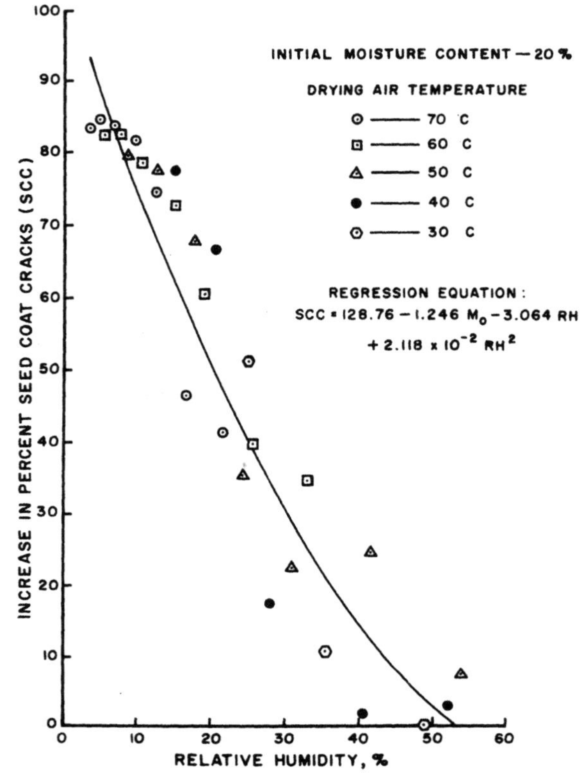

Figure 1. **Increase in percent soybeans with seed coat cracks from drying when initial moisture content is 20%.**

moisture content levels. It is interesting to note that increases in SCC are higher for the lower initial moisture contents.

Measured increases in the percentage of cleavage cracks (CC) were also analyzed in terms of the various drying parameters. CC levels were found to be related to the same factors as SCC. Temperature was also significant (1% level). Preliminary regression analyses indicated that higher order terms would be required to fit the data at the higher relative humidities where little or no increase in CC was observed. These data were removed from the analysis by eliminating CC for readings for:

$$Rh \geq 40\% \text{ for } M_o = 24\% \text{ (w.b.)}$$
$$Rh \geq 30\% \text{ for } M_o = 20\% \text{ (w.b.)}$$
$$Rh \geq 25\% \text{ for } M_o = 16\% \text{ (w.b.)}$$

A regression analysis on the remaining data yielded:

$$CC = 6.054 + 2.785M_o - 3.426Rh + 3.907 \times 10^{-2} Rh^2 \quad \text{(V)}$$

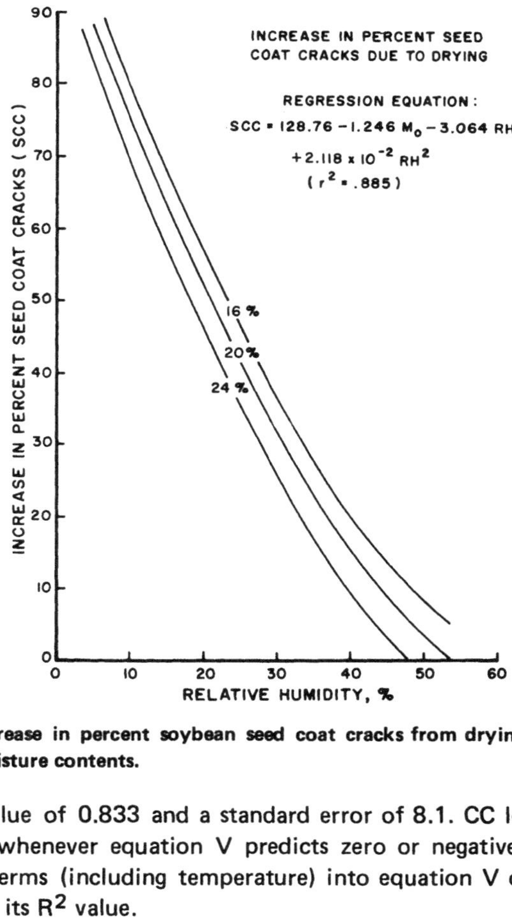

Figure 2. **Increase in percent soybean seed coat cracks from drying at three initial moisture contents.**

with an R^2 value of 0.833 and a standard error of 8.1. CC levels are to be taken as zero whenever equation V predicts zero or negative values. Introducing other terms (including temperature) into equation V did not significantly improve its R^2 value.

Figure 3 shows a plot of equation V for 20% initial moisture along with the experimental data points (including those omitted from the regression analysis). Similar fits were obtained for the 16 and 24% initial moisture contents. The regression and the data indicated that little or no increase in % CC should occur when the drying air relative humidity is above:

(a) 19% for M_o = 16% (w.b.),
(b) 25% for M_o = 20% (w.b.),
(c) 36% for M_o = 24% (w.b.).

A plot of the predicted increase in % CC for 16, 20, and 24% initial moisture contents for various relative humidities is shown in Figure 4.

As can be seen, relative humidity is the most significant factor affecting the production of CC in the Williams cultivar of soybeans when dried with heated air under fully-exposed conditions. Contrary to the results noted with

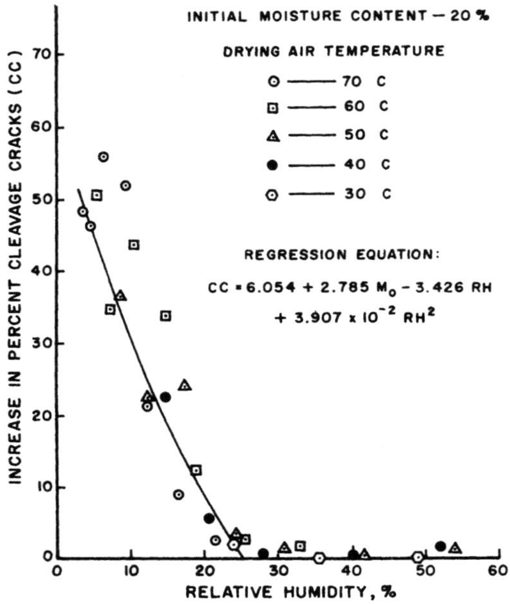

Figure 3. **Increase in percent of soybeans with cleavage cracks from drying from 20% initial moisture content.**

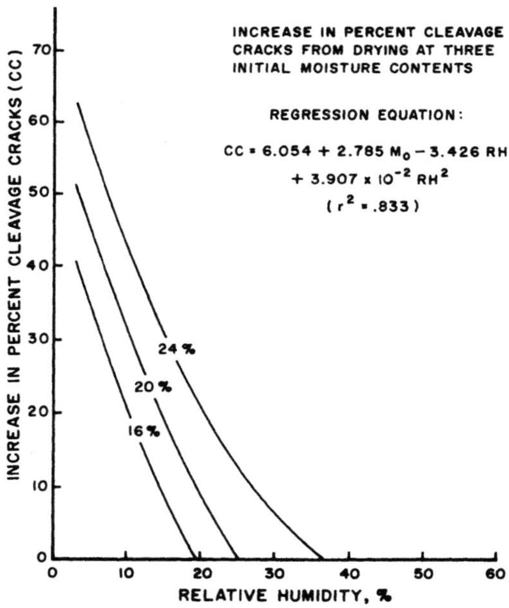

Figure 4. **Increase in percent of soybeans with cleavage cracks from drying at three initial moisture contents.**

SCC the percentage of CC increases with an increase in initial moisture content.

Deep-Bed Drying Damage

Experimental Design and Procedures. In this study a segmented deep-bed soybean drying apparatus was designed so that test samples at selected depths could be removed easily during a drying test in order to establish drying rates and variations in drying damage with time. The levels monitored in most tests were located at the bottom of the column where the drying air entered and at grain depths of 15.24 and 45.72 cm (19).

All drying experiments were designed to determine the time-variation in moisture content and drying damage at different levels within the soybean dryer as affected by different drying parameters. Parameters studied included two soybean initial moisture contents (19 and 25% dry basis), two drying air temperatures (50 and 65 C) and two air flow rates (0.102 and 0.203 $m^3/s \cdot m^2$). Each test was replicated 2 times, making a total of 16 tests. Drying air for all tests was taken from a controlled enviornment chamber maintained at a dew point of 8 C.

Results and Discussion. Results obtained in this investigation included data for each test on temperature, moisture content and drying damage as a function of time for each drying depth considered. Observed levels of soybean damage included the damage which existed in the test beans prior to their use in the drying experiments as well as any damage caused by the drying process. Except as noted otherwise, measured levels of seed coat and cotyledon damage for each sample were combined into a single total damage determination expressed as a percentage of the sample weight.

Figure 5 and 6 are typical of the results obtained from each of the eight treatment combinations (19). In all tests, the drying damage of the soybeans in the bottom layer increased rapidly to a maximum value within 20 minutes after the tests were started. Maximum values for different test conditions were different from each other. The drying damage of soybeans at a depth of 45.72 cm did not increase with time for the air flow rate of 0.102 $m^3/s \cdot m^2$ but did increase with time for the air flow rate of 0.203 $m^3/s \cdot m^2$. Under all drying conditions tested, the drying damage of soybeans at 15.24 cm increased with time.

Figure 7 shows the increase in percent soybean damage resulting from the drying process at three different depths in the dryer for all eight treatments. An analysis of all the data indicated that depth was the most significant factor affecting drying damage. Initial soybean moisture content and drying air temperature were also shown to have a significant influence on drying damage. In general, higher drying air temperatures and flow rates and lower initial moisture contents caused more drying damage. The effects of some interactions were also found to be significant. Lowering the initial

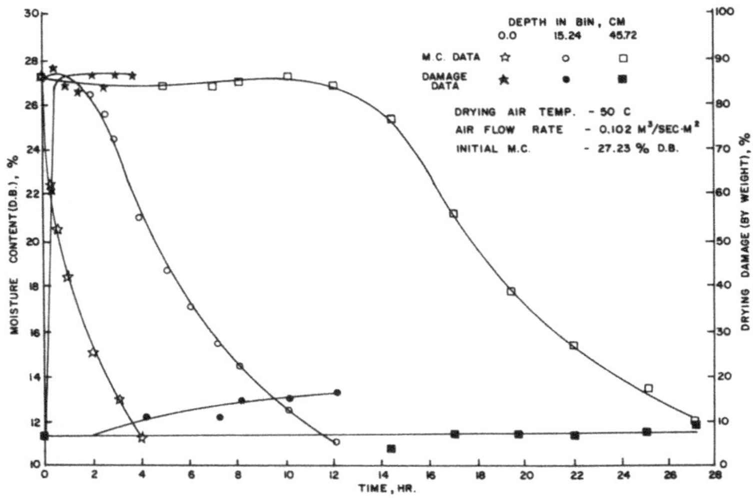

Figure 5. Soybean moisture content and drying damage versus time at different depths for 50 C drying air temperature, 0.102 $m^3/sec \cdot m^2$ air flow rate and 27.2% (dry basis) initial moisture content.

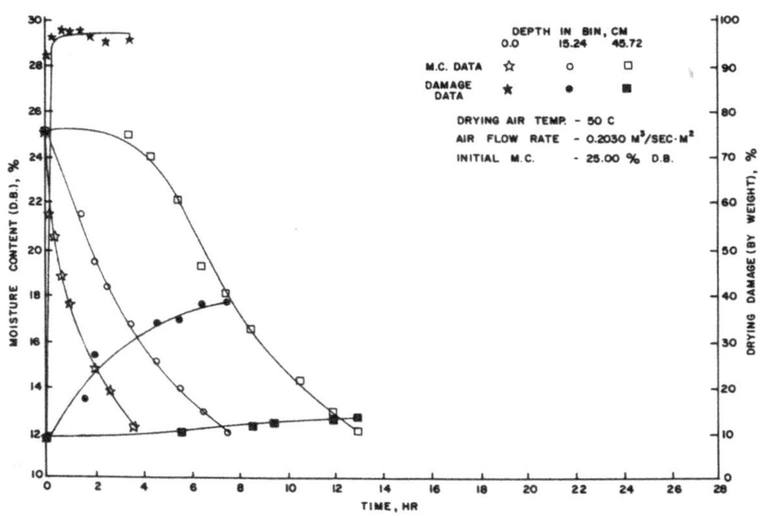

Figure 6. Soybean moisture content and drying damage versus time at different depths for 50 C drying air temperature, 0.203 $m^3/s \cdot m^2$ air flow rate and 25% (dry basis) initial moisture content.

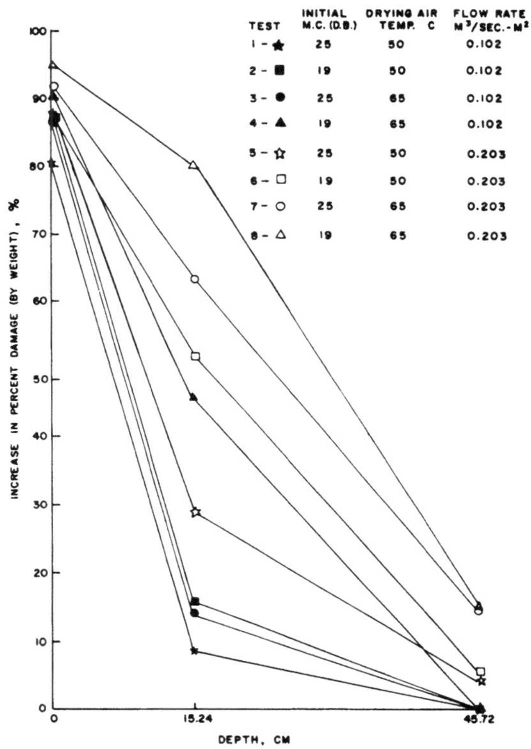

Figure 7. **Increased damage to soybeans dried to 12.36% M.C. (d.b.) at different drying depths.**

moisture content of soybeans caused a significant increase in drying damage in the bottom layer, some increase at the 15.24 cm depth but had no significant effect on soybean damage at the 45.72 cm depth. The soybeans at the depth of 15.24 cm seemed to be more sensitive to changes in the various drying factors than any other layer investigated. The percentage of soybeans with cleavage cracks did not significantly increase at depths of 15.24 and 45.72 cm in any of the tests, but significant increases were found in the bottom layer. The cleavage cracks at this level, unlike seed coat cracks, occurred gradually with time. In general, the percentage of cleavage cracks increased with an increase in drying air temperature and, unlike seed coat cracks, increased with an increase in initial soybean moisture content.

One additional test was conducted to investigate drying damage at other levels within the dryer and to determine the potential increase in damage that might be expected at different depths as the beans dried below 12.36% (dry basis) moisture content. Five different levels in a 60.96 cm column of seeds were monitored during this test. One was at the bottom near where the drying air entered, and the rest where located at 15.24 cm depth increments from the bottom. For this test a 25% (dry basis) initial moisture content was used and the drying air temperature and flow rate were 65 C and 0.203 $m^3/s \cdot m^2$, respectively. The test was terminated when the top layer of seed had been dried to 12.36% (dry basis) moisture content. The time-variation of moisture content and soybean drying damage at five different depths during this test are shown in Figure 8. Unlike the previous tests, the soybeans (except for the layer at 60.96 cm) were dried below 12.36% (dry basis) moisture content. Profiles of drying damage at any time during the drying process were established (Figure 9).

The drying damage curves in Figure 8 show that soybeans of different depths tend to reach different maximum values of drying damage as the drying zone passed by. In other words, when the drying is completed there will still exist a certain profile of drying damage in a deep-bed dryer. The shape of the profile will depend on the existing combination of drying parameters.

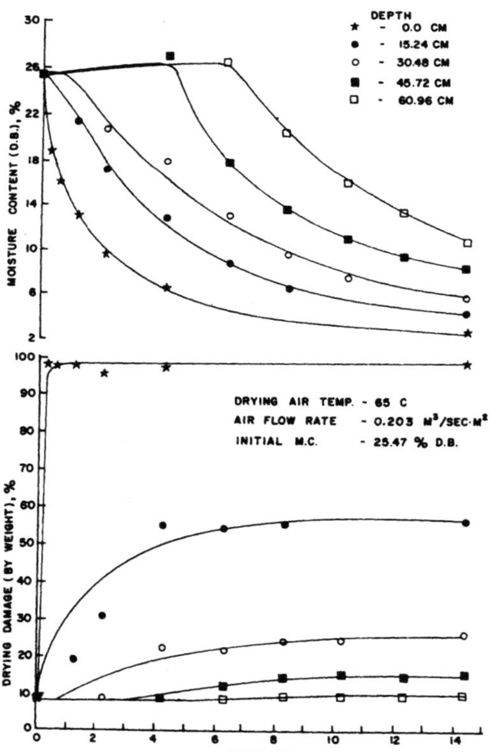

Figure 8. **Time-variation of moisture content (% d.b.) and drying damage of soybeans at 5 different depths.**

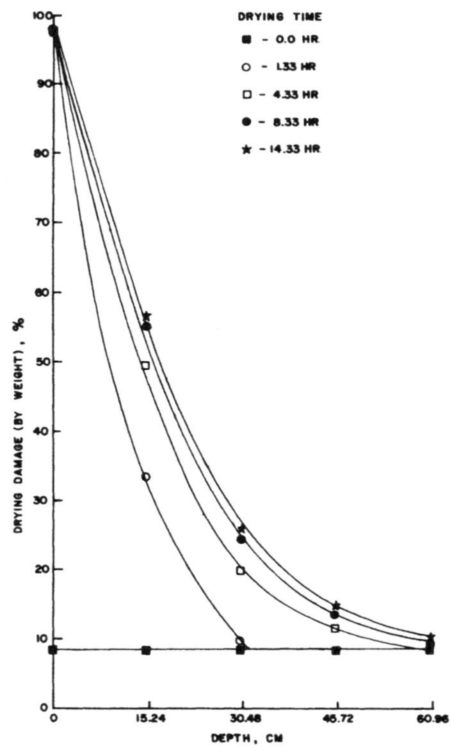

Figure 9. Predicted profiles of soybean drying damage in the bed at different drying times.

RECOMMENDED DRYING PRACTICES

Early harvest of soybeans is limited by the fact they are difficult to harvest on first dry-down until the plant dies from complete maturity or is exposed to a killing frost. In good weather the seed will usually have a moisture content of 15 to 16% when they can first be harvested. If they re-wet in the field after that time they can usually be harvested at somewhat higher moisture contents. High moisture soybeans must be dried before they can be safely stored.

Generally, all conventional grain drying methods are adaptable to soybeans with some restrictions on drying air temperature and relative humidity and on handling equipment and practices. Reasons for the temperature and relative humidity limitations have already been discussed. Handling restrictions are important in that soybeans are more easily damaged from physical abuse than are most grains. They should be handled as little and as gently as possible to prevent seed coat damage and cracked seed. Split and damaged soybeans open the way for fungal growth and can lead to storage problems.

Soybean Seed Drying

The most severe limitations on soybean drying conditions are those imposed when drying planting seed. The seed must be dried to 13% moisture content or less to maintain viability in storage. To maintain viability prior to storage the drying operation must be accomplished within a certain time period depending on seed moisture content and the environment to which it is exposed. Drying air temperature, relative humidity and flow-rate must be such that drying is completed within the prescribed time without reducing seed germination or causing undue physical drying damage. This means that the drying air temperature should be kept below 43 C with a relative humidity greater than 45% and less than 70%. Temperature limitations are usually not a problem if the relative humidity is controlled within the indicated limits.

Natural air drying can be used to dry soybean seed with moisture contents of 16% or less if the air temperature is above 10 C with a relative humidity below 70%. Air flow rates of 2 to 3 m^3/min-t are required. If humidities are higher or if the moisture content is above 16% then a few degrees of supplemental heat and a higher air flow (4 to 6 m^3/min-t) should be used. The maximum allowable temperature rise will usually be fixed by the lower limit on relative humidity. A useful "rule of thumb" is that relative humidity is approximately cut in half for each 11 C temperature rise. For soybean seed moisture contents above 20% Matthes, et al. (9) have suggested heated air drying with the drying air controlled at 40 to 50% relative humidity with air flow rates of 10 to 12 m^3/min-t. At this air flow rate the grain depth should be limited to 1.2 m or less.

Drying for Commercial Use

When drying for commercial use the primary consideration is usually one of economics. Farmers and commercial grain elevator operators are usually willing to accept a certain level of seed coat cracks and other drying damage in their soybeans as long as the degree of such damage is not sufficient to lower the market grade of the soybeans. They are in the position of choosing higher capacity along with higher drying damage on one hand versus lower capacity and higher quality on the other. The choice of operating conditions is usually a compromise between these extremes.

Most recommendations (10,18) suggest a maximum temperature limit of 54 to 60 C. This applies to all types of heated air dryers. At these temperatures, the relative humidity of the drying air will be low and seed coat damage high. Lowering the temperature below these limits will reduce drying capacity but should improve the quality of the dried soybeans. Air flow rates should be essentially the same as those used for other grain crops.

Soybeans which have been dried with heated air should be handled as little as possible since they are more prone to handling damage. Dryers which use recirculators or stirring devices can also result in additional damage and should be avoided.

Low temperature drying is another method for drying commercial soybeans where weather conditions are suitable. It has been used satisfactorily in Illinois (6) and should be a satisfactory drying method in the North Central Region of the U.S. after the mean daily temperature drops below 10 C. In using this type of drying system it is recommended that: (a) air flow rates be 1.1 to 2.2 m^3/min-t with the higher rate applied when soybean moisture contents are 17% or higher; (b) maximum harvest moisture content should be 22%; and (c) the supplemental heater should only be used when the relative humidity exceeds 65% or the ambient air temperatures drops below 5 C.

SUMMARY

This paper has discussed the status of soybean drying research and reviewed current drying practices. It is apparent from the presented material that there is still much to learn about how a soybean dries and how the drying process affects soybean quality.

The effect of drying air temperature on soybean oil quality does not appear to have been investigated very thoroughly. The factors which cause seed coat and cotyledon damage during the drying process have been studied to a greater extent but they still need to be defined more precisely. Methods of predicting drying damage at various locations in a deep-bed soybean dryer are needed so that drying air conditions can be programmed and drying damage minimized.

The economic importance of soybeans as a food crop makes it imperative that we gain a better understanding of all aspects of soybean drying and that we apply this knowledge in the design of practical soybean drying systems.

NOTES

G. M. White and I. J. Ross, Department of Agricultural Engineering and D. B. Egli, Department of Agronomy, University of Kentucky, Lexington, Kentucky 40546.

This paper is published with the approval of the Director of the Kentucky Agricultural Experiment Station and designated Journal Article No. 79-2-3-28.

REFERENCES

1. Alam, A. and G. C. Shove. 1973. Simulation of soybean drying. ASAE Trans. 16(1):134-136.
2. Alam, A. and G. C. Shove. 1973. Hygroscopicity and thermal properties of soybeans. ASAE Trans. 16(4):707-709.
3. Byg, D. M. and W. H. Johnson. 1970. Reducing soybean harvest losses. Ohio Report 55(1):17-18.

4. Haghighi, K. and L. J. Sergerlind. 1978. Computer simulation of the stress cracking of soybeans. ASAE paper no. 78-3560, ASAE, St. Joseph, MI 49085.
5. Henderson, S. M. 1952. A basic concept of equilibrium moisture. Ag. Eng. 33(1): 29-32.
6. Hirning, H. J. 1973. Field experiences with drying soybeans. ASAE paper no. 73-3514, ASAE, St. Joseph, MI 49085.
7. Holman, L. E. and D. G. Carter. 1952. Soybean storage in farm-type bins. Bull. 553, Illinois Ag. Exp. Sta., Univ. of Illinois, Urbana, IL.
8. Matthes, R. K., G. B. Welch and A. H. Boyd. 1974. Heated air drying of soybean seed. ASAE paper no. 74-3001, ASAE, St. Joseph, MI 49085.
9. Matthes, R. K., A. H. Boyd and G. B. Welch. 1973. Heated air drying of soybean seed. Information sheet no. 1246, Mississippi Ag. and Forestry Exp. Sta., Mississippi State Univ., Mississippi State, MS 39762.
10. McKenzie, B. A. 1972. Drying soybeans with heated and unheated air. Publication AE-84, Cooperative Extension Service, Purdue Univ., West Lafayette, IN.
11. Mensah, J. K., G. L. Nelson, F. L. Herum and T. R. Richard. 1978. Mechanical properties related to soybean seedcoat failure during drying. ASAE paper no. 78-3561., ASAE, St. Joseph, MI 49085.
12. Misra, R. N. and J. H. Young. 1978. Finite element analysis of simultaneous moisture diffusion and shrinkage of soybeans during drying. ASAE paper no. 78-3056, ASAE, St. Joseph, MI 49085.
13. Misra, R. N. and J. H. Young. 1978. Finite element procedures for estimating shrinkage stresses during soybean drying, ASAE paper no. 78-3559, ASAE, St. Joseph, MI.
14. Overhults, D. G., G. M. White, H. E. Hamilton and I. J. Ross. 1973. Drying soybeans with heated air,. ASAE Trans. 16(1):112-113.
15. Overhults, D. G., G. M. White, H. E. Hamilton, I. J. Ross and J. D. Fox. 1975. Effect of heated air drying on soybean oil quality. ASAE Trans. 18(5):942-945.
16. Pfost, D. 1975. Environmental and varietal factors affecting damage to seed soybeans during drying. Ph.D. dissertation, the Ohio State Univ., Univ. Microfilms International, Ann Arbor, MI.
17. Rojanasaroj, C., G. M. White, O. J. Loewer, and D. B. Egli. 1976. Influence of heated air drying on soybean impact damage. ASAE Trans. 19(2):372-377.
18. Rodda, E. D. 1974. Soybean drying—seed, food, feed, ASAE paper no. 74-3540, ASAE, St. Joseph, MI 49085.
19. Ting, K. C., G. M. White, I. J. Ross and O. J. Loewer. 1978. Seed coat damage in deep-bed drying of soybeans, ASAE paper no. 78-3006, ASAE, St. Joseph, MI 49085.
20. Walker, R. J. and H. J. Barre. 1972. The effect of drying on soybean germination and crackage, ASAE paper no. 72-817, ASAE, St. Joseph, MI 49085.
21. White, G. M., O. J. Loewer, I. J. Ross and D. B. Egli. 1976. Storage characteristics of soybeans dried with heated air. ASAE Trans. 19(2):306-310.
22. White, G. M., T. C. Bridges, O. J. Loewer, and I. J. Ross. 1978. Seed coat damage in thin-layer drying of soybeans as affected by drying conditions, ASAE paper no. 78-3052, ASAE, St. Joseph, MI 49085.

SOYBEAN SEED STORAGE UNDER CONTROLLED AND AMBIENT CONDITIONS IN TROPICAL ENVIRONMENTS

E. J. Ravalo, E. D. Rodda, F. D. Tenne, and J. B. Sinclair

Favorable results from varietal and agronomic trials at many locations around the world have resulted in wider production of soybeans in tropical and subtropical areas. The maintenance of soybean seed quality during storage in these hot and humid environments has been recognized as a limiting factor in soybean production. High quality seeds are required to obtain adequate stands for profitable soybean production. Soybean seeds must often be imported in the initial stages of production development in a new area. As local seed production becomes possible, a storage technology suitable for the environment and scale of operations must be available.

Soybean seeds do not store as well as seeds of other legumes. The major environmental factors influencing seed deterioration are temperature and moisture (1,2,3,5,8,9,12). Moisture is usually considered in terms of seed moisture content. Since seeds lose or gain moisture from the surrounding air until reaching equilibrium at a given set of conditions, it would also be appropriate to think of moisture in terms of the relative humidity of the storage environment. Storage fungi are a major cause of reduced germination in soybeans, especially when soybeans are stored at moisture contents above 12.5% (3).

Other factors associated with deterioration of soybean seeds in storage are microorganisms, field production history, and storage containers (6,10,11, 13). Various researchers (7,12) reported the association of *Aspergillus* spp. with reduced viability of soybean seeds in storage. Tedia (10) showed that the microorganisms most frequently associated with the decline in vigor of soybean seeds stored at high temperature (35 C) and moisture content (13%) were *Aspergillus* spp. and *Bacillus subtilis*. *Bacillus subtilis* has been shown to be seed-borne in soybeans (11,12).

The primary objective of this project was to study practical and inexpensive methods for storing soybean seeds under tropical conditions. The initial phase of the study was to evaluate tropical storage properties of certified seeds produced in the continental U.S. The final phase of the work was to study the storage properties of soybean seeds produced in Puerto Rice under tropical conditions for various storage environments. Results of these investigations were previsouly reported in ASAE Annual Meeting Papers 77-4057 and 78-6030. This chapter presents a summary of the above reports.

MATERIALS AND METHODS

Research emphasis was directed in developing simple and inexpensive methods for storing soybean seeds using resources generally available in typical small farms in developing countires. The experimental variables investigated were: two storage locates, L_1—Mayaguez Campus, and L_2— Isabela Substation; two storage conditions, S_1—shelf-stored in ambient conditions, and S_2—stored in cardboard cartons with about six inches of rice hulls for insulation; three initial moisture levels, M_1—13%, M_2—10%, and M_3—7%; four containers, C_1—sealed metal can, C_2—metal cans with one mil plastic lining, C_3—fertilizer jute bags with one mil plastic lining, and C_4—cotton cloth bags; and three storage periods, T_1—three months, T_2—six months, and T_3—nine months. Three replications were used for all samples. Sealed samples in cans were also stored in a controlled environment of 3 C and 25 C at Urbana, Illinois. Germination tests, moisture content determinations, and pathological studies were conducted both at the start and at the end of the storage period.

Preparation of Samples for Storage Tests

Four hundry thirty-two 200-g samples were prepared using certified Woodworth seeds from Illinois for the initial phase of the storage studies. Seeds multiplied in Puerto Rico with parent seeds from Illinois were used in the second phase of the storage experiment. The initial moisture content of the seeds was about 10 to 12%. The seeds were reconditioned and allowed to absorb moisture from the surrounding air. This was done by spreading the seeds in a one-inch layer in a cold storage room for several weeks until the moisture content reached 13 to 14%.

The seeds were then dried to each of the three desired moisture levels by using forced air heated to about 33 C. The drying was done by placing all the seeds in the laboratory dryer and withdrawing about one-third of the seeds when the appropriate moisture level was reached. The dried seeds were then sealed in double plastic bags. At the end of the drying period, the seed lots of each moisture content were divided into three approximately equal parts. The seeds were then subdivided into 200 g samples and placed in the different containers.

Preparation of Samples for Germination Tests

The seed samples placed in storage were a mixture of mechanically broken and whole seeds. The percentage of broken seeds ranged from 1 to 2%. Mechanically damaged seeds and seeds damaged by insects were separated from the whole seeds by hand. Samples for germination tests were taken at random from the whole seeds. Four hundred seeds were used for germination tests for each treatment. The seeds were further subdivided into 8 groups of 50 seeds each. Sandbed germination tests were used for determining the seed germination percentage at the end of the storage period.

Constant Temperature Storage Studies at Urbana

Oven moisture content determinations were run on the seeds as received. Quantities of seeds for each test were conditioned to the desired moisture content by allowing moisture absorption in an Aminco chamber or by low temperature drying. Conditioned materials were held at 3 C in double plastic bags until all materials were ready for sample preparation. Each sample contained a minimum of 250 seeds. Tests were run with Illinois-produced seeds, newly harvested Puerto Rico-produced seeds, and the same Puerto Rican seeds held for conditioning in Puerto Rico for ambient storage tests. Container atmosphere CO_2 determinations were made for the 6-month and 9-month samples of the Illinois-produced seed.

The samples were stored in sealed No. 1 cans. The storage conditions studied were as follows: dry basis moisture contents of 7, 10, and 13%; storage temperatures of 3 C (chamber temperature cycle of 0 and 3 C) and 25 C; and for the last series, treatment with thiram, a fungicide. Three replications were run for all samples. Official germination tests were made by the Illinois Crop Improvement Association, Urbana, Illinois.

RESULTS AND DISCUSSION

The results of storage studies of U.S. certified Woodworth and Puerto Rico-produced soybean seed stored under a tropical environment in Puerto Rico and at controlled temperatures at Urbana, Illinois, are presented in Tables 1, 2, and 3, and Figures 1 through 11. All storage variables studied except storage locations had significant effects on the germination of Woodworth soybean seeds stored under different storage conditions.

The apparent influence of storage containers on germination of U.S. certified Woodworth soybean seeds stored in Puerto Rico for different storage periods is presented in Figure 1. Germination was maintained at fairly high levels for three months' storage for all containers. For longer storage periods germination was strongly affected by the storage container, with seed germination dropping to zero after nine months' storage in cloth bag containers. Soybean seeds stored in C_1 sealed and C_2 unsealed tin cans preserved

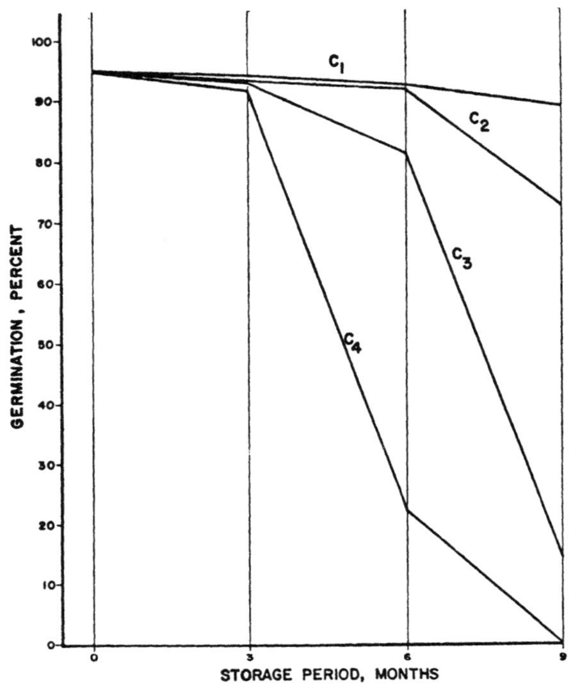

Figure 1. Variation of germination with respect to containers and storage period.

germination at the acceptable level of above 80% after nine months' storage. High germination was associated with containers (C_1 and C_2) which minimized moisture gain of seed while in storage. In contrast low germinations were related to seed stored in containers (C_3 and C_4) which gave little protection against seeds absorbing moisture from the surroundings (Figure 2). These results corroborate the findings of others (2,3) that one of the most important factors affecting the rapid loss of germination in storage is the moisture content of the seeds. It is not only essential to have low initial seed moisture contents, but even more important is the requirement to maintain the low initial moisture content of the seeds throughout the storage periods (Figures 1, 2, 3, and 4). This is particularly important under the hot and humid conditions prevalent in the tropics, which are very favorable for dry seeds to absorb moisture from their surroundings until the seed moisture content is in equilibrium with the atmosphere. The best containers were those that maintained a low initial moisture content (Figures 1, 2, and 4).

Results of storing soybean seeds in sealed tin cans under controlled tempterature and tropical environments in Puerto Rico are presented in Figures 4, 5, and 6). At controlled temperature of 3 C (S_1) which closely resembles conditions of seeds in cold storage, the germination was fairly independent of

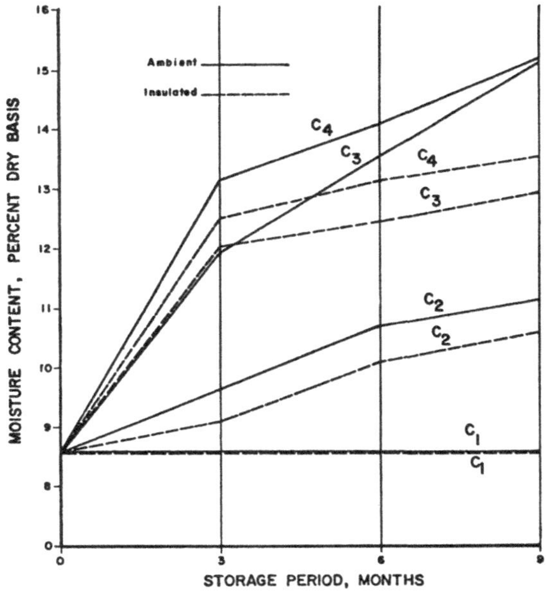

Figure 2. **Change in moisture in storage.**

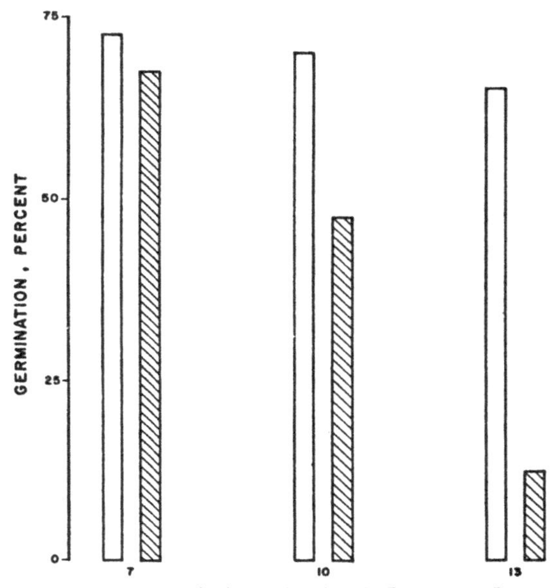

Figure 3. **Mean percentage germination of Illinois- and Puerto Rico-produced Woodworth seed at different moisture contents. □ = Illinois-produced seed, and ▨ = Puerto Rico-produced seed.**

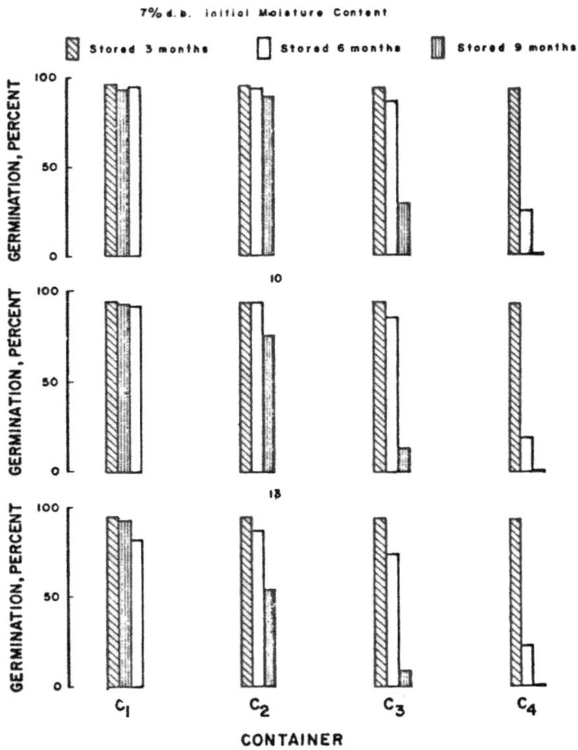

Figure 4. Mean percentage of germination of Illinois-produced Woodworth seed at different storage conditions.

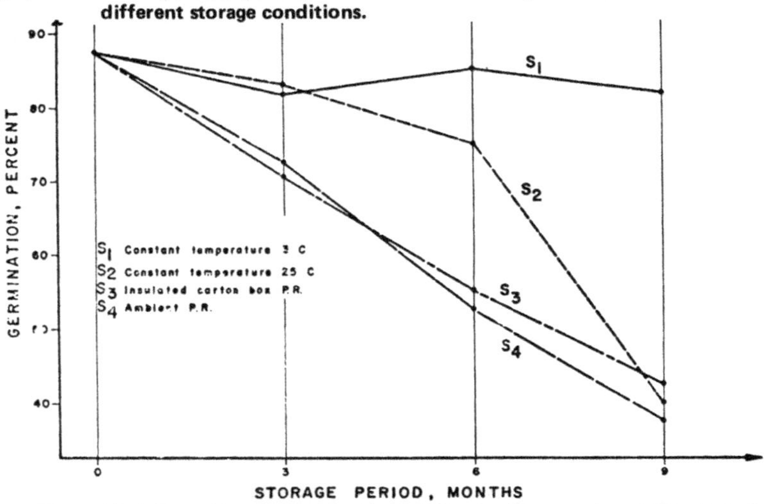

Figure 5. Variation of mean germination of Puerto Rico-produced Woodworth seed in sealed metal cans with respect to storage period and storage environment.

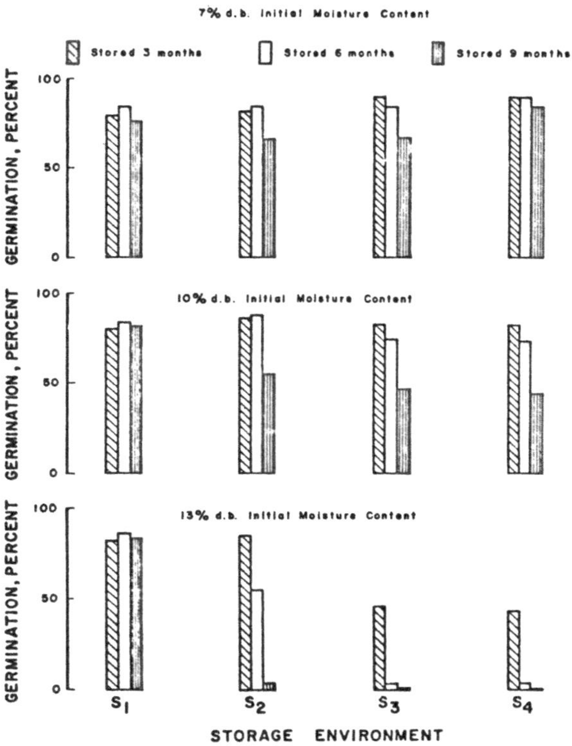

Figure 6. **Mean percentage of germination of Puerto Rico-produced Woodworth seed in sealed cans stored at constant temperature 3 C, S_1; constant temperature 25 C, S_2; insulated box PR, S_3; and ambient PR, S_4.**

storage time. The germination at the end of the nine months' storage period was about the same as the initial germination prior to placing the seeds in storage. The effect of storing soybean seeds in sealed tin cans, in rice hull insulated carton boxes (S_3), and under ambient (S_4) conditions in Puerto Rico was practically identical, with mean germination dropping from the initial germination of about 90% to about 40% after nine months' storage. Insulation had no effect of practical significance (Figure 5). Germination results of the controlled temperature storage at 25 C were similar to those of S_3 and S_4 storage. The drop of mean germination appeared to be related with CO_2 concentration in the tin can atmosphere (Figure 7). Germination was held at a fairly good level of above 80% for CO_2 concentrations below 4%. Above 4% CO_2 concentrations the germination deteriorated rapidly, with germination dropping to zero for CO_2 concentrations of about 11%. Low CO_2 concentration was associated with seeds of low initial moisture content.

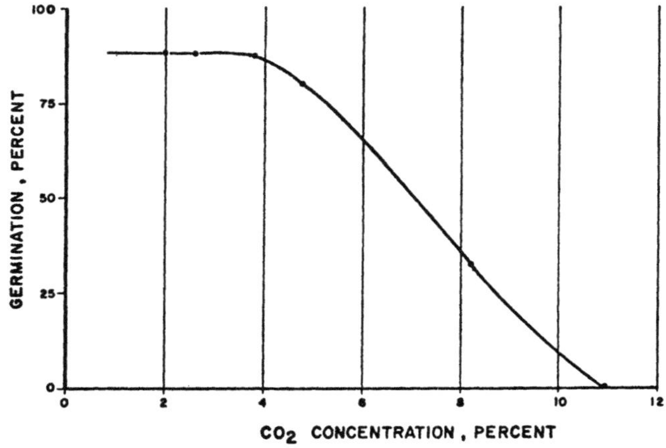

Figure 7. Relationship between mean percentage germination and CO_2 concentration for Illinois-produced seed sealed in metal cans and stored at constant temperature of 25 C.

High CO_2 concentration was related to seeds of high initial moisture content stored for nine months. The rapidly deterioration of seed germination appeared to be related also to *Bacillus subtilis* populations (Figure 8).

Comparisons of mean germination for seed multiplied in Puerto Rico and parent seeds from Illinois are presented in Tables 1 and 2, and Figures 3, 4, 6, and 9. It is quite apparent from these data that seeds produced in Puerto Rico stored much less satisfactorily than the Illinois-produced seeds. Field production history has been shown to be associated with germination properties of the produced seeds (4). The hot and humid conditions prevalent in Isabela were favorable for the development of any internally seedborne microorganism which might be present in the parent materials. Microorganism population, primarily *Bacillus subtilis,* was much higher for the seeds produced in Puerto Rico than for the seeds produced in Illinois (11). *Bacillus subtilis* had been reported as seedborne in soybeans. Treating seeds with thiram in storage did not produce better gemination than the untreated seeds (Table 3, Figures 10 and 11). Ellis et al. (4) have shown that growing pigeon peas, *Cajanus cajan* Millsp., in the less humid condition in Fortuna produced seeds of much higher quality than the seeds produced in the humid conditions in Isabela. Selecting a favorable growing location for seed production with due regard to the foregoing consideration and producing the seeds during the dry season of the year can be expected to be of much help in producing seeds of good germination and storage qualities.

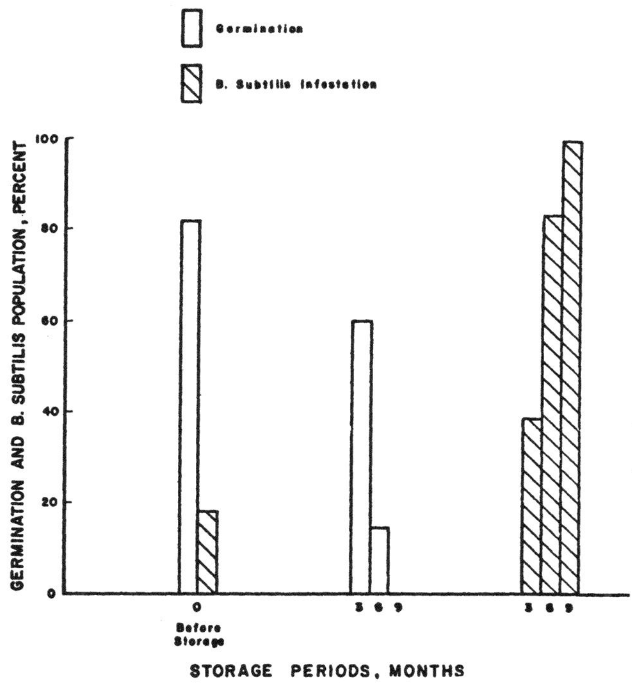

Figure 8. Percent germination and *B. Subtilis* population for Puerto Rico-produced Woodworth seeds in sealed metal cans assayed at 35 C.

Table 1. Mean percentage of germination of Puerto Rico-produced Woodworth soybean seeds in sealed metal cans at various storage environments.

Moisture Content	Storage Periods	Storage Environments: Constant Temperature 3 C	25 C	Ambient PR	Insulated PR
– % –	– *Months* –			– % –	
7	3	80	81	90	90
	6	85	83	86	90
	9	79	67	69	85
10	3	82	86	84	82
	6	84	88	73	74
	9	83	54	47	45
13	3	84	84	45	41
	6	87	54	1	1
	9	85	1	0	0

Table 2. Mean percentage emergence in sand of Woodworth soybean seeds stored in various containers; at three initial moisture levels; either noninsulated (N) or insulated (I) in rice hulls; for 3, 6, and 9 months; either non-treated (NTR) or treated (TR) with 2.08 g/kg thiram, Mayaguez, 1976.

Containers	Initial Moisture	Insulation	3 Months		6 Months		9 Months	
			NTR	TR	NTR	TR	NTR	TR
	– % –				– % –			
C_1 Sealed tin can	7	N	89	91	83	88	62	76
	7	I	91	89	90	89	92	79
	10	N	88	85	70	76	44	50
	10	I	82	82	76	72	60	29
	13	N	38	51	2	0	0	0
	13	I	42	40	2	0	1	0
C_2 Nonsealed tin can	7	N	92	90	86	88	84	83
	7	I	90	92	85	83	90	72
	10	N	82	82	66	63	22	34
	10	I	82	76	54	57	8	20
	13	N	52	33	0	0	0	0
	13	I	34	42	0	0	0	0
C_3 Fertilizer bag (plastic lined)	7	N	92	91	89	81	81	76
	7	I	96	91	87	83	86	63
	10	N	80	85	67	54	23	16
	10	I	84	83	57	52	8	13
	13	N	36	40	0	0	0	0
	13	I	36	35	0	0	0	0
C_4 Cloth bag	7	N	70	57	0	0	0	0
	7	I	38	15	0	0	0	0
	10	N	50	49	0	0	0	0
	10	I	42	29	0	0	0	0
	13	N	51	41	0	0	0	0
	13	I	26	13	0	0	0	0

Table 3. Mean percentage germination of Woodworth soybean seeds stored in sealed cans at three initial moisture levels, at two storage temperatures; for 3, 6, and 9 months; either treated (TR) or non-treated (NTR) with 2.08 g/kg thiram.

Initial Moisture Content	Storage Temperature	3 Months		6 Months		9 Months	
		NTR	TR	NTR	TR	NTR	TR
– % –	– C –			– % –			
7	3	80	83	79	83	78	76
7	25	77	75	71	73	76	53
10	3	75	80	82	75	77	68
10	25	64	77	62	70	36	46
13	3	82	80	78	77	74	61
13	25	55	65	16	17	1	3

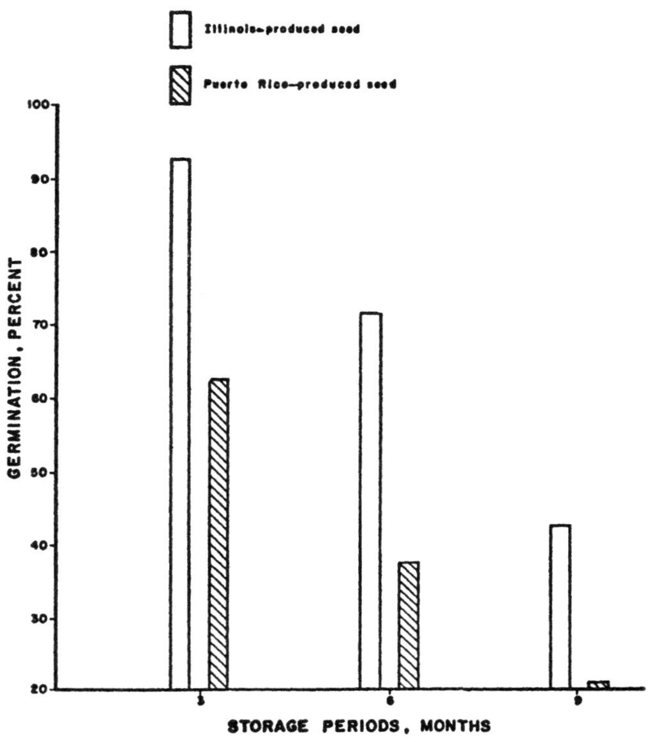

Figure 9. Mean percentage germination of Illinois and Puerto Rico Woodworth seed at various storage periods.

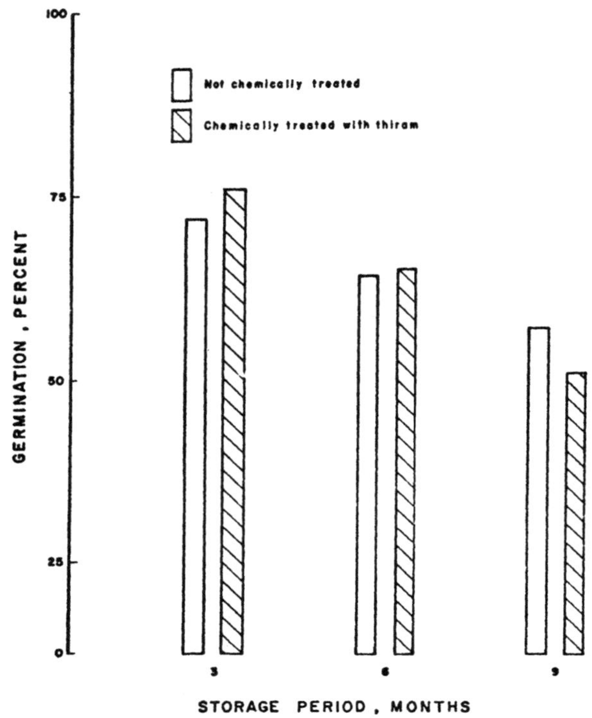

Figure 10. **Mean percentage of germination of Puerto Rico produced Woodworth seed in sealed metal cans either treated (TR) or nontreated (NTR) with 2.08 g/kg thiram.**

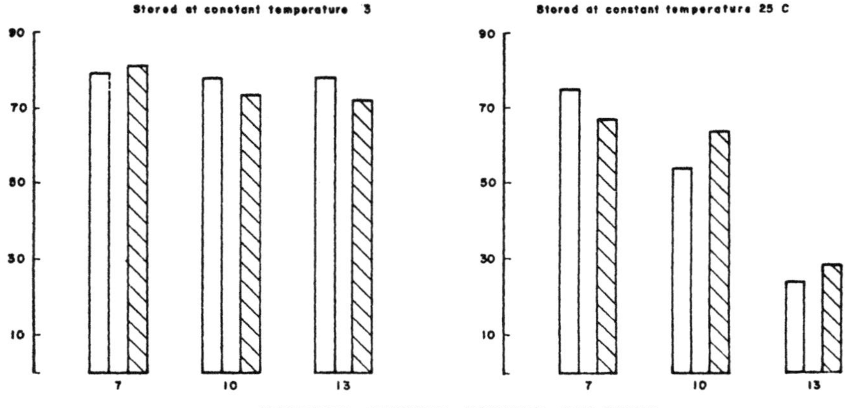

Figure 11. **Mean percentage of germination of Puerto Rico-produced Woodworth seed in sealed metal cans either treated (TR) or nontreated (NTR) with 2.08 g/kg thiram.**

SUMMARY

Storage studies using U.S. certified Woodworth and Puerto Rico-produced soybeans were conducted with the primary objective of developing simple and inexpensive methods for producing and maintaining seed quality during storage in the tropics. Results showed that the viability of high quality seeds declined rapidly to almost zero after nine months of improper storage in a tropical environment. Drying to a low moisture content below 10% and maintaining the low initial moisture content of seeds throughout the storage period will permit satisfactory storage of soybean seeds under tropical environment. Sealed metal containers or plastic bags placed in metal containers with covers offered adequate storage protection for seeds with a low enough initial moisture content. Poor quality seeds did not store well even at low temperature (3 C). Fungicide treatment of seeds in storage did not improve germination. Neither did insulation have any effect of practical significance.

NOTES

Eliodoro J. Ravalo, Department of Agricultural Engineering, University of Puerto Rico, Mayaguez, Puerto Rico 00708; Errol D. Rodda, Department of Agricultural Engineering, University of Illinois, Urbana, Illinois 61801; Frank D. Tenne, formerly of Department of Pathology, University of Illinois, Urbana, Illinois 61801; and James B. Sinclair, Department of Pathology, University of Illinois, Urbana, Illinois 61801.

This study was partially funded by the Rockefeller Foundation and USAID under the INTSOY Program jointly administered by the University of Puerto Rico and the University of Illinois.

A special acknowledgement is due Mr. Larry Gold, Manager of Star Can, Star-Kist Caribe, Inc., for supplying and sealing the cans used in this project.

REFERENCES

1. Baciu, D. 1972. Effect of storage conditions on the quality of soybean seed. Inst. Cercatari Cereale Plants Teh Fundulea C. Amelior. Plant Probl. Genet. Teor. Apl. 38:247-57. Field Crop Astract 4:1822.
2. Byrd, H. W. and J. C. Delouche. 1971. Deterioration of soybean seed in storage. Proc. Assoc. of Official Seed Analysts 61:41-57.
3. Darworth, C. F. and C. M. Christiansen. 1968. Influence of moisture, temperature, and storage time upon changes in fungus flora, germinability, and fat acidity values. Phytopathology 58:1457.
4. Ellis, M. A., E. H. Paschal, E. J. Ravalo, and E. Rosario. 1978. Effect of growing location on internally seedborne fungi, seed germination, and field emergence of pigeon pea in Puerto Rico. Journal of Agr., UPR, 62:(4) 344-555.
5. Foor, S. D. 1977. Germination potential of soybean seed. Ph.D. Thesis, Univ. of Illinois, Urbana, Illinois, 120 pp.
6. Grabe, D. F. 1965. Storage of soybean for seed. Soybean Digest 26(1):14-6.
7. Ramstad, P. E. and W. F. Geddes. 1942. Respiration and storage behavior of soybeans. Minn. Agr. Exp. Sta., Tech. Bul. 156.
8. Ravalo, E. J. and E. D. Rodda. 1977. Tropical storage of U.S. certified soybean seed. ASAE Paper No. 77-4057.

9. Rodda, E. D. and E. J. Ravalo. 1978. Soybean seed storage under constant and ambient tropical conditions. ASAE Paper No. 78-7002.
10. Tedia, M. D. 1976. Effect of storage conditions and environment during maturation of soybean on seed quality and crop performance. Ph.D. Thesis, Univ. of Illinois, Urbana, Illinois, 164 pp.
11. Tenne, F. D. 1977. The relationship of seedborne fungi and *Bacillus subtilis* under temperature and tropical condition. Ph.D. Thesis, Univ. of Illinois, Urbana, Illinois, 98 pp.
12. Tenne, F. D., E. J. Ravalo, J. B. Sinclair, and E. D. Rodda. 1978. Changes in viability and microflora of soybean seed under various conditions in Puerto Rico. The Journal of Agri., UPR, 62:(3)00, 250-255.
13. Toole, E. H. and V. K. Toole. 1946. Relation of temperature and seed moisture to the viability of stored soybean seed. USDA Circular 753.

SIMULATION FOR RESEARCH AND CROP MANAGEMENT

D. N. Baker

Just before the 1920's a methodology called "growth analysis was developed in an effort to describe mathematically the seasonal time course of crop dry matter yield. Gregory (10) defined net assimilation rate (NAR) as: NAR = (1/L)(dW/dt), where L and W are leaf area and plant wt, indicating that NAR is a measure of the excess of photosynthate production over respiratory loss. Blackman (6) pointed out that the rate (R) of increase in dry wt can be regarded as a process of continuous compound interest: R = (1/W)(dW/dt). With the hindsight available to us a half century later we can see that these "models" implicitly contain some conceptual errors and shortcomings. For example, Blackman's compound interest analogy suggests a constant growth rate throughout the season. This implies exponential growth through the season, but we know that crop growth is sigmoid (c.f. Fig. 1). Nevertheless, it is often exponential early in the season when the plant is vegetative and receives an unlimited supply of light, water, and nutrients. So, over the years the principles of growth analysis have been used successfully with forage and root crops. The simulation of fruiting crops is another matter. Beyond the conceptual problem with long-term constancy in the growth rate, we find short-term variations (e.g., those caused by cloudiness, drought or temperature extremes), of much interest (Fig. 2).

By the late 1940's Watson (21) had coined the term leaf area index (LAI), and in the early 1950's the development of canopy light interception-photosynthesis models (13) was begun in an effort to explain in more detail the illumination of leaves in crop stands and the relations between LAI and canopy dry matter accretion. The 1960's and the early part of the present decade saw the empirical description of leaf and crop canopy photosynthesis.

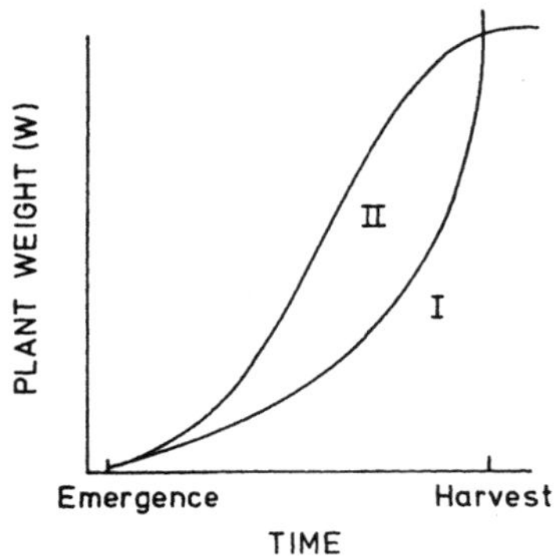

Figure 1. Time courses of plant growth as described by (I) the compound interest law, and (II) the sigmoid pattern found in fruiting crops.

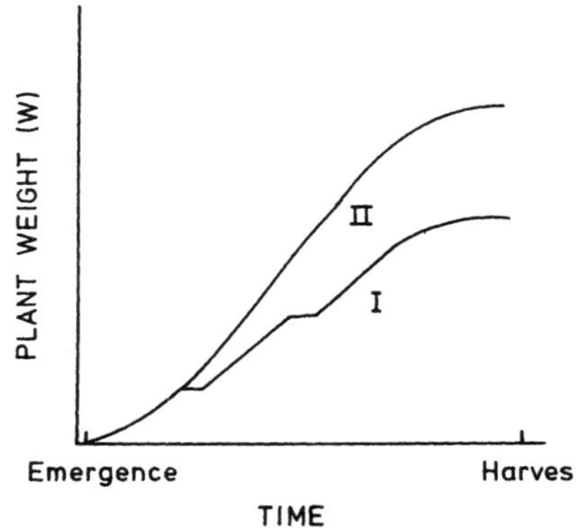

Figure 2. Time courses of plant growth as occurs in the idealized sigmoid pattern (II) and the more typical case (I) where growing conditions have varied or stresses have occurred.

The history of all of this with special reference to the relations between crop architecture, photosynthesis and yield has been reviewed elsewhere (5), tracing the evolutionary process from growth analysis, through the modeling of crop canopy light interception, the modeling and measurement of canopy photosynthesis and respiration, to the first seasonal materials balances, and finally to the development of dynamic crop simulation models. But some of the more important discoveries from the experimental research deserve mention here. The leaves and canopies of well-watered crops are not light saturated at full sun in the field (1,11,17). A four-carbon photosynthetic pathway exists in some species (14), and indeed, this manifests itself in terms of greater efficiency at the crop canopy level (12). The assumption by early growth analysts that canopy dry matter yield was mainly a matter of canopy light capture was confirmed (2). Finally, it was demonstrated that in some species, especially legumes, starch levels in leaves may build up during rapid photosynthesis, thereby reducing photosynthetic efficiency. Through all of this the conceptual understanding of crop photosynthesis was worked out and a data base describing it began to emerge.

This data base and theory would serve as the basis of crop simulation models. Because growth and the collateral physiological processes affecting it do vary diurnally and seasonally and because yield is reckoned in terms of wt (per unit land area), dynamic materials balance models were built and solved iteratively. In other words, the "R" term in Blackman's growth equation is not constant in crops and the scientific effort from that point to the present simulation models has been largely one of elaborating more detailed and better methods of accounting for the physical and physiological factors causing this rate to vary. Initially, the dynamic models accounted only for the effects of temperature and solar radiation on respiration and photosynthesis (3,4). No attempt was made to simulate organ growth and morphogenesis.

In 1968, Stapleton, an agricultural engineer with an interest in the optimization of machinery arrays in cotton production, proposed the development of what we know today as crop simulation models (20). Many of the early efforts here and overseas (8,9,22) can be attributed to the fact that Stapleton brought the possibilities of a new technology to the attention of plant scientists (19). Table 1 contains a representative list of current materials balance dynamic crop modeling efforts. The objectives and the validation methods used by the modelers in this group are almost as varied as their disciplinary backgrounds. Validation efforts range from extensive field testing to simple tests of reasonableness; the former usually is employed where crop management decision making is the objective, the latter when the object is simply to study the interrelations among processes. Some programs (including ours) involve both of these and have the additional objective of system design, i.e., crop breeding and breeding in special combination with cultural practice.

The flow chart in Figure 3 depicts a possible model of soybean growth. The major plant and soil processes are treated as models or subroutines, which are called one or more times per day as needed to predict the progress in that process for the day. This model contains a soil section which describes the movement and distribution of soil. For RHIZOS (15), this would be a two-dimensional model of the soil profile. The remainder of the model concerns the production and distribution of dry matter in the plant. Climate, soil, variety and cultural information are read in at the beginning of the "run." Then each day a main program calls, sequentially, the subroutines in the main line. After PLTMAP has been called and executed, the date is compared with the maturity date (either input or calculated) and if the crop is mature, the requested output is printed and the "run" terminates. CLYMAT massages the input climate data, converts it to the proper units and provides, for each parameter, average values appropriate to the time step being taken, e.g., 3 hr CLYMAT also calls a DATE subroutine which begins with Julian dates and calculates calendar dates, time from emergence, etc., and TMPSOL, which calculates soil temperature at various points in the profile as needed. If a rain has occurred, a RUNOFF model may be called, and then a GRAFLO model may be called to distribute vertically the rain percolating into the profile and to estimate the elution of soluble nutrients down through the profile. An alternative sometimes followed here is to use the capillary flow subroutine CAPFLO to distribute the rainfall in the profile. Evapotranspiration, ET, is calculated, often via a modified Penman approach, and the uptake (UPTAKE) of soluble nutrients and water is calculated. Then, the capillary redistribution of water (Darcian flow) is calculated.

Table 1. Some currently active crop modeling efforts.

Authors	Institutions	Model Name	Species	Processes Treated
Allen, J. and J. H. Stamper	U. of Florida	CITRUSIM	Citrus	Photosynthesis
Arkin, G. F., J. T. Ritchie, and R. L. Vanderlip	Texas A&M U. USDA/ SEA, and Kansas State U.	SORG	Sorghum bicolor	Photosynthesis, respiration, transpiration and evaporation
Baker, D. N., J. R. Lambert, and J. M. McKinion	USDA/SEA at Mississippi, Clemson U.	GOSSYM	Cotton	Photosynthesis, respiration, growth and morphogenesis. Incorporates RHIZOS.
Brown, L. G., J. D. Hesketh, J. W. Jones, and F. D. Whisler	Mississippi State U.	COTCROP	Cotton	Photosynthesis, respiration, transpiration, runoff, drainage, nitrogen uptake, denitrification, leaching, organogenesis, partitioning and growth.
Childs, S. W., J. R. Gilley, and W. E. Splinter	U. of Nebraska	Unnamed	Corn	Photosynthesis, respiration, transpiration, growth, soil evaporation, and soil water flows
Curry, R. B., G. E. Meyer, J. G. Streeter, H. J. Mederski, and A. Eshel	Ohio Agr. Res. & Development Center	SOYMOD OARDC	Soybeans	Photosynthesis, respiration, translocation and evaporation
Duncan, W. G.	U. of Florida, U. of Kentucky	SIMAIZ	Corn	Photosynthesis, processes involved in setting seed number and seed size
Duncan, W. G.	U. of Florida, U. of Kentucky	MIMSOYZ	Soybeans	Photosynthesis, nitrogen fixation, assimilate redistribution, processes for setting seed number and seed size
Duncan, W. G.	U. of Florida, U. of Kentucky	PEANUTZ	Peanuts	Photosynthesis, nitrogen fixation, processes for setting seed number and seed size
Fick, G. W.	Cornell University	ALSIM	Alfalfa	Photosynthesis defined as crop growth rate, and partitioning

Holt, D. A., G. E. Miles, R. J. Bula, M. M. Schreiber, D. T. Doughtery, and R. M. Peart	Purdue University, USDA/SEA	SIMED	Alfalfa	Photosynthesis, respiration, growth, translocation, and soil moisture uptake
van Keulen, H.	Netherlands Agr. U., Wageningen	GRORYZA	Rice	Gross assimilation and respiration
van Keulen, H.	Netherlands Agr. U. Wageningen	ARIDCROP	Natural vegetation in semi-arid regions	Photosynthesis, respiration, transpiration, and water uptake
Loomis, R. S. and E. Ng	U. of California, Davis	POTATO	Potato	Photosynthesis, respiration, transpiration, water uptake, growth, development, and senescence
Loomis, R. S., J. L. Wilson, D. W. Rains, and D. W. Grimes	U. of California, Davis	COTGRO	Cotton	Photosynthesis, respiration, transpiration, water uptake, growth, development, flowering, fruit development, senescence, and soil heat flux
Loomis, R. S., G. W. Fick, W. A. Williams, W. H. Hunt, and E. Ng	U. of California, Davis	SUBGRO	Sugar beet	Photosynthesis, respiration, transpiration, water uptake, growth, development, and senescence
Marani, A.	The Hebrew U. of Jerusalem	ELCOMOD	Cotton (Acala)	Photosynthesis, respiration, growth, morphogenesis, ET, nitrogen uptake, and gravitational soil wetting
Orwick, P. L., M. M. Schreiber, and D. A. Holt	Purdue U.	SETSIM	Setaria	Carbon flow, photosynthesis, respiration, growth, and translocation
Ryle, G. J. A., N. R. Brockington, C. E. Powell, and B. Cross	Grassland Research Institute, Hurley, Berkshire, England	Unnamed	Uniculum barley	Photosynthesis, assimilate distribution, and synthetic and maintenance respiration
de Wit, C. T. et al.	Netherlands Agr. U., Wageningen	PHOTON and BACROS	Any crop	Photosynthesis, respiration, transpiration, reserve utilization, water uptake, and stomatal control

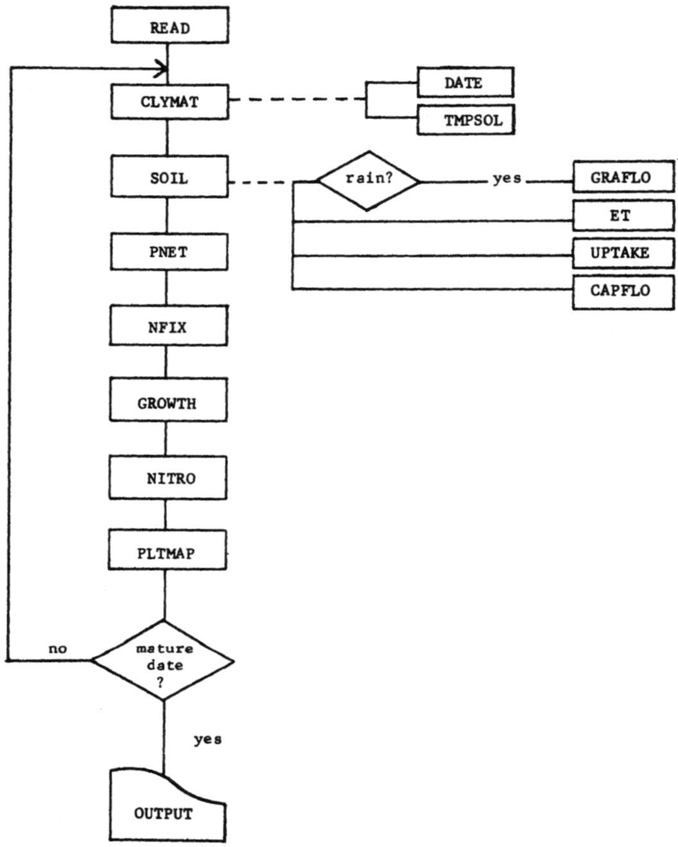

Figure 3. Flow chart of the subroutine structure in a possible soybean model.

The SOIL section provides the overall model with an estimate of soil water potential in the root zone. In the simulation of nonleguminous species, a model such as RHIZOS also determines the daily uptake of nitrogen.

After SOIL the subroutine PNET is called to estimate photosynthate production and respiratory losses. The net photosyntate plus soluble carbohydrates held in reserves are available for growth and nitrogen fixation. Nitrogen fixation is calculated in NFIX. There, the effect of Rhizobial type temperature and soil water potential are accounted for.

Assimilate partitioning is handled in GROWTH. Several strategies are used to distribute dry matter to existing organs. In one, arbitrary priorities and distribution factors are used to decide the fraction of available assimilate going to each organ. This is the so-called "standard plant" concept. In another, the assimilate supply to demand ratio is obtained by calculating the effects of temperature, tissue turgor, and nitrogen supply, on the potential growth of each organ. The product of this ratio and the potential assimilation increments are then used as partitioning factors to calculate the dry wt increases in each of the organs. This approach assumes that over the time step involved, e.g. one day, translocation is not limiting and that no hormone-actuated gating system is involved in assimilate partitioning. Other partitioning strategies assume that translocation is limiting within the plant, and flow resistances together with source-sink relationships are calculated. The appropriateness of the various approaches to assimilate partitioning depends on the anatomy of the plant. Similar rationales are used in estimating the allocation of nitrogen to plant parts in NITRO. The model proposed in Figures 3 and 4 ignores the small amount of dry matter tied up in nodule structures.

The times at which new organs appear on the plant are calculated in PLTMAP, along with the abscission of leaves and fruit. In other words, morphogenetic processes, the effects of insect damage, and physiological stresses are calculated here, and a map of the plant is kept updated in this subroutine.

A simplified conceptual description of the carbohydrate balance section of a typical soybean model is presented in Figure 4. Photosynthesis, respiration and growth are represented. Irregular and rectangular enclosures represent pools of indefinite and definite size, respectively. The valve-shaped

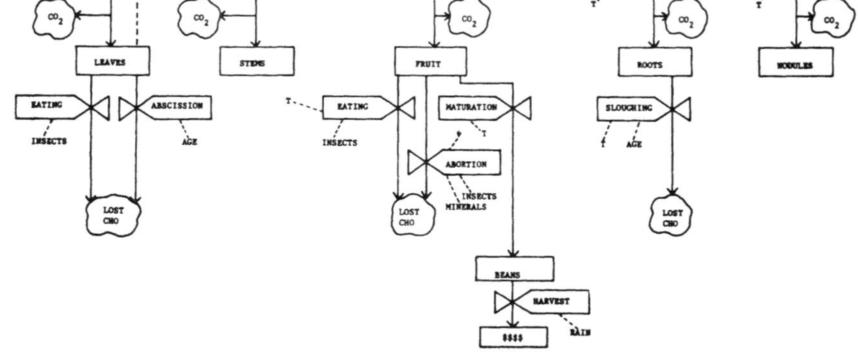

Figure 4. A simplified conceptual model of photosynthate partitioning in the soybean plant.

structures represent flow regulators. Solid and dashed connecting lines depict material and information flows, respectively. The carbohydrate pool receives material via photosynthesis at a rate depending on the amount of shortwave radiant energy absorbed, the leaf temperature and the leaf water potential. Carbohydrate is lost via maintenance respiration at a rate depending on temperature and the biomass of the system. It is also lost via a respiratory process associated with the photosynthesis and temperature of the tissues involved. Another respiratory loss is associated with this growth. Biomass is lost from various organs by several means. Leaf material is lost through insect feeding and senescence. Insect feeding may also take some fruit. In simulating insect damage it is necessary simply to specify daily the age and mass of the organs damaged. Fruit may also be aborted as a result of insect damage or physiological stress. Biomass is also removed from the fruit pool via the formation of mature beans. Depending on soil conditions, roots in certain age categories are lost or sloughed throughout the season.

In the model, the physiological rates are expressed mathematically. Some of the functional relationships suggested in Figure 4 are listed in Table 2. In some cases it is convenient to use several program statements to solve these equations. For example, a curve with a sharp break may be sectioned into two. The growth function is not a simple function of ψ; rather, tissue turgor is a function of ψ, and growth is actually expressed as a function of turgor. Similar lists of functional relationships must be developed for the other plant and soil processes.

Breaking down the system into simple elements identifies experimental research needs. For example, crop canopy photosynthesis must be measured under a range of light intensities, CO_2 levels, temperatures, leaf water potentials and leaf starch levels. The results may simply be expressed in the form of a multiple regression equation. Organ growth, for example, must be measured at various temperatures and tissue turgor levels. If the model strategy is to use a metabolite supply:demand ratio as a partitioning factor, then these growth rates must be potential rates (which are then decreased by the ratio). This implies the measurement of growth under luxury supply conditions. It is ordinarily done at elevated CO_2 levels and bright light. Then, more experiments are done under controlled conditions progressively further from the optimum to develop a data base and equations for decreasing these rates from the potential. Since potential rates of both growth and morphogenesis need to be obtained at elevated CO_2 levels, we (18) have developed a special facility referred to as a SPAR installation. SPAR units contain a meter depth of field soil and the plants grow in a canopy light environment similar to that in the field. Such installations are now in use at several locations where models of the type referred to here are being developed. The SPAR unit represents a physical model of the crop.

Table 2. Some functional relationships in the materials balance of a crop simulation model. Symbols are defined below.[1]

Photosynthesis	$P = f(R_i, CO_2, T, \psi, CHO)$
Maintenance resp.	$R_m = f(T)B$
Light respiration	$R_L = f(T)P$
Growth	$F = f(T, \psi)$
Growth respiration	$G_R = k(G)$
Fruit abortion	$A_B = f(S_p)$
Maturation	$M = f(T)$
Root sloughing	$S_1 = f(T, O_2)$

[1]Where: p = photosynthesis; R_i = incident solar radiation; CO_2 = atmospheric carbon dioxide concentration; T = temperature; ψ = leaf water potential; R_m = maintenance respiration; B = plant biomass; R_L = light respiration; G = growth; k = constant; A_B = fruit abortion, S_P = physiological stress; M = maturation; S_l = root sloughing; O_2 = soil oxygen concentration; and CHO = leaf starch level.

At Mississippi State we have carried modeling a step further. This is represented in Figure 5, taken from McKinion (16). We have tied the computer controlling and logging data from our SPAR units to the computer running our models. The unit on the left in the figure represents the computer system in which the crop simulation model runs. The equipment on the right represents the machine tied to the SPAR units. With this arrangement we can "unplug" a subroutine or several subroutines at a time in the simulator and plug in real data from the SPAR installation of process rates, e.g., photosynthesis or transpiration. This is helpful in identifying sources of error in the remainder of the simulator.

Clearly the modeling of crop systems results in the asking of a characteristic class of questions for experimental research. Not only that, it specifies how experiments shall be done and in some instances how the apparatus or facilities shall be designed. This one might call Phase I research management.

Phase II research management might be that which pertains to the application of a crop simulation model for system design. For example, suppose a breeder wanted to know under what soil moisture conditions it would pay to have a resistance to a certain kind of nematode in a certain crop line. Or, the question may be asked, "In order to get the most benefit from the development of enhanced photosynthetic capability, does fruiting habit need to be modified, or does optimum fertilizer management need to be altered?" Dozens of such questions can be addressed with process-oriented crop simulation models. For the immediate future there is no guarantee that the model predictions will be correct. The model is merely a tool which accounts rationally and quantitatively for as much scientific fact as is available in the

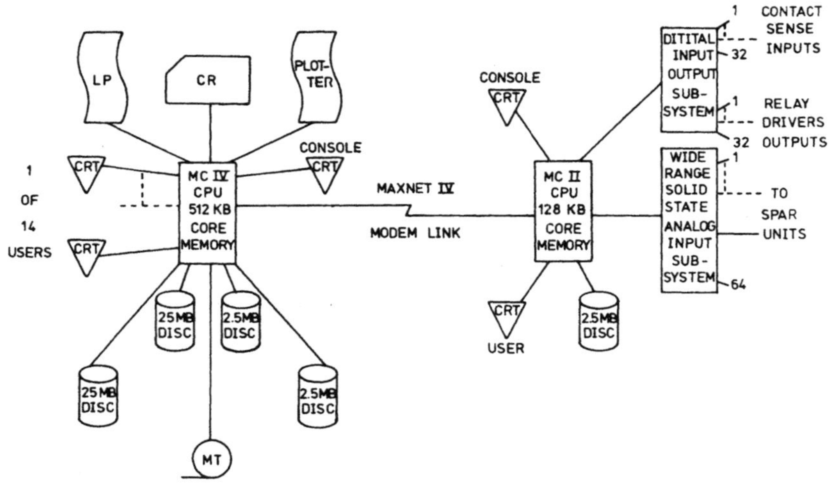

Figure 5. **The computer systems and hardware used in the simulation (Modcomp IV) and physical (SPAR) modeling (Modcomp II) of crops. CRT, LP, and CR represent cathode ray tube terminals, line printers, and card reader, respectively.**

making of such decisions. Such a model should be thought of as an adjunct to, not a substitute for, common sense.

The same or very similar models will soon be used to help crop statisticians or economists sharpen their estimates of the effects of climate, pest and disease disturbances on yield. Initially these applications will focus on picking out and quantifying the trends and relative changes in growth and the resultant effects on yield rather than on absolute yield forecasts. Although our objective is absolute accuracy in yield prediction, we will gain much useful information about crops from these models before such accuracy is achieved.

Farm operators will use these simulation models eventually in conjunction with comparable insect, disease and weed models in making crop management decisions. In fact, one of the main reasons (in addition to the application in breeding) for the process orientation of these models is to provide for the interfacing of insect, weed and disease models.

A simple example of the use of a model in cotton irrigation management is presented in Figure 6. The cotton model GOSSYM gave a nearly perfect simulation of the moisture stress treatments in the rain-out shelter experiments of Bruce and Römkens (7). Their plant population was 50,600 plants/ha—a rather sparse population. We used the model to predict the growth, fruiting and yield of a crop planted at 101,300 plants/ha and grown with their climate and irrigation inputs.

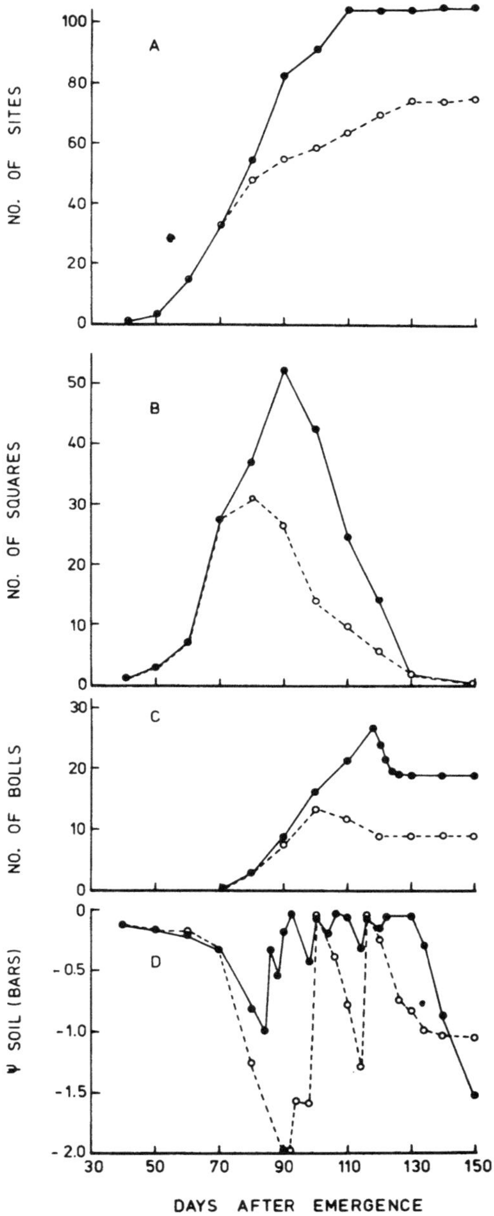

Figure 6. Simulation of soil water potentials and cotton growth under two possible irrigation plans.

The model predictions of the average of soil water tensions at 15-, 30-, and 45-cm depths are graphed in Figure 6-D for two of the treatments. In one treatment soil water potential fell to -1.0 bar on the 85th day from emergence. Then, the crop was irrigated, and the average water potential in the top 45 cm of the profile was maintained at -0.5 bar or greater until near the end of the season. A second treatment was allowed to deplete the soil water in the same region to -2.0 bar on day 90. Then irrigation was begun, returning the soil to field capacity by day 100. In this treatment stresses were again allowed to develop and reached -1.25 bar by day 115 and -1.0 bar by day 135. This (drier) treatment might well represent the usual water stress in the mid-South.

The predicted effects of these treatments on the vegetative growth of the crops are shown in Figure 6-A. The water stresses were allowed to build beginning with first bloom. The model predicts that the mildly stressed crop would produce over 100 fruiting sites/plant and the more severely stressed crop would produce less than 80. Predicted final plant heights were 150 and 118 cm, respectively.

The model predicts that even the small difference in soil moisture between treatments on day 80 would show up as a difference in fruit production and fruit abscission (Fig. 6-B and C). The difference in fruiting resulted in nearly twice as many bolls at maturity for the mildly stressed crop as compared with the more severely stressed one (D), but the model predicted that the bolls would have been somewhat larger, on the average, in the drier crop. A final yield difference of 6.9 vs. 5.4 bales/ha (2.8 vs. 2.2/acre) was predicted. The drier crop would have been earlier, reaching 65% boll open about a week sooner. The farmer would consider this delay in maturity with its increased cost in insect pest management along with the cost of irrigation in deciding whether he could afford to try for the additional yield of 1.5 bales/ha.

Other applications in crop management decision making in tillage and pest control could be presented for various weather scenarios. We believe the soybean farmer, too, will benefit from the use of simulation models in selecting and evaluating culture inputs.

In summary, the quest for mathematical descriptions of crop growth that will be useful in crop management and yield forecasting continues. Growth analysis has evolved into crop modeling and systems analysis. From simple compound interest laws which considered the gross plant attributes of wt and size, and which could only predict growth in the early vegetative stages, we have seen a move toward the study and mathematical description (at first) of plant photosynthesis, respiration, (and later) growth and morphogenesis, and the assembly of these process models, as subroutines, into plant simulation models. The effort to make these models useful over the ecological range of the crops has led to development and incorporation of comparable models (e.g., RHIZOS) of the climate and soil processes, evapotranspiration, soil nitrogen transformation, and the movement of water, nutrients and roots through the soil mass. Further, this effort to make the models useful in the field prompted the development and incorporation of insect, weed and disease models. The research has become truly multidisciplinary and the crop model has become a common medium of communication among scientists. Paralleling the modeling activity has been a great evolution in the controlled environment equipment used to develop rate equations. From the growth cabinets, the field canopy photosynthesis chambers, the weighing lysimeters, phytotrons and rhizotrons, we have moved to the SPAR installation where the above- and below-ground process rates of growth, morphogenesis, photosynthesis and transpiration can all be measured and, to a greater extent, controlled independently. The combination of these physical models of crops with computer models offers the scientist unprecedented analytical power. Even the insect cage gives way to the SPAR unit in some experiments. The evolution has been not so much toward complicated models (they are all very simple in their parts) as toward extensive models to account for the factors determining growth and yield under field conditions. The shift to simulation models was made possible by the availability of large digital computers. The evolution toward the more extensive models needed to simulate and manage crop growth is made practical by the continuing evolution in the computer industry.

NOTES

Donald N. Baker, Crop Simulation Research Unit, USDA, SEA/AR, and Department of Agronomy, Mississippi State University, Mississippi State, Mississippi 39762.

REFERENCES

1. Baker, D. N., and R. B. Musgrave. 1964. Photosynthesis under field conditions. V. Further plant chamber studies of the effects of light on corn (*Zea mays* L.). Crop Sci. 4:127-131.
2. Baker, D. N., and R. E. Meyer. 1966. Influence of stand geometry on light interception and net photosynthesis in cotton. Crop Sci. 6:15-19.
3. Baker, D. N., and D. L. Myhre. 1968. Leaf shape and photosynthetic potential in cotton. Proc. 1968 Beltwide Cotton Prod. Res. Conf., Hot Springs, Ark., Jan. 9-10, pp. 102-109.
4. Baker, D. N., and J. D. Hesketh. 1969. Respiration and the carbon balance in cotton (*Gossypium hirsutum.* L.). 1969. Proc. Beltwide Cotton Prod. Res. Conf., New Orleans, LA, Jan. 7-8, pp. 60-64.
5. Baker, D. N., J. D. Hesketh, and R. E. C. Weaver. 1978. Crop architecture in relation to yield. In U. S. Gupta (ed.) Crop Physiology, Oxford & IBH Publishing Co., New Delhi, Bombay, Calcutta, India, pp. 110-136.
6. Blackman, W. H. 1919. The compound interest law and plant growth. Ann. Bot. 33:353-360.
7. Bruce, R. R., and M. J. M. Römkens. 1965. Fruiting and growth characteristics of cotton in relation to soil moisture tension. Agron. J. 57:135-140.

8. Duncan, W. G. 1972. SIMCOT: A simulator of cotton growth and yield. In C. Murphy (ed.) Proc. Workshop on Tree Growth Dynamics and Modeling. Duke University, Durham, NC, pp. 115-118.
9. Fick, G. W., W. A. Williams, and R. S. Loomis. 1973. Computer simulation of dry matter distribution during sugar beet growth. Crop Sci. 13:413-417.
10. Gregory, F. G. 1917. Physiological conditions in cucumber houses. Third. Ann. Rept., Experiment and Research Station, Cheshunt, England, pp. 19-28.
11. Hesketh, J. D., and R. B. Musgrave. 1962. Photosynthesis under field conditions. IV. Light studies with individual corn leaves. Crop Sci. 2:311-315.
12. Hesketh, J. D., and D. N. Baker. 1967. Light and carbon assimilation by plant communities. Crop Sci. 7:285-293.
12. Kasanaga, A. and M. Monsi. 1954. On the light transmission of leaves and its meaning for the production of matter in plant communities. Jap. J. Bot. 14:304-324.
14. Kortschak, H. P., C. E. Hartt, and G. O. Burr. 1965. Carbon dioxide fixation in sugar cane leaves. Plant Physiol. 40:209-213.
15. Lambert, J. R., D. N. Baker, and C. J. Phene. 1975. Simulation of soil processes under growing row crops. Paper No. 75-2580. Winter Meeting ASAE, Chicago, IL. 12 pp.
16. McKinion, J. M. 1979. Dynamic simulation: A positive feedback mechanism for experimental research in the biological sciences. Agricultural Systems (In press).
17. Moss, D. N., R. B. Musgrave, and E. R. Lemon. 1961. Photosynthesis under field conditions. III. Some effects of light, carbon dioxide, temperature, and soil moisture on photosynthesis, respiration and transpiration of corn. Crop Sci. 1:83-87.
18. Phene, C. J., D. N. Baker, J. R. Lambert, J. E. Parsons, and J. M. McKinion. 1978. SPAR—A soil-plant-atmosphere research system. Trans. ASAE 21:924-930.
19. Stapleton, H. N. 1968. A report of a working seminar on the stimulation of the cotton plant. Sponsored by the Dept. of Agr. Eng., The University of Arizona, Tucson, May 23-24, 55 pp.
20. Stapleton, H. N., D. R. Buxton, F. L. Watson, D. J. Nolting, and D. N. Baker. 1973. Cotton: A computer simulation of cotton growth. Arizona Agr. Exp. Sta. Tech. Bull., Tucson, 123 pp.
21. Watson, D. J. 1947. Comparative physiological studies on the growth of field crops. I. Variation in net assimilation rate and leaf area between species and varieties, and within and between years. Ann. Bot. N. S. 11:41-76.
22. de Wit, C. T. 1970. Dynamic concepts in biology. In Prediction and Measurement of Photosynthetic Productivity. Proc. IBP/PP Tech. Meet., Trebon, Czech. Pudoc, Wageningen, The Netherlands, pp. 17-23.

SIMULATION OF INSECT DAMAGE TO SOYBEANS

W. G. Rudd

It was apparent early in our efforts in the development of integrated pest management programs that a soybean crop growth model that responds correctly to simulated damage by insects and other pests would be required. At the time, research emphasis was on pests that defoliate soybean plants or that feed on pods. A survey of existing soybean plant growth models revealed none in which the emphasis was on this kind of behavior. We therefore set out to construct our own model.

A simple canopy model for soybean above-ground dry matter production has been developed for use in insect pest management modeling research. The model is based on an empirical relationship between leaf area and dry matter accumulation rates. The model outputs include dry wt of stems, leaves and fruits as functions of days after planting. Validation runs show good agreement between model outputs and experimental data for several soybean varieties under different growing conditions. Experiments with simulated mechanical defoliation indicate that some "compensation" occurs if the plant is defoliated heavily after podset. The model agrees well with experimental results when compensation is included. The model also responds correctly to simulated podfeeding and to variations in planting dates. The model is now in use in pest management studies.

MODEL FORMULATION

The model is based upon two primary sets of data. The first is Jensen's thorough study (5) in which Dare soybeans were planted in 122 cm-rows and kept relatively insect free. Periodically, fifteen 61 cm-samples were chosen at random from the 1.48 ha plot. All plants in each 61 cm sample were cut at ground level, placed in polyethylene bags, and carried to the laboratory. Fresh wt, dry wt and counts of leaves, stems, and fruits were taken. Leaf area was measured using a Lambda meter. The other basic datum used to build the model is the dry matter production vs leaf area index (LAI) curve from Shibles and Weber (6).

In Figure 1, bars have been omitted for clarity, but typical coefficients of variation for leaf, stem and fruit dry wt are 10, 22, and 17%, respectively. Three different post-emergence growth patterns are apparent. During vegetative growth (pre-podset), early growth is exponential, followed by linear growth as the canopy closes. The Shibles and Weber data (6) indicate a possible explanation for this behavior. Early in the season, the dry matter production rate is proportional to LAI which is in turn roughly proportional to the total dry wt. Thus the rate of increase is proportional to the wt produced previously. The result is exponential growth. Later on, as the canopy fills, the dry matter production rate approaches a constant value, which we call PTSMAX. A growth process that occurs at a constant rate results in a linear growth pattern. We have refit the Shibles and Weber data (6) to obtain the following equation for the dry matter production rates:

$$f(\mathrm{LAI}) = \begin{cases} (5.4 + 47.78\ \mathrm{LAI} - 6.12\ \mathrm{LAI}^2)\ \mathrm{PTSMAX} & (\mathrm{LAI} < 3.835) \\ [1 - \exp(-.91435\ \mathrm{LAI})]\ \mathrm{PTSMAX} & (\mathrm{LAI} \geq 3.835) \end{cases} \qquad [1]$$

Our arguments above indicate that, during the pre-podfill vegetative growth period, we have:

$$\begin{aligned} dw/dt &= f(\mathrm{LAI}) \\ d\ell/dt &= \mathrm{FRLF} \cdot dw.dt \\ ds/dt &= (1 - \mathrm{FRLF}) \cdot dw/dt \end{aligned} \qquad [2]$$

where: w = total dry wt, ℓ = leaf dry wt, s = stem dry wt, and t = time.

ITEMRG = emergence time
ITLSTP = podset time

The podfill period is more complicated. We assume that all leaf growth stops at podset, provided that defoliation is not too high. In other words, all further dry matter production is devoted toward the filling of the pods. Early in podfill, the pods are too small to absorb all the new dry matter produced each day so that some dry wt must remain in the stems for temporary storage. Later on, the pods grow large enough to assimilate more than the dry wt produced each day. The dry matter stored in the stems can be transferred to the pods. This occurs until the stems have been reduced to their pre-podset wt. This behavior, we feel, is the reason for the "hump" in the stem

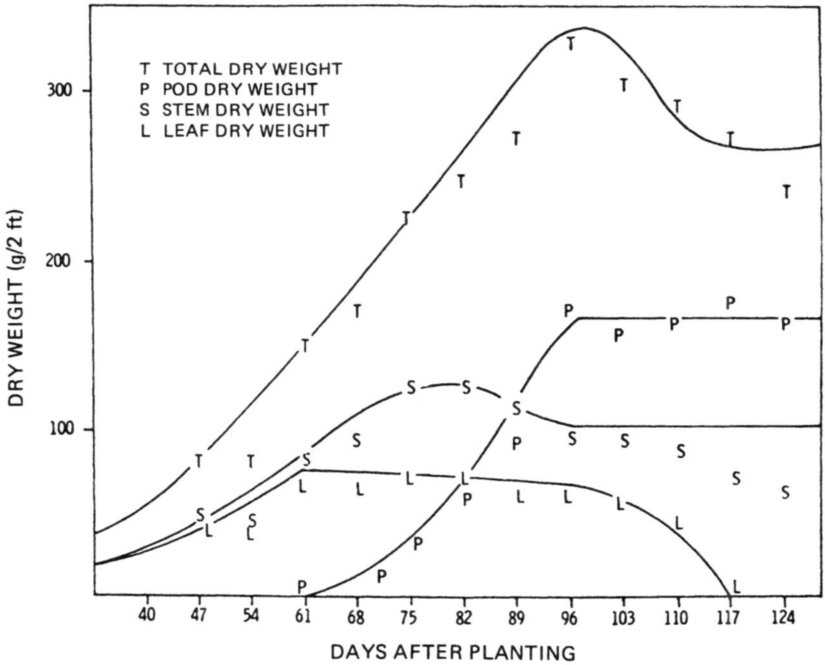

Figure 1. Data (letters) and model results (lines) for Jensen's study (5) of Dare soybean plant growth.

wt in Figure 1. Many other plant growth data sets which we have examined, display this behavior. While other workers have found some evidence to indicate that temporary storage occurs in leaves as well, Jensen's data do not show this to be the case. Nor is this leaf storage process necessary in our model in order to obtain the final observed fruit wt. Our data indicate that some senescence occurs slowly at a constant rate during the podfill period as follows:

$$\begin{aligned} dw/dt &= f(LAI) - K, \\ dp/dt &= Cp, \\ ds/dt &= dw/dt - Cp, \\ d\ell/dt &= -K \end{aligned} \qquad \textit{if}\ \mathrm{ITSLTP} < t \leq \mathrm{ISTOP} \qquad [3]$$

where: ISTOP is the time at which pod filling ends, K is the (constant) rate of senescence, and C is a proportionality constant that determines the rate of fruit growth.

Equation 3 exhibits the correct qualitative behavior. Early in podfill, p is relatively small so that Cp is less than dw/dt and the stems gain in wt. When the fruits become large enough to assimilate more than the total dry matter produced during the day, the stem wt decreases. We have not indicated the stem wt limits in equation 3, but in the computer implementation of the model, the stem wt is not permitted to fall below its level at podfill (t = ITLSTP). One can argue that the plant must keep enough dry matter in the stems to provide structural support.

At time ISTOP, all dry matter production ceases in the model. The leaves senesce rapidly and some dry wt is lost from the stems, presumably because of leaching processes. We have not put a lot of effort into modeling this phase of plant development since insects have little effect on yield during this period with the exception of those that feed directly on pods.

Figure 1 compares the model dry wt results with the original data while Figure 2 shows computed and measured leaf area.

We have found that by changing four parameters—ITLSTP, ISTOP, PTSMAX and FRLF—it is possible to model several other varieties and growth patterns from several geographical locations. To do so, we observe that PTSMAX is (ignoring senescence) the slope of the dry matter production curve during podfill. ITLSTP essentially regulates the amount of vegetative growth prior to podset, while the difference, ISTOP - ITLSTP, determines the length of time fruit are being formed. An easy way to adjust the parameters for a specific cultivar is to notice that, so long as ISTOP - ITLSTP is long enough to permit the stems to be drained of the material temporarily stored there, the final fruit wt is, neglecting a small correction for senescence, simply PTSMAX (ISTOP - ITLSTP).

We have obtained agreement between model and data similar to that in Figures 1 and 2 for several other varieties and growing conditions, including some partial sets of data from Bragg, Pickett, Dare and Lee varieties grown here, three years of data from Henderson and Kamprath (3), and some data on the indeterminate Beeson variety grown in Wooster, Ohio, by Curry (2).

PEST DAMAGE

Since the model is to be used in pest management model studies, it is vital that the model respond properly to simulated pest damage. We have chosen to use the data of Kalton, et al. (4) as representative of the effects of mechanical defoliation on yield. Figure 3 shows a replot of their data.

With no attempt to model any "compensation" for severe defoliation the model seriously overestimates the effects of defoliation early in the season. Since modeled growth is dependent upon leaf area, to add compensation for defoliation in the model it is necessary to have leaf growth occur in a qualitatively different manner when the plant is subjected to high levels of defoliation. In particular, the model should regrow some leaf area lost to defoliation if defoliation is heavy enough to have a significant effect on yield. Before podset, leaf production already proceeds exponentially, which is as

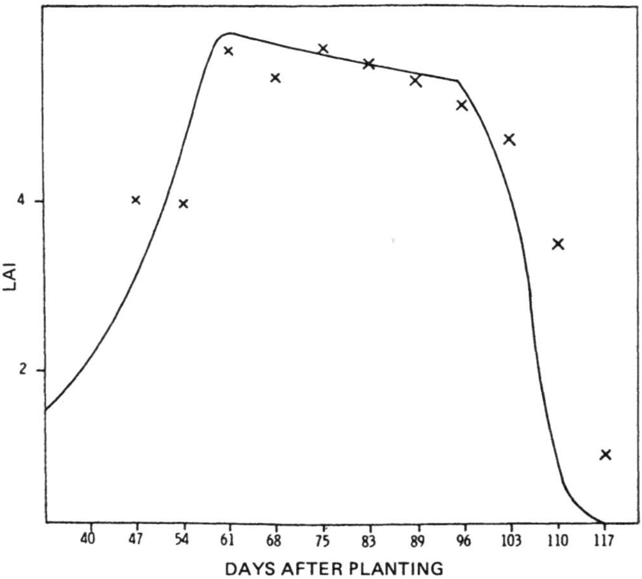

Figure 2. **Leaf area index measured (X) and modeled (–).**

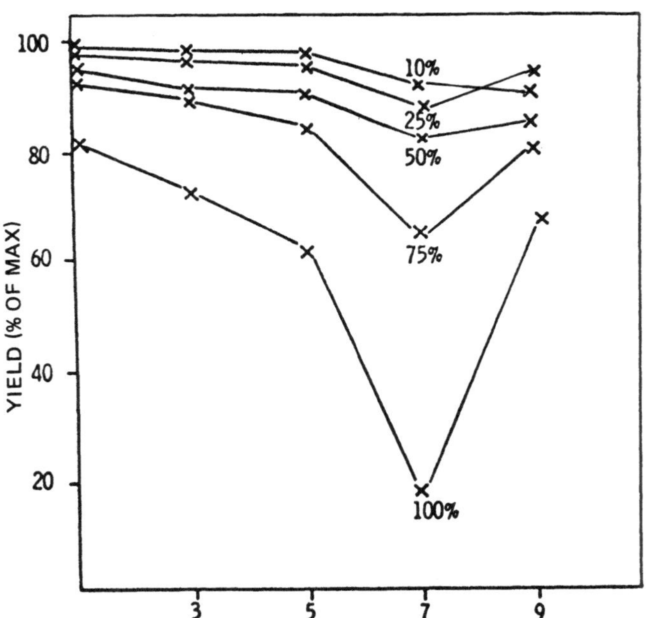

Figure 3. **Data of Kalton, et al. (4) showing dependence of yield loss on mechanical defoliation at various levels at various growth stages.**

fast as is possible. Therefore it makes little sense to try to include compensation during the pre-podfill phase.

We have assumed that the fruit are the only real sink for dry matter after the beginning of podfill. There is reason to believe that, if the plant suffers heavy defoliation, the leaves can become sinks as well, thereby removing some dry matter that normally would go to fruit and using that dry matter to replace lost leaves. In a sense, the modeled plant invests a portion of its anticipated yield in leaf production, hopefully improving future yield by increasing leaf area, thereby increasing productive capacity.

In the computer implementation of the model, we check the leaf area after podset. If the LAI fails below a threshhold value (we use 4), we take a portion (½ appears to work well) of the excess dry matter produced daily and divert it from its normal destination, temporary storage in stems, to leaf wt. This process is repeated daily until the leaf area rises above the threshhold. Once the pods are large enough to absorb all the daily dry matter produced, this compensation procedure stops in the model.

Figure 4 shows a simulation of the experiment of Kalton, et al. At all but the highest defoliation levels, the model agrees well with the experiment.

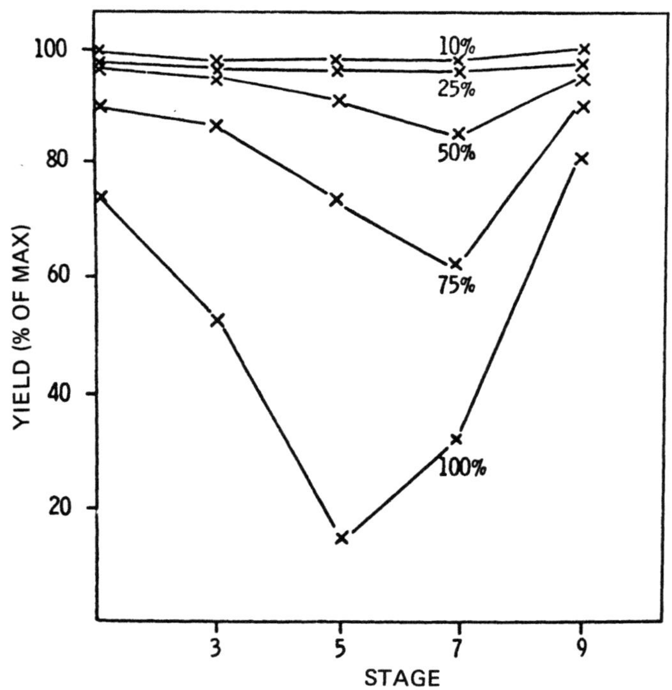

Figure 4. Model simulation of the experiment by Kalton, et al. (4).

We found no special provisions for compensation for pod feeding to be necessary. Figure 5 compares simulated yields after several levels of depodding with experimental results.

By varying the model time between emergence and podset, leaving all other parameters untouched, the effect of changing planting dates on yield from determinate varieties can be determined. Figure 6 compares the results of such a model experiment with field data from Able (1).

CONCLUSIONS

We have developed a simple, efficient soybean plant growth model for use in soybean pest management model studies. The model shows reasonable agreement with several sets of plant growth data and responds much like real systems to simulated stress. Currently, we are using the model for studies such as that in Figure 7, in which we show anticipated yield reduction vs peak soybean looper populations for various maximum undefoliated leaf area indexes. Similar studies permit us to perform model experiments to try various pest management strategies in controlling pod-feeding pests.

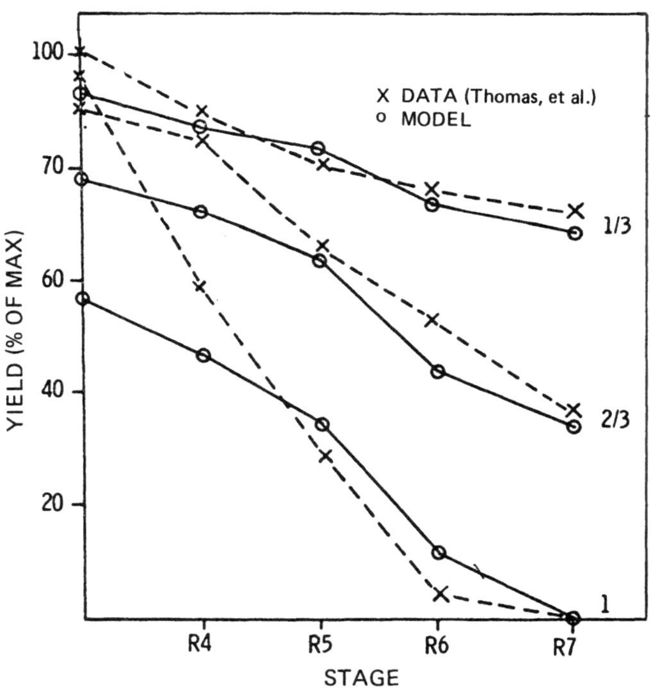

Figure 5. **Comparison of model and data on yield loss resulting from mechanical depodding at various levels at various plant growth stages.**

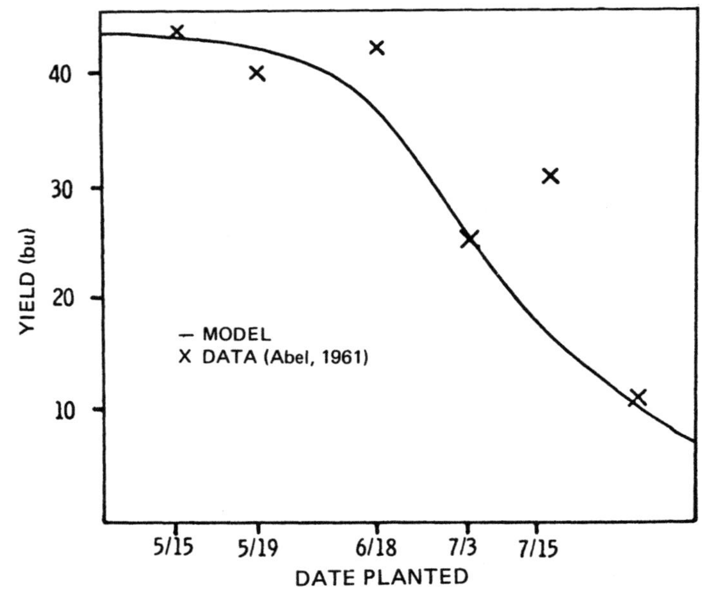

Figure 6. Model and experimental results on the effects of planting date on soybean yield.

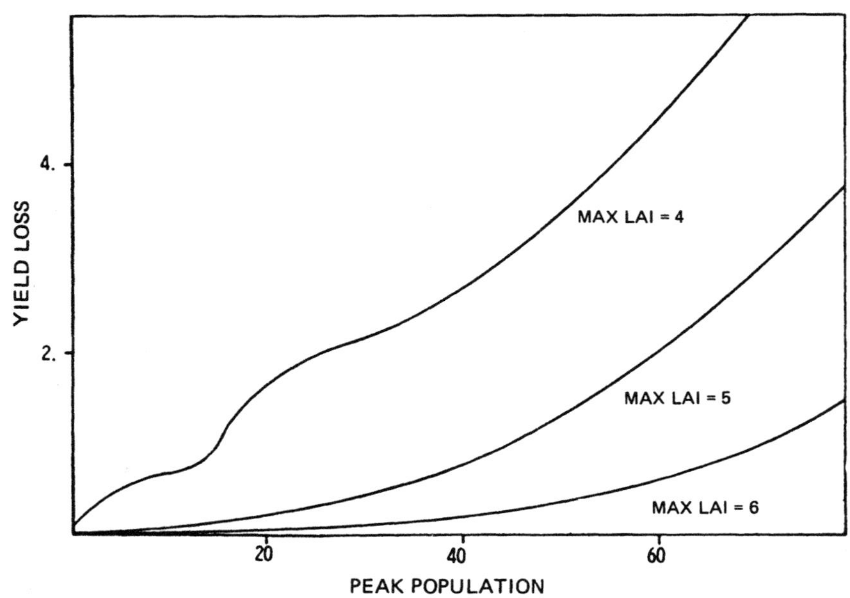

Figure 7. Effect of soybean looper populations on anticipated soybean yield reductions.

Gross effects of defoliation, depodding, and planting dates all can be explained with this simple model in which the only real factor acting is the connection between dry matter production and leaf area. There are no temperature effects or roots in the model and photosynthesis, morphogenesis, translocation, respiration and other processes are not modeled specifically. The fact that a simple leaf-area-driven model can explain several gross observed phenomena of course does not imply that the hypotheses behind the model are correct.

NOTES

W. G. Rudd, Department of Computer Science, Louisiana State University, Baton Rouge, Louisiana 70803.

Research was supported in part by the Graduate Research Council of Louisiana State University and by the National Science Foundation and the Environmental Protection Agency through a grant (NSF GB 34718) to the University of California. The findings, opinions and recommendations expressed herein are those of the authors and not necessarily those of the University of California, the National Science Foundation or the Environmental Protection Agency.

REFERENCES

1. Abel, G. H., Jr. 1961. Response of soybeans to dates of planting in the Imperial Valley of California. Agron. J. 53:95-98.
2. Curry, R. B. 1977. Unpublished data. OARDC, Wooster, Ohio.
3. Henderson, J. B. and E. J. Kamprath. 1970. Nutrient and dry matter accumulation in soybeans. N. C. Ag. Exp. Sta. Tech. Bull. No. 197.
4. Kalton, R. R., C. R. Weber and J. C. Eldridge. 1949. The effect of insury simulating hail damage to soybeans. Iowa Ag. Exp. Sta. Res. Bull. No. 359.
5. Jensen, R. L. 1974. Unpublished data. Louisiana State Univ. Dept. of Entomology.
6. Shibles, R. M. and C. R. Weber. 1965. Leaf area, solar radiation interception and dry matter production in soybeans. Crop Sci. 5:575-577.

SIMULATION OF THE VEGETATIVE AND REPRODUCTIVE GROWTH OF SOYBEANS

R. B. Curry, G. E. Meyer, J. G. Streeter, and H. J. Mederski

In order to present adequately the current state of the art of soybean simulation as requested by the conference planners, it is necessary to cover some of the background on simulation of soybean growth and development. The simulation technique used here refers to application of computer analysis of a system of mathematical equations on a continuous basis to describe the dynamic behavior of a crop plant over a growth season, or portion thereof. Work on simulation of soybean growth and development currently underway at various institutions with which the authors are personally aware will be reviewed. The current status of SOYMOD/OARDC in Ohio will be discussed in some detail as an example of the status of soybean simulation today.

BACKGROUND

Work on soybean growth and development simulation at OARDC dates back to at least 1972. Through the support of the Research Center and several CSRS grants, a team of people have had a part in the evolution of the soybean growth and development simulator. The current version is referred to as SOYMOD/OARDC. Faculty members, R. B. Curry, J. G. Streeter and H. J. Mederski, together with post-doctoral research associates, L. H. Chen, C. H. Baker and G. E. Meyer, represent an interdisciplinary team of agricultural engineers and agronomists who contributed to the modeling effort over the years. In addition, A. Eshel, a post-doctoral research associate, and N. K. Narda, a Ph.D. student, are currently continuing work with the previously-named faculty members on further model development.

About the time our work began here in Ohio, Dr. W. G. Duncan began work on a soybean model at Kentucky. This model was patterned after his corn growth and development model SIMAZ. Numerous exchanges of information and ideas occurred between the two groups.

In 1975, a regional committee was formed in the Southern Region (S-107) to develop a soybean model which would cover the areas of pest control and crop management as well as growth and development. These three areas form the basis for the development of three submodels. Nearly all the Southeastern state experiment stations are involved along with USDA/SEA/AR research workers and the Ohio Agricultural Research and Development Center.

CURRENT RESEARCH

Space does not permit the review of details of models being developed at various locations; therefore, general areas of emphasis and people involved are mentioned in order for interested readers to make further contacts.

Florida—K. Boote, along with J. Jones, and G. Smerage, have begun to work on a soybean model based on experimental data obtained in Florida studies, with emphasis on soil-plant-water relations and defoliation by insects. Their studies contribute to S-107.

Kentucky—As mentioned before, W. G. Duncan has developed a soybean model. Work continues in cooperation with D. Egli and Everett Leggett.

Louisiana—W. Rudd has put together a simplified soybean growth model to generate plant material for the soybean insect simulator he has developed. This model is available. Rudd also has been a major contributor to the insect submodel of S-107.

Nebraska—G. E. Meyer, now on the faculty of the University of Nebraska, has continued to work with SOYMOD/OARDC in cooperation with Ohio with emphasis on adapting it to Nebraska's climate and needs of irrigation management. An interactive version with simplified prompting and help options is being incorporated into AGNET for use in Nebraska.

North Carolina—D. Raper and M. Wann are developing a model of the vegetative growth of soybeans based on phytotron data and patterned after the tobacco model. Their group is continuing to work on the model. Data has been taken on carbohydrate and nitrogen balances for validation. Their work has also been expanded to include study of floral initiation and reproductive development based on an internal aging function.

Ohio—as mentioned before, faculty members have developed a dynamic simulation of soybean growth and development, SOYMOD/OARDC. A brief description is given in the following section. A more detailed history and current status on SOYMOD/OARDC can be found in the literature (3,4,5,8,9).

South Carolina—G. Miles and J. Lambert have provided the coordination for the development of a plant growth submodel under the S-107 Region soybean modeling project. Miles has constructed a systems dynamic type flow chart for soybeans including carbon, nitrogen and water flows and has assembled a set of process rate equations with the help of research personnel at locations in other states.

USDA/SEA/AR—D. Peters, at Illinois, has headed a recently selected team of USDA/SEA/AR researchers in the development of a soybean model. Other members are T. Sinclair and R. Zobel, Cornell; H. Taylor, Iowa; J. Hesketh, Illinois; and D. Baker, Mississippi.

Utah—R. Hill has developed a soybean yield prediction model to be used in irrigation management.

SOYMOD/OARDC

SOYMOD/OARDC was developed to serve as a tool to further the understanding of the growth and development of the soybean plant and as an aid to research on improving the production and management of the soybean. The objectives were to include as much of the known physiological processes as necessary to produce a simulator which will respond to a simulated environment in the same way as field grown soybeans would respond to a real environment. This simulator would then be useful in answering questions of how the plant would respond if we made a particular modification in the plant structure, physiology, or environment. Of course, the accuracy of the predicted responses would need to be checked by field experiments, but the simulator would have narrowed the focus of research.

SOYMOD/OARDC is basically a system of dynamic partial-differential equations describing the mass and energy balance within the soybean plant. The equations describe the soybean as an open system with import, export, and internal control processes. The simulator employs dynamic simulation techniques at a detailed physiological level. There was one fundamental underlying hypothesis used in building the simulator, the processes described are based mathematically on the availability of the most acceptable theory in the literature, the most logical hypothesis based on available experimental data, or observed behavior. These processes were organized into a computer program in such a way that improvements in the simulator could be made easily when new data and improved theories became available.

SOYMOD/OARDC is a modular structured computer program, written in standard FORTRAN IV for execution in either batch or interactive mode. The simulator (Figure 1) consists of a mainline program and 22 subroutines of which 18 describe various plant processes, 1 performs integration, and 3 handle selected output functions. The mainline program acts as an executive system accepting the inputs necessary to run the simulator, queing the various

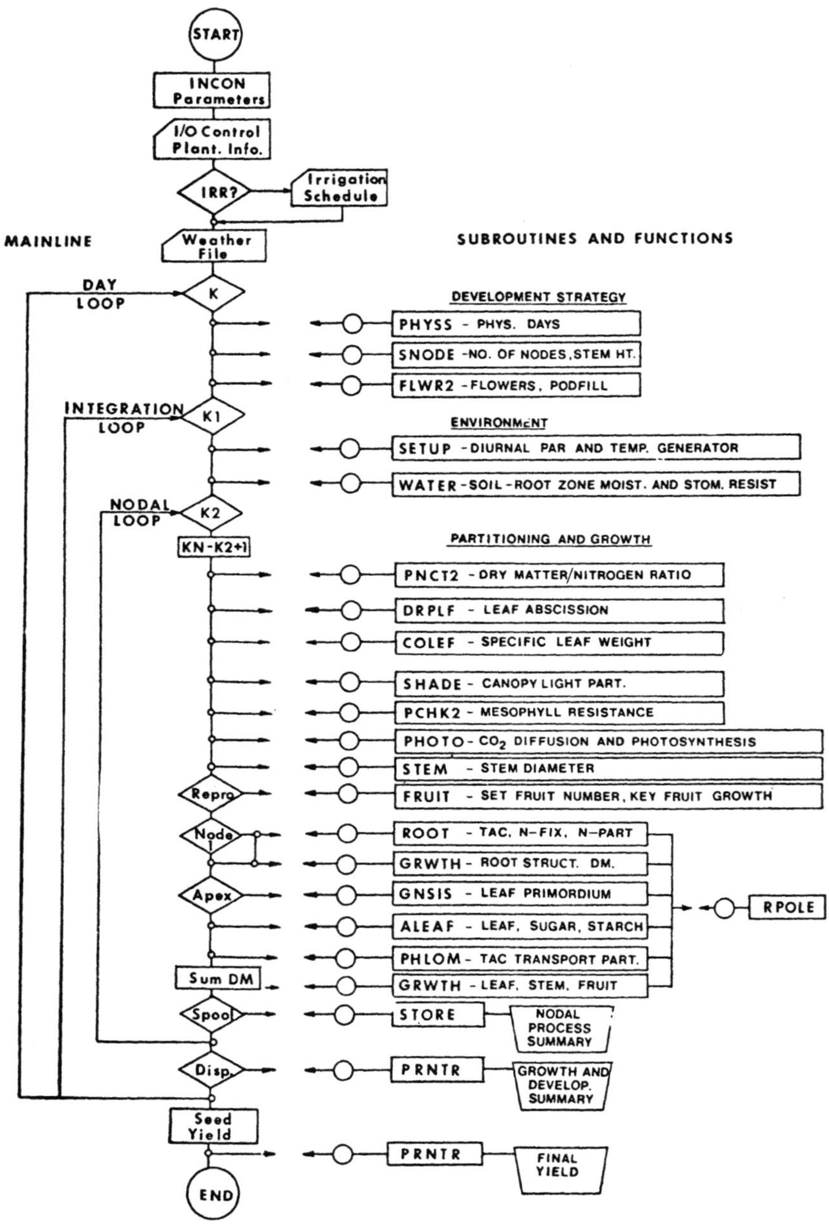

Figure 1. **Simplified flow diagram of SOYMOD/OARDC showing flow of logic and program units.**

process subroutines and managing the selected output. The simulator is being run currently on a Hewlett Packard 3000 Series II computer, but has been run on several other computers including various models of the IBM-370.

Input required and available output are summarized in Tables 1 and 2. Given an emergence date and a length of growing season, a special routine selects the required set of daily weather data from a large multi-year file. One of the subroutines converts the maximum and minimum temperatures and total daily insolation into hourly temperature and insolation. Output forms include chronology and synopsis of crop performance over the growing season, a detailed profile of the soybean canopy on any selected date and an abbreviated weather and crop summary.

The processes simulated include photosynthesis, respiration, carbohydrate translocation and partitioning, nitrogen assimilation and partitioning, leaf senescence and transpiration. Nodal formation, flowering, initiation of podfill and leaf and petiole abscission are considered as discrete events. In order

Table 1. Input data required for SOYMOD

A. For Each Run

1. *Plant Parameters*
   - Variety
   - Emergence date (year, month, day)
   - Row spacing
   - Plant spacing in row
2. *Soil Parameters*
   - Soil type
   - Soil water retention curve
   - Bulk density
   - Hydraulic conductivity
   - Initial soil water content
   - Irrigation? (If yes, can have automatic irrigation or supply irrigation schedule)
3. *Operational Parameters*
   - Choice of output forms
   - Interval between time of printout
   - Special tests (e.g. defoliation, depodding)
   - Maximum number of days simulated

B. Automatically Supplied Climatic Parameters
(From disc storage beginning with date of emergence)

- Daylength
- Maximum air temperature
- Minimum air temperature
- Dewpoint temperature (or minimum temperature)
- Solar radiation
- Rainfall
- Wind speed

Table 2. Output data generated by SOYMOD.

A. *Log* (Chronology of dry matter accumulation and physiological events—*whole plant basis*)

- Dry matter for each plant part (leaf blades, stems + petioles, roots, fruits)
- Plant height
- Maximum stem diameter
- Number of fruits
- Flowering and podfill events
- Leaf area
- Irrigation events
- Leaf abscission events
- Seed yield

B. *Partitioning Summary* (once/day; specify days desired—*each node basis*)

- Leaf area
- Number of fruits
- Total dry matter and dry matter components: nitrogen, available carbohydrate, structural carbohydrate—for each plant part
- Rate of: net photosynthesis, respiration, phloem loading, translocation, and carbohydrate storage
- Growth rate for each plant part

C. *Short Summary* (end of season)

- Location, date, variety, soil type, and planting configuration
- Total rainfall from emergence to maturity
- Total irrigated water from emergence to maturity
- Total solar radiation and heat units from emergence to maturity
- Physiological events: flowering, podfill, senescence, and maturity dates
- Crop summary: number of nodes attained
  - Maximum plant dry matter
  - Total evapotranspiration from emergence to maturity
  - Grams per 100 seed
  - Seed yield
  - Total fruit dry matter per plant

to give the reader a sense of the process formulation included, several of the processes will be described in some detail using Figure 1.

Photosynthesis is described by an equation reported by Lommen (6) and is calculated in subroutine PHOTO. This equation described the C-3 photosynthetic process rate within a leaf with photorespiration and a simplified leaf diffusion resistance network. Stomatal and mesophyll resistances control the flux of CO_2 to and from the site of fixation. Stomatal resistance is controlled jointly by leaf water potential and light intensity. Mesophyll resistances are controlled by the level of starch buildup in the leaf (10). The level of maximum potential photosynthesis is governed by the leaf nitrogen concentration (representing important photosynthetic enzymes) and leaf temperature (1,2, 12,14). Photosynthesis is considered at each node on the shoot where leaves exist, according to the energy available (Photosynthetically Active Radiation) at that nodal layer calculated by subroutine SHADE. The ambient concentration of CO_2 is usually assumed constant at 330 ppm, but can be varied.

A carbohydrate balance is then carried out in the leaves using subroutine ALEAF. This balance includes photosynthate produced by subroutine PHOTO, growth (production of structural carbohydrates), starch synthesis, and export to the phloem. Translocation is calculated using subroutine PHLOM and is based on a mass-flow phloem-transport theory proposed by Münch and described in a book by Nobel (11).

The hydrostatic and dynamic energy available to drive sucrose solution through a set of sieve tubes can be estimated from the concentration of solute within each tube, influx of water from supporting tissue, and influence of gravity. In SOYMOD, water is assumed not to limit the translocation process. Therefore, terms were only developed for the velocity and concentration gradient component of the mass flow.

The use of subroutines PHLOM and STEM allow the total transport through the stem to be limited by the amount of sugar available, temperature, sugar gradient along the stem and the size of the phloem. The process of loading sugar from each leaf at each nodal section into the phloem was described by a second order rate process. The unloading of the phloem was based on the demand at each node by the developing leaf or fruit. Carbohydrate not used at a given node was transported to the next lower node and so forth with the remainder available to the root system (subroutine ROOT).

In order to provide a feedback control system necessary to balance the need for both N and carbohydrate for growth of organs and to provide the necessary energy to produce nitrogen in the nodule system, a concept based on dry matter nitrogen ratio was used. This concept assumes that the plant has an internal regulating system which attempts to maintain a relatively constant dry matter—nitrogen balance in various organs. In order for nitrogen to be available to the shoot tissue, carbohydrate must be translocated from the leaves to the root nodules. The growing tissues demand for nitrogen and carbohydrate is coupled with the consumption of carbohydrate by the nitrogen-fixing mechanism. These demand-supply relationships are described mathematically by a system of coupled mass balance equations (subroutine PNCT2). No growth or photosynthesis occurs below 2% N content in an organ, based on data from Penning de Vries (13). The model assumes that 1 g of glucose will yield .11 g of nitrogen.

Carbohydrate partitioning control is handled in SOYMOD/OARDC by a series of rate equations of the Michaelis-Menten type. These rates are computed for each node to balance the demand for carbohydrate and nitrogen and the supply from photosynthesis and nitrogen fixation. The rate of structural growth of each plant part is calculated using subroutine GRWTH. This series of equations is solved dynamically using the real pole numerical explicit technique described by function RPOLE (5). This method applies a first order solution over the integration interval to a linear equation. The Michaelis-Menten and second order terms of the rate equation were linearized using the first two terms of a Taylor series expansion. These equations were solved sequentially in the computer program.

The timing of reproductive events such as flowering and fruiting were simulated using a set of empirical temperature-daylength iterative regressing equations proposed by Major, et al. (7). This approach seems to work satisfactorily but needs to be tested for other climatic conditions. It would be desirable to replace this empirical approach with physiologically-based internally-controlled reproductive event generation systems when such a step can be developed from physiology theory and verified experimentally.

Verification and validation were handled by comparison of simulation results selecting sets of soybean data from field experiments both in Ohio and elsewhere. Data sets were selected only if they included both the required plant and environmental parameters. A set of data for Beeson soybeans grown at Wooster in 1974 was used in order to determine the parameters in the simulator. The calibrated simulator was then used to simulate soybean plants grown in the field using weather data for 1972, 1976 and 1978, for examples. Validation was then carried out by comparing the simulated results with the actual field data. Figures 2 and 3 show examples of this validation. Table 3 shows simulated and experimental fruit numbers for the 1976 season.

Uses of SOYMOD/OARDC include both tests to further the understanding of the soybean plant and tests on the effect of change in plant structure or management. For example, one can study the effect of planting density. Table 4 shows the variation in yield with planting density for 4 yr. Also shown is the effect of changing moisture regime from year to year.

This set of simulations indicates how a complex nodal canopy structure can be handled rather easily over the growing season using the computer. These results, although not totally validated, indicate the importance of high canopy density for high reproductive yields accepted generally by the agronomic community. The problem of soybean branching and stand compensation is not solved completely in SOYMOD, however.

Another problem of interest to physiologists is the ability of the plant to adjust and compensate for different source-sink relationships. SOYMOD offers a way to study these relationships using the defoliation or depodding options. We have only begun simulations along these lines and this is a wide-open area for future studies. This latter question provides an example of an important test of the simulator: can it respond to a step change? In this example, either the fruit are removed or the leaves are removed.

Table 3. Simulated and actual fruit numbers/plant, summer 1976.[a]

Date	Days After Emergence	Actual No.	Simulated No.
8/5	75	56.0 ± 7.3	58
8/19	85	61.5 ± 6.1	65
9/2	99	44.8 ± 1.9	42
9/17	114	36.8 ± 2.5	39[b]

[a]Beeson soybeans grown in a test plot at Wooster, Ohio. Emerged May 27, 1976. Grown in 76.2 cm (30.0 inch) rows and thinned to an average 5.1 cm (2.0 inch) plant spacing, and irrigated.
[b]Final number.

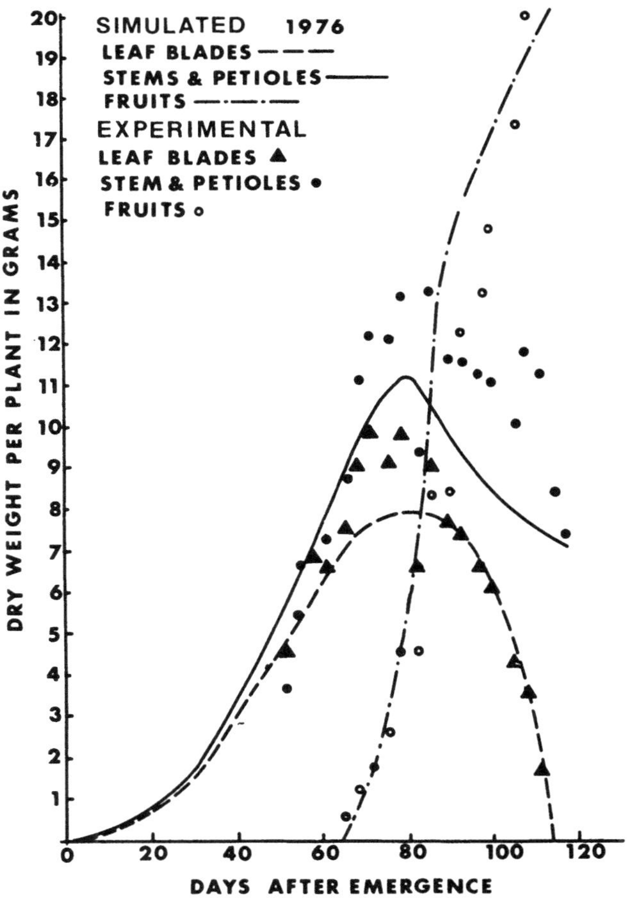

Figure 2. Comparison of simulated results and actual field data for 1976.

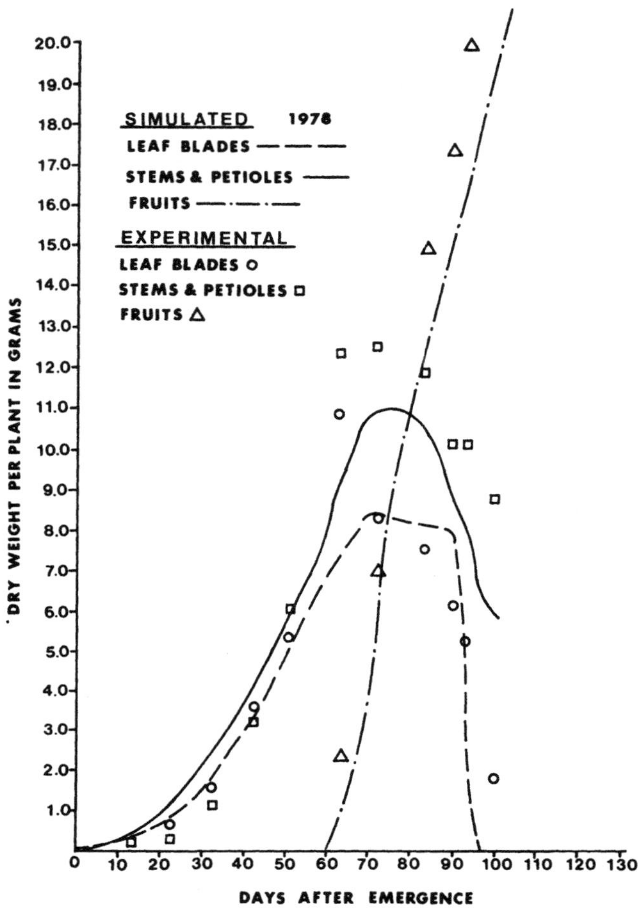

Figure 3. Comparison of simulated results and actual field data for 1978.

In addition to, or in combination with the plant spacing test of SOYMOD mentioned before, the simulator could be used to study the effect of other plant structural changes such as leaf arrangement and size, and light distribution. By testing many different light distributions in a short time and selecting those that showed promise of understanding canopy architecture and its effect on yields, the simulator would reduce the research time and cost necessary to test the best of these combinations in the field.

A reliable and responsive plant growth model is necessary for the development and use of insect and disease infestation. SOYMOD could be used in this way. A combined model would then be useful for predicting the level

Table 4. Simulated soybean performance under various planting configurations and moistured regimes.[1]

Year	Spacing Row by Plant	Density (plant/m^2)	Max. Leaf Area (cm^2/plant)		Fruits Weight (g/plant)		No. of Fruits/Plant		Yield (Kg/ha)	
			Irrigated	Not Irrigated	Irrigated	Not Irrigated	Irrigated	Not Irrigated	Irrigated	Not Irrigated
1975	13x13	62	1154	688	13.8	8.9	22	20	4422	2849
	76x 5	26	1730	1316	15.0	10.4	35	30	2016	1391
	91x15	7	2867	1941	45.4	33.0	71	61	1687	1230
1976	13x13	62	1135	1085	12.2	12.8	20	22	3938	4132
	76x 5	26	1699	1591	19.3	21.1	34	33	2587	2822
	91x15	7	2846	2405	47.2	32.5	67	57	1754	1210
1977	13x13	62	1127	1053	7.5	17.1	13	14	2399	5484
	76x 5	26	1702	1574	13.0	22.5	29	34	1734	3010
	91x15	7	2652	2224	40.6	28.1	70	58	1512	1048
1978	13x13	62	1162	902	15.2	11.7	12	20	4939	3750
	76x 5	26	1721	1338	16.0	12.1	28	34	2144	1626
	91x15	7	2552	1919	43.3	21.6	69	47	1613	806

[1]Based on Wooster, Ohio, weather data, Beeson variety.

of damage on final seed yield, and aid in the management decision to treat or not to treat. As mentioned earlier, some work is going on in other states in developing a plant-pest model.

On a larger scale, another use of SOYMOD may be in an overall crop management model, which would include such management practices as planting, spraying, harvesting, etc. Such a model would be used by a farmer on almost an interactive basis to guide day to day or week to week decision. In order for this to happen, real time weather data would have to be available.

As one can see, a simulator like SOYMOD/OARDC can be an important tool in research and management. SOYMOD/OARDC has been developed to mimic plant growth and development, and primarily as a tool for research, because of the depth of its physiological detail. As computers increase in speed and decrease in cost, SOYMOD could be modified and incorporated in large management models. Also advances in numerical intergration procedures, computation time and cost of simulation have been reduced. Improvements and automation in hybrid computers and advances in parallel processing will also have a dramatic improvement on simulation methodology in the future.

FUTURE NEEDS AND WORK PLANNED

SOYMOD/OARDC needs to be validated using climatic data from other geographic areas. Dr. Meyer is beginning this effort in Nebraska and we will be working on it in Ohio. Related to this need is the necessity to determine if the current structure of SOYMOD/OARDC will respond satisfactorily to determinate variety parameters and the climate of the southern U.S.

With respect to the structure of SOYMOD/OARDC, improvements are needed in the reproductive timing system and the root development-water uptake system. The regression equations developed by Major, et al. (7) have performed satisfactorily to date, but it would be more desirable to have the reproductive timing controlled internally.

The root development-water uptake system used in the present SOYMOD/OARDC is rather elementary and requires modification to include a root generating system which would then provide a water uptake function based on demand and availability of water and location in the soil profile. In order for these systems to work satisfactorily, the soil profile in the model must be layered and grided to show where roots are developing and where water is added and removed.

Finally, work needs to be started to couple SOYMOD/OARDC to management and cultural practice models. This will require modification of the simulator and possible simplification in order to keep the overall models of manageable size.

NOTES

R. Bruce Curry, Professor, Department of Agricultural Engineering, Ohio Agricultural Research and Development Center, Wooster, Ohio; George E. Meyer, Assistant Professor of Agricultural Engineering, University of Nebraska (formerly post-doctoral research associate, OARDC, Wooster); John G. Streeter, Professor, and Henry J. Mederski, Professor, Department of Agronomy, OARDC, Wooster.

This paper was approved for publication as Journal Article No. 24-79 of the Ohio Agricultural Research and Development Center. Research was supported in part by PL88-109 Grant No. 616-15-71 and No. 115-15-116.

REFERENCES

1. Beuerlein, J. E. and J. W. Pendleton. 1971. Photosynthetic rates and light saturation curves of individual soybean leaves under field conditions. Crop Sci. 11:217-219.
2. Buttery, B. R. and R. I. Buzzell. 1977. The relationship between chlorophyll content and rate of photosynthesis in soybeans. Can J. of Plant Sci. 57(1):1-5.
3. Curry, R. B., C. H. Baker, and J. G. Streeter. 1975. SOYMOD I: A dynamic simulator of soybean growth and development. Trans. ASAE 18(5):964-968.
4. Curry, R. B., C. H. Baker, and J. G. Streeter. 1975. An overview of SOYMOD, simulator of soybean growth and development. Proceedings of 1975 Summer Computer Simulation Conferences, San Francisco, CA, pp. 954-960.
5. Keener, H. M. and G. E. Meyer. 1978. Transient solution of differential equations on the digital computer by approximation and solution as first order equations, (In press).
6. Lommen, P.W., C. R. Schwintzer, C. S. Yocum, and D. M. Gates. 1971. A model describing photosynthesis in terms of gas diffusion and enzyme kinetics. Planta. 98:195-220.
7. Major, D. J., D. R. Johnson, J. W. Tanner, and I. C. Anderson. 1975. Effects of daylength and temperature on soybean development. Crop. Sci. 15:174-180.
8. Meyer, G. E., R. B. Curry, J. G. Streeter, and C. H. Baker. 1978. Computer simulation of reproductive processes and senescence in indeterminate soybeans. ASAE paper no. 78-4025, presented at the Summer Meeting of ASAE, Logan, Utah.
9. Meyer, G. E., R. B. Curry, J. G. Streeter, and H. J. Mederski. 1979. SOYMOD/OARDC: A dynamic simulator of soybean growth, development and seed yield. Ohio Research Bulletin No. 113 (In press).
10. Nafziger, E. D. and H. R. Koller. 1976. Influence of leaf starch concentration on CO_2 assimilation in soybean. Plant Physiol. 57:560-563.
11. Nobel, P. S. 1975. Introduction to biophysical plant physiology. W. H. Freeman & Co., San Francisco, CA, Chapter 8, pp. 390-401.
12. Ojima, M., J. Fukui, and I. Watanabe. 1965. Studies on the seed production of soybean II. Effect of three major nutrient elements supply and leaf age on the photosynthetic activity and diurnal changes in photosynthesis of soybean under constant temperature and light intensity. Proc. Crop Sci. Soc. Japan. 33:437-442.
13. Penning de Vries, F. W. T. 1975. The cost of maintenance processes in plant cells. Ann. Bot. 39:77-92.
14. Sato, K. 1976. The growth responses of soybean plant to photoperiod and temperature. I. Responses in vegetative growth. Proc. Crop Sci. Soc. Jap. 45(3):443-449.

APPLICATION OF MODELING TO IRRIGATION MANAGEMENT OF SOYBEANS

J. W. Jones and A. G. Smajstrla

The potential for increasing soybean yields by irrigation is well documented. In Ohio, Mederski et al. (44) reported on experiments in which non-irrigated soybean yields were compared with yields of soybeans irrigated when 20% of available soil water was depleted. They found that yield increases due to irrigation ranged from 5% for a late maturity, high yielding variety to 60% for an early maturing, low yielding variety. In Australia, Constable and Hearn (11) measured about 1000 kg/ha increase in yield when soybeans were irrigated. The smaller seed size of non-irrigated soybean accounted for most of the yield difference.

In a recent study in Florida in which Bragg and Cobb soybean varieties were grown on sandy soil, an extended drought occurred after flowering and soybean yields averaged 976 and 2832 kg/ha for non-irrigated and well irrigated plots respectively (L. C. Hammond, K. J. Boote, J. W. Jones, unpublished data). These data support the findings of Peters and Johnson (56) that about 134 kg/ha of soybeans are produced for each 2.5 cm of water used during July, August and September.

Soybean yield and water use depend on the stage of growth and on available soil water. Results from several experiments (26,61,62) indicate that soybeans are most susceptible to drought when it occurs during the pod filling stage and is least susceptible when it occurs during early vegetative growth. In a phytotron experiment, Sionit and Kramer (63) found that seed yields of Bragg soybean were decreased 16, 11, 34, and 28% when water stress occurred during flower induction, flowering period, early pod formation, and pod fill stages, respectively. Burch et al. (9) found that the ratio of actual evapo-transpiration (ET) to potential evapotranspiration (ET_p) increased from 0.2 early in the season to about 1.2 late in the season for well watered plants. The ET of non-irrigated plants decreased rapidly during a drying cycle.

In arid areas, irrigation is needed to replenish the soil water depleted by ET to avoid reductions in yield. In humid areas, the need for irrigation in soybean production is a more complex issue. Although long-term average rainfall amounts may be sufficient for replenishing the soil water supply, the year-to-year variability in rainfall amounts and the variability in successive days without rain may result in one or more drought periods during a season. Therefore, irrigation of soybeans may be desirable in humid areas, especially if grown on low water holding capacity soils (54). In order to study the need for irrigation as well as irrigation management strategies, models are needed to summarize and operationalize existing knowledge about soybean water use and yield, weather patterns, soil properties and economics into a framework compatible with irrigation objectives.

In this chapter we focus on the use of models in scheduling irrigations. Models have been developed at three levels of sophistication. First, a soil water balance method is described in which estimated ET is subtracted from the soil water in the root zone and irrigation timing is based on replenishment of soil water supply after a critical level is reached. Secondly, a crop yield response model is included with the soil water balance model so that irrigation scheduling for maximizing yield or net profit can be studied. The third level of sophistication involves a dynamic crop growth model sensitive to plant water relations and other weather and management variables. Such a model is coupled with the soil water balance approach to study more complex questions concerning irrigation scheduling and other crop management interactions. Relative advantages and disadvantages of each level are discussed.

WATER BALANCE MODELS

Water balance models for irrigation scheduling were developed as "bookkeeping" approaches to maintaining soil moisture in the root zone within desirable limits. In general, volumes of soil water, defined in terms of the soil moisture characteristics and the root zone of the crop being irrigated, were assumed to be available for crop use. Depletions from this reservoir by ET were made on a daily basis. When the readily available reservoir of water in the plant root zone was depleted, an irrigation was scheduled to replenish that volume.

Water balance models were classified into two categories: (a) Those based on the assumption that water was uniformly available for plant use between the limits of field capacity and permanent wilting point. ET rates were therefore simulated as a function of climatic parameters only. Curve A in Figure 1 illustrates this model. (b) Those based on the assumption that transpiration rates were known functions of soil water potential or water

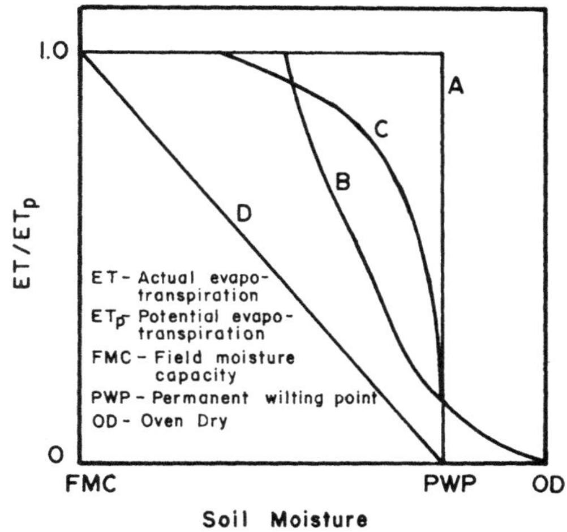

Figure 1. Various proposals for adjusting ET as soil moisture becomes limiting (13).

content. ET rates were simulated to decrease as soil water contents decresed. Curves B, C and D in Figure 1 were all proposed as possible relationships between relative transpiration rates and soil moisture.

The water balance approach as discussed here implied that depletions from the reservoir of available soil water were predictable for specific plant species and climatic conditions. Those depletions were the ET volumes normally predicted on a daily basis by ET models.

Uniformly Available Soil Moisture

Water balance models based on the assumption of uniformly available soil moisture between field capacity and permanent wilting point simulated water use based on climatic variables only. Those simulation models for ET by various crops have been summarized by Jensen (33). For ET predictive techniques to be applicable to irrigation scheduling, they must be capable of predicting water use on a daily, or at least a weekly basis. The techniques must be capable of interpreting short-term climatic variations in terms of ET. Some of the more widely used techniques that meet these criteria are described briefly in this section.

An equation used widely to calculated potential ET for use in irrigation scheduling is the modified Penman (55) equation. This equation is given as:

$$ET_p = \frac{\Delta}{\Delta + \gamma}(R_n + G) + \frac{\gamma}{\Delta + \gamma} 15.36\,(1.0 + 0.0062U_2)\,(e_z^o - e_z) \qquad (1)$$

where U_2 = wind speed at 2 m (km day^{-1}), e_z = vapor pressure at ht z (mb), e_z^o = saturated vapor pressure at ht z (mb), Δ = slope of the saturation vapor pressure-temperature curve (mb C^{-1}), γ = psychrometric constant (mb c^{-1}),

R_n = net radiation (cal cm^{-2} day^{-1}), G = heat flux density from ground (cal cm^{-2} day^{-1}), and ET_p = potential evapotranspiration rate (mm day^{-1}).

The Penman equation predicts potential ET (ET_p), which is that of a well-watered, vegetated surface. Allen et al. (1) determined that ET_p predicted by the Penman equation slightly overestimated measured ET from well watered grasses in lysimeters. Van Bavel (68) modified Equation (I) by the inclusion of a transfer coefficient for water vapor based on a logrithmic wind profile and a surface roughness parameter. This formulation was tested with good results for evaporation from bare soil, open water surfaces, and ET from well watered alfalfa. Ritter and Eastburn (60) determined that the Penman equation overestimated ET from irrigated corn during the entire growing season except for short periods of time during the peak vegetative stage of growth.

To predict actual rather than potential ET a crop coefficient, K_c, was introduced (33) as:

$$ET = K_c \cdot ET_p \qquad \text{(II)}$$

Crop coefficients for specific crops must be determined experimentally. They represent the expected relative rate of ET if water availability does not limit crop growth. The magnitude of the crop coefficient is a function of the crop growth stage.

The Jensen and Haise (34) equation is a second predictive technique sometimes used for irrigation scheduling applications. The recommended minimum time period for application is five days, which may not be a suitable time base in many situations. Their equation estimated potential ET as:

$$ET_p = (0.014T_a - 0.37)\, R_s \qquad \text{(III)}$$

where T_a = average daily temperature computed as $0.5 \cdot (Tmax + Tmin)$ (F), and R_s = solar radiation (mm day^{-1} H_2O). Actual ET was calculated as in Equation (II). The prediction of ET by Ritter and Eastburn (60) for irrigated corn resulted in underestimations during the first year of their study, but accurate estimations during the second year.

Pan evaporation techniques have been used widely as indicators of ET_p. The applicable equation is:

$$ET_p = (C_{et})\,(E_{pan}) \qquad \text{(IV)}$$

where C_{et} = a pan coefficient dependent upon the type of pan used, and E_{pan} = the measured pan evaporation depth on a daily basis (cm). The greatest difficulty encountered with the use of this technique is that the pan coefficient is not a constant, but varies with changing surroundings and weather conditions. However, reasonable approximations of daily ET are attainable if the pan is maintained properly and if care is exercised in the choice of a pan coefficient. Jensen (33) presents a method of estimating the pan coefficient.

A final type of model described here is a modification of the Brown-Rosenberg (8) resistance model. This model was tested with field data for irrigated sorghum by Verma and Rosenberg (69). Hourly and daily model estimates were compared to direct lysimetric measurements with good agreement under moderately advective to non-advective conditions.

In summary, a first approximation of an irrigation scheduling model using a water balance and ET estimated for well watered crop conditions was described. Several techniques for estimating ET were described. To be most useful in a scheduling model, it was required that ET be estimated on a daily or hourly basis. One of the major shortcomings of these models is that they do not account for changes in ET rates due to changing soil moisture levels. Those models will be discussed in the following section.

Limiting Soil Moisture

A number of researchers have developed models to predict actual ET as functions of both climatic demands and soil water potential. This results in a more complex model than those discussed previously, which use climatic indicators only.

Soil moisture levels can reduce ET below potential levels. There are differences of opinion concerning the changes in transpiration rates as affected by the available soil moisture. David and Hiler (13) summarized some of the relationships proposed by various researchers. These are presented in Figure 1.

Norero (52) developed an equation to express ET as a function of soil moisture potential and atmospheric evaporative demand:

$$\frac{d(ET)}{d\Psi_s} = k\left[\frac{ET}{\Psi_s}\right]\left[1 - \frac{ET}{ET_p}\right] \qquad (V)$$

where Ψ_s = total soil moisture potential (mb), and k = a proportionality coefficient. This equation was found to produce results similar to those of several other literature models if proper values of the proportionality coefficient were chosen. Norero et al. (53) verified the suitability of Equation (V) as a predictor of actual ET. They compared simulated and field experimental data for the widely varying available soil moisture levels associated with shallow rooted corn and deep rooted alfalfa with good agreement. They also developed a general relationship between relative ET and the time since the root zone soil moisture was fully replenished. Required information for the use of this technique was the relationship between relative ET and moisture content or potential for at least one evaporative demand intensity.

Norero's model is being used currently with good results by an irrigation scheduling service in the San Joaquin Valley (42). For this application, a dryness coefficient, k_D, was defined. The dryness coefficient was defined in terms of water potentials as:

$$k_D = \frac{ET}{ET_p} = \frac{1}{1 + \frac{\Psi_s}{\Psi_s^*} 2.56/\log(\Psi_{smin}/\Psi_{smax})} \quad \text{(VI)}$$

where Ψ_s^* = water potential corresponding to 0.5 ET_p (mb), $\Psi smin$ = water potential corresponding to 0.05 ET_p (mb), and Ψ_{smax} = water potential corresponding to 0.95 ET_p (mb).

Hanson (22) defined a coefficient of limiting soil water, k_{sw}, as:

$$k_{sw} = \left[\frac{Q - q}{SW - q}\right]^{0.03937\, ET_p} \quad \text{(VII)}$$

where Q = current soil water in the root zone (cm), q = lower limit of soil water in the root zone at the wilting point (cm), and SW = soil water in the root zone at field capacity (cm). This formulation implies that ET decreases below ET_p as soon as the soil water storage decreases below field capacity. As shown in Figure 2, and as contrasted to the single valued functions of Figure 1, the rate of decrease is greater as evaporative demand increases.

Hanson's model was calibrated and verified for pasture ET in the north central U.S. Field and lysimeter data at four sites were used with energy balance and pan evaporation techniques for predicting potential ET. Good agreement between simulated and measured ET was obtained at all sites with both techniques for estimating ET_p, although agreement was better with the energy balance than with the pan evaporation technique.

Kanemasu et al. (38) simulated lysimetric data for soybeans and sorghum with a model developed by Ritchie (58). Ritchie's model separated evaporation and transpiration components of water use. Potential evaporation E_p from a wet soil surface under a row crop (energy limiting) was defined as:

$$E_p = \frac{\tau}{\alpha} ET_p \quad \text{(VIII)}$$

where τ = reduction factor due to crop cover, and α = proportionality constant due to crop and climate.

During the falling rate stage (soil limiting) evaporation rate, e(mm day^{-1}), was defined as a function of time as:

$$E = ct^{1/2} - c(t - 1)^{1/2} \quad \text{(IX)}$$

where c = coefficient dependent upon soil properties (mm day^{-1}), and t = time (days).

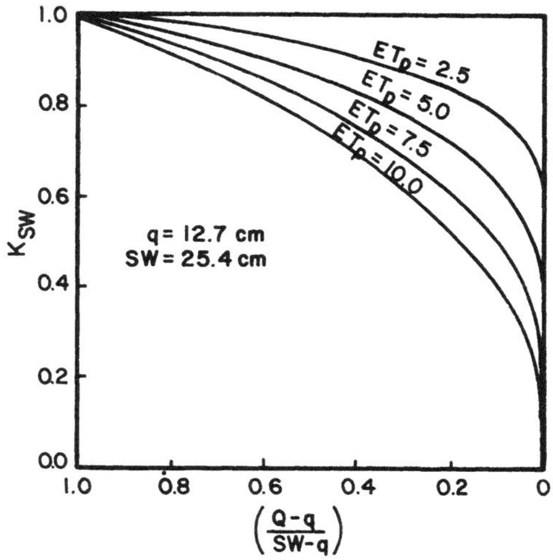

Figure 2. **Relationships between available soil water and the coefficient of limiting soil water as affected by various values of ET_p (22).**

Transpiration rates were calculated separately from evaporation rates. For plant cover of less than 50%, potential transpiration rate, T_p, was calculated as:

$$T_p = \alpha v(1 - \tau)\,[\Delta/(\Delta + \gamma)]\; R_n \tag{X}$$

where R_n = net radiation, and $\alpha v = (\alpha - 0.5)/0.05$.

For greater than 50% crop cover, T_p was calculated as:

$$T_p = (\alpha - \tau)\,[\Delta/\Delta + \gamma)]\; R_n. \tag{XI}$$

This formulation represented transpiration during non-limiting water conditions only. To account for decreasing soil water potential with water content, and effects on transpiration rate, a coefficient of limiting soil water (k_s) was defined by Kanemasu et al. (38) as:

$$k_s = \frac{\theta_a}{0.3\theta_{max}} \tag{XII}$$

where θ_a = average soil water content ($cm^3\ cm^{-3}$), and θ_{max} = water content at field capacity ($cm^3\ cm^{-3}$). At water contents above $0.3\theta_{max}$, transpiration rates were assumed to be controlled by climatic conditions only. Ritchie (59) reported that this model predicted transpiration rates well for sorghum and corn.

Kanemasu et al. (38) modified Ritchie's model by including a term to account for advective climatic effects. The modified model provided daily and seasonal estimates of ET that agreed with lysimetric observations. Daily estimates for soybean and sorghum were both predicted within 2 mm of lysimetric estimates.

In summary, several models for predicting ET rates under both well watered and water stressed conditions were presented. All models adequately predicted ET for the crops and climatic conditions they simulated when calibrated for those conditions. The models presented were all simple approximations of complex dynamic systems. Their simplicity has the advantage of requiring few data inputs, and, therefore, they can be applied with relatively few meterorological, soil or crop measurements taken. However, because of their simplicity, several empirical coefficients are required in each model, and each must be calibrated for specific crops, soil conditions and climatic variables.

YIELD RESPONSE MODELS

A second level of sophistication in the use of models for irrigation scheduling involves crop yield response models. There are two forms: dry matter yield models and grain yield models.

Dry Matter Yield Models

Dry matter yield was calculated as a function of cumulative transpiration, T(cm), by de Witt (14) as:

$$Y = \frac{K_m T}{E_o} \qquad \text{(XIII)}$$

where Y = yield (kg ha^{-1}), K_m = crop factor (kg ha^{-1} day^{-1}), and E_o = average free water surface evaporation rate (cm day^{-1}). The formulation of this equation allowed yield to be expressed as a function of water use. It was assumed that the only limiting factor to maximum productivity was water availability.

For a given crop and year [constant K_m and E_o in Equation (XIII)], Hanks (19) calculated relative yield as a function of relative transpiration:

$$\frac{Y}{Y_p} = \frac{T}{T_p} \qquad \text{(XIV)}$$

where Y_p = potential yield when transpiration is equal to potential transpiration (kg ha^{-1}) and, T_p = cumulative transpiration that occurs when soil water does not limit transpiration (cm). Hanks concluded that this model produced reasonable first order approximations of relative yield. When compared with field data, the model predicted the Y/Y_p versus ET/ET_p seasonal function well, provided that factors were adjusted properly to allow for the separation of evaporation and transpiration for the use of Equation (XIV).

Hanks' work developed and tested a physically-oriented, simple model to predict yield as a function of water use. Many researchers used regression type equations to express relationships between yield and water use. Those relationships are not as useful as physically based models because they are site-specific and crop-specific.

Many investigators (2,20,41,57,70) have shown that dry matter production is related linearly to ET with a high degree of correlation. In all of these studies, no expression is given for grain yields which is the component of major economic importance.

Grain Yield Models

Grain yield has been estimated as a function of soil moisture level during the crop season by various techniques. The most common are simple statistical regression models of seasonal water use versus time or growth stage. Again, as with dry matter regression models, these models are site- and crop-specific. Therefore, they are not as useful as physically based models which are applicable at various locations provided crop water use data can be measured or simulated.

Water production functions, which relate yield to water use, are useful tools for economic analyses of crop production and water use (23). These models commonly assume exponential relationships between yield and individual factors affecting yield.

Other approaches predict yields from physically based models which relate water stresses during various stages of crop growth to final yield, accounting for increased sensitivity to water stress at various stages of growth. An approached used frequently was to interpret ET or T reduction below potential levels as integrators of the effects of climatic conditions and soil water status. Two basic mathematical approaches were taken in the development of these models. One assumed that yield reductions during each crop growth stage were independent. Thus additive mathematical formulations were developed (17,26,49). A second approach assumed interactive effects between crop growth stages. These were formulated as multiplicative models (18,32).

Additive Models. The Stress Day Index model is an additive model presented by Hiler and Clark (26). The model is formulated as:

$$\frac{Y}{Y_p} = 1.0 - \frac{A}{Y_p} \cdot \sum_{i=1}^{n} (CS_i \times SD_i) \qquad \text{(XV)}$$

where A = yield reduction per unit of stress day index, (kg ha^{-1}), SD_i = stress day factor for crop growth stage i, and CS_i = crop susceptibility factor for crop growth stage i. CS_i expresses the fractional yield reduction resulting from a specific water deficit occurring at a specific growth stage. SD_i expresses the degree of water deficit during a specific growth period.

Hiler et al. (27) and Hiler and Clark (26) presented crop susceptibility factors for a number of field crops including soybeans. Yields of several field crops and vegetable crops were predicted well as functions of water stress by the stress day index model.

The stress day index model was utilized to schedule irrigations by calculating the daily SDI value (daily SD x daily CS) and irrigating when it reached a predetermined critical level, SDI_o. This integrated the effects of soil water deficit, atmospheric stress, rooting density and distribution, and crop sensitivity into the plant water stress factor. The critical SDI_o factor for irrigation scheduling is a function of water availability and cost. It is established for the entire growing season. Then, because of varying crop susceptibility factors with growth stage, SDI_o is obtained at lower levels of water stress during critically sensitive periods, and therefore, irrigations are scheduled more frequently.

Moore (49) devised an additive model to predict relative growth during irrigation periods for the crop season as a function of soil water parameters and time. This model was used in an economic analysis of irrigation scheduling, but was not verified with experimental data.

Multiplicative Models. To estimate grain or bean production where conditions during one stage of growth are more critical than others, where nutrients are not limiting, and where effects of water stress at each stage may be interactive, Jensen (32) developed the following model:

$$\frac{Y}{Y_p} = \prod_{i=1}^{n} \left[\frac{ET}{ET_p}\right]_i^{\lambda_i} \tag{XVI}$$

where ET/ET_p = relative evapotranspiration rate during the i-th stage of physiological development, and λ_i = crop sensitivity factor due to water stress during the i-th growth stage. Hanks (19) used this method successfully to predict grain yields of corn grown in Israel and Nebraska under various irrigation regimes. Neghassi et al. (50) reported that this multiplicative model did not predict adequately grain yields of wheat at Nebraska.

Hill and Hanks (28) modified Equation (XVI) by including factors to account for decreased dry matter production due to planting late season crops, and to account for decreased yields due to excess water. Their equation is:

$$\frac{Y}{Y_p} = \prod_{i=1}^{n} \left[\frac{T}{T_p}\right]_i^{\lambda_i} \cdot SYF \cdot LF \tag{XVII}$$

where $[T/T_p]_i$ = relative total transpiration for growth stage i when soil water is not limiting, SYF = seasonal yield factor which approaches 1.0 for adequate dry matter production, and LF = lodging factor which approaches 1.0 for normal soil water conditions and 0.0 for saturated conditions. Because this model relates relative yield to relative transpiration, it was also necessary to predict evaporation rates in order to maintain a water balance. Both E and T were calculated as functions of ET_p using the Jensen-Haise technique (33).

The Hill and Hanks (28) model predicted corn and soybean yields which were in close agreement with field research data. Figure 3 presents the field trial relative yield versus model predicted relative yield for simulation of Missouri soybean field data. Model verification was accomplished by simulating soybean yields and comparing with four separate sets of field data. This yield response model appears to be an excellent simulator of soybean yields as affected by transpiration rates.

Stewart et al. (66) developed a yield model with multiplicative crop sensitivity factors. This model was based on ET deficient, ET_p, and potential yield. The rationale was that each crop variety has a unique genetically determined potential for yield, the level of which is adjusted doward in the real world by environmental factors. The model was presented in two versions. When stages of crop growth were not important the model was given as:

$$Y = Y_p - Y_p B_o (ET_D/ET_p) \qquad \text{(XVIII)}$$

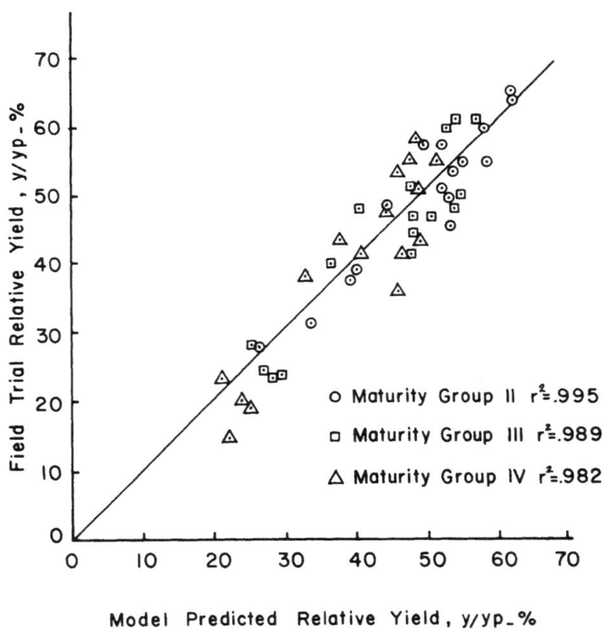

Figure 3. Results of soybean yield response model calibration with Missouri field data (28).

where B_o = slope of the ET_D/ET_p versus Y/Y_p curve, and ET_D = evapotranspiration deficit. In the formulation, B_o was defined as a constant for a given variety and was considered to be independent of location.

When water stress at a given crop growth stage had a greater potential of reducing yield than stress at other stages, a second version of the model was used:

$$Y = Y_p - Y_p(B_V ET_{DV} + B_P ET_{DP} + B_M ET_{DM})/ET_p \qquad \text{(XIX)}$$

where B_V, B_P, B_M = intensity factors for vegetative, pollination and maturity growth stages, and ET_{DV}, ET_{DP}, ET_{DM} = evapotranspiration deficits for the same growth stages. Hanks et al. (21) evaluated this model for corn production at four sites in the western U.S. They concluded that both models [Equations (XVIII) and (XIX)] produced excellent results for dry matter production, and that agreement with field data was also good for grain yield but improvement obtained by using this model was not significantly greater than that of the single stage model.

Hall and Butcher (18) presented a model which related relative crop yield to relative water availability in the plant root zone. Their model was formulated as:

$$\frac{Y}{Y_M} = \left[\frac{W}{W_M}\right]_1^{b_1} \cdot \left[\frac{W}{W_M}\right]_2^{b_2} \cdot \left[\frac{W}{W_M}\right]_3^{b_3} \qquad \text{(XX)}$$

where W = amount of available water at the end of the growth stage (cm), W_M = total available water in the soil profile (cm), and $b_{1,2,3}$ = crop growth coefficients for each growth stage. Hanks et al. (21) analyzed the Hall-Butcher model and obtained good correlations with field data for corn growth at one experimental site.

Minhas et al. (47) proposed another multiplicative model expressed as:

$$\frac{Y}{Y_P} = \prod_{I=1}^{n} \left\{ 1.0 - \left[(1.0 - \frac{ET}{ET_p})^2\right]_i^{\lambda_i} \right\} \qquad \text{(XXI)}$$

where all factors are as previously defined. Howell and Hiler (31) found that it described adequately the yield response of grain sorghum to water stress.

In summary, considerable efforts have been directed toward development of simple models for describing the yield response of crops subjected to water stress conditions. The applicatiion of these models to soybean irrigation management appears tractable and is already beginning (28). The inputs for these models, in addition to soil and weather data needed for E and T calculations, are crop related variables. There exists a paucity of data in the literature for adequate calibration of the crop coefficients. Research efforts should be directed at filling this need.

CROP GROWTH MODELS

Crop growth models are described here as the third level of sophistication in model development for irrigation scheduling. This approach differs from the yield response models described previously in several respects. Final absolute yields are output from crop growth models in contrast to relative yields discussed earlier. Also, the factors considered and the methodology used are very different. Crop growth models provide dynamic estimates of the crop status at any time during a season. The influence of various soil and environmental factors, such as temperature, radiation and soil water status, usually are included.

During the past ten years, considerable progress has been made in the development of dynamic crop growth models (3-6,12,16,24,30,43,46,65). Much of this progress has been brought about by an improved understanding of and techniques for measurement of photosynthesis and respiration. Crop growth models vary in detail and complexity, however, each contains a carbon balance and simulates seasonal growth and production for time varying environmental conditions. In this discussion, generalities are used to describe this overall approach because of a lack of a universally accepted framework for representing crop growth processes and their interrelationships.

Crop growth rate usually is calculated as the difference between photosynthesis and respiration, multiplied by a biomass:CO_2 conversion factor (25):

$$\frac{1}{\phi}\frac{dW}{dt} = (Pg - RoW)/(1 + \phi G_R) \qquad \text{(XXII)}$$

where (dW)/(dt) = biomass growth rate (kg ha^{-1} day^{-1}), ϕ = biomass:CO_2 conversion factor (kg biomass kg^{-1} CO_2), Pg = gross photosynthesis (kg CO_2 ha^{-1} day^{-1}), Ro = maintenance respiration factor (kg CO_2 kg^{-1} biomass day^{-1}), W = biomass (kg ha^{-1}), and G_R = growth respiration factor (kg CO_2 kg^{-1} biomass). The photosynthesis process rate depends on incident photosynthetically active radiation (PAR) and canopy geometry (usually leaf area index). Respiration is based on plant growth rate as well as total plant biomass and is highly sensitive to temperature. Partitioning of biomass into various plant parts is usually dependent on stage of growth or on organ initiation and source-sink relationships (37). Growth of leaves, stems, roots, and fruit are calculated by:

$$\frac{dW_i}{dt} = \alpha_i \frac{dW}{dt} \qquad \text{(XXIII)}$$

where $(dW_i)/(dt)$ = biomass growth rate of leaves (i=1), stems (i=2), roots (i=3) and fruit (i=4), and α_i = partitioning coefficient for leaves, stems, roots and fruit.

In more detailed models, photosynthate carbon first is transferred into a nonstructural carbohydrate pool that is used subsequently for translocation, respiration, and growth (30). In other models, biomass growth is dependent on soil and plant nitrogen balances as well. The important point is that these processes, Pg, Ro, G_R, α_i, are complex and are dependent on crop state and environmental conditions. Usually, daily or shorter time steps are used for calculating the process rates and for integrating them through time for estimating crop biomass in leaves, stems, roots, and fruit.

The sensitivity of a crop to water deficit is described in general terms by Slatyer (64) and by Merva (45). Figure 4 is a modified schematic from Merva showing the effects of water deficits on crop growth processes. The evaporative loss of water from leaves results in reduced turgor and osmotic water potentials in the leaves. The resultant decrease in the total water potential causes an increased water potential gradient between leaves and soil. When soil water content is sufficiently high, root extraction occurs to replenish water lost by the plant. However, when the soil dries, the decrease in soil hydraulic conductivity causes a decreased rate of water uptake by roots and significant internal plant water deficits may occur. The turgor potential may be lowered sufficiently to cause stomatal closure which prevents further leaf dessication. Thus, transpiration is decreased. Furthermore, reduced turgor decreases expansive growth of leaves and photosynthetic rates (7). Overall growth is thus reduced and source-sink relationships may be altered by fruit and leaf abortion. The reduced leaf expansion, over a long-term, will result in decreased canopy size and thus photosynthesis (because of reduced radiation interception), even after water stress is alleviated. Finally, mineral ion availability may be reduced due to decreased N fixation and ion absorption.

For use in irrigation management studies, crop growth models must (a) include a soil water balance to dynamically estimate the water balance components, and (b) be sensitive to internal plant water deficits caused by a failure of the soil-root system to supply water rapidly enough to replace water evaporated from the leaves.

The soil water balances used in crop growth models are esentially the same as those described in the first section of this chapter. The soil water balance must provide an estimate of soil water content and water potential. Decreased rates of water loss by soil evaporation and by transpiration as soil dries are essential features of a soil water balance for crop growth models (as in Figure 2). The ET model developed by Ritchie (58) has been modified and used in several crop growth models (5,36,43).

Potential evaporation and transpiration are calculated separately in the corn growth model described by Childs et al. (10). Potential soil evaporation is a function of leaf area index (LAI), calculated as:

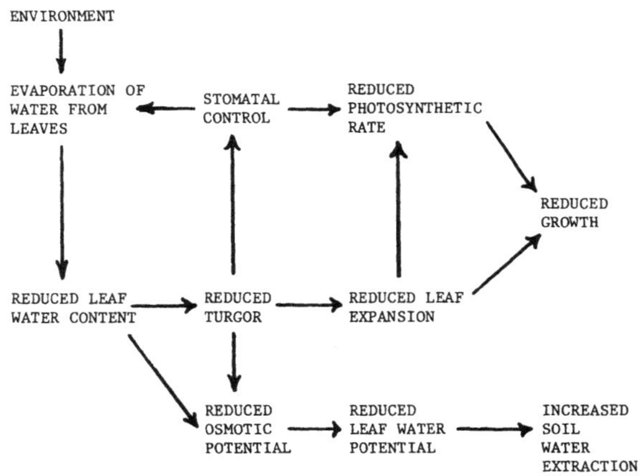

Figure 4. Schematic of the effects of water deficit on crop growth rate and soil water extraction rate.

$$E_p = ET_p\, e^{-\beta(LAI)} \qquad \text{(XXIV)}$$

where β = empirical constant (0.438 was used for corn). Potential transpiration (T_p) depends on intercepted radiation and is estimated by:

$$T_p = ETP(1 - e^{-\beta(LAI)})\ CROPCO \qquad \text{(XXV)}$$

where CROPCO = a crop coefficient. Actual soil evaporation (E) is calculated using a numerical solution to Darcy's law in a one-dimensional soil water model whenever soil evaporation is restricted by soil characteristics. When the soil surface is wet, actual and potential soil evaporation are equal.

Actual transpiration (T) is calculated diurnally by solving a steady-state resistance analogue model similar to that described by Nimah and Hanks (51). The equation solved is:

$$T = (\Psi_1 - \Psi_s)/(r_s + r_p) \qquad \text{(XXVI)}$$

where Ψ_1 = leaf water potential (cm), Ψ_s = soil water potential (cm), r_s = resistance to water flow in soil (cm·hr cm^{-1}), and r_p = resistance to water flow in the plant (cm·hr cm^{-1}). Stomatal closure was assumed to occur at a given minimal leaf water potential, Ψ_{lm}. During the day, the maximum flux of water from the soil to leaves was calculated based on the Ψ_{lm} and soil profile water potentials. Actual transpiration was then the minimum of T and T_p. Thus, actual evapotranspiration was the sum of actual transpiration and soil evaporation values, ET = E + T.

Because of the complex effects of water deficits on crop growth processes, the sensitivity of the crop to internal water deficits has been described by various, simplified means for crop growth models. Characteristic of most crop growth models is some function for reducing the photosynthesis rate depending on the severity of stress. In the corn model described by Childs et al. (10) and Tcheschke and Gilley (67), photosynthesis was decreased in proportion to the average leaf water potential:

$$Pg = Pg' \cdot FWATER$$

where Pg' = photosynthesis rate for crop without water stress (when leaf water potential is above -6000 cm), and FWATER = quadratic function of leaf water potential derived from corn data of Boyer (7). The only effect of water stress on crop growth was a reduction in photosynthesis, yet simulated corn yields compared very favorably with observed yields over a three-year period in Nebraska (Figure 5).

In the corn model described by Barfield et al. (5), water stress was defined as the ratio of water in the soil root zone to the water in the root zone at field capacity. In their model photosynthesis was reduced in proportion to that ratio, and an additional effect was included. If water stress occurred at silking, the maximum ear size was reduced. They simulated corn response to irrigation and compared simulated results with 12 experimental plot-year results. They concluded that the model was an effective tool for estimating long-term benefits from irrigation but that additional evaluation is needed before it is used in irrigation management on a yearly basis.

Other models in various stages of development include the effects of water deficits on other processes, such as changes in leaf area expansion rate, biomass partitioning, organ initiation and abortion rates, and nitrogen uptake rates (36,46).

Inputs to the crop growth models usually include the same climatic data required to estimate ET and calculate a soil water balance because crop water use must be estimated for this approach also. For crop growth processes, air temperature (max and min) and PAR are required. Soil data usually include a soil water retention curve, root zone depth, and unsaturated hydraulic conductivity relationship. Crop parameters must be specified and consist of coefficients used in calculation of process rates.

More detailed models have been developed for simulating soil-plant-water relationships, including transpiration and photosynthesis, for various soil and atmospheric regimes (29,40). To date, these models have been used to study the occurrence and significance of water deficits over sufficiently short time periods that crop parameters are constant. Seasonally, crop morpholoy and physiology are changing continually in response to various environmental factors. Therefore, the effects of water deficits on the

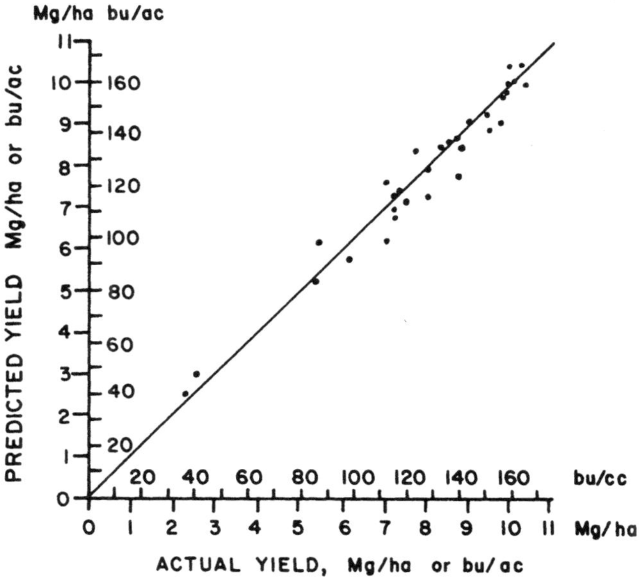

Figure 5. Comparison between simulated and measured corn response to varying water stresses over a three year period in Nebraska using a crop growth model (67).

longer-term growth processes, such as leaf area expansion, root proliferation, and source-sink relations, need to be included in crop growth models. Progress is now being made in adapting the short-term soil-plant-water models for coupling with seasonal crop growth models for more comprehensive studies of the effects of natural weather cycles on crop growth and yield. The use of such models for studying irrigation management strategies is promising.

SIMULATING SOYBEAN WATER REQUIREMENTS

A model was developed for simulating water use and yield response of soybeans grown on well-drained, sandy soils in the humid southeast. The objectives of this model were to provide a research tool for (a) studying seasonal irrigation requirements, (b) estimating the long-term potential for yield increase using irrigation, (c) studying the effects of different irrigation management strategies on water use efficiency and yield of soybeans, and (d) studying the sensitivity of predictions to errors in estimating soil, crop and climatic parameters. A water balance approach was coupled with a yield response model to demonstrate the utility of this methodology.

Model Description

The soil was divided into two zones similar to the approach of Hill and Hanks (28). The upper zone (evaporation zone) was 5 cm deep and the lower zone (transpiration zone) varied in depth depending on root growth. Root growth data of Mitchell and Russell (48) were used to vary the transpiration zone depth during the season to a maximum depth of 152 cm. Evaporative water loss was from the evaporation zone and transpiration water was lost from both zones depending on their respective water contents. Soil properties of a Lakeland fine sand, with a capacity to retain 10% water by volume (71) was used. Due to the high inflitration rates of the soil, all rainfall (and irrigation) was added to the profile until field capacity was reached, and excess water was assumed to drain from the profile.

Potential evapotranspiration (ET_p) was calculated as a fraction of pan evaporation (PEV):

$$ET_p = 0.7\,(PEV). \qquad \text{(XXVIII)}$$

Then, ET_p was partitioned into potential transpiration (T_p) and potential evaporation (E_p) components calculated by:

$$T_p = k_c(ET_p) \qquad \text{(XXIX)}$$

$$E_p = (1 - k_c)\,ET_p \qquad \text{(XXX)}$$

where k_c = crop coefficient based on seasonal ET data of Burch et al. (9), = (0.015) (Days) with $0.2 \leq k_c \leq 1.1$.

Actual E and T were calculated as functions of available water in the two soil zones. Transpiration was calculated as (28):

$$T = (T_p/0.5)\,(SWS/AW) \qquad \text{if } (SWS/AW) < 0.5 \qquad \text{(XXXI)}$$

and

$$T = T_p \qquad \text{if } SWS \geq 0.5 \qquad \text{(XXXII)}$$

where SWS = total root zone soil water storage on the current day (cm), and AW = total root zone soil storage at field capacity (cm).

Evaporation was calculated by:

$$E = \text{minimum}\left\{E_p/(2^{t-1}),\ SWS_{ez}\right\} \qquad \text{(XXXIII)}$$

where t = time from last rainfall or irrigation (days), and SWS_{ez} = soil water stored in the evaporation zone (cm).

A soil water balance was calculated as:

$$\text{WBAL} = \text{SWS}_{ez}\,(\text{I}) + \text{SWS}_{tz}\,(\text{I}) - \text{SWS}_{ez}\,(\text{I}-1) - \text{SWS}_{tz}\,(\text{I}-1) + \text{RAIN} + \text{IRR} - \text{E} - \text{T} - \text{EXCESS}. \qquad \text{(XXXIV)}$$

The yield response model was based on (T/T_p) ratios for four stages of soybean growth. The multiplicative form:

$$Y/Y_p = (T_1/T_{p1})^{\lambda_1}\,(T_2/T_{p2})^{\lambda_2}\,(T_3/T_{p3})^{\lambda_d}\,(T_4/T_{p4})^{\lambda_4} \qquad \text{(XXXV)}$$

was used. The length of each stage and λ_i values were based on the crop susceptibility (CS_i) values of Hiler and Clark (26), Table 1. Since λ_i values were not available, data from a 1978 irrigation experiment on Bragg soybeans (L. C. Hammond, K. J. Boote and J. W. Jones, unpublished data) were used to calculate the A factor for the additive model, Equation (XV) (26). Then the reported CS_i values were used to calculate the (Y/Y_p) with $(T/T_p)_i$ equal to 0.5 for each stage separately. Then, the multiplicative form was used with $(T/T_p)_i = 0.5$ and the calculated (Y/Y_p) to obtain λ_i for each stage. The resulting data are given in Table 1.

Ten years of weather data, rainfall and pan evaporation, for Gainesville, Florida, were used to study the possible long-term water use and yield responses of soybeans grown without irrigation. An emergence date of June 1 was used for all simulations. A maximum root zone depth of 100 cm was used to demonstrate the use of the model. Then, maximum root zone depth was varied from 40 to 152 cm to demonstrate the sensitivity of crop response. Finally, the length of stage 3, the period during which the crop is most susceptible to stress, was varied to demonstrate the importance of an accurate determination of stage lengths.

Simulated Results

Results for the 100-cm root zone are shown in Tables 2 and 3. Actual transpiration was less than T_p during 5, 1, 3, and 4 years of the 10 years for stages 1 through 4, respectively. This implies that water stresses are most

Table 1. Stage lengths and crop sensitivity factors, λ_i, calculated for Bragg soybeans.

Stage	Stage Length, Days	CS_i Values From (26)	Calculated λ_i Values
1. Vegetative	55	0.12	0.24
2. Flowering	30	0.24	0.48
3. Early Pod Fill	30	0.35	0.84
4. Late Pod Fill	25	0.13	0.26

Table 2. Simulated soybean water use and relative yield response for 100-cm root zone depth using Gainesville, Fla., weather data.

Year	Stage 1		Stage 2		Stage 3		Stage 4		Season		
	T	T_p	T	T_p	T	T_p	T	T_p	T	T_p	Y/Y_p
1	9.9	9.9	12.2	12.2	11.7	12.6	8.2	8.2	42.0	42.9	0.94
2	10.2	10.4	11.8	11.8	12.4	13.4	9.1	9.1	43.5	44.7	0.93
3	9.5	9.5	11.2	11.2	12.6	12.6	5.4	9.5	38.7	42.8	0.86
4	9.3	9.7	14.8	14.8	12.1	12.1	7.9	7.9	44.1	44.5	0.99
5	10.2	10.2	11.2	11.2	10.9	10.9	5.8	8.6	38.1	40.9	0.90
6	9.4	9.4	9.9	9.9	11.1	11.1	8.4	8.4	38.8	38.8	1.00
7	8.5	8.5	10.9	10.9	11.2	11.2	7.6	7.6	38.2	38.2	1.00
8	9.5	9.6	10.7	10.7	10.7	10.7	8.5	8.5	39.4	39.5	1.00
9	10.6	12.0	8.3	12.1	11.4	11.4	8.4	8.4	38.7	43.9	0.81
10	8.9	9.8	10.3	10.3	6.6	12.7	1.2	8.1	27.0	40.9	0.34
Average	9.6	9.9	11.1	11.5	11.1	11.9	7.1	8.4	38.9	41.7	0.88
Standard Deviation	0.6	0.9	1.7	1.4	1.7	0.9	2.4	0.6	4.7	2.4	0.20

Table 3. Simulated soybean water use and relative yield response for 100-cm root zone depth using Gainesville, Fla. weather data.

Year	Stage 1		Stage 2		Stage 3		Stage 4		Season	
	ET	ET_p	ET	ET_p	ET	ET_p	ET	ET_p	ET	ET_p
1	18.2	22.7	13.2	13.4	12.2	13.8	8.5	8.9	52.1	58.8
2	15.3	24.2	12.7	13.0	13.0	14.6	9.5	10.0	50.5	61.8
3	16.7	21.4	12.2	12.4	13.0	13.8	5.5	10.4	47.4	58.0
4	14.4	22.5	15.6	16.4	12.7	13.2	8.3	8.7	51.0	60.8
5	16.3	23.3	12.0	12.4	11.4	11.9	6.1	9.4	45.8	57.0
6	16.5	21.9	10.6	10.9	11.8	12.1	8.5	9.2	47.4	54.0
7	15.4	20.3	11.5	12.0	11.9	12.3	7.9	8.3	46.7	52.9
8	14.8	21.3	11.5	11.9	11.3	11.7	8.7	9.2	46.3	54.1
9	15.6	28.0	9.0	13.4	11.8	12.5	8.8	9.2	45.2	63.1
10	14.9	22.5	11.2	11.4	6.8	13.9	1.4	8.8	46.8	56.6
Average	15.8	22.8	12.0	12.7	11.6	13.0	7.3	9.2	47.9	57.7
Standard Deviation	1.1	2.1	1.7	1.5	2.9	1.0	2.4	0.6	2.4	3.5

likely to occur during stage 1 and 4. For maximum root zone depths less than 100 cm, the percentage of years experiencing water stress during the last two growth stages increased. For the 60-cm maximum root zone depth, water stresses occurred 7 of the 10 years, whereas there was no change in the occurrence of stresses during stage 1 when compared with the 100-cm root zone simulated results.

Seasonal average transpiration values were simulated to be 93% of the potential value. Actual evapotranspiration was more affected because of the limited depth of the evaporation zone. In every stage each year, ET was less than ET_p, and averaged only 0.69 of ET_p for stage 1. In general, the variabilities in ET and in T for all the stages were greater than those in ET_p and in T_p, respectively. Simulated ET for the 140-day season was 47.9 cm whereas simulated T was 38.9 cm.

The simulated relative yield, Y/Y_p, averaged 0.88 and varied from 0.34 (1978 weather data) to 1.00 (1974, 1975, 1976 weather data). For comparative purposes, non-irrigated soybean yield in Gainesville was about one-third of the highest yielding, irrigated treatment yield in 1978 (L. C. Hammond, K. H. Boote, and J. W. Jones, unpublished data).

When maximum root zone depth was varied from 40 to 152 cm, the 10-year average Y/Y_p increased from 0.54 to 0.94 (Figure 6). If the well-watered yield (Y_p) was 2600 kg ha^{-1}, then average yield losses of 1200 and 160 kg ha^{-1} would be expected if the root zone was restricted to 40 and 152 cm, respectively. The simulated yield losses varied considerably. For a root zone depth of 100 cm, there was a 20% chance (2 out of 10 simulations) that yield losses would be greater than 500 kg ha^{-1} without irrigation if Y_p was 2600 kg ha^{-1}.

The crop susceptibility to water stress was greatest when stress occurred during the early pod growth stage. Because of the lack of a precisely defined, discrete event to separate stages 3 and 4, a sensitivity analysis was performed to determine the effects of errors in setting the length of stage 3 on predicted yield response to water deficits. For this analysis, the lengths of stages 1 and 2 and the length of the growing season remained constant. The lengths of stages 3 and 4 were varied to study stage 3 lengths of 20, 25, 30, 35 and 40 days for each of 40, 80, 120 and 152-cm root zone depths.

The 10-year average relative yield responses are shown in Figure 7(b) for stage 3 durations of 20 to 40 days. As the length of stage 3 increased, the predicted decrease in Y/Y_p varied from 4 to 6%, depending on root zone depth. These data indicate that, over the long-term, predicted average relative yield reductions are insensitive to small errors in estimating the length of stage 3, regardless of root zone depth. However, depending on seasonal weather conditions, the sensitivity may be greater. For example, predicted Y/Y_p are shown in Figure 7(a) for 1978 weather, which was characterized by

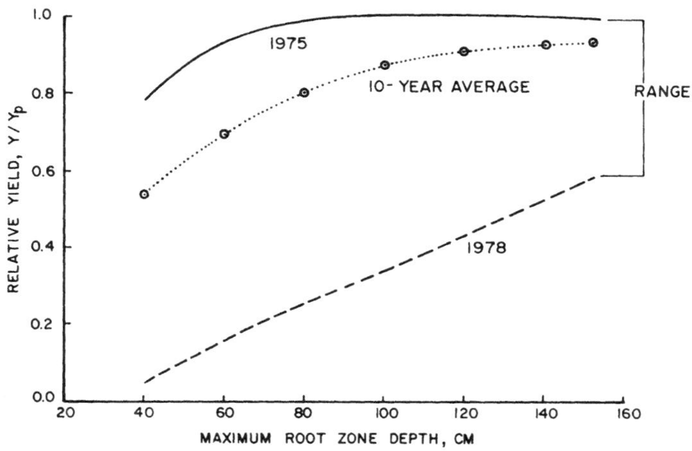

Figure 6. **Simulated relative yield response of soybeans as affected by maximum root zone depth using historical weather data from Gainesville, Florida.**

a long drought beginning in mid-July. For the 152-cm root zone depth, a change in the length of stage 3 from 20 to 40 days resulted in a decrease in estimated Y/Y_p of about 32%. For shallower root zones, the sensitivity of Y/Y_p to changes in stage 3 length decreased.

For long-term results, there was little difference between 120 and 152-cm root zone depths. The 10-year average Y/Y_p values were significatnly affected as root zone depth decreased from 120 to 80 cm. For an extreme year such as 1978, however, the 120-cm Y/Y_p estimates were about 10% lower than those for the 152-cm root zone depth.

In summary, a model was developed for estimating soybean crop water use and yield response to varying climatic conditions. The model was used to study the potential for increasing the yield of soybeans grown on sandy soil in the humid southeast by irrigation. Although unverified, the model indicates that for moderate root zone depths (100 cm), yield increases may average about 14% for well irrigated soybeans, but may vary from no increase to 300%, depending on the year. Results were very sensitive to root zone depth but relatively insensitive, in the long-term, to the length of growth stage 3. The study of irrigation strategies is now possible using the model. Experiments are critically needed to verify the model for local conditions.

SUMMARY

Several modeling approaches have been used for studying crop water use and irrigation requirements. Few researchers have studied soybean irrigation

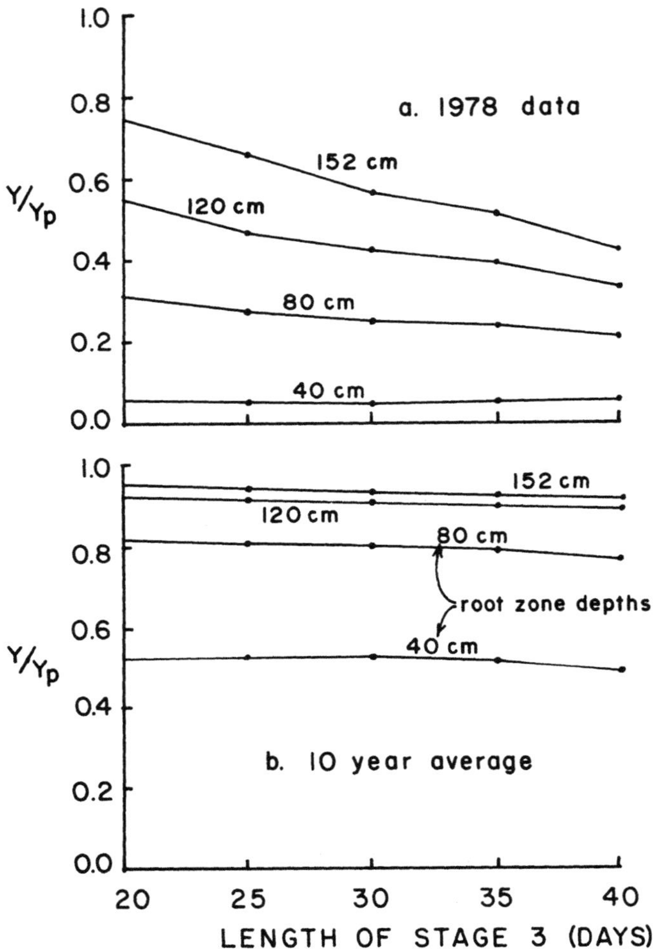

Figure 7. **Relationship of simulated relative yield of soybeans (Y/Y_p) to the length of growth stage 3 and root zone depth for (a) 1978 weather data of Gainesville, Florida, a year having below average rainfall, and (b) average for 10-year historical weather data of Gainesville, Florida.**

scheduling; therefore, this chapter discussed methodologies at several levels of complexity and then developed an example model applicable to soybean irrigation studies.

Basic to each model is a soil water balance. Of the processes that affect the soil water balance, most attention has been placed on ET. Whereas the assumption was made in early models that ET was a function of climatic and

and crop parameters only, more recent soil water balance models separated E and T and included effects of limiting soil moisture on both E and T. Such an approach has been successfully tested for soybeans (38). Irrigation scheduling using a soil water balance model is based on maintaining optimum water levels in the root zone for maximum productivity.

Yield response models have been added to the soil water balance approach to estimate relative yield depending on the occurrence, timing, and degree of plant water stresses. Such models have been used for various crops including soybean. However, data for soybean responses are limited to two studies (26,28) and the differences between the susceptibility factors, λ_i, for the two studies are unresolved. More data are needed to allow this approach to be further developed for soybean.

In this chapter, a yield response model was described and simulated results were discussed. The predicted relative yield (Y/Y_p) was very sensitive to maximum root zone depth, and it is suggested that irrigation management to enhance root proliferation is a viable method to decrease the effects of short drought periods in humid regions.

The use of dynamic crop growth models in irrigation management was discussed. Although several soybean growth models currently are available and more are being developed, their application to studying irrigation management awaits development. Examples were cited wherein corn growth models were used successfully for studying irrigation management. The inputs for such models require careful consideration. Current applications of crop growth models for other than irrigation management are still in a research phase of development.

For practical application to irrigation scheduling, the state of the art appears to be the combination of water balance and yield response models such as the model developed and discussed in this chapter. The formulation of such models is relatively simple although application requires that they be calibrated by determining empricial crop coefficients for specific crops and locations. Outputs of these models provide useful information for irrigation system design and for scheduling irrigation. For research applications, the dynamic crop growth models offer exciting opportunities for studying the effects of limiting water on various crop growth processes and on absolute yield, rather than relative yield, of crops. When fully developed and tested, crop growth models will become useful tools for application to scheduling irrigation.

NOTES

J. W. Jones and A. G. Smajstrla, Agricultural Engineering Department, University of Florida, Gainesville, Florida 32611.

REFERENCES

1. Allen, L. H., J. S. Rogers, and E. H. Stewart. 1978. Evapotranspiration as a benchmark for turfgrass irrigation. Fla. Turfgrass Mgt. Conf. Proc. (In press).
2. Arkley, R. J. 1963. Relationships between plant growth and transpiration. Hilgardia 34:359-589.
3. Baker, C. H. and R. D. Horrocks. 1976. CORNMOD, A dynamic simulator of corn production. Agric. Sys. 1:57-77.
4. Baker, D. N., J. D. Hesketh, and W. G. Duncan. 1972. Simulation of growth and yield in cotton: I. Gross photosynthesis, respiration, and growth. Crop Sci. 12: 431-435.
5. Barfield, B. J., W. G. Duncan, C. T. Haan. 1977. Simulating the response of corn to irrigation in humid areas. ASAE Tech. Paper No. 77-2005. ASAE, St. Joseph, MI 49085.
6. Bowen, H. D., R. F. Colwick, D. G. Batchelder. 1973. Computer simulation of crop production—potential and hazards. Agric. Engr. 54(10):42-45.
7. Boyer, J. S. 1970. Leaf enlargement and metabolic rates in corn, soybean, and sunflower at various leaf water potentials. Plant Physiol. 46:233-235.
8. Brown, K. W. and N. J. Rosenberg. 1973. Shelter effects on micro-climate, growth and water use by irrigated sugarbeets in the Great Plains. Agric. Meterorol. 9: 241-263.
9. Burch, G. J., R. C. G. Smith, and W. K. Mason. 1978. Agronomic and physiological responses of soybean and sorghum crops to water deficits. II. Crop evaporation, soil water depletion and root distribution. Aust. J. Plant Physiol. 5:159-167.
10. Childs, S. W., J. R. Gilley, W. E. Splinter. 1977. A simplified model of corn growth under stress. Trans. ASAE 20(5):858-865.
11. Constable, G. A., and A. B. Hearn. 1978. Agronomic and physiological responses of soybean and sorghum crops to water deficits. I. Growth, development and yield. Aust. J. Plant Physiol. 5:159-167.
12. Curry, R. B., C. H. Baker, and J. G. Streeter. 1975. SOYMODI: A dynamic simulator of soybean growth and development. Trans. ASAE 18:963-968.
13. David, W. P. and E. A. Hiler. 1970. Predicting irrigation requirements of crops. J. Irr. and Dng. Div., ASCE 96(IR3):241-255.
14. de Witt, C. T. 1958. Transpiration and crop yields. Verslag van Landbouwk. Onderzoek. No. 64.4, 88 p.
15. Dougherty, C. T. 1977. Water in the crop model SIMED2. Inf. Ser. N. Z. Dept. Sci. Ind. Res. 126:103-110.
16. Duncan, W. G. 1977. Univ. of Florida. Personal communication.
17. Flinn, J. C. and W. F. Musgrave. 1967. Development and analysis of input-output relation for irrigation water. Aust. J. Agric. Econ. 11(1):1-19.
18. Hall, W. A. and W. S. Butcher. 1968. Optimal timing of irrigation. J. Irr. and Dng. Div., ASCE 94(IR2):267-275.
19. Hanks, R. J. 1974. Model for predicting plant yield as influenced by water use. Agron. J. 60:660-665.
20. Hanks, R. J., H. R. Gardner, and R. L. Florian. 1969. Plant growth evapotranspiration relations for several crops in the Central Great Plains. Agron. J. 61:30-34.
21. Hanks, R. J., J. I. Stewart, and J. P. Riley. 1977. Four-state comparison of models used for irrigation management. J. Irr. and Dng. Div., ASCE 103(IR3):283-294.
22. Hanson, C. L. 1976. Model for predicting evapotranspiration from native rangelands in the Northern Great Plains. Trans. ASAE 19(4):471-477.
23. Heady, E. O. and J. L. Dillon. 1961. Agricultural production functions. Iowa State Univ. Press, Ames, IA, 667 p.

24. Hesketh, J. D. and J. W. Jones. 1976. Some comments on computer simulators for crop growth—1975. Ecol. Modeling 2:235-247.
25. Hesketh, J. D. and J. W. Jones (Eds.). 1979. Predicting photosynthate production and use for ecosystem models. CRC Press, Inc., West Palm Beach, Florida (In press).
26. Hiler, E. A. and R. N. Clark. 1971. Stress day index to characterize effects of water stress on crop yields. Trans. ASAE 14(4):757-761.
27. Hiler, E. A., T. A. Howell, R. B. Lewis, and R. P. Boos. 1974. Irrigation timing by the stress day index method. Trans. ASAE 17:393-398.
28. Hill, R. W. and R. J. Hanks. 1978. A model for predicting crop yields from climatic data. ASAE Tech. Paper No. 75-4030. ASAE, St. Joseph, MI 49085.
29. Hillel, D. 1977. Computer simulation of soil-water dynamics. Int. Develop. Res. Centre, Ottawa, 214 pp.
30. Holt, D. A., R. J. Bula, G. E. Miles, M. M. Schreiber, and R. J. Peart. 1975. SIMED. Res. Bull. 907. Agric. Exp. Sta., Purdue, Univ, 26 pp.
31. Howell, T. A. and E. A. Hiler. 1975. Optimization of water use efficiency under high frequency irrigation—I. Evapotranspiration and yield relationship. Trans. ASAE 18(5):873-878.
32. Jensen, M. E. 1968. Water consumption by agricultural plants, pp. 1-22. In T. T. Kozlowski (Ed.) Water deficits and plant growth 2, Academic Press, New York.
33. Jensen, M. E. (Ed.) 1973. Consumptive use of water and irrigation water requirements. Tech. Comm. on Irrigation Water Requirements. J. Irr. and Dng. Div., ASCE, New York, 215 p.
34. Jensen, M. E. and H. R. Haise. 1963. Estimating evapotranspiration models in irrigation scheduling. Trans. ASAE 21(1):82-87.
36. Jones, J. W., L. G. Brown, and J. D. Hesketh. 1979. COTCROP: A computer model of cotton growth and yield. In J. D. Hesketh and J. W. Jones (Eds.) Predicting Photosynthate Production and Use for Ecosystem Models. CRC Press, Inc., West Palm Beach, FL (In press).
37. Jones, J. W. and G. H. Smerage. 1978. Representation of plant/crop physiology. ASAE Tech. Paper No. 78-4024. ASAE, St. Joseph, MI, 19 p.
38. Kanemasu, E. T., L. R. Stone, and W. L. Powers. 1976. Evapotranspiration model tested for soybean and sorghum. Agron. J. 68:569-572.
39. Lambert, J. R., D. N. Baker, and C. J. Phene. 1975. Simulation of soil processes under growing row crops. ASAE Tech. Paper No. 75-2580. ASAE, St. Joseph, MI 49085.
40. Lambert, J. R. and F. W. T. Penning de Vries. 1973. Dynamics of water in the soil-plant-atmosphere system: A model named TROIKA. In Physical Aspects of Soil Water and Salts in Ecosystems. Springer-Verlag, Berlin.
41. Leggett, G. E. 1959. Relationship between wheat yield, available soil moisture, and available nitrogen in eastern Washington dryland areas. Bull. 609. Washington Agric. Exp. Sta.
42. Lord, J. M., G. D. Jardine, and G. A. Robb. 1977. Operation of a water management ET model for scheduling irrigations. ASAE Paper No. 77-2053. ASAE. St. Joseph, MI 49085.
43. McKinion, J. M., D. N. Baker, J. D. Hesketh, and J. W. Jones. 1975. SIMCOT II: A simulation of cotton growth and yield. In Computer Simulation of a Cotton Production System—Users Manual, ARS-S-52.
44. Mederski, H. J., D. L. Jeffers, and D. B. Peters. 1973. Water and water relations, pp. 239-266. In B. E. Caldwell (Ed.) Soybeans: Improvements, Production and uses. Monograph 16. Amer. Soc. of Agron., Inc., Madison, WI.
45. Merva, G. E. 1975. Physioengineering Principles, The AVI Publishing Co., Inc., Westport, CT, p. 100.

46. Meyer, G. E., R. B. Curry, and J. G. Streeter. 1978. Analysis and computer simulation of flowering and reproductive growth in indeterminate soybeans. ASAE Tech. Paper No. 78-4025. ASAE, St. Joseph, MI 49085.
47. Minhas, B. S., K. S. Parikh, and T. N. Srinivasan. 1974. Toward the structure of a production function for wheat yields with dated inputs of irrigation water. Water Resour. Res. 10:383-393.
48. Mitchell, R. L. and W. J. Russell. 1971. Root development and rooting patterns of soybean [*Glycine max* (L.) Merrill] evaluated under field conditions. Agron. J. 63:313-316.
49. Moore, C. W. 1961. A general analytical framework for estimating production function for crops using irrigation water. J. Farm Econ. 18(4):876-888.
50. Neghassi, H. M., D. F. Heermann, and D. E. Smika. 1975. Wheat yield models with limited soil water. Trans. ASAE 18:549-557.
51. Nimah, M. N. and R. J. Hanks. 1973. Model for estimating soil, water, plant, and atmospheric interrelations: I. Description and sensitivity. SSSA Proc. 37:522-527.
52. Norero, A. L. 1969. A formula to express evapotranspiration as a function of soil moisture and evaporative demands of the atmosphere. Ph.D. Dissertation. Dept. Soils and Meterology, Utah State Univ., Logan, UT.
53. Norero, A. L., J. Keller, and G. L. Ashcroft. 1972. Effect of irrigation frequency on the average evapotranspiration for various crop-climate-soil systems. Trans. ASAE 15(6):662-660.
54. Pendleton, J. W. and E. E. Hartwig. 1973. Management, p. 223. In B. W. Caldwell (Ed.) Soybeans: Improvements, Production, and Uses. Monograph 16. Amer. Soc. of Agron., Inc., Madison, WI.
55. Penman, H. L. 1963. Vegetation and Hydrology. Tech. Committee No. 53. Commonwealth Bureau of Soils. Harpenden, England, 125 p.
56. Peters, D. B. and L. C. Johnson. 1960. Soil moisture use by soybeans. Agron. J. 52:687-689.
57. Powers, J. F., D. L. Grunes, and B. A. Reichman. 1961. The influence of phosphorous fertilization and moisture on growth and moisture absorption by spring wheat. I. Plant growth, uptake, and moisture use. SSSA Proc. 25:209-210.
58. Ritchie, J. T. 1972. Model for predicting evaporation from a row crop with incomplete cover. Water Resour. Res. 8:1204-1213.
59. Ritchie, J. T. 1973. Influence of soil water status and meteorological conditions on evaporation from a corn canopy. Agron. J. 65:893-897.
60. Ritter, W. F. and R. P. Eastburn. 1978. Estimating evapotranspiration rates for corn in Delaware. ASAE Tech. Paper No. 78-2029. ASAE, St. Joseph, MI 49085.
61. Runge, E. C. A. and R. T. Odell. 1960. The relationships between precipitation, temperature, and the yield of soybeans on the Agronomy South Farm, Urbana, Illinois. Agron. J. 52:245-247.
62. Shaw, R. H. and D. R. Laing. 1965. Moisture stress and plant response, pp. 73-94. In Plant Environment and Efficient Water Use. Amer. Soc. of Agron., Inc. Madison, WI.
63. Sionit, N. and P. J. Kramer. 1976. Water potential and stomatal resistance of sunflower and soybean subjected to water stress during various growth stages. Plant Physiol. 58:537-540.
64. Slatyer, R. O. 1967. Plant-Water Relationships. Academic Press, London, 366 p.
65. Stapleton, H. N., D. R. Baxton, R. L. Watson, D. J. Nolting, and D. N. Baker. 1973. COTTON: A computer simulation of cotton growth. Tech. Bull. 206; Agric. Exp. Sta., Univ. of Arizona, Tuscon, 124 p.

66. Stewart, J. E., R. M. Hagan, and W. O. Pruitt. 1974. Functions to predict optimal irrigation programs. J. Irr. and Dng. Div., ASCE, 100(IR2):179-199.
67. Tscheschke, P. and J. R. Gilley. 1978. Modification and verification of Nebraska's corn growth model—CORNGRO. ASAE Tech. Paper No. 78-4033. ASAE, St. Joseph, MI 49085.
68. Van Bavel, C. H. M. 1966. Potential evaporation: The combination concept and its experimental verification. Water Resour. Res. 2(3):455-467.
69. Verma, S. B. and N. J. Rosenberg. 1977. The Brown-Rosenberg resistance model of crop evapotranspiration. Modified tests in an irrigated sorghum field. Agron. J. 69:332-335.
70. Whittlesly, N. K. and L. J. Colyar. 1968. Decision making under conditions of weather uncertainty in the summer fallow-annual cropping area of eastern Washington. Tech. Bull. No. 58. Washington Agric. Sta.
71. Zakaria, A. A. 1976. Simulation of one-dimensional water transport in Lakeland fine sand. Ph.D. Dissertation, Univ. of Florida, Gainesville, 66 p.

A MATHEMATICAL MODEL FOR THE THRESHING AND SEPARATING OF SOYBEANS

V. M. Huynh, T. E. Powell, and J. N. Siddall

A mathematical model is presented to describe the process of threshing and separating soybeans in a conventional cylinder and concave thresher. The probabilistic approach proposed in an earlier work by the same authors (3) is used to derive analytical expressions for the process.

In this model, the detachment of beans from pods, the separation of loose beans through both the straw and the concave grate, and the generation of damaged beans during threshing are recognized as random events. Each is described by a probability density function which can be fitted by an exponential decay function. The parameters characterized by these functions are then related to the process physical quantities such as machine parameters, crop feed rate, operating conditions and crop conditions. The performance of a given thresher is described from these distribution functions. Threshing performance is calculated in terms of threshing loss, separation efficiency, and bean damage. Quantitative relationships for the thresher horsepower demand are also determined from a consideration of the force required to move the compressed straw along the concave. Various graphical trends are obtained to show the characteristics of the proposed model. The influences of different threshing parameters on threshing performance are studied via the model. No attempt, however, is made in this work to confirm these experimentally.

LITERATURE SURVEY

Research on soybean harvesting has so far been concentrated on the combine header loss; and various efforts had been made (1) to improve combine design to reduce this loss. Soybean threshing, however, is usually considered to be of secondary importance and is often neglected in research literature.

Extensive field work in soybean combining was conducted by Lamp et al. (4). Tests were done over a five-year period in mostly high moisture crops. Threshing loss and bean damage were reported as functions of threshing speed and crop moisture level. Similar relationships can be found in the works of Pickett (6) and Singh (7) for navy bean threshing. Pickett indicated that threshing loss was dependent on pod moisture level and that this loss could be reduced by increasing the cylinder speed or decreasing the concave clearance. In a similar work, Singh showed that Pickett's data on mechanical damage could be fitted by an exponential function of both threshing speed and bean moisture level.

Recent experimental work on soybean threshing by rotary combines (5) showed that even though small clearance was required for threshing high moisture crops, this did not have significant influence on damage. Impact is generally considered to be a major cause of seed crackage and split during threshing. In single impact tests, Cain and Holmes (2) found that the bean sample having lowest moisture content was subjected to the most extensive damage and that the percentage of splits and fragments increased rapidly as the impact speed increased.

PHYSICAL MODEL

In this section, the threshing and separating process for soybeans is examined. This leads to the development of a physical model from which quantitative relationships for the process are derived.

Consider the threshing of soybeans in a cylinder and concave thresher shown in Figure 1. The crop mixture, which consists of unthreshed pods, prethreshed beans and straw are fed to the thresher to form a loose mat at the concave entrance. The rasp bar cylinder imparts threshing energy to this mat and accelerates it around the concave. Threshing action takes place along the concave length, releasing beans from the pods. Loose beans are then separated through the straw mat and finally through the concave grate. Unseparated beans, as well as unthreshed beans and straw, are discharged at the rear of the concave. A schematic diagram of the threshing and separating process is shown in Figure 2.

The threshing model is illustrated best by making use of a control volume, a concept derived from fluid mechanics. Consider a control volume defined by the space between the rasp bars and concave and the areas A_1 and A_2 perpendicular to the direction of crop flow (Figure 2). This volume moves around the concave at a velocity $\overline{V}_s$ which represents the average velocity of the crop mixture. For simplicity, it is assumed that crop material does not enter or leave the control volume with the exception of the loose beans at the time of concave separation.

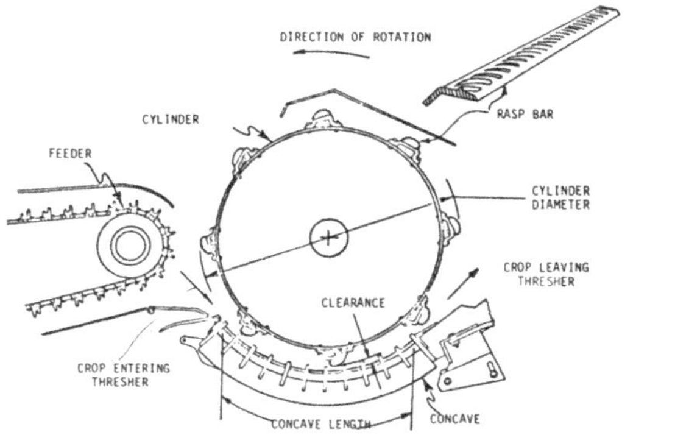

Figure 1. Rasp bar thresher.

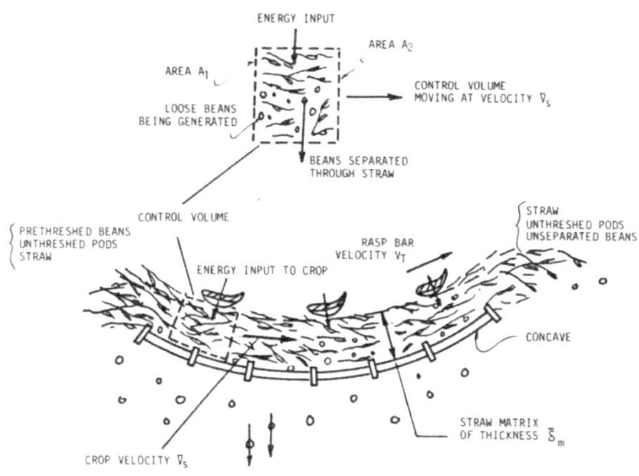

Figure 2. Physical model of threshing.

Since the velocity of the control volume is less than the peripheral speed of the cylinder, this control volume is in contact with the rasp bars a number of times before exiting to the rear. At each contact, energy is transferred to the crop inside the control volume in the forms of impact, compression, and shear forces. Continuous shearing action from the concave is also added to this energy to provide the threshing action to different parts of the crop. Thus, loose beans are generated at random inside the control volume as they are transported along the concave length. After threshing, the loose beans are free to move up and down inside the control volume. The effect of gravity force and centrifugal acceleration is super-imposed on this motion to provide a net downward movement of the beans for separation through the straw.

After penetrating through the straw, the loose beans are moving with the control volume and are in constant contact with the concave grate. Concave separation occurs when a bean drops through an open area in the concave grate and finally leaves the control volume. If it strikes a rod or a bar, the bean bounces off and is carried away by the control volume. This process is repeated until the bean penetrates the grate.

After detaching from the pods, the loose beans are damage prone as they are exposed repeatedly to the impact of the threshing elements. The longer the exposure, the more likely that they will get damaged. High speed impacts, as well as severe shear and crushing, are recognized as the likely causes of seed crackage. In this model, damage is determined by the ratio of the specific energy absorbed by the crop during impact and the facture strength of the beans.

THE PROBABILISTIC MODEL

As indicated in the above description, the threshing and separating of soybeans involves a number of random phenomena, such as: the detachment of beans from pods, the separation of loose beans through the straw, the penetration of beans through the concave grate, and the generation of damaged beans. Each of these can be treated as a statistical event, i.e. the time to occurrence of the event is a random variable. The probability density functions which describe these events are assumed to be of an exponential type:

$$f_i(t) = \frac{1}{\tau_i} e^{-t/\tau_i} \qquad i = 1,2,3,4 \qquad \text{(I)}$$

where t is the time to occurrence of the event such as threshing of a bean, and τ_i is the mean time to occurrence of the event. Thus, τ_i could represent the average time elapsed between the entry of the crop to the thresher and the threshing of a bean. The basic assumption implied in the above expression is that the event is likely to occur at any time, e.g. a bean has an equal chance to be threshed anywhere along the concave length. Thus, τ_i as defined above is a constant and is independent of the time variable t. Equation (I) is an exponential decay function illustrated in Figure 3. Probabilistic theories are applied to derive analytical expressions for the threshing process. Details of the derivation are available (3).

The threshing loss or the fraction of unthreshed beans is given by:

$$R_1 = e^{-\lambda_1 t_c} \qquad \text{(II)}$$

where t_c is the dwell time of the crop in the threshing zone, λ_1 is equal to $1/\tau_1$ and is the mean threshing rate in kernels per second. The separation efficiency of a thresher, given in fraction of bean yield, is predicted by:

$$\eta = 1 - \frac{\lambda_2\lambda_3(\lambda_2 - \lambda_3)e^{-\lambda_1 t_c} + \lambda_3\lambda_1(\lambda_3 - \lambda_1)e^{-\lambda_2 t_c} + \lambda_1\lambda_2(\lambda_1 - \lambda_2)e^{-\lambda_3 t_c}}{(\lambda_1 - \lambda_2)(\lambda_2 - \lambda_3)(\lambda_3 - \lambda_1)} \qquad \text{(III)}$$

where λ_2 = mean separation rate of beans through the straw (= $1/\tau_2$), and λ_3 = mean separation rate of beans through the concave grate (= $1/\tau_3$).

There is a different separation rate for each bean depending on the distance from the concave at which it was threshed. In an earlier paper (3), this was modelled by obtaining a mean overall separation efficiency. Numerical trials have indicated that a simpler but adequate approximation is simply to use a mean separation rate of beans through the straw as defined above.

The fraction of damaged beans produced by the threshing process is derived as:

$$D_a = 1 - e^{-\lambda_4 t_c} \qquad \text{(IV)}$$

where λ_4 is the damage infliction rate on the beans (= $1/\tau_4$).

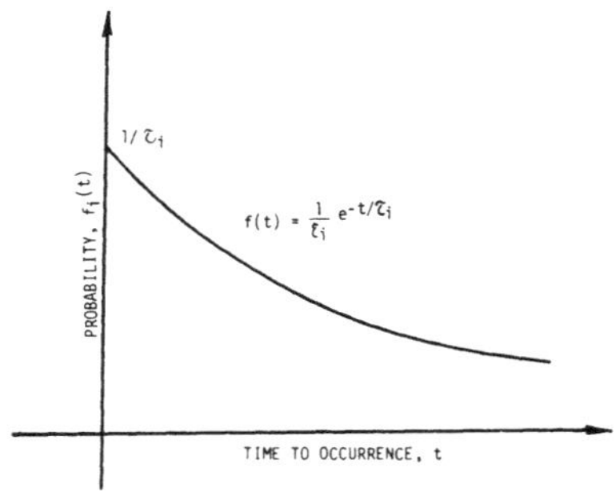

Figure 3. Probability density function describing the probability of occurrence of a random event.

PROCESS CHARACTERISTIC VARIABLES

Equations (II) to (IV) describe the threshing performances in terms of five characteristic variables: the mean rates of occurrence; λ_1, λ_2, λ_3, and λ_4 and the crop dwell time, t_c. In this section, these variables are related to physical quantities of the threshing and separating process.

Model of Crop Flow

A crop motion model is developed to define the flow of crop in the threshing area. High speed motion studies have indicated that crop velocities are non-uniform across the concave clearance and are different for different crop particles depending on their physical properties. For a given particle, this velocity is also a time varying quantity. Particle instantaneous speeds, for most crop material, range from zero to slightly greater than the rasp bar peripheral speed. In general, the highest particle speed occurs near the rasp bar and lowest near the concave.

To simplify the model, the crop particle speeds are assumed to be uniform and constant with time. The equivalent crop speed is given by:

$$\bar{V}_s = \frac{1}{2}(k_f V_o + k_r V_T) \tag{V}$$

where V_o and V_T are the feeder chain speed and the rasp bar peripheral speed, respectively, and k_f and k_r are the slip factors. For a given feeder chain geometry and a particular rasp bar and concave design, these slip factors are considered to be dependent on crop properties only.

Equation (V) defines the average speed of the beans and straw in the threshing area. The time duration for the crop to travel the full length of the concave at this speed is given as:

$$t_c = \frac{L}{\bar{V}_s} \tag{VI}$$

where L is the concave length. The average free mat thickness of the straw matrix according to this model of crop flow is:

$$\bar{\delta}_m = \frac{Q_s}{\rho \bar{V}_s W} \tag{VII}$$

where Q_s is the MOG mass feed rate, ρ is the free density of the MOG, and W is the thresher width.

Threshing Parameter

Soybean threshing involves the application of energy to the crop in order to: a) break open the pod sufficiently for the bean to escape, and b) remove the bean from the open pod. In the cylinder and concave thresher, this energy is supplied in the form of impact action, tension, compression, and shear. Among these, shear and direct impact are recognized as the main mechanisms responsible for bean threshing under normal operating conditions.

For a given bean pod, threshing is achieved when the specific energy absorbed by the pod is greater than the energy requirement cited in the last paragraph. This energy requirement is dependent on the mechanical properties of the crop which are affected by crop maturity, crop moisture content, and crop variety.

In this analysis, the threshing rate is assumed to be of the form:

$$\lambda_1 = k_T \frac{\rho V_T{}^2 W D}{Q_s C M_C} \qquad \text{(VIII)}$$

where Q_s is the MOG feed rate, C is the concave clearance, M_C is the crop moisture content, D is the cylinder diameter, and k_T is a threshing factor. In the above expression, the threshing rate is assumed to be proportional to the inverse of crop moisture content and proportional to the specific energy absorbed by the crop in shear and during impact. For a given thresher, the threshing factor is dependent only on crop variety and is considered constant in this analysis.

Separation of Beans Through the Straw

During threshing, the soybean straw, in the form of a loose mat, acts as a damper to reduce damage inflicted on the loose beans. However, the presence of the straw mat also poses some resistance to bean separation. The centrifugal force and the acceleration due to gravity induce the necessary motion for the bean to overcome this resistance to penetration of the mat.

To describe the motion of a bean through the straw mat, a linear resistive force model (3) is used to represent the resistance to motion of the crop mat. In this model, the motion resistance per unit thickness of a mat is assumed to be proportional to the applied acceleration. According to this model, the time duration for a bean to pass through a given mat thickness δ is given by:

$$\tau = K_m \left(\frac{\delta}{G + \dfrac{2\bar{V}_s{}^2}{D}} \right)^{1/2} \qquad \text{(IX)}$$

where G is the acceleration due to gravity, D is the cylinder diameter, and K_m is a proportional constant.

Since the beans originate from different mat depths, the time for each bean to cross the mat to reach the concave is different depending on the relative position of the bean within the thickness of the mat. To take this into consideration, the straw mat of thickness $\bar{\delta}_m$ is divided into n number of discrete thin layers. Within each layer, the beans possess the same mean time, τ'_j, for mat crossing. Thus, the overall mean time for the loose beans to penetrate the mat, $\bar{\delta}_m$, is the average sum of τ'_j for all layers, i.e.:

$$\tau_2 = \frac{k_m}{m} \sum_{j=1}^{m} \left(\frac{j\frac{\bar{\delta}_m}{m}}{G + \frac{2\bar{V}_s^2}{D}} \right)^{1/2} \quad \text{(X)}$$

The mean rate of bean separation through the straw is, therefore, given by:

$$\lambda_2 = \frac{1}{\tau_2} \quad \text{(XI)}$$

Passage of Beans Through Concave Grate

After passing through the mat, the loose beans are now sliding across the concave surface at an assumed velocity V_s. Consider a cell formed by rods and bars on the concave whose geometry is shown in Figure 4. When a bean is presented to this cell, there are three possibilities: a) the bean passes cleanly through the cell, b) the bean hits a rod or a bar and after a series of reflections it drops through an opening in the cell, and c) the bean bounces off from an obstruction and is carried toward the next cell. In this analysis, the effect of bean reflection is neglected and concave separation is assumed to occur when the projection of the bean is within the open area of the cell. Thus, the probability of passage for the bean through a cell is:

$$p = (a_1 - a_2 - d)(b_1 - b_2 - d)/(a_1 b_1) \quad \text{(XII)}$$

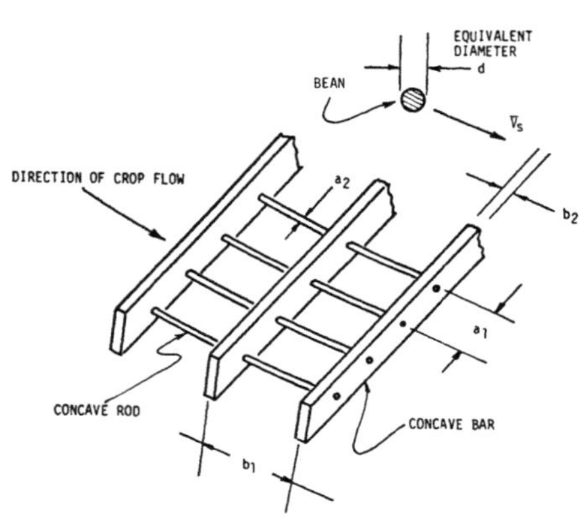

Figure 4. Concave geometry.

where a_1 and b_1 are the centerline distances between the rods and between bars, respectively; a_2 is the rod diameter; b_2 is the width of the bar; and d is the bean size which is assumed to be spherical.

It is also assumed that if a bean does not separate through a given cell, it will move to the next cell at a velocity $\overline{V}_s$. Thus, the amount of time available for each bean to penetrate through any given cell is the same. The probability that bean will pass through a cell during this time is given by:

$$p = \lambda_3 \frac{b_1}{\overline{V}_s} \qquad \text{(XIII)}$$

where λ_3 is the mean rate of concave separation and $b_1/\overline{V}_s$ is the crop dwell time within the cell.

Combining Equations (XII) and (XIII), one obtains:

$$\lambda_3 = \frac{1}{\tau_3} = \frac{\overline{V}_s(a_1 - a_2 - d)(b_1 - b_2 - d)}{a_1 b_1^2} \qquad \text{(XIV)}$$

Bean Damage

Mechanical damage inflicted on soybeans during threshing includes both visible and invisible damage. Visible damage takes the form of seed coat cracks, splits, and fragments. Invisible damage relates to the failure of the embryo or any internal structure which renders the bean incapable of germinating. In the threshing area, loose beans are subjected to damage as they are exposed to the multiple impacts of the threshing elements. Damage is determined by the energy input to the crop and the bean strength, or the ability to absorb this energy without failure. The energy input to the crop in a single impact is proportional to the square of impact velocity, and the bean strength is related to the bean moisture content.

In this model, the mean damage rate for soybeans is proposed to be proportional to the ratio of the specific impact energy and the minimum energy to cause bean failure, i.e.:

$$\lambda_4 = K_d \frac{V_T^2}{Q_s M_C} \qquad \text{(XV)}$$

where K_d is a damage factor. For a given thresher, this factor is dependent on crop variety and is considered constant.

THRESHING POWER REQUIREMENT

During threshing, the cylinder horsepower is expended in impacting, compressing, and shearing the crop mat and in transporting it around the concave. This power is also dissipated in bearing friction and aerodynamic forces created by the rotating cylinder. This part of the power, however, constitutes only a small fraction of the total power demand. In this analysis, the cylinder horsepower is estimated from a consideration of the cylinder torque required to overcome the motion resistance of the crop mat which is subjected to a compression between the rasp bars and the concave surface.

Soybeans are considered to be compressible materials. The static compression curves for most crops follow a power law, i.e.:

$$p_c = k_p\left(\frac{h_o}{h}\right)^n \qquad \text{(XVI)}$$

where h and h_o are the current and original thicknesses of the crop mat, respectively, and k_p and n are constants. Typical compression curves for soybeans are shown in Figure 5. The constant n is about 6.

Consider the threshing area shown in Figure 6. The straw mat of thickness $\bar{\delta}_m$ is compressed to a thinner mat of thickness C, the average rasp bar to concave clearance. The average pressure on this mat is given by:

$$p_r = k_p\left(\frac{\bar{\delta}_m}{C}\right)^{n'} \qquad \text{(XVII)}$$

where n' is a constant related to straw dynamic compression. In general, "n'" in this expression is slightly less than "n" given in Equation (XVI). This is because in dynamic compression, part of the mat spring force is lost

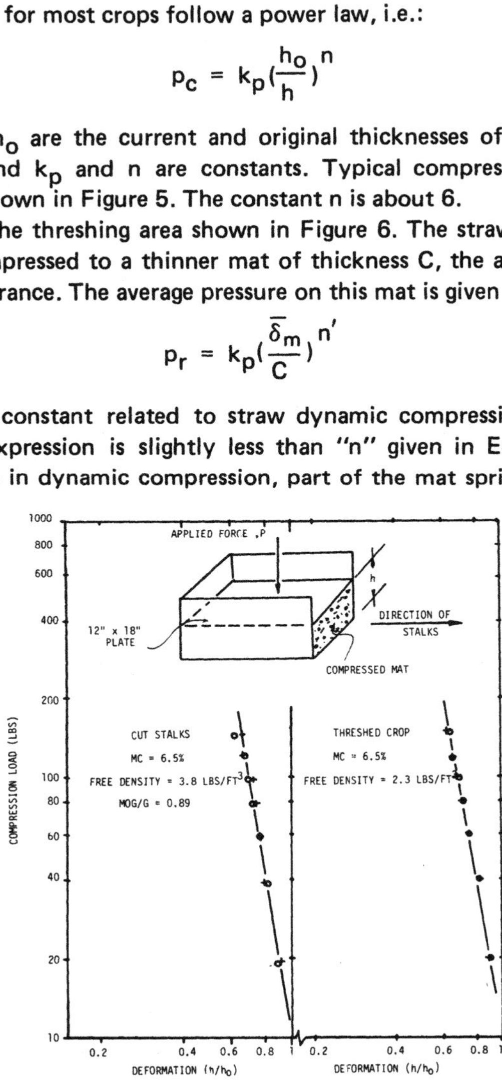

Figure 5. Soybean compression curves.

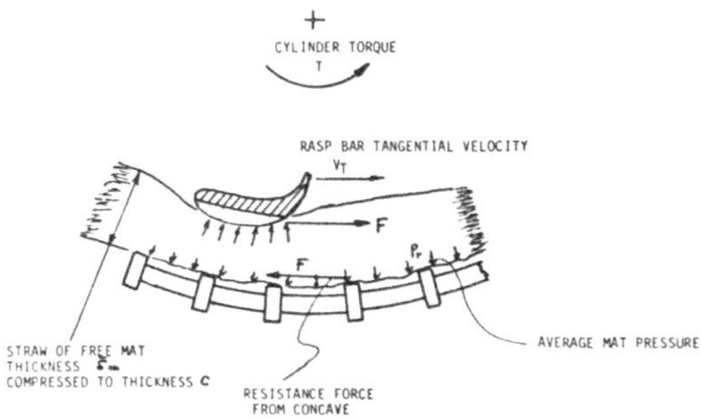

Figure 6. Forces acting on the straw mat in threshing area.

due to the shear failure and the crushing of straw particles. For simplicity, "n′" is considered the same as "n" in this analysis.

To estimate the force to move the mat around the concave, a coefficient of motion resistance is assigned to the mat. This coefficient reflects the sliding friction, shearing, and plowing resistance of the crop mat. The resistance force is computed by:

$$F = \mu p_r W L \qquad \text{(XVIII)}$$

where μ is the coefficient of motion resistance of the straw. For a given thresher design, this coefficient is dependent on the crop moisture content.

The threshing horsepower demand is derived from the cylinder torque required to overcome this resistance, and is given by:

$$P_o = \tfrac{1}{2} NFD + P_N \qquad \text{(XIX)}$$

where N is the cylinder RPM, and P_N is the idle power of the thresher.

PREDICTED RESULTS

Equations (II) through (XIX) constitute a mathematical model of the process of threshing and separating soybeans. This model can be used to predict the performances of a given thresher. In this section, general trends are established for the threshing performances to indicate the behavior of the model. Quantitative results in terms of numerical values, however, are not obtained in this primary part of the study. These could be calculated by the use of a computer program. The procedure for calculating the threshing performances in such a program would be as follows: (a) Determination of Input Data: These data include the cylinder and concave dimensions, shown in Table 1, and the crop and operating conditions, Table 2. (b) Determination of Model Constants: These constants are determined from exeriments are are listed as shown in Table 3. (c) Calculation of Crop Flow Parameters:

Table 1. Cylinder and concave data.

Cylinder Diameter, D
Cylinder Width, W
Concave Length, L
Concave Width, W
Centerline Distance Between Adjacent Concave Rods, a_1
Concave Rod Diameter, a_2
Centerline Distance Between Adjacent Concave Bars, b_1
Width of Concave Bars, b_2

Table 2. Crop conditions and operation conditions.

MOG Feed Rate, Q_s
MOG/G
MOG Free Density, ρ
Moisture Content, M_C
Coefficient of Motion Resistance, μ
Bean Equivalent Diameter, d
Average Cylinder to Concave Clearance, C
Feeder Linear Speed, V_o
Cylinder Peripheral Speed, V_T

Table 3. Model constants.

Constant	Equation
n	(XVI)
k_f	(V)
k_r	(V)
k_T	(VIII)
k_m	(X)
k_d	(XV)
k_p	(XVI)

The velocity of crop flow around the concave and the thickness of the crop mat for a given MOG feed rate are determined using the crop motion model. From this, the crop dwell time in the threshing area is also calculated. (d) Threshing Loss Calculation: The threshing rate is determined for a given set of operating conditions. The threshing loss is then calculated via Equation (II). (e) Bean Damage Calculation: Similar calculations are performed for the damage rate and the damage level. The damage level is evaluated by Equation (IV). (f) Determination of Separation Rates: The straw mat is divided into a number of layers and the mean separation rate through the straw mat is determined. The probability of passage through the concave grate is also evaluated from the given concave dimensions to give a mean rate of concave passage. (g) Calculation of Separation Efficiency: From the threshing rate and the rates of passage, the concave separation efficiency is determined via Equation (III); and (h) Horsepower Calculation: The average mat pressure is calculated from the mat compression equation. The crop motion resistance is then defined, and the cylinder horsepower subsequently is evaluated.

Distribution of Beans Collected Under Concave

A typical distribution of separated beans is established to show the inherent characteristics of the model. This distribution is determined in terms of the fractions of beans collected in a number of boxes placed along the concave length (Figure 7). Greatest separation occurs near the concave entrance, and the fraction of separated beans decreases exponentially as the crop moves towards the concave exit (Figure 7). This figure also shows the effect of MOG

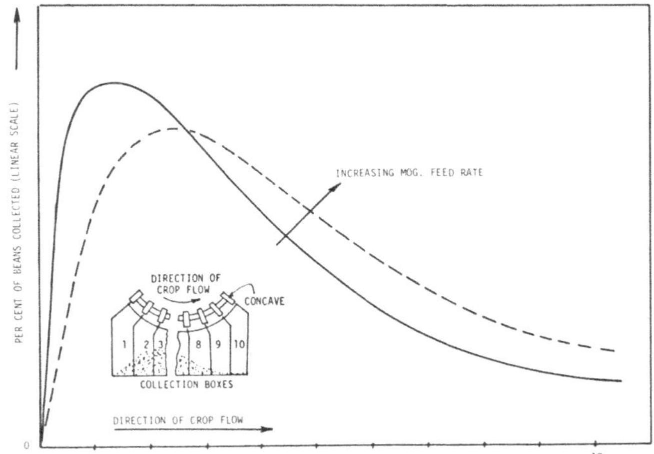

Figure 7. Distribution of beans collected under concave.

feed rate on the distribution curve. As MOG feed rate increases, threshing is retarded and the separation rate through the straw is reduced. Thus, the distribution curve becomes flatter and more beans are separated near the rear of the concave.

Threshing Loss

Mature beans are fairly easy to thresh especially at low moisture level. Threshing loss, however, can be excessive when harvesting wet crops.

Effect of Concave Length. To illustrate the behavior of the threshing model, the threshing loss for similar threshers of different concave lengths was determined. The main influence of increasing concave length is that the crop dwell time in the threshing area is increased. The threshing loss decreases exponentially as the concave length increases. This relationship is presented by a straight line in a semi-log paper (Figure 8). As the concave reaches a certain length, the economic return of a small reduction in threshing loss does not justify any further increase in the concave length. This is shown as point D in the loss curve plotted in linear scales on the graph insert. There is a certain amount of bean prethreshing before the crop reaches the concave. This prethreshing would cause the loss, curve 1 in Figure 8 to shift to a lower level, curve 2, Figure 8. In the same figure, threshing losses for crops at different moisture contents are shown as concurrent lines of different slopes. The line of greater slope is for low moisture crops, e.g., line 3

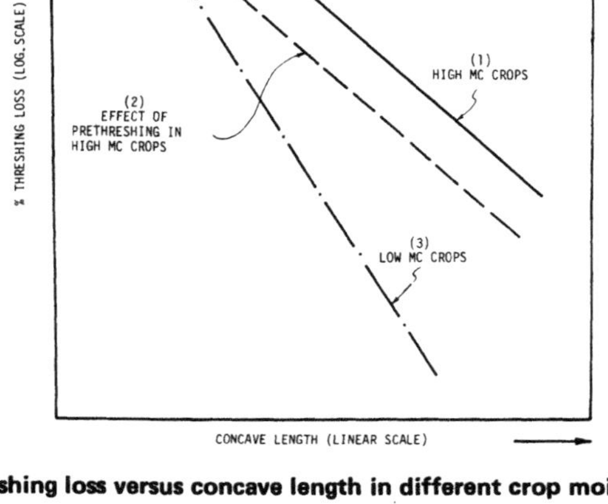

Figure 8. Threshing loss versus concave length in different crop moistures.

in Figure 8. Thus, the threshing loss is reduced when harvesting low moisture beans.

Threshing Speed. Threshing loss decreases exponentially as the impact energy increases (Figure 9). For a given crop throughput, this impact energy is proportional to the square of the cylinder speed. For most difficult-to-thresh crops, doubling the cylinder speed over the normal speed range is sufficient to reduce the threshing loss to an acceptable level. More energy is required to thresh crops at high moistures. The model shows that the impact energy should be increased in the same proportion to the rise in crop moisture in order to keep the threshing loss at the same level (Figure 9).

Concave Clearance. Threshing loss can be reduced to a certain extent by decreasing the concave clearance. As the clearance decreases, the shearing action of the rasp bars and concave increases, thus providing more threshing energy to the crop. The effect of concave adjustment on threshing loss at various crop moisture levels is shown in Figure 10.

MOG Feed Rate. For a given threshing speed, the threshing energy per unit crop input decreases as the amount of MOG feed rate increases. Thus, the threshing loss is greater at high MOG feed rates (Figure 11). An increase in crop moisture content causes the loss curve to swing upward.

Bean Damage

Soybean threshing is normally limited by the susceptibility to damage of the seeds. Threshing effort is usually kept at a minimum level to reduce damage.

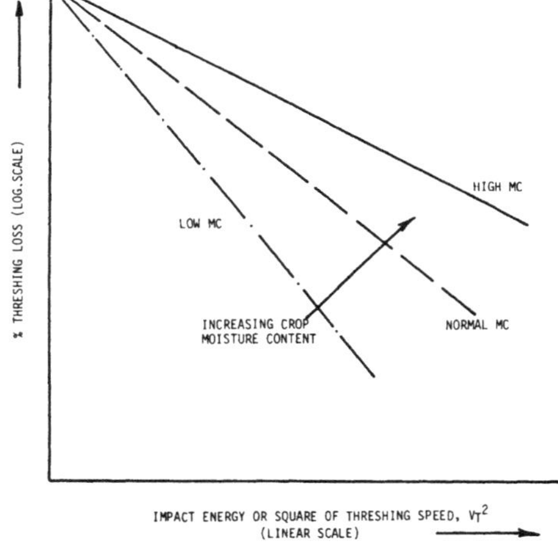

Figure 9. Effect of cylinder speed on threshing loss.

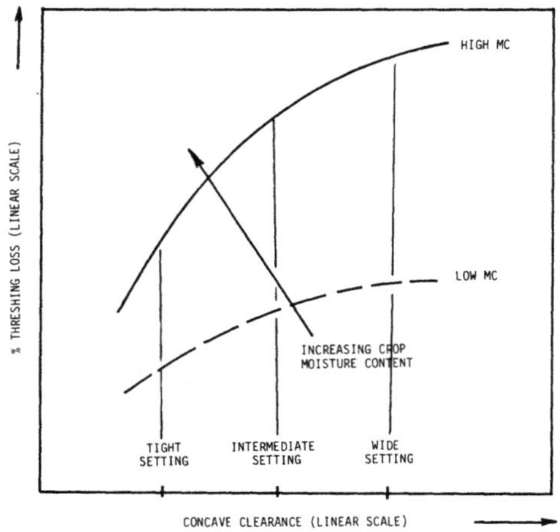

Figure 10. Effect of clearance on threshing loss.

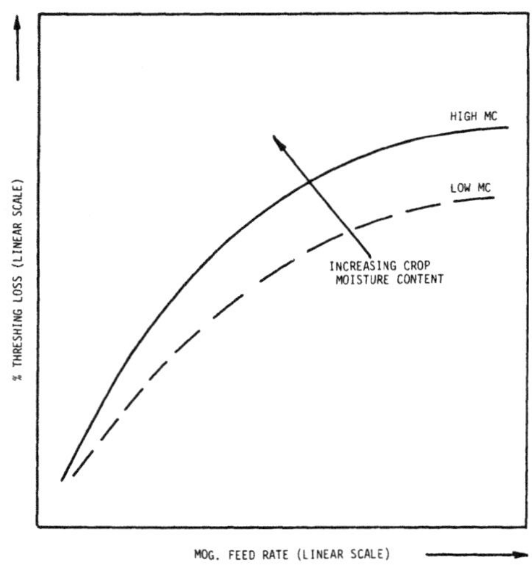

Figure 11. Effect of MOG feed rate on threshing loss.

Threshing Speed. Soybean damage is related to the energy absorbed by the crop during impact from the rasp bars. The model shows that as the impact energy increases, the portion of intact beans decreases exponentially. This portion is also greater for high moisture crops as the beans at this moisture are more tolerant to mechanical impact. This effect is presented in Figure 12. The damage vs. speed curves are also shown in linear scales in the graph insert.

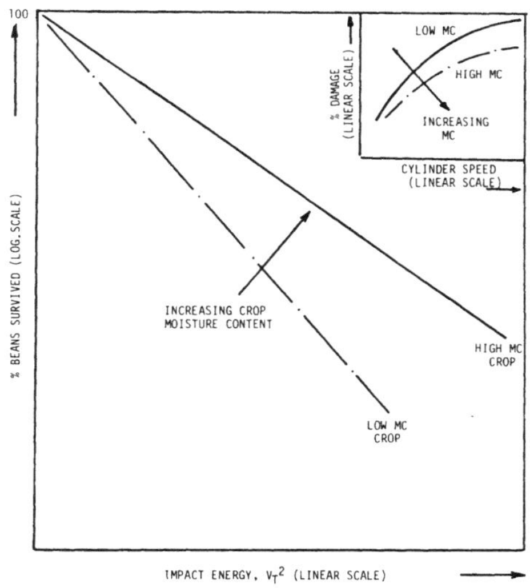

Figure 12. Effect of threshing speed on damage.

MOG Feed Rate. The presence of the soybean straws in the threshing area affects the seed crackage in the following manner: (a) the straw prevents the loose beans from getting into direct contact with the threshing elements where severe damage is likely to occur, and (b) the straw mat gives a cushioning effect to crop impact thus reducing the energy transmitted to the beans. Bean damage level, therefore, is reduced when the straw or MOG feed rate is increased (Figure 13).

Threshing Horsepower

Effect of Throughput. Figure 14 illustrates the relationship between the thresher horsepower and crop feed rate. As the crop mat pressure is proportional to throughput to the power "n", ($n > 1$), the cylinder torque increases rapidly as MOG feed rate increases. Consequently, the thresher power demand is much greater at high feed rates. The crop moisture content, as well as the presence of green weeds in a soybean crop affects the resistance to motion of the crop around the concave. Figure 14 shows that for a given MOG feed rate, the horsepower demand is greater for crops at higher moisture level. The horsepower curves in this figure form parallel straight lines when replotted on log scales (see graph insert). The slopes of these lines are proportional to "n", the constant in the straw mat compression equation.

Effect of Cylinder Speed. For a given feed rate, the cylinder torque is determined by the crop mat thickness in the threshing area. An increase in the cylinder speed causes a decrease in the mat thickness. This results in a

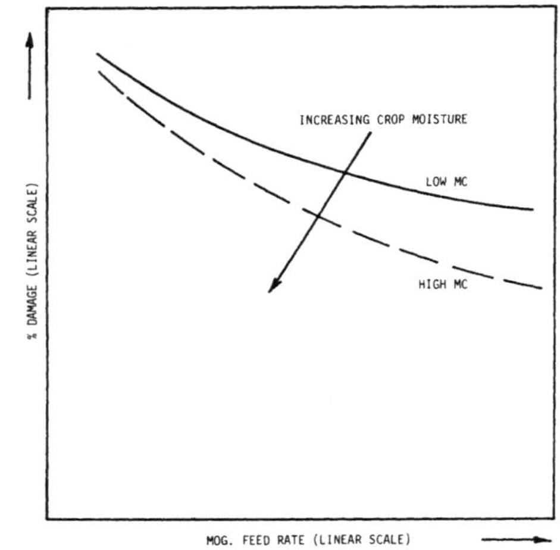

Figure 13. Effect of crop feed rate on damage for a given threshing speed.

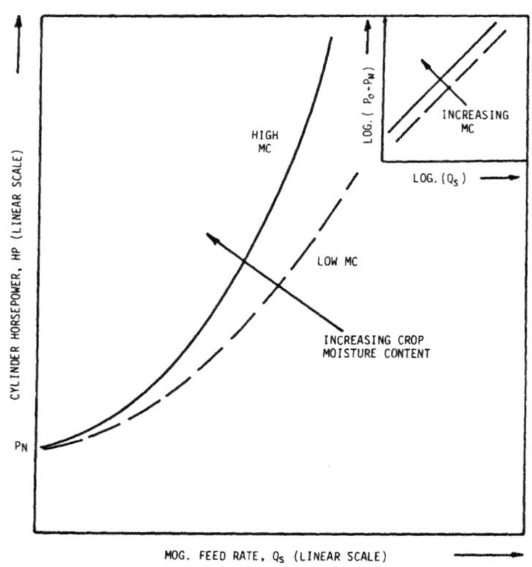

Figure 14. Cylinder horsepower vs. MOG feed rate.

large reduction in the cylinder torque and a slight decrease in the threshing horsepower. This effect is shown in Figure 15.

DISCUSSION

In this section, some of the assumptions used in the derivation of the model are examined.

The Probabilistic Model

Exponential probability density functions are used in this work to describe the phenomena involved in the threshing and separating process. It is assumed that these phenomena occur at random and, thus, the rates of occurrence of these are independent of time. This type of distribution function gives a reasonable starting point for the analysis and is quite convenient for mathematical manipulation. It should be noted, however, that the rates of occurrence of these phenomena may deviate from being constant. For example, the beans may require less energy to thresh further along the concave, the beans may get damaged more easily closer to the concave rear because of the cumulative effect. Also, the beans can penetrate the mat sooner near the concave exit because of a thinner mat, etc. Experimental investigations in this area may provide useful information.

Crop Flow Model

The present model for crop motion utilizes constant slip factors for crop velocity in relation to the feeder speed and rasp bar speed. These factors are

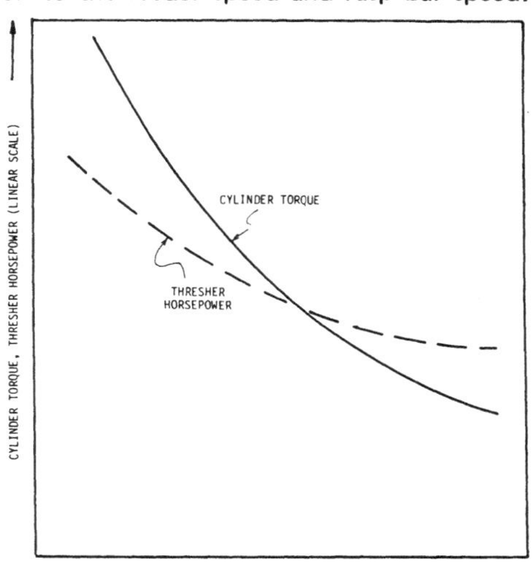

Figure 15. **Effect of cylinder speed on cylinder torque and hrosepower for a given crop throughput.**

assumed to be dependent on thresher design and crop properties. It is recognized that these factors may also depend on the material feed rate, feeding speed, drum speed, and concave clearance. A constant crop average velocity is also used in the model to determine the crop dwell time and the mat thickness in the concave area. In practice, the crop material is fed into the thresher at low speed and it is accelerated to the rear of the thresher. This non-uniformity of the crop speed would affect the overall thresher performances predicted by the model.

Threshing Rate and Damage Rate

The threshing and damage rates in this model are assumed to be inversely proportional to the minimum energy required to thresh and the minimum energy to cause damage, respectively. These energy quantities are described as simple functions of crop moisture content. It is known, however, that these quantities are complex functions of bean and pod moisture contents, crop variety, crop threshing speed, and clearance. Further studies in this area may be necessary to improve the model.

Passage of Beans Through Concave Grates

In modelling the separation of the threshed beans through the concave, the passage of beans through the concave opening by particle reflection is neglected and the loose beans are assumed to travel at a constant speed in the threshing area. In actual fact, the reflection of beans by the concave rods or bars increases the probability of concave passage and also causes a reduction in the bean travelling speed. Thus, actual concave separation would be greater than that predicted by the model.

SUMMARY

A probabilistic method of approach has been proposed for the mathematical modelling of the threshing and separating of soybeans in a conventional combine thresher. The parameters for the model were defined in terms of process physical quantities. General trends were established for the threshing performances using the model. The effect of various crop threshing variables were also studied via the predicted trends.

NOTES

V. M. Huynh, T. E. Powell, Engineering Department, White Farm Equipment, Brantford, Ontario, Canada; and J. N. Siddall, Mechanical Engineering Department, McMaster University, Hamilton, Ontario, Canada.

REFERENCES

1. Byg, D. M. 1975. Research on soybean harvesting equipment, pp 197-214. In L. D. Hill (Ed.) Proceedings World Soybean Research Conf., Interstate Printers, Danville, IL.

2. Cain, D. F. and R. G. Holmes. 1977. Evaluation of soybean impact damage. ASAE Paper No. 77-1552.
3. Huynh, V. M., T. Powell, and J. N. Siddall. 1977. Threshing and separating process—A mathematical model. ASAE Paper No. 77-1556.
4. Lamp, B. J., W. H. Johnson, and K. A. Harkness. 1961. Soybean harvesting losses—Approaches to reduction. ASAE Trans. 4(2):203-205.
5. Newbery, R. S., M. R. Paulsen, and W. R. Nave. 1978. Soybean quality with rotary and conventional threshing. ASAE Paper No. 78-1560.
6. Pickett, L. K. 1972. Mechanical damage and processing loss during navy bean harvesting. ASAE Paper No. 72-640.
7. Singh, B. and D. E. Linvill. 1977. Determining the effect of pod and grain moisture content on threshing loss and damage of navy beans. ASAE Trans. 20(2):226-231.

SIMULATION APPLICATIONS IN SOYBEAN DRYING

J. H. Young and R. N. Misra

In order to obtain maximum profits, soybeans must be harvested within a short period of time at which maximum yield and quality may be attained and must be maintained at a moisture content safe for storage. Hunt and Harper (12) developed harvesting timeliness factors for two varieties of soybeans (Shelby and Harosoy) and concluded that there were "only a couple of days when the monetary loss is at a minimum" (Figure 1). Harvesting prior to the time at which the bean moisture content reaches 13% wet basis (W.B.) results in penalty costs for the marketing of wet beans while later harvests incur increased harvesting losses in addition to weight loss due to the reduction in moisture content below the market level of 13%.

A number of investigators have studied soybean field losses and various changes which may be made in harvesting equipment to minimize losses (5,6,10,11,15,21-23,25,26,29,30,32,34). Although improvements in equipment can reduce harvest losses, the optimum period for harvest remains relatively short requiring high capacity harvesting equipment which can be used for only a short period of time. Hunt and Harper (12) showed that a significant portion of the field losses occur prior to harvest due to field aging. These losses cannot be reduced by machinery modifications. Byg and Johnson (5) showed that field losses increased as the moisture content decreased. An 8% shatter loss was found when beans were combined at 10% (w.b.) moisture compared with only a 2% loss when combined at 17% moisture. These observations suggest that the length of the harvesting season may be increased by harvesting prior to the time at which beans reach the 13% market level of moisture and completing the drying artificially to a safe storage level. Artificial drying may also be advisable in wet seasons to prevent

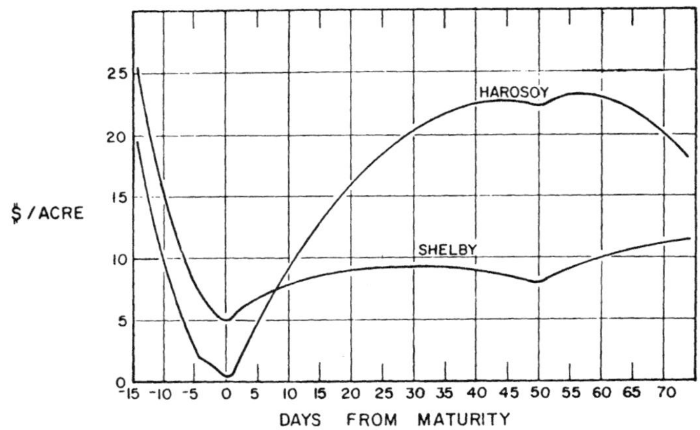

Figure 1. **Monetary loss with time for two soybean varieties [evaluated by Hunt and Harper (12) assuming a yield of 40 bu/a and a price of $2.40/bu].**

losses in seed viability due to field aging which according to Moore (19) is dependent mainly on the moisture and temperature changes in the seed and the duration and frequency of exposure to moisture and temperature changes.

In order for atificial drying of soybeans to be practical, the drying process must be accomplished at a cost which is less than the additional field losses which would occur in field drying and there must be no significant deterioration or cracking of the seed. It is the purpose of this chapter to review studies which have been conducted to improve drying efficiency and quality of the dried product, to discuss simulation procedures which offer promise for the design of better drying systems, and to indicate priorities for future soybean drying simulation.

REVIEW OF SOYBEAN DRYING RESEARCH

A review of literature on soybean drying research reveals an evolvement of efforts which may be categorized as follows: (a) Experimental investigations of drying curves and damage caused by various drying conditions, (b) Development of simulation procedures for predicting drying curves and the determination of required mass transfer properties, and (c) Development of simulation procedures for predicting both drying curves and stress development and the determination of required physical properties.

Walker and Barre (31) conducted thin-layer drying tests on two soybean varieties (Harosoy 63 and Chippewa 64) at temperatures ranging from 32 to 66 C and relative humidities ranging from 20 to 60%. They then evaluated damage due to seed coat cracks and reduced germination. They concluded that seed coat cracks can be prevented by using relative humidities above 40%

and that drying air temperature above 54 C will cause reductions in germination.

Matthes and Welch (13) investigated the drying of 22% moisture content Dare soybeans in bins. 1.1 m in depth. Tests were conducted at relative humidities of 42 and 55% with air flow rates ranging from 0.02 to 0.16 m^3/m^3s and resulting seed germination was evaluated. They concluded that it is feasible to dry high moisture seeds (up to 22%) with heated air and obtain a product of acceptable quality, that the relative humidity of the drying air can range from 40 to 50%, and that the air flow rate should not be less than 0.14 to 0.16 m^3/m^3s.

White et al. (33) and Rojanasaroj et al. (27) dried soybeans in thin layers at air temperatures ranging from 24 to 74 C and evaluated the effects on storage characteristics and susceptibility to impact damage. It was concluded that soybeans dried at high temperatures were more prone to mold development, had reduced germination, had higher free fatty acid contents, and were more susceptible to cracking or splitting due to impact in handling. Since relative humidity was not controlled in their studies it is not clear whether the damage was actually due to the high temperatures or to the low relative humidities which would result from heating ambient air.

Several mathematical models have been developed within the past few years for simulating the moisture removal during soybean drying. Alam and Shove (1) modified a corn drying model of Bloome and Shove (4) and successfully simulated the drying of 20.7% (w.b.) soybeans in a 1.83-m column using ambient air at a rate of 1.58 m^3/m^2 min. Relationships to define equilibrium moisture content, specific heat, latent heat of vaporization, and bulk density of soybeans were developed and included in the simulation model.

Overhults et al. (24) investigated the use of an exponential drying relationship for describing the moisture removal from thin-layers of soybeans. They obtained a good description of experimental results when they used the relationship:

$$MR = \frac{M - M_e}{M_o - M_e} = \exp\left[-(kt)^n\right] \qquad (I)$$

where MR = moisture ratio, M = moisture content (dry basis) at any time (t), M_o = initial moisture content (dry basis), M_e = equilibrium moisture content (dry basis), k = drying rate parameter (hr^{-1}), t = time (hr), and n = constant for a particular temperature. The drying parameters, k and n, were found to vary with temperature as follows:

$$k = a + b/T_a \quad \text{and} \quad n = \beta_o + \beta_1 T \qquad (II)$$

where T_a = absolute temperature, T = drying air temperature, and a, b, β_o and β_1 = constants. The thin layer drying simulation model was evaluated using experimental data obtained for soybeans of the Cutler variety ranging in initial moisture content from 20 to 33% (w.b.). Drying air temperatures used in the study ranged from 37.8 to 104.4 C with a constant dew point temperature of 7.8 C. Severe physical damage to the soybeans was noted at the higher temperatures and moisture contents.

Sabbah et al. (28) used the Michigan State Univ. drying model developed by Bakker et al. (3) to simulate the drying of soybeans in fixed beds with periodic air direction reversals. A thin-layer drying model was developed in which the drying parameters were rather complicated relationships of the grain shape, drying air state, equilibrium moisture content of the grain, and initial moisture content of the grain.

The simulation models of Alam and Shove (1), Overhults et al. (24) and Sabbah et al. (28) all consider only the moisture removal curves for soybeans with no attempt toward describing the bean shrinkage and resulting stresses which may cause crackage. More recently simulation models have been developed by Misra (16) and Haghighi (7) which use finite element procedures for simulating the moisture transfer and stress development phases of soybean drying. These models both give adequate simulations of the moisture transfer function. However, there are basic differences in model assumptions and neither is presently adequate for explaining the grain crackage which occurs during drying. Differences in assumptions and simulation results will be discussed in a later section of this chapter.

REASON FOR SIMULATION

Morey et al. (20) have reviewed the present status of grain drying simulation and have suggested two answers to the question, "why simulate?". They are: (a) The process of model development and testing contributes significantly to the understanding of the mechanisms and processes involved in the drying of grains, and (b) simulation models help to predict the performance of dryers and dryer systems so that they may be used to improve dryer designs and to aid in the operation and management of drying systems.

At this time, the cracking phenomena in soybeans are not well understood and the efforts of Misra (16) and Haghighi (17) to simulate these processes mathematically have helped to evaluate some theories concerning crackage. If these models can be developed to the point that grain stresses (and thus failures) can be predicted for various drying conditions, then they will be quite important tools for drying design or management.

COMPARISON OF THE MISRA AND HAGHIGHI MODELS

The models of both Misra and Haghighi consider the soybean to be a spherical body which is divided into finite elements for numerical analysis.

Misra has assumed a homogeneous sphere while Haghighi has assumed an inner kernel surrounded by a spherical shell (the seed coat). Based on the work of Arora et al. (2) Misra chose to neglect the heat transfer problem within the kernel and the thermal stresses which may result from temperature gradients while Haghighi chose to solve both moisture and heat transfer problems and assumed the material shrinkage to be affected by both moisture and temperature gradients. Misra chose to simulate the moisture transfer and soybean shrinkage problems simultaneously while Haghighi first solved the heat and moisture equations neglecting shrinkage and then super-imposed the shrinkage as a function of temperature and moisture gradients. Misra assumed that the moisture diffusion coefficient varied with moisture content and material density (which changes due to shrinkage) while Haghighi assumed the diffusion coefficient to be a constant for a given drying temperature. Misra assumed that the elastic modulus of the soybean decreased according to a power function of the moisture content while Haghighi assumed constant elastic modulus values. Both models assumed the soybean to be a Hookean material.

Misra (16) conducted experimental drying tests at four dry-bulb temperatures (35,55,75, and 95 C) and four dew point temperatures (10,15,20, and 25 C). These data were used for parameter evaluations. Haghighi based his parameters on values obtained from the literature and compared his simulated drying curves with those of Sabbah et al. (28) and Overhults et al. (24). Figure 2 is a reproduction of simulated and experimental drying curves from Misra and Young (17) at a dew-point temperature of 10 C. Figure 3 compares simulated drying curves of Haghighi and Segerlind (8), Overhults et al. (24) and Sabbah et al. (28). It is apparent that with proper parameter evaluation either model can simulate the moisture removal.

Figure 4 illustrates the predicted radial and tangential stresses as determined by Misra et al. (18) while Figure 5 shows the stresses determined by Haghighi and Segerlind (8). Each model predicts compressive radial stresses throughout the soybean while tangential stresses are compressive in a large center core of the sphere but become tensile near the surface. The large jump in tensile stresses near the surface in Figure 5 are due to the assumption by Haghighi of stiffer material properties for the seed coat.

Figure 6 illustrates the tangential stress at the surface predicted by Misra et al. (18) for four dry-bulb temperatures at a dew-point temperature of 10 C. Note that the peak values are approximately the same for all drying temperatures but the time at which they occur is later for the lower temperatures. Since a number of investigators including Misra et al. (18) have observed more cracking at the higher temperatures, the predicted maximum stress values cannot be used to predict crackage. It appears, however, that an assumption of viscoelastic material properties (rather than

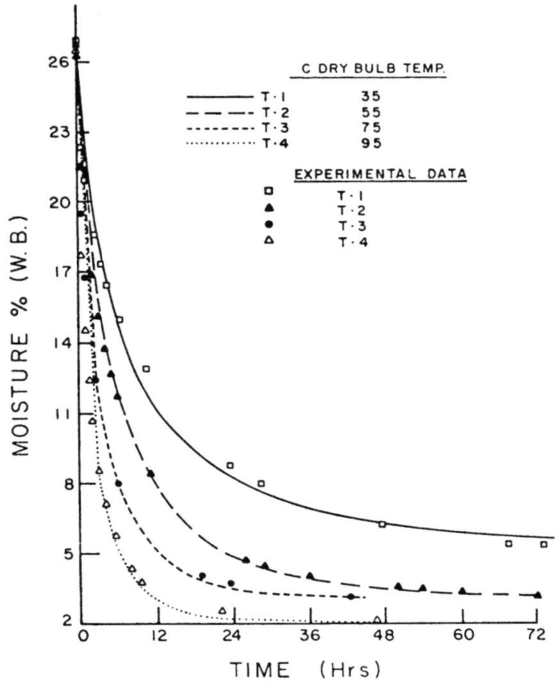

Figure 2. **Experimental observations and simulated drying curves for rewetted Ransom soybeans at a dew-point temperature of 10 C (17).**

elastic) would have allowed more stress relaxation at the lower temperatures such that peak values may then be an indicator of crack susceptibility.

Figure 7 shows the tengential stress curves predicted by Haghighi and Segerlind (8) for the seed coat when soybeans are dried at 35, 55, and 75 C. Note that the peak stresses increase with temperature and always occur after approximately the same drying time. Haghighi suggests that the increased stress levels at higher temperatures as due to his inclusion of temperature gradients in his shrinkage relationship. However, higher temperatures of the drying air should cause expansion of the outer layers (not contraction as results from moisture removal) and thus thermal stresses in the seed coat would counteract rather than add to the hydral stresses. There is apparently some as yet undetermined difference in the moisture removal relationships of the two simulation models which is responsible for the drastic difference in predicted stress reactions to different temperature levels.

Figure 8 is a plot of crackage data presented by Misra et al. (18) for four dry-bulb and four dew-point temperatures. Note that the percentage of

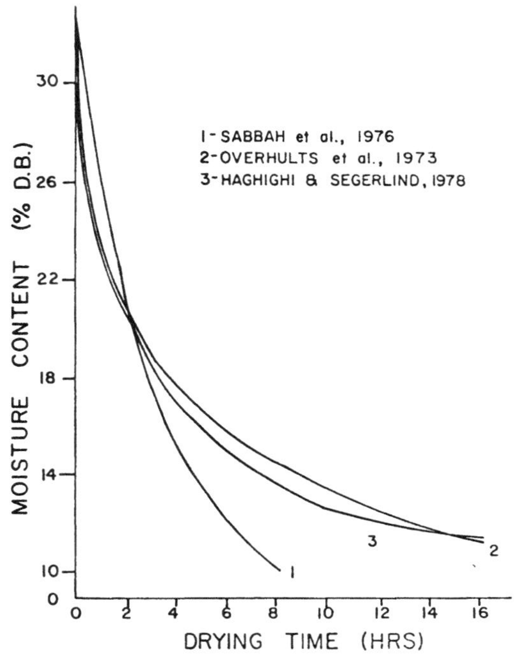

Figure 3. Comparison of drying curves simulated by (28), (24), and (8).

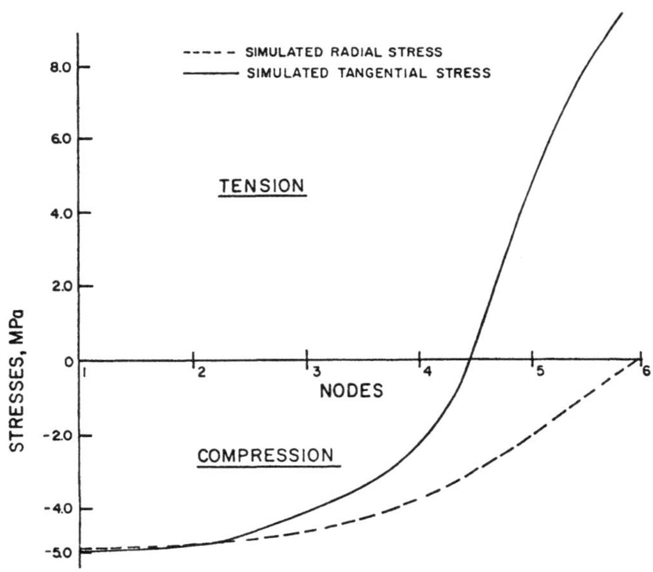

Figure 4. Simulated radial and tangential stresses of Misra et al. (18) for soybeans drying at 35 C dry-bulb and 10 C dew-point temperature after 3.5 hr.

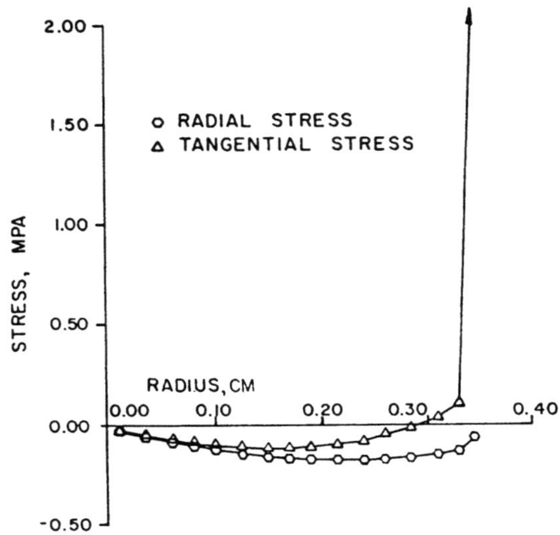

Figure 5. Simulated radial and tangential stresses of Haghighi and Segerlind (8).

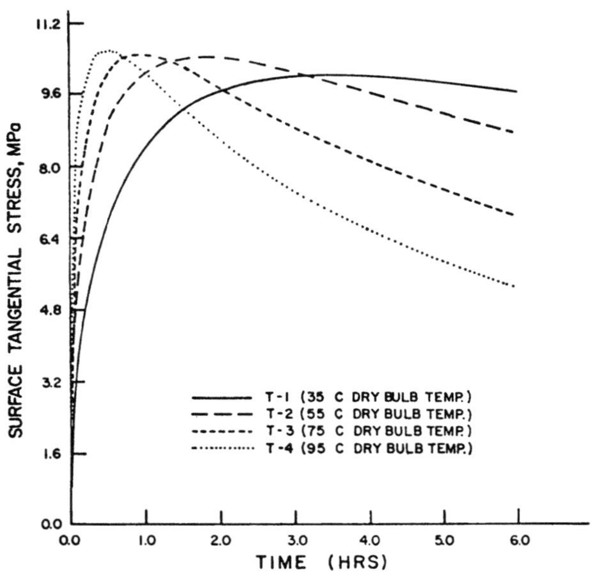

Figure 6. Simulated tangential stresses at the surface for soybeans drying at a 10 C dew-point temperature (18).

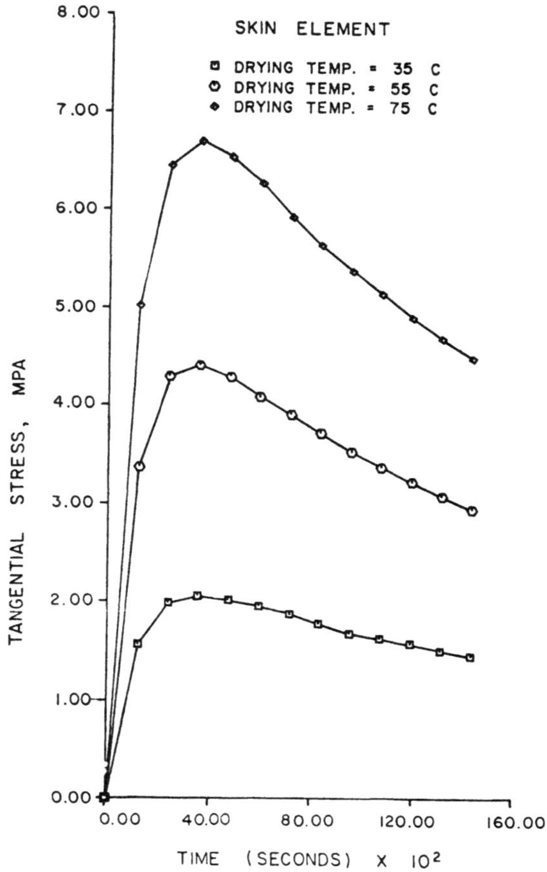

Figure 7. Simulated tangential stresses at the surface (8).

cracked beans increases with drying temperature at each of the dew-point temperatures. A primary question is whether the increased crackage at the higher temperatures is due to the temperature gradients or to the larger moisture gradients which result from the faster drying. A replot of the data (Figure 9) as percent crackage versus relative humidity of the drying air seems to suggest that the later is the primary cause. Note that above approximately 40% relative humidity there was no crackage as had been found previously by Walker and Barre. Though the data is too limited for definite conclusions, indications are that at a particular relative humidity the effect of higher temperatures is to lower the percent cracks. This would support the theory that cracks develop primarily due to moisture gradients.

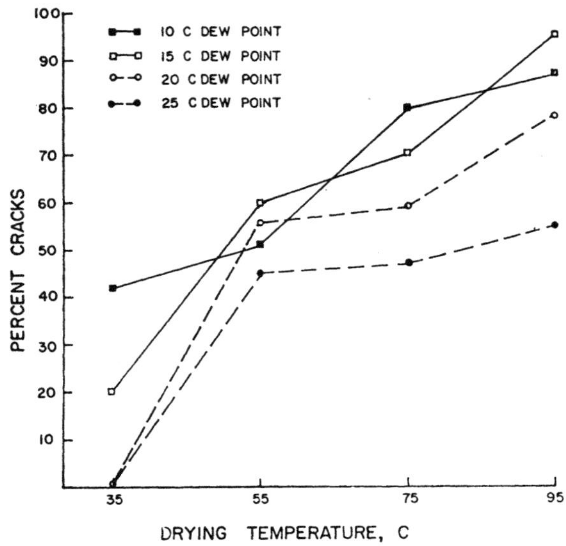

Figure 8. Percent cracks versus dry-bulb temperature at four dew-point temperatures.

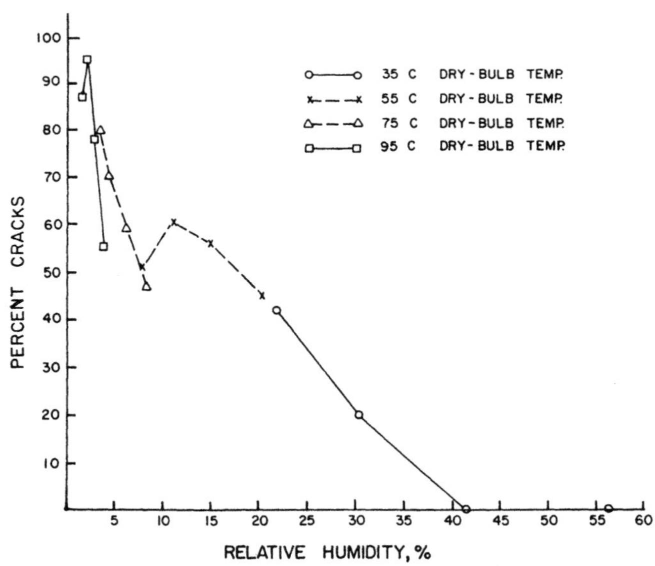

Figure 9. Percent cracks versus relative humidity for four different dry-bulb temperatures.

SUMMARY

During the past ten years a great deal of information has been accumulated on the moisture removal curves of soybeans in different drying environments and on the physical damage which may result. There are a number of models which give satisfactory simulation of the moisture removal. These models may be used for studying the performance of drying systems and as aids in the proper management of the systems. However, they require additional information on grain quality variations in order to optimize the drying processes.

The models of Misra (16) and Haghighi (7) attempt to simulate both the moisture transfer and the stress development in the soybeans. They offer promise for use in design of systems to optimize grain quality as well as moisture removal efficiency. However, further study of the models is needed in order to explain discrepancies in predicted stresses and in order to relate predicted stresses to observed physical damage.

Both Misra (16) and Haghighi (7) have noted that viscoelastic properties of the soybean are needed for better simulation models. Herum et al. (9) and Mensah et al. (14) have reported work in the area of evaluating these properties. Further work toward incorporating the viscoelastic property data in the drying models is needed.

NOTES

J. H. Young, Department of Biological and Agricultural Engineering, N. C. State University, Raleigh, North Carolina 27650; and R. N. Misra, Agricultural Engineering Department, Ohio Agricultural Research and Development Center, Wooster, Ohio 44691.

REFERENCES

1. Alam, A. and G. C. Shove. 1973. Simulated drying of soybeans. Trans. ASAE 16(1):134.
2. Arora, V. K., S. M. Henderson, and T. H. Burkhardt. 1973. Rice drying cracking versus thermal and mechanical properties. Trans. ASAE 16(2):320.
3. Bakker-Arkema, F. W., L. E. Lerew, S. F. DeBoer, and M. G. Roth. 1974. Grain dryer simulation. Research Report 224, The Michigan State Univ. and Agric. Exp. Sta., East Lansin, MI.
4. Bloome, P. D. and G. C. Shove. 1971. Near equilibrium simulation of shelled corn drying. Trans. ASAE 14(9):709.
5. Byg, D. M. and W. H. Johnson. 1970. Reducing soybean harvest losses. Ohio Report. 55(1):17-18.
6. Dunn, W. E., W. R. Nave, and B. J. Butler. 1973. Combine header component loses in soybeans. Trans. ASAE 16(6):1032.
7. Haghighi, K. 1978. Finite element formulation of the thermo-hydro stress problem in soybeans. ASAE Paper No. 78-3560. ASAE, St. Joseph, MI 49085.
8. Haghighi, K. and L. J. Segerlind. 1978. Computer simulation of the stress cracking of soybeans. ASAE Paper No. 78-3560. ASAE, St. Joseph, MI 49085.

9. Herum, F. L., J. K. Mensah, H. J. Barre, and K. Majizadeh. 1976. Viscoelastic behavior of soybeans due to temperature and moisture content. J. Article No. 114-76, The Ohio Agric. Res. and Dev. Center, Wooster, OH.
10. Hoag, D. L. 1972. Properties related to soybean shatter. Trans. ASAE 15(3):494.
11. Hoag, D. L. 1975. Determination of the susceptibility of soybeans to shatter. Trans. ASAE 18(6):1174.
12. Hunt, D. and R. W. Harper. 1967. Timeliness of soybean harvesting. ASAE Paper No. 67-615. ASAE, St. Joseph, MI 49085.
13. Matthes, R. K. and G. B. Welch. 1974. Heated air drying of soybean seed. ASAE Paper No. 74-3001. ASAE, St. Joseph, MI 49085.
14. Mensah, J. K., G. L. Nelson, F. L. Herum, and T. R. Richard. 1978. Mechanical properties related to soybean seedcoat failure during drying. ASAE Paper No. 78-3561. ASAE, St. Joseph, MI 49085.
15. Miller, L., R. K. Matthes, and W. J. Drapala. 1971. A study of variables affecting the threshing of soybeans. ASAE Paper No. 71-624. ASAE, St. Joseph, MI 49085.
16. Misra, R. N. 1977. Failure and stress analysis in soybeans subjected to a temperature and moisture gradient. Ph.D. Thesis, North Carolina State Univ., Raleigh, NC.
17. Misra, R. N. and J. H. Young. 1978. Finite element analysis of simultaneous moisture diffusion and shrinkage of soybeans during drying. ASAE Paper No. 78-3056. ASAE, St. Joseph, MI 49085.
18. Misra, R. N., J. H. Young and D. D. Hamann. 1978. Finite element procedures for estimating shrinkage stresses during soybean drying. ASAE Paper No. 78-3559. ASAE, St. Joseph, MI 49085.
19. Moore, R. P. 1960. Soybean germination. Seedman's Digest 11(2):12.
20. Morey, R. V., H. M. Keener, T. L. Thompson, G. M. White, and F. W. Bakker-Arkema. 1978. The present status of grain drying simulation. ASAE Paper No. 78-3009. ASAE, St. Joseph, MI 49085.
21. Nave, W. R., D. E. Tate, and B. J. Butler. 1972. Combine headers for soybeans. Trans. ASAE 15(4):632.
22. Nave, W. R. and D. L. Haog. 1975. Relationship of sickle and guard spacing and sickle frequency to soybean shatter loss. Trans. ASAE 18(4):630.
23. Nave, W. R. and R. R. Yoeger. 1975. Use of air-jet guards to reduce soybean harvesting losses. Trans. ASAE 18(4):626.
24. Overhults, D. G., G. M. White, H. E. Hamilton, and I. J. Ross. 1973. Drying soybeans with heated air. Trans. ASAE 16(1):112.
25. Quick, G. R. 1970. Laboratory analysis of the combine header. ASAE Paper No. 70-630. ASAE, St. Joseph, MI 49085.
26. Quick, G. R. and W. F. Buchele. 1974. Reducing combine gathering losses in soybeans. Trans. ASAE 17(6):1123.
27. Rojanasaroj, C., G. M. White, O. J. Loewer, and D. B. Engli. 1976. Influence of heated air drying on soybean impact damage. Trans. ASAE 19(2):372.
28. Sabbah, M. A., G. E. Meyer, H. M. Keener, and W. L. Roller. 1977. Reversed-air drying for fixed bed of soybean seed. Trans. ASAE 20(3):562.
29. Tate, D. E. and W. R. Nave. 1971. Air-conveyer header for soybeans. ASAE Paper No. 71-607. ASAE, St. Joseph, MI 49085.
30. Tunnell, J. C., W. R. Nave, and R. R. Yoerger. 1973. Reducing soybean header losses with air. Trans. ASAE 16(6):1020.
31. Walker, R. J. and H. J. Barre. 1972. The effect of drying on soybean germination and crackage. ASAE Paper No. 72-817. ASAE, St. Joseph, MI 49085.
32. Weeks, S. A., J. C. Wolford, and R. W. Klies. 1975. A tensile testing method for determining the tendency of soybean pods to dehisle. Trans. ASAE 18(3):471.

33. White, G. M., O. T. Loewer, I. J. Ross, and D. B. Engli. 1976. Storage characteristics of soybeans dried with heated air. Trans. ASAE 19(2):306.
34. Williams, M. M. and C. B. Richey. 1973. A new approach to gathering soybeans. Trans. ASAE 16(6):1017.

APPLICATION OF VON BERTALANFFY'S EQUATION TO SOYBEAN GROWTH MODELING

R. Guarisma

The emergence of soybeans [*Glycine max* (L.) Merr.] in the field is a function of the interaction of the seeds and the soil environmental factors. Stress during the emergence period of soybean production results in either inadequate numbers of plants for profitable production or a highly diverse plant population which makes crop management difficult. Important environmental factors include soil temperature, soil moisture, aeration, soil insects, and soil mechanical impedance. Gilman et al. (7) have found that less than optimum levels of temperature or moisture may weaken the seed and render it more susceptible to other factors of the environment and, consequently, reduce emergence. Nonuniform stands increase the difficulty in determining the optimum time of performing subsequent production operations. Poorly timed operations usually reduce profits, and for the grower, this means that the crop must be managed more effectively. Efficient management requires an understanding of the system which the grower is attempting to control. The ultimate understanding of a system can be represented by a model whereby the system can be simulated. Modeling can be very beneficial, even if the initially developed models are not sufficient for accurate simulation (3). Modeling and computer simulation of plant growth has increased greatly in recent years. Hatfield and Egli (9) developed prediction equations for soybean hypocotyl elongation as a function of soil temperature and depth of planting (2) and Wanjura et al. developed equations which describe the rate of shoot and hypocotyl elongation in corn and cotton, respectively. The equations developed by Blacklow (2) and Wanjura (13) were incorporated into computer models which made it possible to predict the time of emergence as a function of soil temperature. Von Bertalanffy (12) has developed a growth function for animals, and Richards (10) has shown that when limitations imposed by its theoretical background are discarded, it may have wide applications in empirical botanical studies, facilitating comparisons between sigmoid curves of different shapes.

The objective of the present study was to determine whether Von Bertalanffy's equation could be adapted to simulate soybean hypocotyl growth during seedling emergence as influenced by the soil physical environment, moisture, temperature, and mechanical impedance. In order to evaluate the suitability of Von Bertalanffy's equation for simulating soybean hypocotyl elongations we chose to use a set of average hypocotyl elongation data developed from growth chamber data.

EXPERIMENTAL PROCEDURE

Description of Data

We worked with a set of average soybean hypocotyl elongation data developed by Bowen (4) under a wide range of steady-state soil environmental conditions. He germinated seeds of soybean 'Forest 1975' cultivar for 18 hr at 30 C in vermiculite, selecting those with uniform 3 mm-radicle lengths, which were planted in plywood boxes containing soil. Hypocotyl growth readings were taken through the transparent side of the box.

Four replications of 15 seeds each of the 48 combinations of temperature, moisture, and mechanical impedance were observed. The statistical arrangement was a factorial design. For each replication, the order of preparation of the boxes and position of the boxes in the growth chamber were each randomized.

DESCRIPTION OF VON BERTALANFFY'S EQUATION

The four parameter form of Von Bertalanffy's equation used in this study was derived from the version presented by Richards (10). This equation has the advantage that the four parameters define any one curve just as effectively as do the original constants of the growth equation (12), and in a way which conveys physiological information more easily.

The equation is given by:

$$Y^{1-m} = E^{1-m} [1 \pm B \exp(-KT)] \qquad [I]$$

In equation [I], the + sign applies when $m > 1$ and the - sign applies when $m < 1$. The hypocotyl length (cm) is represented by Y at any time T (days); E represents the maximum hypocotyl length (cm) possible under prevailing environmental conditions; m determines the proportion of the final size, E, at which the inflexion point occurs; and B usually has no biological implication and can be eliminated from the equation by adjustment of the time scale.

Richards (10) formulated universally comparable average rate parameters among the family of curves; thus R = EK/2 m + 2 represents a weighted average absolute elongation rate throughout development (cm/day), S = K/2 m + 2 has a similar interpretation in relation to 'proportional elongation' rates (day^{-1}), and W = K/m is the actual relative elongation rate at the point of inflexion on the growth curve, where the absolute rate is maximal (day^{-1}). If m = 0, equation [I] reduces to the mono-molecular function, and if m = 2 to the autocatalytic function. When m = 1 exactly, equation [I] cannot be solved, but it may be shown (10) that as unity is approached the growth curve approaches more closely the Gompertz curve, which is indeed the limiting form. Values of m intermediate between 0 and 1 result in a family of curves ranging in shape from the mono-molecular to the Gompertz, and the autocatalytic.

RESULTS

Calculation of Von Bertalanffy's Parameters

Best-fit parameters were derived for each of the 48 sets of mean hypocotyl elongation values over time by means of an iterative, non-linear least squares program (NLIN, Statistical Analysis System, NCSU). Time was expressed in days in order to speed up exponential calculations. The E (asymptote) value was determined from observations of elongation - time curves based on the actual data for each combination of soil environmental factors.

This approach was applied to the 48 sets of mean hypocotyl elongation data using three different values for m (0.7, 0.9, and 1.5). The results obtained from the fitting process for the three values of m showed that the best fit is obtained when m = 1.5. Table 1 shows the best-fit parameters for a fixed value of m = 1.5 along with the R^2 values for goodness of fit for this particular case and for the 48 treatment combinations. Observation of the R^2 values shows that 41 treatments were fitted with R^2 values higher than 0.80 and 21 treatments with values higher than 0.90. The lowest R^2 values are found in those treatments at 15 C temperature and 10 bars of soil moisture tension and in one case at 20 C and 1.0 bar of moisture tension.

The best fit values for parameter E agreed closely with the observed mean hypocotyl length at the end of the growth period.

Influence of Soil Environmental Conditions on The Best-Fit Parameters

Each of the 48 sets of hypocotyl elongation data now has associated with it a best-fit estimate for each of the three parameters (E, B, and K). Average rate parameters were formulated as combinations of the best-fit parameters

Table 1. **Best-fit values for Von Bertalanffy's parameters and R^2 values for good-of fit of equation (II) for 48 observations.**

OBS	E	B	K	T[a]	M[a]	P[a]	R	S	W	Z	R^2
1	7.90	6.52	0.31	15	10.00	0.23	0.49	0.06	0.21	16.13	0.64
2	2.09	3.49	0.50	15	10.00	1.12	0.21	0.10	0.33	10.00	0.72
3	2.05	5.16	0.64	15	10.00	2.24	0.26	0.13	0.43	7.81	0.75
4	1.77	2.09	0.57	15	10.00	3.36	0.20	0.11	0.38	8.77	0.77
5	10.00	3.71	0.29	15	1.00	0.23	0.58	0.06	0.19	17.24	0.91
6	2.67	4.79	0.52	15	1.00	1.12	0.23	0.10	0.35	9.62	0.92
7	2.63	4.00	0.41	15	1.00	2.24	0.22	0.08	0.27	12.20	0.90
8	2.95	1.90	0.31	15	1.00	3.36	0.18	0.06	0.21	16.13	0.88
9	10.20	7.96	0.53	15	0.33	0.23	1.08	0.11	0.35	9.43	0.97
10	4.15	5.30	0.47	15	0.33	1.12	0.39	0.09	0.31	10.64	0.94
11	2.74	1.85	0.37	15	0.33	2.24	0.20	0.07	0.25	13.51	0.83
12	3.44	1.39	0.22	15	0.33	3.36	0.15	0.04	0.15	22.73	0.93
13	8.74	6.14	0.42	20	10.00	0.23	0.73	0.08	0.28	11.90	0.85
14	3.63	5.99	0.49	20	10.00	1.12	0.36	0.10	0.33	10.20	0.93
15	3.13	3.11	0.51	20	10.00	2.24	0.32	0.10	0.34	9.80	0.89
16	3.02	2.38	0.38	20	10.00	3.36	0.23	0.08	0.25	13.16	0.93
17	8.19	5.52	0.53	20	1.00	0.23	0.87	0.11	0.35	9.43	0.86
18	5.88	4.14	0.41	20	1.00	1.12	0.43	0.08	0.27	12.20	0.82
19	4.28	1.94	0.31	20	1.00	2.24	0.27	0.06	0.21	16.13	0.94
20	2.59	2.28	0.49	20	1.00	3.36	0.25	0.10	0.33	10.20	0.63
21	10.10	6.63	0.76	20	0.33	0.23	1.54	0.15	0.51	6.58	0.97
22	4.09	3.46	0.56	20	0.33	1.12	0.46	0.11	0.37	8.93	0.93
23	4.34	2.09	0.37	20	0.33	2.24	0.32	0.07	0.25	13.51	0.96
24	3.54	3.40	0.58	20	0.33	3.36	0.41	0.12	0.39	8.62	0.94
25	7.55	4.15	0.50	25	10.00	0.23	0.75	0.10	0.33	10.00	0.88
26	3.11	4.33	0.66	25	10.00	1.12	0.41	0.13	0.44	7.58	0.98
27	3.04	3.29	0.67	25	10.00	2.24	0.41	0.13	0.45	7.46	0.93
28	2.32	2.36	0.63	25	10.00	3.36	0.29	0.13	0.42	7.94	0.85
29	9.92	5.99	0.80	25	1.00	0.23	1.59	0.16	0.53	6.25	0.94
30	4.45	3.91	0.64	25	1.00	1.12	0.57	0.13	0.43	7.81	0.91
31	3.06	4.04	0.75	25	1.00	2.24	0.46	0.15	0.50	6.67	0.86
32	2.68	1.15	0.43	25	1.00	3.36	0.23	0.09	0.29	11.63	0.83
33	8.20	8.40	1.21	25	0.33	0.23	1.98	0.24	0.81	4.13	0.92
34	4.45	1.85	0.52	25	0.33	1.12	0.46	0.10	0.35	9.62	0.96
35	3.41	1.74	0.59	25	0.33	2.24	0.40	0.12	0.39	8.47	0.81
36	3.05	1.27	0.50	25	0.33	3.36	0.30	0.10	0.33	10.00	0.93
37	9.78	8.60	0.51	30	10.00	0.23	1.00	0.10	0.34	9.80	0.90
38	2.86	2.79	0.58	30	10.00	1.12	0.33	0.12	0.39	8.62	0.89
39	2.45	2.28	0.60	30	10.00	2.24	0.29	0.12	0.40	8.33	0.88
40	2.09	4.13	1.01	30	10.00	3.36	0.42	0.20	0.67	4.95	0.71
41	9.90	3.37	0.51	30	1.00	0.23	1.01	0.10	0.34	9.80	0.89
42	3.55	3.53	0.78	30	1.00	1.12	0.55	0.16	0.52	6.41	0.91
43	2.80	7.51	1.27	30	1.00	2.24	0.71	0.25	0.85	3.94	0.97
44	2.76	1.26	0.56	30	1.00	3.36	0.31	0.11	0.37	8.93	0.90
45	9.82	6.29	1.13	30	0.33	0.23	2.22	0.23	0.75	4.42	0.97
46	4.51	1.91	0.60	30	0.33	1.12	0.54	0.12	0.40	8.33	0.97
47	3.83	6.82	1.20	30	0.33	2.24	0.92	0.24	0.80	4.17	0.76
48	3.21	1.19	0.59	30	0.33	3.36	0.38	0.12	0.39	8.47	0.95

[a]T = temperature, C; M = moisture, bars; and P = mechanical impedance, kg/cm^2.

and, the relationship between these rate parameters (R, S, and W) and corresponding levels of soil temperature, soil moisture tension, and soil mechanical impedence was examined by multiple linear regression analysis. The influence of soil environment on each parameter was explained satisfactorily by three different regression models shown in Tables 2, 3, and 4. The R^2 values for fit of each regression model to parameters E, B, and K were 0.97, 0.75 and 0.82, respectively. For all temperatures, the elongation parameter (E) decreases sharply with an increase in mechanical impedance at any fixed level of soil moisture tension. The rate of change of this parameter was very slow in the mechanical impedance range 2.24 to 3.36 kg/cm^2 . Parameter R (weighted average absolute elongation rate) decreases with an increase in mechanical impedance at any given level of temperature. The maximum value for this parameter is attained at 30 C, 0.33 bars, and 0.24 kg/cm^2 and its minimum value is attained at 15 C, 10 bars, and 3.36 kg/cm^2. This parameter increases with an increase in moisture level at any given temperature.

Parameter S decreases when the temperature goes down but seems to be insensitive to mechanical impedance or soil moisture tension. The weighted mean relative elongation rate W shows sensitivity to temperature but not to moisture or mechanical impedance. Its highest value is attained at 30 C, 1.0 bars, and 2.24 kg/cm^2 and its lowest value is found at 15 C, 0.33 bars, and 3.36 kg/cm^2. One more parameter considered in this study is parameter Z which is the reciprocal of S and represents the time required for the major part of hypocotyl elongation. This parameter shows high sensitivity to temperature and increases with a decrease in soil temperature.

USE OF THE MODEL IN SIMULATION

We next developed a model to simulate average soybean hypocotyl elongation under fluctuating conditions of the soil environment. This involved substitution of the predicted values for parameters E, B, and K (obtained from regression models) into the basic Von Bertalanffy's equation, i.e.,

$$\hat{Y} = \hat{E}\ [1 + \hat{B} \exp(-\hat{K}T)] \qquad [II]$$

and thus obtaining a predicted average hypocotyl length ($\hat{Y}$) for any time under any combination of soil conditions for which the model is valid.

Simulations of hypocotyl elongation under fluctuating conditions of the soil environment were carried out using the step-input approach of Guarisma (8). Each hourly growth rate was calculated as the difference between the predicted hypocotyl length under the soil conditions prevailing at time T + 1 hr and predicted hypocotyl length at time T. Figures 1 and 2 are the result of simulations during a 10-day period for some particular combinations of soil environmental conditions showing the effect of temperature and mechanical impedance on the predicted hypocotyl elongation.

Table 2. **Multiple linear regression model to predict Parameter E in Von Bertalanffy's equation.**

Parameter	Estimate	SE
Intercept[a]	- 6.027	11.0
T (temperature, C)	3.066	1.51
M (moisture, bars)	- 0.217	0.08
P (impedance, kg/cm^2)	-26.747	4.93
T^2	- 0.160	0.06
T^3	0.002	0.001
TP	1.290	0.37
TM	- 0.009	0.01
T^2P	- 0.027	0.006
T^2P^3	0.001	0.0004
TP^3	- 0.083	0.013
P^2	6.335	1.31
TP^2	- 0.025	0.06
M^2P	0.001	0.001
M^3T	0.0001	0.0001

[a]Mean = 4.72; SD = 0.585; DF = 14; R^2 = 0.97; and n = 48.

Equation [II] was implemented by a computer program which carries out the simulation of hypocotyl elongation during a 10-day period and handles changes in environmental conditions within the range of the data use in the experiment. Input to the program consists of soil temperature (T) in C degrees, soil moisture tension (M) in bars, and soil mechanical impedance (P) in kg/cm^2. Output consists of daily elongation (cms) and elongation rate in mm/day during a 10-day growth period.

DISCUSSION

The results obtained in this study show that at any fixed level of mechanical impedance and soil moisture tension, average hypocotyl elongation rate shows an optimal response to soil temperature with higher elongation rates occurring at 30 C (Figure 1). These results agree with the findings by Hatfield and Egli (9) and by Gilman et al. (7) whose investigations have shown that soybean emergence occurred in the least amount of time when the soil temperature was in the range of 25 to 35 C. The exact temperature optimum could not be determined in the present study because of problems encountered during data collections at 32 and 35 C.

The dampening effect of soil mechanical impedance on hypocotyl elongation (Figure 2) agrees with results obtained by Arnold and Alston (1) and Garner and Bowen (3). Wanjura (13) and Vaughan (12) have observed that

Table 3. Multiple linear regression model to predict Parameter K in Von Bertalanffy's equation.

Parameter	Estimate	SE
Intercept[a]	3.82	5.60
T (temperature, C)	-0.57	0.79
M (moisture, bars)	-0.20	0.81
P (impedance, kg/cm^2)	0.54	2.73
T^2	0.03	0.03
MT	-0.04	0.09
PT	0.07	0.35
TMP	-0.005	0.003
M^2	0.04	0.04
P^2	-0.81	0.67
M^2P	0.33	0.13
M^2T	-0.20×10^{-3}	0.001
P^2M	-0.07	0.04
P^2T	0.05	0.02
T^2M	0.20×10^{-2}	0.0004
T^2P	-0.01	0.01
$M^2T^2P^2$	0.31×10^{-5}	0.36×10^{-5}
T^3	-0.54×10^{-3}	0.53×10^{-3}
P^3MT	0.54×10^{-4}	0.35×10^{-3}
P^3	0.88×10^{-2}	0.14
T^3M	-0.35×10^{-4}	0.6×10^{-4}
T^3P	0.25×10^{-3}	0.22×10^{-3}
P^3T^2	-0.36×10^{-3}	0.15×10^{-3}
M^3P	-0.31×10^{-1}	0.012
P^3T	0.44×10^{-2}	0.01×10^{-2}
P^3M	0.91×10^{-2}	0.1×10^{-1}

[a]Mean = 0.587; SD = 0.156; DF = 22; R^2 = 0.82; and n = 48.

the hypocotyl with attached cotyledons generates greater resistance to movement through the soil mass and consequently, shows higher sensitivity to soil compaction when compared to the root.

The low R^2 values obtained under high stress conditions are the result of the difficulty encountered with the reading of hypocotyls whose growth was taking place away from the plexiglass side of the box. This situation is expected to improve with more replications under the same conditions and with the use of x-rays in those cases in which the hypocotyls cannot be seen from the outside. Our observations show that, for treatments at 15 C and 10 bars, the best-fit value for parameter E in Von Bertalanffy's equation tended to exceed the observed data and did not represent a "reasonable" asymptotic value in a physiological sense. We investigated the predictions of equation [II] under high stress condition using a value of m = 0.7 and found a significant improvement of the R^2 values as compared to those obtained when m = 1.5. This suggests the possibility of using a combined model to improve

Table 4. Multiple linear regression model to predict Parameter B in Von Bertalanffy's equation.

Parameter	Estimate	SE
Intercept[a]	19.36	57.37
T (temperature, C)	- 1.72	8.09
M (moisture, bars)	- 8.71	8.37
P (impedance, kg/cm^2)	5.54	28.00
T^2	0.09	0.36
MT	0.58	0.95
PT	0.02	3.59
TMP	- 0.05	0.03
M^2	0.43	0.44
P^2	- 8.34	6.92
M^2P	2.31	1.33
M^2T	-0.91×10^{-2}	0.01
P^2M	- 0.29	0.42
P^2T	0.55	0.29
T^2M	- 0.02	0.04
T^2P	- 0.05	0.15
$M^2T^2P^2$	0.12×10^{-4}	0.36×10^{-4}
T^3	-0.16×10^{-2}	0.54×10^{-2}
P^3MT	0.26×10^{-2}	0.36×10^{-2}
P^3	0.86	1.51
T^3M	0.29×10^{-3}	0.62×10^{-3}
T^3P	0.12×10^{-2}	0.23×10^{-2}
P^3T^2	0.21×10^{-2}	0.16×10^{-2}
M^3P	- 0.21	0.12×10^{-2}
P^3T	-0.19×10^{-1}	0.92×10^{-1}
P^3M	-0.13×10^{-1}	0.10

[a]Mean = 3.90; SD = 1.60; DF = 22; R^2 = 0.75; and n = 48.

the predictive ability of the present one making it a more reliable mathematical model to simulate soybean hypocotyl elongation under stress conditions.

CONCLUSIONS

In the present study we have shown that Von Bertalanffy's equation can be adapted as a mathematical model to simulate average soybean hypocotyl elongation of cultivar 'Forest 1975' under a wide range of steady-state soil environmental conditions. The equation shows sufficient flexibility for accurate simulation and the results outlined here have exciting potential and could be extended easily to other cultivars. Furthermore, the parameters of Von Bertalanffy's equation contain important physiological information that is conveyed very conveniently.

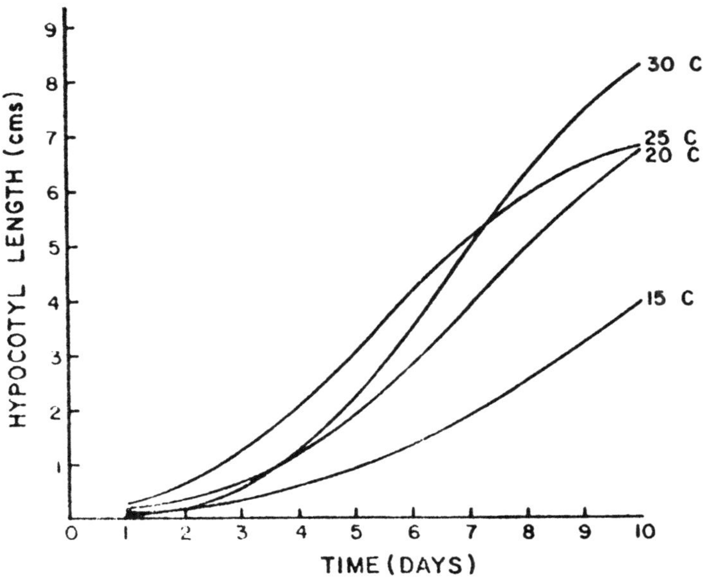

Figure 1. Growth time response for 10 bars moisture and 0.23 kg/cm^3 mechanical impedance illustrating temperature effect.

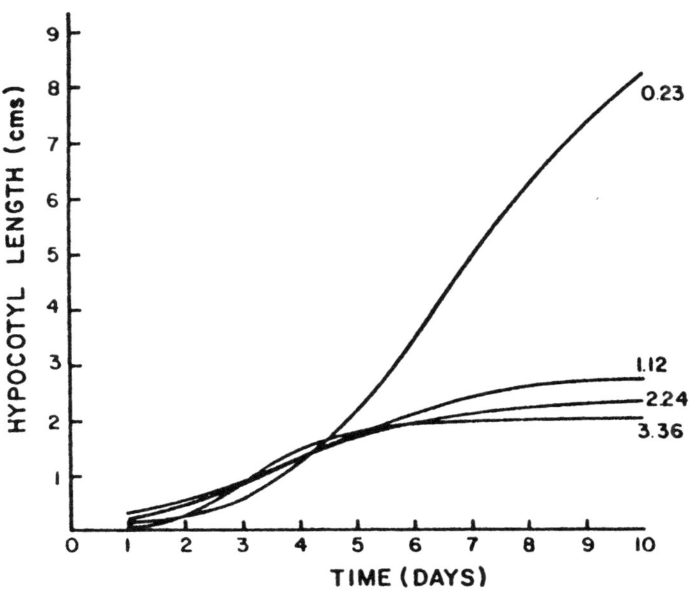

Figure 2. Growth time response for 30 C temperature and 10 bars moisture illustratting mechanical impedance effect.

NOTES

Raul Guarisma, Universidad Central de Venezuela, Maracay, Venezuela.

REFERENCES

1. Arnold, A. and R. E. Alston. 1961. Certain properties of hypocotyl of *Impatiens balsmina* reflecting physiological complexity. Plant Physiol. 36:650-655.
2. Blacklow, W. M. 1972. Influence of temperature on germination and elongation of the radicle and shoot of corn (*Zea mays* L.) Crop Sci. 12:647-650.
3. Bowen, H. D., R. F. Colwick, and D. G. Batchelder. 1973. Computer simulation of crop production: potentials and hazards. Agr. Eng. 54(10):42-45.
4. Bowen, H. D. and J. W. Hummell. 1979. Critical factors in soybean seedling emergence. In this volume.
5. Day, P. R., C. Van Bavel, V. Jaminson, H. Kohnle, J. Lutz, R. Miller, J. Page, and T. Peele. 1956. Report of the Committee on Physical Analyses, 1954055. Soil Sci. Soc. Amer. Proc. 20:167-169.
6. Gerner, T. H. and H. D. Bowen. 1966. Plant mechanics in seedling emergence. Amer. Soc. Agr. Eng. Trans. 9:650-653.
7. Gilman, D. F., W. R. Fehr, and J. S. Burris. 1973. Temperature effects on hypocotyl elongation of soybeans. Crop Sci. 12:246-249.
8. Guarisma, R. 1978. Use of the Gompertz equation for prediction of soybean hypocotyl elongation during emergence. Ph.D. Thesis. North Carolina State Univ., Raleigh, N.C.
9. Hatfield, J. L. and D. B. Egli. 1974. Effect of temperature on the rate of soybean hypocotyl elongation and field emergence. Crop Sci. 14:423-426.
10. Richards, F. J. 1959. A flexible growth function for empirical use. J. Exp. Bot. 10(29):290-300.
11. Vaughan, D. H. 1974. A simulation of cotton radicle growth during emergence. Ph.D. Thesis, North Carolina State Univ., Raleigh, N.C.
12. Von Bertalanffy, L. 1957. Quantitative laws in metabolism and growth. Quart. Rev. Bio. 32(3):217-229.
13. Wanjura, D. F., D. R. Buxton, and H. N. Stapleton. 1970. A temperature model for predicting initial cotton emergence. Agron. J. 62:741-743.

SIMULATION OF SOIL COMPACTION UNDER TRACTOR WHEELS

H. D. Bowen and H. Jaafari

World-wide there is a keen sense of urgency in finding a simple and reliable way of predicting the amount of compaction that may be expected from operation of heavy-wheeled and tracked vehicles on agricultural soils. The pressure of gaining labor efficiency is increasing the demand for more powerful and heavier tractors at a time when there is no practical way to determine whether damage is being done to the soil structure by traffic until damage has been done. Soil compaction increases draft required for tillage, limits root growth where excessive due to high mechanical impedance, reduces infiltration of water, increases time to drain the soil to field capacity, and when excessive reduces yields of all seed, forage, root, and tree crops.

Soil compaction is not a new phenomenon. Soil compaction problems have been limiting natural pasture production for centuries. The cow path or any place where animals congregate are seldom productive. Soil compaction on tilled lands was a well recognized problem in farming before the advent of the tractor. Tractor powered farming with heavy machinery has aggrevated compaction problems, but it has also made it possible to alleviate compaction by tillage.

Tractor wheel traffic is the worst offender in causing excessive compaction in modern power farming and results when pressures on the soil exceed the shear strength of the soils. The risk of excessive compaction increases as the wheel loading and draft increases, the tire pressure increases, the number of passes increases and the soil moisture increases. The larger the tractor the deeper the compaction.

Although these generalizations are helpful guidelines they are not precise enough for the farm producer-manager who must maximize labor efficiency, while slowing down the rate of accumulation of excessive compaction. The farm manager can make a wise decision on whether to purchase the larger of two tractors only if he knows the compaction resulting from operation of each of the two tractors. In the daily decisions of crop production he may have to decide whether to go to the field to harvest when the field is a little too wet or risk losing a crop from a predicted weather front that is moving into the area. He needs to know what the consequence of his decision will be. The availability of an easy-to-use, inexpensive computer simulation for calculating the compaction by different size tractors operating at various moisture contents on his soil would allow him to sharpen his ability to make decisions involving compaction.

The purpose of this chapter is to review briefly the history of past attempts to develop a practical simulation of soil compaction under wheel traffic and assess the prospects for obtaining a satisfactory simulation in the near future.

HISTORY OF SIMULATION ATTEMPTS

In the early 1950's a German engineer, Walter Söehne, made the first comprehensive study of soil compaction under tractor tires. In that study Söehne brought to bear on the problem the approaches of soil mechanics developed by civil engineers for use in predicting the soil pressures under circular and strip footings for buildings and dams. In his investigations in Germany, Söehne made tests under a variety of conditions and found that Boussinesq's equations, modified by Fröelich for different soil conditions, gave a reasonable estimate of the major principal stress, σ_1, occurring under wheel traffic in many situations (11).

Söehne made the assumption that soil compaction under wheel traffic is uniquely determined by σ_1. That assumption was justified by the soil mechanic and foundation engineering literature of the time. The relation of σ_1 to soil compaction was determined by compacting the soil in a confined compression test, which consisted of obtaining the porosity of the soil as a function of the average pressure developed by a close-fitting piston compressing soil in a cylindrical vessel.

When the piston has a small clearance with the cylinder, the soil is confined on all sides and shear stresses and lateral strains are assumed to be prevented. In this apparatus all resulting axial strain is presumed to be due to the axial stress, σ_1, imposed upon the soil by the moving piston. The intermediate and minor principal stresses, σ_2 and σ_3, are equal and are due to passive reaction of the vessel walls to the soil lateral stress generated by the axial stress, σ_1. From laboratory tests on each soil type, it is possible to construct semi-log plots of the porosity versus $\log_{10}$ pressure for several levels of soil moisture and obtain a family of curves similar to Figures 1 and 2.

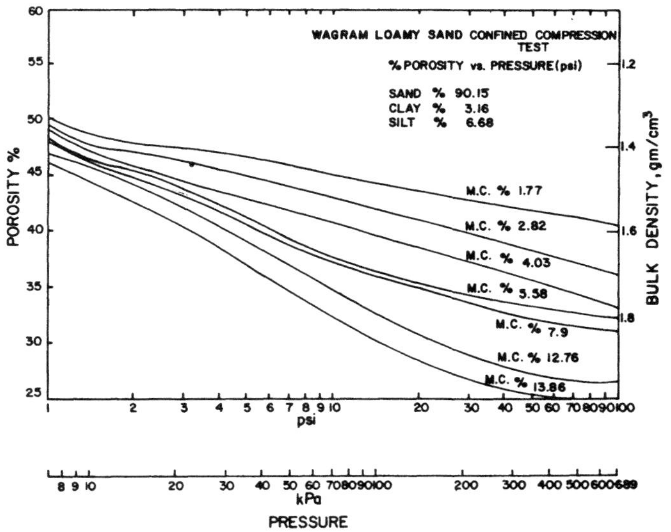

Figure 1. Semi-log plot of confined compression test of a Wagram loamy sand (8).

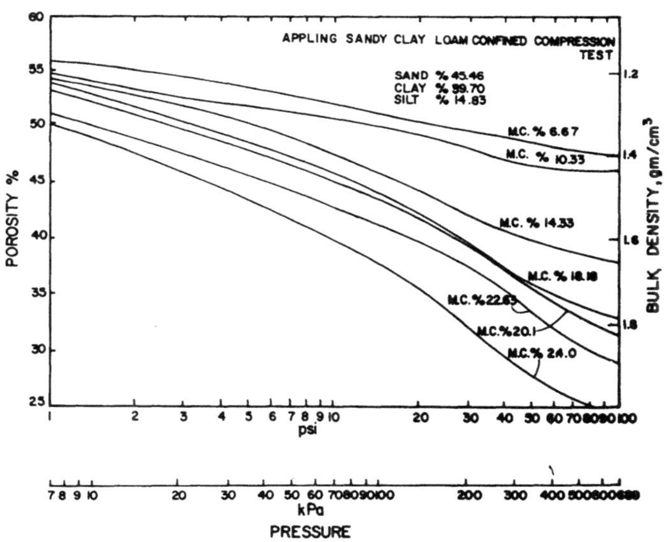

Figure 2. Semi-log plot of confined compression test of an Appling sandy clay loam (8).

The curves of porosity versus $\log_{10}$ pressure are roughly straight lines over a wide range of pressure for most soils at pressures greater than about 40 kPa. so that porosity at each moisture level is written as:

$$\eta = -A \log_{10} p + C \qquad (I)$$

where η is the soil porosity of the soil, p is the pressure from the confined compression test, A is the slope of the semi-$\log_{10}$ plot of porosity vs. p, and C is the intercept of the curve at p = 1.0. Thus σ_1 can be calculated from the wheel loading with Boussinesq's equation and related to soil compaction through the confined compression tests at any place where σ_1 uniquely determines compaction.

From his own investigations Söehne found that the Boussinesq equation provided a reasonable estimate of soil compaction at deeper depths, but gave too high values from the tire-soil interface to some distance below the interface. Söehne published his approach to simulation of soil compaction in 1958 based on his laboratory and field test results in Germany (11). In the late 1950's he came to the U.S. to work with several well known agricultural engineers and soil scientists in an effort to confirm his assumptions and approach in the best laboratories and test facilities in the U.S.

A complete confirmation of Söehne's theory of soil compaction never came, but tests showed that there were pressure distributions similar to those he was proposing. The porosity vs. $\log_{10}$ pressure relation were generally satisfactory, but several investigators were not satisfied with the assumption that σ_1 uniquely determined soil compaction. Research workers recognized the pioneering job Söehne had performed in organizing the problem into a rational simulation approach, but could not accept it as a reliable method for predicting soil compaction for agricultural purposes without further clarification.

Throughout the late 1960's and early 1970's engineers and soil scientists continued to investigate various aspects of the mechanics of soil compaction. VandenBerg (13) of the National Tillage Machinery Laboratory proposed a theory and performed tests that related soil compaction to the mean normal stress, σ_m, and the maximum natural shear strain $\overline{\gamma}_{max}$ rather than to the principal normal stress, σ_1, as assumed by Söehne. VandenBerg's equation for the bulk density, BD, was:

$$BD = A \log(\sigma_m + \sigma_m \overline{\gamma}_{max}) + B \qquad (II)$$

where A is the slope of the curve and B is the intercept of the curve on the BD axis. The mean normal stress, σ_m was given by $\sigma_m = 1/3(\sigma_1 + \sigma_2 + \sigma_3)$, and $\overline{\gamma}_{max}$ the maximum natural shear strain was determined from tests using triaxial apparatus. Dunlap and Weber (6) showed that under certaining loading paths σ_2 rather than σ_1 or σ_m controlled compaction.

Jaafari (8) and Söehne et al. (12) reported that the maximum vertical strain under a tire does not occur at the tire-soil interface where the maximum σ_1 occurs, but rather at a depth approximately half the tire width below the tire-soil interface on the centerline below the tire track. At greater depths along the centerline Boussinesq's equation gave a satisfactory pressure. Jaafari showed that for no-draft operation of three tractors on a Wagram loamy sand and an Appling sandy clay loam Boussinesq's equation could be used with a multiplier factor for each depth to predict the bulk density from the tire soil interface to the subsoil.

Chancellor and Schmidt (4) have shown that when the depth to width ratio of the tire rut is small as from operation of a wide low pressure tire on medium strength soil, the maximum vertical strain occurs at a shallower depth than when the depth to width ratio of the rut is large as from a high pressure narrow tire operating on soft soil.

Bodman and Rubin (1) have shown that confined compression tests with shear strain superimposed during compaction results in additional compaction. Davies et al. (5) and Raghavan and McKyes (9) showed that draft loads and slippage in the field results in greater compaction than when no draft and no slippage occurs.

Chancellor (3) showed that soils vary in sensitivity to slippage and shear strains. He indicated that the sensitive soils can be identified either from soil texture or from triaxial tests and developed a computer program to compute soil compaction that accounts for sensitivity of the soil to shear strain.

Bowen (2) developed a computer program to give a visual display of bulk density resulting under a tire. He used Söehne's approach with Boussinesq's equation for calculating for the pressure, σ_1, under the tire and substituted this into Equation (I) to obtain the porosity, η. The program then calculated dry bulk density BD from:

$$BD = 2.65\,(1 - \eta/100) \tag{III}$$

Figure 3 shows a computer printout of the visual display and bulk density code from a hypothetical soil. The inputs required for the simulation are indicated and except for the soil factors are readily available.

THE SIMULATION OUTPUT

What would a soil compaction simulation show you that you do not already know? A simulation would give an output similar to that in Figure 3 wherein by looking at the code printed out one can see the predicted distribution of bulk density under the tire for the first pass over a tilled soil. The simulation program will plot graphs of porosity and bulk density vs. moisture content for the soil, similar to Figures 4 and 5 for the Wagram loamy sand and the Appling sandy clay loam respectively. By selecting the bulk density

```
                SOIL PROFILE - BULK DENSITY DISTRIBUTION

FFFFFFFFFFFFFFFFFFFFFFFFFFFFFFFFFFFFFFFFFFFFFFFFFFFFFFFFFF
FFFFFFFFFFFFFFFFFFFFFFFFFFFFFFFFFFFFFFFFFFFFFFFFFFFFFFFFFF
FFFFFFFFFFFFFFFFFFFFFFFFFFFFFFFFFFFFFFFFFFFFFFFFFFFFFFFFFF
FFFFFFFFFFFFFFFFFFFFFFFFFFFFFFFFFFFFFFFFFFFFFFFFFFFFFFFFFF
LLLLLLLLLLLLLLMMMMMMMMMLLLLLLLLLLLLLLLLLLLLLLLLLLLLLLLLLLL
FFFFFFFFFFGHIKLMNNNNNNMLKIHGFFFFFFFFFFFFFFFFFFFFFFFFFFFFFF
FFFFFFFFFGHIJKLMNOOONMLKJIHGFFFFFFFFFFFFFFFFFFFFFFFFFFFFFF
FFFFFFFFFGHIJKLMMNONMMLKJIHGFFFFFFFFFFFFFFFFFFFFFFFFFFFFFF
FFFFFFFFGHHIJKLIMMNMMLIKJIHHGFFFFFFFFFFFFFFFFFFFFFFFFFFFFF
FFFFFFFFGHIIJKKLIMMMLIKKJIIHGFFFFFFFFFFFFFFFFFFFFFFFFFFFFF
FFFFFFFGGHIIJJKKKKLMLKKKJJIIHGGFFFFFFFFFFFFFFFFFFFFFFFFFFF
FFFFFFGHHIIJJJKKKKKKKKKJJJIIHHGGFFFFFFFFFFFFFFFFFFFFFFFFFF
FFFFFFGGHHIIIJJJKKKKKJJJIIIHHGGFFFFFFFFFFFFFFFFFFFFFFFFFFF
FFFFFFGGHHIIIJJJJJJJJJJIIIHHGGFFFFFFFFFFFFFFFFFFFFFFFFFFFF
FFFFFFGGHHHIIIJJJJJJJJJIIIHHHGGFFFFFFFFFFFFFFFFFFFFFFFFFFF
FFFFFFGGHHHIIIIJJJJJJIIIIHHHGGFFFFFFFFFFFFFFFFFFFFFFFFFFFF
FFFFFFGGHHHHIIIIIIIIIIIIIHHHHGGFFFFFFFFFFFFFFFFFFFFFFFFFFF
FFFFFFGGHHHIIIIIIIIIIIIIIHHHGGGFFFFFFFFFFFFFFFFFFFFFFFFFFF
FFFFFFGGGHHHHIIIIIIIIIIIHHHHGGGFFFFFFFFFFFFFFFFFFFFFFFFFFF
FFFFFFGGGHHHHHIIIIIIIIIHHHHHGGGFFFFFFFFFFFFFFFFFFFFFFFFFFF
FFFFFFGGGGHHHHHHHIIIHHHHHHHGGGGFFFFFFFFFFFFFFFFFFFFFFFFFFF
FFFFFFGGGGHHHHHHHHHHHHHHHHHHHGGGGFFFFFFFFFFFFFFFFFFFFFFFFF
FFFFFFGGGGGHHHHHHHHHHHHHHHHHGGGGGFFFFFFFFFFFFFFFFFFFFFFFFF
```

Figure 3. **Visual printout and code for left half profile of simulation of soil compaction produced by a tractor and disk harrow operating on a deep tilled soil.**

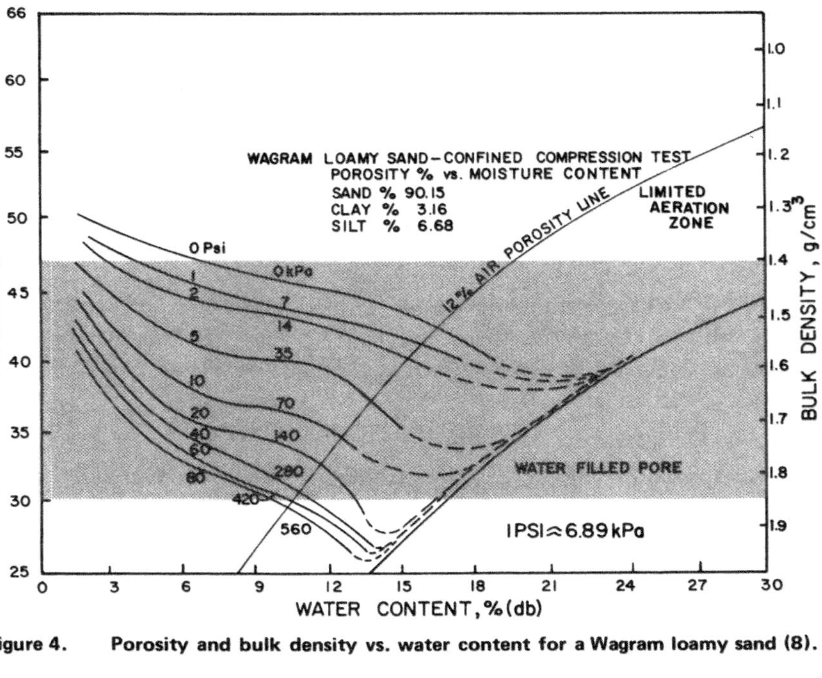

Figure 4. **Porosity and bulk density vs. water content for a Wagram loamy sand (8).**

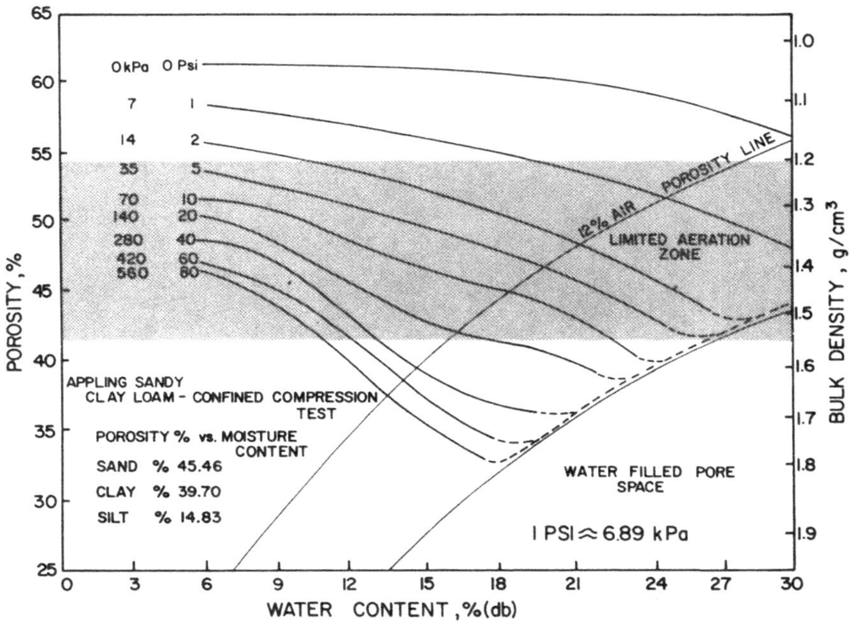

Figure 5. **Porosity and bulk density vs. water content for an Appling sandy clay loam (8).**

under a tire one can see from Figure 4 or 5 how the Wagram loamy sand or the Appling sandy clay loam is compacted. In addition curves like those in Figure 6 will be plotted showing the effect of numbers of passes of any given size tractor on soil bulk density.

Soil compaction will reduce root and top growth either by an excess of mechanical impedance or a lack of aeration. Penetrometer readings with a cone index CI, greater than 1033 to 2112 kPa (150 to 350 psi) on soil at field capacity moisture will limit root growth. From the graphs of Figures 4 or 5 one can determine how near the soil is approaching a limiting compaction. The shaded area represents the range of bulk density encountered for plant growth. Some clay soils will have a bulk density following tillage of less than 1.0. The bulk density of some sands may be as low as 1.2 following tillage but available moisture will be limiting. Compaction to 1.2 or more on clay soils and to 1.4 or more on sandy soils usually increases available moisture and does not seriously limit root impedance or aeration. However, increasing the bulk density over the range 0.8 to 1.6 reduced available water to one-third for several of silty soils in England and Wales (10). Air capacity of silty soils was limiting at bulk densities greater than about 1.2 to 1.3. Thus it

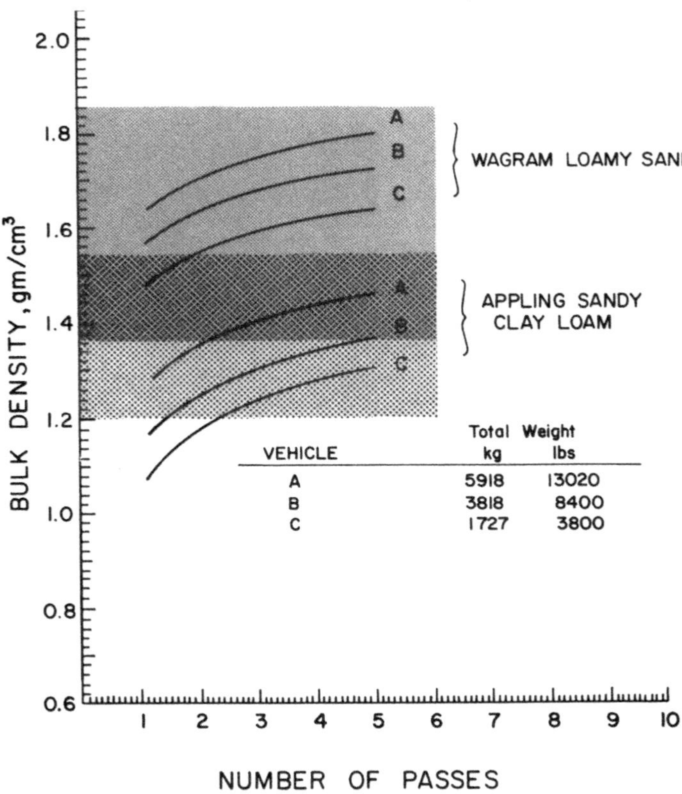

Figure 6. **Maximum bulk density in center of wheel track vs. number of passes for three tractors on two soils at near field capacity moisture (8).**

would seem that the silty soils would respond the most to management schemes of essentially zero traffic, while clays and sands benefit from moderate compaction and can tolerate relatively severe compaction if only part of the soil is compacted and hard-pans are not continuous.

The bottom edge of the shaded area of Figures 4 and 5 indicate a porosity and bulk density that will severely limit plant growth from both aeration and mechanical impedance or an interaction of them. In general it was found on Swedish soils (7) that when a soil was compacted to 200 kPa (29 psi) with one trip of a wheel across a tilled soil at 100 cm of moisture tension, about 30 to 40% of the macropore space (>30 μm equivalent diameter) was destroyed. The soil will still grow crops at this level of compaction. However a loading causing 800 kPa (116 psi) in the soil at 100 cm moisture tension resulting from one pass of the wheel was sufficient to destroy all of the macropores in most soils and result in no aeration, even though there may have been a considerable porosity in pores <30 μm equivalent diameter. The pores <30 μm would not have drained at 100 cm moisture tension.

In Figure 6 the bulk density is a linear function of the $\log_{10}$ (number passes of the tractor). If this relation holds for 100 or more passes then we can reasonably expect to account for the accumulated compaction resulting from traffic from any mix of tractor sizes and passes. For instance, the simulation of the operation of a standard two-wheel drive tractor vs. a four-wheel drive tractor of the same weight shows almost identical total compaction, but the four-wheel drive has the zone of maximum compaction at a lesser depth because the total load on each drive wheel is less and the tire is narrower. This assumes that there is about half the loading on each drive wheel of the four-wheel drive as for the two-wheel drive after weight transfer, and that tire pressures are the same. Unfortunately, there are not enough data relating soil compaction to number of passes at the 100 or more pass level to validate this relation. This is only one possible way to account for accumulated compaction.

Heavy wheel loads compact more and to deeper depths than light loads as shown by the data of Figure 7. Here we can see that tractor A weighing about 3.4 times as much as tractor C has a penetrometer cone index, CI,

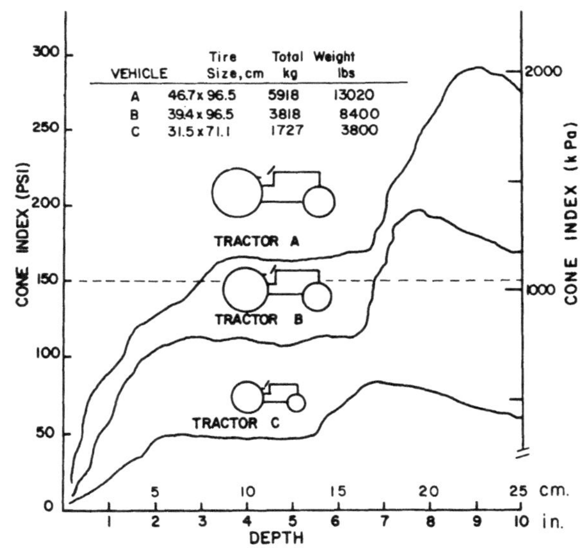

Figure 7. **Penetrometer Cone Index, CI, vs. depth in center of the wheel tracks of three tractors of different weight after one pass on an Appling sandy clay loam near field capacity (8).**

which is about 3 times that of tractor C. Also we can see clear evidence of the peak resistance being at greater depths for the larger tractors.

ASSESSMENT OF PROSPECTS FOR A SIMULATION FOR GENERAL USE

Technically we can provide the computer programs to develop the kind of information shown in this chapter at a reasonable cost and its use is well within the computer facilities available to farmers in most areas even now. However, there is still the problem of implementing a program that will provide the soil parameters that are needed.

The computer simulation is only as good as the soil compaction data provided and the most useful simulations would come from making a confined compression test on the soils on each farm. However, it seems likely that availability of confined compression data on a few dozen soil types with appropriate criteria for matching such as (a) a mechanical analysis of the soil, (b) known sensitivity of soil to packing, and (c) the presence of an easily compacted layer, would allow simulations that would greatly improve our ability to manage compaction of our soil.

The primary objective of the plant-soil dynamics subcommittee of the American Society of Agricultural Engineers ASAE PM 45/5 is to make available a soil compaction simulation with confined compression test data on some 50 soils in the U.S. and world. This would be published in the ASAE yearbook. The ASAE PM 45/5 subcommittee is composed of agricultural engineers and soil scientists who are keenly interested in providing a valid simulation. A reasonable estimate of the normal time to assemble and document such a program would be 5 years, but with the national emphasis on energy-saving and concern for structure damage in the soil this program may possibly be made available in less time. In the meantime if sufficient interest is shown much of the simulation could be made available to research and extension personnel within a year of this publication.

NOTES

H. D. Bowen, Department of Biological and Agricultural Engineering, N. C. State University, Raleigh, North Carolina 27650; and H. Jaafari, University of Technology and Agricultural Production, Isfahan, Iran.

REFERENCES

1. Bodman, G. and J. Rubin. 1948. Soil Puddling. Soil Sci. Amer. Proc. 13:27-36.
2. Bowen, H. D. 1975. Simulation of soil compaction under tractor-implement traffic. ASAE Paper No. 75-1569, St. Joseph, MI.
3. Chancellor, W. J. 1966. Combined hypotheses for anticipating soil strains beneath surface impressions. Trans. ASAE 9(6):887-892, 895.
4. Chancellor, W. J., and R. H. Schmidt. 1962. Soil deformation beneath surface loads. Trans. ASAE, 5:240-246, 249.

5. Davies, D. B., J. B. Finney, and S. J. Richardson. 1973. Relative effects of tractor weight and wheelslip in causing soil compaction. J. Soil Sci. 24(3):399-409.
6. Dunlap, W. H., and J. A. Weber. 1971. Compaction of an unsaturated soil under a general state of stress. Trans. ASAE 14(4):601-607, 611.
7. Ericksson, J., I. Hakansson, and B. Danfors. 1975. The effect of soil compaction on soil structure and crop yields. Swedish Institute of Agric. Eng. Bulletin 354. Uppsala, Sweden. (English translation by J. C. Aase, USDA, ARS, Sidney, Montana.)
8. Jaafari, H. 1978. Simulation of soil compaction under tractor-traffic. Ph.D. Thesis, N. C. State Univ., Raleigh, NC.
9. Raghavan, G. S. V., and E. McKyes. 1977. Laboratory study to determine the effect of slip-generated shear on soil compaction. Can. Agr. Eng. 19(1):40-42.
10. Reeve, M. J., P. D. Smith, and A. J. Thomasson. 1973. The effect of density on water retention properties of field soils. J. Soil. Sci. 24(3):355-367.
11. Söehne, W. 1958. Fundamentals of pressure distribution and soil compaction under tractor tires. ASAE Journ. 39:276-281, 290.
12. Söehne, W., W. J. Chancellor, and R. H. Schmidt. 1962. Soil deformation and compaction during piston sinkage. Trans. ASAE 5:235-239.
13. Vanden Berg, G. E. 1962. Triaxial measurements of shearing strain and compaction in unsaturated soil. ASAE Paper No. 62-648, St. Joseph, MI.

RAW MATERIAL AND SOYBEAN OIL QUALITY

T. L. Mounts

The current U.S. grades of soybeans (Table 1) give little attention to the quantity or quality of the oil and protein obtained from the crop, although grade levels of splits, damaged kernels, and foreign material measure factors that reduce quality. In both domestic and export trade, the most common grade of bean is the U.S. #2. This derives strictly from economic factors with little regard for the quality of products.

Our food system flows through three sectors—inputs (such as land), farm operations, and product marketing—to the fourth and final sector, the consumer. The third sector can be more completely described as encompassing all facets of post-harvest handling, storing, processing, and distribution. However, there is some overlap from those immediately preceding and following sectors. Pre-harvest factors in production of the crop certainly affect the quality of the crude oil and processing efficiency. At the same time, the price, availability, and quality (esthetic and organoleptic, as well as nutritive) of food for the consumer are functions of post-harvest technology.

PRE-HARVEST EVENTS AFFECT OIL QUALITY

While our many farm crops are growing, a multitude of factors play a role in the successful harvest of the crop. A dramatic example occurred in 1971 throughout much of the southeastern portions of the United States. Because wet weather delayed harvest, mature soybeans remained in the fields for prolonged periods, and much of the crop was damaged. When the damaged soybeans were placed in storage at high moisture levels, the damage became progressively worse.

Table 1. **Grades and requirements for soybeans.**[a]

U.S. Grade	Test Wt, lb/Bushel	Moisture, %	Splits, %	Damaged Kernels, % Total	Damaged Kernels, % Heat Damaged	Brown-Black Beans, %	Foreign Material, %
1	56	13	10.0	2.0	0.2	1.0	1.0
2	54	14	20.0	3.0	0.5	2.0	2.0
3	52	16	30.0	5.0	1.0	5.0	3.0
4	49	18	40.0	8.0	3.0	10.0	5.0

[a]Official grain standards of the United States (1).

Robertson et al. (2) reported that as the severity of damage increased, the percent free fatty acid (FFA) and Lovibond color increased while the phospholipid (P) content of the oil decreased. The authors suggested that the phospholipids were probably destroyed during storage by enzymatic action. Refiners reported that oil extracted from field-damaged beans contained non-hydratable phospholipids that were not easily removed by degumming or refining. Such oil formed stable emulsions during refining, and high refining losses were experienced. The high FFA content of the oil from damaged beans was of greatest concern to processors. Oil refining losses rose from a normal of 1 to 1½% to over 4%.

Finished salad oils prepared from field-damaged beans had poor flavor stability, and the residual phospholipids were suspected of contributing to the flavor deterioration. However, subsequent research (3), indicated that increased iron (Fe) content of the oil caused the flavor deterioration. Iron acts as a catalyst for the oxidation of the oil. The increased iron content, which parallels an increase in FFA content caused by glyceride hydrolysis in the damaged beans, arises from natural iron in the bean and from attack of the fatty acids on iron surfaces and on particulate iron. As shown in Table 2, the oxidation value of the crude oil from damaged soybeans was increased substantially over that of oil from sound beans (4). Peroxide value (PV) gives a measure of hydroperoxides present in the oil, while the anisidine value (AV) is a measure of secondary oxidation products. The AV plus two times the PV yields the oxidation value of the oil.

Comparative studies were performed to determine the effects of processing upon the anisidine reactive material in sound and field-damaged beans (Fig. 1). Relatively little increase in AV occurs during degumming or refining of either sound or field-damaged soybean oils. Bleaching raised AV in both oils but to a lesser extent in sound oil. Deodorization had a markedly different effect on the two types of oils. Anisidine values always decreased during deodorization of sound oils, whereas they sometimes increased during deodorization of field-damaged oils.

Table 2. Effect of iron on oxidation of crude soybean oil.

Bean Condition	Iron Content, ppm	Peroxide Value	Anisidine Value	Oxidation Value
Sound	1.2	0.5	0.4	1.4
Sound	1.5	0.1	0.5	0.7
Sound	2.8	0.0	0.4	0.4
Sound	2.8	0.8	0.5	2.1
Damaged	2.0	0.2	3.4	3.8
Damaged	3.9	1.2	0.4	2.8
Damaged	4.0	1.0	3.8	5.8
Damaged	6.0	1.5	1.8	4.8

The final measure of a finished salad oil, of course, was its flavor. Processed oils were evaluated organoleptically by a 20-member analytical type taste panel. Panelists were asked to score both the odor and flavor intensity of the oils on a 1 to 10 scale. Specific flavors or odors were scored on a 1 to 3 scale. Freshly deodorized soybean salad oils from sound beans usually score 7 to 8 on overall intensity. Although this analytical or difference type panel cannot indicate consumer acceptance, a rating below 6 is generally indicative of poor quality.

Damaged oils were processed to finished products by laboratory simulations of commercial procedures (5). Alternative refining parameters were evaluated to determine which were the most effective for yielding improved

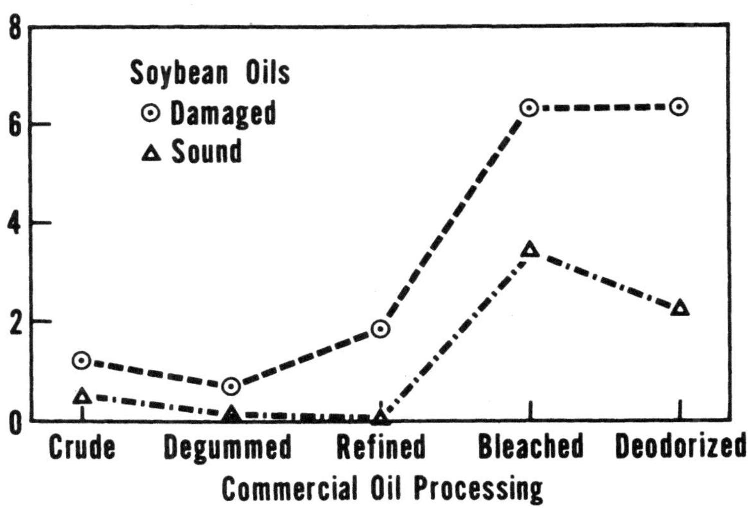

Figure 1. Effect of processing on anisidine values (4).

oils. Crude oils were degummed with water, acetic anhydride, or phosphoric acid; degummed oils were single or double refined; refined, washed oils were vacuum bleached with activated clay or activated carbon; deodorization was performed at 210 C, 230 C, or 260 C. Results of the most significant experiments are shown in Table 3. Deodorization of the oils at 210 C yielded products having low flavor scores. Increasing the temperature of deodorization to 260 C gave oils with flavor scores above 7. The results of these studies showed that elevated deodorization temperature was the most important refining factor improving the quality of salad oils from damaged beans.

POST-HARVEST HANDLING AFFECTS OIL QUALITY

Recent studies (6) indicate that post-harvest handing of the bean can affect the extracted crude oil quality. Shipments were sampled at the origin and destination. Individual samples were combined after grading, and the combined samples were fractionated to give composite, whole bean, and split bean fractions for each sampling site. Crude oil was extracted from each fraction by laboratory simulations of commercial practice. Results of critical oil quality determinations (Fe, FFA, PV, and P) for a shipment of soybeans from New Orleans, La., to Tilbury, England, are presented in Table 4. The analyses of extracted oils for Fe, FFA, and PV show that the composite and split bean fractions deteriorated during shipment. There was no in-transit deterioration in these oil quality factors with the whole bean fraction. Oil extracted from split bean fractions showed significant differences in oil quality factors relative to whole bean oil. Analysis of the oil extracted from whole and split bean fractions of soybeans sampled from local elevators (Table 5) showed significant differences in the results for Fe, FFA, and PV and confirmed the findings with the exported beans. Split beans appear to influence greatly the quality of oil, as reflected by increased levels of Fe, which is a catalyst for oxidation of soybean oil, and of FFA.

Table 3. Mean flavor scores compared by degumming method and deodorization temperature.

Degumming Method	Mean Flavor Scores and Sig[a] Deodorization Temperature 210 C	260 C
Water	4.80	7.17
Acetic anhydride	5.59	7.11
Phosphoric acid	4.82	7.16
Mean	5.07	7.14

[a]LSD (least significant difference) between degumming methods = 0.45 flavor score unit. LSD between deodorization temperatures = 0.78 flavor score unit.

Table 4. Soybean oil quality evaluation[a] during post-harvest handling.

Fraction	Fe, ppm	FFA, %	PV, meq/kg	Phosphorus (ppm)	
				Crude	Degummed
New Orleans (barge-train)					
Composite	0.4 ± 0.04[b]	0.7 ± 0.3[b]	0.9 ± 0.0[b]	442 ± 11.4[b]	37 ± 1.1[b]
Wholes	0.7 ± 0.4	1.0 ± 0.3	1.2 ± 0.1		
Splits	1.2 ± 0.5	1.1 ± 0.6	2.2 ± 1.0		
New Orleans (vessel)					
Composite	0.5 ± 0.4	1.1 ± 0.5	0.7 ± 0.1	500 ± 10.0	40 ± 1.1
Wholes	0.5 ± 0.2	0.7 ± 0.5	1.3 ± 0.5		
Splits	1.9 ± 0.4	1.4 ± 0.0	1.7 ± 0.9		
Tilbury					
Composite	2.5 ± 0.9	1.7 ± 0.3	2.8 ± 1.0	359 ± 11.0	183 ± 63.9
Wholes	0.4 ± 0.1	0.9 ± 0.7	1.8 ± 0.7		
Splits	2.5 ± 0.2	1.7 ± 0.2	1.9 ± 1.0		

[a]All values are the average of triplicate analyses.

[b]Standard deviation.

Table 5. Oil quality evaluation[a] of soybeans from local elevators.

Sample	Iron, ppm	FFA, %	Peroxide Value
Trivoli, Illinois			
Wholes	0.4 ± 0.2[b]	0.6 ± 0.3[b]	0.3 ± 0.3[b]
Splits	1.1 ± 0.6	1.0 ± 0.4	2.8 ± 0.6
Edwards, Illinois			
Wholes	0.3 0.1	0.6 ± 0.2	1.0 ± 0.2
Splits	1.4 ± 0.3	1.9 ± 0.9	3.2 ± 0.9
Farmer City, Illinois			
Wholes	0.9 ± 0.1	0.7 ± 0.2	1.5 ± 0.4
Splits	2.8 ± 0.4	1.2 ± 0.2	3.6 ± 0.9

[a] All values are the average of triplicate analyses.

[b] Standard deviation.

The phosphorus content of the crude oil (Table 4) from the exported beans sampled at the origin was normal. Removal during water degumming (approximately 90% removal of phosphatides) was adequate. Oil from the beans at destination showed deterioration as indicated by the low phosphorus content of the crude oil. Degumming of the oil failed to lower the phosphorus content below 183 ppm. Thus, as with the field-damaged beans discussed earlier, nonhydratable phosphatides were formed during export shipment of the beans. Foreign processors have long contended that they could not achieve adequate phosphatide removal by simple water hydration. The evidence of this study indicates that phosphatides deteriorate during shipment of the beans and explains the poor phospholipid removal experienced by foreign processors.

Investigations are continuing to ascertain what specific factors of export shipment contribute to phosphatide deterioration and formation of nonhydratable phosphatides. The data on oil from split beans, however, clearly show that improved methods of handling to minimize bean breakage could contribute to improved oil quality and lower refining losses to the processor.

NOTES

T. L. Mounts, Northern Regional Research Center, Agricultural Research, SEA, USDA, Peoria, Illinois 61604.

The mention of firm names or trade products does not imply that they are endorsed or recommended by the U.S. Department of Agriculture over other firms or similar products not mentioned.

REFERENCES

1. Official Grain Standards of the United States, Revised to 1970, USDA Consumer and Marketing Service, Grain Division, Washington, D.C.

2. Robertson, J. A., W. H. Morrison, III, and D. Burdick. 1973. Chemical evaluation of oil from field- and storage-damaged soybeans. JAOCS 50:443.
3. Evans, C. D., G. R. List, R. E. Beal, and L. T. Black. 1974. Iron and phosphorous contents of soybean oil from normal and damaged beans. JAOCS 51:444.
4. List, G. R., C. D. Evans, W. F. Kwolek, B. K. Boundy, and J. C. Cowan. 1974. Oxidation and quality of soybean oil: A preliminary study of the anisidine test. JAOCS 51:17.
5. List, G. R., C. D. Evans, K. Warner, R. E. Beal, W. F. Kwolek, L. T. Black, and K. J. Moulton. 1977. Quality of oil from damaged soybeans. JAOCS 54:8.
6. Mounts, T. L., G. R. List, and A. J. Heakin. 1979. Post-harvest handling of soybeans: Effects on oil quality. JAOCS (In press).

NUTRITIONAL ASPECTS OF SOYBEAN OIL UTILIZATION

E. A. Emken

A total of 17.5 billion lbs of fat and oil was consumed in the U.S. during 1976 (1). This amount equals a daily per capita intake of 0.35 lbs (157 g). The contribution of vegetable oil to the total dietary fat consumed has continually increased from 16.8% in the early 1900's to 43.3% in 1976. Visible fat alone contributed 12.4 billion lbs to the American diet in 1976. The contribution of various fats to this total is given in Figure 1, which shows that soybean oil provides almost 60% or 7.3 billion lbs of visible fat (20). The contribution of partially hydrogenated soybean oil to the total yearly per capita consumption is estimated to be 24.1 lbs (Figure 2) (33). This estimated value also includes a small amount of other hydrogenated vegetable oils.

Since 40 to 50% of the total calories in the American diet are contributed by fats, the dietary and nutritional importance of soybean oil and partially hydrogenated soybean oil are great. Partially hydrogenated soybean oil supplies 10 to 15% of our daily calories. Soybean oil is high in polyunsaturated fatty acids and has a high vitamin E content. In a properly refined and deodorized oil, there are no problems with contaminants such as herbicides, pesticides, microorganisms, or heavy metals.

One possible area for concern is the reported toxicity of cyclic monomers, dimers, and oxidation products formed in heated soybean oil (8,44,45). However, only severely abused oils, which no longer have good flavor and odor characteristics, contain appreciable quantities of these dimers and oxidation products.

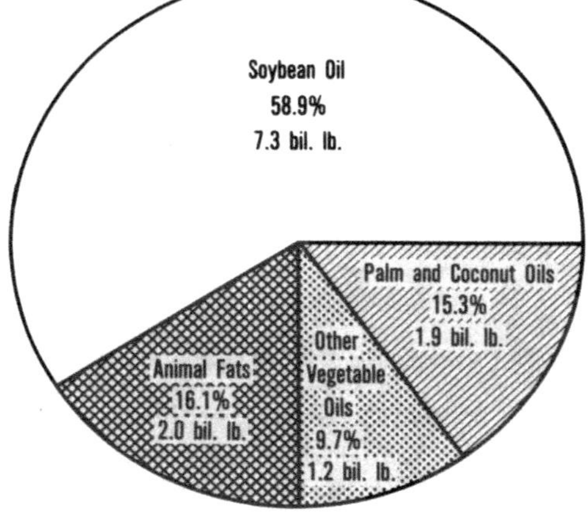

12.4 billion pounds total fat

ERS/USDA Statistical Bulletin 627, May 1976

Figure 1. Sources of visible dietary fat and oil consumed in the U.S. during 1976.

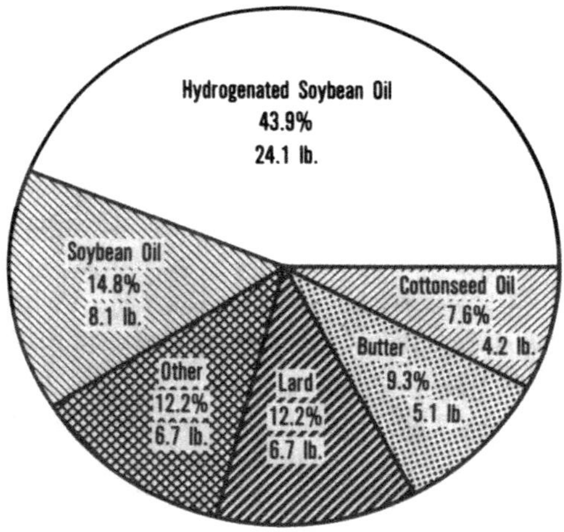

Total visible fat consumption 54.9 lb./capita

Fats and Oil Situation, FOS-281, ERS/USDA, February 1976

Figure 2. Sources of visible dietary fat consumed in lbs/capita/year in the U.S. during 1974.

NUTRITIONAL ROLE OF FATS IN METABOLIC DISORDERS

A variety of common and rare disorders known to involve fats and fat metabolism are listed in Table 1. In many cases, dietary fats are not properly absorbed, utilized, oxidized, or removed from cell tissue. These problems are normally the result of inherited metabolic defects that cause a multitude of clinical symptoms. In some diseases, the problem can be traced to a missing enzyme, or alternation of lipid-controlling feedback mechanisms. In any case, these symptoms illustrate the importance of the role fats play in our general well-being.

Non-genetic disorders such as essential fatty acid deficiency are non-existent in the normal population. Essential fatty acid requirements are easily met by the American diet, partly because soybean oil is a rich source of linoleic acid.

The possible involvement of fats and, in particular, hydrogenated soybean oil in coronary heart disease and cancer has caused much concern and resulted in many conflicting scientific reports (11, 13, 18, 21, 24, 32, 38, 43, 48). However, at present, the question of whether hydrogenated soybean oil contributes to the etiology of diseases such as cancer or coronary heart disease cannot be answered with any certainty.

FATTY ACID ISOMERS PRESENT IN PARTIALLY HYDROGENATED SOYBEAN OIL

The controversy concerning hydrogenated soybean oil is the result of the isomeric fatty acids formed during hydrogenation. Both geometric (*trans*) and positional isomers are found in hydrogenated soybean oil. The general isomeric fatty acid content of hydrogenated soybean oil is shown in Table 2 (27).

Table 1. Disorders that involve fat metabolism.

Abeta lipoproteinemia	Farber's disease
Acrodermatitis enteropathica	Fucosidosis
Allergis enteritis	Gaucher's disease
Atherosclerosis	Hepatosplenomegaly
Bilary disease (cholelithiasis)	Hyperlipoproteinemia
Cancer	Kinky hair disease
Colliac disease (sprue)	Krabbe's disease
Coronary heart disease	Metachromatic leukodystrophy
Cystic fibrosis	Multiple sclerosis
Diabetes mellitus	Niemann-pick disease
Diarrhoea	Obesity (adiposity)
Diverticula	Steatorrhoea
Essential fatty acid deficiency	Tay-sachs disease
Fabry's disease	

Table 2. Fatty acid composition (%) of soybean oil (SBO) and partially hydrogenated soybean oils[a].

Fatty Acid[b]	SBO	HSBO-1[c]	HSBO-2[c]	HSBO-3[c]
16:0	11	11	11	11
18:0	4.1	4.3	7	10.5
c-18:1[d]	22	29	33	18
t-18:1[d]	--	12	12	51
9c,12c-18:2	54	31	22	--
9c,12t- and 9t,12c-18:2	--	4	6	--
9t,12t-18:2	--	--	--	--
9c,13t- and 8t,12t-18:2	--	2	4	9
Conjugated isomers	--	2	0.5	--
18:3 (all isomers)	7.5	2.3	2.0	--
>20:0	1	1	1	1

[a]Data from Ref. 27.
[b]Geneva numbering system used for fatty acid abbreviations.
[c]HSBO = partially hydrogenated soybean oil.
[d]Includes positional octadecenoic acid isomers.

Figures 3 and 4 give a detailed analysis of the distribution and relative amounts of the positional monoene isomers in commercially hydrogenated soybean oil products (42, 49).

ABSORPTION OF FATTY ACID ISOMERS

Soybean oil and hydrogenated soybean oil are well over 90% absorbed in normal human subjects (12, 15, 17). The data in Table 3 indicate that the melting point of the fat is the main factor that effects its absorption. Melting points greater than 50 C are accompanied by decreased digestibility. Tristearin, for example, has an mp of 70 C and a digestibility of 19%.

Recent studies with deuterated tri-*trans*-12-octadecenyl glycerol (mp 52 C), tri-*cis*-12-octadecenyl glycerol, tri-*trans*-9-octadecenyl glycerol, and tri-*cis*-9-octadecenyl glycerol showed that all four fats were equally well absorbed by humans provided the high melting *trans*-12-octadecenyl glycerol was first melted and then emulsified with a mixture of sugar, sodium caseinate, and water to inhibit crystal formation (14,15,17).

TRANS FATTY ACID CONTENT OF HUMAN TISSUES

Analysis of human lipids confirms that the *trans* fatty acid isomers are absorbed and incorporated (23, 25, 29, 31, 40, 46). The *trans* fatty acid contents of various human tissue are summarized in Table 4.

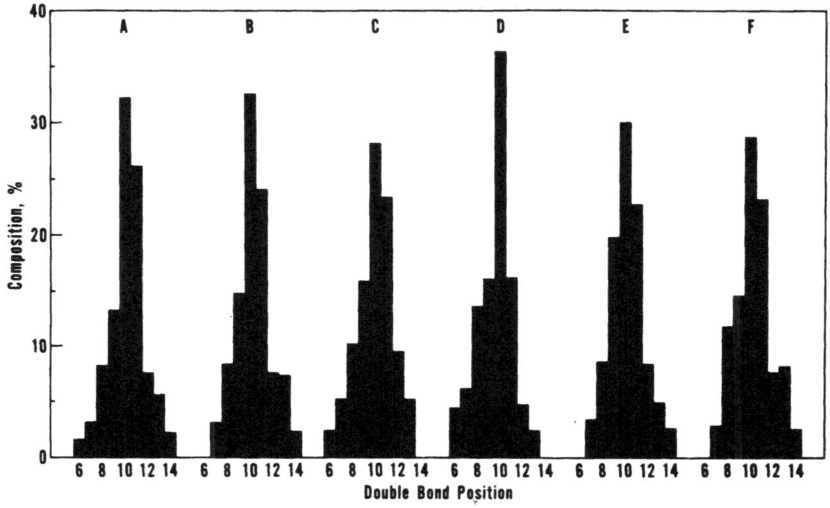

Figure 3. Distribution of positional isomers in the *trans*-octadecenoic acid fraction from commercial samples of margarine and shortening.

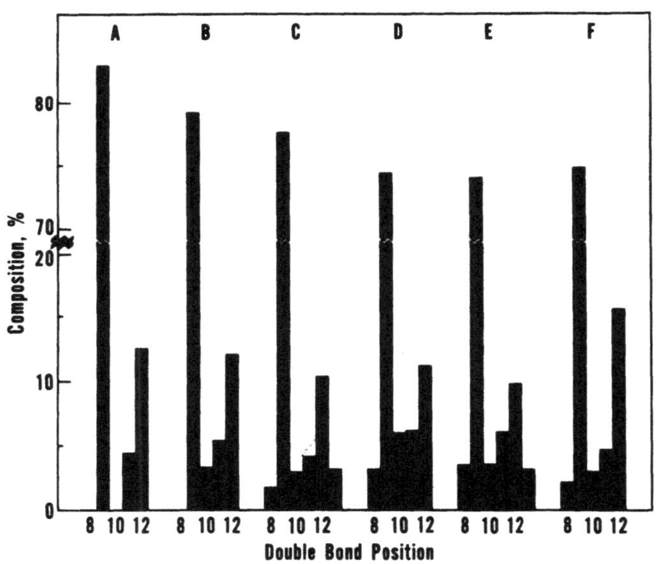

Figure 4. Distribution of positional isomers in the *cis*-octadecenoic acid fraction from commercial samples of margarine and shortening.

Table 3. Digestibility of fats[a].

Fat or Oil	Melting Point, C	Coefficient of Digestibility	
		Human	Rat
Mutton	50	88.0	84.8
Hydrogenated cottonseed	54	--	68.7
Hydrogenated cottonseed	46	94.9	83.8
Hydrogenated peanut	50	92.0	--
Hydrogenated peanut	52.4	79.0	--
Hydrogenated corn	43	95.4	--
Hydrogenated soybean	35	96.7	97.0
Soybean	--	--	98.5
Lard	37	--	96.6
Butter	34.5	--	90.7
Tripalmitin	66.5	--	27.9
Tristearin	70	--	18.9

[a]Data from Ref. 12.

Table 4. Percent *trans* in human tissues.

Tissues	*trans* (%)
Adipose	2.4-12.2
Liver	4.0-14.4
Fetal liver	0.0-0.5
Heart	4.6-9.3
Heart	1.0-8.2
Aorta	2.3-8.8
Aorta	2.0-8.0
Aorta	0.67-1.53
Maternal fat	1.5-6.8
Myocardium	0.27-1.53
Jejunum	0.38-1.21
Sebum	1.0-1.7
Milk	2-4
Serum	2.7
Erythrocytes	3.1
Serum sphingomyelin	4.6

Considering that the *trans* fatty acid intake has been estimated at 8% of the total fat (34), it is surprising that higher *trans* fatty acid levels are not found. Certainly, there is no evidence for the accumulation of *trans* isomers. This observation is consistent with studies on humans feed deuterated fatty acid isomers (14-17). In these studies, both deuterium-labeled elaidic acid and oleic acid were completely removed from essentially all of the plasma lipids within 48 to 72 hr after feeding. Also similar turnover rates were found for 12t-18:1 and 12c-18:1, which were completely removed from plasma lipids within 48 hr after feeding.

STUDIES WITH PARTIALLY HYDROGENATED SOYBEAN OIL

Multiple generation feeding studies in rats have been conducted by several laboratories to assess the long-term consequence of partially hydrogenated soybean oil (3, 22, 54). The data in Table 5 are typical of the results using this approach. This study with partially hydrogenated soybean oil containing 35% *trans* fatty acids showed no statistically significant effect on the growth, weight, reproduction, and longevity of rats. Pathological examination showed no identifiable difference between control and experimental rats.

Partially hydrogenated soybean oil has been fed to a variety of other species including humans in numerous experiments. In these studies, the goal was to correlate increases in serum cholesterol, triglyceride or phospholipid levels with the *trans* fatty acid content of the diet. The results have been variable and seem dependent upon the species as well as the experimental design. Results from several studies involving partially hydrogenated vegetable oil fed to humans are shown in Table 6 (2, 5, 6, 7, 9, 10, 19, 26, 28, 36, 37, 39, 52, 53). Where normal diets are fed, the results indicate essentially no difference between partially hydrogenated and non-hydrogenated oil.

Table 5. Multigeneration experiment with rats fed fats containing 35% *trans* fatty acids[a].

		Wt at 90 Days (g)	
Fat Level (%)	Generation	Males	Females
11.2	40	290	200
	75	310	212
16.0	5	283	219
	25	278	209
Purina rat chow		280	210

[a]Data from Ref. 3.

Table 6. Effects of *trans* acids or partially hydrogenated oils on human serum lipid levels.

Hydrogenated Oil or Isomeric Fat	Change in Serum Level		
	Triglyceride	Phospholipid	Cholesterol
Peanut	N.D.[a]	N.D.	Increase
Cottonseed	Increase	Increase	Increase
Corn	Increase	Increase	Increase
Corn	N.D.	N.D.	Increase
Sunflower	Increase	No change	Increase
Margarines (2 brands)	N.D.	N.D.	Increase
Margarines (6 brands)	N.D.	N.D.	No change
Soybean	N.D.	N.D.	No change
Soybean	No change	No change	No change
t, t- and c, t-18:2	Increase	N.D.	Increase
44% *trans* fatty acids	No change	N.D.	No change
34% and 37% Elaidic acid	N.D.	N.D.	Increase

[a]N.D. = not determined.

An insight into the metabolism and utilization of isomeric fatty acids can be obtained by determining the distribution of isomeric fatty acids in tissue from animals fed diets containing known amounts of *trans* or isomeric fatty acids. This approach is used to determine if the *trans* or isomeric fatty acid is distributed selectively into specific tissues or lipid fractions in comparison to the normal *cis* fatty acids. A refinement of this approach is to use enzymatic studies in order to understand why fatty acid selection occurs and what is the likely consequence of this action. Many investigations using these approaches have shown very definite differences in the enzymatic and metabolic steps which control the utilization of the various isomeric fatty acids.

ISOMERIC FATTY ACID INCORPORATION INTO TISSUE LIPIDS

Selectivity values for the distribution of elaidate vs. oleate into human plasma lipid fractions are given in Table 7 (15,16). Selectivity values calculated from chicken, rat, and rabbit studies are in general agreement with values from human studies, although there is some variation (41, 50).

Relative selectivities for the preferential incorporation or discrimination of elaidate vs. oleate into tissue lipids are calculated by dividing the elaidate/oleate ratio fed by the elaidate/oleate ratio found in the tissue. Selectivity values less than one indicate discrimination against elaidate; conversely, values larger than one indicate a preference for elaidate.

In general, phospholipid fractions show a preference for incorporation of elaidate. A particularly strong selectivity was found for elaidate incorporation into the 1-acyl position (PC-1) of phosphatidyl choline. This general

Table 7. Selectivity values for elaidate vs. oleate incorporation into lipid fractions.

Lipid[a]	Human (Plasma)	Chicken (Yolk)	Rat (Liver)	Rabbit (Serum)
TG	0.90	0.69	0.93	1.31
FFA	1.31	--	--	--
CE	0.19	--	--	0.61
PE	1.96	--	1.96	--
PS	1.39	--	1.49	--
PC	1.34	1.13[b]	1.80	3.06[b]
PC-1	3.68	--	--	--
PC-2	0.68	--	--	--
SM	1.99	--	--	--
LPC	0.83	--	--	--

[a]TG = triglycerides; FFA = free fatty acids; CE = cholesteryl esters; PE = phosphatidyl ethanolamine; PS = phosphatidyl serine; PC = phosphatidyl choline; PC-1 = 1-acyl phosphatidyl choline; PC-2 = 2-acyl phosphatidyl choline; SM = sphingomyelin; LPC = lyso phosphatidyl choline.
[b]Total phospholipids.

preference was also noted in laying hen, rat, and rabbit phospholipids. The cholesteryl ester fractions from human and rabbit studies show a strong discrimination against elaidate incorporation. These data suggest that the fatty acid composition of tissue lipid fractions is regulated in order to control their physical and biochemical properties. The data in Table 7 are also in agreement with in vitro experiments and suggest that elaidate is utilized similarly to a saturated fatty acid (35). The discrimination against incorporation of elaidate into cholesteryl esters suggests that elaidate may not have a major effect on cholesteryl ester fatty acid composition.

The data in Table 8 list the selectivity values for the 12t-18:1 and 12c-18:1 positional isomers compared to 9c-18:1 (14). The strong selectivities for incorporation of these isomers were found in both human plasma and rat liver phospholipids. The low selectivities for the triglyercide fractions probably reflects the preferential removal of the isomers from the triglyceride pool for use in the phospholipids. The similarity of the human selectivity values to rat, rabbit, and laying hen data in Tables 7 and 8 indicates results from previous animal studies are applicable to human subjects. Probably because of the difference in species, the magnitude of some of the selectivity values varies considerably. Thus, in spite of differences, animal studies provide useful guidelines, but they should be confirmed by human studies.

ESSENTIAL FATTY ACID REQUIREMENTS

Studies of human populations have shown that relatively small amounts of fats are necessary to satisfy nutritional requirements (30). Fat intake by an

Table 8. Selectivities for 12t-18:1 and 12c-18:1 vs. 9c-18:1 incorporation into human plasma and rat liver lipids[a].

Lipid Fraction	12 t-18:1		12c-18:1	
	Human (Plasma)	Rat (Liver)	Human (Plasma)	Rat (Liver)
TG	0.55	0.64	0.71	0.42
FFA	0.73	--	0.83	--
CE	0.23	--	0.73	--
PE	1.04	4.58	1.30	2.25
PS	0.93	2.83	1.58	1.73
PC	1.98	3.74	3.65	2.58
PC-1	5.38	--	3.24	--
PC-2	0.14	--	4.53	--
SM	1.18	--	2.85	--
LPC	0.63	--	4.13	--

[a]See Tables 2 and 7 for abbreviations.

aboriginal tribe in India was estimated at only 5 g/day/person. Previously, the essential fatty acid requirement was considered to be 4 to 5 g/day, which equals about 2% of the daily calorie intake for a 2000 kcal diet (51). Recently, recommendations for essential fatty acid have been revised to 3% of the total calorie intake for normal human adults and infants to 4.5% during pregnancy and to 5 to 7% during lactation. An essential fatty acid intake of 10 to 13% of calories is recommended as a protection against cardiovascular disease (47). However, subjects have been fed less than 2 g linoleic acid/day for 6 months without exhibiting any adverse effects.

trans Fatty acids inhibit elongation of linoleic acid which may accentuate the essential fatty acid requirement (4). If essential fatty acid deficiency exists, the intake of isomeric fats may not be desirable. However, hydrogenated soybean oil is still a good source of essential fatty acids and would prevent EFA deficiency in spite of its isomeric fatty acid content.

RECOMMENDATIONS

Americans consume more than enough fat; the problem is one of too much dietary fat, which overwhelms the body's capacity to utilize it. Soybean oil (unhydrogenated and partially hydrogenated) is a major contributor to the total dietary fat intake, but it is also an excellent source of polyunsaturated fat. It would thus seem prudent to concentrate on reducing the intake of saturated fats rather than soybean oil, if one is concerned about coronary heart disease and is trying to lower serum cholesterol levels.

The many feeding studies with partially hydrogenated soybean oil do not provide convincing evidence that its intake should be reduced in preference to other fats. The problem is that isomeric fats formed during hydrogenation are utilized differently from the usual Δ9 *cis* fatty acids. The implications of these differences are currently unknown. Man has been consuming fatty acid isomers in ruminant fats for many centuries and hydrogenated soybean oil has been used for 30 yrs. During this time, no short- or long-term effects have been clearly correlated with isomeric fatty acids. At present, *trans* isomers such as elaidate appear to be similar to stearate in their nutritional value.

The basic concept of limiting consumption of specific food items to moderate amounts while maintaining desirable body wt is probably the best advice that can be offered. Before more specific recommendations are possible, much more needs to be learned about the nutritional effects of the isomeric fats in partially hydrogenated soybean oil.

NOTES

E. A. Emken, Northern Regional Research Center, Agricultural Research, SEA, USDA, Peoria, Illinois 61604.

REFERENCES

1. Agricultural Statistics 1977, p. 140, U.S. Government Printing Office 1977, cat. no. AI.47:977.
2. Ahrens, E. H., J. Hirsch, W. Insull, T. T. Tsaltas, R. Blomstrand, and M. L. Peterson. 1957. The influence of dietary fats on serum-lipid levels in man. Lancet 1: 943-953.
3. Alfin-Slater, R. B., A. F. Wells, L. Aftergood, and H. J. Deuel. 1957. Nutritive value and safety of hydrogenated vegetable fats as evaluated by long-term feeding experiments with rats. J. Nutr. 63:241-261.
4. Anderson, R. L., C. Fullmer, Jr., and E. J. Hollenbach. 1975. Effects of the *trans* isomers of linoleic acid on the metabolism of linoleic acid in rats. J. Nutr. 105: 393-400.
5. Anderson, J. T., F. Grande, and A. Keys. 1961. Hydrogenated fats in the diet and lipids in the serum of man. J. Nutr. 75:388-394.
6. Antonis, A., and I. Bersohn. 1961. The influence of diet on serum-triglycerides in South African White and Bantu prisoners. Lancet 1:3.
7. Antonis, A., and I. Bersohn. 1961. The influence of diet on serum lipids in South African White and Bantu prisoners. Am. J. Clin. Nutr. 10:484-499.
8. Artman, N. R. 1969. The chemical and biological properties of heated and oxidized fats. In R. Paoletti and D. Kritchevsky (eds.) Advances in Lipid Research, 7:245-330, Academic Press, NY.
9. Beveridge, J. M. R., and W. F. Connell. 1962. The effect of commercial margarines on plasma cholesterol levels in man. Am. J. Clin. Nutr. 10:391-397.
10. Bronte-Stewart, B., A. Antonis, L. Eales, and J. F. Brock. 1956. Effects of feeding different fats on serum-cholesterol level. Lancet 1:521-526.
11. Carroll, K. K. 1975. Experimental evidence of dietary factors and hormone-dependent cancers. Cancer Res. 35:3374-3383.

12. Deuel, H. J., Jr.. 1955. The digestibility of fats. In The Lipids, Volume II, The Interscience Publishing Co., NY, p. 195-246.
13. Ederer, F., P. Leren, O. Turpeinen, and I. D. Frantz, Jr. 1971. Cancer among men on cholesterol-lowering diets. Lancet 2:203-206.
15. Emken, E. A., W. K. Rohwedder, H. J. Dutton, W. J. DeJarlais, R. O. Adlof, J. F. Mackin, and J. M. Iacono. 1979. Incorporation and distribution of deuterium labeled elaidic acid in man: Neutral blood lipids. Metabolism 28:575-583.
16. Emken, E. A., W. K. Rohwedder, H. J. Dutton, W. J. DeJarlais, R. O. Adlof, J. F. Mackin, R. Dougherty, and J. M. Iacono. 1979. Incorporation of deuterium labeled *cis* and *trans*-9-octadecenoic acids in humans: Plasma, erythrocyte and platelet phospholipids. Lipids 14:547-554.
17. Emken, E. A., W. K. Rohwedder, H. J. Dutton, R. Dougherty, J. M. Iacono, and J. Mackin. 1976. Dual labeled technique for human lipid metabolism studies using deuterated fatty acid isomers. Lipids 11:135-142.
18. Enig, M. G., R. J. Munn, and M. Keeney. 1978. Dietary fat and cancer trends—a critique. Fed. Proc. 37:2215-2220.
19. Erickson, B. A., R. H. Coots, F. H. Mattson, and A. M. Kligman. 1964. The effect of partially hydrogenated dietary fats, of the ratio of polyunsaturated to saturated fatty acids, and of dietary cholesterol upon plasma lipids in man. J. Clin. Invest. 43:2017-2025.
20. ERS/USDA Statistical Bulletin 627, May 1976.
21. Fredrickson, D. S., and R. I. Levy. 1972. Familial hyperlipoproteinemia. In J. B. Stanbury, J. B. Wyngaarden, and D. S. Fredrickson (eds.) The Metabolic Basis of Inherited Disease, McGraw-Hill Co., 3rd Edition, p. 545-614.
22. Gottenbos, J. J. and H. J. Thomasson. 1965. The biological action of hardened oils. Bibl. Nutr. Dieta. Fasc. 7:110-129.
23. Heckers, H., M. Korner, T. W. L. Tuschen, and F. W. Melcher. 1977. Occurrence of individual *trans*-isomeric fatty acids in human myocardium, jejunum and aorta in relation to different degrees of atherosclerosis. Atherosclerosis 28:389-398.
24. Heyden, S. 1975. Epidemiological data on dietary fat intake and atherosclerosis with an appendix on possible side effects. In A. J. Vergroesen (ed.) The Role of Fats in Human Nutrition, Academic Press, NY, p. 44-113.
25. Hirvisalo, E. L., and O. Renkonen. 1970. Composition of human serum sphingomyelins. J. Lipid Res. 11:54-59.
26. Horlick, L. 1960. The effect of artificial modification of food on serum cholesterol level. Can. Med. Assoc. J. 83:1186-1192.
27. Houtsmuller, U. M. T. 1978. Biochemical aspects of fatty acids with *trans* double bonds. Fette. Seifen. Anstrichm. 80:162-169.
28. deIongh, H., R. K. Beerthuis, C. den Hartog, L. M. Dalderup, and P. A. F. Van der Spek. 1965. The influence of some dietary fats on serum lipids in man. Bibl. Nutr. Dieta 7:137-152.
29. Johnston, P. V., O. C. Johnson, and F. A. Kummerow. 1957. Occurrence of *trans* fatty acids in human tissue. Science 126:698-699.
30. Jones, R. J. 1974. Role of dietary fat in health. J. Am. Oil Chem. Soc. 51:251-254.
31. Kaufman, H. P., and G. Mankel. 1964. Uber das Vorkommen von trans-Fettsauren. Fette. Seifen. Anstrichm. 66:6-13.
32. Keys, A. 1975. Coronary heart diease—the global picture. Atherosclerosis 22:149-192.
33. Kromer, G. W., and S. A. Gazelle. 1976. U.S. edible fats and oils refining capacities, 1975, Fats and Oil Situation, FOS-281, ERS/USDA, p. 18-32.
34. Kummerow, F. A. 1975. Lipids in atherosclerosis. J. Food Sci. 40:12-17.

35. Lands, W. E. M. Selectivity of microsomal acyltransferases. In The Essential Fatty Acids, Proceedings of the Miles Symposium '75, The Nutrition Society of Canada and Miles Laboratories, Ltd., Miles Laboratories, Ltd., Rexdale, Ontario, Canada, p. 15-26.
36. Mattson, F. A., E. J. Hollenbach, and A. M. Kligman. 1975. Effects of hydrogenated fat on the plasma cholesterol and triglyceride levels of man. Am. J. Clin. Nutr. 28:726-731.
37. McOsker, D. E., F. H. Mattson, H. B. Sweringen, and A. M. Kligman. 1962. The influence of partially hydrogenated dietary fats on serum cholesterol levels. J. Am. Med. Assoc. 180:380-385.
38. Milne, M. D. 1974. Hereditary disorders of intestinal transport. In D. H. Smyth (ed.) Biomembranes, Plenum Press, NY, p. 985-992.
39. Mishkel, M. A., and N. Spritz. 1969. The effects of *trans* isomerize trilinolein on plasma lipids in man. In W. L. Holmer, L. A. Carlson, K. Paoletti (eds.) Adv. Exp. Med. Biol., Volume 4, Plenum Press, New York, NY, p. 355-364.
40. Morello, A. M., and D. T. Downing. 1976. *trans*-Unsaturated fatty acids in human skin surface lipids. J. Invest. Dermatol. 67:270-272.
41. Mounts, T. L., E. A. Emken, W. K. Rohwedder, and H. J. Dutton. 1971. Metabolism of labeled isomeric octadecenoates by the laying hen. Lipids 6:912-918.
42. Parodi, P. W. 1976. Composition and structure of some consumer-available edible fats. J. Am. Oil Chem. Soc. 53:530-534.
43. Pearce, M. L., and S. Dayton. 1971. Incidence of cancer in men on a diet high in polyunsaturated fat. Lancet. 1:464.
44. Perkins, E. G. 1976. Chemical, nutritional and metabolic studies of heated fats. I. Chemical aspects. Rev. Fr. Corps Gras 23:258-262.
45. Perkins, E. G. 1976. Chemical, nutritional and metabolic studies of heated fats. II. Nutritional aspects. Rev. Fr. Corps Gras 23:313-322.
46. Picciano, M. F., and E. G. Perkins. 1977. Identification of the *trans* isomers of octadecenoic acid in human milk. Lipids 12:407-408.
47. Report of the joing FAO/WHO Expert Consultation on the Role of Dietary Fats and Oils in Human Nutrition. 1977. Rome, Italy, p. 21-31.
48. Rogers, A. E., and P. M. Newberne. 1975. Dietary effects on chemical carcinogenesis in animal models for colon and liver tumors. Cancer Res. 35:3427-3431.
49. Scholfield, C. R., V. L. Davison, and H. J. Dutton. 1967. Analysis for geometrical and positional isomers of fatty acids in partially hydrogenated fats. J. Am. Oil Chem. Soc. 44:648-651.
50. Schrock, C. G., and W. E. Conner. 1975. Incorporation of the dietary *trans* fatty acid (c 18:1) into the serum lipids, the serum lipoproteins and adipose tissue. Am. J. Clin. Nutr. 28:1020-1027.
51. Soderhjelm, L., H. F. Wiese, and R. T. Holman. 1970. The role of polyunsaturated acids in human nutrition and metabolism. In R. T. Holman (ed.) Progress in the Chemistry of Fats and Other Lipids, Volume 9, Pergamon Press, p. 555-585.
52. Vergroesen, A. J. 1972. Dietary fat and cardiovascular disease: Possible modes of action of linoleic acid. Proc. Nutr. Soc. 31:323-329.
53. Vergroesen, A. J., and J. J. Gottenbos. 1975. The role of fats in human nutrition: An introduction. In A. J. Vergroesen (ed.) The Role of Fats in Human Nutrition, Academic Press, NY, p. 1-32.
54. Vles, R. O., and J. J. Gottenbos. 1972. Long-term effects of feeding butterfat, coconut oil, and hydrogenated or non-hydrogenated soybean oil. 2. Life-span experiments in rats. Voeding 33:455-465.

FOOD AND INDUSTRIAL USES OF SOYBEAN LECITHIN

B. F. Szuhaj

Commercial soybean lecithin is one of the most complex and least understood products of the soybean. To most chemists, lecithin refers to the phospholipid phosphatidylcholine. To many food chemists, lecithin is an ingredient used for its multi-functional purposes in their food systems. The next time you have an opportunity at the supermarket, check ingredient statements for lecithin. You may be surprised to find that lecithin is on many ingredient lists.

Lecithin has many food uses. If you had margarine for breakfast this morning, you had lecithin. That coffee creamer, donut, hot chocolate or instant breakfast you ate probably contained lecithin.

In the non-food area lecithin is widely used as a process aid. The concrete which you walked or drove here on might have had lecithin used as a release and curing compound. The shoe polish on your shoes might have been formulated with lecithin. Even the eye shadow, lipstick or shampoo you might have used could have contained lecithin.

Yes, you have used lecithin today, you have seen the benefits of lecithin indirectly. I would like to provide more information on lecithin for you by developing how lecithin is manufactured, its physical/chemical properties, its functionality, and the different types of products and their applications.

LECITHIN MANUFACTURE

Commercial soybean lecithin is obtained from crude unrefined soybean oil (10). The process is a simple separation of the soybean phosphatides from crude oil by means of hydration and centrifugation. Live steam or warm water is added to crude soybean oil to hydrate the lecithin which agglomerates and then is separated from the oil. The resulting lecithin and water emulsion is batch or film dried from a moisture of about 25 to 50% to about 0.5 to 2% or lightened in color, then dried.

PHYSICAL-CHEMICAL PROPERTIES

As with most food ingredients, commercial soybean lecithin can be quantified and qualified by its physical/chemical properties. The following lists the major chemical and physical analysis performed on lecithin products. Many are AOCS official methods.

- Acetone Insolubles (AI)
- Acid Value (AV)
- Benzene Insolubles (BI)
- Moisture (H_2O)
- Peroxide Value (PV)
- Iodine Value (IV)
- Viscosity (CPS)
- Phosphorus (P)
- Nitrogen (N)

Acetone Insolubles (AI)

Acetone insolubles (1) are used as a rapid quantitative method for determining the phosphatide content of commercial soybean lecithin. This test is a direct result of lecithin's very low solubility in cold acetone. This value will range from 50 to 95%.

Acid Value (AV)

The acid value (2) is a direct measure of the total acidity of the lecithin product. It is not directly related to free fatty acid content or pH but rather the phosphatide acidity. Acid values are usually between 15 and 35.

Benzene Insolubles (BI)

This is an official AOCS method (3) for determining the insolubles (soy fiber) in lecithin. With today's EPA requirements benzene is being replaced by kerosene or hexene in most laboratories.

Moisture (H_2O)

Moisture is determined by one of three methods: Karl Fischer moisture (8), toluene distillation (14), and oven drying. The method of choice depends on the lecithin product and the range of accuracy. Moisture levels are between 0.5 and 2%.

Color (Gardner)

Color is determined visually rather than spectrophotometrically. The Gardner color scale (5) is used in the Gardner Hellige varnish color comparator as the industry standard. These values range between GH 11 and GH 17.

Peroxide Value (PV)

The peroxide value (11) of a lecithin product is usually a measure of residual bleach rather than active oxidation within the lecithin product. Most lecithin products have a PV of less than 5 but may have peroxide values exceeding 100 depending upon the method of modification .

Iodine Value (IV)

The iodine value (6) has been used to characterize lecithin by the amount of unsaturation. However, it is not recommended as a good measure of phosphatide purity. This is a classical analysis independent of function. Crude natural lecithin has a IV of 95 to 100.

Viscosity (CPS)

The viscosity of lecithin is usually measured in centipoise with a Brookfield viscometer or by the bubble time method (4). Liquid lecithins exhibit Newtonian flow characteristics. Low viscosity lecithin have viscosities near 1000 cps.

Phosphorus and Nitrogen

The phosphorus (13) and nitrogen (9) content of soybean lecithin are used for identifying and characterizing lecithin and its fractions. Crude lecithin contains approximately 2% P and 1% N.

COMPOSITION

There are several major classes of compounds present in commercial soybean lecithin (7,10,12,14,15). These include phospholipids, triglycerides, sterols, tocopherols, sterylglycosides, phytoglycolipids and carbohydrates. These compounds have been identified through two dimensional thin layer chromatography and further chemical analysis by UV, IR and NMR techniques. The composition of commercial soybean lecithin will not be addressed in this chapter. It should suffice to say that the major phosphatides are phosphatidyl choline, phosphatidyl ethanolamine and phosphatidyl inositol. The fatty acid composition is very similar to soybean oil: C 16:0, 16%; C18:0, 5%; C18:1, 15%; C18:2, 56% and C18:3, 8%.

PRODUCT CLASSES

Commercial soybean lecithin products are available in many color grades, viscosities, and degrees of functionality (13). I would like to give you a broad descriptive classification to better follow these products. The four major classes of commercial soybean lecithin would fall into the following: (1) crude-natural lecithin, (2) refined lecithin, (3) blended lecithin, and (4) modified lecithin.

Crude-Natural Lecithin

These lecithin products are the most basic classification. These are the original lecithin products that reflect changes only in color and viscosity.

Refined Lecithin Products

The refined lecithin products are the result of further processing of the natural lecithin. The refining process may be the removal of soybean oil and providing a product that is oil free or the addition of other types of oils to change the physical/chemical properties as well as improve stability and functionality.

Blended Lecithin Products

The blended products have been formulated to meet both handling and functional needs of the user. Blending may be used to meet viscosity requirements through the addition of edible oils or cosurfactants to enhance the functionality.

Modified Lecithin Products

Modified lecithin products are those that have been chemically modified through selective chemical treatment. The modification that results is intended to improve or enhance the compatibility of lecithin in certain systems and provide further functionality.

NSPA Lecithin Specifications

From a reference standpoint, the National Soybean Processors Association soybean lecithin specifications (16) cover only the crude or natural products previously discussed and the major analysis include AI, moisture BI, acid value, color and viscosity. The represent the different color classifications of natural or unbleached, single bleached and double bleached lecithin classified as fluid and plastic (Table 1).

GRAS Status of Lecithin

In general, all lecithin products are intended for food application. That is to say, lecithins are modified or upgraded with kosher and/or food grade materials. Currently the GRAS status of lecithin is being reviewed by the FDA. All indications are that lecithin will maintain its GRAS status. For further definition of lecithin, one should consult the Food Chemicals Codex.

FUNCTIONALITY

Lecithin is a unique food ingredient because of its multi-functional nature. Lecithin may perform in one system as an emulsifer and a release product and in another system as a viscosity modifier and a stabilizer.

Table 1. National Soybean Processors Assn. Soybean Lecithin Specifications for 1975-76.[1]

Analysis	Grade		
	Fluid Natural Lecithin	**Fluid Bleached Lecithin**	**Fluid Double-Lecithin**
Acetone Insoluble, Min	62%	62%	62%
Moisture, Max	1%	1%	1%
Benzene Insoluble, Max	0.3%	0.3%	0.3%
Acid Value, Max	32	32	32
Color, Gardner, Max	10	10	10
Viscosity, poises, @ 77 F, Max	150	150	150
	Plastic Natural Lecithin	**Plastic Bleached Lecithin**	**Plastic Double-Lecithin**
Acetone Insoluble, Min	65%	65%	65%
Moisture, Max	1%	1%	1%
Benzene Insoluble, Max	0.3%	0.3%	0.3%
Acid Value, Max	30	30	30
Color, Gardner, Max	10	10	10
Penetration, Max	22 mm	22 mm	22 mm

[1]Year Book and Trading Rules—National Soybean Processors Association (1975-76).

Commercial soybean lecithin products are employed for their surface active effects. Since lecithin products contain both a lipophilic and a hydrophilic group within the same molecule, lecithin tends to be found at the boundary between immiscible materials. Therefore, lecithin's mode of action is through a surface modifying effect. A classic example of lecithin as a emulsifier is in the area of margarine. Margarine is a water-in-oil emulsion that contains minute droplets of water with any hardened vegetable oil base. Lecithin orients itself on the surface of the water droplets with its hydrophilic portion in water and its lipophilic portion in the oil. This orientation lowers the water surface tension with respect to the oil. Lecithin assists in the emulsification and stabilization of the water within the oil base.

In general, the major functions of lecithin as surfactants are in the following categories: (1) colloidal dispersion, (2) wetting, (3) lubrication and release, (4) crystallization control, and (5) complexing.

Colloidal Dispersion

Lecithin products function as colloidal dispersing agents in: (1) emulsification, (2) solubilization, and (3) suspension.

Emulsification

Commercial soybean lecithin acts as a water-in-oil emulsifier in systems such as margarine and chocolates. Modified lecithins tend to function in both water-in-oil and oil-in-water emulsion systems.

Solubilization

Modified commercial lecithins are employed to produce oils that are "soluble" in aqueous systems since their particle size is extremely small and approaches that of a true solution. For example, flavor oils and oil soluble colors can be made water soluble using this process.

Suspensions

Lecithin helps to redisperse materials that separate during storage. Lecithin is often used in conjunction with other stabilizing agents to facilitate pigment dispersing in paints. Lecithin coats the individual particles and prevents agglomeration.

Wetting

Lecithin products are effective wetting agents. When plated on a powder lecithin will tend to instantize it by aiding in the dispersion and ultimate wetting.

Lubrication and Release

Lecithin is an excellent lubricant and release agent. The mechanism of this action is not fully understood; however, when lecithin is applied as a thin film over a surface, it promotes release from surfaces such as cooking utensils or plastic molds.

Crystallization Control

Lecithin controls the crystallization of sugar in fat systems. It promotes more stable and smoother products through crystal structure modification. Lecithin also serves to coat fat-containing particles preventing migration and recrystallization. The coating of cocoa fibers to inhibit chocolate bloom is a good example.

Complexing

Lecithin is believed to be involved in the starch complexing process associated with staling in baked goods. Lecithin tends to retard this crystallization of starch.

Multiple Functionalities

In the areas previously mentioned lecithin was described as being effective in one dimension. However, lecithin is usually involved in many functional aspects of a finished product. Lecithin is a multi-functional ingredient providing high versatility in a surface active agent. Where there is a surface boundary of any sort, it can be modified by a lecithin product.

APPLICATIONS OF COMMERCIAL SOYBEAN LECITHIN

In this last area I will cover the food and industrial applications of commercial soybean lecithin. As previously discussed, lecithin is used because of its surface active effects. The major categories of applications are as follows: food processing, cosmetics, pharmaceuticals, dietary supplement, coatings, plastic and rubber, glass and ceramics, and paper and printing.

Food Processing

Margarine. Lecithin is used as an emulsified and anti-spattering agent. It is added directly to the oil during batching.

Confection and Snack Foods. In chocolates, lecithin is used for crystallization control while in caramels lecithin is used for viscosity control. In the coatings area lecithin is used for its anti-sticking properties. Chewing gums contain lecithin as a softener. In popcorn glazes it is used to decrease tackiness. In breakfast cereals and breakfast bars it is used to prevent agglommeration.

Instant Foods. Lecithin is used as a wetting and dispersing agent and emulsifier in cocoa powders, instant drinks, instant cocoa, instant coffee, protein drinks, coffee whiteners, milk replacers, cake mixes, puddings, instant toppings, and instant soups.

Commercial Bakery Products. Lecithin is used as a starch complexing agent for crystallization control; it is also used as an emulsifier, a wetting agent, and a release agent (internal and external). The areas of use include: breads, rolls, donuts, cookies, cakes, pasta products, pies, and crusts.

Cheese Products. In cheese products, lecithin is used as an emulsifier and a release agent in natural and imitation cheeses.

Meat and Poultry Processing. Lecithin is used as a surfactant, sealant, browning agent, phosphate dispersant in meat, and in poultry it is used in glazes and basting compounds. For pet foods and bacon products lecithin is a dietary supplement and release agent as well as an emulsifier.

Dairy and Imitation Dairy Products. In dairy and imitation dairy products, lecithin is used as an emulsifier, wetting and dispersing agent, an anti-spatter agent and release agent. Product areas include: infant milk formulas, milk cream replacers, egg replacers and imitation eggs, whipped toppings, ice cream, flavored milks, flavored butters, and basting butters.

Miscellaneous Products. Lecithin is used for crystallization control and as an emulsifier in this miscellaneous category of peanut spread, salad products, and flavor and color solubilization.

Package Aid. Lecithin is employed indirectly as a packaging aid for its function as a release agent and sealant. The polymer package area include can interior coatings, sausage casing coatings, and stocking nets for hams and rolled meats.

Process Equipment. Lecithin is used on processing equipment for internal and external application as a release agent, a lubricant and an anti-corrosive agent. The processing equipment includes: frying surfaces, extruders, conveyors, broilers, dryers, blenders, and evaporators.

Cosmetics

Hair Preparations. Lecithin is used as an emulsifier, emollient, softening agent, texture controller, lubricant, and anti-corrosive agent in shampoos and shaving creams.

Makeup Preparations. In makeup preparations, lecithin is used as a pigment wetting and dispersing agent and emollient and an emulsifier in eye color cream and lipsticks.

Creams and Oils. Lecithin is an emulsifier, emollient, and moisturizer when used in hand creams and oils, body creams and oils and cream bases.

Pharmaceuticals

Dietary Supplement. Lecithin is used as a direct dietary supplement for man and farm animals.

Intermuscular Injections. Lecithin functions as a dispersing and emulsifying agent for intermuscular injections.

Vitamins. Lecithin is used to disperse and emulsify vitamins for easier dispensing.

Creams and Ointments. Lecithin is a dispersing and emulsifying agent and emollient in creams and ointments for pharmaceutical products.

Coatings Manufacture and Finishing

Paints. Lecithin is used as an emulsifier, pigment dispersing and wetting aid, pigment grinding aid, and viscosity modifier in water emulsion paints as well as alkyd and oil base paints.

Magnetic Tape Coatings. Lecithin is used to disperse and wet particles for magnetic tape coatings.

Waxes and Polishes. Lecithin is an emulsifier as well as a particle wetting and dispersing agent for waxes and polishes.

Wood Coatings. Lecithin is used as a preservative in wood coatings.

Plastic and Rubber Industry

Injection Die Molding. Lecithin is incorporated as an external and internal release agent as well as a plasticizer for injection and die molding of polyethylene, polypropylene, nylon, and rubber.

Tire Manufacture. Lecithin is applied in the tire manufacturing process as an external release agent, pigment dispersant, and wetting agent.

Natural Latex and Rubber Products. Lecithin is used in the rubber glove industry as a release agént, pigment dispersing agent, anti-caking agent, and processing aid.

Polymer Extrusion. In polymer extrusion, it is used as a release agent and an anti-static compound.

Polymer Coloration and Pigmentation. Lecithin is employed in polymer paints and coatings to disperse pigments and act as a wetting agent.

Toy Manufacture. Lecithin is used as a release agent in plastic and rubber moldings. Lecithin is also used in the pigmentation of modeling clay.

Glass and Ceramics Processing

Lecithin is used as a release agent for the manufacture of glass and ceramic products as well as for pigment and particle dispersing properties.

Paper and Printing

Printing Inks. Lecithin is used as a pigment wetting and dispersing agent and emulsifier in typewriter ribbons as well as carbon and transfer papers.

Paper Manufacture. Lecithin is a dispersing agent and softening agent in paper manufacture.

Masonry and Asphalt Products

Lecithin is used in concrete, shingling, floor tiles, asphalt, surface sealants, internal and/or external form release agents, particle dispersing and wetting agent, and as an emulsifier.

Petroleum Industry

Specialty Greases and Oils. Lecithin is used in drilling oils, cutting oils, and marine lubricants for its properties as a lubricant and detergent action.

Metal Processing

Cutting. In the processes of threading, milling, and turning metals, lecithin is used as a lubricant, a flushing aid, and an anti-corrosive agent.

Welding. In welding, lecithin is used as a slag release agent.

Wire Drawing. Lecithin is used as a lubricant when drawing wire.

Aluminum Sheet and Foil Rolling. Lecithin is also a lubricant to facilitate the rolling of aluminum foil.

Casting. Lecithin is employed as a release agent in the casting of metals.

Polishing. Lecithin is used as a particle dispersant in compounds for polishing of finished metals.

Miscellaneous

Pesticides. Lecithin is used in the pesticide industry as an emulsifier and wetting and dispersing agent.

Adhesives. Lecithin is also used in the adhesive area for its wetting and dispersing properties.

Textiles and Leathers. Lecithin is used as a softening agent, a penetrant, and lubricant for textiles and leathers. It is used in the dyes for textiles and leather pigmentation as a solubilizing agent.

CONCLUSION

In this brief article I have tried to give an overview of commercial soybean lecithin, its manufacture, its physical/chemical properties, product classes, its functionality, and most important of all, its applications. Lecithin is indeed a multi-functional ingredient for both food and industrial use. Despite the multitude of applications, lecithin is one of the most complex and versatile ingredients from the soybean.

Future of Commercial Soybean Lecithin

The newer market areas for soybean lecithin will probably be in the non-food area. The future of commercial lecithin depends on the awareness of its availability and utility by the technologists.

1. As an *emulsifier,* lecithin has been competing against the more exotic but less complex surfactants. Lecithin, however, is quite unique. It is truly a natural food grade amphoteric surfactant.

2. Lecithin is an outstanding wetting agent and therefore has a future in *instant products.* Coating powder will facilitate the dispersion and suspension of beverages, desserts, toppings, and sauces. The potential in non-food areas is relatively unexplored.

3. The *release and lubrication* properties will continue to expand the application of lecithin into food and non-food systems. Lecithin is a safe food grade product that can be applied directly onto a surface or be incorporated into a formula to permit internal release properties.

4. The *health and nutritional* benefits of soybean lecithin have yet to be explored. More research is needed in basic biochemistry and clinical studies in a host of disorders. The most recent interest in lecithin is in its involvement in neurological and nervious disorders and also in the aging process.

NOTES

Bernard F. Szuhaj, Food Research, Research and Engineering Center, Central Soya Co., Inc., Fort Wayne, Indiana.

REFERENCES

1. Acetone Insolubles. Amer. Oil Chem. Soc. Official Method. Ja 4-46.
2. Acid Value. Amer. Oil Chem. Soc. Method. Ja 6-55.
3. Benzene Insoluble. Amer. Oil Chem. Soc. Method. Ja 3-55.
4. Bubble Time Method. Amer. Oil Chem. Soc. Method. Tg 1a-64.
5. Gardner Varnish Scale. Amer. Oil Chem. Soc. Method. Td 1a-64.
6. Iodine Value of Lecithin. Amer. Oil Chem. Soc. Method. Cd-1-15.
7. Jacini, G., and G. de Zotti. 1957. Composition of the Phosphates of Soybean, Inds. Parfum et Cosmet., 12:389.
8. Karl Fischer Moisture Method. Amer. Oil Chem. Soc. Method. Ca-2e-55.
9. Kjeldahl Method. Amer. Oil Chem. Soc. Method. Aa 5-38.
10. Markley, K.S. 1951. Soybeans and Soybean Products, Volume II, Interscience Publishers, Inc., New York, p. 593-601.
11. Peroxide Value. Amer. Oil Chem. Soc. Method. Cd 8-53.
12. Singh, H. and O. S. Privett. 1970. Glycolipids and phospholipids of immature soybeans. Lipids 5:692.
13. Thaler, H. and E. Just. 1944. Determination of the phosphatide content of fats, Fette u. Seifen. 51:55.
14. Toulene Distillation Procedure. Amer. Oil Chem. Soc. Method. Ja 2-46.
15. Wittcoff, H. 1951. The Phosphatides, Reinhold Publishing Corp., New York, p. 219-223.
16. Yearbook and Trading Rules. 1975-76. National Soybean Processors Association.

INDUSTRIAL USES OF SOYBEAN OIL

K. T. Zilch

Whenever a research chemist contemplates the use of soybean oil as a raw material for the manufacture of industrial chemicals, he takes into consideration its molecular structure relative to sites of reactivity, its fatty acid composition and that, upon hydrolysis, glycerol is produced as a by-product. With these facts in mind he develops products that will satisfy various end use applications.

Since the 1940's, one of the principle uses of soybean oil, as a non-food grade material, has been in the manufacture of alkyds which are incorporated into various coating formulations and sold in the marketplace as house paints, bridge paints, air dry automotive finishes, industrial paints, etc. Soybean oil satisfies this end use application because it contains a polyester configuration that can be rearranged in the presence of glycerol and phthalic acid to yield a flexible polymer. Upon exposure to air this polymer oxidizes to a higher molecular weight polymer having film properties. It is possible for oxidative polymerization to occur because a high percentage of the fatty acids in soybean oil contain olefinic linkages.

With a decline in the sales of soybean oil for use in alkyds from a higher of 242 million lb/yr in 1953 to the present-day level of approximately 154 million lb/yr, the research chemist began to conjure up ways of modifying alkyd resins which resulted in improved performance properties. The advances served to better satisfy the coatings industry as well as helping to develop new chemicals that could find utilization in more lucrative markets or in new end use applications. This is the subject which will be addressed in the remainder of this chapter.

It might be interjected at this point that other natural oils and/or fatty acids such as safflower oil, sunflower oil, cottonseed oil, rapeseed oil, and tall oil fatty acids can also be utilized in the same way as soybean oil to satisfy the same markets. Therefore, soybean oil, to hold its share of the market, must be priced competitively or have a higher percentage of the desired chemical configuration for a particular end use application.

MODIFIED WATER BASED ALKYD COATINGS

As stated in the Chemical Economics Handbook (5), the overall consumption of alkyd surface coatings is expected to decrease about 2 to 3% annually through 1981. The major reason for this diminished demand is government antipollution regulations regarding solvent use and emissions. However, this decrease in alkyd resins would be far greater if it were not for the research chemist inventing ways of modifying these resins to improve their performance properties and at the same time shifting away from solvent-based coatings to water-based systems. Fortunately the chemistry of alkyd systems is such that they can be tailored to meet a variety of end-use requirements through the choice and ratio of reactants and/or modifiers.

For example, this past September, a patent (6) was issued on a polyurethane-modified water-soluble alkyd wherein the resin was prepared by reacting soybean oil unsaturated fatty acids and phthalic anhydride with an excess of polyol along with a co-reactant dimethyol-propionic acid, and subsequently reacted with toluene diisocyanate to yield the polyurethane derivative. Water solubility was obtained by neutralizing the resins with ammonia or an amine.

Since alkyd-urethane coatings in the past several years have been emerging as tough, wear-resistant materials that can be specially formulated to coat many types of surfaces, alkyd modifications of this type are of interest to the company or chemist attempting to expand the use of soybean oil as a raw material.

EPOXIDIZED SOYBEAN OIL

Another use for soybean oil is in the manufacture of epoxidized soybean oil. It is made commercially by reacting soybean oil with peracetic acid as illustrated in Figure 1. The other R alkyl groups shown can also be epoxidized if they contain carbon to carbon couble bonds.

Epoxidation of soybean oil converts it into a more compatible system for a variety of resins such as polyvinyl chloride wherein it acts both as a plasticizer and a stabilizer. When incorporated into the resin it is relatively permanent to volatility and extraction of much the same order as the conventional polymeric plasticizers.

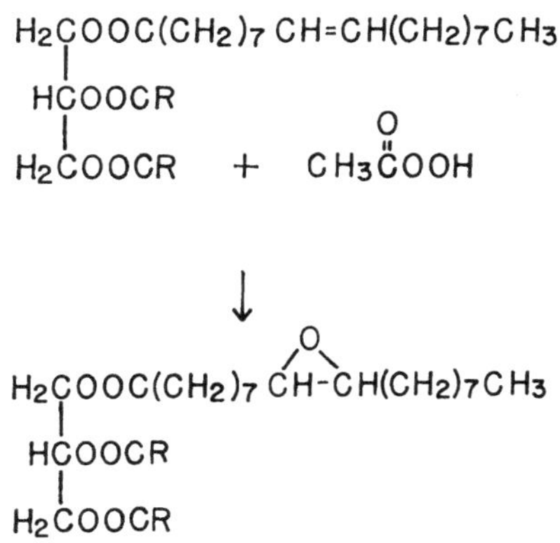

Figure 1. **Reaction of SBO with peracetic acid.**

Because it is acceptable to the Food and Drug Administration as a plasticizer/stabilizer system in resins that come in contact with food, it enjoys a large share of the market for this type of product. However, in certain applications it must compete with other epoxized esters such as epoxidized octyl tallate, alkyl epoxystearates, epoxidized glycol dioleates, and other esters of dibasic acids.

According to government statistics (14), the U.S. production of epoxidized soybean oil has decreased from a higher of 127 million lb in 1974 to 92 million lb in 1977. However, in the same period of time there was a decrease in the production of all epoxy ester plasticizers from 154 million lb in 1974 to 120 million lb in 1977.

POLYMERIZATION OF SOYBEAN OIL UNSATURATED FATTY ACIDS

The unsaturated fatty acids isolated from the hydrolysis of soybean oil can be polymerized thermally or catalytically to yield both dimer and trimer polybasic acids which are commercially utilized in the manufacture of corrosion inhibitors, polyamide resins, lubricants, fuel additives, antiwear agents, coating modifiers, etc.

In the thermal polymerization of soybean oil fatty acids or esters, the polyunsaturated acids are principally polymerized to yield polybasic acids have mono-, di- and tricyclic configurations (3,8,10,11,16). Figure 2 shows the polymerization of methyl linoleate to yield one of the several monocyclic isomers found in the reaction product.

In the catalytic polymerization of linoleic acid, a higher percentage of monocyclic dimer acid is formed, whereas in thermal polymerization a greater amount of polycyclic products is formed.

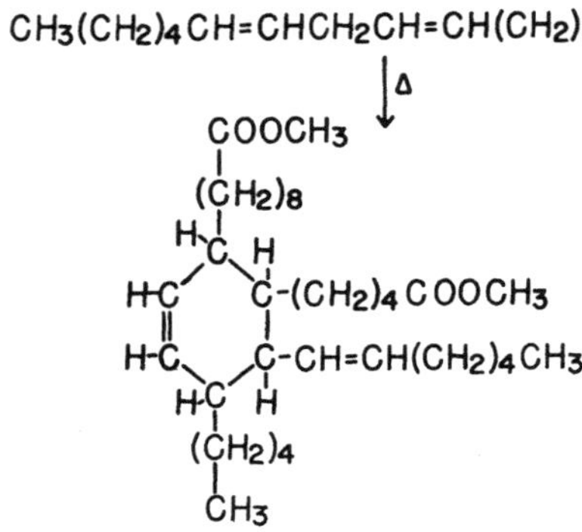

Figure 2. **Polymerization of methyl linoleate.**

It should also be noted that in both the thermal and catalytic polymerization of unsaturated fatty acid and esters, monobasic acids represent a portion of the yield which are removed by distillation. Also the dimer acids or esters can be fractionated from the higher polymeric acids or esters by means of molecular distillation.

Corrosion Inhibitors

One of the most widely used applications of dimer acids is for corrosion inhibition. The fact that dimer acids contain two polar carboxyl groups and a large non-polar hydrocarbon component makes it an ideal acid for its adsorption onto metal surfaces. The adsorbed complex gives a hydrophobic surface which protects against penetration by water and other polar agents. Dimer acid formulations are used to protect down-well parts in oil rigs against corrosion from agents such as acetic acid, H_2S, and CO_2 (7,12). Dimer acids are also incorporated into jet fuels as a stabilizer against thermal degradation, and in gasoline as a detergent, forming easily removed granular deposits.

Lubricants and Lubricant Additives

In the area of lubrication, dimer acids are utilized as stabilizers in the cold rolling of metals to inhibit the degradation of petroleum products. Also esters of dimer acids have found use as lubricants and lubricant additives. For example, they are utilized as base fluids in the formulation of synthetic lubricants or added as anti-squawk agents in transmission fluids. When dimer esters are added to lithium soap greases they provide superior quality products.

Polyamide Resins

Polyamides of dimer acids are also of commercial significance and are derived through the reaction of dimer acids with polyfunctional amines such as ethylene diamine illustrated in Figure 3.

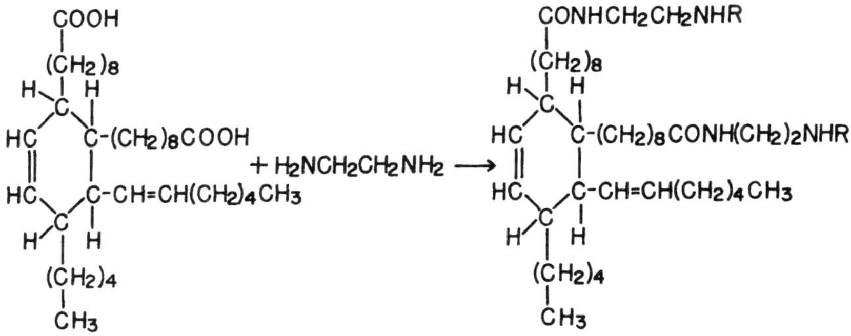

Figure 3. Reaction of dimer acid with ethylene diamine. R = hydrogen or dimer acid group.

Dimer acid polyamides are widely used as the resin binder in flexographic printing inks and are especially suited for printing on metallic foil laminates and plastic films because they adhere so tenaciously to these surfaces. Recently, polyamide resins suitable for use in flexographic printing have been developed in which water is used as the primary solvent (17). Aqueous systems are preferred because water is inexpensive, non-flammable, and does not present pollution problems.

Dimer acid polyamides having relatively sharp melting points, are suitable for use as hot melt adhesives. The major application for these polyamides is in the shoe industry where their excellent adhesion permits the bonding of sole to uppers without the necessity of stitching. Also because of their excellent adhesion properties, they can be used as welding cements in the formation of cans.

Many more end use applications of dimer acids and their derivatives can be cited which are of industrial significance; however, because of time, may I suggest that those individuals interested in this subject refer to a recent publication in dimer acids (6) which deals with this subject in a general way but also contains many valuable references. In addition, commercial literature is available which lists abstracts on dimer acids and their projected uses.

DEFOAMERS

A number of natural oils, including soybean oil, have found utilization in PVC dispersions, spinning baths, textile baths, and fermentation processes because of their antifoaming properties when incorporated into aqueous systems.

One of the most critical problems associated with fermentation is that of foam caused by the large quantity of air and high degree of agitation used as well as the composition of the fermentation broth itself. Normally excessive foaming is associated with a decrease in yield of the desired product. Any defoaming agent added to a fermentation process must be completely nontoxic and effective in small quantities.

Soybean oil has been shown to be effective as a defoaming agent in the manufacture of penicillin. In addition, it has been reported that soybean oil stimulates penicillin production in tank and shake-flask fermentations, but the effect is not interpreted as one of surface action (2).

SOYBEAN OIL GLYCEROL

In the hydrolysis of soybean oil, frequently referred to as "splitting," one of the co-products produced along with the fatty acids is glycerine as shown in Figure 4. The glycerol/water (sweet-water) product obtained from the process is concentrated to crude glycerine or treated in various ways to yield refined glycerine. Both crude and refined glycerol are used in a number of end use applications.

In many medical and pharmaceutical preparations it is used as a humectant or as a vehicle to impart smoothness. In cosmetics and toiletries it is frequently used as a humectant and as a lubricant. One of the largest end uses in this category is in the manufacture of toothpaste. It is estimated that some 63 million lb of glycerine were used in 1977 for the manufacture of alkyd resins (9). With a gradual decrease in the production of alkyd resins, as indicated above, this market is expected to decrease at a rate of 1%/yr through 1980. Glycerine finds use as a plasticizer in cigarette papers and cigarette filters as its triacetate derivative. In cigarette tobacco it is incorporated as a humectant to help retain moisture. This market has remained relatively steady since 1967.

Mono- and diesters of glycerol are prepared by reacting fatty acids or fatty glycerides with an excess of glycerol. These esters find their greatest use as emulsifiers in the food industry. They are also used in the preparation of cosmetics, synthetic elastomers, and textiles.

Polyglycerols prepared by the alkali condensation of glycerol and polyglycerol esters of fatty acids are finding increasing use in foods, particularly in shortenings and in margarines. The polyglycerol esters offer a wide range

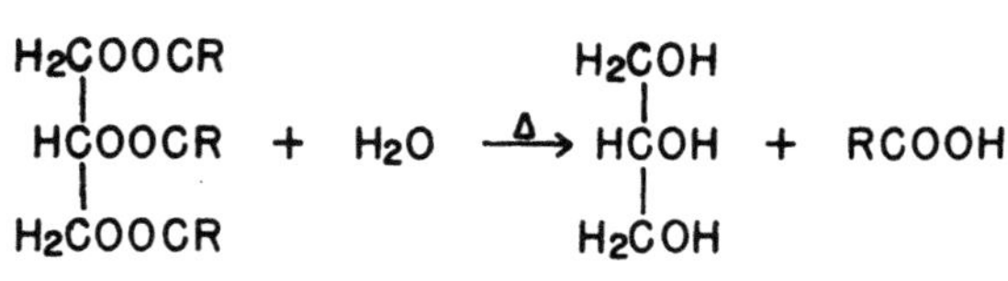

Figure 4. **Hydrolysis of soybean oil. R = palmitic, stearic, oleic, linoleic, linolenic acid moieties.**

of hydrophilic and lipophilic properties. The most rapidly growing market for glycerol is in the manufacture of polyether polyols which in turn are used in the manufacture of polyurethane foams. The projected annual growth for this end use application is 8%/yr, through 1980. Polyether polyols are prepared by condensing propylene oxide with glycerol in a base catalyzed reaction as shown in Figure 5. Further reaction of the polyether with toluene diisocyanate produces resins which are used today in all major applications for flexible polyurethane foams.

DETERGENT FUEL ADDITIVES

As we all know, many changes are taking place in the design of internal combustion engines to meet new federal standards relative to exhaust gas emissions. For example, prior to the early 1970's crankcase gases were vented to the atmosphere whereas today they are recycled to the intake air supply of the carburetor. Another change involves the recycling of a portion of the exhaust gases to the combustion zone of the engine. Both recycled gases contain substantial quantities of deposit-forming substances which promote the formation of deposits in the carburetor leading to inefficient operations of the engine and causing an increase in harmful exhaust emissions. Therefore, there is a need for a carburetor detergent in fuels to prevent the formation of these deposits.

Although the market for carburetor detergents is small, it has the potential for tremendous growth. Because of this potential, more and more patents are appearing in the literature describing the preparation of such detergents. One such patent (13) describes the preparation of an aminimide derivative of tallow, coconut, and soybean fatty acids by reacting the corresponding ethyl ester with the condensation product of ethylene or propylene oxide with dimethyl hydrazine as shown in Figure 6. The fatty acid moiety of the

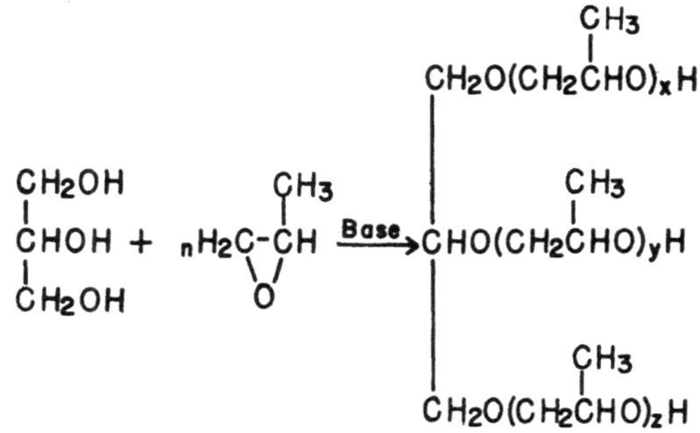

Figure 5. Polyether polyol of glycerol.

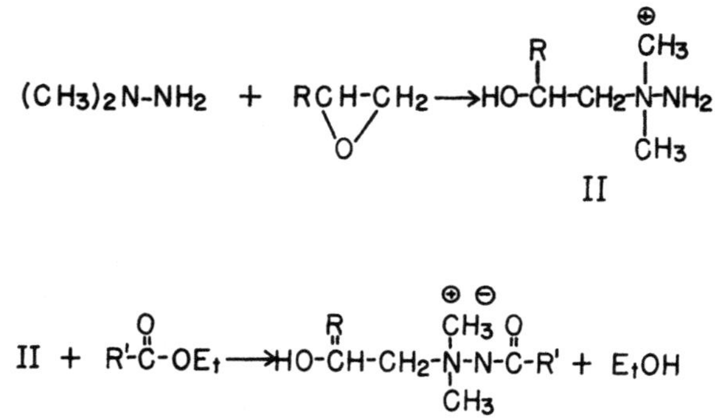

Figure 6. Preparation of fatty acid aminimide derivative.

aminimide derivative provides the lipophilic property to make this detergent compatible with the fuel system.

CONCLUSION

There are many more industrial uses for soybean oil which could be cited. However, these are some of the latest and more interesting developments in this field.

NOTES

Karl T. Zilch, Emery Industries, Cincinnati, Ohio 45232.

REFERENCES

1. Abstracts of dimer acid use-patents and journal references, Vol. I. Tech. Bull. 118 (1976) and Vol. II, Tech. Bull. 109 (1978), Emery Industries, Cincinnati, OH.
2. Abu-Shady, M. R., F. M. El-Beih, and S. S. Radwan. 1976. Effect of pure lipids and natural oils on the production of antibiotics by microorganisms. Fett. Seifen. Anstrichmittel 78:478-480.
3. Bradley, T. F. and W. B. Johnston. 1941. Ind. Eng. Chem. 33:86.
4. Bradley, T. F. and D. Richardson. 1940. Ind. Eng. Chem. 32:802, 963.
5. Connolly, E. M. and J. C. Dean. 1977. Alkyd surface coatings. In Chemical Economics Handbook, 592.5821 E, Stanford Research Institute.
6. Harris, R. R. and W. J. Pollack. 1978. U.S. P. 4,116,902.
7. Leonard, E. C. 1975. The dimer acids, Humko Sheffield Chemical, Memphis, TN.
8. Nimerick, K. H. 1968. U.S. Pat. 3,378,488.
9. Oosterhof, D. 1976. Glycerin. In Chemical economics handbook, 662.5021 E, Stanford Research Institute.
10. Paschke, R. F., L. E. Peterson, and D. H. Wheeler. 1964. Dimer acid structures. The thermal dimer of methyl 10-*trans,* 12-*trans,* linoleate. J. Amer. Oil Chem. Soc. 41:723.
11. Sen Gupta, A. K. 1968. Untersuchungen uber die struktur dimerer fettsauren. III. Zur frage der Diels-Alder reaktion bei der thermischen polymerisation von linolsaure methylester. Fette. Seifen. Anstrichmittel 70:153, 267.

12. Standord, J. F. 1968. U.S. pat. 3,412,024.
13. Sung, R. L., P. Dorn, W. P. Cullen, and R. C. Schlicht. 1978. U.S. Pat. 4,078,901.
14. U.S. International Trade Commission. Synthetic Organic Chemicals, United States Production and Sales 1974 (Pub. 776) and 1977 (Pub. 833).
15. Wheeler, D. H. and J. J. White. 1967. Dimer acid structures. The thermal dimer of normal linoleate methyl 9-*cis*, 12-*cis* octadecadienoate. J. Amer. Oil Chem. Soc. 44:298.
16. Whyzmuzis, P. D. and C. W. Wilkus. 1974. U.S. Pat. 3,786,007.

ANTI-NUTRITIONAL FACTORS AS DETERMINANTS OF SOYBEAN QUALITY

I. E. Liener

Ever since the soybean was first introduced into this country in the early part of this century, its value as a rich source of protein in the diet of animals and man has been found to be vastly improved by heat treatment. Implicit in this observation is the realization that there are heat-labile factors present in soybeans which interfere with the utilization of its protein. In addition to those factors which are inactivated by heat, other heat-stable factors are known to be present which can also detract from the nutritional quality of soybean protein, albeit to a relatively minor extent and only under rather special circumstances. In Table 1 a compilation of the heat-labile and heat-stable anti-nutritional factors known to be present in soybeans is presented. Each of these will be discussed in turn and an attempt will be made to evaluate their nutritional significance, particularly in the human diet, wherever possible.

HEAT LABILE FACTORS

Trypsin Inhibitors

The trypsin inhibitors are probably the best known, and certainly the most studied, of all the anti-nutritional factors known to be present in soybeans. From the observation that the thermal inactivation of these inhibitors is accompanied by a marked enhancement in the nutritive quality of the protein it has been generally concluded that the trypsin inhibitor is the main cause of the poor growth of animals fed inadequately heated soybeans. There are, however, several lines of evidence which indicate that the trypsin inhibitors are only partially responsible for the poor nutritive value of raw soybeans.

Table 1. Anti-nutritional factors in soybeans.

Heat-labile	Heat-stable
Trypsin inhibitors	Saponins
Hemagglutinins	Estrogens
Goitrogens	Flatulence factors
Anti-vitamins	Lysinoalanine
Phytate	

For example, if rats are fed heated soybeans to which have been added the isolated inhibitors so as to provide the same level of antitryptic activity which is present in raw soybeans, the reduction in growth falls short of that of raw soybeans (27). An investigation of a large number of different varieties of soybeans with respect to the growth-promoting quality of their protein in rats (PER) and their trypsin-inhibitory activity revealed no correlation between these two parameters (21). Finally, if the trypsin inhibitory activity of a crude extract of soybeans is removed selectively by affinity chromatography with Sepharose-bound trypsin, the resulting extract is still capable of causing growth inhibition and pancreatic hypertrophy (Table 2). It may be estimated from the data shown in Table 2 that the trypsin inhibitor accounts for about 40% of the growth inhibition and pancreatic enlargement observed with raw soybeans.

These findings raise the question as to what is responsible for the remaining 60% of the growth-retarding and pancreatic effects of raw soybeans. A possible clue came from experiments in which the crude soybean protein extract from which the inhibitor had been removed selectively was subjected to digestion with trypsin *in vitro.* Heat treatment of such a preparation of soybean protein produced an increase in the digestibility of the protein over and above the digestibility of a similar sample from which the inhibitor had not been removed (19). This observation suggests that native, undenatured soybean protein is in itself refractory to enzymatic attack unless denatured by heat,

Table 2. Contribution of soybean trypsin inhibitors to the growth inhibition and pancreatic hypertrophy observed with raw soybeans (19).

Source of Protein	PER	Wt of Pancreas g/100 g Body Wt
Soy protein, unheated	1.4	0.71
Soy protein, heated	2.7	0.57
Soy protein minus inhibitor[a]	1.9	0.65
% change due to removal of inhibitor	+38	- 41

[a]Trypsin inhibitors were removed by passage of unheated soy flour extract through a column of Sepharose-bound trypsin.

and this may very well account for the growth depression seen with the inhibitor-free soybean protein. Since the level of active trypsin in the intestine is believed to control the size of the pancreas by a negative feedback mechanism (15), the undenatured protein may in fact, act as a competitive inhibitor of trypsin and thus serve to reduce the level of active trypsin in the intestines, causing an enlargement of the pancreas.

Many soybean products on the market today are made from soy isolates which may contain as much as 30% of the trypsin inhibitor activity of raw soybean meal. An examination of the trypsin inhibitor activity of several textured meat analogs during various stages of their fabrication from soy isolates reveals that, although the protein isolate may be rich in antitryptic activity, the latter is reduced to very low levels in the final product (Table 3). Household cooking of such products would be expected to reduce these levels even further. Of particular concern to the pharmaceutical industry is the possibility that infants fed soy milk manufactured from soy isolates may be more sensitive to the physiological effects of trypsin inhibitors. Churella et al. (6), however, have shown that the heat treatment involved in the processing and sterilization of soy formulas prepared from inhibitor-containing isolates reduced the trypsin inhibitor content to less than 10% of the original activity. This residual level of activity did not cause any weight reduction or pancreatic hypertrophy in rats. These observations are consistent with the findings of Rackis et al. (35) who found that only 70 to 80% of the trypsin inhibitor activity need be destroyed in order to achieve maximum weight gains and PER with rats, and only 40 to 70% destruction is needed to eliminate pancreatic hypertrophy.

Assuming for the moment that processing conditions may not have been sufficient to reduce the level of trypsin inhibitor activity below that of the threshold level established for rats, would the residual trypsin inhibitor activity still pose a risk in the human diet? Trypsin inhibitor activity is

Table 3. Trypsin inhibitor activities of textured meat analogs made from soy isolate.

Product	Antitrypsin Activity TIU[a]/g Dry Solids x 10^{-3}	% of Soy Flour
Soy flour (unheated)	86.4	100
Soy isolate	25.5	30
Soy fiber	12.3	14
Chicken analog	6.9	8
Ham analog	10.2	12
Beef analog	6.5	7

[a]TIU = trypsin inhibitor units as determined by the method of Kakade et al. (20).

invariably measured *in vitro* on the basis of the ability of soybean preparations to inhibit bovine or procine trypsin since these sources of trypsin are readily available commercially in a highly purified state. Human trypsin is known to exist in two forms, a cationic species, which constitutes the major component, and an anionic species, which accounts for about 10 to 20% of the total trypsin activity of human pancreatic juice (12,36). While the latter is inactivated completely by the soybean inhibitor, the cationic form of trypsin which comprises about 80 to 90% of the total trypsin of human pancreas is inhibited very weakly. This observation, of course, casts considerable doubt as to whether the soybean trypsin inhibitors are of any nutritional significance as far as humans are concerned.

Hemagglutinins

It has been recognized for many years that soybeans and most other legumes contain hemagglutinins or lectins which have the unique property of binding carbohydrate-containing substances (26). Ever since the time of Ehrlich, it has been known that some of these lectins, such as ricin from the castor bean, are extremely toxic. Little is known even now, however, concerning the extent to which these substances might affect the nutritive value of the more common legumes which constitute a rich source of protein in the diet of man and animals. When the isolated soybean hemagglutinin was fed to rats, the results obtained were somewhat ambiguous (25). As long as the animals were allowed free access to their food, there was a significant depression in growth. However, since this growth depression was accompanied by a concomitant decrease in food consumption, it was not clear whether the failure of the animals to grow was a consequence of lowered food intake or whether the lower food consumption was the result of depressed growth. When the food intake was equalized, however, the soybean hemagglutinin had little effect on growth. This negative effect was subsequently corroborated when it was found that rats fed soybean extracts from which the hemagglutinin had been removed selectively by affinity chromatography grew just as poorly as those receiving the original crude soybean extract (Table 4). It would appear, therefore, that the soybean hemagglutinin does not play any major role as a determinant of the nutritional quality of soybean protein.

Goitrogens

Unheated soybeans have been reported to cause marked enlargement of the thyroid gland of the rat and chick, an effect which could be counteracted by the administration of iodine (as KI) or partially eliminated by heat (2,32). A number of cases of goiter have also been reported in human infants fed soybean milk (16,43), a situation which could likewise be alleviated by iodine supplementation. Presumably the heat treatment employed for sterilizing

Table 4. **Effect of removing soybean hemagglutinin (SBH) on the growth-promoting activity of raw soybean extracts (41).**

Protein Component of Diet	Hemagglutinating Activity Units/g Protein x 10^{-3}	PER
Original soy extract	324	0.91
Original soy extract minus SBH[a]	29	1.13
Original soy extract, heated	6	2.25
Raw soy flour	330	1.01
Heated soy flour	13	2.30

[a]SBH was removed from an aqueous extract of soybeans by passage through a column of Sepharose-bound concanavalin A.

some soybean milk infant formulas may, in some instances, be insufficient for the complete destruction of the goitrogenic principle. Iodine fortification has therefore been recommended for soybean infant formuls (11). A goitrogenic principle has been isolated from soybean whey and has been characterized as a low molecular weight oligopeptide of two or three amino acids or a glycopeptide containing one or two amino acids and a sugar residue (23,24).

Antivitamins

Vitamin D. The inclusion of unheated soybean meal, or the protein isolated therefrom, in the diet of chicks may cause rickets unless higher than normal levels of vitamin D_3 are added to the diet (4). This rachitogenic effect can also be eliminated by autoclaving or by supplementation with calcium and phosphorus (18). It has been suggested that the rachitogenic properties of soy protein may be due to phytic acid (18), although the evidence on this point is not conclusive.

Vitamin E. Anti-vitamin E activity has been reported in isolated soy protein (13) as measured by growth, mortality, exudative diathesis, and encephalomalacia. The identity of this anti-vitamin factor has not been established, although it has been suggested that it might be tocopherol oxidase (31).

Vitamin B_{12}. Not only is the soybean lacking in vitamin B_{12}, but it has also been reported to contain a heat-labile substance that increases the requirement for this vitamin (10) and causes an increased excretion of metabolites associated with enzymes that require vitamin B_{12} as a coenzyme (9). This increased requirement for vitamin B_{12} in rats fed raw soy flour has been attributed to a decreased availability of the vitamin produced by the intestinal flora and to an increased turnover of the absorbed vitamin.

Phytate

Phytic acid is known to interfere with the biological availability of various minerals including zinc, manganese, copper, molybdenum, calcium, magnesium, and iron. Most of these same effects are observed in diets containing soy protein isolates (33) and can be eliminated effectively by autoclaving or by adding chelating agents. The effect of heat is most likely due to the fact that it favors the interaction of phytic acid with protein and thus diminishes its ability to bind with minerals. Since phytase activity has been reported to be present in the intestinal mucosa of man (1), it may be that this enzyme could negate some of the deleterious effects of phytate in the human diet.

HEAT-STABLE FACTORS

Saponins

Although saponins from some plants have an adverse effect on animal growth, it would appear that the saponins of soybeans are relatively innocuous to chicks, rats, and mice even when fed at levels three times greater than the levels found in soy flour (0.5%) (17). Saponins are hydrolyzed by bacterial enzymes in the lower intestinal tract, but neither saponins nor their aglycones (sapogenins) can be detected in the blood of test animals. It is probably safe to say that saponins should be removed from the list of anti-nutritional factors in soybeans.

Estrogens

Substances exhibiting estrogenic activity, namely genistein daidzein, and coumestrol are present in soybeans, and can in fact inhibit growth (28,29,30) and interfere with reproductive performance (5) when fed to rats at high enough levels. However, in order to attain such high levels from soybeans themselves, the latter would have to be the sole constituent of the diet. It is unlikely, therefore, that the estrogens present in soybeans would constitute a health hazard to man as part of a normal varied diet.

Flatulence Factors

One of the important factors limiting the use of legumes in the human diet is the production of flatulence associated with their consumption, and the soybean is no exception in this regard. The principle offenders appear to be low molecular weight oligosaccharides containing α-galactosidic and β-fructosidic linkages, namely raffinose and stachyose (34). Thus, flatus activity in humans has been noted mostly with soybean products from which the carbohydrate has not been removed, such as full-fat and defatted soy flours (34,37). As shown in Table 5, when soy flour is extracted with 80% ethanol to produce a concentrate, the flatulence effects are reduced considerably.

Table 5. Effect of soy products on flatus in man (34).

Soy Product[a]	Flatus Volume[b] ml/hr
Defatted flour	71
Protein concentrate	36
Whey solids	300
Alcohol extract	240
Protein isolate	13

[a]All soy products were toasted with live steam at 100 C for 40 min and fed at a level equivalent to 146 g defatted soy flour/day.
[b]Average of 4 subjects in each soy product.

Flatus activity resides namely in the soy whey solids and in the alcohol extract which contain the low molecular oligosaccharides. Protein isolates, and products prepared therefrom, and fermented soy preparations such as tempeh (3) are virtually devoid of flatus activity.

Flatulence is generally attributed to the fact that man is not endowed with the enzyme (α-galactosidase) necessary for hydrolyzing the α-galactosidic linkages of raffinose and stachyose so as to yield readily absorbably sugars (14). Consequently, the intact oligosaccharides enter the lower intestine where they are metabolized by the microflora producing such gases as carbon dioxide, hydrogen, and, to a lesser extent, methane. It is the production of these gases which are responsible for nausea, cramps, diarrhea, abdominal rumbling, and to the social discomfort generally associated with the ejection of rectal gases. It should be emphasized that there is considerable variability in individual response to the flatus-producing effects of beans. Many individuals are completely unaffected by the ingestion of beans, but the exact reason for this variable response is not understood completely, although it seems reasonable to assume that it is probably related to individual differences in the microbial population of the lower intestines.

Since the flatus-producing factors in soybeans are heat-stable, attempts have been made to eliminate these factors by enzymatic hydrolysis. Although treatment of soybeans with mold enzymes virtually eliminated stachyose and raffinose (3,40), there was no significant reduction in flatus activity in human subjects (3). A significant reduction in the oligosaccharide content of soybeans can also be achieved by a combination of soaking, germination, and resoaking (22), although in this case it is difficult to assess how much of this reduction is due to a leaching out of the oligosaccharides or to autolysis by endogenous enzymes.

Lysinoalanine

Alkaline extraction of soybeans which is used frequently to prepare protein isolates is known to lead to a reduction in the nutritive value of the protein, attributable, at least in part, to the destruction of cystine (7). One of the decomposition products of cystine is dehydroalanine which can interact with the ε-amino group of lysine to form lysinalanine. Alkali-treated soybeans have been shown to produce kidney lesions in rats, an effect which can be reproduced by the administration of free lysinoalanine (44). Inconsistent and variable results, however, have been reported by other workers in this field (8,39,42), and it now appears that the response to lysinoalanine depends on the species of test animal employed (even among strains of the rat differences have been noted), the composition of the basal diet, and whether the lysinoalanine is peptide-linked or not.

Sternberg et al. (38) have shown lysinoalanine to be widely distributed in cooked foods, commercial food preparations, and food ingredients, many of which had never been subjected to alkaline treatment. Many of these foods had levels of lysinoalanine which were considerably higher than those found in commercial samples of soy protein isolate. The wide distribution of lysinoalanine among commonly cooked foods would tend to indicate that this is neither a novel or serious problem since humans have long been exposed to proteins containing lysinoalanine with apparent impunity. Its presence in soy protein can hardly be considered a serious problem for man.

NOTES

Irvin E. Liener, Department of Biochemistry, College of Biological Sciences, University of Minnesota, St. Paul, Minnesota 55108.

REFERENCES

1. Bitar, K. and J. G. Reinhold. 1972. Phytase and alkaline phosphatase activities in intestinal mucosas of rat, chicken, calf, and man. Biochem. Biophys. Acta 268: 442-452.
2. Block, R. J., R. H. Mandl, H. W. Howard, C. D. Bauer, and D. W. Anderson. 1961. The curative action of iodine on soybean goiter and the changes in the distribution of iodoamino acids in the serum and in thyroid gland digests. Arch. Biochem. Biophys. 93:15-24.
3. Calloway, D. H., C. A. Hickey, and E. L. Murphy. 1971. Reduction of intestinal gas-forming properties of legumes by traditional and experimental food processing methods. J. Food Sci. 36:251-255.
4. Carlson, C. W., H. C. Saxena, and L. S. Jensen. 1964. Rachitogenic activity of soybean fractions. J. Nutr. 82:507-511.
5. Carter, M. W., G. Matrone, and W. G. Smart, Jr. 1955. Effect of genistin on reproduction of the mouse. J. Nutr. 55:639-645.
6. Churella, H. R., B. C. Yao, and W. A. B. Thompson. 1976. Soybean trypsin inhibitor activity of soy infant formulas and its nutritional significance for the rat. J. Agr. Food Chem. 24:393-397.
7. DeGroot, A. P. and P. Slump. 1969. Effect of severe alkali treatment of proteins on amino acid composition and nutritive value. J. Nutr. 98:45-56.
8. DeGroot, A. P., P. Slump, L. van Beek, and V. J. Feron. 1977. Severe alkali treatment of proteins, pp. 270-283. In C. E. Bodwell (ed.) Evaluation of Proteins for Humans, AVI Publishing Co., Westport, CT.

9. Edelstein, S. and K. Guggenheim. 1970. Changes in the metabolism of vitamin B_{12} and methionine in rats fed unheated soya bean flour. Brit. J. Nutr. 24:735-740.
10. Edelstein, S. and K. Guggenheim. 1970. Causes of the increased requirements for vitamin B_{12} in rats subsisting on an unheated soybean flour diet. J. Nutr. 100: 1377-1382.
11. Federal Register. 1971. Label statements relating to infant food. 36:23555.
12. Figarella, C., G. A. Negri, and O. Guy. 1975. The two human trypsinogens. Inhibition spectra of the two human trypsins derived from their purified zymogens. Eur. J. Biochem. 53:457-463.
13. Fisher, H., P. Griminger, and P. Budowski. 1969. Anti-vitamin E activity of isolated soybean protein for the chick. Z. Ernahrungswiss. 9:271-278.
14. Gitzelmann, R. and S. Auricchio. 1965. The handling of soya alpha-galactosidases by a normal and a galactosemic child. Pediatrics 36:231-235.
15. Green, G. M. and R. L. Lyman. 1972. Feedback regulation of pancreatic enzyme secretion as a mechanism for trypsin inhibitor-induced hypersecretion in rats. Proc. Soc. Exp. Biol. Med. 140:6-12.
16. Hydowitz, J.D. 1960. Occurrence of goiter in an infant on soy diet. N. E. J. Med. 262:351-353.
17. Ishaaya, I., Y. Birk, A. Bondi, and Y. Tencer. 1969. Soybean saponins. IX. Studies of their effect on birds, mammals, and cold-blooded organisms. J. Sci. Food Agr. 20:433-436.
18. Jensen, L. S. and F. R. Mraz. 1966. Rachitogenic activity of isolated soy protein for chicks. J. Nutr. 88:249-253.
19. Kakade, M. L., D. Hoffa, and I. E. Liener. 1973. Contribution of trypsin inhibitors to the deleterious effects of unheated soybeans fed to rats. J. Nutr. 103:1772-1778.
20. Kakade, M. L., N. Simons, and I. E. Liener. 1969. An evaluation of natural vs. synthetic substrates for measuring the antitryptic activity of soybean samples. Cereal Chem. 46:518-526.
21. Kakade, M. L., N. Simons, I. E. Liener, and J. W. Lambert. 1972. Biochemical and nutritional assessment of different varieties of soybeans. J. Agr. Food Chem. 20: 87-90.
22. Kim, W. J., C. J. B. Smit, and T. O. M. Nakayama. 1973. The removal of oligosaccharides from soybeans. Lebensm.-Wiss. u. Technol. 6:201-201.
23. Konijn, A. M., S. Edelstein, and K. Guggenheim. 1972. Separation of a thyroid-active fraction from unheated soya bean flour. J. Sci. Food Agr. 23:549-555.
24. Konijn, A. M., B. Gershon, and K. Guggenheim. 1973. Further purification and mode of action of a goitrogenic material from soybean flour. J. Nutr. 103:378-383.
25. Liener, I.E. 1953. Soyin, a toxic protein from the soybean. I. Inhibition of rat growth. J. Nutr. 49:527-539.
26. Liener, I. E. 1976. Phytohemagglutinins (Phytolectins). Ann. Rev. Plant Physiol. 27:291-319.
27. Liener, I. E., H. J. Deuel, Jr., and H. L. Fevold. 1949. The effect of supplemental methionine on the nutritive value of diets containing concentrates of the soybean trypsin inhibitor. J. Nutr. 39:325-339.
28. Lookhart, G. L., B. L. Jones, and K. F. Finney. 1978. Determination of coumestrol in soybeans by high-performance liquid and thin-layer chromatography. Cereal Chem. 55:967-972.
29. Magee, A. D. 1963. Biological responses of young rats fed diets containing genistin and genistein. J. Nutr. 80:151-156.
30. Matrone, G., W. W. G. Smart, Jr., M. W. Carter, V. W. Smart, and H. W. Garren. 1956. Effect of genistin on growth and development of the male mouse. J. Nutr. 59:235-241.

31. Murillo, E. and J. K. Gaunt. 1975. Invesitgations on alpha tocopherol oxidase in beans, alfalfa, and soybeans. The First Chemical Congress of the North American Continent, Mexico City. Abstract no. 155.
32. Patton, A. R., H. S. Wilgus, Jr. and G. S. Harshfield. 1939. The production of goiter in chickens. Science 89:162.
33. Rackis, J. J. 1974. Biological and physiological factors in soybeans. J. Am. Oil Chem. Soc. 51:161A-174A.
34. Rackis, J. J., D. H. Honig, D. J. Sessa, and F. R. Steggerda. 1970. Flavor and flatulence factors in soybean protein products. J. Agr. Food Chem. 18:977-982.
35. Rackis, J. J., J. E. McGhee, and A. N. Booth. 1975. Biological threshold levels of soybean trypsin inhibitors by rat bioassay. Cereal Chem. 52:85-92.
36. Robinson, L. A., W. J. Kim, T. T. White, and B. Hadorn. 1972. Trypsins in human pancreatic juice—their distribution as found in 34 specimens. Two human pancreatic trypsinogens. Scand. J. Gastroenterol. 7:43-45.
37. Steggerda, F. R., E. A. Richards, and J. J. Rackis. 1966. Effects of various soybean products on flatulence in the adult man. Proc. Soc. Exptl. Biol. Med. 121:1235-1239.
38. Sternberg, M., C. Y. Kim, and F. J. Schwende. 1975. Lysinoalanine: presence in foods and food ingredients. Science 190:992-994.
39. Struthers, B. J., R. R. Dahlgren, and D. T. Hopkins. 1977. Biological effects of feeding graded levels of alkali-treated soybean protein containing lysinoalanine (N^{ϵ}-DL-[2-amino-2-carboxyethyl]-L-lysine) in Sprague-Dawley and Wistar rats. J. Nutr. 107:1190-1199.
40. Sugimoto, H. and J. P. Van Buren. 1970. Removal of oligosaccharides from soy milk by an enzyme from *Aspergillus saitoi*. J. Food Sci. 35:655-660.
41. Turner, R. H. and I. E. Liener. 1975. The effect of the selective removal of hemagglutinins on the nutritive value of soybeans. J. Agr. Food Chem. 23:484-487.
42. Van Beek, L., V. J. Feron, and A. P. DeGroot. 1974. Nutritional effects of alkali-treated soy protein in rats. J. Nutr. 104:1630-1636.
43. Van Wyk, J. J., M. B. Arnold, J. Wynn, and F. Pepper. 1959. The effects of a soybean product on thyroid function in humans. Pediatrics 24:752-760.
44. Woodward, J. C., D. D. Short, M. R. Alvarez, and J. Reyniers. 1975. Biologic effects of N^{ϵ}-(DL-2-amino-2-carboxyethyl)-L-lysine, lysinoalanine, Pt. 2, pp. 595-618. In M. Friedman (ed.) Protein Quality of Foods and Feeds, Vol. 1, Marcel Dekker, New York, NY.

SOY PROTEIN PRODUCT CHARACTERISTICS

M. F. Campbell

Edible soy protein is one of the world's least expensive and highest quality protein sources that is available in large quantities. The worldwide usage of soy proteins as food ingredients has been increasing gradually during the last 50 years. Developments of new technology in recent years have led to substantially greater rates of usage. For maximum utilization of soy protein in human foods, many different soy protein products have been developed.

Since there are so many soy products with different characteristics and properties, I would like to familiarize you with basics of these products. Soy protein products are used in foods for nutritional, functional and economic purposes. However, unless the finished food product is delicious, the other advantages become very difficult to sell. The selection of the appropriate soy protein ingredient is critical to meet the consumer's demands. My goal is to describe basic food ingredients made from soy protein.

BASIC SOY PRODUCTS

The major commercially available soy proteins fall into three categories. These are: (1) soy flour (less than 65% protein), (2) soy protein concentrate (65 to 89% protein), and (3) soy protein isolate (90% and higher protein). The categories and protein percentages are those defined in the FDA's Tentative Final Regulation—Common or Usual Names for Vegetable Protein Products (2). All three categories of products may be texturized by one of several methods to yield various forms of textured vegetable protein.

Soy Flour

Soy flour is the basic form of edible soy protein. Soy flour or grits are made by grinding and/or screening of defatted soy flakes. Defatted soy flakes are prepared by solvent extraction of full fat soy flakes to remove the oil. The basic steps for production of defatted flakes have been described adequately in the literature (5). During the desolventizing process, heat is used to inactivate antinutritional factors found in raw soybeans. Various levels of heat treatments or toasting are used to prepare soy flour or grits for specialized applications. The level of heat treatment is usually specified by Nitrogen Solubility Index (NSI) or Protein Dispersibility Index (PDI).

Soy flour products are available in: (1) various particle sizes (flours are fine; grits are coarse), (2) solubility (measured by NSI), and (3) fat levels (most are sold as defatted products, but some refatted or lecithinated products are available). A typical analysis for defatted soy flour is shown in Table 1. The protein level is 50% or higher while various carbohydrates make up 32 to 34% of the flour. The carbohydrate composition of soy flour is shown in Table II (4). Soy flour is the least expensive soy protein. The cost is presently in the range of $.14 to .16/lb.

Significant amounts of soy flour are used in bakery products as substitutes for non-fat dry milk. Soy flour may be used alone or in combination with dairy whey. A limited amount of soy grits are used in various types of extended meat patties.

Table 1. Defatted soy flour (typical analysis).

	%
Protein (N x 6.25)	51
Fat	1
Fiber	3.2
Ash	5.8
Carbohydrates	32-34
Moisture	5-10

Table 2. Carbohydrate composition of soy flour.

Constituent	% of Flour
Pollysaccharide (Insoluble), Total	15-18
Acidic polysaccharides	8-10
Arabinogalactan	5
Cellulosic material	1-2
Oligosaccharide (Soluble), Total	15
Sucrose	6-8
Stachyose	4-5
Raffinose	1-2
Verbascose	Trace

Soy Protein Concentrate

Soy concentrates contain 70% protein and are made by removing the soluble sugars from soy flour. The three basic processes for removal of the soluble sugars include extraction with (1) 60 to 80% aqueous alcohol, (2) water acidified to about pH 4.5, and (3) water, if the defatted flakes have been heat-treated to totally denature the protein (3). The yield of concentrate via these processes is about 60 to 70% of defatted flake or flour wt (3).

As shown in Table 2, the carbohydrates of soy flour are divided into two categories: polysaccharides (insoluble) and oligosaccharides (soluble). The polysaccharide fraction is primarily nondigestible and provides a dietary fiber type of carbohydrate. However, the oligosaccharide fraction presents a problem. The stachyose, raffinose and verbascose are responsible for flatulence in humans (4). A primary reason for making a soy protein concentrate is the removal of sugars that cuase flatulence, flavor improvement and functionality control. The typical analysis of soy protein concentrate (Table 3) shows the protein level at about 70% (dry substance basis) and carbohydrates at 20.8%. The carbohydrates are primarily the insoluble polysaccharide fraction. The level of stachyose and raffinose is less than 10% of the amount present in soy flour.

Depending on the process used, soy protein concentrate may be produced in flour and/or grit forms. Recent advances in technology have led to some breakthroughs in the production of improved products. Flavor improvement has been the primary result of such efforts. Soy protein concentrate products with low flavor profiles can be used in a wider variety of products than concentrates made via traditional processes. The functional properties of soy protein concentrate make it an ideal protein ingredient for processed meat products. This includes the use of grits in coarse ground meats and fermented sausages, and flour in meat emulsion products, such as bologna-type products. Functional characteristics such as moisture absorption and fat

Table 3. Typical analysis of soy protein concentrate.

	%
Protein (as is)	64.9
Protein (dsb)	70.5
Fat	0.3
Carbohydrates + Fiber[1]	20.8
Moisture	8.0
Ash	6.0
pH	6.8

[1]Primarily dietary fiber, majority non-digestible plus about 1.2% soluble sugar (stachyose, 0.4%; raffinose, 0.1%; and sucrose, 0.7%).

binding make soy protein concentrate an ideal ingredient in almost any meat application.

Some of the major food products in which soy protein concentrate can be used include bakery products, cereal and convenience foods, snack products, baby foods and dry mixes, and pasta products.

At present, concentrates sell for about \$.40/lb. They do hold three times their wt in water and therefore cost about \$.10/lb in hydrated form when used in many meat applications.

Soy Isolates

Soy isolates are 90% protein (dsb) and are made by removing both the soluble and insoluble carbohydrate fractions from the soy flour (3). This is a more complicated process than the soy protein concentrate process and is much more expensive. It does result in protein fractionation and protein losses. The yield of soy protein isolate may vary from 30 to 40% of the starting defatted flake wt (3). In addition to the 90% protein, isolates are composed of ash, fiber and carbohydrates. Commercial isolates are produced in two forms. One is the isoelectric form (pH 4.5) which is nondispersible and the second is the neutralized form (pH 6.7 to 7.3) which is water dispersible (3).

Isolates are the most expensive soy protein products (\$.70 to .90/lb). In the past, isolates have been used primarily for functional properties because of the low flavor level. With the development of low flavor soy protein concentrate, many food processors do not believe it is necessary to use isolates except in a few food applications where the extra functionality is worth the extra cost. The development of functional, low flavor soy flour and soy concentrate products has eliminated the need for isolates in many processed meat applications.

The major applications for isolates are as replacers for dairy protein in dairy type items such as whipped toppings, liquid coffee whiteners and frozen desserts. Instant breakfast preparations and powdered protein supplements are examples of beverage products that contain soy protein isolate.

Textured Protein Products

Soy flour (50% protein), soy concentrate (70% protein) and soy isolate (90% protein) are the three primary soy protein products. Each of these can be texturized by various processes to yield textured soy products. The best known example is the texturization of soy flour by thermoplastic extrusion. This yields the lowest price textured product. Generally the finished products sell for \$.30 to .35/lb.

Texturization of soy flour yields the textured vegetable protein products that have received much publicity in recent years, especially when hamburger prices increase rapidly. Textured soy flour is viewed by some of us in the industry as the first generation product. Textured soy protein concentrates made by the same thermoplastic extrusion process are the second generation of practical, economic textured protein products. The lower flavor profile, lack of flatulence and improved texture properties make textured soy protein concentrate an attractive ingredient for the processed meat industry.

Textured soy protein concentrate is hydrated with 3 parts water to yield a hydrated product that is 17 to 18% protein. Because of the absence of the beany or grassy flavors, textured soy protein concentrates can be used at replacement levels of 30 to 40% in ground meat applications. Procedures have been developed to use concentrates in fabricated meat products, such as beef and pork rolls and similar new fabricated meat foods.

Textured soy protein concentrate products sell for about $.50 to .55/lb. On a hydrated ready-to-use basis, this is $.12 to .14/lb. This is only a few cents per pound above the cost of textured soy flour. The flavor and functional advantages of textured soy protein concentrates allow more meat to be replaced; therefore, the economic advantages become attractive very quickly.

The next class of textured products is spun soy isolates. Although this technology has been available for many years, the products have never been accepted to any great extent. The processing costs plus the cost of isolates make this a very expensive product. Another disadvantage is that this product cannot be dried because it will not rehydrate. Therefore, these products are sold in the frozen state so more costs are incurred in storage and transportation.

Modified Soy Isolates

The seventh and last class of soy products is modified soy isolates. The Staley Company, through its Gunther Products Division, is the only manufacturer of these products in the U.S. Enzymes are used to modify soy isolates to develop specific functional properties. Gunther proteins are used to provide whipping, aeration and foaming properties. Applications include bakery products, confectionery, frozen desserts and bar mixes.

SELECTION OF SOY PRODUCTS AS FOOD INGREDIENTS

The problem of which soy protein product to use is often confusing. Factors to consider include: (1) Product application—does the soy protein work or function in the specific food system? (2) Nutrition—does the soy protein offer the required nutritional factors to the food system? (3) Cost—does the particular soy protein offer the best overall quality and benefit from the cost? (4) Flavor—does the soy protein product have low enough flavor profile so that it doesn't detract from the finished product's quality? (5) Flatulence—does the soy product contain soluble sugars which cause flatulence? This may not be a problem at low usage levels, but is a problem at the high usage levels that are sometimes necessary for optimum economy. (6) Convenience—is the product available in convenient, easy-to-use form? Does it require special storage conditions, such as refrigeration or freezing?

The best approach to selecting a soy product is first to decide on the base product, that is, soy flour, concentrate or isolate. Then decide on particle size or textural properties. The protein quality of soy protein products as measured by the Protein Efficiency Ratio test varies somewhat from product to product. However, there is growing evidence to show that in human studies, soy protein is as good in quality as animal protein. Other nutritional advantages of the soy protein products described include the low level of lipid material and the lack of cholesterol. The low level of fat allows the formulation of food products with the desired fat composition. In addition to the absence of cholesterol, recent research has shown that a diet composed of soy protein reduced the plasma cholesterol level in young, healthy, normolipidemic women (1). Soy flour and soy protein concentrate do contribute dietary fiber from the insoluble carbohydrate fraction of the product. Soy protein concentration contains 7 to 8% neutral detergent fiber.

Since soy protein contains a good balance of amino acids, they are a natural for use with animal proteins. Lysine is present in a larger amount in soybeans than many other cereal grains. This helps provide a synergistic effect of blending soybeans with cereal to yield a well-balanced economical protein.

THE FUTURE

I have tried to describe briefly the basic soybean protein products. However, to make a conference such a this a success, we need to make plans for the future. In this regard, I believe there are several areas where scientists at this meeting can help improve the utilization of edible soy protein. Most of the recommendations deal with changing soybean composition via breeding. These goals are difficult to achieve but they would help increase the consumption of soy protein as food. These recommendations include: (1) increased level of sulfur-bearing amino acids, (2) reduced soluble sugars—no flatus factors, (3) reduced anti-nutritional factors, (4) improve flavor and color, and (5) lower linolinate content.

We are making progress in improving these characteristics of soy proteins via new processing techniques, but giant strides could be made if we could grow soybeans with the desired characteristics. Changing these characteristics by processing is feasible, but does add significantly to the cost. If beans were available with some of these improvements, there would be many more opportunities for increased consumption of soybean protein as human food on a worldwide basis.

NOTES

M. F. Campbell, A. E. Staley Manufacturing Company, Research and Development, Decatur, Illinois 62525.

REFERENCES

1. Carrol, K. K., P. M. Giovannetti, M. W. Huff, O. Moase, D. C. K. Roberts, and B. M. Wolfe. 1978. Hypocholesterolemic effect of substitutiong soybean protein for animal protein in the diet of healthy young women. Am. J. Clin. Nutr. 31:1312-1321.
2. Food and Drug Administration. 1978. Tentative food regulation—common or unusual names for begetable protein products. Federal Register 43 (136):30472-30491.
3. Meyer, E. W. and L. D. Williams. 1976. Soy protein concentrates and isolates, p. 904-917. In L. D. Hill (ed.) Proceedings of the World Soybean Research Conference. Interstate Printers and Publishers, Inc., Danville, IL.
4. Rackis, J. J. 1974. Biological and physiological factors in soybeans. J. Am. Oil Chem. Soc. 51:161A-174A.
5. Smith, A. K. and S. J. Circle. 1972. Soybeans: Chemistry and technology, p. 316-319. In A. K. Smith and S. J. Circle (eds.) Proteins, Vol. 1, AV Publishing Co., Westport, CT.

ANALYSIS OF SOY PROTEIN IN MEAT PRODUCTS

A. C. Eldridge and W. J. Wolf

Textured soybean protein products are finding increased use as meat extenders. They were used initially in institutional feeding programs, and this utilization was given additional impetus in 1971 when the USDA Food and Nutrition Service permitted addition of textured soy products to meat dishes in the National School Lunch Program.

In Europe, only three countries allow soy products in meats. Ten other countries would permit the addition of soy products if adequate procedures were available for measuring the various forms of soybean protein products that might be used, such as flours (50% protein), concentrates (70% protein), and isolates (90% protein). A further complication is that all three of these commercially available products can also be obtained in texturized forms that, additionally, may be artificially colored and may be fortified with vitamins and minerals.

Research effort on methods for the determination of soy products therefore increased in the last few years, particularly in Europe. A summary of more promising techniques will be presented along with their merits and weaknesses. New procedures such as the measurement of stable isotopes and fluorescent techniques will also be discussed.

MICROSCOPY AND HISTOLOGICAL METHODS

Until recently, microscopy and histological methods have been mainly qualitative rather than quantitative. This is because the techniques often detect nonprotein materials such as the carbohydrates accompanying soy protein in soy flour. However, when soy flour or flakes are fractionated, the

the nonprotein components are partially removed and, as a result, the carbohydrate content varies with the type of soy product. That is, a defatted soy flour has 29% carbohydrate, a soy concentrate has 16%, and an isolate may contain only 2% carbohydrate (9). Consequently, it is necessary to know what type of soy product is present and the techniques are not applicable when more than one type is added to the meat item.

Probably, the oldest and the best known microscopy method is inspection for characteristic hour-glass and palisade cells in the residue remaining after extracting with potassium hydroxide (35). More elaborate methods now exist that use different stains that enable the measurement of not only carbohydrates but also proteins. Specifically, detection and even quantitative approximation can be made of textured soy flour (TSF). Smith (41) suggests four useful stains: toluidine blue, iodine, periodic acid/Schiff reagent, and acridine orange. Flint et al. (7,15) quantitated TSF added to meat products by using a toluidine blue stain; they measured TSF with a standard deviation of 1.85% at the 45% level. They report that an experienced person can analyze one or two samples per day, a rate too slow for routine screening of many samples. However, with automation, the analysis could be speeded up, but their staining technique is applicable only to TSF. Concentrates or isolates cannot be determined because the amount of carbohydrate present in these products is variable.

Bergeron and Durand (6), using several stains, developed a histological technique that is reported to be rapid and capable of detecting as little as 1% soybean protein in meat products. They reported satisfactory results with fresh, heated, or putrefied meat containing soy flours, concentrates, or isolates. Parisi et al. (37) in Italy claim that they can detect soy flours, concentrates, and isolates in commercial meat products by the periodic acid-Schiff base reaction.

IMMUNOCHEMICAL TECHNIQUES

Theoretically, immunological techniques should be the best procedures for the determination of nonmeat proteins in meat products because of the high specificity of antibodies and the sensitivity of the antigen-antibody reaction. By having a series of different antibodies, the researcher should be able to determine which commodity has been added to a meat-soy blend and also how much was added. Several authors have demonstrated the potential value of the immunochemical approach. The procedures used by investigators vary considerably, and the immunochemical technique can be applied after electrophoresis in gel media, in solution, or other methods.

Probably the major problem to be dealt with is whether antibodies prepared from an undenatured protein will react with a denatured protein. Several authors have reported that heat treatments (cooking) are detrimental to quantitation of foreign proteins (20,22,25).

West German researchers (5,17,18) reported recently that after heating soy protein to 120 C for 50 min, antigenicity was lost. However, the heated proteins or isolated polypeptide chains can be conjugated with a carrier protein, and corresponding antibodies can be prepared. These antibodies may then be applied to detect heated soy protein in various products.

In 1978, Koh (24) published a very interesting article describing identification and quantitation of the amount of soy protein added to both cooked and uncooked beef mixtures. Koh cooked products to 71 C internal temperature and analyzed the mixtures by rocket immunoelectrophoresis. The author used an unusual technique of preparing antibodies from a "renatured" protein. This approach needs to be investigated further, and the quantitation should be studied thoroughly.

ELECTROPHORESIS

Although many papers have described the use of gel electrophoresis for the determination of nonmeat proteins in meat products, only a few of the more promising methods will be discussed. In 1972, Parsons and Lawrie (38) identified an electrophoretic protein band unique to soy and a protein band characteristic of meat. They quantitated soy added to meat products by measuring the areas of the two bands densitometrically. Essentially this same procedure has been used by several other researchers (19,27,28). Parsons and Lawrie reviewed their most recent research at a symposium on the detection of soy in meat products (39).

Persson and Appleqvist (40) in Stockholm used polyacrylamide gel electrophoresis with dodecylsulphate in the buffer instead of urea, which was employed by Parsons and Lawrie. Thus, they were able to measure the amount of soy protein added to hamburger type foods. Persson and Appleqvist used a different set of protein bands for quantitation than did earlier workers (19,27,28,38).

Another approach has been reported by Fischer and Belitz (11), who identified at least two bands characteristic of soy protein. In a later publication (12), they isolated "band 2" and characterized it by its amino acid composition, N-terminal amino acid, isoelectric point and molecular weight. All procedures that involve the use of electrophoretic bands, especially in the presence of either urea or sodium dodecylsulphate, appear to be very effective systems for determining soy in meat-soy blends; they deserve more effort and collaborative studies.

NOVEL APPROACHES

Several new approaches to the determination of soy in soy-meat have been described recently. Bailey et al. (2-4,31), isolated and characterized a unique penta-peptide (Ser-Glu-Glu-Ala-Arg), which they obtained by trypsin hydrolysis of the major soy protein—the 11S globulin. This method has the advantage that the "tag" is unique (assuming that other plant proteins do not yield the same peptide on hydrolysis). Being covalently bound, the tag is not lost when concentrates or isolates are prepared from soy flour.

Some authors have suggested measuring the amount of canavanine in the soy-meat blend to indicate the amount of soy in the mixture (13), while others recommend measuring the amount of meat present by determining creatine (23) or 3-methyl histidine (21,26).

An approach that may have some promise to estimate the amount of non-meat protein in meat blends is based on computer comparisons of the total amino acid pattern, as described by Lindqvist et al. (29). These Swedish authors prepared a mixture containing 54% milk protein, 36% whey protein, and 10% soy protein. After the amino acid analysis of the blend and a step-wise multiple-regression analysis with a computer, they concluded that their mixture contained 50% milk protein, 37% whey protein, and 13% soy protein. Another group of researchers in the Netherlands is also trying this technique (36).

A method that we have investigated (10) depends on the fact that materials in soy flours, concentrates, and isolates fluoresce at 440 nm when excited at 360 nm. The method involved dissolving the sample in 6M guanidine hydrochloride and measuring the fluorescence of the solution. However, this procedure does not work on cooked products; apparently, additional fluorescent materials form during cooking.

The determination of the ratio of stable isotopes in meat-soy blends (16) has also been investigated. The authors used $^{13}C/^{12}C$ ratios to differentiate corn-fed animal protein (C_4 plant) from soy protein (C_3 plant). The error in this procedure appeared to be rather large, and additional analyses are required in order that a judgement can be made on its merits.

MISCELLANEOUS

Attempts have been made to determine components added to meats by density gradient centrifugation. De Hoog et al. (8) claimed to detect 0.1% nonmeat protein in the presence of starch and polyphosphate by this procedure. In an elaborate paper, Montag (34) reported the densities of 63 substances used in the meat industry. He noted that defatted soybean flour had a band with a density between 1.3 to 1.5 in a bromoform:benzene mixture. Soybean protein occurs at the interface of a carbon tetrachloride:toluene mixture that has a specific gravity of 1.42 (1).

One of the proteins of soy, the 7S protein, has been studied and found to contain ca. 4% of mannose that is bound covalently to the protein. The analysis of soy protein for mannose was indeed successful (9), but the error in measuring a small amount of this sugar was too large for any quantitative work. Carbohydrate analysis of soy products did not indicate any unusual carbohydrates, except pinitol (9). However, measurement of pinitol will not be of any value, because it is not bound covalently and its concentration decreases during fractionation of soy flour to concentrates and isolates.

Isoelectric focusing has been tried by the Laboratory of the Government Chemist in London (14,30,32). Five percent added soy protein isolate (Promine D) in fresh pork sausage could be determined with a relative accuracy of ±20%. Below this level of addition, results were even less accurate. The technique of isoelectric focusing could not be applied to products that had been heated extensively during manufacture. The London researchers also examined high-performance liquid chromatography (14,33) as a means of measuring soybean protein isolates added to pork sausage. After examining column packings and extraction solvents, the authors claimed a relative accuracy of ±15%, but encountered problems of short column life. The analysis was unsuccessful for cooked products.

CONCLUSION

This review reveals the difficulties encountered in attempting to quantitate soy protein added to meat products. Analytical chemists have responded to this challenging and important problem in various ways. At present, it is safe to say that the problem has not been solved satisfactorily. Clearly much more research is needed to develop a method or series of methods applicable to the variety of soy proteins now available as well as to the large number of processed meat products in the U.S. and Europe that are potential candidates for extension and fortification with soy proteins.

NOTES

A. C. Eldridge and W. J. Wolf, Northern Regional Research Center, Agricultural Research, Science and Education Administration, USDA, Peoria, Illinois 61604.

The mention of firm names or trade products does not imply that they are endorsed or recommended by the U.S. Department of Agriculture over other firms or similar products not mentioned.

REFERENCES

1. Anon. 1977. Study and the determination of casein and soybean protein in meat products. Riv. Soc. Ital. Sci. Aliment. 6:143-147.
2. Bailey, F. J. 1976. A novel approach to the determination of soya proteins in meat products using peptide analysis. J. Sci. Food Agric. 27:827-830.
3. Bailey, F. J. and C. Hitchcock. 1977. A novel approach to the determination of soya protein in meat products using peptide analysis. Ann. Nutr. Aliment. 31: 259-260.
4. Bailey, F. J., J. W. Llewellyn, C. H. S. Hitchcock, and A. C. Dean. 1978. The determination of soya protein in meat products using peptide analysis and the characterization of the specific soya peptide used in the calculations. Chem. Ind. 477-478.

5. Baudner, S., H. O. Günther, and A. Schweiger. 1977. Detection of soya protein in canned meat heat treated at 120 C. Mitt. Geb. Lebensmittelunters. Hyg. 68:183-185.
6. Bergeron, M. and P. Durand. 1977. Histological identification of different forms of soya in meat products. Ann. Nutr. Aliment. 31:261-270.
7. Coomaraswamy, M. and O. Flint. 1973. The histochemical detection of soya novel proteins in comminuted meat products. Analyst (London) 98:542-545.
8. De Hoog, P., S. Van den Reek, and F. Brouwer. 1970. Detection and determination of hydrolyzed milk protein and soybean protein in meat products. Fleischwirtschaft 50:1663-1666.
9. Eldridge, A. C., L. T. Black, and W. J. Wolf. 1979. Carbohydrate composition of soybean flours, protein, concentrations, and isolates. J. Agr. Food Chem. 27:799-802.
10. Eldridge, A. C. and L. G. Holmes. 1979. Evaluation of a fluorometric technique for quantitative determination of soy flour in meat-soy blends. J. Food Sci. 44:763-764.
11. Fisher, E.-H. and H.-D. Belitz. 1971. Detection of soybean-protein in meat products by electrophoresis. Z. Lebensm.-Unters.-Forsch. 145:271-272.
12. Fischer, K.-H. and H.-D. Belitz. 1976. Studies of the detection of soybean protein in meat products. Isolation of a characteristic protein zone. Z. Lebensm.-Unters.-Forsch. 162:231-233.
13. Fischer, K.-H., H.-D. Belitz, and G. Kloos. 1976. Studies of the detection of soybean protein in meat products. Detection of canavanine. Z. Lebensm.-Unters.-Forsch. 162:227-229.
14. Flaherty, B. 1975. Current research in the laboratory of the government chemist. Chem. Ind., June 21, 495-497.
15. Flint, F. D. and M. V. Meech. 1978. Quantitative determination of textured soy protein by a stereological technique. Analyst (London) 103:252-258.
16. Gaffney, J. S., A. P. Irsa, L. Friedman, and E. A. Emken. 1979. $^{13}C/^{12}C$ Analysis of vegetable oils, starches, proteins and soy-meat mixtures. J. Agr. Food Chem. 27:475-478.
17. Günther, H. O. 1977. Detection of plant proteins in meat products. I. Methods and results of routine work. Ann. Nutr. Aliment. 31:179-182.
18. Günther, H. O., A. Schweiger, and S. Baudner. 1977. Detection of plant proteins in meat products. II. Recent developments. Ann. Nutr. Aliment. 31:229-230.
19. Guy, R. C. E., R. Jayaram, and C. J. Willcox. 1973. Analysis of commercial soya additives in meat products. J. Sci. Food Agric. 24:1551-1563.
20. Herrmann, C., C. Merkl, and L. Kotter. 1977. The problem of serological reactions for detecting foreign proteins in heated meat products. Ann. Nutr. Aliment. 31: 153-156.
21. Hibbert, I. and R. A. Lawrie. 1972. Technical note: the identification of meat in food products. J. Food Technol. (London) 7:333-335.
22. Kamm, L. 1970. Immunochemical quantitation of soybean protein in raw and cooked meat products. J. Assoc. Off. Anal. Chem. 53:1248-1252.
23. Khan, A. W. and D. C. Cowen. 1977. Rapid estimation of muscle proteins in beef-vegetable protein mixtures. J. Agr. Food Chem. 25:236-238.
24. Koh, T. Y. 1978. Immunochemical method for the identification and quantitation of cooked or uncooked beef and soya proteins in mixtures. Can. Inst. Food Sci. Technol. J. 11:124-128.
25. Koie, B. and R. Djurtoft. 1977. Changes in the immunochemical response of soybean proteins as a result of heat treatment. Ann. Nutr. Aliment. 31:183-186.
26. Lawrie, R. A. 1977. Meat identification in fresh and processed foods. Food Technol. Aust., September 1977, pp. 347-349.

27. Lee, Y. B., D. A. Rickansrud, E. C. Hagberg, E. J. Briskey, and M. L. Greaser. 1975. Quantitative determination of soybean protein in fresh and cooked meat-soy blends. J. Food Sci. 40:380-383.
28. Lee, Y. B., D. A. Rickansrud, E. C. Hagberg, and R. H. Forsythe. 1976. Detection of various nonmeat extenders in meat products. J. Food Sci. 41:589-593.
29. Lindqvist, B., J. Ostgren, and I. Lindberg. 1975. A method for the identification and quantitative investigation of denatured proteins in mixtures based on computer comparison of amino-acid patterns. Z. Lebensm.-Unters.-Forsch. 159:15-22.
30. Llewellyn, J. W. 1977. Detection of soya protein by isoelectric focusing. Proc. Div. Chem. Soc. 14:75-76.
31. Llewellyn, J. W., A. C. Dean, R. Sawyer, F. J. Bailey, and C. H. S. Hitchcock. 1978. Technical note: The determination of meat and soya proteins in meat products by peptide analysis. J. Food Technol. (London) 13:249-252.
32. Llewellyn, J. W. and B. Flaherty. 1976. The detection and estimation of soya protein in food products by isoelectric focusing. J. Food Technol. (London) 11:555-563.
33. Llewellyn, J. W. and R. Sawyer. 1977. Application and limitation of isoelectric focusing and high performance chromatography in the estimation of soya proteins in meat products. Ann. Nutr. Aliment. 31:231-232.
34. Montag, A. 1973. Application of density-gradient centrifugation for investigation of auxiliary material used in meat processing. Z. Lebensm.-Unters.-Forsch. 153: 213-223.
35. Official Methods of Analysis. 1975. Assoc. Off. Anal. Chem., 12th Ed., p. 424, method 24.045.
36. Olsman, W. J. 1979. Methods for detection and determination of vegetable proteins in meat products. J. Am. Oil Chem. Soc. 56:285-287.
37. Parisi, E., A. Maranelli, and A. Giordano. 1974. Histological detection of soybean protein in cooked meat products. Vet. Ital. 25:384-391.
38. Parsons, A. L. and R. A. Lawrie. 1972. Quantitative identification of soya protein in fresh and heated meat products. J. Food Technol. (Lond.) 7:455-462.
39. Parsons, A. L. and R. A. Lawrie. 1977. Quantitative identification of plant proteins in food products. A procedure based on laser densitometry of proteins extractable in 10 M urea and resolved by thin-layer polyacrylamide electrophoresis. Ann. Nutr. Aliment. 31:201-206.
40. Persson, B. and L. A. Appleqvist. 1977. The determination of non-meat proteins in meat products using polyacrylamide gel electrophoresis in dodecylsulphate buffer. Ann. Nutr. Aliment. 31:225-228.
41. Smith, P. R. 1975. Detection and estimation of vegetable protein in the presence of meat protein. Inst. Food Sci. Technol. (U.K.), Proc. 8:154-162.

ORIENTAL SOYBEAN FOODS

D. Fukushima and H. Hashimoto

Soybeans have been an important source of protein, fat, and flavor for Oriental people for thousands of years. A large variety of foods which were developed from soybeans can be classified into two groups; fermented and non-fermented foods. Recently, several reviews have been written in English on these soybean products (1,3-7). The main fermented soybean foods are shoyu (soy sauce), miso (fermented soy paste), sufu (Chinese soybean cheese), tempeh (fermented soybean cake), and natto (fermented whole soybeans). The development of fermented foods, which depends on the use of rather sophisticated microbiology, was a remarkable achievement in the early history of China.

The ancestor of shoyu (soy sauce) and miso (fermented soybean paste) is "chiang" which originated in China approximately 2,500 years ago. Chiang was introduced in Japan during the 7th century by Buddhist priests, who taught the people to avoid eating meat and fish. Therefore, this new seasoning from soybeans gradually took the place of the salted seasoning prepared from meat or fish which had been previously used. During the following centuries, chiang was transformed into the present Japanese shoyu and miso, which are now quite different from their Chinese counterpart.

Sufu (Chinese soybean cheese) also originated in China in ancient times. Tempeh (fermented soybean cake) is used mostly in Indonesia. On the other hand, natto is the fermented soybean food originating in Japan in ancient times, and it is manufactured and consumed only in Japan. In Japan, about 439,000 tons of soybeans and defatted soybeans are used per year to make these fermented soybean foods. As shown in Table 1, however, 586,000 tons are used for the traditional non-fermented soybean foods.

Table 1. Consumption of soybeans and defatted soybeans in Japan, 1976

	Whole Soybean	Defatted Soybean Grits	Total
		– ton –	
For Foods			
Fermented			
Shoyu (soy sauce)	10,000	165,000	175,000
Miso (fermented soypaste)	190,500	5,000	195,500
Natto (fermented whole soybeans)	69,000	0	69,000
Non-Fermented			
Tofu and Aburage	411,500	55,000	466,500
Kori-tofu	29,000	0	29,000
Others	16,000	75,000	91,000
For Feeds	30,000	1,950,000	1,980,000
Total	756,000	2,250,000	3,006,000

The main non-fermented soybean foods are tofu (soy mild curd), kori-tofu (dried tofu), yuba (coagulant film of soy milk), kinako (roasted soybean powder), and moyashi (soybean sprout). Among these non-fermented soybean foods, tofu and its derivatives are the largest. The amount of soybeans used for the traditional soy milk curd, tofu, is about 466,500 tons/yr in Japan. However, as tofu is perishable and difficult to transport for long distances due to its high moisture content, the scale of tofu production is fairly small, in spite of its large demand. There are over 35,000 tofu plants in Japan.

During the last two decades, the technology and engineering on these traditional foods have made great progress, particularly in the areas of soy sauce and miso in Japan. In this chapter, the manufacturing techniques and their recent progress on these products are described.

FERMENTED SOYBEAN FOODS

Shoyu (Soy Sauce)

Shoyu is the Japanese type of soy sauce with a salty taste and sharp flavor which is made by fermenting soybeans, wheat, and salt (6,7). Shoyu is the only traditional fermented product that has become well-known in the cookery of Western countries, and is often referred to as soy sauce. It is used widely throughout East Asia and many other countries with different designations: chiang-yu in China, tao-yu in Indonesia, tayo in the Philippines, and kanjang in Korea.

There are five kinds of shoyu recognized by the Japanese government at present, as shown in Table 2. Koikuchi-shoyu is a representative Japanese shoyu which forms 85.4% of the annual production of shoyu in Japan. It is

Table 2. Typical composition of different kinds of shoyu.

Kind	Be	NaCl	Total Nitrogen	Formol Nitrogen	Reducing Sugar	Alcohol	pH	Color	Annual Production
			– % (W/V) –			– % (V/V) –			– % –
Koikuchi	22.5	17.6	1.55	0.88	3.8	2.2	4.7	Deep Brown	85.4
Usukuchi	22.8	19.2	1.17	0.70	5.5	0.6	4.8	Light Brown	11.7
Tamari	29.9	19.0	2.55	1.05	5.3	0.1	4.8	Dark Brown	2.2
Saishikomi	26.9	18.6	2.39	1.11	7.5	Trace	4.8	Dark brown	0.3
Shiro	26.9	19.0	0.50	0.24	20.2	Trace	4.6	Yellow to Tan	0.4

an all-purpose seasoning agent used in the preparation of foods as well as a table condiment, and it is characterized by a strong aroma and deep brown color. Usukuchi-shoyu, which forms 11.7% of the total production, is characterized by its lighter brown color and milder flavor. It is used mainly for cooking when one wishes to preserve the original flavor and color of the foodstuff itself. Tamari is the Chinese type shoyu and is made up of only soybeans or a mixture with a higher percentage of soybeans than wheat. Tamari, saishikomi, and shiro-shoyu are produced and consumed only in certain localities for special uses.

Shoyu manufacturing consists of three major processes: the koji making process, brine fermentation, and refining (Figure 1). Traditionally, whole soybeans are used for koji making, however, in recent years defatted soybean grits or flakes have taken its place.

The cooking of soybeans is carried out in a batch-type cooker. Recently, however, a continuous cooker has been developed by which a high pressure short-time cooking of soybeans is carried out.

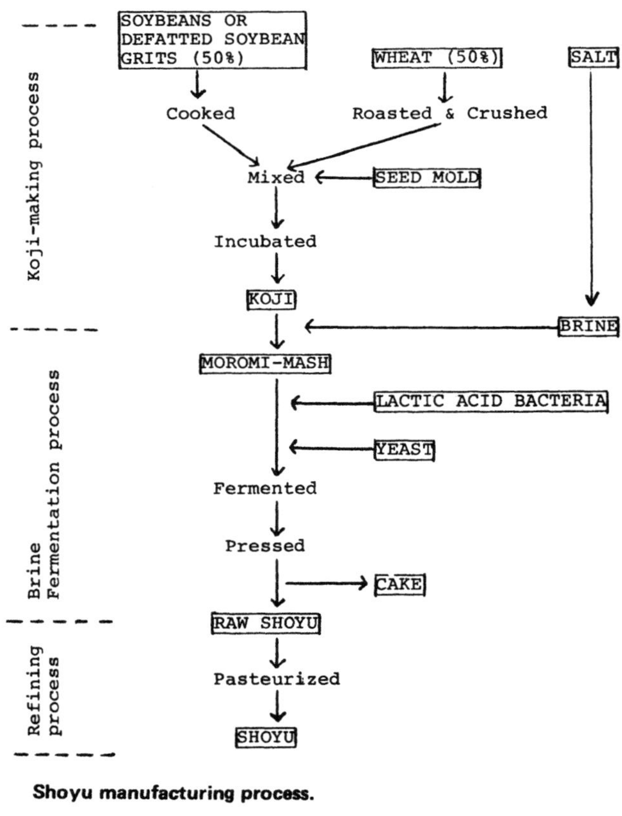

Figure 1. Shoyu manufacturing process.

The wheat is roasted in a continuous roaster and then cracked into pieces. Next, equal amounts of cooked soybeans and wheat are mixed and then inoculated with a pure culture of *Aspergillus sojae* or *Aspergillus oryzae.* Traditionally, this mixture is put into wooden trays. During koji making, the mixture in each tray is stirred by hand three times to cool. In recent years, however, automatic koji making processes have been developed to replace this traditional way. The substrate mixture is put into a large shallow perforated vat and the temperature and moisture controlled air are circulated through the mass to give the proper conditions for mold cultivation and enzyme formation. After two or three days, the moldy material which is called "koji" is harvested. The harvested koji is transferred to deep fermentation vessels with about a 25% salt brine. The resulting mixture is called moromi and moromi mash.

During the brine fermentation for six to eight months, enzymes dervied from koji hydrolyze most of the proteins of the materials into peptides and amino acids. Much of the starch of the wheat is converted to simple sugars, which are fermented primarily to lactic acid, alcohol, and carbon dioxide. The pH drops from an initial value of 6.5 to 7.0 down to a pH of 4.7 to 4.8. The high salt concentration, around 18%, effectively limits growth to a few desirable types of microorganisms. At the first stage of moromi mash, *Pediococcus halophilus* is grown and it produces lactic acid which lowers the pH. At the second stage, *Saccharomyces rouxii* is grown and as a result, a vigorous alcoholic fermentation occurs. Recently, the pure cultured *Pediococcus halophilus* and *Saccharomyces rouxii* are often added to the moromi mash to accelerate the lactic acid and alcoholic fermentations, respectively. At the last stage of moromi fermentation, *Torulopsis* are grown, which are a group of salt resistant yeasts. These strains produce a mature aroma which is composed of phenolic compounds that are important as aroma compounds of soy sauce.

Usually, moromi is fermented in indoor vessels. Recently, however, closed outdoor moromi fermenters have been developed and are available commercially in Japan.

The final process is refining which includes filtering and pasteurizing. The aged moromi mash is put into a cloth and the liquid part of the mash is separated from the residue with a hydraulic press until the moisture content of the residue becomes about 25%. The filtrate of an aged mash is heated at 70 to 80 C by a plate heater. This heating is necessary to develop the color and aroma, and to inactivate most enzymes. Heating of the raw shoyu causes the inactivated enzyme proteins to coagulate and form a sediment which reaches 5 to 10% of the volume of the finished product. The sediment could occur mainly through hydrophobic interactions among the denatured protein molecules (2). After removing the sediment by filtration, clear refined shoyu is bottled and packaged. Sometimes, sodium benzoate is added to shoyu as a preservative, but the recent tendency is for asceptical bottling without adding any synthetic preservatives. More than 50% of the soy sauce on the market in Japan does not contain any preservatives.

There were two marked developments in shoyu technology during the recent two decades. One was the improvement of the treating method of soybeans and the other was the mechanization of the processes. Thus, the yield of shoyu production on the basis of total nitrogen increased from 65 to almost 90% during the past 25 yr. Furthermore, the quality of shoyu has been very much improved.

Miso (Fermented Soy Paste)

Miso is also a fermented product of soybeans and cereals in the presence of salt, which is used widely throughout East Asia. The progenitor of miso is referred to as "chiang" developed in China long before the Christian era, and transformed into "jang" in Korea, "miso" and "shoyu" in Japan, "tao-tjo" in Indonesia and Thailand, and "tao-si" in the Philippines.

There are many kinds of "chiang" in China. Actually, the term "chiang" includes a very wide range of foods. Most of chiang in China is prepared at home, just as the Western people make their own jams and pickles. On the other hand, miso in Japan is now manufactured commercially in a modernized factory on a large scale.

There are many varities of miso in Japan as there are chiang in China based on the ratio of substrates, salt concentration, and the length of fermentation and aging. In appearance, most of the miso in Japan is a paste resembling peanut butter in consistency and smoothness in texture. Its color varies from a creamy yellowish white to a very dark brown. Generally speaking, the darker the color the stronger the flavor. The product is typically salty and has a distinctive pleasant aroma.

Miso can be classified into three major types on the basis of the raw materials as shown in Table 3. Rice miso is made from rice, soybeans and salt; barley miso is made from barley, soybeans, and salt; and soybean miso is made from soybeans and salt. These types are further classified by taste into three groups: sweet miso, semi-sweet miso, and salty miso. Each group is further divided by color into white-yellow miso and red-brown miso groups. Among these miso, rice miso is the most popular one, which forms 81% of total miso consumption.

In miso manufacturing, methods differ from variety to variety, but basically the process is the same. The production process of rice miso which is the most popular is outlined in Figure 2. There are two basic differences between miso and shoyu manufacturing, though both are very much alike. One is in koji making. The koji of shoyu is made by using all the raw materials, that is, the mixture of soybeans and wheat, whereas the koji of miso is

Table 3. Chemical compositions of major types of miso.

Classification			Aging	Chemical Composition					Annual Production	
Material	Taste	Color	Time	Protein	Fat	Carbohydrate	Ash	Moisture		
						– % –			– Ton	– % –
Rice miso									467,000	81
	Sweet	Yellowish white	5-20 days	11.1	4.0	35.9	7.0	42.0		
		Reddish brown	5-20 days	12.7	5.1	31.7	7.5	43.0		
	Semi-sweet	Bright light yellow	5-20 days	13.0	5.4	29.1	8.5	44.0		
		Reddish brown	3- 6 months	11.2	4.4	27.9	14.5	42.0		
	Salty	Bright light yellow	2- 6 months	13.5	5.9	19.6	14.0	47.0		
		Reddish brown	3-12 months	13.5	5.9	19.1	14.5	47.0		
Barley miso									63,000	11
	Semi-sweet	Yellowish to reddish brown	1- 3 months	11.1	4.1	29.8	13.0	42.0		
	Salty	Reddish brown	3-12 months	12.8	5.2	21.1	15.1	46.0		
Soybean miso									46,000	8
	Salty	Dark reddish brown	5-20 months	19.4	9.4	13.2	13.0	45.0		

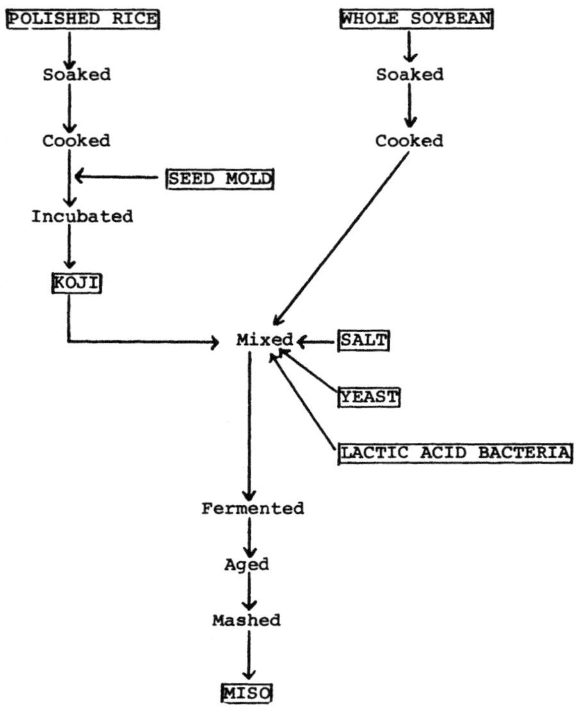

Figure 2. Rice miso manufacturing process.

made by using only carbohydrate material, that is, rice or barley. The soybeans are used in miso making without the inoculation of koji mold, except on soybean miso. The other difference is that miso is a solid paste and, therefore, the making process has no need for filtration process which has a very large influence on the cost in shoyu making. The fungus and yeast used in miso manufacturing are very similar and sometimes the same as in shoyu manufacturing.

Recently, a revolution has occurred in the manufacture of miso, which has taken the form of automated equipment and continuous processing. Particularly, the use of a rotary fermenter is effective in the preparation of rice or barley koji. Cooked and mold inoculated rice is put into the large trommel of this rotary fermenter in which air is circulated at a controlled temperature and humidity. The trommel is rotated several times to prevent agglomeration of the rice during the fermentation. After the completion of the fermentation, the resulting koji is mixed with salt, cooked whole soybeans, pure cultured yeasts and lactic acid bacteria, and water, and then kept for an appropriate period for the second fermentation. The resulting aged mixture is mashed and packed as miso.

Sufu (Chinese Soybean Cheese)

Sofu is a soft cheese type product which is made from soy milk curd by the action of microorganisms. Sofu which originated in China in the fifth century has been consumed widely as a relish by Chinese people. However, sufu is not consumed in Japan.

The sufu making process consists of three major process which are outlined in Figure 3. The first step, the soy milk curd making process, is substantially the same process as the tofu making process. However, the water content of tofu is reduced to less than 70% when making sufu. The normal water content of tofu is about 90%.

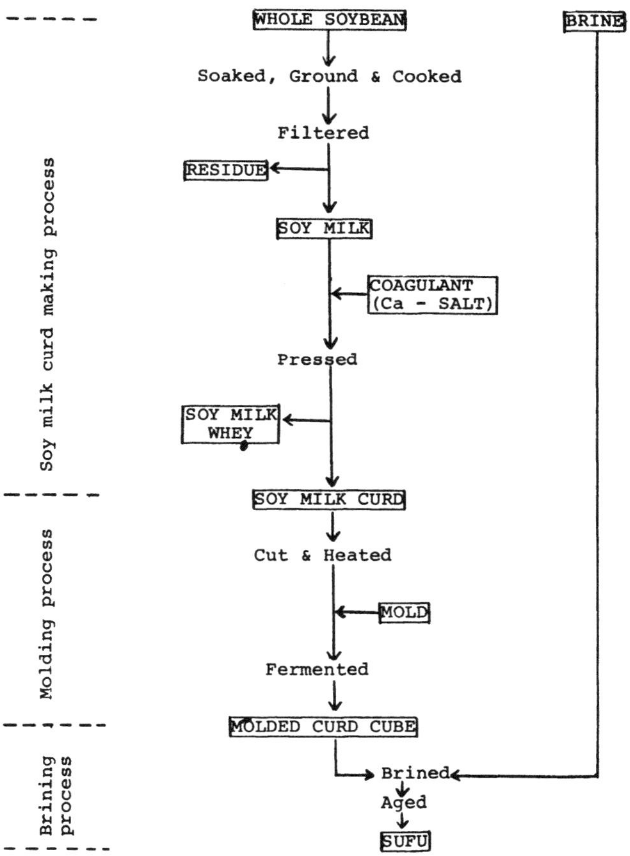

Figure 3. Sufu manufacturing process.

The second process of sufu making is the molding process. After the hard-made tofu is cut into 3 cm cubes, the cubes are heated for pasteurization and for lowering the water content of the cube surface, and then the mold is inoculated onto it. In obtaining good quality sufu, the mold for sufu fermentation must have a white or yellowish-white mycelium to warrant an attractive appearance. The texture of mycelial mat should also be dense and thick so that a firm film will be formed over the surface of the fermented tofu cubes to prevent any distortion in its shape.

The molds belonging to the genus of *Mucor* or *Actinomucor* are used usually, but the molds belonging to the genus of *Rhizopus* are also sometimes used. For instance, *Actinomucor elegans, Mucor hiemalis, Mucor silvaticus, Mucor praini, Mucor subtilissimus,* and *Rhizopus chinensis* are used for the fermentation. The time of mold fermentation differs depending upon the varieties of mold. It takes about 7 days at 12 C in *Rhizopus chinensis,* 3 days at 20 C in *Mucor hiemalis* and *Mucor silvaticus,* and 2 days at 25 C in *Mucor praini.*

The last step in making sufu is brining and aging. The freshly molded cubes are placed in various types of brining solutions depending on the flavor desired. The usual brining solutions are fermented rice mash with salt, soy sauce moromi mash, fermented soy paste, or a 5% salt solution containing rice wine amounting to about 10% ethyl alcohol. The time of aging ranges from 1 to 12 months depending upon the variety of the brining solution. After aging, the product is bottled with brine, sterilized, and marketed as sufu. Because sufu is a cream cheese type product and has a mild flavor, Western people would find it suitable as a cheese-type dish.

Tempeh (Fermented Soybean Cake)

Tempeh is one of the most important fermented soybean foods originating in Indonesia and is consumed widely in Indonesia, Malaysia and the Philippines. Unlike most of the other fermented soybean foods, which are usually used as flavor agents or relishes, tempeh serves as a main dish in these countries. Tempeh is not produced and consumed in Japan. Tempeh is a cake-like product made by fermenting soybeans with *Rhizopus.* A typical process for making tempeh is outlined in Figure 4A.

The soybeans are soaked in water, dehulled and boiled for 1 hr with excess water. The cooked soybeans are spread out for surface drying and then mixed with a previous batch of tempeh. The inoculated soybeans are wrapped with banana leaves and fermented at room temperature for 1 day. During the fermentation, the soybeans are covered with white mycellium of mold and bound together by the mycellium as a cake. The cake of tempeh is sliced and fried, or if preferred, dipped in brine or spices before frying. It can also be baked or added to soup. By itself, tempeh has a bland flavor. Recently, in a

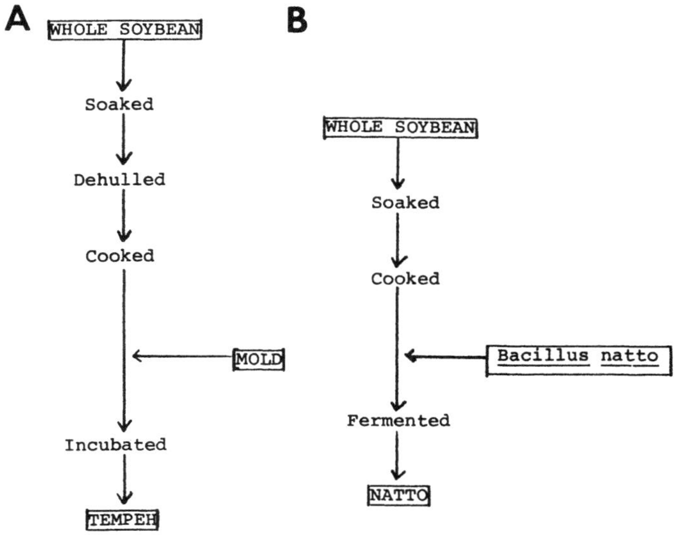

Figure 4. (A) Tempeh manufacturing process, and (B) natto manufacturing process.

larger scale fermentation of tempeh, the pure cultures of molds such as *Rhizopus oligosporus, Rhizopus oryzae, Rhizopus arrhizus,* and *Rhizopus stolonifer* have sometimes been used.

Natto (Fermented Whole Soybeans)

Natto is a traditional fermented soybean food originating in the Northeastern parts of Japan about 1,000 yr ago. There are two major types of natto. One is itohiki-natto and the other is hama-natto. Hama-natto is a food fermented by *Aspergillus* species and rather resembles soybean miso in taste. The production scale of hama-natto is very small, as it is produced only in certain localities. On the other hand, itohiki-natto is very popular and produced in large amounts, particularly around Tokyo and the Eastern parts of Japan. Therefore, the word natto usually means itohiki-natto.

Natto is a unique soybean product fermented by *Bacillus natto.* In this product, the shape of cooked whole soybean particles remains intact, and the surface of the particles is covered with a very viscous substance which consists of glutamic acid polymers produced by *Bacillus natto.* The manufacturing process is outlined in Figure 4B.

The manufacturing process of natto is very simple and the time of fermentation is short. A small quantity of the inoculated cooked whole soybeans is put into a small plastic tray with cover and packed. The resulting packed tray is kept at 40 C for the fermentation by *Bacillus natto.* After 14 to 18 hr the packed tray is cooled to 2 to 7 C and then shipped to the market.

Natto is a very cheap and nutritious protein food and its annual production is about 124,000 tons. Natto is tasty and does not have much odor, but its viscous fluid is distinctive. Natto is served with shoyu and mustard.

Besides these traditional fermented foods, a new fermented soybean product appeared on the marked recently in Japan. It is a soy milk drink fermented by lactic acid bacteria. This product and others are now being researched for possible future use of soybeans.

NON-FERMENTED SOYBEAN FOOD

Tofu (Soy Milk Curd)

Tofu is a traditional non-fermented soybean food which has been consumed for thousands of years in Oriental countries. Due to a high content of protein and fat (Table 4), tofu has made a substantial contribution to nutrition as well as fish and meat.

To make tofu, soybeans are washed, soaked in water overnight, ground with a continuous addition of small amounts of water and heated at 100 C for several minutes with a definite amount of water. The final ratio of water to the original weight of soybeans is about 10 to 1. A soy milk of 5 to 6% solid is obtained by filtration. The calcium sulfate solution is added to the milk at a ratio of 2 to 3% of the soybeans at 70 to 80 C. After the resultant coagulated curd has been kept for 10 min, the curd mixture is poured into a coarse cloth set in a wooden box and pressed to remove the excess whey. Then the curd is taken out from the wooden box and placed in water. It is then cut in thick slices of about 300 to 500 g and sold while fresh.

The above-mentioned tofu is called momen-tofu or ordinary tofu. The process of ordinary tofu is outlined in Figure 5A. In the process as shown in Figure 5A, when the concentration of soy milk is higher, say around 10%, the soy milk is coagulated as a whole without separating any whey. Therefore, a different type of tofu, called kinugoshi-tofu is made. This product is rather smooth and softer than ordinary tofu, and it holds soluble whey components such as soluble protein, sugar and vitamin B1, in itself.

Table 4. Chemical composition of traditional non-fermented soybean foods.

	Moisture	Protein	Fat	Soluble Carbohydrate	Fiber	Ash
			– % –			
Tofu	88.8	6.0	3.5	1.9	0	0.6
Aburage	44.0	18.6	31.4	4.5	0.1	1.4
Kori-Tofu	10.4	53.4	26.4	7.0	0.2	2.6
Yuba	8.7	52.3	24.1	11.9	0	3.0
Kinako	5.0	38.4	19.2	29.5	2.9	5.0
Soybean sprout	93.8	3.0	0.1	2.2	0.6	0.3
Soybean	12.0	34.3	17.5	26.7	4.5	5.0

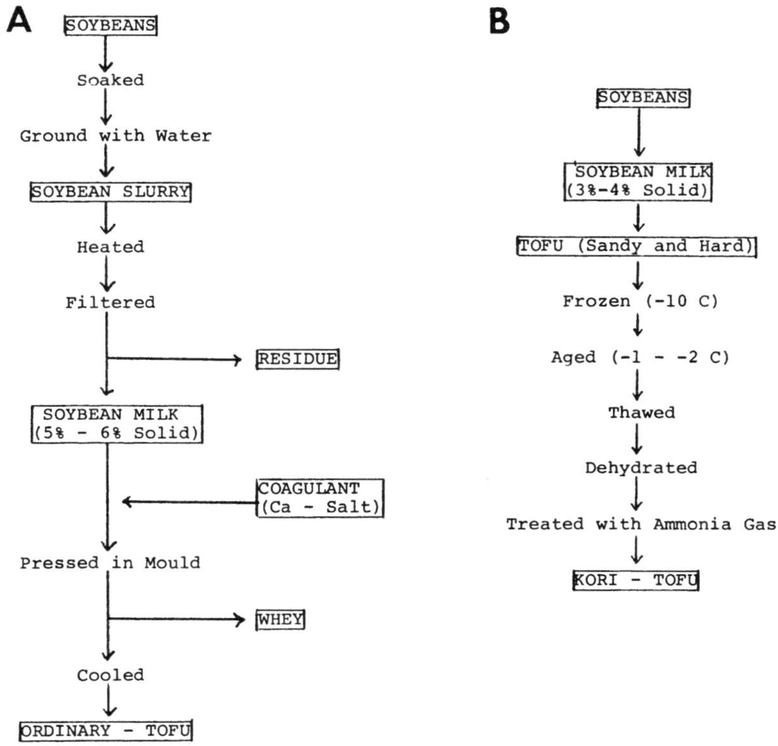

Figure 5. (A) Ordinary-tofu manufacturing process, and (B) kori-tofu manufacturing process.

As shown in Table 4, tofu contains 88% moisture, 6% protein, and 3.5% oil. It is perishable, fragile, and difficult to transport for long distances.

Recently, a new and improved product, packaged tofu, has appeared. Packaged tofu is made by applying the process of kinugoshi-tofu manufacturing in which all soy milk components are retained. After being cooled, soy milk of a higher solid concentration (about 10%) than ordinary tofu is poured into a plastic (poly-styrene or polypropylene) film container with calcium sulfate or gluconodelta-lactone. The container which is sealed tightly, is heated in a water bath at 90 to 95 C for about 50 min and then cooled. The milk coagulates gradually as a whole without separating the whey. Packaged tofu is easy to transport and somewhat more preservable than ordinary and kinugoshi-tofu.

There are several traditional tofu derivatives such as aburage (deep-fried tofu) and yaki-tofu (baked tofu). Recently, packaged tofu and fried tofu have been manufactured on a large scale due to easy transportation and good preservation qualities.

Kori-Tofu (Dried Tofu) and Yuba (Coagulant Film of Soy Milk)

Tofu holds a great deal of water in a thick gelatinous texture, and it is very hard to make a dried product without any damage to the constituents. However, we can see how elegantly this difficult problem is solved in the traditional kori-tofu making process. The kori-tofu making process is outlined in Figure 5B.

Tofu is frozen rapidly to -10 C within a 3 hr period and it is kept at -1 to -2 C for 2 to 3 weeks for aging. Aging gives the tofu a sponge-like quality after thawing. This characteristic texture is brought about by the molecular interaction in the locally condensated protein solution among the ice crystals. The bonds responsible for the polymerization of soy milk protein causing the characteristic texture might be hydrophobic bonds and S-S bonds.

After being dehydrated by pressing, tofu is dried by forced hot air and exposed to ammonia gas to make kori-tofu large and soft when soaked in hot water at the time of cooking. As shown in Table 4, the protein content of kori-tofu is 53% or more, and the oil content is 25% or more. These values are much higher than those of the original soybeans. As kori-tofu is preservable and easy to transport, its production scale is much larger than that of tofu. The consumption of soybeans for kori-tofu is about 30,000 tons/yr in Japan.

There is one more traditional non-fermented soybean food, yuba, which is produced by heating soy milk at a temperature just below the boiling point in a flat pan; the resulting coagulant film is scooped up successively by a fine stick and dried. Formation of the film may be also explained by the occurrence of polymerization through hydrophobic interaction and S-S bonding among soy milk protein monomers or oligomers. Yuba is manufactured on a small scale in certain localities of Japan. Kori-tofu and yuba are served after cooking with seasonings or other ingredients.

Kinako (Roasted Soybean Powder) and Moyashi (Soybean Sprout)

Kinako is the powder of roasted soybeans. The flavor and digestability of soybeans are improved by roasting and powdering. Chemical composition of kinako is similar to that of soybeans (Table 4). This simple food, kinako, is eaten by adding sugar and then sprinkling on rice cake or cooked rice by the Japanese people.

Moyashi is the sprout of soybeans. It is made by soaking the soybeans in water and then the moistened soybeans are moved to a dark room with a temperature of 22 to 23 C. The beans begin to sprout within 24 hr. After 1 week the srpouts are harvested. They are approximately 10 cm long. For moyashi manufacturing, some other beans such as green beans are used as well as soybeans. Moyashi is rich in vitamin C and is one of the most popular vegetables in the Oriental countries.

CONCLUSION

The fermented and non-fermented soybean foods described in this chapter are mostly traditional foods in Oriental countries. Traditional foods are closely related to the regional food cultures and therefore some of their flavors might be unacceptable for the people who are not involved in those food cultures. These traditional foods are the products which have been developed through thousands of years and have survived the long experiences of the people in Oriental countries. It is possible these flavors would be acceptable for most people even though it might be hard to accept on the first trial. Consider the fermented soy sauce (shoyu) in the U.S. It had first been consumed only by the Oriental people in the U.S. Today, however, the fermented soy sauce is penetrating rapidly into the population as a whole, indicating that a traditional flavor in a certain area can also be accepted by most of the people in other areas. This might be mentioned with regard to most of the other soybean foods if the quality is superior. Considering the important food value of soybeans, it might be of great significance to take a further look at these traditional foods by using scientific methods and developing new food technology in order to appeal to world population.

NOTES

Danji Fukushima and Hikotaka Hashimoto, Kikkoman Foods, Inc., P. O. Box 69, Walworth, Wisconsin 53184.

REFERENCES

1. Fukushima, D. 1978. Proc. Int. Soya. Protein Foods Conf., Singapore, p. 39.
2. Hashimoto, H. and T. Yokotsuka. 1974. J. Ferment. Tech. 52:747.
3. Hesseltine, C. W. and H. C. Wang. 1972. Soybeans: Chemistry and Technology, Vol. 1, AVI Publishing Co., Westport, Connecticut, p. 389.
4. Watanabe, T., H. Ebine, and M. Okado. 1974. New Protein Foods, Vol. 1A, Academic Press, New York, p. 415.
5. Wantanabe, T. 1978. Proc. Int. Soya Protein Foods Conf., Singapore, p. 35.
6. Yang, F. M. and B. J. B. Wood. 1974. Advanced Appl. Microbiology, Vol. 17, Academic Press, New York, p. 157.
7. Yokotsuka, T. 1960. Advanced Food Res. 10:75.

FOODS FROM WHOLE SOYBEANS

A. I. Nelson, L. S. Wei, and M. P. Steinberg

This chapter discusses research on home and village use of soybeans. Some of the information concerning this subject was published as an INTSOY bulletin (6). The objectives of this research were to develop whole soybean preparation concepts that would result in highly acceptable products, encompass short preparation times and require modest amounts of energy (fuel) during preparation. This study was directed towards developing concepts suitable for home and village use in developing countries. However, use of these concepts in highly developed countries is just as appropriate and could be carried out with greater ease due to such kitchen devices as blenders, motorized grinders, a greater assortment of cooking utensils and more effective cooking facilities.

Whole soybeans have been used for human food in the Orient since recorded history. However, in other parts of the world, very few whole soybeans are used for food. Almost the entire world soybean crop is used for oil extraction and high protein meal production. The oil is used for human food while about 97% of the oil-free meal is sold as a protein supplement for animal feed. The remaining 3% is used mainly for preparing textured vegetable protein, soy concentrates, and isolates for human food use. Thus, sound concepts for using whole soybeans for human food appear to offer much promise for solving world protein and calorie needs.

The diet of many millions of people in various parts of the world is deficient in protein and calories. At this time one of the best answers to this shortage is whole soybeans. This statement seems to imply that if whole soybeans are available and, if added to the diet, the problem of protein malnutrition would be solved immediately. Incorporation of a new crop into the diet of a local people is rarely easy. However, introduction of a new cereal or vegetable crop is relatively simple as compared to soybeans. There are several reasons for this situation. In many parts of the world, over a period of many years, soybeans have been tried and rejected by local people. One very important reason for this reaction was that many of the foods that were prepared had "beany" or "painty" flavors and odors which were considered unacceptable in most parts of the world. Another serious drawback to the use of soybeans was the long cooking time required to achieve acceptable tenderness (4,10). Cooking time for whole soybeans might be as long as 3 hr and this discouraged housewives. Still another and increasingly important factor is the fuel required for such long cooking periods. Thus, whole soybean use at the home and village level has been beset by problems which were usually not associated with many other new plant food introductions.

The previous statement has considered some of the problems encountered when efforts are made to incorporate whole soybeans into the human diet. Answers to the problems may be offered readily but practical solutions are at best formidable. The material to be presented in this chapter offers an answer to one important question: acceptability of the finished foods. The products described here have been sampled by hundreds of people from all parts of the world. Acceptance has been highly positive with some individuals suggesting that certain local ingredients, including spices and flavors, should be added to some products to obtain a wider acceptance in their country. Thus, it is believed that essentially all product concepts can be used, following certain local recipe additions or deletions, in a manner which will make the foods highly attractive.

This chapter is limited to the development of sound concepts for preparing highly acceptable soyfoods for home or village use. However, it would be remiss not to mention the final obstacles that must be overcome if soybeans are to be accepted and used on a regular and continuing basis.

In developing countries a highly palatable soybean food can be prepared and served for a gathering of local people. Although this food may be received with enthusiasm, it is doubtful that any of the families in attendance will start a regular incorporation of soybeans in their diet. Acceptance of soybeans as a staple food item will only occur after a long and continued program which is presented regularly to the families. This may be accomplished by outreach extension programs which might be sponsored by the government or international agencies such as CARE or UNICEF. These programs must continue for at least several years to firmly establish the worth and develop a use pattern of soybean foods for home and village.

In developed countries extension workers using publications, radio statements, television demonstrations as well as on-the-spot preparation at clubs and fairs should be effective. Again the program must continue for a substantial period of time if the soybean is to achieve status as a staple food item. Since only 3% of solvent extracted soybean meal is used for human food, use of whole soybeans for human foods circumvents the expense of oil and meal separation and offers a new and valuable food source.

The economics of direct use of whole soybeans for human foods is most favorable. Animal protein is increasing in cost rapidly and is far too expensive for many peoples of the world. Conversion of vegetable protein to animal protein is highly inefficient; for example, in case of the highly efficient broiler chickens, 5 lb of vegetable protein is consumed to produce 1 lb of animal protein. Needless to say, the cost of vegetable protein is but a very small fraction of the total cost of animal protein.

It is interesting to note that the soybean plant will grow and produce a satisfactory crop over a wide range of the earth's cultivatable land. This extends from a tropical climate to the north and south temperature zones. In many small countries it may not be possible to produce enough soybeans to operate a solvent extraction plant economically. However, even in small countries, soybeans can be produced in adequate amounts for use in home, village or small commercial operations.

Nutritional Value

The nutritional value of soybeans has, over the years, been discussed thoroughly by many authors (3,5,6). Mature, dry soybeans contain about 40% protein which is very high when compared to most other foods. Cows' milk contains about 3.5% while the common red bean is 22% protein. This high protein content is characterized by a good balance of amino acids. In fact, the amino acid balance of soybeans is better than that of any other common vegetable source of protein and approaches the FAO dietary standards for protein (3). Cereals, such as rice or wheat, are good sources of the sulfur-bearing amino acids and can be blended with soybeans to give a combined protein that meets the FAO standard. The protein quality of a soybean-cereal food combination is substantially better than that of products prepared solely from either soybeans or cereals and should be as good as cows' milk casein. It is fortunate that the acceptability of soy-cereal combinations are generally very good.

Mature, dry soybeans contain about 20% oil which is very high as compared to that of most cereals and vegetables. Soybean oil is about 85% unsaturated and is cholesterol-free which makes it highly desirable in the human diet. In many developing countries, calorie content of the diet is too low, which results in protein being used for calories instead of tissue building. Oil provides over twice the calories of carbohydrates and proteins, which enhances dietary use of soybeans. Other nutritional aspects of the soybean should not be disregarded. The soybean is a good source of minerals and vitamins, with the exception of vitamin C. Carbohydrate in soybeans, although not all utilizable by the human system, adds to the total caloric contribution.

A CONCEPT FOR PREPARATION OF WHOLE SOYBEANS FOR HUMAN FOOD USE

As discussed previously, the use of soybeans for human foods presents acceptability problems if the products are beany or painty in flavor and tough and chewy in texture. The flavor problem is caused by action of an off-flavor producing enzyme system in raw soybeans that is known as lipoxygenase. This enzyme must be inactivated at the proper time to prevent the off-flavors from developing (6,7). In addition, raw soybeans contain a number of antinutritional factors which, if not inactivated, drastically reduce the protein nutritional value of the finished food.

Inactivation of the Lipoxygenase System

In whole, sound, dry soybeans, the sites of the lipoxygenase enzyme and the substrate are located separately in the cell so that off-odors and off-flavors are not present in sound dry beans. Breaking or damaging of the cell tissue exposes the enzyme and substrate sites, but as long as the tissue remains dry the enzyme does not function. However, addition of water to broken or damaged tissues results in instant off-odor and off-flavor development (6).

Sound, dry, whole soybeans can be soaked in about a 5:1 water: soybean ratio and the beans will double in weight in about 6 hr. Hydrated sound beans do not develop beany or painty odors and flavors because the enzyme and substrate remain separated. However, hydrated beans are very susceptible to damage which will allow the enzyme and substrate to interact in the presence of water. Therefore, the hydrated beans must be handled very carefully to prevent tissue damage.

The lipoxygenase enzyme can be inactivated readily by cooking in water as is practiced in blanching vegetables for freezing or canning. Soaked whole soybeans can be dropped directly into boiling water which inactivates the enzyme in several minutes. Dry whole soybeans or dehulled cotyledons can also be dropped directly into boiling water for simultaneous hydration and enzyme inactivation. However, hydration and blanching time for dry soybeans will be much longer.

If it is desirable to produce a soy food with reduced fiber content, the soybeans can be dehulled. If dehulling is practiced it is accomplished most effectively on the dry whole soybeans. However, even the best dehulling equipment will cause some damage to the cotyledon tissue. Thus, the dehulled cotyledons must not be soaked prior to blanching. The dry cotyledons should be dropped directly into boiling water to achieve hydration and instant blanching of exposed cell tissue. Cotyledons are readily hydrated and blanched with this technique.

Use of Sodium Bicarbonate Blanch

One of the drawbacks encountered in using whole soybeans for human food is the long cooking time, up to 3 hr, needed to attain acceptable texture. These long cooking times are wasteful of energy; long soaking and cooking times discourage use of soybeans. Tenderization of soybean tissue can be achieved more rapidly by the addition of sodium bicarbonate (baking soda) to the cooking or blanching water (5,6). Sodium bicarbonate is added to cooking water in the range of 0.05 to 0.5%. Concentration of bicarbonate used depends upon the size of the particle being blanched; thus, whole beans require a high concentration and cotyledon pieces a low concentration. The cooking time to a given bean tenderness is usually reduced by about 50% when a desirable level of bicarbonate is added to the cooking medium.

The use of optimum levels of sodium bicarbonate in the cooking water invariably improves flavor of the cooked product. However, use of bicarbonate tends to perceptibly darken the beans. Fortunately, this color deterioration is slight and is not noticed unless direct comparisons are available.

Inactivation of Anti-Nutritional Factors

Anti-nutritional components include trypsin inhibitor, hemagglutinin, phytic acid as well as other minor factors. Because of their importance in animal nutrition, proteinase inhibitors in legumes, including soybeans have been studied thoroughly (9). Very little research relates the effects of proteinase inhibitors to human nutrition. At present it is generally accepted that inactivation treatments which are adequate for animal nutrition will also be suitable for human nutrition. Trypsin inhibitor can be destroyed in soybeans by adequate heat treatment and it is known that this factor is the most heat resistant of all the anti-nutritional components. Several interesting facts are known regarding trypsin inhibitor inactivation. This anti-nutritional factor is eliminated in hydrated soybean tissues more rapidly than in dry tissues. An alkaline medium, such as sodium bicarbonate blanch, is a factor in a more rapid destruction.

Trypsin inhibitor requires longer cooking or blanching to inactivate than does the lipoxygenase enzyme system. It is quite fortunate that both of these factors are eliminated when beans are cooked long enough to develop suitable tenderness or texture in the beans. This would usually be about 20 to 30 min for whole soybeans.

Effects of Quality and Variety of Soybeans for Direct Food Use

Dry whole soybeans must be free from foreign material such as pods, stems, and stones. Crushed beans and splits should be removed to as great an extent as possible. However, it is extremely important to use mold-free soybeans. Moldy beans affect flavor and color drastically and may contain dangerous toxins. Thus, soybeans for direct food use should be free from all types of defects and damage and should be golden or yellow in color. Field varieties grown for oil extraction are suitable for preparation of soybean food products. Vegetable varieties of soybeans have not been found to be superior to general field varieties.

Use of the Concept

The concept for preparation of acceptable soy foods is followed easily and results in a bland, cooked soybean component that is suitable for incorporation into many foods. Specific examples such as soy milk and soy breakfast and patty foods will be described in detail. It is suggested that these concepts be generally followed using flavors and ingredients that are accepted by the local people.

HOME PREPARATION OF SOYMILK

Introduction

Since time immemorial the contribution of cows' milk to human nutrition has been of incalculable value. However, it is now clearly documented that some people are intolerant to cows' milk. A substantial percentage of the world's population has hypolactosia (low level of lactase in the intestinal mucosa) which results in lactose intolerance. Lactose intolerance ranges from about 1% of the people in Europe to as high as 100% in Africa, Asia, Latin America and the Middle East (8). The cause for lactose intolerance is due to low consumption of cows' milk after weaning of the child.

Symptoms of lactose intolerance usually follow a few hours after ingestion of milk or milk-containing products. These include abdominal pain, diarrhea, bloating, cramps, abnormalities in the stool, and flatulence. In extreme cases, deficiency of β-galactosidase is also present and results in a metabolic disorder called galactosemia which results in high galactose blood levels which can be fatal.

Although lactose is the major problem, some people are allergic to the protein in cows' milk (12). Thus, soymilk is sorely needed for those people who are intolerant to cows' milk.

There is a very real need for soymilk that can be prepared in the home using simple equipment and techniques. Quite obviously such a product should be bland in flavor and free from beany or painty off-flavors. The development of the concept for preparation of such a soymilk follows.

Experimental

In the initial studies, Wayne, Bragg, Clark 63, Corsoy and Bonus varieties were used. However, for the main work the Bonus variety was selected and 100 g of raw material were used for each sample. The soymilks were prepared by several different extraction methods. However, with all methods the raw material was cooked or blanched in the first operation to prevent off-flavor development. Sodium bicarbonate was incorporated into both blanching and extraction water. After substantial initial experiments four methods were established as potentially suitable.

Method 1. One hundred grams of cleaned dry whole soybeans were dropped into 500 ml of boiling water containing 0.05% sodium bicarbonate and boiled for 5 min. This blanch water was discarded. The partially blanched beans were dropped into one liter of boiling water containing 0.04% sodium bicarbonate and the blanching was continued for an additional 5 min. The blanched beans were ground into a slurry in which the second blanch water was included. Grinding was accomplished by blending at high speed for 1 min. The slurry was then cooked for 20 min and filtered.

Method 2. Same as Method 1 except that the slurry was filtered immediately after blending. The filtrate or extract was then cooked (simmered) for 20 min.

Method 3. One hundred grams of hammer-cracked or coarsely broken whole soybeans were used as in the preparation procedure described in Method 1.

Method 4. Raw, dry whole soybeans were ground with a hand grinder into a coarse flour fine enough to pass through a 20 mesh screen. One hundred grams of soy flour were sprinkled into 500 ml boiling water containing 0.05% sodium bicarbonate. This slurry was simmered for 5 min. At this time 500 ml boiling water containing 0.04% sodium bicarbonate were stirred into the slurry and the entire mixture was simmered for another 15 min. The slurry was then filtered.

Six folds of cheesecloth were used for filtering all slurries. After all the slurry had filtered freely, the corners of the filter cloth were brought together and the residue was pressed by twisting the cloth as much as possible. This procedure forced considerable liquid through the filter cloth. The filtrate or milk produced by each method was formulated using 0.2 g salt, 4 g sugar (sucrose) and 17.6 μl artificial cream flavor (Fritzsche, N.Y.) for every 100 ml of soymilk.

The volume, protein, solids content, viscosity and pH of all soymilks were determined before addition of salt, sugar and flavorings. The volume was measured with a graduated glass cylinder. The pH was determined with a Beckman Research pH Meter. Nitrogen content of all the samples was analyzed by the Micro-Kjeldahl AOAC method (2) and the protein was calculated as N x 6.25. Solids were determined by drying, first in an air oven and then in a vacuum oven, for a total of 48 hr. Total and free fat contents of all the samples were determined by chloroform-methanol extraction and hexane extraction, respectively. Ash was determined by Method 14.007 A.O.A.C. (2) and calcium by the wet ashing method (G. Stone, personal communication.

Results and Discussion

The quality and acceptability of soymilk is related to factors such as blanching time, wholeness of beans, use of sodium bicarbonate and incorporation of blanch water. For example, when whole, raw beans were blanched for 5, 7, 10 and 20 min, the final soymilk was quite different in overall acceptability. Soymilk prepared from 5 min blanched beans had a total solids of 6.6% while the product from 7 min blanch contained 6.2% total solids. Both soymilks contained 2.9% protein but developed a highly objectionable off-odor and off-flavor which was reminiscent of cut grass. The 10 min blanch produced a product free from all off-flavors and off-odors. This soymilk contained 2.87% protein and 6.4% total solids. The soymilk produced from 20 min blanched beans contained only 1.37% protein and 4.76% total solids. It was very evident that 20 min had exceeded the optimum length of blanching time. Thus, 10 min was clearly superior and was used for all further studies.

The quantity of water used for blanching had essentially no effect on total protein extracted as long as total blanch time was limited to 10 min. The 10 min blanch was performed in two separate 5 min steps. The first blanch water was discarded. The objective was to clean the beans thoroughly and to discard higher sugars and some pigments that were leached. The water from the second blanch was incorporated into the final product; because it was prepared with sodium bicarbonate, it imparted a slight alkalinity to the milk.

In Method 1, the slurry was cooked and then filtered, while in Method 2 the slurry was filtered before final cooking. However, the final products from both methods were smooth in texture, bland in flavor and generally high in acceptability. In Method 3, cracked or broken soybeans, including hulls, were used as the raw material. The dry, raw whole beans were either hand-cracked with a hammer or in a loosely-spaced Buhr mill. General acceptability was rated as inferior to products from Methods 1 and 2. Method 4 started with a raw soybean flour passing a 20 mesh screen. The drawback with this method was a very objectionable cooked-cereal flavor in the final product.

Table 1 presents the proximate analyses of soymilk prepared by the four methods. It is interesting that the milk prepared with soy flour (Method 4) contained the highest protein (3.3%) and total solids (7.32%). No doubt this was the result of a high percentage of suspended material and no protein was lost during blanching. In descending order, the total solids and protein were 7.25 and 2.82%, respectively, for soymilk prepared from whole beans when

Table 1. Composition of soymilk prepared from different forms of soybeans.

Component	Method 1 Whole Bean Slurry Cooked	Method 2 Whole Bean Filtrate Cooked	Method 3 Cracked	Method 4 Soy Flour
		– % –		
Water	94.36	92.75	95.12	92.68
Total solids	5.64	7.25	4.88	7.32
Protein	2.59	2.82	2.04	3.33
Total fat	0.92	1.62	1.21	1.51
Free fat	0.67	0.68	0.96	1.14
Calcium	0.14	0.16	0.11	0.15
Ash	0.68	0.73	0.61	0.66
Carbohydrate (by difference)	1.31	1.92	0.91	1.67

the slurry was filtered before cooking (Method 2) and 5.64 and 2.59% in milk when the slurry was cooked for 20 min and then filtered (Method 1). Soymilk prepared from cracked soybeans (Method 3) contained 2.04% protein and 4.88% total solids.

It is interesting to note, Table 1, that filtrate-cooked soymilk (Method 2) had a substantially higher fat content, 1.62%, as compared to 0.92% for the slurry-cooked product (Method 1). When these values were corrected for final yield of soymilk, as was done for protein yields, Table 2, the difference between the two values was slightly reduced. Thus it is evident that filtering of the slurry prior to final cooking substantially increases the extraction of lipid materials.

The data presented in Table 2 represent six replications. It should be noted that with all four methods the starting relationship of water was ten times the weight of the raw material used. In order to compare the real protein yields the final soymilk was corrected to a 600-ml yield. The data are presented in Table 2. On a corrected final yield basis, the soymilk prepared from flour (Method 4) contained 3.24% protein, whole bean soymilks (Methods 1 and 2) each contained 2.74% protein while soymilk from cracked beans (Method 3) contained 2.04% protein. Thus, there was no advantage in protein recovery for either Method 1 or 2 but Method 2 was definitely superior regarding lipid extraction.

A panel of seven members of the Food Science Laboratory organoleptically evaluated the products for color, flavor, off-flavor, chalkiness, bitterness, mouth-drying and viscosity using a nine point hedonic scale. A score of 9 indicates the product to be excellent, 5 acceptable and 1 entirely

Table 2. Yield, total solids and protein contents of soymilk prepared by four methods.

	Method	Yield	Total Solids in Soymilk	Protein Liquid Sample	Protein Corrected to 600 ml
		– ml –		– % –	
1	Soymilk from whole beans (slurry cooked)	635	5.64	2.59	2.74
2	Soymilk from whole beans (slurry filtered and filtrate cooked)	585	7.25	2.82	2.74
3	Soymilk from cracked beans	600	4.88	2.04	2.04
4	Soymilk from soy flour	590	7.32	3.33	3.24

unacceptable. The results are shown in Table 3. A statistical analysis of variance showed that only color and viscosity gave significant differences (5% level of confidence) within treatments; full-fat flour gave superior color and cracked beans resulted in inferior viscosity. All other organoleptic characteristics showed some variation in scores; however, these differences were not significant. It is interesting and important that all organoleptic scores for all treatments and product characteristics were well above the acceptable level of 5.0; the lowest mean score was 6.2.

Soy milk, Table 3, prepared from whole beans was rated superior to milks made from cracked beans or full-fat soy flour in all organoleptic characteristics except color. The ratings for many of the characteristics such as bittnerness, chalkiness and mouth-drying averaged about 8.0 for the whole bean product. These ratings are very high for soy milks and suggest that the soy milk procedure using whole beans was superior to other methods.

Conclusions

Cooking or heating is the most important step in preparation of soymilk. Raw soybeans must be partially hydrated and cooked before grinding into a slurry to prevent lipoxygenase activity which results in beany and painty flavors. Analyses of samples prepared by the four methods were free from trypsin inhibitor. Use of sodium bicarbonate in both blanching and grinding waters results in a higher extraction of protein in soymilk.

Soymilk prepared from cracked beans resulted in the lowest protein extraction and viscosity when compared to other methods. Soymilk from soy flour was characterized by a cooked-cereal flavor which resulted in lowest acceptability. Soymilk from whole soybeans contained a high percentage of

Table 3. **Organoleptic evaluation of soymilk prepared from three different forms of soybean and formulated with three percent sugar.**

	Mean Organoleptic Scores For						
Treatment	Color	Flavor	Off-flavor	Chalkiness	Bitterness	Mouth drying	Viscosity
Whole beans	6.2	7.4	7.1	7.8	8.3	8.0	7.2
Cracked beans	6.2	6.9	6.7	7.7	8.0	7.6	6.6
Full-fat soy flour	7.1	7.0	6.4	7.5	8.0	7.6	7.0

protein and was highly acceptable as regards all organoleptic characteristics. The residue remaining after filtering contains about 50% of the original protein and fat. This is a highly nutritious and valuable material and is quite suitable for use in preparing weaning foods, patties and many other high protein foods (6).

Recommendations

The following recommendations should be considered: (a) Care must be exercised in measuring ingredients and especially in controlling cooking times to exact recommendations. (b) Dry clean and sort whole soybeans, drop 100 g beans (2/3 cup whole raw soybeans) directly into 500 ml (2 cups) boiling water containing 0.05% sodium bicarbonate (1/8 tsp) and simmer for 5 min. (c) Discard this blanch water and drop partially blanched beans into 1,000 ml (4 cups) boiling water containing 0.04% sodium bicarbonate. (d) Simmer the beans for an addition 5 min and grind in the blanch water using a blender for 1 min at high speed. (A hand grinder which will produce a finely ground material can be used in place of a blender.) (e) Stir the slurry and filter through six-fold finely woven, moist cheesecloth. Squeeze out as much extract as possible. (f) Cook the filtrate at simmer for 20 min. (g) Flavor with 0.2 g salt, 4 g sugar, and 17.6 μl artificial cream flavor/100 ml milk. (Other flavors such as vanilla, chocolate, coffee, coconut, etc., can be used in place of cream flavor.) Equivalent household measures for flavoring 1 cup of soymilk are 2¼ tsp. sugar, 1/8 tsp. salt, and 1 drop of cream flavor. (h) Dissolve all ingredients, pour hot into bottle and cool. Store in refrigerator and use as fresh milk; and (i) Follow directions carefully. Boiling times longer than recommended will reduce the amount of protein in final product.

SOYBEAN BREAKFAST AND PATTY FOODS

Introduction

The problems relating to lipoxygenase enzyme activity and anti-nutritional factors in raw soybeans apply to all food products prepared from whole soybeans. Thus, the general concepts already discussed apply here and must be followed carefully to assure final products of high quality. The soy foods described in this section need no presoaking of the raw soybeans, require short preparation and cooking times and have very good texture and flavor.

Raw Bean Treatments

One of the main considerations in the development of these foods was a short preparation and cooking time which would result in a highly acceptable texture. This constraint encouraged initial studies on breaking or crushing of whole dry beans prior to cooking. Initial studies showed clearly that this technique was very effective in reducing cooking time to a given bean tenderness or texture. For the initial studies the dry beans were broken with a hammer into about 20 pieces. Later studies showed that the beans could be broken satisfactorily in hand-operated or power-drive Buhr-type mills.

The Wayne cultivar of soybeans was used throughout these studies. However, later trials indicated that any good quality, fully mature field or vegetable variety would be suitable. The beans must be dry cleaned and hand-sorted to remove foreign material, mold and other defects.

Hydration

Preliminary cooking studies were made using crushed and whole soybeans, with and without sodium bicarbonate and varying the lengths of cooking time. The beans were dropped directly into boiling water and simmered for as long as 120 min. When cooking times greater than 30 min were used, 100 ml of cooking medium were added for each additional 30 min to replace the evaporated cooking medium.

Results (Figure 1) showed that two important factors were involved in water absorption; these were length of cooking time and whether beans were whole or crushed. Soybeans absorb more cooking medium with longer cooking times up to about 60 min. From 60 to 120 min, some samples showed slightly higher water absorption. Crushed beans always absorbed more cooking medium than whole beans (Figure 1). Increased water absorption by crushed beans may be due to three factors. First, the intact hull is a barrier to moisture penetration into whole beans (11). Second, crushed beans have a very much greater surface area per unit weight that is available for water absorption. Third, breaking or crushing enhances the water permeability into the bean meats during cooking and increases the rate of heat penetration into the bean tissue. Increased hydration and heat penetration should lead to more rapid inactivation of enzymes and inhibitors. These points were generally verified in earlier studies (1).

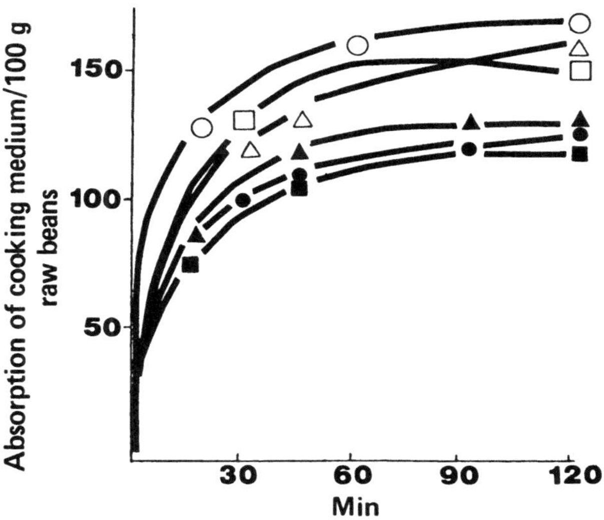

Figure 1. Soybean absorption of cooking medium after various treatments. ● = Whole, tap water; ■ = Whole, .05% sodium bicarbonate; ▲ = Whole, .5% sodium bicarbonate; ○ = Crushed, tap water; □ = Crushed, .05% sodium bicarbonate; and △ = Crushed, .5% sodium bicarbonate.

Tenderization

A study was carried out to show whether crushed beans were more tender than whole beans after cooking. Texture or tenderness was measured using a Lee-Kramer shear press (6). Figure 2 shows the effects of cooking time, composition of cooking medium and wholeness of beans on tenderness. Texture values decreased with increase in cooking time, with crushing of beans and with addition of either 0.5 or 0.05% sodium bicarbonate to the cooking medium. As expected, the use of 0.5% sodium bicarbonate was more effective than 0.05% sodium bicarbonate or tap water for tenderization. Crushed soybeans cooked in tap water were less tender than whole beans cooked in 0.5% sodium bicarbonate. Thus, the effect of the use of sodium bicarbonate is a powerful factor in tenderization during cooking (Figure 2). However, the combination of crushing with sodium bicarbonate cooking medium was by far the most effective method for rapid tissue tenderization.

Trypsin Inhibitor Inactivation

Two replications using four samples were made to determine the cooking time required for trypsin inhibitor inactivation. Ten minutes cooking time eliminated all trypsin inhibitor in the first replication and 94% inhibitor

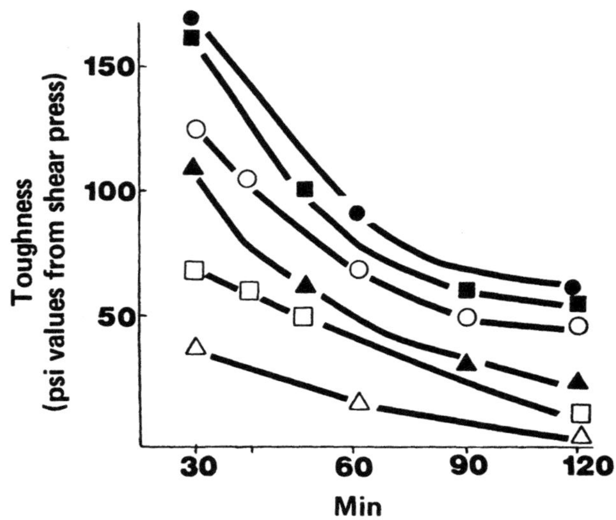

Figure 2. Tenderness of beans determined after various cooking treatments. ● = Whole, tap water; ■ = Whole, .05% sodium bicarbonate; ▲ = Whole, .5% sodium bicarbonate; ○ = Crushed, tap water; □ = Crushed, .05% sodium bicarbonate; and △ = Crushed, .5% sodium bicarbonate.

in the second. However, when the samples were cooked for an additional 15 min or for a total time of 25 min all treatments showed 100% elimination of trypsin inhibitor.

Soybean Foods

The preliminary studies showed clearly that crushed or broken whole soybeans, including hulls, could be cooked in a low concentration of sodium bicarbonate solution to achieve optimal tenderness in no more than 30 min. This encouraged development of a concept using crushed soybeans and requiring short cooking times for substantial fuel savings. Two food types were considered, a weaning or breakfast food and a fried patty that could serve as a meat substitute. Combinations of soy:cereals or vegetables on approximately a 1:1 solids ratio were selected arbitrarily for concept development. Since the 0.05% sodium bicarbonate solution was adequate for tenderizing tissue, this concentration was used in all development work.

Many experiments, not included here, were required to develop the final formulations as well as other procedures. After thorough experimentation and statistical analyses of the data, it was determined that a 10-min initial cook of crushed beans using a solution containing 0.05% sodium bicarbonate followed by 15 min additional cooking with other ingredients yielded the highest organoleptic scores for appearance, flavor and texture. Thus a total cooking time of 25 min was used for weaning and breakfast foods.

Initial cooking procedures for patty-type soy foods were similar to that used for weaning or breakfast products. Grinding the cooked mixture is critically important and the usual food chopper or grinder is not recommended. The most desirable hand grinder was a Molino Corona, Landers, Mora and CIA Ltd., Colombia. With this grinder, the food is augered between two round serrated plates which can be placed under considerable tension to increase the grinding action. This type of hand grinder is available in most developing countries and is generally available in the U.S. at health food stores.

WEANING OR BREAKFAST FOODS

Soy-Whole Wheat

Drop 100 g dry, crushed soybeans directly into 750 ml boiling 0.05% sodium bicarbonate cooking medium. Simmer for 10 min and add ¼ tsp. salt, 2 tbsp. molasses, 100 g ground whole wheat flour and 25 g brown sugar. Continue simmering this mixture for another 15 min with continual stirring. Adjust the hand grinder for high tension between plates and grind the slurry. Heat the ground product to the simmering point and serve. The final heating will eliminate undesirable contamination by microorganisms that may occur during the grinding step.

Soy-Corn

Drop 100 g dry, crushed soybeans directly into 750 ml boiling 0.05% sodium bicarbonate cooking medium. Simmer for 10 min and add ¼ tsp. salt, 2 tbsp. molasses, 100 g cornmeal and 25 g brown sugar. Continue simmering this mixture for another 15 min with continual stirring. Pass the cooked material through a hand grinder adjusted for high tension. Heat the ground product to the simmering point and serve.

Soy-Sweet Potato

Drop 100 g dry, crushed soybeans directly into 750 ml boiling 0.05 % sodium bicarbonate cooking medium. Simmer for 10 min and add 300 g raw, peeled, and finely diced sweet potatoes. Continue simmering this mixture for another 15 min with occasional stirring. Pass the cooked material through a hand grinder adjusted for high tension. Heat the ground product to the simmering point and serve.

These formulations were evaluated by a nine member taste panel which represented a broad contrast in food preferences. Individual organoleptic ratings ranged from good to excellent with a very good average. Since completion of these studies, at least 300 visitors from many countries have sampled these products and their reactions and comments have been highly favorable.

FRIED PATTY MEAT SUBSTITUTE

Soy-Potato

Drop 100 g dry, crushed soybeans directly into 800 ml boiling 0.05% sodium bicarbonate cooking medium. Simmer for 10 min and add 450 g diced (3/8 in. or less) raw white potatoes and ¼ tsp. salt. Continue simmering this mixture for another 15 min. with occasional stirring. During this cooking, remove cover from pan as needed to allow most of the cooking medium to evaporate. Adjust the hand grinder for high tension between plates and grind the thick slurry. Add chopped green peppers and onions to taste. Form patties from the mixture and fry until brown (about 15 to 20 min). Serve the finished patties while hot.

Soy-Rice

Drop 100 g dry, crushed soybeans directly into 1,000 ml boiling 0.05% sodium bicarbonate cooking medium. Simmer for 10 min and add 100 g dry white long grain rice and ¼ tsp. of salt. Continue simmering this mixture for another 15 min. with continual stirring. During this cooking, remove the cover from the pan as needed to allow most of the cooking medium to evaporate. Adjust the hand grinder for high tension between plates and grind the thick slurry. Add chopped green peppers and onions to taste. Form patties from the mixture and fry until brown (about 15 to 20 min). Serve the finished patties while hot.

Soy-potato and soy-rice products were prepared for organoleptic evaluation. For comparison, patties were also prepared from potatoes or rice alone; soybeans were not included in these patties but the prescribed amount of salt, chopped green pepper and onion was added. The four samples, as shown in Table 4, were evaluated by a nine-member taste panel. Organoleptic ratings were made on a scale of 1 to 9. The average organoleptic scores are presented in Table 4. A statistical analyses of variance showed that the only signficant difference was in appearance. Soy-rice was significantly superior at the 5% confience level to the rice patty. All other treatments and organoleptic characteristics were not significantly different from one another. However, the soy-potato and soy-rice products had an overall average score of 7.0 while potato and rice patties averaged 6.7. Flavor and texture scores were about the same for products with and without soy.

Conclusions

Concepts for highly acceptable soybean with cereal or vegetable requiring short preparation and cooking times were developed. The recipes presented here were generally rated as very good but should not be considered as final formulations. These recipes should be altered to suit local tastes as

Table 4. Organoleptic evaluation of fried patty meat substitute.

	Potato		Rice	
Characteristic	**With Soy**	**Without**	**With Soy**	**Without**
		– Mean Scores –		
Appearance	7.2	6.5	7.8	6.6
Texture	6.6	6.7	6.7	6.0
Flavor	6.7	6.3	7.0	7.0

well as use of local raw materials. These concepts may be applied to the development of many new products such as soups or porridges or adapted to local recipes. The amount of final cooking volume will vary depending on intensity of boiling and whether the cooking pot is covered. Slight adjustments may be necessary in total cooking solution used and intensity of cooking flame.

NOTES

A. I. Nelson, L. S. Wei, and M. P. Steinberg, Department of Food Science, University of Illinois, Urbana, Illinois 61801.

REFERENCES

1. Albrecht, W. J., G. C. Mustakas, and J. E. McGhee. 1966. Rate immersion studies on atmospheric steaming and immersion cooking of soybeans. Cereal Chemistry 43:100.
2. A.O.A.C. 1970. Official methods of analysis. 11th Ed. Association of Official Agricultural Chemists, Washington, D.C.
3. FAO/WHO. 1973. Energy requirements and protein requirements. WHO Tech. Rep. Ser. No. 522, p. 63.
4. Mueller, D. C., B. P. Klein, and F. O. Van Duyne. 1974. Cooking with soybeans. University of Illinois, College of Agriculture Cooperative Extension Service, Circ. No. 1092.
5. Nelson, A. I., M. P. Steinberg, and L. S. Wei. 1976. Illinois process for preparation of soymilk. J. Food Sci. 41:57-61.
6. Nelson, A. I., M. P. Steinberg, and L. S. Wei. 1978. Whole soybeans for home and village use. INTSOY Series No. 14, International Agriculture Publications, Univ. of Ill., Urbana.
7. Nelson, A. I., L. S. Wei, and M. P. Steinberg. 1971. Food products from whole soybeans. Soybean digest 31(3):32.
8. PAG Bulletin. 1972. Vol. II, No. 2
9. Rackis, J. J. 1972. Biologically active components, pp. 158-203. In A. K. Smith and S. J. Circle (Eds.) Soybeans: Chemistry and Technology. Volume 1, Proteins. AVI, Westport, Conn.
10. Rice, W. 1973. Treasure beneath the surface of the fair skinned soybean. Washington Post (May 10).

11. Spata, J. A., A. I. Nelson, and S. Singh. 1974. Developing a soybean Dal for India and other countries. World Crops. 26(2):82-84.
12. Woodruff, C. W. 1976. Milk intolerance. Nutrition Review 34(2):33-37.

SOYBEANS IN WORLD TRADE

R. E. Bell

Soybeans are one of the fastest growing items in world trade. The volume of world trade in soybeans and soybean products amounted to 35.6 million metric tons in 1977-1978. This was 66% higher than 5 yr earlier. The value of world trade in soybeans and soybean products was worth more than $8.5 billion last year.

Prospects are for a further expansion in world trade in soybeans and soybean products this year, as well as in the foreseeable future. World trade in soybeans and soybean products is being encouraged by an absence of import barriers in key importing countries, particularly in Japan and the European Community (e.g., the European Common Market).

As a result of their rapid growth in world trade, world consumption of soybeans increased 45% during the past 5 yr. World consumption was a record 77 million metric tons in 1977-1978, with 85% of the soybeans being crushed for oil and meal. Only in East Asia is there a large food consumption of soybeans as whole beans.

The growth in world consumption of soybeans in recent years is the result of a rapid growth in world demand for high-protein feed, particularly in North America, Europe, the Soviet Union and East Asia, and an expansion in world demand for edible vegetable oil, especially in developing countries. The regions of largest growth in soybean consumption in recent years have been North America and Western Europe, although consumption has been also growing at a rapid rate in the Soviet Union and Eastern Europe.

The recent growth in world trade in soybeans and soybean products would not have been possible without a sharp expansion in soybean production in Brazil. Brazilian soybean production increased 13-fold during the past decade, and nearly 80% of the increase went into export as beans, meal and oil, with most of the exports going to Western and Eastern Europe.

The U.S. also expanded its soybean production during the past decade, but not to the same extent as Brazil. Soybean production in the U.S. was more than 50 million metric tons (50 mt) in 1978, and U.S. exports of beans, meal and oil will total a record 28 million tons in 1978-1979.

As a result of the rapid growth in world trade in soybeans and soybean products, prices received by farmers for soybeans more than doubled during the past 5 yr. Taking inflation into account, the increase was about 45% in real terms. This increase compares very favorably with prices received in real terms for corn and wheat during the same period.

The higher prices for soybeans in recent years have encouraged an expansion in world soybean acreage, especially in the U.S., Brazil and Argentina. World soybean acreage increased nearly 25 million acres during the past 5 yr.

The growing world demand for soybeans during the past 5 yr has been met primarily by the expansion in soybean acreage. If the consumption trends of the past 5 yr are to continue, it will be necessary to achieve a break-through in average yields per acre for soybeans.

If recent trends in crop acreages continue, soybeans will soon replace corn as the leading U.S. crop in terms of harvested acreage. Soybeans have already replaced both corn and wheat as the top cash crop in the U.S.

Government policies and the spread of soybean production technology have been important contributors to the recent increases in world production, trade and consumption of soybeans. Agricultural policy in the U.S. has encouraged soybean production, as have government policies in Brazil.

There has never been any government restrictions on U.S. soybean production. Production has been guided by market prices and new technologies. This is in contrast to the situations for wheat, feed grains and cotton, which at times have been partially restricted in their production by government policies.

Government policies have actively encouraged soybean production in Brazil. The Brazilian Government has also protected Brazil's soybean crushing industry through a combination of tax incentives, export subsidies and export controls on raw soybeans.

In Western Europe, government policies have encouraged greater consumption of soybeans, particularly in the European Community (EC). The EC's policy of restricting imports of corn and other grains through its system of variable import levies has spurred soybean consumption in EC countries. This trend has been further encouraged by heavy imports of cassava from Southeast Asia for use in mixed feeds for livestock. Cassava contains no protein and therefore requires larger quantities of protein meals to balance livestock rations when it is used in place of grains.

Government policies aimed at raising living standards by increasing meat consumption in the Soviet Union and Eastern Europe have also spurred soybean consumption in recent years. International policies limiting offshore fishing have encouraged more livestock production in Japan and consequently increased the demand for soybean meal for livestock feeding.

Two decades ago the European Community undertook a commitment during a world-wide international trade negotiation not to impose any import charge on soybeans imported into EC countries. This concession is the most important one ever granted an agricultural product in international trade negotiations and is in large part responsible for the large growth in soybean consumption in Western Europe in recent years.

Japan will make a similar concession on soybeans in the international trade negotiations soon to be concluded in Geneva. This will mean that the world's two largest importers of soybeans will be committed by international treaties to the duty-free entry of soybeans. Japan needs to follow the commitment on soybeans with similar ones on soybean products as Europe has done.

A very significant development in the world soybean market in 1978-1979 was the emergence of the Soviet Union as a net importer of edible fats and oils for the first time in history. Current estimates indicate the Soviet Union will be a net importer of edible fats and oils by 70,000 mt in 1978-1979. As recently as 5 yr ago, the Soviet Union was a net exporter of edible fats and oils by 550,000 t. The shift of the Soviet Union from a net exporter to a net importer during this period is equivalent to 3.5 mt or 125 million bu of soybeans. There is no indication that the trends in the Soviet Union of the last 5 yr will reverse in the foreseeable future. In fact, the trend toward a larger net importer is likely to accelerate rather than diminish.

The rapid increase in soybean production in Brazil of the past decade appears to have stagnated during the past 2 yr. Drought sharply reduced the crops of 1978 and 1979. Future gains in Brazilian soybeans will be more difficult than those of the early 1970's. Earlier gains were the result of replacing other crops and pasture with soybeans. Over 80% of the growth in soybean acreage in the three largest soybean producing states of Rio Grande do Sul, Parana and Sao Paulo came from the replacement of other crops with soybeans.

Most of the future growth in Brazil's soybean acreage will need to occur in the Central West states of Mato Grosso, Goias, and Minas Gerais. The expansion in these areas will likely be much slower than was the case in traditional soybean producing states. Most observers of the Brazilian farming scene believe Brazilian soybean production may reach 15 mt by 1982. How much farther production can go beyond this level will depend on how expansion proceeds in the Central West. The speed and degree of this expansion will depend in large part on the willingness of the Brazilian government to subsidize the expansion.

Argentina has also increased its soybean production in recent years, but its production is still relatively small compared to that in the U.S. and Brazil. This year's harvest in Argentina will likely be around 3.3 mt. A large portion of the Argentine crop is exported, however, and the size of the Argentine crop is therefore important to world trade in soybeans.

Any rapid increase in world exports of soybeans during the next several years will need to come from the U.S. The U.S. has the capacity to further increase its production and exports of soybeans and it can make these increases fairly quickly by replacing corn acreages with soybeans in the U.S. Corn Belt. The U.S., for example, has the capability of increasing its soybean production in 1979 to 2 to 5 mt over last year. Depending on weather conditions, the U.S. should easily be able to produce between 52 and 53 mt of soybeans in 1979. A harvested acreage of 66.5 million acres and an average yield of 29 by/acre would product a crop of 52.5 mt. A crop of 55 mt could be produced with favorable weather and an average yield of a little more than 30 bu/acre.

A crop of this size may be needed to prevent unstable soybean prices sometime during the next 18 months, particularly if the current rate of inflation continues in the U.S. A period of volatile prices might occur this summer if this year's Brazilian harvest is below the present consensus estimate of 11.5 mt. The Brazilian government has committed 1.1 mt of soybeans for export out of the 1979 Brazilian crop, and another 1.0 mt will need to be saved for seed or will be lost through shrinkage and waste. This leaves only 9.4 mt to be crushed by domestic crushers, and this is about 25% below Brazil's crushing capacity.

As was the case last year, Brazil will probably try to crush and ship as much of its 1979 crop as possible before the 1979 harvest begins in the U.S. next September. About 70% of Brazil's annual crush will likely occur between April and September of this year. This means the U.S. will have the world market mostly to itself during the period from October 1979 through March 1980. This should give U.S. producers strong prices for their 1979 crop if world demand for soybeans holds up, and there is little reason, if any, to believe it will not.

It is possible that the U.S. carryover of soybeans next August 31 will be only 100 to 125 million bu (2.7 to 3.4 mt). Although this would double the U.S. stocks of 60 million bu on hand on August 31, 1973, it is also important to remember that world consumption of U.S. soybeans is now 65% higher than it was at the time soybean prices became so volatile in the early summer of 1973. For the longer run, it is imperative that a breakthrough be achieved in soybean yields per acre if price volatility is not to occur periodically.

NOTES

Richard E. Bell, Executive Vice President, Riceland Foods, Stuttgart, AK 72160.

SOYBEAN PROTEINS IN HUMAN NUTRITION

W. J. Wolf

Soybeans are classified as oilseeds because they were first processed for oil and, today, processing is still centered around the hexane extraction process, which yields both oil and meal. Although the meal was at first considered a byproduct, its well-balanced amino acid composition and its high nutritional value soon changed the evaluation and established the meal as our most important protein resource for animal feeds. At present, we are at the beginning of another transition, namely, the small but growing use of soy proteins in foods. The increasing use of soy proteins as extenders for traditional animal proteins is generating a great deal of interest in their nutritive value for humans. Before discussing the current situation on human nutrition of soy proteins, it is desirable to review briefly soybean composition and the contributions that soy proteins make to our food chain.

SOYBEAN COMPOSITION

From a compositional viewpoint, the soybean could logically be called a protein seed rather than an oilseed. The average proximate composition is: 40% protein, 21% oil, 34% carbohydrate, and 5% ash (10). Because the soybean contains twice as much protein as oil, there is good justification for emphasizing protein rather than oil. Seed composition also determines the material balance that results when soybeans are processed by hexane extraction (8). Products from the hexane extraction of 1 bushel (60 lb) of soybeans are meal (43.3 lb), oil (11 lb), hulls (4.2 lb), and shrinkage (1.5 lb). The yield of defatted meal with a protein content of about 50% is almost four times that of the oil, thereby giving additional emphasis to the importance of the protein.

CONTRIBUTION TO U.S. DIET

Although only a small amount of soybean protein is at present consumed directly in foods, our soybean crop is a vast protein reserve used mainly for conversion to animal proteins and for export. The size of this protein reserve is seen more clearly in Table 1, which shows the supply of soybeans that was available in the U.S. in 1977. The crop plus the carryover amounted to 1.819 billion bushels, which contained 19.8 million metric tons of protein. This becomes a more comprehensible figure when it is expressed as 249 g of available protein/capita/day. By comparison, the recommended dietary allowance of protein/day for a 70-kg man is 56 g (16), and the daily per capita protein in the U.S. food supply is estimated to be 100 g (17). Consequently, soybeans in 1977 had the potential of supplying 4.5 times the recommended dietary allowance and 2.5 times the amount of protein provided by the usual food supply.

Let us now consider how the soybean supply was disposed of in 1977. Table 2 shows that 935 million bushels, or just over one-half of the supply, were processed, which provided 128 g of protein/capita/day or more than twice the dietary allowance of 56 g/day. Nearly 40% of the supply was exported, and the remaining 10% comprised the amount used for seed and carryover.

Table 1. U.S. soybean supply for 1977[a].

		Protein	
	Soybeans, Million bu	Million Metric Tons	Available Protein, g/Capita/Day
Production	1,716	18.7	235
Carryover	103	1.1	14
Total	1,819	19.8	249

[a]Fats and Oils Situation (3).

Table 2. Disposition of soybeans in U.S. in 1977[a].

	Million bu	% of Supply	g Protein/Capita/Day
Processing	935	51.4	128
Export	700	38.5	96
Seed	59	3.2	8
Carryover	125	6.9	17
Total	1,819	100.0	249

[a]Fats and Oils Situation (3).

Direct Food Uses

For direct food uses of soy protein, defatted flakes are converted into flours and grits, concentrates, and isolates. Table 3 shows the growth in the amounts of these products for the decade 1967 to 1977. Production figures for flours and grits include the amounts of these materials converted into concentrates and isolates as well as the textured flours. Consequently, for estimating per capita consumption only the flours and grits are considered. The 1 billion lb of edible flours and grits produced in 1977 is equivalent to 3 g of protein/capita/day as compared to the available supply of 249 g of soybean protein (Table 1). The amount of soy protein actually consumed in the U.S. per capita is less than 3 g/day, because some of the flours and grits are exported (e.g., Public Law 480 programs), and there is some loss of protein when flours and grits are converted into concentrates and isolates. A more realistic figure for daily consumption is 2.5 g/capita, which is 4.5% of the dietary allowance and 2.5% of total daily food protein supply. By contrast, in Japan where soybean foods have a long tradition of use, the daily per capita intake of soybean protein in 1975 was 10 g or 13% of the total protein consumption of 79 g (23).

Indirect Food Uses

As shown above, direct food use of soybean protein is still very small. Nonetheless, soybean protein makes a major contribution to our dietary protein supply since it is fed to livestock and poultry and converted into the traditional proteins of meat, milk, and eggs. Table 4 shows how soybean meal was used in feeds in 1977. Of the total meal, 75% is fed to poultry and swine; poultry consume 42% and swine 33%. The remainder is divided equally between dairy and beef cattle plus miscellaneous uses.

In terms of the daily per capita supply, the soybean meal used in feeds is equivalent to 87 g of protein. This represents 35% of the total soybean protein available in 1977 and is equal to 87% of the total daily food protein

Table 3. Production estimates for edible soybean protein products, 1967 to 1977[a].

	Annual Production, Million lb		
Product	1967	1972	1977
Flours and grits	105-110	325-500	1,000
Concentrates	17-30	40-50	70
Isolates	22-35	35-40	75
Textured flours		35-40	140
Textured concentrates			5

[a]From Wolf (25).

Table 4. U.S. soybean meal usage in 1977[a].

Use	Million Metric Tons[b]	%	g Protein/Capita/Day
Poultry	6.6	42.3	37
Broilers	2.8	17.9	16
Hens and pullets	2.7	17.3	15
Turkeys	1.1	7.1	6
Swine	5.2	33.3	29
Dairy cattle	1.6	10.3	9
Beef cattle	1.6	10.3	9
Other	0.6	3.8	3
Total	15.6	100.0	87

[a]Feed Situation (4).
[b]On a 44% protein basis.

supply per person. As pointed out earlier, the amounts of soy flour and grits that are diverted to food uses contain only 3 g of protein/capita/day, thus the ratio of feed to direct food uses is 29:1. Clearly, feeding of soybean meal to poultry and livestock is the dominant use of this protein resource at present.

NUTRITIONAL QUALITY OF SOYBEAN PROTEINS

As just discussed, the average per capita consumption of soybean protein per day is still small. Nonetheless, certain segments of the population are consuming higher than average levels because soy proteins are being used as meat extenders and meat analogs in institutional feeding programs, such as in schools and nursing homes. As a result of increasing use and significance of these proteins in foods, there is growing concern about their nutritional properties. Questions which are being asked are (a) What are the nutritional qualities of soybean proteins as compared to traditional animal proteins such as meat, milk, and eggs which are considered as being of high quality? and (b) Does processing remove or inactivate antinutritional factors known to be present in raw soybeans?

Animal Assays

The traditional approach to evaluation of quality of soybean proteins has been to feed them to experimental animals. The rat has been the most widely used animal, and the Protein Efficiency Ratio or PER test is commonly employed for such purposes. The PER value is the wt gain divided by the wt of protein consumed under standardized conditions; results of such tests show consistently that soybean protein ranks lower than casein. Some PER values from several protein sources are casein, 100; rapeseed, 93; beef (defatted), 82;

soybean, 74 to 82; oat flour, 66; rice, 57; wheat flour, 28; peas, 23; and wheat gluten, 6 (14).

Although the PER test is an official method of the Association of Official Analytical Chemists (2) and is used for regulatory purposes, it and other rat assay procedures have come under criticism because they yield information that may not be directly applicable to humans (6,21). Hopkins and Steinke (7), for example, recently reviewed the various rat assays employed for evaluating protein quality and called attention to several factors that affect test results. A major criticism is that the estimated amino acid requirements of man and rats are not the same (Table 5). They also point out that the amino acid requirements for man are not well known. Amino acid requirements for man have been estimated by two expert groups, the Food and Agriculture Organization of the United Nations (FAO/WHO) (1) and the Food and Nutrition Board (FNB) of the National Academy of Sciences (24). The important differences between the requirements for man and the rat are in lysine, methionine, and arginine. Man does not require a dietary source of arginine, but the growing rat does. As far as soybean protein is concerned, it is the methionine plus cystine level that is limiting in meeting the essential amino acid requirements of the rat.

The higher requirement of certain amino acids by the rat as compared to humans is attributed to: (a) its faster growth rate; (b) more hair, which

Table 5. Amino acid requirements of man and the rat[a].

	Man		Rat
Amino Acid	FAO/WHO[b] %	FNB[c] %	NRC-NAS[d] %
Lysine	5.5	5.1	7.5
Threonine	4.0	3.5	4.2
Methionine + cystine	3.5	2.6	5.0
Tryptophan	1.0	1.1	1.2
Valine	5.0	4.8	5.0
Leucine	7.0	7.0	6.2
Isoleucine	4.0	4.2	4.6
Phenylalanine + tyrosine	6.0	7.3	6.7
Histidine[e]	--	1.7	2.5
Arginine	--	--	5.0
Total essential A.A.	36.0	37.3	47.9
Total nonessential A.A.	64.0	62.7	52.1
Total	100.0	100.0	100.0

[a]Expressed as percent of dietary protein.
[b]Food and Agriculture Organization/World Health Organization (1).
[c]Food and Nutrition Board, National Academy of Sciences (24).
[d]National Research Council-National Academy of Sciences (15).
[e]Essential only for infants.

increases the requirement for sulfur amino acids; and (c) less recycling of amino acids (7). Nutritionists are now recognizing that the rat may not be an appropriate model for measuring protein quality in humans and are pointing out that more rapid and less expensive techniques such as in vitro tests are needed by the food industry and regulatory agencies (7).

Human Assays

The ultimate answer to protein quality in humans must come from studies with humans. However, such experiments are expensive and difficult because of practical and ethical problems of working with human subjects. Nonetheless, several groups are actively performing protein quality measurements with infants, young children, and adults. Some of these studies are reviewed here.

Studies with Infants. Investigations of soybean protein quality in infants are of considerable importance, because a number of infants are allergic to cow's milk and must therefore be placed on formulas in which soybean protein isolates may be the sole source of dietary protein up to 6 months of age. Fomon and coworkers (5) have conducted numerous studies with infants in which they compared soy protein isolate fortified with methionine to cow's milk. Nitrogen balance studies indicate that methionine-fortified soy protein isolate is equal to cow's milk protein in promoting retention of nitrogen. Results of one study are shown in Figure 1 where nitrogen retention is plotted against nitrogen intake for soy isolate and cow's milk. There was no statistically significant difference between the two protein sources.

Fomon and Ziegler (5) have also conducted growth studies on infants with methionine-fortified soy protein isolate, cow's milk, and human milk. Table 6 summarizes and compares results from three studies. They found no statistically significant differences in growth between the three sources of dietary protein. In still another study, soy protein isolate was fed with and without supplementation with L-methionine. Food intake and growth showed no significant differences, but serum concentrations of urea were lower in infants receiving the methionine-supplemented diets. The latter suggests that protein quality was improved by methionine addition, although nitrogen retention measurements failed to show any benefits from methionine.

Studies with Growing Children. Torun and associates (22) have conducted tests with two protein isolates in children 19 to 44 months old. Milk was used as the reference protein. The isolates were well tolerated and their digestibilities were comparable to that of milk. Nitrogen balance studies indicated that the isolates gave nitrogen retentions similar to those obtained with milk. It should be emphasized that these results were obtained without methionine supplementation. Furthermore, they are contary to what one

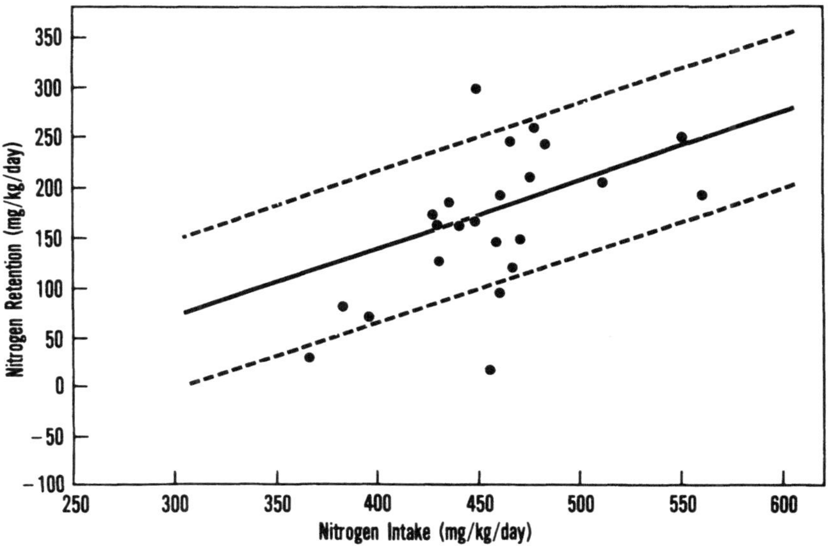

Figure 1. **Retention of nitrogen as a function of nitrogen intake by normal infants 8 through 120 days of age receiving diets supplying 8 to 11% of energy from protein. Each dot corresponds to a 3-day metabolic balance study with protein provided by soy protein isolate. Solid curve is regression for balance studies with cow's milk and dashed lines are 95% confidence levels. From Fomon and Zeigler (5).**

would expect with rat bioassays, such as the PER test, which show that soy protein ranks lower than casein, the major protein fraction in milk.

Studies with Adults. Several investigators have evaluated nutritional properties of soybean proteins in adult humans. For example, Kies and Fox (11) conducted nitrogen balance studies and found that the subjects were in negative balance at an intake of 4 g of beef nitrogen/day. With the same intake of nitrogen from textured soy flour, the subjects became more negative in nitrogen balance. Supplementation with methionine improved nitrogen retention and indicated that methionine was limiting at the low protein intake. However, at a more adequate intake of 8 g of nitrogen from textured flour or 50 g of protein/day, nitrogen balance was positive and the same as with beef.

Zezulka and Calloway (26) determined the minimum amount of soy protein isolate with and without methionine supplementation needed to meet the amino acid requirements of adult men. Their results (Fig. 2) indicated that 3 g of soy protein nitrogen gave a negative balance; but when soy protein isolate was supplemented with methionine to bring the total sulfur amino acids to 900 mg/day as recommended by FAO/WHO (1), the subjects were in

Table 6. Mean wt gains for normal human female infants between 8 and 112 days of age[a].

Protein Source	Number of Subjects	Wt Gain: Age 8 to 42 Days g/Day	Age 8 to 42 Days g/100 kcal	Age 42 to 112 Days g/Day	Age 42 to 112 Days g/100 kcal	Age 8 to 112 Days g/Day	Age 8 to 112 Days g/100 kcal
Soybean + methionine	13	30 ± 6	7.1 ± 1.2	23 ± 4	4.3 ± 0.4	25 ± 4	5.1 ± 0.5
Cow's milk	10	30 ± 7	7.1 ± 1.0	22 ± 5	4.2 ± 0.8	24 ± 4	5.1 ± 0.7
Human milk	116	34 ± 9	--	22 ± 6	--	26 ± 5	--

[a]Fomon and Ziegler (5). Soy protein isolate provided 6.5% of calories whereas in cow's milk-based formula protein supplied 6.0% of calories.

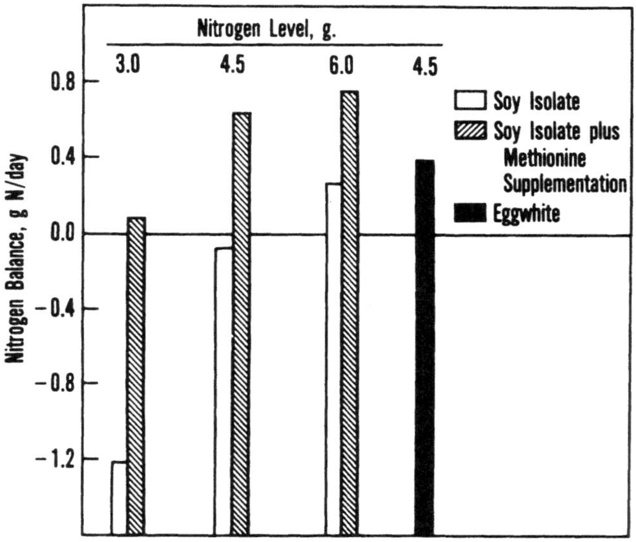

Figure 2. **Nitrogen retention in young men receiving soy protein isolate with and without methionine in comparison with egg white protein. Nitrogen intake is expressed as g/day. In the supplemented diets, methionine was added to raise the total sulfur amino acid content of the diet to 900 mg/day. From Zezulka and Calloway (26).**

balance. An intake of 4.5 g of soy protein nitrogen brought the subjects nearly into balance; methionine supplementation clearly put the subjects into positive balance. At intakes of 6.0 g of soy protein nitrogen or higher, the men were in positive balance without methionine supplementation although methionine addition still gave a positive effect. A level of 6.0 g of soy protein nitrogen was equivalent to 4.5 g of egg white nitrogen (Fig. 2). The results indicate that methionine is limiting at low levels, but at daily protein intakes of 38 g and higher, the sulfur amino acid requirements were met.

Scrimshaw and Young (21) have compared egg protein and soy protein isolates using nitrogen balance techniques (Fig. 3). The nitrogen balance curve for egg lies above the curve for soy isolate, and the values for the two proteins at zero nitrogen balance of body nitrogen equilibrium indicate that soy isolates are 80% as effective as egg protein in maintaining nitrogen balance.

Scrimshaw and Young (21) also examined the effect of methionine supplementation on nitrogen balance when isolates were fed at a level of 0.51 g protein/kg/day (Fig. 4). At this low level of isolate intake, supplementation with 1.1% methionine gave a positive response that was equal to that obtained

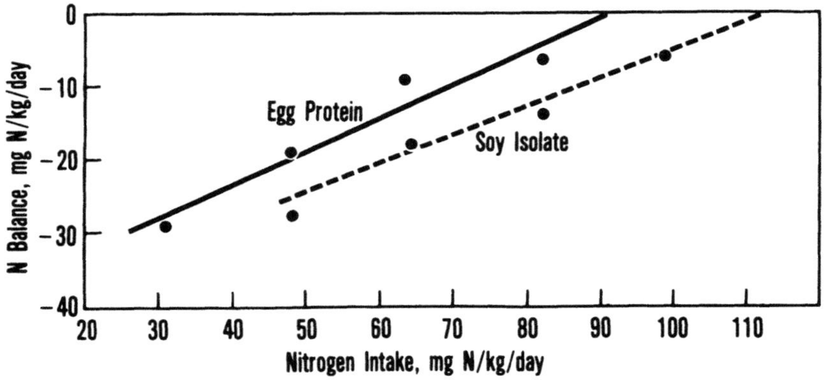

Figure 3. Nitrogen retention in young men receiving graded intakes of egg or soy protein isolate. From Scrimshaw and Young (21).

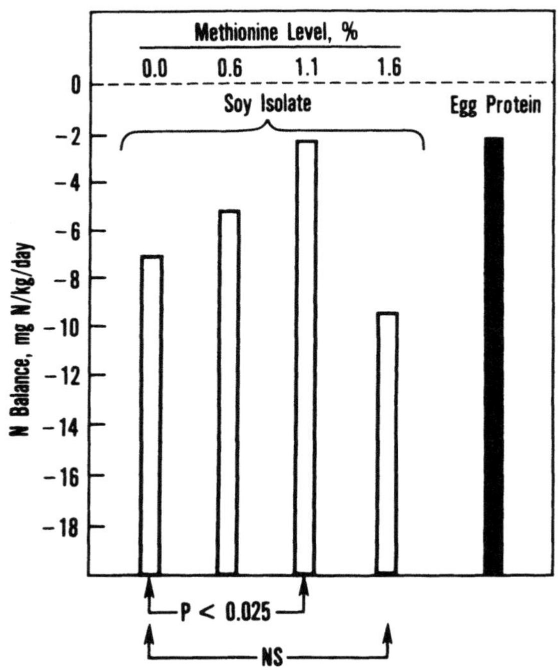

Figure 4. Nitrogen retention in young men receiving soy protein isolate plus varying levels of L-methionine. Supplementation is expressed as the percent of total protein intake (0.51 g protein/kg/day). From Scrimshaw and Young (21).

with egg protein. However, addition of 1.6% methionine resulted in a lower nitrogen balance than with zero supplementation. On the other hand, when the level of soy isolate was increased to 0.8 g/kg/day to meet the dietary allowance for total protein, additional methionine had no significant effect. On the basis of these and other studies, Scrimshaw and Young (21) concluded that under normal usage in adults, methionine supplementation of properly processed soy protein products is unnecessary and probably undesirable.

TRYPSIN INHIBITORS

Among the various factors in raw soybeans that affect nutrition, the trypsin inhibitors have received the most attention. Their primary effect is on the pancreas; in rats they cause the organ to enlarge and they also cause poor growth. However, these effects can be readily overcome by moist heat treatment (Fig. 5). Thus, steaming for 10 min inactivates about 80% of the trypsin inhibitor activity and the protein efficiency ratio reaches a maximum. Enlargement of the pancreas is abolished after even shorter times of steaming; only about 50% inactivation of the inhibitor activity is necessary to eliminate this effect (19).

Because trypsin inhibitors affect the pancreas in animals, there is concern about their possible effects when they are ingested by humans. Moist heat treatment will reduce inhibitor activity of flours and grits to low levels (Fig. 5), but not all edible soy protein products are processed in this manner. Analyses

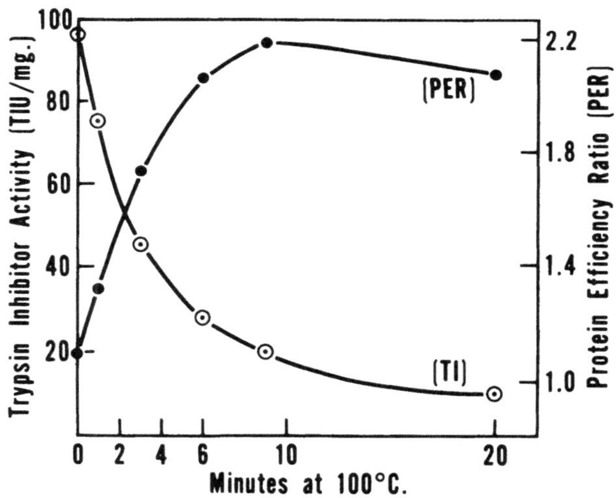

Figure 5. Effect of steaming time on trypsin inhibitor activity and protein efficiency ratio (PER). From Rackis (18).

of commercial products reveal that varying inhibitor levels remain in them (Table 7). Toasted and textured flours have inhibitor activities about 10% of those found in raw soy flour. Concentrates and isolates may also be this low in activity, although some range as high as 20 to 30% of the value for a raw flour.

Edible soy protein products are rarely consumed directly, but rather they are used as ingredients in foods that receive additional heat treatment either during processing or in cooking before consumption. Consequently, additional inactivation of trypsin inhibitors may occur as noted in cooked meat products (12), with the result that only low levels of inhibitor are likely to be ingested. Nonetheless, questions remain of possible effects of long-term ingestion of residual trypsin inhibitor activity in soy products. A recent cooperative study (20) between the Northern and Western Regional Research Centers involved feeding commercial flour, concentrate, and isolate to rats for 285 days (approximately one-half of the normal life-span). Detailed examination of the pancreas revealed no abnormal effects. Additional studies corresponding to the complete life-span of the rat are planned to clarify this important question.

SUMMARY

If the U.S. population consumed soybeans and soybean proteins directly in foods, the annual crop would supply more than four times our protein needs without any other source of protein in the diet. In actual practice, however, we use soybean proteins directly in foods to provide less than 5% of our requirements. We process about one-half of the crop into oil and meal and export the rest. The meal obtained by processing enters our food chain primarily as animal feed for conversion to meat, milk, and eggs. Recent studies with humans indicate that still greater diversion of soybean protein into direct food use in the future would present no major problems, particularly with adults, insofar as methionine limitations are concerned.

Table 7. Trypsin inhibitor activities of soy protein products.

Product	Trypsin Inhibitor Activity	
	mg/g Sample	% of Raw Soy Flour
Raw soy flour[a]	56	100
Toasted soy flour[a]	5	9
Protein concentrate[a]	6	11
Protein isolates[a]	5-10	9-18
Protein fiber[b]	7	13

[a]Kakade et al. (9).
[b]Liener (13).

However, questions are being raised about potential physiological responses of humans to the residual trypsin inhibitor activity in presently available protein products. Preliminary long-term studies with rats have revealed no problems, but more research is needed to determine whether low levels of trypsin inhibitor are of any consequence in human diets.

NOTES

W. J. Wolf, Northern Regional Research Center, Agricultural Research, SEA, USDA, Peoria, Illinois 61604.

The mention of firm names or trade products does not imply that they are endorsed or recommended by the USDA over other firms or similar products not mentioned.

REFERENCES

1. Anonymous. 1973. Energy and protein requirements. Report of a joint FAO/WHO *Ad Hoc* Expert Committee, FAO Nutrition Meetings Report No. 52, Food and Agriculture Organization of the United Nations, Rome, 117 p.
2. Association of Official Analytical Chemists. 1975. Official Methods of Analysis, 12th Edition.
3. Fats and Oils Situation. 1978. Economics, Statistics, and Cooperatives Service, USDA, Washington, D.C., FOS-292, 8.
4. Feed Situation. 1978. Economics, Statistics, and Cooperatives Service, USDA, FdS-271,26-27.
5. Fomon, S. J., and E. E. Ziegler. 1979. Soy protein isolates in infant feeding. In H. L. Wilcke, D. T. Hopkins, and D. H. Waggle (eds.) Soy Protein and Human Nutrition, Academic Press, New York, p. 79-99.
6. Hegsted, D. M. 1974. Assessment of protein quality. In Improvement of Protein Nutriture. National Academy of Sciences, Washington, D.C., p. 64-88.
7. Hopkins, D. T., and F. H. Steinke. 1978. Updating protein quality measurement techniques. Cereal Foods World 23:539-543.
8. Horan, F. E. 1974. Soy protein products and their production. J. Am. Oil Chem. Soc. 51:67A-73A.
9. Kakade, M. L., J. J. Rackis, J. E. McGhee, and G. Puski. 1974. Determination of trypsin inhibitor activity of soy products: A collaborative analysis of an improved procedure. Cereal Chem. 51:376-382.
10. Kawamura, S. 1967. Quantitative paper chromatography of sugars of the cotyledon, hull, and hypocotyl of soybeans of selected varieties. Kagawa Univ. Fac. Tech. Bull. 18:117-131.
11. Kies, C., and H. M. Fox. 1971. Comparison of the protein nutritional value of TVP, methionine enriched TVP and beef at two levels of intake for human adults. J. Food Sci. 36:841-845.
12. Kotter, L., A. Palitzsch, H. -D. Belitz, and K. -H. Fischer. 1970. The presence and significance of trypsin inhibitors in isolated soya protein intended for use in the manufacture of meat products which are to be heated to high temperatures. Fleischwirtschaft 50:1063-1064.
13. Liener, I. E. 1975. Effects of anti-nutritional and toxic factors on the quality and utilization of legume proteins. In M. Friedman (ed.) Protein Nutritional Quality of Foods and Feeds, Part 2, Marcel Dekker, Inc., New York, p. 523-550.

14. McLaughlan, J. M. 1976. The relative nitrogen utilization method for evaluating protein quality. J. Assoc. Off. Anal. Chem. 59:42-45.
15. National Research Council. 1972. Nutrient Requirements of Laboratory Animals, Second Revised Edition, National Academy of Sciences, Washington, D.C.
16. National Research Council. 1974. Recommended Dietary Allowances, 8th Revised Edition, National Academy of Sciences, Washington, D.C.
17. Phipard, E. F. 1974. Protein and amino acids in diets. In "Improvement of Protein Nutriture, National Academy of Sciences, Washington, D.C., p. 167-183.
18. Rackis, J. J. 1974. Biological and physiological factors in soybeans. J. Am. Oil Chem. Soc. 51:161A-174A.
19. Rackis, J. J., J. E. McGhee, and A. N. Booth. 1975. Biological threshold levels of soybean trypsin inhibitors by rat bioassay. Cereal Chem. 52:85-92.
20. Rackis, J. J., J. E. McGhee, M. R. Gumbmann, and A. N. Booth. 1979. Effects of soy proteins containing trypsin inhibitors in long-term feeding studies in rats. J. Am. Oil Chem. Soc. 56:162-168.
21. Scrimshaw, N.S., and V. R. Young. 1979. Soy protein in adult human nutrition: A review with new data. In H. L. Wilcke, D. T. Hopkins, and D. H. Waggle (eds.) Soy Protein and Human Nutrition. Academic Press, New York, p. 121-148.
22. Torun, B. 1979. Nutritional quality of soybean protein isolates: studies in children of preschool age. In H. L. Wilcke, D. T. Hopkins, and D. H. Waggle (eds.) Soy Protein and Human Nutrition, Academic Press, New York, p. 101-119.
23. Watanabe, T. 1978. Traditional nonfermented soybean foods in Japan. Proc. Int. Soya Protein Food Conf., Singapore, American Soybean Association, p. 35-38.
24. Williams, H. H., A. E. Harper, D. M. Hegsted, G. Arroyave, and L. E. Holt, Jr. 1974. Nitrogen and amino acid requirements. In Improvement of Protein Nutriture, National Academy of Sciences, Washington, D.C., p. 23-63.
25. Wolf, W. J. 1978. New soy protein food products in the U.S. Proc. Int. Soya Protein Food Conf., Singapore, American Soybean Association, p. 59-65.
26. Zezulka, A. Y., and D. H. Calloway. 1976. Nitrogen retention in men fed varying levels of amino acids from soy protein with or without added L-methionine. J. Nutr. 106:212-221.

PRICING SOYBEANS ON THE BASIS OF CHEMICAL CONSTITUENTS

N. J. Updaw and T. E. Nichols, Jr.

The value of each bushel of soybeans to a processor is determined by the quantities of oil and meal to be extracted and their respective wholesale prices. It has been documented that the quantities of oil and meal obtained through the crushing process, as well as the protein content of the meal, are associated with the oil content and protein content of the soybeans to be processed (2). Until recently it has not been possible for market participants to measure protein content and oil content at time of sale; however, the development of near-infrared reflectance instruments will permit soybean buyers to obtain simultaneous estimates of the oil content, protein content and moisture content of the beans available to the market.

The purpose of this chapter is to report the highlights of a study which attempted to document the economic effects of the addition of such measurements to the grading standards for soybeans sold in the U.S. The results pertain to an environment in which each bushel of soybeans receives a price that reflects the market value of its oil and protein content. Profit margins for crushers and handlers are assumed to remain constant in this analysis, so that attention is centered on the response of soybean producers and the consumers of oil and meal. The results indicate which groups in our economy may benefit from the addition of chemical measurement and which groups may be hurt, in addition to the practical difficulties inherent in the inclusion of these instrument readings to the grading procedures for soybeans.

BACKGROUND

Since the end of World War II the production of soybeans in the U.S. has increased to such an extent that only corn and wheat now exceed soybeans in total annual harvested acreage (8). The increase in soybean production and the relatively high soybean prices which have prevailed in recent years have resulted in revenue to farmers that is greater than that of any other crop except corn. All of the soybeans produced each year are sold in domestic and foreign markets or are stored so that they may be sold at a later date. The ultimate purchaser of most of the crop is a processor who crushes soybeans to obtain crude soybean oil and soybean meal.

Since soybeans are used as an input in the production of soy oil and meal, the demand for soybeans is a derived demand for the chemical constituents of the commodity. Processors who strive to maximize profits will purchase soybeans at the prevailing market price only if they anticipate that the value of the oil and meal obtained from the beans will exceed the purchase price plus the cost incurred in processing the beans. On average, American processors have been able to extract approximately 10.8 lb crude oil and 47.3 lb meal from each 60.0 lb bu of soybeans (2). Therefore, for every one cent increase in the wholesale price of soybean oil, processors can be expected to be willing to pay up to an additional 10.8 cents/bu for soybeans, all other things remaining constant. Similarly, for every one cent increase in the price of soybean meal, processors should be willing to pay up to an additional 47.3 cents/bu, all other things remaining constant. Undoubtedly, there are instances in which the price paid for soybeans does not reflect exactly any changes in the value of the products obtained in a crush, but competition among processors to obtain soybeans for their mills will tend to maintain a positive correlation between the demand for soybeans and the value of the products obtained from their use (4).

Soybeans, like other agricultural products, are a heterogeneous commodity. The chemical components (moisture, oil, protein, fiber, hull, foreign matter) tend to vary from bushel to bushel. All other things equal, purchasers prefer to buy soybeans which have relatively large portions of oil and protein since these components have positive economic value which is captured when the beans are crushed. Sellers, on the other hand, would like to receive the highest possible price for the products which they produce without regard for their end use. Given these conflicting interests and the practical difficulties inherent in the sale of large quantities of an agricultural commodity, market participants have established grading standards, sampling procedures and measurement techniques to apply at every market transaction. This has involved the definition of a standard soybean (US No. 1 yellow) according to weight, moisture content, the proportion of each bushel comprising beans that are damaged, diseased or discolored, and the proportion of each bushel which is composed of foreign material. The purpose of the grading system is to provide systematic guidelines for classifying products into more uniform categories of quality and to determine the difference in price by which each bushel may be sold in an effort to establish a fair return for each trader's effort. Its use has improved the marking system by reducing the amount of time and effort expended in the determination of quality and by providing objective measurement of soybean value and price for producers, grain merchants and processors.

The grading system presently used provides information primarily on physical characteristics and grade tolerances for soybeans. However, it fails to provide needed information on levels of oil and protein. Processors attach considerable importance to physical characteristic information but they also need additional information concerning the oil and protein content to assure themselves of a profitable crush. During those years of high oil prices and low meal prices, such as the early 1950's, the greater part of the value of each bushel of soybeans was attributed to the oil which could be extracted. Naturally, processors were most interested in the oil content of the soybeans that they were crushing since the quality of oil extracted is directly related to the oil content of the inputs. More recently, worldwide increases in the demand for protein has raised the meal value of each bushel of soybeans to a level exceeding the oil value; hence, processors have exhorted producers to take steps to increase the protein content of their beans. The data in Table 1 provide an illustration of the changes in oil value and meal value of a typical bushel of soybeans in recent years. These data demonstrate how the economic interests of soybean buyers can change in response to developments in the markets for oil and meal. As a rough approximation, the oil value of a bushel of soybeans will dominate the meal value whenever the relative price ratio of the products (P_m/P_o) falls below 0.228. Given the present supply-demand relationships of protein and oil, it appears likely that the meal value of each bushel will be of greater interest to processors than the oil value for some years to come. It is conceivable, however, that emphasis will return to oil content some time in the future.

Chemical analyses can provide processors an estimate of the constituents of the soybeans they have been purchasing. The standard laboratory methods for estimating protein and oil levels in soybeans are by Kjeldahl and Soxhlet pet ether extraction, respectively. Results from these tests are very reliable; however, the cost associated with running the samples together with the 2 to 3 day delay in getting the results make the use of these tests impractical in the procurement of soybeans. Because of these problems, processors have in the past followed the practice of analyzing only a few representative samples early in the season. Geographic areas identified as producing low quality beans were avoided whenever possible. If low quantity beans were purchased, they were usually held in segregated storage for blending with high quality beans purchased from other areas to yield a protein level needed in processing.

In recent years the concerns of processors for more precise information on the constituent values have led several American firms to develop machines

Table 1. **Relative oil value and meal value of one bushel of soybeans yielding 10.8 lb of oil and 47.3 lb of meal.**

Year	Average Price of Oil[a]	Oil Value	Average Price of Meal[b]	Meal Value	Total Value	P_m/P_o
	– ¢/lb –	– $/bu –	– ¢/lb –	– $/bu –	– $/bu –	
1950	13.7	1.48	4.0	1.89	3.37	.292
1955	11.6	1.25	2.6	1.23	2.48	.224
1960	8.8	.95	3.0	1.42	2.37	.341
1965	11.2	1.21	4.1	1.94	3.15	.366
1970	12.0	1.30	4.0	1.89	3.19	.333
1975	25.4	2.74	7.4	3.50	6.24	.291

[a]Average wholesale price of soybean oil, Midwestern Mills, Agricultural Statistics, U.S. Department of Agriculture.

[b]Average wholesale price of 44% soybean meal, Midwestern Mills, Agricultural Statistics, U.S. Department of Agriculture.

which are capable of providing simultaneous measurements of oil, protein and moisture content of soybeans. By using near-infrared (NIR) instruments, it is now possible to analyze each load of soybeans as they are purchased. With the aid of these machines it is also possible to incorporate oil and protein into the routine inspection and grading procedure. Furthermore, with the use of a premium-discount schedule it is possible to adjust soybean prices by value differences in oil and protein levels similar to that being used for moisture, foreign material and other quality factors.

The Federal Grain Inspection Service (FGIS) is currently testing these new measurement devices for reliability and is considering the merits of adding oil and protein measurements to existing grade standards. If the machines are proven to be reliable and the changes in standards are adopted, soybeans could be priced according to the value of the chemical constituents rather than by present factors.

In order for oil and protein to be included as factors in a grading system, there must be widespread acceptability that such determination improves pricing and marketing efficiency. From an economic viewpoint, the benefits accruing to all segments of the marketing system must offset the costs generated by changing present marketing methods if efficiency is to be increased. Incorporation of oil and protein as factors in grade determination and component pricing will have an impact on producers, handlers, processors and consumers. Economic theory provides a framework in which empirical estimates may be made of the value of component pricing to each group and of the changes in market conditions that are likely to come about in response to component pricing. Each group will be analyzed separately and the results summarized to provide some conclusions concerning the desirability of the component pricing of soybeans for society as a whole.

EFFECTS OF COMPONENT PRICING ON SOYBEAN PROCESSORS

If it is assumed that the introduction of component pricing of soybeans does not confer upon buyers the opportunity to exploit soybean producers through some monopsonistic power, then the major benefit to processors from the measurement of chemical constituents is to reduce the risk of purchasing soybeans at a price which exceeds the value of the products extracted in a crush. It is conceivable that those processors and grain merchants who first incorporate oil and protein measurements in their purchasing patterns will earn excess profits or rents through their ability to avoid those beans expected to yield below-average crushing margins; however, if the profits of those processors and merchants who do not screen soybeans deteriorate because their competitors draw away the superior soybeans, all competitors will find it desirable to measure oil and protein components as well. Eventually, crushing margins for the indsutry would be expected to return to a level commensurate with a normal rate of return on investment.

For ready acceptance and adoption of the NIR technique by processors, it is necessary to demonstrate how estimates of oil content and protein content of soybeans by this method can reduce the risks ordinarily incurred in the operation of a processing mill. The quantities of soybean oil and soybean meal produced in a crush are determined by the oil and meal content of the soybeans that are processed. In general, the larger the oil content of the soybeans which are crushed, the greater the quantity of oil to be extracted. Since the material which remains after the oil has been extracted is used in the production of soybean meal, the greater the oil content of the soybeans to be crushed, the lesser the quantity of meal to be obtained. The relationships which exist among oil content and the quantities of recoverable oil and meal are linear. They have been expressed in the form of an equation (2):

$$O_i = -0.62 + 60.72\ X_i \qquad \text{(I)}$$

$$Y_i = 59.34 - 69.00\ X_i \qquad \text{(II)}$$

where O_i = number lb soy oil obtained from a bushel containing X_i oil content; X_i = oil content expressed as a decimal; and Y_i = lb soybean meal obtained from a bushel containing X_i oil content. These relationships apply to soybeans with 13.0% moisture content. It is assumed that the average weight of the soybeans that are to be crushed is 60 lb/bu.

Equations (I) and (II) may be helpful to the oilseed processor as he attempts to fulfill his daily production requirements. If, for example, he plans to process No. 1 soybeans that have an average oil content of 18.0%, the estimated quantities of soy oil and meal to be extracted would be 10.31 lb/bu and 46.92 lb/bu, respectively. The remaining 2.77 lb of the 60 lb bu of beans represents the milling loss incurred in the crushing process.

Of course, knowledge of the quantities of oil and meal to be extracted from a bushel of soybeans is not sufficient to allow the processor to estimate the crushing margin he might earn; he must be capable of estimating the protein content of the meal to be produced, as well. This is due to the fact that processors are penalized whenever the soybean meal they produce does not contain the stated protein content. The protein content of meal is determined by the oil content and the protein content of the soybeans that are crushed, according to the following equation:

$$Z_{ij} = -0.1343 + 0.6712\, X_i + 1.3203\, X_j \qquad \text{(III)}$$

where Z_{ij} = protein content of soybean meal produced by soybeans that have X_i oil content and X_j protein content; and X_j = protein content of soybeans to be crushed, expressed as a decimal. If the protein content equals or exceeds the standard (44% protein meal), the processor can compute the crushing margin that will be earned from every batch of beans, provided that he can measure the oil content and protein content of the soybeans to be crushed.

Since the grading procedure now in use ignores oil and protein content, it is very possible that different soybeans will be purchased at the same price but will yield quantities of oil and meal which will vary in value. Grading discounts provide soybean producers with an incentive to alter the moisture content and the foreign matter of the beans they bring to market but do very little to induce farmers to change oil content, protein content, or hull and fiber. Therefore, oil content and protein content may vary in a random fashion or may be correlated with the location in which the soybeans are produced. Whichever is the case, these fluctuations in oil and protein content become reflected in the crushing margins earned by processors who are unaware of the chemical constituents of their beans. Table 2 illustrates this point. The market values depicted are those that prevailed during one day in the summer of 1974 in a North Carolina market. All of these observations were made on the same day in a single market. Crushing margins range from - $.05/bu for beans coming from Piedmont counties to + $.26/bu for beans produced in the Coastal Plains area. The relative prices of soybean oil and meal which prevailed at this particular time $.25/lb and $140/ton, respectively) made those beans containing relatively high oil content most valuable to the buyer.

While the ability to estimate crushing margins is very valuable to the processors, it is uncertain that this feature alone would be sufficient in attracting their support for oil and protein measurements. It is necessary to demonstrate that a system of soybean discounts and premiums that are acceptable to all market participants may be devised and implemented at a reasonable cost, so that the reduction in crushing margins associated with

Table 2. Market price of U.S. No. 1 soybeans and value of chemical constituents in North Carolina, July 1974.

County in Which Soybeans Were Grown	Oil Content	Protein Content	Price of Beans	Value of Components[a]	Crushing Margin
	– % –	– % –	– $/bu –	– $/bu –	– $/bu –
Duplin	18.0	38.0	7.68	7.87	.19
Pitt	19.0	37.0	7.68	7.93	.25
Harnett	19.0	37.0	7.68	7.93	.25
Granville	21.0	34.0	7.68	7.94	.26
Alamance	19.0	34.0	7.68	7.64	-.04
Randolph	17.0	37.0	7.68	7.63	-.05

[a]Based on oil prices at $.25/lb and 44% soybean meal at $140/ton.

component pricing is at least as valuable to the processing industry as the additional cost of screening soybeans.

COMPONENT DISCOUNTS AND PREMIUMS AS THEY MIGHT BE APPLIED IN THE MARKET

If the grading of soybeans is to include oil and protein content measurements, it is necessary that market participants accept the measurements provided by infrared analyzers and that there be some set of premiums and discounts available so that both buyer and seller will be able to agree upon a price. The question of the reliability of machine readings is under investigation at this time. A paper published by Rinne et al. (7) indicates that machine measurements of oil content and protein content in the laboratory compare quite favorably with those of conventional laboratory tests when the machines are maintained in a controlled environment. While further study of the machines operating under field conditions may be needed before conclusions concerning their accuracy can be reached, we will proceed under the assumption that machine readings of oil and protein components are accurate on average with insignificant deviations. In this case, it becomes possible to anticipate acceptance of component pricing by all market participants, provided that a set of discounts is established which is not likely to distribute wealth from one group of market participants (sellers) to another (buyers). This implies that soybean producers who sell soybeans that contain chemical constituents which provide above-average crushing margins would receive a premium price, and those selling soybeans which yield below-average crushing margins would receive a discounted price. It is inconceivable that producers would accept a discount scheme which not only stablized crushing margins but also raised them on average. To keep rates of return on investment constant for all groups in the marketing chain, discounts and premiums must maintain processor crushing margins at their customary levels. Such a discount scheme has been developed by Bullock, Nichols and Updaw (2). It is included in this chapter to illustrate the steps that must be taken to construct a table of discounts and premiums.

To begin, it is necessary to define the "standard" soybean according to oil and protein content. In this example we have assumed that the standard bean is capable of being processed into 49% protein meal since this is an important product for processors who serve the poultry industry. We have chosen 18.0% as the oil standard and then used equation (III) to determine the protein content required to make 49% meal; this happens to be 35.86%. Therefore, the standard soybean is defined as one which would now be graded as US No. 1 and is composed of 18.0% oil and 35.86% protein.

Since oil content and protein content both vary from bushel to bushel, it is conceivable that any bushel might deserve a discount or premium for both constituents, or a discount for one and a premium for the other. Therefore, discounts should be applied for each component separately. Since the oil value of any bushel equals the quantity of extractable oil times the wholesale price of oil, oil content discounts (premiums) should be assessed according to the following relationship:

$$D_i^o = -P_o\, 60.72\, (X_i - .18) \qquad \text{(IV)}$$

where D_i^o = oil discount or premium assessed to soybeans containing oil content X_i, in cents per bushel; and P_o = wholesale price of crude soy oil, in cents per lb. Equation (IV) reflects the difference in value of recoverable oil extracted from the "standard" bushel and any bushel brought to market. If measured oil content exceeds the standard, the discount for oil will be negative, which means the soybeans will receive a premium for oil. Conversely, if measured oil content falls short of the standard, the oil discount will be positive. In either case, the oil discount will fluctuate with the prevailing wholesale price of crude soy oil.

Discounts and premiums for soybean protein content are somewhat more complicated than those for oil content because two factors must be taken into account: (a) quantity of meal obtained from each bushel is determined by its oil content, and (b) the protein content of the meal produced is determined by protein content and oil content of the soybeans processed. Since processors are penalized whenever the meal sold by them does not meet stated protein content, soybean protein discounts that are based solely on the quantity of meal expected to be produced will overestimate the protein value of soybeans that have relatively low protein content. Discount penalties for protein-deficient meal usually require a 1% reduction in price for each 1% below standard in the protein content of soybean meal. Thus, the meal value of every bushel of soybeans may be approximated by the following equation:

$$M_{ij} = \frac{P_{44}}{.44}(59.34\text{–}69.0\,X_i)(-0.13 + 0.67\,X_i + 1.3203\,X_j) \qquad \text{(V)}$$

or

$$M_{ij} = \frac{P_{44}}{.44}(-7.71 + 48.73\,X_i + 78.33\,X_j - 46.23\,X_i^2 - 91.08\,X_iX_j)$$

where M_{ij} = meal value of 1 bu of beans containing oil content X_i and protein content X_j, in cents/bu; and P_{44} = price of 44% protein meal, wholesale, in cents/lb. On the basis of this relationship and the component standards, the protein discount (premium) for any bushel of soybeans becomes:

$$D_{ij}^p = \frac{-P_{44}}{.44}\{48.73\,(X_i - .18) + 78.33\,(X_j - .3586) - 46.23\,(X_i^2 - .0324) - 91.08\,[\,X_iX_j - (.0645)]\} \qquad \text{(VI)}$$

Component discounts (premiums) for each bushel of soybeans will equal the sum of the oil and protein discounts (premiums).

Equations (IV) and (VI) may appear so imposing that market participants would hesitate to use them. However, they may be converted rather easily to tables such as that given in Table 3. Growers bringing soybeans to market with only 15% oil and 28% protein would receive discounts for both oil and meal totaling \$1.26/bu (upper left corner). However, growers producing beans containing an average 25% oil and 38% protein might receive a premium of \$1.21/bu for their product (lower right corner). Market participants could consult a table of discounts that would apply at varying prices of oil and 44% protein meal to determine component discounts and premiums. As long as measurement devices provide tolerably accurate readings of oil content and protein content, there should be no technical difficulties which would prevent the component pricing of soybeans from being put into practice.

EXPECTED COSTS AND RETURNS TO PROCESSORS FROM THE COMPONENT PRICING OF SOYBEANS

If component pricing is perceived by processors to be a profitable proposition, it will over time become an accepted practice. The long-run effects on the indsutry should be nil since any competitive advantages earned by those firms who adopt the new technology early dissipate once other firms accept the new marketing procedure. Crushing margins may decrease slightly if processors are unable to pass along the additional measurement cost to consumers of oil and meal. This is consistent with a long-run equilibrium in the industry, however, as long as component pricing reduces the risk of doing business.

The expected cost associated with component pricing would vary from firm to firm. In a recent study conducted by the USDA, Niernberger (6)

Table 3. Per bushel discounts and premiums for alternative levels of oil and protein content when the price of soybean oil is $.25/lb and the price of 44% soybean meal is $140/ton.[a]

Protein Content	Type of Discount	Oil Content (%) 15	17	19	21	23	25
– % –		– ¢/bu –					
28	Oil	- 45.54	- 15.18	+15.18	+45.54	+75.90	+106.26
	Meal	- 83.25	- 78.46	- 76.26	- 74.65	- 73.63	- 73.20
	Total	- 126.79	- 93.64	- 61.08	- 29.11	+ 2.27	+ 33.06
30	Oil	- 45.54	- 15.18	+15.18	+45.54	+75.90	+106.26
	Meal	- 60.67	- 58.46	- 56.84	- 55.81	+55.37	- 55.52
	Total	- 106.21	- 73.64	- 41.66	- 10.27	+20.53	+ 50.74
32	Oil	- 45.54	- 15.18	+15.18	+45.54	+75.90	+106.26
	Meal	- 40.09	- 38.46	- 37.42	- 36.97	- 37.11	- 37.83
	Total	- 85.63	- 53.64	- 22.24	+ 8.57	+38.79	+ 68.43
34	Oil	- 45.54	- 15.18	+15.18	+45.54	+75.90	+106.26
	Meal	- 19.51	- 18.46	- 18.00	- 18.13	- 18.85	- 20.15
	Total	- 65.05	- 33.64	- 2.82	+27.41	+57.05	+ 86.11
36	Oil	- 45.54	- 15.18	+15.18	+45.54	+75.90	+106.26
	Meal	+ 1.07	+ 1.54	+ 1.42	+ 0.71	- 0.58	- 2.47
	Total	- 44.47	- 13.64	+16.60	+46.25	+75.32	+103.79
38	Oil	- 45.54	- 15.18	+15.18	+45.54	+75.90	+106.26
	Meal	+ 21.65	+21.54	+20.84	+19.56	+17.68	+ 15.21
	Total	- 23.89	+ 6.36	+36.02	+65.10	+93.58	+121.47

[a]Relative to the standard soybean containing 18% oil and 35.86% protein.

found that there were economies of scale to the implementation of infrared analyzers, which cost about $20,000 each, that would give larger processors a competitive edge over small firms. For the average oilseed crusher, the cost of measuring oil and protein content was estimated at approximately 0.5 cents/bu crushed per year. To this cost must be added the value of the additional time spent determining discounts and premiums once measurements have been taken. This apparently nominal sum can become quite substantial when calculated for the entire annual crush; nevertheless, it is conceivable that processors would be willing to incur this expense if they can be assured more stable year-to-year profits.

EXPECTED EFFECTS ON SOYBEAN PRODUCERS

While there may not be any long-run effects on processors or grain merchants from the introduction of component pricing, there are reasons to expect changes in the activities of soybean producers and the wholesale consumers of soybean oil and meal. If the existence of constituent measurements and compensation generate incentives for producers to alter the chemical composition of their soybeans, there will be changes in the relative supplies of soy oil and meal to leave the processing mills. These effects will be permanent if soybean producers have not been growing beans whose characteristics reflect adequately the relative demands for the two soy-derived products. In other words, if processors are correct in their complaints that the soybeans produced today are deficient in protein, then component pricing will lead to extraordinary profits for producers who grow beans that possess above-average protein content. This will create incentives for producers to change their production practices so that the protein content of the average soybean is increased. In addition, those growers who possess locational advantages for the production of high-protein beans will be given an incentive to expand production at the expense of producers who have an advantage in the production of high-oil soybeans. These changes will be reflected in the markets as a decline in the price of meal and an increase in the price of oil, all other things equal, when compared to prices which would have prevailed in the absence of component pricing.

In the following section of this chapter, empirical estimates of the reactions of producers will be provided and expected effects in the markets for oil and meal will be presented also. The extent of these effects will determine the merits of constituent measurements since they represent the long-run economic adjustments of the proposal. If producers have the capability of changing the chemical characteristics of soybeans while maintaining overall profitability, then the prospects for component pricing may be good. If producers are unable to alter oil content and protein content profitably, then it would be unlikely that measurements for oil and protein would be attractive to them.

REGIONAL ACREAGE RESPONSE TO COMPONENT PRICING OF SOYBEANS

The establishment of component pricing would shift the economic penalties associated with variations in oil content and protein content from processors to producers. Soybean production would become more risky than it is presently because expectations concerning the price to be received by each grower would have to be adapted to include traditional chemical characteristics.

It has been alleged by some soybean processors that there exists a predictable relationship between the chemical composition of soybeans and the region of the country in which they are grown. Specifically, it is believed that as one proceeds from the Canadian border to the Gulf of Mexico, the oil content of soybeans increases and the protein content decreases. If this is true, the addition of oil and protein measurements to the grading of soybeans is expected to shift acreage from the Corn Belt and Great Lakes regions to the South and the Mississippi Delta states. In an attempt to assess the regional variation of chemical composition of soybeans, the authors have obtained data collected by agronomists who conduct experiments at research station sites located throughout the soybean-producing states. Oil content and protein content of several varieties (on a dry-matter basis) have been recorded and published by the USDA (9). Locations have been grouped into regions as indicated in Table 4. An analysis of vairance of oil content and protein content by regions has been conducted, and the resulted summarized in Table 5.

By setting the Corn Belt as the standard region and comparing all other regions to it, the results indicate that during the early 1960's there was no significant difference in the average protein content of soybeans, no matter where they were grown. Soybeans grown in the Delta region had a significantly higher oil content than that of the Corn Belt, and those produced in the Lakes region had an average oil content that was significantly lower than that of the Corn Belt. If the chemical composition of soybeans grown at research station sites is highly correlated with all soybeans grown in the surrounding region, then soybeans grown in the Delta region during the early 1960's should have been the most profitable to be crushed, those grown in the Lakes region the least profitable, and those grown elsewhere would fall somewhere in between.

An analysis of soybean samples taken in the early 1970's produced slightly different results (Table 6). Findings indicated that there were no significant differences in average oil content between the Corn Belt and all other regions, but there were significant protein differences. The average protein content of soybeans grown in the South and Atlantic regions was significantly higher than the protein content of those grown in the Corn Belt and the average protein content of beans grown in the Lakes region was significantly lower than that of the Corn Belt. If these samples were representative of the area, one could conclude that soybeans grown in the South and Atlantic regions should have been more profitable for processors than those grown in the Corn Belt; those grown in the Lakes region should have been less profitable than soybeans produced in the Corn Belt; and those grown in all other regions should have been equally profitable, on average, as those grown in the Corn Belt.

This analysis provides little support for the contention that southern soybeans tend to possess above-average oil content and Northern soybeans possess above-average protein content. If the relationships which exist at research station sites hold true in the fields of soybean producers, component pricing would enhance the price of soybeans in the Delta, South and Atlantic

Table 4. Regional grouping of major soybean producing states.[a]

Region	States Contained Within Region
South	Florida, Georgia, Alabama, Tennessee, Oklahoma, Texas
Plains	North Dakota, South Dakota, Kansas, Nebraska
Atlantic	Maryland, Virginia, North Carolina, South Carolina
Lakes	Minnesota, Wisconsin, Michigan
Delta	Mississippi, Arkansas, Louisiana
Corn Belt	Ohio, Indiana, Illinois, Iowa, Missouri
North	New Jersey, Kentucky, Delaware

[a]Derived from Houck and Subotnik (4).

Table 5. Average oil content, protein content, and 95% confidence intervals for soybeans grown at research station sites by regions, 1962-65.

Years	Region	Average Oil Content	Confidence Intervals, Oil	Average Protein Content	Confidence Intervals, Protein
				– % –	
1962-65	Corn Belt	21.13	20.63 - 21.63	40.09	39.40 - 40.78
1962-65	Atlantic	21.20	20.77 - 21.63	39.79	39.09 - 40.49
1962-65	Delta	22.14	21.66 - 22.62[a]	39.09	38.21 - 39.97
1962-65	South	21.23	20.55 - 21.91	40.06	38.82 - 41.30
1962-65	North	20.68	20.11 - 22.11	41.19	40.23 - 42.15
1962-64	Lakes	19.65	19.05 - 20.25[b]	40.17	39.03 - 41.31
1962-64	Plains	21.11	20.11 - 22.11	38.57	37.50 - 39.64

[a]Denotes significantly higher than Corn Belt.
[b]Denotes significantly lower than Corn Belt.

Table 6. Average oil content, protein content, and 95% confidence intervals for soybeans grown at research station sites by regions, 1972-76.

Years	Region	Average Oil Content	Confidence Intervals, Oil	Average Protein Content	Confidence Intervals, Protein
				– % –	
1972-76	Corn Belt	21.43	21.22 - 21.64	40.69	40.44 - 40.94
1972-76	Atlantic	21.57	21.29 - 21.85	41.32	41.00 - 41.63[a]
1972-76	Delta	22.26	21.56 - 22.96	40.41	39.93 - 40.89
1972-76	South	21.91	21.58 - 22.24	41.49	41.17 - 41.81[a]
1972-76	North	22.13	21.50 - 22.76	41.74	40.91 - 42.57
1972-76	Lakes	21.70	20.95 - 22.45	38.97	38.03 - 39.91[b]
1972-76	Plains	21.36	21.00 - 21.72	40.67	40.27 - 41.07

[a]Denotes significantly higher than Corn Belt.
[b]Denotes significantly lower than Corn Belt.

regions vis-a-vis the rest of the country because the total amount of valuable product obtainable from these beans exceeds that of the others. It must be the case that soybeans grown in the Delta, South and Atlantic regions possess lower levels of hull and fiber than those grown in the north.

In addition to regional acreage response, soybean producers may be able to alter, within limits, the chemical composition of their beans through changes in agronomic practices. For the purposes of this study, such responses will be grouped in two categories: (a) changes in fertilizer applications and (b) changes in the variety of seed planted. Continuing efforts by agronomists make any definitive statements tenuous in nature; however, there is considerable evidence that farmers have the capability to increase protein content at the expense of a reduction in oil content. While it is beyond the scope of this chapter to review the entire literature, a review of the study published by the N.C. Agricultural Extension Service (5) might outline the possibilities. In this study, it was determined that lime applications had the effect of increasing protein content, increasing yield and decreasing oil content. Furthermore, it was detected that varieties which possessed above-average oil content had below-average protein content; the converse was true, as well. The variety effects are shown in Figure 1. On the face of it, producers can be expected to have the capability to react to premiums or discounts assessed on the basis of oil and protein measurements.

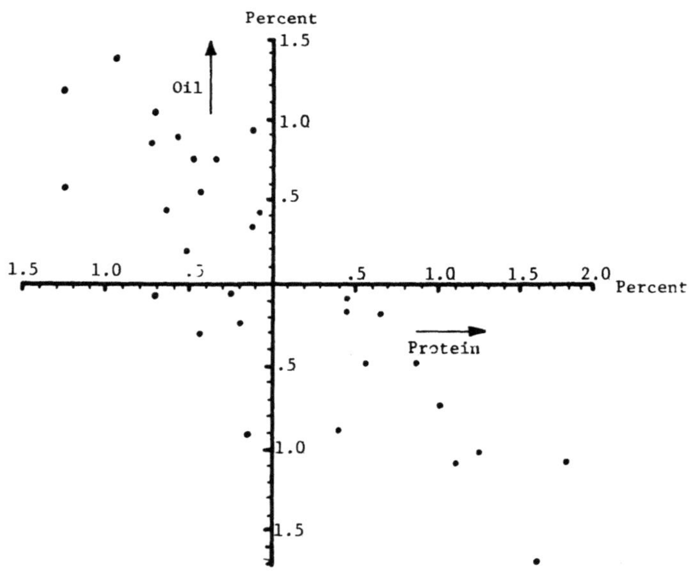

Figure 1. **Varietal deviations from average oil and protein content of soybeans, estimated from pooled data at 33 locations in N.C. (5).**

While producers may have the ability to alter the protein content and oil content of soybeans through the application of nitrogen, fertilizer and lime, it is not clear that they would find it profitable to do so. One of the conclusions of the study cited above was that additional applications of nitrogen and potash above certain levels increased protein but not yield and the value of the increased protein was less than the cost of the fertilizer applied. If one is to assess the reaction of producers to component pricing, it is necessary to develop an economic model which will identify the most profitable alternatives made available to producers. In an analysis of component pricing, it is necessary to convert observations of yield, oil content and protein content to per acre production of oil and protein to assess alternatives. If producers are paid for the value of the products to be obtained from their beans, rather than the beans themselves, they may no longer base their management decisions strictly upon expected yield.

To determine the economic viability of changing seed variety in response to component pricing, observations of average yield, oil content, and protein content for 20 varieties were drawn from *Uniform Soybean Tests* over the period 1972 to 1976 (10). Observations were converted to lb of oil/acre and lb of protein/acre so that an analysis of variance could detect systematic variation attributable to differences in location, year of observation, and variety. When varietal differences were isolated, there was no significant difference in per acre output of oil and protein for any variety when compared with the standard variety, Amsoy 71 (Table 7). The scatter of output variation attributable to differences among varieties is shown in Figure 2. This evidence indicates that conventional soybean varieties are too similar in production capabilities to allow soybean producers to change variety as the market discounts and premiums for oil and protein would change. The yields of those varieties that appear above-average in either oil content or protein content are much too low compared to those of the average variety to produce any long-run effects among soybean producers in their choice of varieties. If companies' pricing is to become effective, additional research by breeders must be directed toward the generation of varieties which have higher yields and a wider range of oil and protein contents than presently exists. If those varieties possessing average levels of oil and protein continue to provide higher yields than those that are above average in either component, producers will likely find it more profitable to plant conventional varieties even if they bear substantial discounts.

Additional research is also needed to determine whether producers would find it profitable to change fertilizer practices in response to component pricing of soybeans. With the exception of the possibility of some changes in regional acreage, the evidence to date is that oil and protein measurements would not induce farmers to alter the average chemical

Table 7. **Output of oil and protein per acre by variety (10).**

Variety	Oil/Acre	Protein/Acre
	– lb –	
Amsoy 71	530	940
Beeson	493	969
Bonus	483	1040
Bragg	342	838
Clark 63	423	859
Corsoy	515	950
Dare	435	894
Essex	449	1009
Hark	465	906
Hodgson	533	880
Hutton	314	858
Kent	458	937
L66-1359	528	986
Lee 68	404	939
M65-69	508	869
Mack	455	939
Pickett 71	357	838
Ransom	403	821
SL 11	452	972
Williams	529	1040

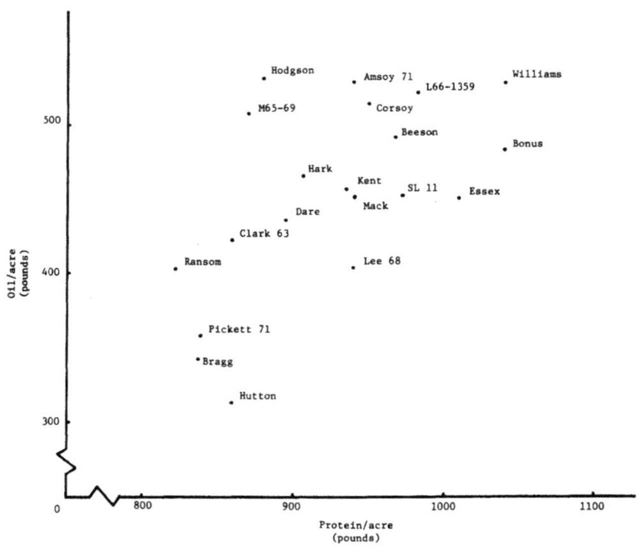

Figure 2. **Pounds of oil and protein per acre produced from different soybean varieties.**

constituents of the soybeans that they now produce. If this conclusion is correct, component pricing will confer no long-run benefits upon either producers or the consumers of oil and protein. It is not at all clear that any major group of participants in the markets for soybeans, soybean oil or meal will derive any significant, permanent economic gains from the implementation of component pricing. Tentative findings indicate that component pricing under present conditions would create a net social cost of $8.1 million/yr (10). Based on average annual expenditures of oil and protein in the U.S. if the predicted social costs were borne entirely by consumers it would be expected to raise the price of oil and protein by about .03 cents/lb.

SUMMARY

The purpose of this chapter has been to highlight the economic aspects of the inclusion of oil content and protein content measurements to the grading standards for soybeans in the U.S. The results of this study indicate that the proponents of component pricing may find it difficult at this time to convince the buyers and sellers of soybeans to accept protein and oil content measurements as a part of the grading standards. Until soybean breeders can develop new varieties that exhibit high yields and unusually high levels of oil or protein, it is unlikely that soybean producers will possess the tools that will allow them to alter chemical composition without suffering a decline in profits. Unless producers change the oil content and protein content of their soybeans, consumers of soybean oil and meal will face the same relative price ratio for these two goods that would have prevailed in the absence of component pricing. Without widespread acceptability of component pricing by most market participants, it is difficult to envision how the inclusion of oil and protein measurements will occur.

One additional complication is the possibility that the measuring devices that would be employed at soybean buying stations would provide unreliable readings of oil content and protein content. Unless producers, grain merchants and processors are convinced that infrared analyzers give consistent, accurate readings under various rates of use or differing environmental conditions, it is unlikely that market participants would press for their use; bad information may be worse than no information at all.

Despite all of these problems, it may become feasible for producers and consumers to gain additional economic welfare from the introduction of component pricing. For this to occur, it is necessary that the manufacturers of analyzers sell devices that are acceptable to all segments of the soybean complex. Furthermore, it is necessary for soybean breeders to develop varieties which will permit producers to alter the chemical composition of soybeans which they produce in response to changing prices for oil and meal. As long as producers do not possess the technological capability to alter the levels of protein and oil within soybeans profitably, the measurement of these chemical constituents can be of little value to society; its costs will exceed its benefits.

That grim fact may paradoxically give reason for hope. Pressure to develop new varieties that have higher yields and a wider range of oil and protein contents is surfacing among plant breeders. Additional research is underway to develop varieties which would permit producers to respond directly to market incentives in the production of more oil or more protein per acre. With the ability by producers to vary output in response to relative price changes without a reduction in profit and the development of better devices to measure these chemical constituents, it is conceivable that component pricing will be adopted by farmers and processors. Acceptability of a component pricing with a premium-discount system based on oil/protein content, however, will vary greatly among producers and handlers and processors.

Processors would have the greatest incentive for widespread adoption of the practice once reliable instruments for quick determination of oil and protein are available. Handlers would be less enthusiastic about changing the present practices unless handling margins were widened sufficiently to cover the cost of testing and handling.

Farmer acceptance would depend largely on the quality of beans produced and the premium-discount schedule applied. Growers producing high protein-low oil beans would undoubtedly favor adoption of component pricing when protein prices are high relative to oil. On the other hand, farmers producing high oil-low protein beans would not favor any change in the present pricing system.

While there would be gainers and losers, depending upon the quality of products marketed, widespread adoption of component pricing in soybeans could result in a net benefit to society. However, this does not appear to be the situation at the present time. Recent studies indicate that component pricing of soybeans would create a net social cost of $8.1 million/yr. If the entire cost was passed on to consumers, it would be expected to raise the price of oil and meal by .03 cents/lb, respectively. Only through profitable changes in oil/protein production at the farm level can the component pricing of soybeans be expected to improve social welfare in the future.

NOTES

N. J. Updaw, Department of Agricultural Economics, Oklahoma State University, Stillwater, OK 74024; and T. E. Nichols, Jr., Department of Economics and Business, N.C. State University, Raleigh, NC 27650.

REFERENCES

1. American Soybean Association. 1975. Soybean Digest Blue Book. Hudson, IA.

2. Bullock, J. B., T. E. Nichols, and N. J. Updaw. 1976. Pricing soybeans to reflect oil and protein content. Economics Research Report No. 37, N.C. Agric. Expt. Station, Raleigh, NC.
3. Houck, J. P., and J. S. Mann. 1968. Domestic and foreign demand for U.S. soybeans and soybean products. Tech. Bul. No. 256, Agric. Expt. Station, Univ. of Minnesota, Minneapolis, MN.
4. Houck, J. P., and A. Subotnik. 1969. The U.S. supply of soybeans: Regional acreage functions. Agric. Econ. Res. 21:99-108.
5. Nichols, T. E., J. G. Clapp, and R. K. Perrin. 1975. An economic analysis of factors affecting oil and protein content of soybeans. Economics Information Report No. 42, N. C. Agric. Extension Service, Raleigh, NC.
6. Niernberger, F. F. 1978. Near-infrared reflectance instrument analysis of grain constituents: a cost study. ESCS-20, USDA, Washington, DC.
7. Rinne, R. W., S. Gibbons, J. Bradley, R. Seif, and C. A. Brim. 1975. Soybean protein and oil percentages determined by infrared analysis. ARS-NC-26, USDA, Washington, DC.
8. U.S. Department of Agriculture. 1978. Agricultural Statistics. U.S. Government Printing Office, Washington, DC.
9. U.S. Department of Agriculture. 1979. Uniform soybean tests. U.S. Government Printing Office, Washington, DC.
10. Updaw, N. J. 1979. Producer response to the component pricing of soybeans. Ph.D. dissertation. Dept. of Economics and Business, N.C. State Univ., Raleigh, NC.

PROTECTIONISM IN TRADE – EXPANSION IN OILSEEDS

S. J. Hauck

It is a pleasure to be here in the South, addressing a conference whose scope is worldwide. The trading world has been made even wider recently by new relations between the U.S. and the People's Republic of China. One terrible fear I have about getting involved with a country that has one billion people is that President Carter will get carried away and say, "y'all come!"

Occasionally, though, what happens in Congress is not all unbelievable nonsense, and the B.S. begins to stand for Believable Substance. One issue which is firmly in that category, and which now faces our Government squarely in the face, is the prospective new international trade agreement that has emerged from the many years of negotiations in Geneva, under the General Agreement on Tariffs and Trade.

It is not my purpose today to discuss in any detail the specific impact of this agreement on U.S. soybean trade. Our exports to Japan and some other countries will be benefited, but not by as much as we felt was right. Nor is it my purpose to emphasize the importance of congressional approval of this agreement, although it is important, since its rejection by the United States would stymie global implementation of its reciprocally reduced trade barriers and the more disciplined trading procedures it creates.

What I would like to say about this latest round of trade negotiations is that they signify a remarkable and shared determination among the economic leaders of the world to maintain the momentum behind wider free trade in principle. This principle is as old as the Industrial Revolution itself, but free trade is not only important in theory. Indeed, it was the British prime minister Disraeli who said, in 1843, "free trade is not a principle, it is an expedient."

In fact, the most significant expansion of free trade has come in times of prosperity when the demand for goods exceeds the supply, and the opportunity for profit is compromised by limits on trading volume—thus the cry for reducing those limits. But Disraeli was only partially right. When prosperity at home has appeared threatened by cheaper competition abroad, domestic pressures for protection of native industries have sharpened. However expedient on an international scale, free trade then seems a liability to workers within nations, especially when those nations are experiencing inflation.

Such a juncture in the international economy occurred in the thirties, when frightened governments raised trade barrier after trade barrier. Our knowledge of the history of that time demonstrates the long-term folly of seeking short-term protection. At such times, it's not unwise to cling to free trade as a principle whose wisdom is more real than apparent.

That's the task of the Congress today, and it's no mean task. It never has been easy. In 1824, the English historian, Lord Macaulay said, "Free trade, one of the greatest blessings which a government can confer on a people, is in almost every country unpopular."

Among those of us in the agricultural community, however, free trade has been popular traditionally, and especially so among those of us intimately involved in promoting the economic strength of U.S. soybeans and soybean products. American soybean processors, and the association which represents them, have consistently believed that the most efficient economic environment for expanding oilseed markets is trade unfettered by tariffs, non-tarrif barriers, or government export subsidies that undermine free competition. We've maintained that belief because of our fundamental confidence in the future growth of demand for our industry's products.

Yet the demand for soybeans and soybean products is in large part a function of worldwide prosperity, which in turn is dependent on a stable system of trade and monetary exchange. If that system stands to be improved by the agreement emerging from the Tokyo Round of the multilateral trade negotiations, those of us who wish to preserve and extend the position of U.S. oilseed trade should pay close attention to how the agreement is approved or amended by the Congress.

Fortunately, the President's Special Trade Representative, Robert Strauss, has special influence on Capitol Hill, and the new trade agreement will have stout advocates. Those of us who support the agreement can join their ranks, but as a group we should be unabashed about focusing on our special interest. That seems to be the scramble in Washington these days, and those who stay out will no doubt get left out.

The National Soybean Processors Association has been engaged historically in the decisionmaking process in Washington, notably in the case of issues in international trade. While the top policy makers are generally attentive to the overall framework of expanding trade, they expect us to remind them of how the specific issues and cases that affect our crop and our industry can be fairly resolved within that framework.

I'd like to outline six cases of action by the Nation's soybean processors in the past few years, which illustrate how an industry committed to expanding trade can take practical steps to improve its prospects.

CASE NUMBER ONE: PALM OIL

In 1976, NSPA observed that international bank loans, for which the U.S. provides capital, to palm oil producing countries were effectively subsidizing 45% of all world palm oil production, which had increased by 87% in the previous 6 yr. Palm oil exports rose by 185% in that period displacing soybean oil. Palm oil was challenging the ability of domestic edible oil industries to respond with accommodations of price and supply in the free market, and it was forcing the protein fraction to carry the increased cost of producing oilseeds.

The response of our industry was to urge the U.S. Government to withdraw its support for international financing arrangements which were subsidizing palm oil production. After acquiring vocal allies in the Congress, this point of view ultimately prevailed, and in 1977, President Carter instructed U.S. delegates to World Bank committees to refrain from voting for more palm oil loans.

The *principle* at stake here was free trade, which was violated by artificially lowering the cost handicap on producing one particular oil-bearing crop. Free trade stems from the merits of a product, the ingenuity behind its promotion and its inherent characteristics of cost and quality. But if a government, or international bank, gives a product an effective discount, which competing products don't receive, the competition is neither equal nor free.

CASE NUMBER TWO: NON-FAT DRY MILK

In 1976, the European Community decided to try and get rid of a million-ton surplus of non-fat dry milk, which its subsidized dairy farmers had produced. The method it chose was to require European importers of protein, mainly U.S. soybean meal, to pay deposits on it, unless and until they bought an equal amount of skim milk powder, which could also be compounded in animal feeds. The result was the loss of at least 200,000 tons annually of U.S. soybean meal exports to Europe.

The response of NSPA was to secure the cooperation of the American Soybean Association and file a formal complaint under the Trade Act of 1974, demonstrating that the European import deposits were a violation of the zero-duty bindings on soybeans and meal that were long a part of an existing GATT agreement. The U.S. Government took our case before the GATT, and a GATT panel essentially vindicated the principle of our position.

CASE NUMBER THREE: BRAZILIAN EXPORT SUBSIDIES ON SOYBEAN OIL AND MEAL

In the last six month period before the effects of last year's drought in Brazil, Brazilian exports of soybean meal increased 61% while U.S. meal exports rose 16%. In that same October 1977 to March 1978 period, Brazilian soybean oil exports rose 177%, while U.S. oil exports increased 31%. These figures represent a trend begun several years ago. Indeed, the share of the world meal export market represented by *non*-American soybean meal equivalent trade increased from 4% in 1968 to 25% in 1978.

As the dimension of this trend became sharply evident in 1976, NSPA concluded that the reason was to be found in the export subsidies granted by the Brazilian government to their domestic crushers, enabling them to undersell both U.S. and European processors in the big European meal market. Internal Brazilian taxes on income, merchandise circulation and futures trading had the effect of subsidizing processors' exports of oil and meal.

NSPA began discussions with the U.S. Government which led to a bilateral agreement between the U.S. and Brazil under which the export subsidies on oil have been phased out. But not so the subsidies on meal as yet. Last year, NSPA formed a special Brazilian Trade Policy Committee to develop comprehensive documentation of the evidence of the anti-competitive impact of Brazilian export subsidies, and to press the OSTR, USDA and the Department of the Treasury and other U.S. agencies to seek their elimination.

American soybean processors don't shrink from meeting overseas competition with the best products we have to offer. But just as we emphatically opposed the U.S. Government's embargo on soybeans in 1973, we've protested foreign governments' intervention in the world system of oilseed supply. Free trade is not only compromised when a nation's consumers demand a foreign product and the nation's government slaps a tariff on it to limit consumption; that is a form of demand-interference. But when governments contrive lower prices by subsidizing exports, that's a form of supply-interference, and it's equally a compromise of free trade. Just remember that free trade is infringed by governments to the advantage of industries in their countries. To defend free trade, the industry whose competitive position is damaged has to appeal through its government to restore free competition by new agreements governing the trading community.

CASE NUMBER FOUR: THE AUSTRIAN OILSEED TAX

In the spring of 1978, NSPA learned that the Austrian Government was considering imposing a regulatory tax on imported oilseed products, and we made strong representations to U.S. officials to protest the Austrian idea. A letter to Secretary Bergland about the issue noted: "If the general goal of freer trade is to be realized gradually, specific barriers to current trade must not be tolerated." The State Department seconded our opposition, and after the Austrian Government decided to shelve the proposed tax, Secretary Bergland wrote back to NSPA saying that the government "appreciated the advice and information supplied to us by your organization."

CASE NUMBER FIVE: LOWER JAPANESE IMPORT BARRIERS

This year, U.S. negotiators obtained from their Japanese counterparts an agreement for a zero-duty binding on soybeans to be part of the new GATT accord. When we heard that, we wrote to Ambassador Strauss and urged him to obtain a zero-duty binding on soybean meal. We noted the significant foreign exchange benefit to the U.S. that could result, and we also recommended that an effort be undertaken to secure a zero-duty binding on soybean oil.

CASE NUMBER SIX: COMMUNICATION WITH EUROPE

Now I have saved for last what is perhaps the most significant export-market action of the soybean processing industry in recent years. Clearly, the palm oil, milk powder and Brazilian subsidy problems have all been of serious magnitude. But to resolve them, we have no choice but to work through the U.S. Government and its trade negotiators and policymakers.

There is another trading condition that might not yield so readily or appropriately to government action, and it has been the source in the past few years of concern. That is the question of how efficiently our U.S. meal export system has operated at a time when price-subsidized Brazilian soybean meal has displaced a growing share of the European market. Some were heard to say that a problem existed in the quality of soybean meal reaching Europe. To determine whether there was any fire behind this smoke, NSPA formed last year a Soybean Meal Export Development Committee. Its members, including top executives of several soybean processing companies, made two intensive trips to Europe in 1978, the first to talk to meal importers, consumers and trade associations, and the second to inspect port facilities and procedures for unloading and sampling meal. The Committee also made a detailed inspection of U.S. Gulf Coast meal exporting facilities. The result included amendments to NSPA Meal Trading Rules, and the development of new Meal Export Trading Rules, to insure new barge shipment, meal blending and sample procedures that will remove any cause for concern about the quality of soybean meal which reaches our foreign customers.

NSPA has also engaged in a continuing series of meetings with government and industry representatives to explain these changes, and to disseminate accurate information about the complexities of export trade in soybean meal. We hosted a Soybean Meal Seminar in Amsterdam last fall, for USDA ag attaches and ASA-European representatives, and another such seminar in Washington earlier this month for USDA officials and ASA representatives state-side. In addition, we have stationed a special NSPA consultant in Europe to maintain the personal liaison with our customers and other U.S. representatives abroad which we have found so essential in clearing away the misinformation, and providing reliable information about U.S. soybean meal exports.

One key reason these actions are significant is that they represent the initiative of the industry, unprompted by the intrusion of government. There is no substitute when you want action for taking action yourself—and there is nobody who can be trusted better to know your business than your own people. That has been the basis for our sustained effort over the past eighteen months to extend the U.S. share of the European protein market, and to obviate the need for government regulations which would be far less likely to be sensitive to the realities of the export trade.

The NSPA meal export development project continues the same spirit typified by our previous involvement with our Government in protesting foreign subsidies of competing products and foreign barriers to U.S. products. That spirit is the determination to insure a system of trade which is free enough to carry the volume of traffic that the future demand for our products seems to promise. Given general economic stability, the future of the world oilseed market appears sound. The demand for oil and meal which is at the heart of that market can only be satisfied through a trading system free of all forms of protectionism, the protection of tariffs, and also the reverse protection of subsidies. The soybean processing industry intends to continue its action on behalf of that principle, because it's also expedient.

NOTES

Sheldon J. Hauck, National Soybean Processors Association, 1800 M Street, N.W., Washington, D. C. 20036.

DEMAND FOR SOYBEANS AND SOYBEAN MEAL IN THE EUROPEAN COMMON MARKET

H. C. Knipscheer

Traditionally the West European market has been an important outlet for U.S. soybeans and soybean products. Since the entrance of the U.K., Denmark and Ireland into the European Economic Community (E.E.C.) in 1973, the E.E.C. represents the largest market for U.S. soybeans. This expansion of the market and the potential for growth increased the importance of the E.E.C. to U.S. soybean producers as well as to non-U.S. soybean growers. Effects of shifts in E.E.C. policy have now become increasing important.

This chapter focuses on the effect of the E.E.C. agricultural policies on the demand for soybeans and soybean meal. The impact of changes in E.E.C. policies on the demand for soybeans and soybean meal are quantified with the help of an econometric model that accurately describes the demand relationships. The study is confined to the demand for soybean meal in whatever form it may appear on the market, either as soybean meal or still in the form of soybeans. Thus, the demand for soybeans and soybean meal is intended to mean the demand for "aggregated" meal, which is the quantity of soybean meal consumed plus the quantity of soybean consumed in meal equivalents.

The same econometric model can be used to make predictions about future demand for soybean meal, in whatever form, assuming a number of expected changes in the values of the determinants of the model.

IMPORTANCE OF THE E.E.C. SOYBEAN MEAL IMPORTS

The exports of soybeans and soybean products are one of the main components of the U.S. agricultural exports. The importance of the E.E.C. market for U.S. soybeans can be measured in terms of soybean equivalents (1 bu soybeans = 47.5 lb of soybean meal). Since the analysis of the demand for soybeans is confined to the soybean meal component, the soybean complex is reduced to its meal content, and soybeans are converted into soybean meal equivalents. Figure 1 shows the flow of soybean production from U.S. growers to the different markets in 1973.

Thus, expressed in soybean meal equivalents, about 50% of the soybean crop ultimately found its way to the export market.

The most important single market for U.S. agricultural products, and soybeans in particular, is formed by the nine countries that are presently united in the E.E.C. In the 1973-1974 season, more than 60% of the total soybean meal exports, and 47% of the total soybean exports, were shipped to the countries of the Common Market. This implies that in terms of soybean equivalents, more than 25% of all soybeans grown by U.S. farmers is imported, in one way or another, by the E.E.C. Despite the extraordinary high price level in the calendar year 1973, the 1973-1974 season seems to be representative for the relative quantity of U.S.-E.E.C. trade in other years (Table 1).

The important of the U.S. soybean products exported to the E.E.C. may also be expressed in dollar values. During the fiscal year 1976, the value of these exports to the Common Market amounted to nearly two billion dollars ($1,944.5 million)—a significant contribution to U.S. farm income.

E.E.C. Soybean Imports and E.E.C. Livestock Farmers

Imports of soybeans into the E.E.C. occur only in the form of soybeans and soybean meal. Imports of soybean oil have become negligible. Domestic production of soybeans within the E.E.C. is also very modest.

For livestock farmers in the countries of the Common Market, soybean imports have become increasingly important as a protein source for animal feeds. These imports accounted for 38% in 1965, 45% in 1969, and 58% in 1972, of the total crude protein in concentrates. Broken down into the two main protein components, methionine and lysine, soybeans supplied about 28% of the methionine in 1965, 33% in 1969, and 45% in 1972, and even more importantly, about 46% of lysine in 1965, 50% in 1969, and 64% in 1972 of the lysine used in animal feeds in the E.E.C. (9). Note that these percentages reflect the composition of protein concentrates used in animal feeds. The bulk of the total protein supply for West European livestock still comes, of course, from pastures and meadows. In 1972 pastures and meadows supplied 60% of the total protein supply for livestock feeding.

Figure 1. Flow of soybean production in soybean meal equivalents, 1973.

Table 1. Relative importance of U.S. exports of soybean meal to the nine countries that are *now* members of the E.E.C.[a] Sources: American Soybean Association, 1975; U.S. Fats and Oils Statistics, 1970-75; U.S.D.A., Statistical Bulletin No. 560; and FATUS, U.S.D.A., 1978.

	1971	1972	1973	1974	1975	1976
U.S. - E.E.C. exports of soybeans (1000 bu)	189,853	193,435	222,161	235,280	211,157	264,419
U.S. - E.E.C. exports of soybeans in meal equivalents (1000 MT)	4092.5	4169.8	4789.0	5071.8	4551.8	5699.9
U.S. - E.E.C. exports of soybean meal (1000 MT)	3263	2686	4866	3454	2834	3107
U.S. - E.E.C. exports total in meal equivalents (1000 MT)	7355.5	6855.8	9655.0	8525.8	7385.8	8806.9
U.S. - world exports of soybeans (1000 bu)	423,967	440,653	485,774	512,200	459,159	563,359
U.S. - world exports of soybean meal (1000 MT)	4504	3990	4866	5361	4170	5360
U.S. - world exports total in meal equivalents (1000 MT)	13643.2	13488.9	15337.5	16402.2	14067.8	17504.0
U.S. soybean production (million bu)	1176.0	1270.6	1547.2	1214.8	1547.4	1287.6
U.S. production in meal equivalents (1000 MT)	15350	17390	33352	26187	33356	27756
E.E.C. imports as % of U.S. production	29	25	29	33	22	32

[a]The entry of U.K., Ireland and Denamrk was effective in 1973. In this table their imports were included for all 6 yr.

The dependence of West European livestock farmers on U.S. soybean meal became particularly visible in 1973 when the U.S. government imposed an embargo on soybean exports. Since that year the Common Market has given much attention to the supplies of protein concentrates. The protein crisis which hit European husbandry so hard in 1972-1973 is perhaps merely a warning signal of a situation, which, when repeated, might force the Community to reorientate its husbandry in the near future or even the whole of its agriculture. Vachel et al. (9) mention a study which was completed for the Commission of the E.E.C. in the year immediately following the crisis.

NEED FOR AN E.E.C. MODEL

Several economic analysts have constructed models intended to describe the demand for soybeans and their products (4,5,6,7,10,11). However, most of these analysts have treated the demand for soybeans and soybean products by the E.E.C. as a proportional part of the aggregate world demand for U.S. soybean exports. One may question the correctness of such an approach since the usefulness of an aggregate world model for only a part of the world depends on identical demand and supply conditions for soybeans and soybean products throughout the world. Clearly, the agricultural policy of the E.E.C. has resulted in supply and demand conditions in the Common Market that are unique and different from the conditions in the rest of the world, and therefore warrant analyses designed specifically to evaluate these conditions. It may be necessary to include "typical E.E.C. factors" in any model of the market for U.S. soybeans and their products.

The U.S. literature provides two basic approaches for the demand of soybeans and soybean products: the three market approach and the two market approach. The three market approach was developed by Houck and Mann (4). Their model carries three important characteristics: (a) soybean meal and soy oil are joint products of the crushing industry; (b) there exist three distinct markets for soybean oil, meal and beans; and (c) there is more than one market for each product, such as domestic and export markets. This is their basic analytical framework.

This model is still very popular in the U.S. Essentially the same model was used by Matthews, Womack and Hoffman to generate forecasts for the U.S. soybean economy (8). Matthews repeated the forecasts again in 1973.

The two market approach was developed by Vandenborre. A first most important observation Vandenborre made was that for soybeans in the U.S. as well as for soybeans at the export market there was no final demand for soybeans (10). Thus there are only two export markets: one for oil, shipped under whatever form (oil or beans) and one for meal under whatever form (oil or beans). Thus Europe imports soybeans to the extent that oil and meal cover each other. One should be aware that this approach is basically different from the three market model of Houck and Mann. It is precisely for this reason that in this study the demand for soybeans is not analyzed separately.

Although in some studies an attempt is made to distinguish several export regions, of which the E.E.D. is one, none of the models incorporates specific Common Market policy factors, such as European price levels and/or production surpluses. Therefore these models are in no way appropriate to determine the effects of West European agricultural policies, and their usefulness in predicitng changes in the demand for soybean products in the E.E.C. will depend considerably on factors that are deleted in these models.

Once the E.E.C. policy factors are clear, the models developed by Houck, Vanderborre and others, together with demand analyses for other grains such as was done by W. Y. Mo, K. W. Meinken, A. Womack, and S. Roy et al become very helpful in selecting the major factors and determinants that underlie the E.E.C. demand for soybean meal.

E.E.C. POLICY

In order to analyze the effect of E.E.C. policies on the demand for soybean meal in the countries of the E.E.C., the most important apsects of the Common Agricultural Policy (CAP) should be mentioned: (a) objectives of agricultural policy, (b) price structure for agricultural products, (c) surplus sectors, (d) oilseed sector, and (e) meat sector.

Objectives of the Agricultural Policy

The European Economic Community was founded by the Treaty of Rome, signed by Belgium, France, Italy, Luxenburg, the Netherlands, and West Germany. It became effective on January 1, 1958, although a common market was not created immediately at that date, but over a period of ten years. The United Kingdom, Denmark and Ireland joined the Community on January 1, 1973.

The basic objectives of the common agricultural policy of the E.E.C. set in the treaty were: (a) to increase agricultural productivity, (b) to ensure a fair standard of living for the agricultural population, (c) to stabilize the market, and (d) to assure supplies to consumers at reasonable prices. These objectives are promoted by a set of common market regulations. One of the basic elements up to now has been the three price system consisting of target prices, intervention prices and threshold prices, combined with levies on imported agricultural products to absorb the differences between domestic and world market prices, and export subsidies.

Price Structure of Agricultural Products

Basic price regulations for cereals, pork, poultry and some other groups of products were adopted in 1962 and applied uniformly after July 1, 1967.

For grain, the target price (the producers' price of a commodity) is set in the community's largest deficity area, Duisburg, West Germany. Regional target prices are set in relation to the Duisburg prices minus the costs of transportation. The intervention price (the guaranteed prices for community crops) is tied to the target price, usually set at 90 to 95% of the target prices. The national intervention authorities are committed to accept during the entire marketing season grain harvested in the community and offered to them, provided it meets the minimum qualities described in the E.E.C. regulations. The seller is paid the price applying to one of the three nearest intervention centers, plus the value added tax (v.a.t.). The price is "free store, undischarged," i.e., the seller pays the carriage to storage. A refund for storage costs is paid only after the intervention agency has accepted the seller's offer.

Threshold prices are the minimum import prices at the border, derived by subtracting from the domestic target prices the transportation, handling and other costs incurred between port of entry and marketing center. The threshold price acts as a barrier and protects the interior market against the normally very much lower world markets (Table 2). The levy amounts to the difference between the threshold price and the world market price (c.i.f. Rotterdam). The levy is in fact a sliding-scale tariff which fully reflects any price fluctuations. The commission reviews daily the levy for all kinds of grain whose price has changed.

It is clear that the price relations within the Common Market differ greatly from those on the "world market." One of the first modifications to be made in the soybean demand model seems therefore to be the inclusion of one or more price factors, which will allow us to determine the effect of changes of target prices set by the E.E.C. authorities. In this regard, Houck et al. (4) mention the substitutability of grains and soybean meal, which become feasible at appropriate price relationships. Another goal of the E.E.C. price policy is to develop workable price-relations between the various kinds of grains. The administration upholds the limit prices (the target and intervention prices) for each cereal partly with the aim to keep supply equal to demand, but also with the aim to base the minimum prices (intervention prices) on the nutritional value of each feed grain. This is the basic idea that underlies the so-called "silo-cathedral system" (Figure 2).

This sytem intends to facilitate the incorporation and the consumption of wheat and other cereals in the community in order to achieve a more rational usage of community cereals, and to permit the development of a natural price scale on the market reflecting the (nutritional) value of the grains.

E.E.C. Surpluses

The surpluses of milk (butter and skimmed-milk powder) are one of the most persistent headaches for the commission in Brussels. Few expect these

Table 2. **E.E.C. prices of certain agricultural products in % of world market prices.**

Product	1975-1976	1974-1975	1973-1974	1972-1973	1971-1972	1970-1971	1969-1970	1968-1969	1967-1968
					– % –				
Soft wheat	124	107	79	153	209	189	214	195	185
Barley	117	107	96	137	185	146	203	197	160
Corn	128	106	98	143	176	141	159	178	160
White sugar	109	41	66	127	186	203	298	355	438
Oilseeds[a]	127	80	77	131	147	131	155	203	200
Olive Oil		113	96	125	153	165	167	173	166
Butter	320	316	320	249	171	481	613	504	397
Skimmed-milk powder	266	139	156	145	112		380		
Beef	158	162	110	112	133	140	147	169	175
Pork[b]	113	109	131	147	131	134	137	135	147
Eggs[b]	131	164	111	159	162	201	151	137	132

[a]Soybeans and soybean meal imports are duty free.

[b]Calendar year figures (first of split year).

Figure 2. **E.E.C. grain prices 1977-1978 (silo-cathedral system). Source: Commission of the European Communities.**

Maximum: target price
Minimum: intervention price

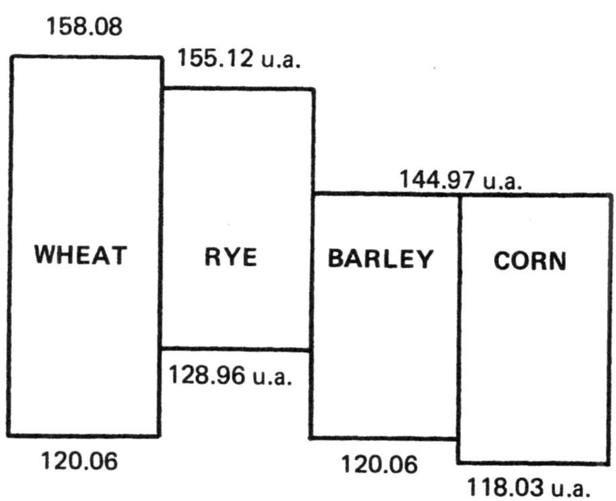

surpluses to decrease in the near future. The degree of self-sufficiency for skimmed-milk powder during 1973-1975 has been between 123 and 158%. The degree of self-sufficiency is the usable production (or usable production from indigenous base material) expressed as a percentage of the domestic uses.

$$\text{Degree of self-sufficiency} = \frac{\text{Domestic production x 100}}{\text{Domestic uses}}$$

The skimmed-milk powder is used mainly as a source of crude protein for feedstuff, especially for young animal feed such as "calve's milk" and also for sows. Presently, the commission stimulates the reduction of intervention stocks by a conditional subsidy on the use of skimmed-milk powder in "calve's milk." The feedstuff industry is allowed to obtain the subsidy only when they use more than 60% skimmed-milk powder for the preparation of "calve's milk." Despite these measures, only a rather small amount of milk products is absorbed yearly by the feedstuff industry. Table 3 shows the quantities of soybean meal and milk products used in animal feed, as well as the total E.E.C. milk production in crude protein units.

When we realize that most of the skimmed-milk powder (1974, 75%; 1975, 61%) is used on the basis of minimum nutritional requirements and therefore at "free" market prices, the amount of reduction of the demand for soybean meal, because of surpluses in skimmed-milk powder stocks, becomes rather small.

Oilseed Sector

Along with fishmeal, oilseed products are the main sources for commercial animal feed protein. The most important oilseeds are cottonseed, groundnut seed, palm-kernel, copra, linseed, rapeseed, sesame seed, sunflower seed, and, above all, soybeans. The increasing dominance of soybean meal on the oilseed meal and cake market becaomes clear in Figure 3. This is due partly to the high content of lysine and methionine in soybean meal, the so-called

Table 3. Milk products for animal feeds, compared with the use of soybean meal. Source: Eurostat, "Feed Balance Sheets: Resources," 1976.

Product	1970-1971	1971-1972	1972-1973	1973-1974
		– 1000 MT crude protein units[a] *–*		
Soybean meal	3,513	3,612	3,949	4,195
Milk products (total)	1,086	1,164	1,152	1,150
Skimmed-milk powder	291	332	382	382

[a]The crude protein unit, which gives the crude protein content (proteins, amino acids, non-protein nitrogen) of a feedstuff is determined by multiplying the total amount of nitrogen measured by the Kjeldahl method by 6.26.

Figure 3. Development of E.E.C.-9 consumption of protein concentrates in protein equivalents (1).

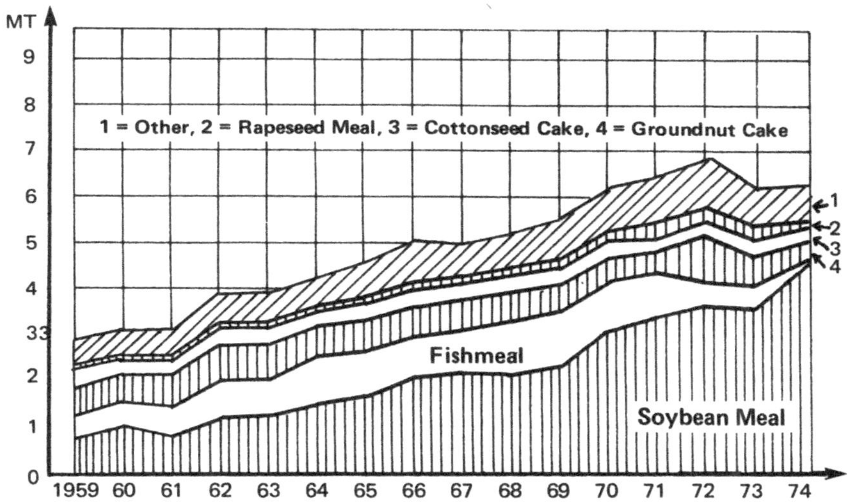

"highly favorable amino-acid profile" of soybean meal. The amino-acid profile of a protein concentrate is especially important for the feeding of monogastric animals (hogs and poultry), which are unable to transform other amino-acids.

Another reason for the prominence of soybeans among oilseeds is their relatively high meal content (and low oil content). With the help of Table 4, it is possible to calculate quantities of different meals and oilseed under a common denominator: protein equivalents. Protein equivalents are the simple results of multiplying the quantities of products with their relative protein content.

For most of the oilseeds, fats and meals, the degree of self sufficiency is small. In the case of oilseeds, the import dependence is even more than 90%. Since areas available for the cultivation of rapeseed, sunflower seed and other oilseeds are limited, there seems little chance of an increase to this degree of self-sufficiency.

Meat Sector

The meat consumption per capita in the E.E.C. has been steadily rising over the last two decades. Table 5 illustrates the patterns in meat consumption over the last three years. The total meat figures comprise meat from cattle, pigs, sheep and goats, equines, poultry, other meat and edible offal. A number of observations can be made: (a) the importance of pork, (b) the stagnancy of beef consumption, and (c) the difference in consumption patterns in the different member countries.

Table 4. Protein content of protein concentrates according to different sources.

Protein Source	Literature Source		
	Beyer (1)	Houck (7)	Hoffmeyer (3)
— meal or cake —		*— % crude protein by wt —*	
Soybean	45	42-50	45
Cotton seed	39	36-43	40
Groundnut	49	45-56	50
Sunflower seed	40	37-38	42
Linseed	34	32-39	35
Copra	21	22	21
Palm-kernel	19	23	20
Rapeseed	33		33
Sesame seed	40		
Fishmeal	65	60-73	65
Animal meal			50-65
Skimmed-milk powder			33

ECONOMETRIC MODEL

Based on economic principles, the analysis of the effects of the CAP, and experience in the past, the following model was developed:

$$QDSM = f(PSM/PECC, ECPL, APFU, APS \text{ and } T),$$

Where the consecutive symbols stand for quantity soybean meal demanded (QDSM), price of soybean meal (PSM), E.E.C. price of cereals (PECC), E.E.C. profitability index for the livestock sector (ECPL), animal protein feed units (APFU), availability of protein substitutes (APS) and a time trend (T). The most important characteristics of this model are: (a) the use of E.E.C. prices for cereals as well as for the calculation of the profitability index, (b) the use of animal protein feed units instead of livestock numbers, and (c) the inclusion of skimmed-milk powder production in the availability of protein substitution. Animal Protein Feed Units are livestock production quantities multiplied by the amount of crude protein that is required for this production based on the nutritional requirements as formulated by nutrition experts. This protein intake can be derived directly from production figures (milk, meat and eggs), without going into complicated livestock number computations. The group of protein substitutes is limited to the commercial protein products such as fishmeal, oilseed products other than soybean meal, animal meal and skimmed-milk powder. Under animal meal we comprise meat-meal, blood meal, offals, feathermeal and other animal meal. With the help of Table 4 it was possible to compare the produced quantities with their relative protein content.

Table 5. Consumption of livestock products in the E.E.C. (kg/capita). Source: Eurostat, Supply Balance Sheets, 1976.

Products	Year(s)	E.E.C.	W. Germany	France	Italy	Netherlands	Bellux	U.K.	Ireland	Denmark
Eggs	1966-70	--	15	12	10	12	14	15	15	12
	1972	14	17	13	12	11	13	16	13	11
	1973	14	17	13	12	11	12	15	12	11
	1974	14	17	13	11	11	12	15	13	11
	1975	14	17	13	11	11	11	12	13	11
Beef and veal	1966-70	--	24	30	23	22	26	21	18	20
	1972	25	23	29	25	20	27	23	20	16
	1973	24	23	28	28	21	28	21	19	15
	1974	25	23	30	25	22	31	24	22	15
	1975	25	23	30	23	22	31	25	29	16
Pork	1966-70	--	43	30	11	29	32	--	28	30
	1972	32	49	33	15	32	36	28	31	34
	1973	32	49	33	16	31	38	27	32	35
	1975	32	51	34	18	35	39	23	27	39
Sheep and goat meat	1966-70	--	0	3	1	0	1	--	11	1
	1972	3	0	3	1	0	1	10	11	0
	1973	3	0	4	1	0	1	8	11	0
	1974	3	0	3	1	0	1	8	11	0
	1975	3	1	4	1	0	1	8	11	0
Poultry meal	1966-70	--	8	14	10	5	8	9	9	4
	1972	11	9	15	12	6	10	12	10	5
	1973	12	9	14	15	7	10	12	13	7
	1974	12	9	14	15	7	9	12	11	7
	1975	12	9	14	16	7	10	11	11	9
Total meat	1966-70	--	80	93	50	62	77	--	--	--
	1972	80	87	94	61	67	84	78	87	63
	1973	79	86	95	66	66	88	72	87	63
	1974	81	89	97	65	71	92	74	92	64
	1975	82	90	99	65	72	90	73	101	70

Table 6. Annual soybean meal consumption/country. Source: FEDIOL, Statistical Report, 1963-1977; F.A.O. Annual Trade Statistics, 1962-1976.

Year	Germany	France	Italy	Netherlands	Bellux	U.K.	Ireland	Denmark
				– 1000 MT's –				
1961	561	217	161	194	110	287	19	320
1962	927	414	266	297	170	514	23	368
1963	934	419	263	174	139	434	28	365
1964	1290	538	311	334	180	425	28	425
1965	1271	569	467	383	201	474	52	439
1966	1971	713	547	403	241	441	49	482
1967	1870	726	638	488	278	363	49	331
1968	1704	776	688	533	283	381	66	382
1969	1996	834	710	579	337	199	63	457
1970	2035	1192	931	1031	463	492	92	544
1971	2193	1313	1004	1098	465	524	97	555
1972	2258	1416	1122	1090	540	569	93	607
1973	2120	1528	1163	930	558	748	80	614
1974	2328	1941	1521	1391	760	878	127	663
1975	2774	1820	1492	1322	706	766	122	685
1976	2960	2115	1712	1436	774	985	159	754
Total	29212	17531	12996	10683	6205	8680	1147	8091

The model was tested by pooling the time-series data of the E.E.C. member countries. Belgium and Luxenbourg were counted as one economic unit. The data for Ireland were incomplete so this country was deleted from the analysis. Table 6 shows that this deletion was of minor consequence for the E.E.C. total figures since the soybean meal demand for Ireland has been very small. Table 6 also shows the differences in total soybean meal consumption per country, due mainly to country size. In order to overcome this problem of differences in country size, the consumption of soybean meal per animal protein feed unit (QDSM/APFU) was taken as dependent variable in the model.

This econometric model was tested by the application of three different stochastic models, each based on a different set of assumptions: the Ordinary Least Squares Model (OLS), the Error Components Model (ECM), and the Classical Factor Analysis Regression Model (CFAR). The econometric model performed well in each of these stochastic forms, although some differences appeared (Table 7).

The estimates of the Error Component Model were preferred to the other estimates. Not only are the assumptions underlying the model preferred to those of the OLS model, but estimates seem also more likely to be true in view of the additional estimated values by the Classical Factor Analysis Regression Model.

Effects of E.E.C. Policies on the Demand for Soybean Meal

With the help of the estimated model a number of effects of changes in the Common Agricultural Policy can be explored. The estimated changes in E.E.D. soybean demand are shown in Table 8.

Table 7. Estimates of regression coefficients from tests by three stochastic models.

Explanatory Variables	Estimated Value of the Regression Coefficients		
	ECM	OLS	CFAR
PSM/PECC (price)	-0.1226	-0.0833	-0.1449
T (time)	0.0313	0.0294	0.0509
ECPL (profit)	0.0146	0.0014	0.0104
AOS (substitutes)	$-2.9118 \cdot 10^{-4}$	$-2.1950 \cdot 10^{-4}$	$-11.7 \cdot 10^{-4}$
$R^2 =$	.68	.85	.82

Table 8. Effects of changes in the CAP on the yearly demand for soybean meal (1961-1976).

CAP Changes	Soybean Demand Changes	
	Relative Change	Absolute Change
	– % –	*– 1000 MT –*
- 10% cereal prices	- 2.7	- 162
+10% meat prices	+1.05	+ 63
+10% skimmed milk powder production	- 0.8	- 48

In respect to the estimated elasticities three observations can be made. (a) The value of the price elasticity is considerably lower than any estimation to date. Although Vandenborre (10,11) obtained an elasticity of - .28 in his first study, he arrived at 0.40 in a later one. Houck and Mann (4) obtained an elasticity of - 0.33. All these estimates however are computed for the U.S. market. Moreover, in each of these studies the writer quotes other studies that have resulted in higher estimates. Houck, Ryan and Subotnik (7) applied a series of models to the E.E.C.-6, covering the period 1951-1967 and obtained price elasticities varying from - 0.67 to - 1.64. In the most recent study, Houck and Ryan (5,6) developed a U.S. export model that used European livestock numbers. This model provided a price elasticity for the demand for U.S. soybean meal exports of - 0.34 (own calculations).

There are several explanations for the low value of the price elasticity that is obtained in this study. First, the presence of a profit variable, not used in any of the above models, may have reduced the estimated effects of price changes somewhat. Secondly, it can be explained by the price and income stabilization policy of the E.E.C. Because of the CAP it is likely that West European compound feed manufacturers and livestock farmers are less sensitive to soybean meal prices than their colleagues in other parts of the world.

(b) The elasticity of the demand for soybean meal with respect to the price of livestock products is also very low. Again, it resembles the effects of the CAP that cause profitability variations to remain within the limits of "fair" income and "fair" consumers prices. (c) Although it is often argued that the E.E.C. skimmed milk-powder production has a considerable negative effect on the demand for soybean meal, this negative effect seems rather small and possibly negligible if one realizes that this decrease in meal demand is offset partly by an increase in soybean meal utilization required for the production of this surplus.

In addition to the above estimated effects of the E.E.C. policy, possible changes in E.E.C. monetary policy can be simulated. Presently, there is no uniform price level within the E.E.C. because of the frequent changes in currency exchange rates. For agricultural products the price level is highest in West Germany, and lowest in the U.K. It is the aim of E.E.C. policy makers to arrive at one common price level. However, the question is which price level will prevail. The relevance of this question in respect to the demand for soybean meal is analyzed by hypothesizing two different levels of common E.E.C. prices: the West German and the U.K. price levels. The analysis shows that the difference in soybean meal demand between these two price levels would amount to about 2.3% of the total soybean meal consumption. In view of the annual increases in soybean meal consumption, the difference in demand at the two extreme price levels is small.

This leads to the conclusion that differences in regional price levels can be neglected in case of middle and long term predictions. This conclusion facilitates the use of the model for prediction purposes as it appears safe to insert aggregate E.E.C. prices directly into the model.

Prediction of Soybean Meal Demand in 1980 and 1985

The accuracy of the estimation of soybean meal demand depends not only on the reliability of the estimated regression coefficients, but also on the hypothesized values of the explanatory variables. For these values of the explanatory variables, this part of the study leans heavily on the recent work of Boddez et al. (2). Boddez et al. completed a set of price hypothesis in preparation for an extensive study about the demand and supply of agricultural products in the E.E.C. for 1985-1986.

The implementation of these prices in the demand model, combined with a number of reasonable assumptions about the quantity variables, leads us to believe that the total soybean meal consumption in the countries of the E.E.C. between 1975 and 1980 will increase by about 21.5%. Soybean meal consumption in 1975 (average 1974-1976) was about 10 million MT (Ireland included). In 1980, a total consumption of 12.2 million MT is expected.

For 1985 the predicted increase in soybean meal consumption is 41.2%. Expressed in an absolute figure, this amounts to a 14.1 million MT demand for soybean meal in 1985. Both figures mean an average yearly growth of about 4% per year over a ten-year period. This is a lower rate of growth than occurred over the last ten and/or five years. Since domestic soybean production in the E.E.C. is unlikely during this period, the implications for the U.S. exports of "aggregated" soybean meal are favorable. The E.E.C. market remains a prime market for U.S. soybean and soybean meal exporters for many years to come.

For E.E.C. policy makers, the projection proves an increasing dependence on foreign protein sources. This dependency may encourage protectionist measures about which form, size and effects predictions are very difficult, if not impossible, to make. However, the low price elasticity indicates that even a rather high levy on the imports of oilseed meals may have only a small effect on the total E.E.C. demand for soybean meal.

NOTES

Hendrik C. Knipscheer, Agricultural Economist, International Institute of Tropical Agriculture, Ibadan, Nigeria.

This study was completed at the University of Illinois, Urbana, IL 61801.

REFERENCES

1. Beyer, V. 1977. Der weltmarkt fur eiveissfuttermittel, Agrarmarkt-Studien Heft 25, Verlag Paul Parey, Hamburg.

2. Boddez, G. 1978. Prognose van de vraag naar en het aanbod van landbouwproducten—1985-86, Deel I. C.L.E.O. Catholic University of Louvain, Heverlee.
3. Hoffmeyer, M. 1977. Die voraussichtliche entwicklung der internatinalen vorsorgung mit landwirtschaftlichen erzeugnissen und ihre folgen fur die gemeinschaft, mitteilungen uber die landwirtschaft Nr. 36, Commission of the European Communities.
4. Houck, J. P., and J. S. Mann. 1968. An analysis of domestic and foreign demand for U.S. soybeans and soybean products. Technical Bulletin 256. Ag. Exp. Sta., Univ. of Minnesota.
5. Houck, J. P., and M. Ryan. 1976. A study of U.S. exports of soybeans and soybean meal. Technical Bulletin 309. Ag. Exp. Sta., Univ. of Minnesota.
6. Houck, J. P., and M. Ryan. 1976. Export of Minnesota soybeans. Minnesota Agr. Econ. Ag. Ext. Service, Univ. of Minnesota.
7. Houck, J. P., M. Ryan and A. Subotnik. 1972. Soybeans and their products, Univ. of Minnesota Press., Minneapolis, Minn, p. 72.
8. Matthews, J. L., A. W. Womack, and R. G. Hoffman. 1973. Formulation of market forecasts for the U.S. soybean economy with an econometric model. FOS-260, U.S.D.A., Washington, D.C.
9. Vachel, J. P. 1974. Use of substitute products in livestock feeding. Internal Information on Agriculture, No. 130. Commission of the European Communities.
10. Vandenborre, R. L. 1967. An econometric analysis of the market of soybean oil and soybean meal. Bulletin 723, Ag. Exp. Sta., Univ. of Illinois, Urbana, Ill.
11. Vandenborre, R. J. 1970. Econometric analysis of relationships in the international vegetable oil and meal sector. Dept. of Agr. Econ., AERR No. 106, Univ. of Illinois, Urbana,

IMPROVING GRADES AND STANDARDS FOR SOYBEANS

L. D. Hill

PURPOSES OF STANDARDS

Grades and standards for grain have three purposes in a market economy: (a) to classify all grain into a few homogeneous categories to facilitate trade, (b) to permit market transactions on the basis of description, and (c) to enable buyers to identify relative value for various end uses. The first two purposes are met by almost any set of descriptive factors, so long as the buyers and sellers accept the standards as the basis for their transactions. The third purpose can be met if the descriptive characteristics determining grade can be related to the quantity and quality of end products derived from the grains.

The ideal grain standard would enable an end user to estimate product yield from the description of the grain characteristics provided by the standard. Differences in end use value could then be converted into price differentials attached to each grade factor. Under competitive market prices these premiums or discounts would be reflected to each handler in the market, to each producer of grain and even to suppliers of seed. At each point in the market channel, price differentials provide the incentive for making changes in breeding, production, harvesting, storing, and marketing practices, that would improve grain quality consistent with cost.

The present system of grades and standards for soybeans meets the first two purposes of grading but is less than ideal with respect to the third, having developed from practices of grain merchandisers rather than from research on end use characteristics. Although soybeans were introduced into the United States around 1912 and U.S. production had increased to over 44 million bushels (17,000 metric tons) by 1935, official soybean standards were not established until 1940, when an amendment to the U.S. Grain Standards Act of 1917 provided for the inspection and grading of soybeans, requiring official inspection on all soybeans moving in interstate and export trade. The grade factors used were similar to those already in use for other grains: test weight, moisture, total damaged kernels, heat damaged kernels and foreign material. Factor limits for splits and color were included as additional important indicators of quality beyond the requirements for corn and wheat.

STANDARDS AND DISCOUNTS

Grade standards do not determine price or value. They only provide a method for describing selected characteristics of each lot of grain. The market then assigns value to these grade factors in the form of discounts or premiums. These price differentials are not limited to the grade factors nor is the market required to differentiate price on all grade factors. For example, premiums for protein in wheat have been used for many years even though protein is not a factor in the determination of numerical grade. Test weight for soybeans is often ignored by the country elevator buying from farmers. Price differentials for various qualities are based on supply and demand, on cost of changing the grade, and on estimated value in end use. Consequently, in a market system free to adjust to changing conditions, quality discounts may fluctuate over time and between locations.

CURRENT DISCOUNTS

A survey of a small sample of Illinois elevators and processors shows typical discounts on the various grade factors and illustrates the degree of variability among individual firms. Price discounts are established generally for each factor and applied whenever a sample of grain fails to meet the quality of the base price grade. This base is No. 1 for soybeans but No. 2 for corn. Each elevator manager in the survey was asked to identify the method used for compensating for the moisture content above 13% in soybeans purchased in 1977. The responses were grouped into four adjustment procedures (Table 1). Use of a shrink formula or factor was reported by 46% of the respondents. This method reduces the quantity of grain purchased by the weight of water above the 13% base. Shrink tables are available showing an exact weight adjustment, but many elevators prefer to use a shrink factor that simplifies the calculation.

A second alternative is to reduce the base price by a fixed amount for each percent above 13. This method is seldom used (only 4% in this survey) because changes in base price require compensating changes in discount rates. The third method avoids this complexity by setting the price discount

Table 1. Percent of Illinois grain firms reporting use of different types of moisture discounts, 1978.

Type of Moisture Discount	Percentage of Firms Reporting
Shrink formula or factor	43
Discount as a percent of price	26
Discount of __ cents per bushel per point	4
Combination of shrink and discount	23
No discount used	4

equal to a percent of base price. Increases in base price automatically increase the discount. The discount system, reported by 30% of the respondents, is established usually to compensate for differences in weight due to excess water and to incorporate costs of drying or local demand and supply conditions. It can be demonstrated that this method is equivalent to the shrink factor. Several managers (23%) used a combination of shrink plus a price discount. This permits adjustment to a base moisture while allowing market forces to influence the value of different moisture levels.

The various moisture discounts are clearly attempts to equate the value of different qualities and to compensate for costs of upgrading (drying costs). Fluctuations in these discounts over time suggest that supply and demand of wet beans relative to dry beans also influence the size of the discount.

In the case of foreign material the discount has become very standardized. Although 10% of the respondents were not using any discount for foreign material, all the rest reported a reduction in the purchase weight by the weight of foreign material above 1%, identified by the term "dockage." This implicitly suggests that the first 1% of foreign material has the same value as good soybeans; any FM above 1% has zero value.

The grade factor of splits provided an interesting variation in the discount procedures. The grade standards allow only 10% splits for No. 1 soybeans. However, 67% of the grain firms reported no discounts were used for the factor of splits. The remaining 33% applied discounts only if splits exceeded 20% of the total sample, the factor limit for No. 2 soybeans, not No. 1. Test weight also indicates variability in the discounts reported. Nearly one-fourth of the respondents reported using no discounts for test weight. Where discounts were used, they varied from ½ cent/bu/lb below 54 lb to 2 cents/bu/lb below 54 lb (Table 2).

VALUE-BASED DISCOUNTS

It is evident even from this limited survey of grading and discounting procedures that the industry does not consider all grade factors to be equally important in measuring quality and value. Some factors are ignored frequently in establishing value, others are given widely different discounts by different

Table 2. Percent of Illinois grain firms reporting use of different test wieght discounts.

Type of Test Weight Discount	Percent of Firms Reporting
No discount	24
½ cent/bu/lb below 54 lb	47
1 cent/bu/lb below 54 lb	12
2 cents/bu/lb below 54 lb	5
Graduated scale for different values below 54 lb	12

firms, while others (especially moisture) are discounted on a relatively consistent and uniform basis. These variations may raise questions but they provide no factual information on the relationship between grades, discounts and value. Although more information is needed on the relationship between grade factors and end use value, to establish equitable discounts, a procedure for calculating discounts can be illustrated using the grade factor of foreign material (FM).

Much of the FM in soybeans is actually small pieces of soybeans with a low yield of protein and oil but it has value for a processor. If the foreign material screenings are assigned a value (e.g., 50% of the value of whole beans), the value of soybeans containing any percent of FM can be calculated from the formula:

$$V_B = [P_B\ (\%\ WB) + P_{FM}\ (\%\ FM)]\ Q_B$$

where: V_B = value of soybeans containing a given percent of foreign material, P_B = price of beans, P_{FM} = price of foreign material, Q_B = quantity of beans including foreign material, % WB = percent of whole beans in Q_B (i.e. 1 - % FM) and % FM = percent of foreign material in Q_B.

For purposes of illustration, set the price of beans at \$6/bu, the price of FM at \$3/bu and % FM at 5%. The formula then becomes:

$$V_B = \$6.00\ (.95) + \$3.00\ (.05)$$
$$V_B = \$5.70 + \$0.15$$
$$V_B = \$5.85 \text{ value of beans with 5\% FM}$$

The calculated value of the beans (V_B) can be compared with the discounted price based on the accepted practice of the trade of subtracting the weight of FM above 1%. In selling one bu of beans with 5% FM, the producer would be paid on the basis of 1 - .04 = .96 bu of beans. At the assumed price of \$6/bu the discounted price would be \$5.76 in contrast to V_B = \$5.85. The price paid for soybeans containing FM above 1% will be less than its value as a result of treating FM as dockage above 1%. The difference between calculated value and the discounted price will be greater given higher prices for soybeans, higher percentage FM and higher value of FM relative to whole beans. Only at FM levels near or below 1% does the dockage system of present standards return more than the computed value formula. Since the standards allow 1% FM at no discount, it is to the advantage of the seller to incorporate up to 1% of FM in any lot of beans sold.

LIMITATIONS OF CURRENT STANDARDS

There have been relatively few changes in the grade limits for soybean standards since 1940. They have proven satisfactory as a basis for trade and meet the first two objectives of standards. Recently, however, questions have been raised concerning the ability of the standards to reflect value in processing. There is little evidence that No. 1 soybeans will yield more oil or soybean meal than No. 2 or No. 3. Neither is there a correlation between the quality of oil or meal and the numerical grade of the beans being processed.

The preceding discussion of discounts has suggested already some major problems in soybean grades. The current standards have five deficiencies as a basis for trade in the modern, sophisticated world market: (a) the most important quality factors in terms of end products (protein and oil) are excluded from the grade factors, (b) not all the factors now in the standards have economic importance to end users, (c) foreign material includes any material passing through an 8/64-inch screen, and is generally subtracted as dockage from the weight being purchased even though it contains mostly pieces of soybeans, (d) maximum or minimum limits on each factor in the numerical grades encourage blending of beans of different quality levels to include the maximum allowable moisture, foreign material, and damaged kernels, and (e) differences between grades on the factor of heat damage are too small to differentiate using the current sample size of 250 g.

Exclusion of Economic Factors

The exclusion of oil and protein content from the soybean standards is in part the result of a lack of equipment for rapid and accurate measurement. As testing equipment is further refined, these become candidates for inclusion in future revisions. Meanwhile, some processors in the domestic markets find it advantageous to identify production areas where they can purchase beans with higher oil and protein content. Importers are in general less able to make such selection.

Irrelevant Factors

Very few of the factors in soybean grades are based on research data justifying their relevance in determining value or quality. The presumption has been that the original factors will be retained until proven useless. Removal of any factor will therefore require extensive research for justification. The storability and product yield of splits is not well documented in current research publications. Heat damage at levels below .5% is also of questionable value as a measure of end product yield or quality. Preliminary research relating test weight to oil and protein content shows no significant relationships.

The grain inspection department of Champaign, Illinois, provided 104 soybean samples which were selected at random from deliveries to processors during the month of January 1979. An additional 10 samples were selected from the University of Illinois to incorporate extreme values of oil and protein. These 114 samples were graded and all grade factors recorded. The Northern Regional Research Laboratory of the U.S. Department of Agriculture, Peoria, Illinois, determined oil and protein content of each sample using an infrared reflectance type analyzer. The sample was ground as received for analysis rather than sorting out the whole beans, in order to maintain those characteristics of the sample causing test weight differences. The results of this analysis are summarized in Table 3. None of the differences are statistically significant. Protein shows a decline as test weight increases from 55.0 to 57.5 lb/bu but the highest protein content is found at 58.0 lb/bu. Oil content appears to increase with test weight between 55 lb/bu and 57.5 lb/bu, but again, the trend reverses at 58.0 lb/bu.

The correlation between grade factors and oil and protein was very low as shown in Table 4. Individual grade factors clearly have very little explanatory power with respect to the two chemical properties of greatest interest to the processor. Additional research is needed to provide rapid reliable and inexpensive methods for determining value in end use or to relate chemical properties to readily measured physical properties.

Definition of FM

Foreign material in soybeans includes any material passing through an 8/64-inch screen. Dirt, weed seeds, and other grains are not differentiated from broken pieces of soybeans. Consequently, the percent of FM does not provide accurate information on the value of a lot of soybeans containing FM. Treating FM as dockage underestimates the value of soybean screenings in soybeans. Where FM is primarily pieces of soybeans the grade system should encourage payment consistent with value.

Table 3. Relationship between test weight and protein and oil content of soybeans, Illinois, 1979.

	Test Weight (lb/bu)							
	55	55.5	56.0	56.5	57.0	57.5	58.0	58.5
Average protein (%)	41.97	41.56	41.22	41.28	41.10	41.00	42.6	41.75
Average oil (%)	19.56	19.67	20.05	20.02	20.13	20.28	19.65	19.75
Number of samples	7	8	20	31	29	15	2	2

Table 4. Correlation between soybean grade factors and oil and protein percent.

	Protein	Oil
Test weight	.077	.016
Moisture	.112	-.096
Total damage	.022	-.095
Foreing material	-.055	-.199
Splits	.055	-.102
Multiple regression (R^2)	.06	.05

Incentive for Blending

Numerical grades encourage blending. Blending and commingling of sub-lots is an essential operation in a large volume grain industry. However, a grading and pricing system that provides a premium for incorporating a small percentage of screenings, splits, other grains, and damaged kernels is not serving the long-run best interests of the entire grain industry, from producers through consumers. Most export beans are sold on a contract calling for grade No. 2. This means that the beans may contain 3% damaged kernels and 2% FM. The exporter delivering on this contract will not be a good business-man if he does not do his best to find enough moldy beans to blend with this cargo to make the 3%. This provides a good market for a farmer with a bin full of moldy grain, but it does not turn bad grain into good grain just because it meets No. 2 standards.

Statistical Requirements

Statistical principles were ignored largely in the development of the grain standards. Heat damage of .2 or .3% cannot be differentiated statistically using the required sample of 250 g. This difference is equivalent to one bean showing heat damage. Random chance and sampling variability are greater than the value for discriminating between grades No. 1 and 2 on the factor of heat damage.

Arbitrary Limits

A final illustration of the arbitrariness of the grade limits can be found in the limits for test weight. It would appear that soybean standards adopted in 1940 were developed out of experience with the standards for corn. They generally include the same factors, and have similar values for damaged kernels. The values for test weight, however, are identical. Despite the difference in end uses of processing, test weight limits are 56, 54, 52 and 49 lb/bu for the first four grades of corn and soybeans. This becomes even more surprising when related to the legal weights per bushel. Corn has a legal weight of 56 lb/bu as a basis for price, and, logically, it also has 56/lb/bu as the test weight standard for No. 1 corn. Soybeans, with a 60 lb legal weight, have a 56 lb test weight limit for No. 1 grade. Corn and soybeans have different legal weights but identical test weights as indicators of quality.

These and other limitations to the present U.S. grades and standards for soybeans demonstrate the need for a thorough examination of the standards in light of the three objectives outlined in the opening paragraph of this paper. A lack of research on the economic implications of quality hampers recommendations for change. Changes should not be undertaken until the research results are available to demonstrate that the revision will increase the efficiency of the market and produce value in excess of the costs.

THE BASIS FOR UNIFORM INTERNATIONAL STANDARDS

Increased soybean production in many countries of the world must be accompanied by the development of a more sophisticated marketing system. In most of these markets a system of grades and standards will be needed to facilitate trade by description. The standards selected should meet the three purposes stated in the first paragraph of this paper. In general, the current U.S. grade standards for soybeans should not be adopted in these new markets without first conducting a careful economic analysis of the alternatives. As additional countries enter the world markets, as either buyers or sellers of soybeans, the need for a uniform basis for trade increases. A diversity of contracts, grade descriptions, and quality specifications adds to the cost of marketing. A uniform terminology and quality standard would improve communication between buyer and seller and would facilitate the transfer of buyer preferences back to producers and plant breeders.

If the objectives of grading and standardization stated in the opening paragraph of this paper can be accepted by the soybean traders of the world, then uniform international standards would be possible. The relationship between quality characteristics and product yield is fairly consistent, and independent of country of origin and use. It only remains for research to find those descriptive, measurable qualities that are reliable indicators of end use properties. An international research conference is an ideal forum in which to initiate the discussion of international soybean standards. this topic will be on the agenda of the Third International Soybean Research Conference.

NOTES

Lowell D. Hill, Department of Agricultural Economics, University of Illinois, Urbana, Illinois 61801.

ROLE OF SOY PROTEIN PRODUCTS IN NATIONAL DEVELOPMENT

F. H. Schwarz and J. K. Allwood

If there is a role for soy protein products in the process of national development we should be able to describe it and its ramifications and state specific examples along with the implications to public policy. Many economic advantages, both macro and micro, have accrued from these products in developing countries over the last several years.

In order that we might see the role of soy protein products in a developing economy I would like to concentrate my comments on three points: a description of soy protein products, the relevant aspects of national development (principally those of the food and agricultural sector), and the role of these products in national development.

AN OVERVIEW OF SOY PROTEIN PRODUCTS

Soy protein products are economical, nutritional and functional food ingredients used principally by the food processing industry. In today's commercial terms they are not found per se as consumer products. The technological development of the soy protein industry has far passed the day when these products were considered only in terms of high protein low cost foods. There are a variety of different kinds of soy protein products from the pure refined protein powders termed isolated soy proteins down to edible grade soybean meal known as soy flour and grits. Each have their own specific use and their own economic and technical advantages. It is our experience that isolated soy proteins are the most versatile soy protein ingredient. Functional and organoleptic deficiencies of less pure forms of soy proteins limit their application in food systems. Because of the major differences which exist between forms of soy protein products, comments which follow will try to distinguish when needed between soy proteins.

Soy protein products can provide a significant economic advantage in food processing as they are less expensive per pound of protein content, than animal proteins such as meat, milk, and eggs. Animal proteins originate from the biological conversion of vegetable proteins. The production of animal proteins is, therefore, less efficient than the direct use of a vegetable protein as a food ingredient. In fact, we have found in the last several years, in several developing countries that even compared to cattle grown on grass isolated soy proteins can offer a lower cost source of protein.

In addition to being economical, soy proteins are nutritious for human consumption. Recent investigations by the Massachusetts Institute of Technology, the Institute for Nutrition in Central America and Panama and Tokushima University in Japan have compared the nutritional quality of isolated soy protein with meat, milk, eggs and fish for human consumption. This subject is treated very professionally in a book published by the Academic Press (6). Without further detail here the point I would like to make is that isolated soy proteins, for example, when used in meat, fish and milk products need not alter the protein quality of the original product.

There are numerous uses for soy protein products. The first category is the use to supplement animal protein supplies at a lower cost per unit of protein. For example isolated soy proteins can be used in combinations with meat, fish, or milk to produce processed products. Typical products would include sausages and canned meats. An isolated soy protein solution can be injected into hams, other whole meat cuts, poultry and fish to increase thêir quantity while maintaining their quality. These same isolated soy proteins with water and a fat source can be used to extend the supply of milk to produce dairy-like products. Because of the relatively large carbohydrate fraction of edible soybean flours and grits, their application in this area is limited.

In the past, a popular view of soy proteins was the manufacture of meat-like products without any meat content. Commercialization of this concept has been slow due to consumer resistance. In contrast to this, the important business today and for the future involves the combination of soy protein products with animal protein resulting in acceptable protein forms. One hundred kilograms of beef being used in sausage production can be combined with 25 kg of isolated soy protein and water to be equivalent to 125 kg of meat. Thus, with the existing supply of meat an increase of 25% can be realized. In the production of hams yield increases of 20 to 30% can be realized. Fish can be injected with isolated soy protein solutions to increase yields by 15 to 20%.

Two points must be emphasized. First, soy protein products are being used to supplement the existing, albeit growing, supply of meat, milk, and fish. Second, the nutritional and organoleptic characteristics of the final food product need not be affected by the addition of isolated soy proteins. By organoleptic characteristics I mean taste, color, odor and texture. Thus, the final food product does not require significant changes in consumer preferences. This was not an easy task. It has, in fact, been a great challenge. However, technology developed over the last three to four years has now made it possible. Further technological improvements are being actively pursued.

The second category of use for soy protein products that I would like to discuss is the fortification of such basic products as cereals, bread and cookies. There are other newer products in this category of fortification such as drink mixes, sandwich spreads, and other commonly accepted foods. Because protein sources are more expensive than carbohydrate sources, fortified products on a per unit basis generally cost more than their non-fortified counterparts. But they are less expensive than providing animal protein products to supplement the protein deficiencies in the basic carbohydrate sources. The principal use of these products in the context of this chapter are in institutional feeding programs such as schools, hospitals, and to the labor forces of large plantation operations. Most of these institutions are concerned with budgetary constraints and the use of soy protein products offers a more economical form of delivering the daily protein requirement. Here too, soy protein products are used to supplement the existing food supply to produce highly acceptable food products without altering the final product's organoleptic characteristics.

To summarize the description of soy protein products we can say that they are economical food ingredients which when used properly can be used to supplement existing food supplies in the production of final food products without changing their nutritional and organoleptic qualities. Proper use involves both choosing the correct form of soy protein for the intended application and then managing its use at levels so as not to radically alter the characteristics of traditional foods.

RELEVANT ASPECTS OF NATIONAL DEVELOPMENT

Let us turn to our second point, that of national development. I would like to limit my remarks to the characteristics of development found in those countries, in the Rostovian sense, that have moved out of the traditional stage and have not as yet achieved the stage of full industrialization. The problems facing national development, especially those that relate to the food and agricultural sector, are numerous. The first that always comes to mind is that of rising population. In these countries population is increasing at more than 2% per annum (7).

Second, the people of these countries are experiencing rising but still inadequate incomes. Within the developing world this increase is reported to run from 0.4 to 2.8% annually in real terms (3). While food supplies are available from either domestic or foreign sources it is a lack of purchasing power that most significantly affects adequate food intake. One of the many conclusions of the World Food Conference was, "It is obvious . . . that an adequate income in cash or kind tends to ensure the availability of an adequate amount of food for the family and therefore reasonable energy and protein intakes" (3).

Third, it should be emphasized that food preferences are a very traditional part of any culture. This does not mean that food preferences do not change. What it does mean is that they can be expected to change through rather predictable traditional patterns. Economists refer to this change when it accompanies rising income, as the income elasticity of demand. The higher the elasticity, the higher the preference being exhibited. In developing countries this elasticity has been shown to be 0.22 for all food calories versus 0.56 for animal proteins. This contrasts with an elasticity of 0.28 for animal proteins in industrialized nations (4). Thus it shows clearly the preference towards more meat products in the developing world.

Fourth, these countries are experiencing a growing rate of urbanization. A recent World Bank report dramatizes this point. They project that, "By the year 2000, at least 22 cities in the world will have populations of 20 million or more—most in barely industrialized nations. Today's largest urban area, New York, has only 16 million persons" (5). This is critical because the urban poor suffer more from malnutrition than rural poor and the income effect on food consumption is most noticeable among urban dwellers (3).

As we look at these characteristics, rising populations, increasing but inadequate incomes, traditional but changing food preferences and increasing urbanization, we must focus on their implications. First, rising populations with increasing but inadequate incomes will demand more food at the most economical cost. Second, the pattern of traditional but changing food preference indicates that these people will create new demands for the specific foods they prefer and will be slow to accept just any food because it is nutritional and has economic benefits. If one traces the experience of such consumer products as the high protein low cost foods and their commercial position in the world today, we find that food preferences have significantly limited their universal consumption.

Third, increasing urbanization forces basic structural changes in the agricultural sector. As more people move off the farm reliance on subsistance agriculture decreases. With the complexity of marketing and distribution channels for food products to urban populations more processed foods will be necessary. Processing may involve such basic developments as increasing the capacity of abbitoirs, milk processing plants, and the canning of foods. Storage facilities and the preservation and packaging of food will need to be improved. Food processing industries will need to be developed to meet these specific needs of a more urban population.

These three basic implications are economic demands. They are: the need for more food at more economical costs, the need for preferred food products such as meat, fish and milk and the need for more processed foods. Having considered the demand side, let us now consider the supply side.

There are limitations to increasing food supplied to meet these changing demands at economical prices. These limitations include land, capital, energy management, technology, education and foreign exchange. The problems facing public policy-makers include the priorities for the best use of limited resources, how the process of agricultural development can best be encouraged and whether the rate of development will be sufficient to meet the needs of the people.

The challenge of developing food production is enormous. In the last decade the rate of increase in food production in the developing world has nearly matched that of idustrialized nations. However, on a per capita basis, food production in the developing world has remained nearly constant (2). I do not wish to be so presumptuous as to make any recommendations regarding specific national plans for agricultural development. We can only assume that each country will do its utmost to meet its own goal.

What I would like to present here is a simple option to supplement the supply, and thereby increase the rate of development, of more food, more economical foods, preferred foods, and processed foods, with the resources available. Having described the situation I would like to recommend two major roles for soy protein in the context of this development.

First, there is a role of soy proteins, more specifically isolated soy proteins, as food ingredients to be used in combination with meat, fish, and milk. They can be used as ingredients in processed meat products such as sausages and canned meats. They can be used to augment cuts of meat and fish. They can be used to supplement the supply of milk and the manufacture of dairy-like products.

Regardless, of the supply of meat, fish, and milk and regardless of the increase in their supply, soy protein products have a major role. If national demand for these traditional foods is not being met soy protein isolates can be used to help meet this demand. If the national demand is being met they can be used to increase food product production for export. If the cost of these traditional food products is not within the purchasing power of the entire population, soy protein isolates can be used to reduce the cost of these foods. Without reiterating the points made above, I believe that with the examples given it can be seen that the supply of more food, more economical foods, the preferred foods, such as meat products, and more processed foods can all be improved rapidly through the use of soy protein products.

The second major role that I can recommend for soy protein is their use to fortify existing foods in institutional feeding programs. In this situation the problem is one of meeting the daily nutritional requirements with acceptable foods at the lowest possible cost. Soy protein products can be used as food ingredients in a variety of such foods and provide increased protein supplies at more economical costs than animal protein products. Through the cooperation of the management of these programs with the local food processing industry and the suppliers of soy protein products many of the desired food products, with the higher protein contents can be made available.

Having described soy protein products, the relevant aspects of national development and the role of soy protein products in national development and having made two specific recommendations regarding this role, three questions remain. Should each country develop its own protein industry? If not, what is the effect on foreign exchange? And last, what public policy should be adopted with regard to soy proteins?

In reply to the first question, should each country develop its own soy protein industry, I would answer no. The production of soy protein products can be accomplished most economically the closer it is to the supply of raw material—soybeans. The efficient production of soybeans is limited by climatic conditions and is not as well suited to the tropical and sub-tropical environments, in which most of the developing world are located, as it is to the temperate climates of such countries as the U.S. and Brazil. To develop such an industry would again involve the use of limited resources which would otherwise be used to improve the rate of development of the agricultural sector and of the food processing industry.

Soy protein products are available. The technology of their use in food products can be provided by existing soy protein suppliers. The programs to increase their use in each national food supply, as indicated above, can be implemented.

Given this response, let us now turn to the question of the use of foreign exchange. Rather than dedicating areas to soybean production this land can be used to produce more traditional crops. In this form of agriculture the experience, technology and infrastructure exist and the rate of development can be more rapid. Such crops can be used to substitute for current imports and/or be used for export thus saving foreign exchange which can be used for the importation of soy proteins.

As a second example I would like to refer to a paper in which Bray describes the development of animal agriculture and the resulting need for feedstuffs (1). He deals with the case in which feedstuff production is being developed to its fullest and further feedstuff requirements need to be imported. The question then is the decision to import feedstuffs to produce meat for food products versus importing soy proteins to be used as ingredients in meat products. Using pork as an example he shows that 9.8 kg of feedstuffs are necessary to produce one kilo of boneless pork. In contrast,

200 g of isolated soy protein in combination with 800 g of water can be used in a meat product to accomplish, in combination with the existing meat supply, the same increase. Thus, one kilo of imported isolated soy protein is equivalent to importing 48.9 kg of feedstuffs. He concludes there is a dramatic savings in foreign exchange with this comparison.

A third but less common case in foreign exchange involves the country that may currently be importing carcass meat. With the economic advantages described above it is obvious that less foreign exchange would be necessary to import an isolated soy protein than to import carcass meat.

SOY PROTEIN POLICY RECOMMENDATIONS

Our final question then is what public policies should be adopted with regard to soy proteins. I would like to suggest that the maximum benefits of soy protein products can be realized if the following policies are adopted.

First, permit soy protein products for use in all food products to be maximum levels technically achievable. This varies considerably depending on the form of soy protein being used. Food laws in some countries were written many years ago. The new technology available today encompassing the use of isolated soy protein products was not considered years ago by those who wrote these regulations. Where these cases exist the regulations should be revised. The Codex Alimentarius Commission of the Food Agriculture Organization and the World Health Organization of the U.N. are now taking a leadership role in this area and the experience and information needed to properly rewrite regulations are available.

Second, encourage the use of soy protein products in institutional feeding programs. The technological development of isolated soy protein products has been so rapid that it has been difficult to disseminate the latest information throughout the world. With the encouragement and education of the leaders of institutional feeding programs great benefits can be realized more quickly.

Third, in general, improve the awareness of the benefits of these products through all appropriate levels of government, academia, the food industry and consumers. Fourth, allow imported soy protein products access to national markets by reducing the resistance to imports currently institutionalized through such means as restrictions on foreign exchange and customs tariffs. The case for using foreign exchange has, I believe been made. It now needs to be developed into actual practice.

In conclusion, I believe it has been shown that there is a role for soy protein products in national development. What is needed now is the total cooperation of government officials, academic leaders, the food processing industry, the management of institutional feeding programs and the suppliers of food proteins to increase the consumption of these products and thus provide benefits to the consumer from the advantages they offer.

NOTES

F. H. Schwarz, International Business Development, Protein Division, and J. K. Allwood, Economic Research Department, Ralston Purina Company, St. Louis, Missouri 63188.

REFERENCES

1. Bray, J. 1977. The foreign exchange and other economic variables affecting the decision to import protein food ingredients. Stanford Research Institute, Menlo Park, CA.
2. Population food supply and agricultural development. 1974. Monthly Bulletin of Agricultural Economics and Statistics, FAO, UN, Rome 23(9):2.
3. Preliminary assessment of the world food situation present and future. 1974. World Food Conf., U.N., Rome.
4. Study of trends in world supply and demand of major agricultural commodities. 1976. Organization for Economic Co-operation and Development, Paris.
5. U.S. News and World Report, February 12, 1979, p. 16.
6. Wilcke, H. L., D. T. Hopkins, and D. H. Waggle (Eds.). 1979. Soy protein and human nutrition. Academic Press, New York, NY.
7. World population: 1975, recent demographic estimates for the countries and regions of the world. ISP-WP-75. International Statistical Programs Center Bureau of the Census. U.S. Depart. of Commerce, Washington, June 1976.

PRINCIPAL DETERMINANTS OF VARIATIONS IN SOYBEAN PRICES

R. McFall Lamm, Jr.

Over the last twelve years soybean prices have varied from a low of \$2.30/bu in October 1969, to a high of \$10.84/bu in June 1973. Soybean oil prices have ranged from 7.3 cents/lb in October 1968 to a high of 43.3 cents/lb in August 1974. Soybean meal prices peaked at \$412.50 in June 1973, following a low of \$69.80 in January 1969, on a per ton basis for Decatur 44% protein. Price variations of this magnitude are significant when compared with price variations of nonagricultural products, and complicate decision-making and planning for both consumers and producers of soybeans and soybean products.

If significant variations in soybean and soybean product prices are fully anticipated, decision-making and planning are not complicated. For example, a producer who knows that the price of soybeans will be \$6.00/bu at harvest can easily determine whether to plant soybeans; production costs can be estimated prior to planting and expected profit can be determined. In actuality, soybean prices are not known prior to harvest and producers must make plans on the basis of expected prices. If expectations are not realized, soybean production may not yield a normal profit, forcing producers to leave the industry or produce other commodities.

This chapter discusses the principal determinants of variations in soybean prices, focusing on the interaction between the demand for soybean products and the supply of soybeans as the price-determining mechanism. The discussion is both retrospective, in that the major causes of price variation in recent years are reviewed, and prospective, in that the potential for future price variation is examined.

RECENT PRICE BEHAVIOR AND INFLATION

Prior to 1972, nominal soybean and soybean product prices remained fairly stable on an annual basis (Table 1). In 1972, however, there was an explosion in soybean meal prices, principally as a consequence of large exports of soybean meal in that year. The results were record-high soybean prices. In 1973, reduced production of commodities on a world-wide basis resulted in further increases in soybean and soybean product prices. A drought and other events helped to maintain nominal prices at record levels again in 1974, roughly double 1972 prices.

Although not generally acknowledged, higher nominal soybean and soybean product prices in 1973 and 1974 were attributable partly to higher inflation rates in the U.S. in those years. In the late 1960's inflation rates were relatively low, averaging 4% per year. In the 1970's inflation rates accelerated, reaching more than 10% in 1974. These increases resulted in higher soybean and soybean product prices as adjustments were made to increases in the money supply.

Because of the importance of inflation in pushing up nominal prices, it is instructive to consider both nominal and real soybean and soybean product prices for analytical purposes. Table 1 presents nominal and real soybean and soybean product prices from 1967 to 1978. Real prices are derived using the implicit price deflator and are reported in 1972 dollars. Quite interestingly, even after stating prices in constant dollars, it is apparent that there was a substantial increase in soybean and soybean product prices in 1973 and 1974. The real increase, however, is much less than the implied nominal increase. In addition, stating soybean and soybean product prices in real dollars indicates that soybean prices appear to be higher in recent years than in the later 1960's. The same is true for soybean product prices. For this reason, it is apparent that soybean and soybean product prices have increased faster than the general price level over the last decade.

MARKET STRUCTURE

Soybeans and soybean products are major international commodities traded world-wide. The U.S. and Brazil are the leading producers and exporters of soybeans. The European Economic Community (EEC) and Japan are the major soybean importers. The U.S., Brazil, and the EEC are major producers and exporters of soybean oil; the U.S. and Brazil are the dominant exporters of soybean meal. Many countries import soybean oil and meal.

The markets for soybeans and soybean meal are essentially free markets internationally. According to Lamm (2), however, free trade in soybean oil is restricted by Japan, the EEC, and the U.S., which impose tariffs on soybean oil imports. Japan and the EEC impose tariffs on soybean oil essentially to

Table 1. Soybean and soybean product prices, 1967 - 1978.[a]

Year	Soybean Price		Soybean Oil Price		Soybean Meal Price	
	Nominal	Real	Nominal	Real	Nominal	Real
			— dollars —			
1967	2.62	3.32	9.61	12.16	80.83	102.33
1968	2.49	3.02	8.19	9.89	81.88	99.19
1969	2.43	2.80	9.07	10.45	79.07	91.16
1970	2.60	2.85	11.95	13.07	84.35	92.34
1971	2.94	3.06	12.60	13.12	83.67	87.16
1972	3.30	3.30	10.56	10.56	111.09	111.09
1973	6.50	6.15	19.84	18.67	244.97	232.03
1974	6.42	5.52	35.80	30.77	148.27	127.85
1975	5.24	4.13	25.39	20.03	132.54	104.25
1976	5.58	4.17	18.82	14.04	172.41	128.62
1977	6.82	4.84	23.78	16.84	202.55	143.60
1978	6.26	4.04	25.37	16.37	180.59	116.51

[a]Soybean price is the average received by farmers; soybean oil price is per 100 lbs for crude, tank cars, fob, Decatur; and soybean meal price is per ton for bulk, 44% protein, Chicago. Data for 1978 are preliminary. Source: U.S. Department of Agriculture.

protect their oilseed processing industries, according to Parris and Ritson (5). The U.S. tariff on soybean oil was imposed prior to U.S. domination of soybean and soybean product markets.

Because the U.S. market for soybeans and soybean products is so large relative to similar markets in other countries, the U.S. price of soybeans and soybean products is considered generally to be the world price. Market arbitrage would be expected to eliminate any spatial disparity which might occur. However, information presented in Table 2 indicates that this may not always be the case on an annual basis. For example, in some years the Netherlands' export price of soybean oil is less than that of the U.S. Similarly, in 1972 and 1973 the Netherlands' export price for soybean meal was lower than that of the U.S. These discrepancies are consequences of short-run demand and supply conditions in the EEC, however, and would be expected to occur periodically in an international market.

Further examination of the data presented in Table 2 indicates that the U.S. price of soybean oil has been historically lower than the European export price. Once the EEC 10% ad valorem tariff and transportation costs are added to the U.S. price, only in 1965, 1972, 1975, and 1976 does U.S. soybean oil become competitive with soybean oil produced in the EEC. In this regard, a recession of the EEC tariff on soybean oil would likely stimulate U.S. soybean oil exports.

Table 2. Representative international soybean and soybean products prices, 1964 - 1977.[a]

Year	Soybeans			Soybean Oil			Soybean Meal	
	U.S. Wholesale Price	U.S. Export Price	U.S. Import Price	U.S. Wholesale Price	Netherlands Export Price	European Import Price	U.S. Wholesale Price	Netherlands Export Price
				– *dollars/100 kg* –				
1964	9.2	10.3	11.1	20.3	21.5	22.9	8.1	9.7
1965	9.6	10.7	11.6	24.7	28.4	27.0	8.4	9.7
1966	10.8	11.8	12.7	25.8	26.6	26.2	9.7	10.3
1967	9.6	10.6	11.4	21.2	23.5	21.6	8.9	10.4
1968	9.2	10.1	11.2	18.1	17.7	17.8	9.0	10.1
1969	8.9	9.8	10.7	20.0	18.4	19.8	8.7	10.2
1970	9.6	10.7	11.9	26.3	24.4	28.9	9.3	10.2
1971	10.8	11.9	13.1	27.8	29.5	30.4	9.2	11.2
1972	12.1	13.4	14.4	23.3	27.3	24.0	12.2	12.1
1973	23.9	27.4	22.3	43.7	40.7	43.9	27.0	21.5
1974	23.6	25.6	27.4	78.9	71.5	83.2	16.3	20.7
1975	19.3	20.7	22.3	55.9	72.2	56.3	14.6	18.1
1976	20.5	21.9	23.4	41.5	46.7	43.8	19.0	20.5
1977	25.1	26.8	28.3	52.4	56.2	57.4	22.3	28.5

[a]Product definitions are the same as those given in Table 1; U.S. export price is fob gulf ports. Source: *Fats and Oils Situation* and *Production Yearbook.*

DEMAND FOR SOYBEAN PRODUCTS

The demand for soybean products and the supply of soybeans interact to determine market equilibrium prices and quantities. Because the real prices of soybeans and soybean products are now generally higher than a decade ago, and since consumption of these commodities is now greater, it is apparent that increases in demand have dominated supply increases over the last decade. Whether this dominance will continue requires a careful analysis of the factors underlying demand and supply changes in the markets for soybeans and soybean products.

Soybean demand is principally a derived demand—it depends inherently on the demand for soybean products. An increase in the demand for soybean oil or soybean meal is passed through directly to the soybean market as an increase in demand. In this way, higher soybean oil or meal prices normally imply higher soybean prices, given supply.

Five factors determine the final demand for any commodity. These include: (a) the income or wealth levels of consumers of the commodity, (b) consumer tastes, (c) the prices of substitutes for the commodity, (d) the population level, and (e) the inflation rate of the currency in which price is reported. Changes in any of these factors affect absolute demand, and can increase or decrease prices depending on the direction of change. Each of these factors has been important in determining the demand for soybean products.

Higher income levels over the last decade, both domestically and internationally, have resulted in increased demands for soybean oil and soybean meal. The importance of income as a determinant of the demand for soybean oil and other vegetable oils is illustrated in Table 3, which presents per capita income, vegetable oil, and animal fat consumption levels for selected developed and developing countries. In high-income developed economies, vegetable oil and fat consumption are extremely large, amounting to more than 70 lb per capita in some countries. In low-income developing economies, vegetable oil and animal fat consumption are substantially lower. Hence, as income levels of developed and developing countries have increased in recent years, the demand for soybean oil has been stimulated.

Higher incomes affect the demand for soybean meal through the demand for meat. Higher income levels increase demands for meat. Greater demands for meat imply greater demands for feeds used in meat production. Since soybean meal is an important ingredient in feeds produced commercially, higher incomes stimulate greater demands for soybean meal.

From 1963 to 1965 per capita meat consumption in the U.S. averaged 197 lb (beef, pork, and poultry). Slightly more than a decade later, over the period 1974 to 1976, per capita meat consumption averaged 223 lb, an increase of more than 13%. Over the same time period, real national income

Table 3. Per capital income, vegetable oils and fat consumption for selected countries.[a]

Country	Income	Oil									
		Soybean	Palm	Groundnut	Cottonseed	Sunflowerseed	Coconut	Olive	Rapeseed	Lard	Butter
	dollars					*– pounds –*					
Developed Countries											
Belgium-Luxembourg	8053	11.1	6.0	3.2	.0	2.1	7.5	.2	.9	--	20.9
France	7176	3.7	2.5	10.3	.0	4.3	3.1	.9	1.7	--	--
West Germany	8413	13.3	6.5	1.1	.9	4.4	8.4	.1	2.8	--	14.6
Italy	3473	5.9	2.0	2.5	.0	.9	1.7	24.6	8.0	--	2.8
Netherlands	7695	21.4	9.4	4.3	.0	2.8	11.8	.2	1.7	--	9.9
United Kingdom	4366	3.2	5.5	4.6	.6	3.9	2.9	.1	1.9	7.5	15.6
Japan	6065	11.1	2.9	4.8	.9	--	1.7	--	6.2	--	1.2
Spain	3186	14.7	.4	.0	.7	8.9	--	23.9	.1	--	--
U.S.	8624	34.2	2.8	1.1	2.5	.0	4.9	--	--	3.7	4.7
Developing Countries											
India	145	.4	.1	5.8	.6	--	.9	.0	--	--	1.8
Pakistan	183	3.1	4.3	.0	2.9	--	.1	.0	--	--	6.7
Nigeria	398	.0	16.1	3.3	2.9	--	.2	.0	--	--	.2
Venezula	2776	2.9	.0	4.0	7.2	--	3.4	.0	--	.7	1.1
Brazil	1446	16.4	.2	1.3	3.8	--	.1	.2	--	2.5	1.3

[a]Income is in 1976 dollars. Consumption is for 1975 or 1976, depending on data availability.

increased approximately 37%, implying a positive relationship between income and meat consumption. Soybean meal consumption increased more than 55% over the period, implying a positive relationship between meat and soybean meal consumption. By inference, higher income levels have had a positive effect on the demand for soybean meal.

Changes in tastes, the second factor affecting demands for soybean products, are difficult to quantify. In the U.S. in recent years there has been increased concern about the consumption of highly saturated fats. Although there has been a decline in consumption of highly saturated fats, particuarly of butter and lard, and an increase in the consumption of less saturated vegetable oils, particularly of soybean oil, it is not clear whether this is a consequence of changes in relative prices or changes in tastes. A similar situation exists in southern Europe where historically there has been a preference for olive oil over other vegetable oils. In recent years, however, more soybean oil has been consumed in southern European countries. Again, it is not clear whether this increased consumption is a result of changes in tastes or a consequence of lower soybean prices relative to olive oil prices.

Fairly conclusive evidence exists regarding the role of substitute prices, the third factor which may affect the demand for soybean products. Table 4 presents selected vegetable oil prices from 1963 to 1976. From the early 1960's to the mid-1970's the price of soybean oil declined with respect to the prices of cottonseed oil, peanut (groundnut) oil, sunflowerseed oil, and olive oil. Over the same period, the price of soybean oil increased with respect to the price of rapeseed oil, coconut oil, and palm oil. Relative increases in the prices of the former group of oils have resulted in increases in the demand for soybean oil, while decreases in the prices of the latter group have decreased the demand for soybean oil.

With respect to soybean meal, the relative prices of most substitutable feeds have increased in recent years (Table 5). From the early 1960's to the mid-1970's, the prices of cottonseed meal, groundnut meal, corn, barley, and sorghum all increased relative to soybean meal price. This implies a direct increase in the demand for soybean meal as a result.

Population growth, the fourth factor which increases the final demand for commodities, has also stimulated the demand for soybean oil and soybean meal. In 1965, the U.S. population was 194 million. In 1975, U.S. population was 214 million, an increase of almost 10%. Population growth in other areas of the world has been even more substantial, further contributing to increased demand for soybean oil and meal.

Accelerating rates of inflation, the final factor which can increase the demand for commodities, stimulates demand for commodities as stores of value. In periods of rapid inflation, the values of currencies decrease in proportion to the rate of inflation. To avoid real losses incurred by holding

Table 4. Selected vegetable oil prices, 1963 - 1976.[a]

Year	Soybean	Cottonseed	Groundnut	Sunflowerseed	Rapeseed	Coconut	Palm	Olive
				– *1972 cents/lb* –				
1963	8.9	10.2	12.1	10.5	9.4	11.7	9.9	45.0
1964	9.2	10.3	14.2	11.5	11.4	12.5	10.5	26.5
1965	11.2	11.6	14.7	13.2	11.9	14.6	12.3	30.1
1966	11.7	14.1	13.4	11.8	11.1	11.2	10.8	29.9
1967	9.6	11.7	12.8	9.6	9.4	12.9	8.8	31.2
1968	8.2	13.0	12.2	7.7	7.3	16.1	7.5	31.2
1969	9.1	10.8	15.0	9.6	9.1	13.3	8.5	36.1
1970	11.9	13.4	17.1	15.1	12.8	15.5	11.6	31.7
1971	12.6	15.2	20.3	17.0	13.2	13.6	12.0	32.6
1972	11.4	11.5	19.2	14.8	10.3	9.7	9.8	40.9
1973	19.8	19.5	24.7	21.9	18.6	24.2	17.0	61.8
1974	35.8	38.1	49.8	44.7	34.8	50.2	30.4	92.5
1975	25.4	27.2	42.3	33.9	25.0	18.1	19.5	115.9
1976	18.8	23.3	33.0	26.7	18.8	19.1	18.4	100.4

[a]Source: *Monthly Bulletin of Agricultural Economics and Statistics* and *Fats and Oils Situation.*

Table 5. Selected vegetable meal and feed grain prices.[a]

	Meal			Feed Grain		
Year	Soybean	Cottonseed	Groundnut	Corn	Barley	Sorghum
	– 1972 dollars/ton –			*– 1972 dollars/bu –*		
1963	107.0	94.7	94.7	1.67	1.70	2.77
1964	100.8	82.9	84.9	1.56	1.67	2.88
1965	102.1	80.9	94.3	1.60	1.69	2.74
1966	114.7	99.9	108.3	1.71	1.76	2.74
1967	102.3	97.5	100.8	1.45	1.52	2.65
1968	99.2	90.9	100.6	1.22	1.41	2.19
1969	91.2	76.1	89.0	1.34	1.49	2.37
1970	92.3	80.8	89.8	1.41	1.40	2.38
1971	87.1	75.1	78.4	1.26	1.48	2.31
1972	111.1	89.5	93.4	1.17	1.50	2.12
1973	232.0	161.2	166.7	2.15	2.42	4.07
1974	127.9	107.1	120.7	2.72	2.70	4.67
1975	104.3	94.2	99.5	2.19	2.20	3.80
1976	128.6	124.6	118.4	2.00	2.05	3.29

[a]Source: *Agricultural Statistics.*

currency as a store of value, consumers and business instead hold other goods, such as commodities. In this respect, an increasing rate of inflation would be expected to stimulate the demand for soybeans.

As a simple test of whether inflation may have stimulated holdings of soybean stocks in the U.S., consider the two periods 1970 through 1972, and 1974 through 1977 (1973 is omitted because government controls were in effect during that year). In the former period, the rate of inflation averaged 3.1%; over the latter period, the inflation rate averaged 5.5%. September 1 stocks of soybeans as a percentage of total U.S. soybean production averaged 11.0% in the former period and 12.3% in the latter. Hence, the higher inflation rate from 1974 through 1977 is positively correlated with higher carry-over stocks of soybeans. This indicates that higher inflation rates may have stimulated greater holdings of soybean stocks over the period considered.

SOYBEAN SUPPLY

Five factors determine the supply of any commodity. These include (a) the state of technology, (b) costs of resources used in production, (c) the prices of alternative products which could be produced, (d) weather, and (e) government activity. Changes in any of these factors can increase or decrease the supply of a commodity depending upon the directions of change.

Technology change occurs when more of a commodity can be produced at a later point in time using the same quantity of resources required to produce the commodity at an earlier point in time. For example, technology change would occur with the introduction of a new seed variety which, when planted in the same quantity as the old seed variety, would yield more soybeans, all other things constant. Improvements in planting and harvesting equipment, fertilizers, pesticides, and labor efficiency would all be considered changes in technology. Technology improvements increase supply, leading to lower prices for a given demand level.

Although soybean production has been characterized by some technological advances in recent years, the nature of technology change in the soybean industry has been evolutionary rather than revolutionary. There has been no single major technological breakthrough, but a collection of minor ones. These gradual changes in technology have increased the supply of soybeans, moderating soybean price increases in recent years. Changes in the costs of resources used in soybean production, the second factor which can increase or decrease soybean supply, have not been significant in recent years. The basic costs of soybean production consist of land rent, equipment depreciation, labor, fuel, seed, fertilizer, pesticides, charges for storage, and marketing costs. Although the prices of some of these items have increased faster than general price indices, there has been no overwhelming increase in any component. In this respect, increases in the costs of inputs used in soybean production have not affected supply significantly. The one major exception to this statement was a large increase in fuel costs in 1973 as a consequence of the Arab embargo on petroleum exports to the United States.

Increases or decreases in soybean supply also occur as a consequence of changes in the prices of crops which could be produced as an alternative to soybeans. This is the third factor which can affect the supply of soybeans. Although wheat, cotton, peanuts, and other crops can be produced as alternatives to soybeans in different regions of the country, corn is the one major crop which can be substituted for soybeans in most regions. Since the price of corn has increased relative to the price of soybeans over the last decade, a decrease in soybean supply is implied, all other things constant. When the relative price of corn increases, producers grow more corn and fewer soybeans. The magnitude of increase in corn price relative to soybean price has been small, however, so there has really been little negative effect on soybean production as a consequence of relative increases in corn prices.

Perhaps the most important factor affecting soybean supply is weather variation (the fourth factor which can affect commodity supply). Droughts or too much moisture are the major causes of damage which can reduce supply. High prices of soybeans in 1974 were partly attributable to bad weather which drastically reduced yields. From 1971 to 1973, soybean yields were 27.5, 27.8, and 27.7 bu/acre. In 1974, soybean yields decreased to 23.2 bu/acre, mostly as a consequence of drought in the Midwest. From 1975 to 1977, yields were 28.9, 26.1, and 30.6 bu/acre. This indicates, at least partially, the effect weather conditions can have on soybean production.

One other factor affecting soybean supply is governmental activity. For most agricultural commodities, governmental activity takes the form of support and loan programs. Since the early 1970's government activity in the soybean market has been minimal, except for the imposition of an export embargo on soybeans, soybean oil, and soybean meal in the summer of 1973. The embargo was imposed to reduce exports at a time when domestic prices were at record highs. According to Lamm (4), the embargo imposed substantial costs on producers, but lowered prices of soybeans and soybean products to consumers.

OUTLOOK

In the future, there are several developments which may have significant effects on the soybean market. On the demand side, income levels will continue to increase, preferences for vegetable oils over animal fats are likely to become more pronounced, the prices of some substitutes for soybean products will likely increase, population will expand, and inflation is forecast to accelerate—all leading to greater demands for soybean products and higher prices. However, massive plantings of oil palms in Malaysia and Indonesia in recent years will result in increased palm oil supplies and lower relative palm oil prices in the future, according to Lamm and Dwyer (3). Lower palm oil prices will decrease demand for soybean oil. In addition, expanded sunflowerseed oil and meal production in the U.S. may decrease demand for soybean oil and meal according to Doty (1).

On the supply side, it is not clear whether the prices of crops which could be produced as alternatives to soybeans will increase or decrease. Costs of resources used to produce soybeans are not likely to change drastically, although fuel, fertilizer, and pesticide prices may increase. In addition, there is likely to be little government intervention in soybean markets, except during crises. These facts indicate that weather variation will continue to be the major determinant of changes in soybean supply, much as it has been in the past.

SUMMARY

This chapter has presented a review of the principal determinants of variations in soybean prices. Both demand and supply-side factors are important in this determination. Over the last decade, increases in soybean demand have led to higher soybean prices. Weather variation on the supply-side has been the major cause of price instability in the short run. In the future, it is not clear what events will contribute most to changes in soybean demand. Weather variation will continue to be the major determinant of changes in soybean supply.

NOTES

R. McFall Lamm, Jr., Economist, Commodity Economics Division of the Economics, Statistics, and Cooperatives Service, U.S. Department of Agricultural, Washington, D.C.

Opinions presented do not necessarily represent those of the Economics, Statistics, and Cooperatives Service or the U.S. Department of Agriculture.

REFERENCES

1. Doty, H. 1978. Future of Sunflowers As An Economic Crop in North America and the World. In J. F. Carter (ed.) Sunflower Science and Technology. American Society of Agronomy, Madison, WI.
2. Lamm, R. 1979. Oilseeds and Oilseed Products: International Marketing Policies and Trade. CED Working Paper, ESCS, USDA, Washington, D.C.
3. Lamm, R. and J. Dwyer. An Analysis of the Effects of U.S. Policies on the World Palm Oil Market Using a Quarterly Econometric Model. Malayan Economic Review, 23:37-48.
4. Lamm, R. 1977. A Policy Analysis of the U.S. Vegetable Oilseeds, Oils and Oil Products Industry Using a Monthly Econometric Model. Research Division Bulletin 130, Virginia Polytechnic Institute and State University, Blacksburg, VA.
5. Parris, K. and C. Ritson. 1977. EEC Oilseed Products Sector and the Common Agricultural Policy. Wye College, University of London, London, England.

SOY OIL UTILIZATION—CURRENT SITUATION AND POTENTIAL

D. R. Erickson and R. A. Falb

In this crop year of 1978/1979 it is projected that worldwide production of fats and oils will total approximately 55 million tons. Of this total, soy oil is projected to account for 12.2 million tons or about 22% of total world fats and oils production. Of that 12.2 million tons, the U.S. is expected to produce 8.2 million tons or 67% of the total world soy oil production. Only animal fats (which include tallow, lard and butter) exceed soy oil production (14.6 million tons). On a world basis, sunflowerseed was projected as third in production with 4.9 million tons or 8.9% of total production.

On a world basis, expansion is anticipated in the production of palm, rapeseed and sunflowerseeds. This will generate large quantities of edible oils as these all have a high ratio of oil. This will produce increased competition to soy oil from these oils in major markets. If we project a linear trend from 1979 to 1985, total world fats and oils production will reach a projected 64 million tons from the 55 million tons in 1979. This will probably outrun projected demand by a million tons. With the state of oil processing technology, in at least the most developed nations and especially in the U.S., the various vegetable fats and oils and certain animal fats have become virtually indistinguishable in end products such as shortening. It is inevitable that in these markets at least, increasing substitution will cause the price to become a more important factor.

Per capita consumption of total edible fats and oils increases or decreases in answer to increasing or decreasing income rather than any fluctuations in price. Therefore, demand would not grow necessarily with increasing large surpluses of oil which would engender decreasing prices overall. It is possible then, that faced by increasing competition, soy oil imports into major markets may decline as a share of total world imports of edible oils. Growth of U.S. soy oil imports may not continue at the same pace as in recent years unless expanded markets can be developed in the third world. Many areas of the world have very low oil consumption relative to the developed countries of Western Europe and North America. These areas offer potential for increasing consumption of fats and oils as incomes in these areas increase. However, future developments in these areas are very uncertain. Financing of oil imports could be a problem in many of these areas because of small cash reserves and an emphasis on priorities other than edible oil consumption.

In the U.S., projected production of soy oil in 1979 is 8.1 million tons. This is 58.3% of the projected total production of 13.9 million tons of all fats and oils. Soy oil is the number one edible oil in the U.S. This holds true even if you throw in all fat and oil sources. This dominance is predicted to continue and actually to increase to 65% in 1985. This prediction is not especially surprising when you remember that soy oil is a liquid oil and the trend in the industry, for various reasons, is toward greater and greater use of liquid oils as opposed to solid fats. However, in the long run in the U.S., competition could come from sunflowerseed oil. The trade is predicting that 10 million acres will be planted to sunflowers within 5 years. Whatever the predictions for future potential competition may be, soy oil's overwhelming dominance of the U.S. edible oil market is an irrefutable fact. Just as irrefutable is the fact that there is very little general knowledge of soy oil.

Three different studies have been conducted which point this out. Two focus group studies were conducted in conjunction with a general market research study on the situation in the edible oil market. This indicated a low level of knowledge and a mixed attitude toward the concept of an edible oil made from soybeans. The second was a consumer panel questionnaire survey conducted by a leading consumer-type magazine which, to a degree, began to quantify the knowledge level and attitudes toward soy oil. The third was a national probability sample of 1,200 female heads of households (better known to most of us as the woman of the house) which was designed to give us quantifiable results on knowledge of, and attitude toward, edible oils and soy oil in particular. We needed this to set a base from which we could, in future years, measure changes in attitude and knowledge of the American consumer.

The research study was conducted by Burke Market Research using a complete National Probability sample. Female heads of the household were contacted who had used either a liquid cooking oil or solid shortening within the three months prior to being contacted. Respondents were asked about their cooking habits, their awareness of various plant oils, ratings of certain oils, importance of certain oil attributes, oil purchase, reasons for purchase of certain oils, main uses and frequency of use of various oils, recognition of oils in certain products, and demographics. Half of the respondents were asked to rate (if aware) and give purchase reasons (if purchased) for soybean, palm and corn oils; the other half rated and gave reasons for purchase for soybean, palm and sunflowerseed oils. The interviewing was conducted by phone from March 27 to April 4, 1978.

In summary, three major findings and conclusions were reached by analyzing the results of the study. (a) Eight-three % of the people asked "What oils can you think of?" did not mention soy oil, indicating a very low awareness of soy oil. (b) When soy, corn, palm and sunflowerseed oils were compared on a variety of characteristics, soy oil placed third, below corn and sunflowerseed. We would particularly like you to remember this and compare it to the results of the product placement study on which we will report later. (c) Soy oil was rated most favorably with respect to the amount of cholesterol it contained. This was surprising in that there was no advertising, promotional or informational campaign linking soy oil with no cholesterol, while there has been such a campaign with corn oil.

Only 7% of 1,200 people reported having purchased soy oil in the past 6 months. This, of course, was not in agreement with market actualities where soy oil has nearly 60% of the total edible fats and oils market in the U.S. Only 4% stated that soy oil was in the last liquid cooking oil they purchased. The fact is that soy oil accounts for 75% of the oil sold as liquid cooking/salad oil. Only 5% said soy oil was in the last margarine they purchased, although market data shows soy oil as having 80% of the market for edible oils in margarine. When people were asked to give reasons for purchasing a particular oil, more of them make comments which fell into a Health/Nutrition category than any other. In a government consumer survey between 1973 and 1975, those food shoppers who selected "making sure that my family and I get the nourishment we need" as their primary concern about food, jumped from 41% to 50%.

In considering the unaided recall of various plant oils, the response "vegetable oil" was considered as a mere repetition of the question by the respondent. Therefore, although almost 20% of the respondents claimed unaided awareness of vegetable oil, this reply was treated as a "don't know" in the comparisons of awareness. Unaided awareness values were: corn, 52%; peanut, 36%; safflower, 23%; soy, 17%; olive, 15%; cottonseed, 4%; sunflower, 3%; coconut, 2%, sesameseed, 2%; and palm, 1%.

As you can see, corn retained its number one position. This is a prime example of the effect of a major advertising/education program for a commodity. The halo effect caused respondents to put corn oil into products it is not used in, merely because it has become so well known and advertised as an edible fats and oils product.

The shift of olive oil from fifth to second place in unaided and aided recall probably reflects the fact that olive oil was not classified in many respondents' minds as a "plant" oil. Total awareness values were: corn, 97%; olive, 94%; peanut, 90%; safflower, 73%; soy, 67%; cottonseed, 47%; coconut, 46%; sesameseed, 42%; sunflower, 42%; and palm, 13%.

The rankings of the means of the oils as far as six attributes are presented in Table 1. The lower the mean, the better the oil was considered to be. From a comparative viewpoint, soy is viewed as superior to sunflowerseed and corn oils in terms of the amount of cholesterol it contained. It is seen as coming between these two oils with respect to greasiness and expense and it is considered to be the worst with respect to odor, aftertaste, and smoking/burning. You have to realize that these are perceptions or attitudes regarding oils about which the respondents had very little actual knowledge. You will see in the report on the Placement and Recall study that under actual use conditions, there was little differentiation between these oils on most of the attributes listed in Table 1.

On a three-point scale, all the attributes were rated as somewhat important, with all ratings over two. Importance of attribute ratings were: (a) grease left on food, 2.82; aftertaste, 2.81; smoking, 2.82; odor, 2.60; amount of cholesterol, 2.45; and expensiveness, 2.39. When examining the reasons for purchase, there were so few people purchasing palm oil (only 7 of 1,200) or sunflowerseed oil (only 16 of 1,200) that it was not included in the comparisons with soy and corn. Half of the 16 people who bought sunflowerseed oil bought it for health reasons and 40% indicated taste was an important factor. For both soy and corn, health and nutritional reasons (40%) were indicated as the chief reason for purchase, while "uses" was the next most important reason. However, for soy the third most important reason for purchase was price/economy while for corn it was product attributes (Table 2).

Although 93% of the respondents claimed to have purchased a liquid cooking oil in the past 6 months, about 48% indicated they were not certain what kind of oil that product contained. The oil purchased last by most of those who could name a kind of oil was corn (34%) followed by soy and sunflowerseed at about 4% each. The lack of real knowledge of the consumer is evident when respondents were questioned about purchase of solid shortening. About 64% had purchased but 57% indicated they did not know what kind of oil their shortening contained. The highest percentage (4%) who claimed they know, said corn (there is no corn oil in shortening) and 1% said

Table 1. The rankings of means of oils on six attributes.

Oil	Cholesterol	Greasiness	Odor	Expensiveness	Smoking/ Burning	Aftertaste
Soy	2.18	2.22	1.90	2.87	2.31	1.82
Sunflower	2.21	2.39	1.97	2.94	2.38	2.02
Corn	2.28	2.40	2.62	3.52	2.58	2.51

Table 2. Reasons for respondent purchase of a liquid cooking oil.

Base—Total Asked Reasons	(106) Soy	(385) Corn
All Mentions of Cholesterol	14%	27%
All Mentions of Polyunsaturated	11	12
Health/Nutrition (net)	41	40
Uses (net)	31	32
Price/Economy (net)	24	12
Product Attributes (net)	21	25
Taste/Flavor (net)	18	21
Texture/Consistency (net)	12	8
Ease/Convenience (net)	–	5

soy. Most respondents had purchased margarine (92%) in the past six months of which 49% indicated they didn't know the oil it contained. Of those who claimed to be aware, most said corn oil (37%) followed by soy (5%) and safflowerseed (1%).

A second market research study was conducted by the American Soybean Association during the period from June 5 through July 25, 1978. This we refer to as the Soy Oil Placement and Recall Study. In it we used unidentified bottles of soy, corn and sunflowerseed oils. The purpose of the study was to get respondents' opinions of these oils under actual use conditions without the psychological factors of product or brand identification to influence these opinions. The U.S. was split up into 6 regions and a city was selected in each region. These cities were: Atlanta, Georgia; Boston, Massachusetts; Dallas, Texas; Davenport, Iowa; Oakland, California; and Portland, Oregon. We used a paired comparison design with half of the respondents using a soy/corn combination and half a soy/sunflowerseed combination. The products were rotated on a "used first" basis to eliminate any bias. The respondents used the first product for 2 weeks after which time they were called to obtained their opinions and reactions. They were then sent the second product to use for 2 weeks, and were again called back.

In summary, the major findings and conclusions were: (a) Soy was statistically significantly preferred over corn oil at the 95% level of confidence. (b) In terms of overall preference and ratings, soy oil was not significantly different from sunflowerseed oil. (c) The main difference between soy oil and corn oil was that soy oil is perceived as the lighter of the two oils. (d) There were no other statistically significant differences in response to questions concerning likes, dislikes and reasons for preference between soy and corn or between soy and sunflowerseed. (e) The oils were perceived as virtually the same on a variety of attribute measures; and (f) The oils were used in the same way. These overall results would seem to be a positive indication that although consumers know almost nothing about soy oil, it is as well liked as its market share would seem to indicate.

Most respondents had used liquid oil in the week prior to placement with the test product. Those who had used corn oil knew it; those who had used soy oil apparently did not. Respondents used the test oils mostly for frying, salads, and baking.

For each oil, respondents were asked if it left food greasy, had an odor, left an odor in the home, had an aftertaste, or smoked. If the person interviewed did think an oil possessed one of these characteristics, she was asked to indicate the degree of the attribute the oil had (a lot or a little). In both tests, there were no differences between the oils with respect to the percentages of respondents agreeing an oil had an attribute, or reporting the degree to which the oil had it. In other words, all oils were the same on all these attributes. It would seem that the differences which do exist between these liquid oils have no appreciable affect on most measures which pertain to home usage of the oils. The overall rating of any given oil (liked it a lot plus liked it somewhat) does not differ from any other. Ratings were as follows: (a) liked soy, 84%; (b) liked corn, 83%; and (c) liked sunflowerseed oil, 84%. Approximately the same percentage of respondents made favorable comments for each oil in each test (Table 3). The only exception was the texture/consistency comments in the soy/corn test where a significantly higher percentage of people made favorable comments of this nature about soy oil (36%) than about corn oil (24%).

In overall preference, respondents preferred soy oil to corn oil 52 to 39%. This difference was statistically significant at the 95% level of confidence. Not statistically significant was the difference between sun and soy of 43 to 39%. It is interesting to note that a higher percentage (18%) had no preference in the sunflowerseed/soy test as opposed to 9% who had no preference in the soy/corn test.

Table 3. Consumer opinions of vegetable oils under use conditions.

Response	Soy/Corn Test Group		Soy/Sun Test Group	
	Soy	Corn	Soy	Sun
Gave favorable comment	91%	88%	94%	89%
Product attribute	64	59	65	60
Taste/flavor	56	55	58	56
Uses	50	53	54	51
Texture/consistency	36	24	31	36
Odor	21	16	24	21
Ease/convenience	7	8	9	4

Regional differences are listed in Table 4. Bases varied from city to city and are small. In only one city on one test was the preference difference statistically significant, and that was in Portland on the soy-corn test. There were no statistically significant differences between the oils on any of the specific attributes about which respondents were asked (Table 5).

Soy oil is unique among the common vegetable oils in that it has a significant percentage of linolenic acid. This is one of its main problems and one which is and will be receiving a lot of attention. The key to the growth of soy oil in the edible oil market has been the reduction of linolenic acid by partial hydrogenation. The market in the U.S. has been able to absorb the costs of this special processing which makes soy oil comparable to all other oils in terms of stability and acceptance. However, even with this special processing, soy oil is less tolerant of mishandling and special care must be observed in the processing. This special care adds to the cost of manufacturing quality soy oil products.

The problem of the linolenic acid content and the necessity of special processing is one that stands out as begging for a solution. In many countries, this factor presently limits its potential because of this instability of the unhydrogenated oil or because of regulations which do not allow it to be hydrogenated. The added cost of further processing also tends to limit its potential in some areas.

Table 4. **Influence of geographic location on consumer ratings of vegetable oils under use conditions.**

	Soy/Corn Preference				Soy/Sun Preference			
City	Base	Prefer Soy	Prefer Corn	No. Pref.	Base	Prefer Soy	Prefer Sun	No. Pref.
Atlanta	(51)	49%	37%	14%	(46)	33%	50%	17%
Boston	(50)	44	50	6	(48)	35	44	21
Dallas	(45)	42	47	11	(44)	50	32	18
Davenport	(52)	58	39	4	(47)	45	43	13
Oakland	(40)	55	35	10	(40)	28	45	28
Portland	(54)	61	30	9	(53)	40	45	15

Table 5. **Consumer ratings of vegetable oils under use conditions.**

	Soy/Corn Attribute Rankings		Soy/Sun Attribute Rankings	
Attribute	Soy	Corn	Soy	Sun
Left food greasy	17%	22%	18%	17%
Had an odor	13	15	10	9
Left odor in home	10	10	8	10
Had an aftertaste	9	13	8	10
Smoked	15	13	7	10

We are presently reassessing the genetic approach to the problem of the linolenic acid content. The elimination or reduction of linolenic acid content of the crude oil could eliminate the necessity of hydrogenation and reduce the cost of processing soy oil significantly. This could expand greatly the potential for use of soy oil on a worldwide basis.

In many countries, oil utilization is the key to increased potential market for soybeans. Many countries now are prime markets for protein from soybean meal but for various reasons, cannot utilize the oil because of cost or food regulations. This leaves these countries with the alternative of processing the beans and exporting the oil, or importing the soybean meal. Neither of these alternatives are as economically attractive as processing the beans and utililizing both the oil and the meal.

The elimination of the linolenic acid content in soy oil, without expensive processing, would be a large step in solving this problem. It could also broaden the base of use for soy oil in the developed countries, such as the U.S., in which it presently is used extensively. However, we realize that this approach of trying to eliminate the linolenic acid content in the raw bean is a long-term and rather involved research effort that will take a major commitment to accomplish.

NOTES

David R. Erickson and Richard A. Falb, American Soybean Association, P. O. Box 27300, St. Louis, Missouri 63141.

RISK MANAGEMENT IN MARKETING SOYBEANS

D. E. Kenyon

Soybean price variability has increased substantially during 1973-78 compared to 1960-72 (Table 1). The coefficient of variation of monthly within year prices averaged .052 during 1960-72 but averaged .179 during 1973-78, more than three times the previous level. Increased price variation is the result of changed government programs, reduction in stock levels, increase in exports, the devaluation of the dollar, and variation in world production. Increased market risks along with increased credit needs have substantially increased the financial risks of most agricultural producers (1).

Table 1 contains coefficients of variation computed from annual yields and deflated annual prices. Price variation is four times that of yield. Prior to 1973, the variation in prices was not markedly greater than in yields. Thus, the increased financial risks since 1972 are largely the result of increased price variation. The market events since 1972 have greatly increased the producer's demand for information on methods of reducing price risk. Producers have become keenly aware of the potentially large returns from "better" marketing strategies, and thus have sought both public and private sources for information on managing market risks.

Producers have sought market information in basically two areas. First, they have sought basic supply and demand information in an attempt to more accurately forecast future prices. Second, they have sought information on forward pricing. Forward pricing is any method used to establish a price prior to delivery. The two methods analyzed in this study are cash contracting and hedging via futures.

This chapter outlines briefly the theory behind evaluating alternative marketing strategies, presents the results of some recent empirical studies, and

Table 1. Coefficient of variation for monthly soybean prices and annual mean prices and yield, 1960-78.

Year	U.S. Average Farm Soybean Price[a]	Standard Deviation of Monthly Prices	Coefficient of Variation of Monthly Prices	U.S. Average Soybean Yield	Average Farm Price Deflated by WPI
	– $ –	*– $ –*		*– bu –*	*– 1967 = 100 –*
1960	1.98	0.029	.015	23.5	2.07
1961	2.48	0.265	.107	25.1	2.62
1962	2.32	0.042	.018	24.2	2.45
1963	2.48	0.065	.026	24.4	2.62
1964	2.49	0.116	.047	22.8	2.63
1965	2.62	0.194	.074	24.5	2.71
1966	2.93	0.259	.088	25.4	2.94
1967	2.62	0.123	.047	24.5	2.62.
1968	2.49	0.086	.035	26.7	2.43
1969	2.43	0.116	.048	27.4	2.41
1970	2.60	0.164	.063	26.7	2.36
1971	2.94	0.108	.037	27.5	2.58
1972	3.30	0.255	.077	27.8	2.77
1973	6.50	1.719	.264	27.7	4.83
1974	6.42	1.046	.163	23.2	4.01
1975	5.24	0.575	.110	28.8	3.00
1976	5.58	0.939	.168	26.1	3.05
1977	6.85	1.440	.210	30.6	3.48
1978	6.03	0.958	.159	28.3	2.94
Average				26.1	2.74
Standard deviation				2.1	0.887
Coefficient of variation				0.080	0.324

[a]Average of unweighted calendar year monthly prices.

draws implications for future research in this area. The empirical section is divided between pre-harvest and post-harvest pricing strategies. The emphasis is on comparing the results of forward pricing strategies with traditional cash market sales. Only price risk is considered. The interrelationship between price, yield, and financial risk is discussed briefly.

THEORETICAL FRAMEWORK

Theory suggests a producer would choose among marketing alternatives with outcomes expressed in probabilities by selecting the alternative that maximized utility. His choice depends upon the expected mean and variance of each outcome and his personal attitude toward risk. The procedure for evaluating different risky alternatives is presented in Figure 1, and is referred

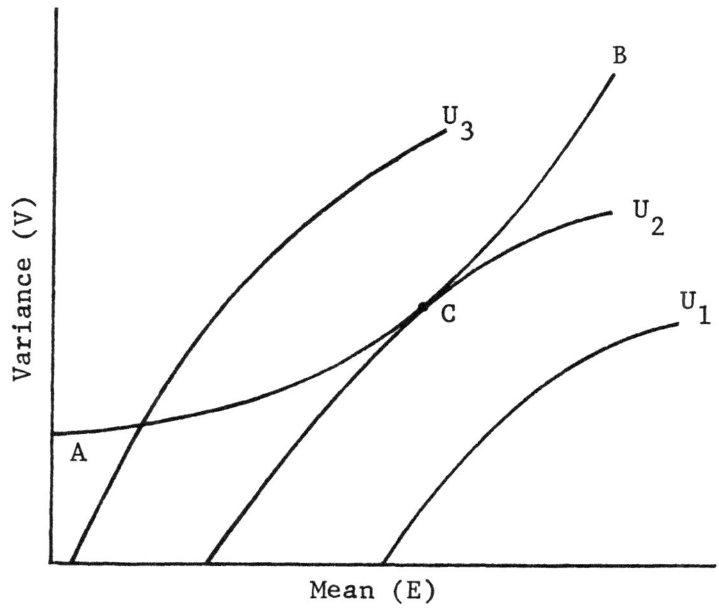

Figure 1. **Illustration of (E-V) analysis ($U_1 > U_2 > U_3$).**

to as E-V analysis. For each alternative, the expected mean price (E) is computed and the variance of price (V) is estimated based on historical data. The problem is to find the marketing alternative that maximizes the producer's utility. Curve AB depicts those efficient alternatives in that they constitute alternatives having maximum E for given V or minimum V for given E. Any alternative above AB has an alternative on AB with greater utility.

The optimal marketing alternative is the point on AB that maximizes the producer's utility. The risk averter attaches greater disutility to losses than to gains when the magnitude of gains and losses are equal in amount and likelihood. This risk-averse behavior is shown with isoutility curves, U_1, U_2, and U_3. Point C on the efficient set gives the maximum utility.

Increasing price risk since 1972 means to maintain expected price, the producer must tolerate more variable prices. Or conversely, to keep risk constant, the producer must sacrifice expected price levels. These adjustments to increasing price variability reflect higher costs of risk bearing. Thus producers seek market information that update and improve their expectations of future events. This is evidenced by renewed interest in outlook information, and expansion of various market information services since 1972.

In the empirical section that follows, the mean and variance or standard deviation of alternative pricing strategies will be presented. When strategies have similar means, those with a lower variance will be preferred. When strategies have both higher means and variances, the preferred strategy would depend on the individual producer's risk aversion.

PRE-HARVEST PRICING STRATEGIES

Nichols and Ikerd (6) suggest a procedure of forward pricing to establish profits to reduce price risk. Little published research is available analyzing such a strategy. Nichols and Ikerd (6) start by determining an asking price. Asking price is total production costs plus a return for management and profit. When the futures market or cash contracting offer a price greater than the asking price, the producer should price his crop. Several variations of this marketing strategy plus one strategy to allow producers to respond to new market information were analyzed.

Since soybean price variation increased dramatically starting with 1973, the 1973-78 production seasons were chosen for analysis. Planting was assumed to occur on May 1. Harvest occurred during the month of October. The harvest prices and basis estimates are for the Northern Neck Region of Virginia. An asking price was calculated for each year. The management return was calculated at 7% of gross sales/acre. Profits were computed at the rate of 10 and 20% of total production costs/acre. Yields were assumed constant at 26.5 bu/acre. All production was sold at harvest.

The following marketing strategies were analyzed: (a) Sell November futures starting on May 1 when the futures price minus basis is greater than the asking price. The asking price includes 10% profit and the entire expected production is sold. (b) Same as strategy (a) except 20% profit included in asking prices. (c) Sell 1/4 of expected production in June, 1/4 in July, 1/4 in August and 1/4 in September. Sell when the adjusted futures price or cash contract price is above the asking price *and* the 5 and 10-day moving averages give a sell signal. If no sales are made in June, and a sell signal occurs in July, sell 1/2 of the expected production, and (d) Starting in June, sell when November futures minus basis is greater than asking price and the 5 and 10-day moving averages give a sell signal. Buy back the futures contract when moving averages give a buy signal. Sell and buy back according to the moving average signals until October 1.

Strategy (c) was included to attempt to avoid hedging when prices were trending up. Strategy (d) was analyzed to determine if farmers could increase average price by responding to new market information. The 5 and 10-day moving average is only one of many trend following methods. They were chosen for two reasons. First, preliminary investigations of several moving averages indicated that the 5 and 10-day average performed reasonably well. Second, the 5 and 10-day average is relatively simple to calculate and understand, an important feature for farmers who allocate little time to marketing and are relatively untrained in marketing skills. A sell signal is given when the 5-day cuts through the 10-day average from above. A buy signal is given when the 5-day cuts through the 10-day average from below. An analysis of past November soybean futures prices from January to October during 1973-78 indicates the highest price in each year but one occurred in June or later. Therefore, straegies (c) and (d) begin on June 1.

The results of these strategies are presented in Table 2. Strategies (A) and (B) reduce the average price $.41 and .19/bu, respectively, without any reduction in price variation when compared to cash sales at harvest. However, notice that under strategy (A) the price received always permitted a 10% profit/bu. Selling the entire crop at harvest resulted in prices substantially below the asking price in 1975 and 1977. For producers with low equity financing, protection against prices below cost of production may be crucial to long-run survival.

Strategy (C) increases the average price by $.15/bu and reduces price variation by 75% compared to cash sales at harvest. The improvement in average price is a result of spreading the pricing decision over four months and only selling when the moving averages indicated declining prices.

Table 2. **Prices received under various pre-harvest pricing strategies for soybeans, 1973-78.**

Year	Oct. Cash Price[a]	Asking Price[b]	Strategy			
			A	B	C	D
			– *$/bu* –			
1973	5.30	3.57	4.17	4.17	6.20	4.90
1974	7.70	4.36	4.35	4.35	6.33	8.01
1975	4.91	5.33	5.65[c]	6.03[c]	5.69	5.83
1976	6.12	5.45	5.86	6.33	6.35	6.68
1977	4.98	5.73	7.07	7.07	5.72	5.66
1978	6.58	5.95	6.04	6.50	6.18	6.00
Average	5.93	5.07	5.52[d]	5.74[d]	6.08[d]	6.18[e]
Standard Deviation	0.99	--	1.00	1.09	0.27	0.97
Coefficient of Variation	0.17	--	0.18	0.19	0.04	0.16

[a] Northern Neck, Virginia.

[b] Production cost plus 7% of gross sales for management and 10% of production costs for profit.

[c] During October and November 1974, $8.00/bu prices could have been established for part of the 1975 crop.

[d] If hedging instead of contracting is used, subtract $.03/bu.

[e] Hedging costs of $.03/bu/trade have been subtracted. On average, this strategy had 5 trades/yr.

Strategy (D) gives the highest average price, but has a coefficient of variation similar to cash sales at harvest. Comparing strategy (C) and (D), only farmers indifferent to risk would prefer strategy (D) over strategy (C). In addition, relatively few farmers have the market knowledge, discipline, or time to implement strategy (D). On the average, strategy (D) involved selling and buying five futures contracts between June and October. The results of strategies (A - C) can be obtained by contracting as well as hedging.

PRE- AND POST-HARVEST PRICING STRATEGIES

Bolen, Baker, and Hinton (3) investigated corn and soybean marketing strategies for farmers with specified financial and risk characteristics. A model 600 acre farm in central Illinois with historical time-series data for 1965-74 was used. Yield risk was not considered. Twelve strategies involving cash sales, contracting, and hedging prior to and after harvest were evaluated. Strategy results are shown in Figure 2. The details of each strategy can be found in Bolen, et al. (3).

The boxed strategies are dominant, i.e., for a given mean price no strategy would yield a smaller variance, or for a given variance no strategy would yield a higher mean price. In this sense, the boxed strategies are "best." The strategies SII-2, SIV-1, SIV-3, SIV-4, and SV-1 survived the dominance test. These strategies are: (SII-2) Price 1/6 of expected production April, May, June, July, August, and September. Deliver all at harvest. (SIV-1) Store at harvest; price and deliver in March. (SIV-3) Store at harvest, sell and deliver for cash, 1/3 in January, 1/3 in March, and 1/3 in May. (SIV-4) Store at harvest; sell and deliver for cash, 1/4 in March, 1/4 in April, 1/4 in May, and 1/4 in June. (SV-1) Forward price 1/4 of the crop in May for harvest delivery, 1/4 in July for harvest delivery. Store the remainder at harvest and sell 1/4 for cash in January and 1/4 for cash in March.

Strategy results are generally consistent with theoretical expectations. The higher the average price, the greater the risk. With the exception of SIV-2, all the strategies with average prices above $3.40/bu do not include contracting or hedging. The lower variance strategies involving hedging (SIII-1, SIII-2, SIII-3) as compared to cash sales at harvest (SI-1) reduce the price risk 50% by giving up $.15/bu in average price. Individual farmers could determine which strategy is "best" based on their aversion to risk and credit restraints.

All strategies in set (B) (Figure 2) involve selling at least part of the crop after harvest. All strategies in Set A involve pricing the crop before or at harvest, although part of the crop may be sold after harvest. The two highest average price strategies, SIV-3 and SIV-4, involve spreading out cash sales from storage over several months from January to June.

Using the dominant strategies and assuming various levels of risk aversion by farmers, Bolen, et al. (3) determined optimal marketing strategies. The

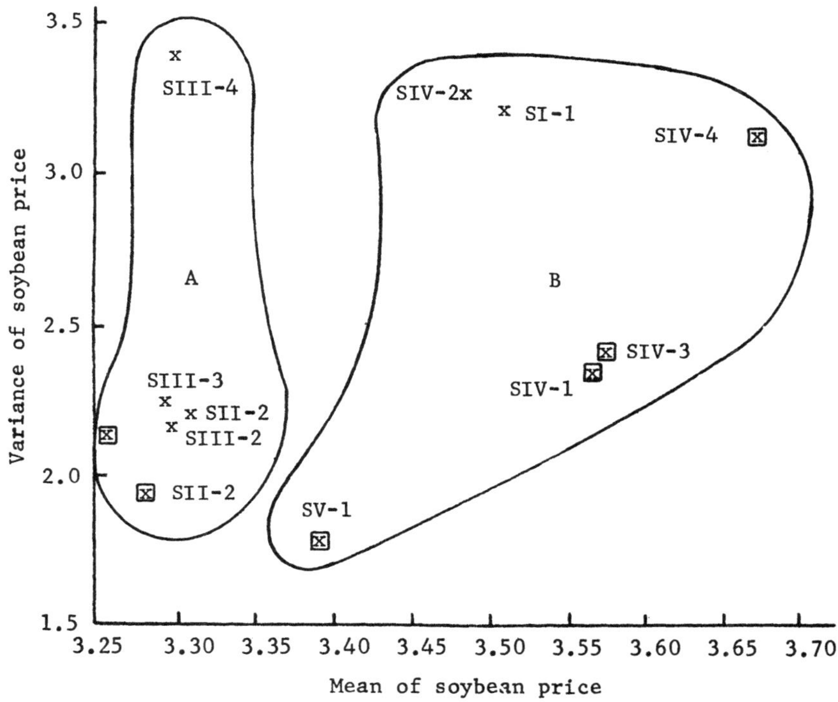

Figure 2. **Means and variances soybean marketing strategies, 1965-74.**

conservative marketer used a combination of strategies including pricing during production via contracting and cash sales from storage. Hedging was not included. The riskier marketer did not price prior to harvest, stored his entire crop and spread cash sales over the months January through June. Hedging or contracting was not used. Net returns were $90,359 and $93,931 and standard deviation of returns were $20,567 and $28,382 for the conservative and risky marketer, respectively. Thus, the risky marketer gained relatively little additional income while increasing risk substantially.

Commodity futures were not included in the optimal strategies. The advantage of futures over contracting is the flexibility to change the pricing decision. The inclusion of yield risk or new marketing information during the production or marketing season might make hedging more attractive relative to contracting and cash market selling.

Barry and Williams (2) found banks would make larger loans when 1/3 to 2/3 of the expected production was contracted compared to contracting all or none of the expected production during the growing season. Barry and Williams conclude " . . . that when credit is valuable, optimal growth plans will include contracting even for managers with little or no aversion to risk even though profit possibilities based on expected market prices appear more favorable with open market sales. Hence, it is clear that research and counseling programs with agricultural producers and their lenders on the merits of contracting as a market instrument should include some treatment of the credit effects."

This statement takes on added importance if the marketing strategy involves hedging. The lender should understand the cash flow needs related to margin calls, and the necessity of making margin calls, to avoid premature exit from a hedging strategy. In addition, the lender should develop a system to monitor the futures market positions of the borrower to guard against speculation. But with either contracting or hedging, the additional credit made available because of reduced price risk, may make these strategies very desirable even though they result in lower average prices.

HARVEST AND POST-HARVEST PRICING STRATEGIES

Since 1972, the highest prices for soybeans have occurred during summer months where producers had small quantities of soybeans to sell. This has led to producer interest in pricing strategies for marketing soybeans from storage. Lutgen (4) analyzed seven marketing strategies for soybeans over the years 1971-76. Storage costs of $.02/bu/month were assumed. Interest was charged at 7%. Commission charges were $.0084/bu. This cost was calculated per bushel even though soybean futures can only be traded in 1000 and 5000-bu units. Harvest cash price was the average of the last five days of October and the first five days of November. The cash and futures prices for the other designated selling months were the average prices for the first 10 days of the month. The seven marketing strategies analyzed were: (a) Sell all production at harvest. (b) Sell 1/3 at harvest, 1/3 in February, and 1/3 in April. (c) Store and sell 1/3 in February, 1/3 in April, and 1/3 in May. (d) Sell 1/3 at harvest, hedge remaining 2/3 at harvest and sell 1/2 in February, and 1/2 in April. (e) Hedge entire crop at harvest, sell 1/3 in February, 1/3 in April, and 1/3 in May. (f) At harvest, if July futures price is greater than harvest cash price plus $.50 basis and storage costs until May, hedge. If not, sell entire crop at harvest. Lift hedge in February, April, or May when basis is equal to or less than $.50. If hedge not removed by May, lift in May regardless of basis level, and (g) Assume farmer at harvest has perfect knowledge of cash prices from harvest through May. The last strategy is used for comparison purposes to indicate the value of perfect information.

The results of these seven marketing strategies are presented in Table 3. Strategy (e) increases gross income $6/acre without increasing risk compared to selling cash at harvest. At 30 bu/acre, average price is increased

Table 3. Nebraska adjusted gross income/acre by marketing strategy, 1971-76 (4).

Strategy	1971	1972	1973	1974	1975	1976	Average	Standard Deviation
a	$72.95	$ 80.78	$140.59	$210.42	$130.11	$154.08	$131.49	$50.64
b	77.20	116.89	137.02	162.77	114.27	186.39	132.42	38.66
c	79.83	154.02	130.42	131.76	107.56	212.05	135.94	44.99
d	73.65	85.36	145.38	133.39	131.98	151.39	120.19	32.55
e	73.62	104.14	148.37	217.80	132.62	150.04	137.77	48.87
f	72.95	80.78	148.62	240.68	134.02	149.59	137.78	60.52
g	88.78	200.90	149.20	241.00	134.02	230.99	174.15	59.89

$.20/bu. Strategy (e) involves hedging using July soybean futures and selling 1/3 in February, 1/3 in April, and 1/3 in May. Selling 1/3 in each of these months without hedging generates slightly less income per acre [strategy (c)], but also has a smaller standard deviation. Thus, in this study, hedging increased returns without increasing risk. (With such small samples, the standard deviations are not significantly different in a statistical sense. However, to include years prior to 1971 with little price variation would not give a representative indication of the likely results of these pricing strategies during the volatile prices of 1972-78.)

A comparison of strategies (a-f) with strategy (g) indicated a substantial difference. Lutgen (4) states, "...this difference indicates farmers should not blindly follow any one marketing strategy year after year. If the producer is going to achieve better than average income through marketing, he will have to be more aware of outlook predictions, and be willing to change his marketing strategy not only year to year, but also within the year as market information becomes available. The producer will need to know what outlook information to watch and how they relate to market prices."

Lutgen's (4) seven marketing strategies were analyzed using Virginia prices, the addition of the 1977-78 marketing year, and the modification of strategy (f) and the addition of strategy (h). The same assumptions concerning storage costs and futures market commissions were used. Strategy (f) was changed to hedge if July futures minus $.20 historical May basis minus storage cost to May was greater than harvest price. If a hedge was placed, it was held until May when the futures were bought back and the soybeans sold. If a hedge was not indicated, the entire crop was sold at the harvest price. Strategy (h) assumes all production is stored at harvest, and 1/6 of the quantity stored is sold in each month from January through June in the cash market.

The results of analyzing these eight strategies over the years 1971-77 are shown in Table 4. The results are similar to Lutgen's (4). Strategy (h)

Table 4. **Northern Neck, Virginia, average annual price and standard deviation for selected post-harvest marketing strategies, 1971-77.**

	Strategy[a]							
Year	a	b	c	d	e	f	g	h
1971	2.95	2.92	3.04	2.83	2.88	2.95	3.04	2.99
1972	3.21	4.81	6.37	3.40	3.38	3.21	6.46	6.46
1973	5.30	5.40	5.34	5.35	5.45	5.46	5.46	5.36
1974	7.70	5.95	5.08	8.18	8.63	8.69	8.69	5.13
1975	4.91	4.41	4.43	4.83	4.94	4.97	4.97	4.60
1976	6.12	7.30	8.29	5.95	5.95	6.12	8.29	7.95
1977	4.98	5.57	6.20	5.05	5.18	5.05	6.20	6.11
Average	5.02	5.19	5.54	5.08	5.20	5.21	6.16	5.51
Standard Deviation	1.63	1.36	1.65	1.75	1.88	1.92	1.94	1.56

[a]All prices are net of storage costs and futures market commissions.

compares favorably to strategy (c) with almost identical averages and standard deviations. Both of these strategies involve spreading cash sales over the months following December of the harvest year. Strategy (e) increased average price of $.18/bu, compared to an estimated $.20/bu in the Lutgen study (4). However, the standard deviation over the period 1971-77 increased compared to cash sales at harvest. In the Lutgen study (4), the standard deviation remained about constant compared to cash sales at harvest.

Strategy (f) generated the same or higher prices as selling at harvest each year. Therefore, although it has a larger standard deviation than strategy (a), it would be preferred over strategy (a) because it always generated prices as high or higher than strategy (a). None of the other strategies, except strategy (g) which assumes perfect information, generated prices equal to or greater than harvest cash prices each year.

CONCLUSIONS AND SUGGESTIONS FOR FURTHER RESEARCH

The pre-harvest and post-harvest strategies analyzed were generally consistent with theoretical expectations. Pricing strategies with higher average prices were associated usually with higher variances of price. Pre-harvest pricing strategies involving contracting or hedging all of expected production at planting reduced average price $.20 to $.40/bu without much effect on price variance. However, these strategies generated profits each year during 1973-78, whereas selling for cash at harvest resulted in losses in two years. The use of 5 and 10-day moving averages in timing the placement of the hedge in conjunction with an asking price improved average price and reduced price variation substantially. These results, combined with the likelihood of more lenient credit policies for producers who forward price, make these strategies attractive to producers with low equity positions.

The Bolen, et al., study (3) indicated that pricing before harvest during 1965-74 reduced returns approximately $.10 to $.40/bu, depending upon the strategy used. Hedging was not included in any of the optimal strategies, although contracting was used by a representative conservative marketer.

In terms of post-harvest or storage pricing strategies, spreading cash sales over the months January through June gave the highest returns without any increase in price variation compared to selling at harvest. A storage hedge using July futures and selling in May increased returns $.18/bu. The standard deviation of price also increased, but the prices generated under this strategy were always equal to or greater than those obtain by selling at harvest.

Based on the pre- and post-harvest pricing strategies, the strategy of spreading the pricing decision over several months improved returns. Pricing 1/4 of expected production in June, July, August, and September when a price greater than the asking price could be established increased the average price and reduced price variation compared to selling all production at harvest [strategy (c), Table 2]. Selling 1/6 of total production from storage each month from January through June [strategy (h), Table 4] increased the average price without increasing price variation compared to selling all production at harvest.

The pre-harvest pricing strategy results presented should be interpreted with caution. None of the reported results include yield risk. McKinnon's (5) theoretical model provides the following rules for minimizing variance of income: the greater the output variability is relative to price variability, the smaller will be the optimal forward sale; and the more highly negatively correlated are price and output, the smaller will be the optimal forward sale. Since price and yield are negatively correlated, risk minimization would lead to pricing less than expected production.

The impact of yield variation on pricing strategies needs to be investigated. The McKinnon model does not estimate the tradeoff between average price and variance of price for alternative pricing strategies under conditions of variable yields. Other important questions are: should a producer forward price a larger percentage of his crop at planting if substantial profits can be established based on expected yields? What is the relationship between various marketing strategies, yield risk, and cash flow requirements? These are questions farmers are asking for which there is little empirical research results to base policy recommendations.

NOTES

David E. Kenyon, Department of Agricultural Economics, Virginia Polytechnic Institute and State University, Blacksburg, Virginia 24061.

REFERENCES

1. Barry, P. J. and D. R. Frazer. 1976. Risk management in primary agricultral production: Methods, distribution, rewards, and structural implications. Am. J. Agr. Econ. 58(2):286-295.
2. Barry, P. J. and D. R. Williams. 1976. A risk-programming analysis of forward contracting with credit constraints. Am. J. Agr. Econ. 58(1):62-70.
3. Bolen, K. R., C. B. Baker, and R. A. Hinton. 1978. Marketing corn and soybeans under conditions of market risk. Ill. Agr. Econ. 18(2):12-19.
4. Lutgen, L. H. 1978. An analysis of marketing strategies for soybean producers. Dept. of Ag. Econ. Rpt. No. 86, Univ. of Nebraska, Lincoln, Nebraska, pp. 1-13.
5. McKimmon, R. J. 1967. Futures markets, buffer stocks, and income stability for primary producers. J. Pol. Econ. 75:844-861.
6. Nichols, Jr., T. Everett and J. E. Ikerd. 1973. Three steps to a successful hedge, N. Carolina Agr. Ext. Serv. Circular 573:1-23, Raleigh, North Carolina.

COMPETITIVE POSITION OF NEW OILSEED SUNFLOWERS WITH SOYBEANS

H. O. Doty, Jr.

Sunflowers are second in importance to soybeans as an oilseed crop in the world. World production is about 12 million mt. The USSR is the leading sunflower producer with about half of the total. The U.S. is the second largest producer with over a tenth of the total, followed closely by Argentina and Romania.

Two kinds of sunflowers are grown in the U.S., the oilseed type and the non-oilseed or confectionery type. The non-oilseed type is used for human food and birdfeed and plantings are relatively stable at about 225 thousand acres. Non-oilseed sunflowers account for about 6% of total U.S. sunflower production.

Oilseed sunflowers in the U.S. are no longer a curiosity, an experimental crop, or a fad. They have become an established crop in certain sections of the U.S. This season about 15 mills are processing sunflowerseed in the U.S. Two-thirds are cottonseed mills in the South processing sunflowers part-time. U.S. oilseed sunflower production has increased greatly in recent years (Table 1). This trend will no doubt continue as new technical developments promise increased productivity.

Although sunflower is a plant native to the Americas, sustained production of the oilseed type began in the U.S. only 13 yr ago. Since then, production acreage expanded very rapidly. In 1978, over 3 million acres were planted to oilseed sunflowers. Farmers' 1979 planting intentions for sunflowers in the 4 major producing states place the expected acreage at over 5 million acres, an 87% increase. Production of sunflowers is expanding at a much faster pace than did soybeans after their U.S. introduction. It wasn't until 1938 that soybeans harvested for seed surpassed 3 million acres. Last

Table 1. U.S. Sunflower seed and oil: Estimated supply and disappearance, 1977-78 and 1978-79 (1).

Item	Sunflowerseed		
	1977-78	1978-79	Change
	– Kmt –		– % –
Beginning stocks, Sept. 1	23	76	
Production	1,322	1,810	37
Imports	3	4	
Total supply	1,348	1,890	40
Domestic crush	219	385	76
Non-oil usage	100	110	10
Planting seed	4	5	
Exports	949	1,305	37
Total disappearance	1,272	1,805	42
Season average price received by farmers $229/mt in 1977-78 $227/mt in 1978-79			
	Sunflowerseed Oil		
Production (Oct.-Sept.)	88	155	76
Domestic disappearance	54	115	113
Exports	34	40	18
Total disappearance	88	155	
Price of crude, Minneapolis $705/mt in 1978-79			

year, there was more acreage planted to sunflowers than planted to oilseed crops of flax, safflower, or peanuts, or the small grain crops of rye or rice.

TECHNICAL DEVELOPMENTS RESPONSIBLE FOR SUNFLOWER EXPANSION

Sunflowers recently have benefitted from two major technical breakthroughs. The first of these was a genetic breakthrough by Russian plant breeders which doubled the oil content from 20% to over 40%. This doubling of oil content made the crop economically competitive with other U.S. crops. U.S. farmers started growing these open-pollinated oilseed sunflower varieties in 1966. The second breakthrough was the production of hybrid sunflowers made possible by discoveries of French and U.S. scientists. Hybrids in the U.S. were the first planted on a major scale in 1977 (about 90% of the crop). These hybrids increased yields about 25%, making oilseed sunflowers an even sounder crop economically.

According to sunflower plant breeders, a large reservoir of genetic material is available with which to further improve sunflowers. Sunflowers are now at about the same stage of development as corn in the early 1930's—before widespread use of hybrid seed. Improved sunflower hybrids promise better disease and insect resistence which should increase seed production per acre. Increased oil content of the seed is also anticipated. Both result in increased oil output per acre and would make sunflowers a stronger competitor (Table 2).

Soybeans have not benefitted from an increase in oil content such as that which took place in sunflowers. Soybean oil yield per bushel increased from 7.4 lb in 1924 to 11.0 lb in 1953, but most of this increase was due to further improved extraction techniques which recovered more of the oil in the beans. Soybean oil yield per bushel is about the same today as it was in 1953. There is no commercial production of hybrid soybeans and none seems likely in the near future.

CONSEQUENCES OF TECHNICAL DEVELOPMENTS IN SUNFLOWERS

As a direct result of these technological developments for sunflowers, unit costs of producing sunflowers have been greatly reduced and new markets for sunflower oil and meal have been developing rapidly. These developments have put sunflowers in direct competition with other crops, including soybeans, for farm acreage. It has also put sunflowers in competition with soybeans for oilseed markets and markets for oilseed products—oil and meal.

Production Costs and Returns

Generalizations on production costs and returns of sunflowers are difficult to make because of differences in management practices, land, and weather. Production costs change quickly and soon become outdated because of changes in the various cost components. However, production costs and

Table 2. Value of four oil crops/acre for selected countries, 1971-75 averages (2).

Crop	Country	Yield/Acre Oil	Yield/Acre Meal	Value/Acre[a] Oil	Value/Acre[a] Meal	Value/Acre[a] Total
		– lb –			– $ –	
Oil palm	W. Malaysia	3,475[b]	500	638[b]	27	665
Soybean	U.S.	288	1,272	62	100	162
Sunflower	USSR	526	658	138	41	179
Peanuts	U.S.	739	972	225	71	296
Peanuts	Nigeria	202	244	61	18	79

[a]Using 1971-75 average prices for Europe.
[b]Includes 3,100 lb palm oil and 375 lb palm kernel oil.

yield estimates can show differences in costs and expected returns at a particular time for sunflowers compared with other crops.

Recent production costs and returns comparisons between sunflowers and other crops in the many different production areas of the country are limited, particularly between sunflowers and soybeans. The following are some recent studies on this subject (4,8).

Sunflower competes with flax and small grains in the northern production areas. Estimated 1978 production costs and yields in east central North Dakota, under better than average management, are shown in Table 3. Breakeven prices for sunflower were $7.77/cwt, wheat $2.72/bu, barley $1.84/bu, and flax $5.53/bu (4). With these relationships many farmers in this area switched to sunflowers in 1978.

Similar estimated 1978 production costs and yields on the High Plains of Texas, for better than average management, are shown in Table 4. Under dryland conditions, the breakeven prices were: sunflower $12.00 cwt, cotton

Table 3. **Estimated 1978 production cost and yield for sunflower, wheat, barley, and flax for East Central North Dakota (4).**

Production Item	Sunflower	Wheat	Barley	Flax
		– $/a –		
Direct costs				
Seed	5.25	6.60	6.00	6.00
Fertilizer	6.60	10.80	11.75	2.65
Herbicide	7.70	3.10	3.10	4.40
Insecticide	1.25			
Machinery repair	6.65	5.55	5.65	5.40
Fuel and lubricants	4.20	3.05	3.20	2.90
Crop insurance	3.00	2.10	2.40	2.60
Drying	1.65			
Custom hire	.80	1.20	1.20	
Interest on operating capital	1.40	1.30	1.35	.95
Total direct costs	38.50	33.70	34.65	24.90
Labor	7.75	6.10	6.40	5.75
Indirect costs				
Machinery replacement	11.85	10.30	10.45	10.10
Machinery interest and insurance	6.90	5.05	5.20	4.90
Land charge	30.00	30.00	30.00	30.00
Overhead	7.35	7.35	7.35	7.35
Total indirect costs	56.10	52.70	53.00	52.35
Total production costs	102.35	92.50	94.05	83.00
Expected yield	1,318 lb	34 bu	51 bu	15 bu
Break-even price	7.77/cwt	2.72/bu	1.84/bu	5.53/bu

Table 4. Estimated 1978 production cost and yield for sunflower and competing crops for Texas high plains (4)[a].

Production Item	Dryland			Irrigated		
	Sunflower	Cotton	Sorghum	Sunflower	Cotton	Sorghum
			– $/a –			
Direct costs						
Seed	7.50	3.81	1.50	9.00	6.60	5.00
Fertilizer	--	--	--	6.00	10.40	27.20
Herbicide	7.00	5.38	--	7.00	7.00	3.85
Insecticide	10.00	--	--	10.00	--	5.00
Crop insurance	--	5.54	--	--	13.20	--
Fuel, lub., and repair	12.04	12.82	11.88	12.69	12.63	13.13
Irrigation	--	--	--	18.36	18.36	35.86
Harvest expense	10.25	28.15	11.88	18.00	60.50	31.90
Interest on opr. cap.	1.45	.88	.29	2.03	2.33	2.59
Total direct costs	48.24	56.58	25.55	83.08	131.02	124.53
Labor at $4/hr	13.05	21.02	12.88	32.33	32.33	36.98
Indirect costs						
Machinery	15.67	15.18	15.45	16.05	16.01	15.99
Irrigation	--	--	--	20.04	20.04	34.10
Land	23.00	23.00	23.00	36.00	36.00	36.00
Overhead	3.27	4.00	3.22	8.09	8.09	9.25
Total indirect costs	41.94	42.18	41.67	80.18	80.14	95.34
Total production costs	103.23	119.78	80.10	195.59	243.49	256.85
Expected yield (lb)	860	260	1,780	2,010	575	6,670
Break-even price	12.00/cwt	.41/lb[b]	4.50/cwt	9.73/cwt	.38/lb[b]	3.85/cwt

[a]Source: (M. Sartin, personal communication).
[b]Value of cottonseed at $60/t is deducted from total cost to obtain break-even price of cotton.

$0.41/lb, and sorghum $4.50/cwt. Under irrigation, the breakeven prices were: sunflower $9.73/cwt, cotton $0.38/lb, and sorghum $3.85/cwt (4).

Estimated 1979 costs of production and breakeven prices under South Carolina conditions are shown in Table 5. Under the higher yield assumption, the breakeven prices were: soybeans $5.48/bu, corn $1.99/bu, sunflowers $9.00/cwt, and cotton $0.67/lb (8). These cost comparisons do not take into account how well crops fit together in diversified farming. Equipment particularly plays a major role. Soybeans, sunflowers, and corn offer flexibility and in combination reduce overall investment requirements.

Competition for Acreage

Present sunflower production in the U.S. is concentrated in North Dakota, Minnesota, South Dakota, and Texas. Also, some commercial

Table 5. Estimated 1979 break-even prices for sunflower and competing crops for South Carolina (8).

Item	Soybeans		Corn		Sunflowers		Cotton	
	21 bu	30 bu	80 bu.	100 bu	1500 lb	1750 lb	450 lb[a]	600 lb[b]
				– $ –				
Variable cost	86.77	110.33	135.09	137.16	110.05	110.50	245.10	294.57
Fixed cost	19.57	20.18	25.82	25.82	18.75	18.75	51.19	53.99
Land and overhead	31.94	33.83	35.81	35.97	33.59	33.59	47.59	53.57
Total costs	138.28	164.34	196.72	198.95	162.39	162.84	343.83	402.13
Break-even	4.13/bu	3.68/bu	1.69/bu	1.37/bu	.07/bu	.06/lb	.54/lb	.49/lb
Total costs	6.58/bu	5.48/bu	2.46/bu	1.99/bu	.11/lb	.09/lb	.76/lb	.67/lb
Estimates of average prices for 1979 crop planning	6.25/bu		2.25/bu		10.5/lb		.64/lb [c]	

[a]Nine insecticide applications.
[b]Twelve insecticide applications.
[c]Does not include value of cottonseed at 6.0/lb.

production is found in most midwestern states, in the southern states of South Carolina, Georgia, Alabama, and Mississippi, and in the West Coast states of California and Oregon.

In the main, northern production area plantings were first made on acreage which had been devoted to flax. In recent years, sunflower has continued to replace flax and has also taken over acreage formerly planted to wheat, barley, and other small grains. This trend is expected to continue in this region during the next few years.

When cotton's acreage allotments were reduced several years ago, cottonseed crushing mills were left with excess crushing capacity and processors were looking for other oilseeds to crush. Sunflowers grown on some of this former cotton land provide oilseeds for the mills and give farmers more diversity and another cash crop.

Competition between sunflowers and soybeans for acreage will intensify as sunflower production continues to expand. Some expansion in sunflower acreage is espected on the fringe of present soybean, corn, and cotton growing areas where yields of these crops are relatively low. In areas where double cropping is practiced, sunflowers are competing well with soybeans because they have a shorter growing season and usually do better under drought conditions.

In the future, sunflowers will be exerting more pressure on soybeans in their competition for acreage because oilseed sunflower yields will be increasing faster than soybeans both in oil content and production per acre. Farmers will be obtaining higher sunflower yields as they learn how to handle the crop. Improved sunflower varieties will also give higher yields.

Competition for Markets

Sunflower is an oilseed crop grown primarily for its oil. Roughly 75% of sunflower's value is obtained from the oil and only 25% from its high protein meal. With the anticipated increase in oil content of sunflower, the proportionate value coming from oil should increase. In contrast, soybean is an oilseed crop grown primarily for its high protein meal. In recent years, roughly 60% of the value of soybeans has come from the high protein meal and 40% from the oil. On a per lb basis, the oil is more valuable than the meal. Oilseed sunflowers yield over 40% oil, whereas soybeans yield about 18%. Sunflowers may become known in the U.S. as the oil crop and soybeans as the protein crop.

Oilseed. The largest outlet for oilseed sunflower is the export market. The oilseed sunflower export market is similar to that of soybeans in that most sales are of seed rather than products, and are destined for Western Europe. In recent years, between 70 to 80% of the U.S. oilseed sunflower crop (oil equivalent) was exported. In 1977-78, exports were three-fourths of U.S. production. Exports of U.S. soybeans and products (oil equivalent) in recent years were about 50% of U.S. production.

Oil. Sunflower and soybean oils are both edible and complete for use in food fat products. Soybean oil presently is the dominant food fat in the U.S. It comprises about 60% of total fats and oils and about 80% of the vegetable oil used in food products.

Sunflower oil is a premium quality oil. It is very stable, has a high smoke point, is light in color, bland in flavor, and requires less refining than soybean oil (6). When produced in the north, it also is very high in polyunsaturates (Table 6) (3). For these reasons, it sells at a premium over soybean and cottonseed oils.

A sizable market has developed in recent years for fat products high in polyunsaturation due to concerns about high blood cholesterol and heart attacks (6). Blood cholesterol levels can be reduced by decreasing fat intake and substituting polyunsaturated fatty acids for saturated ones. If it were firmly established that high blood cholesterol levels cause heart attacks, the polyunsaturated food fat market could expand rapidly. Northern sunflower oil is much higher in polyunsaturation than soybean oil. It contains 64% linoleic acid compared to 51% for soybean oil (Table 6) (3).

Sunflower oil is particularly desirable for use in making high quality margarine and cooking and salad oils. In the premium grade fat products' market it competes primarily with corn and safflower oils. Sunflower oil is higher in polyunsaturates than corn oil and is much more stable than safflower oil. Since sunflower has major advantages over these oils, it could take over a large share of the premium market in the next few years. Blends of sunflower and soybean oil are expected to compete with soybean and other oils for the medium priced fat products' markets. Soybean oil's primary advantage over sunflower oil is its lower price. This gives soybean oil an advantage in some major fat product markets such as vegetable shortening and lower priced vegetable oil margarine. When sunflower oil has become well established in oil products, it will start competing more directly with soybean oil.

Europeans are familiar with sunflower oil and many prefer it over other oils. Responding to this demand, we have exported most of our sunflower seed production to Europe. There is now a dependable U.S. supply of oilseed sunflowers for crushing. In 1978, three large American manufacturers of oil products began producing sunflower oil-based margarine and cooking and salad oils. Both 100% sunflower oil-based products as well as blended sunflower and soybean oil products are on the market. Consumers have indicated a strong demand for sunflower oil products. Due to the very successful introduction of sunflower oil-based products, test markets are being expanded for some products, others are going into national distribution,

Table 6. Fat content and major fatty acid composition of selected fats and oils in decreasing order of linoleic acid content. After (3) except where noted.

Fats and Oils	Total Fat	Saturated Fatty Acids[b]	Unsaturated Fatty Acids[a]	
			Oleic[c]	Linoleic[d]
		– % –		
Safflower (high linoleic)	100	9	12	73
Sunflower (high linoleic) Northern grown, USA	100	10	21	64
Corn	100	13	25	57
Soybean	100	15	23	51
Cottonseed	100	26	18	50
Sunflower (high oleic) Southern grown, USA	100	11	34	49
Sesame	100	15	39	40
Peanut (7)	100	18	47	29
Chicken fat	100	33	40	17
Lard	100	40	41	10
Palm	100	48	38	9
Qlive	100	14	72	8
Tallow, beef	100	48	36	4
Palm kernel	100	81	11	2
Coconut	100	86	6	2

[a]Total is not expected to equal "total fat."
[b]Includes fatty acids with chains from 6 through 20 carbon atoms.
[c]Monounsaturated.
[d]Polyunsaturated.

and several additional manufacturers are planning to introduce sunflower oil-based products in 1979.

Sunflower appears to be a "natural" for advertising and promotion. This gives it an advantage over other oils. Sunflowers have grown wild in country areas for a long time and many consumers admire than as beautiful large yellow and light brown flowers grown widely in gardens or as a delicious edible kernel. The colorful blossom on attractive advertising and packaging of sunflower products has eye appeal and makes it easy to identify the products.

Industrial oil uses are not a major market for sunflower and soybean oils. Both are used in drying oil products. High linoleic acid sunflower oil has good drying oil properties and finds use in white and pastel shades of paint because of its non-yellowing characteristic. Soybean oil is used in making paints, resins, and plastics because it is a low priced semi-drying oil.

Meal. Competition between sunflowers and soybeans also takes place as their meals compete for use in livestock feeds. Sunflower seeds processed without dehulling produce a 28% protein meal, and when partially dehulled a 44% protein meal is produced. Further removal of hulls is possible by screening the meal, thus producing a 51% protein meal (5). In comparison, soybean meal is available in 44% and 49% protein products.

Sunflowers have one decided advantage over many of the oilseeds. They contain no growth-inhibiting or toxic elements. Cottonseed, rapeseed, and soybeans all contain toxic components. Heat must be used to deactivate the enzymes in soybeans, making the meal suitable for feeding. Sunflower meal's main disadvantage is its rather high fiber content. This can be reduced by removing the hulls either before or after processing. Sunflower meal produced under good processing conditions has a nutritive value equivalent to soybean meal (5). However, it contains a different amino acid makeup. Its mineral and energy content compare favorably with other oilseed meals (5,6).

Numerous animal feeding studies have shown that sunflower meal is an excellent, high quality, protein ingredient in livestock feeds. In feeding tests, steers fed sunflower meal performed equally well compared to those fed soybean meal. Researchers found that up to 50% of the protein concentrate in rations for laying chickens could be replaced with sunflower meal. In swine rations, a rate of gain similar to soybean meal was obtained with up to 25% of the protein from sunflower meal. If care is taken to assure adequate levels of lysine and energy, sunflower meal can be fed in any nonruminant ration as well as in ruminant rations (5,6).

Meal, a naturally bulky product, is costly to ship and usually is consumed by livestock within a relatively short distance from the crushing plant. Processors have experienced no difficulty in finding markets for sunflower meal. This is not surprising since there is a strong demand for protein in the world. Sunflower meal has found sufficient markets in or close to production areas. It has replaced flaxseed, cottonseed, and soybean meals in some livestock rations in these areas. Sunflower meal use has been mainly in ruminant rations and this will probably be its main market.

Although sunflower flour is edible and contains no toxic elements, it will not displace edible soybean flour to any extent at the present time. This is because sunflower protein contains a constituent, chlorogenic acid, which oxidizes and in so doing causes a progressive color change from white to beige, to green, and then to brown (6). This is undesirable in food products where color is an important factor. Removal or deactivation of chlorogenic acid in sunflower protein will be necessary before it can compete successfully with soybean protein in most edible markets. However, sunflower protein flour has some properties which make it more desirable in food supplements than soybean protein flour. It has a desirable bland flavor and is low in sugars, which produce the flatulence (gases) in the digestive tract. Soybean flour, on the other hand, has an objectionable beany flavor, and when eaten produces undesirable flatulence (6).

SUMMARY

Sunflowers and soybeans are competing in oilseed markets. They compete for export sales and their oils compete in domestic markets for use in margarine and in cooking and salad oils, while their meals compete in livestock feeds. In the future, sunflower will vie for acreage on the fringe of present soybean and corn growing areas where yields of these crops are relatively low. Competition between sunflowers and soybeans will intensify in the years ahead.

NOTES

Harry O. Doty, Jr., National Economics Division, Economics, Statistics, and Cooperatives Service, U.S. Department of Agriculture, Washington, D.C. 20520.

REFERENCES

1. Anonymous. 1979. Fats and oils situation. Economics, Statistics, and Cooperatives Services, USDA, FOS-294, p. 13.
2. Boutwell, W., H. O. Doty, D. Hacklander, and A. Walter. 1976. Analysis of the fats and oils industry to 1980—With implications for palm oil imports. ERS, USDA, ERS-627, p. 15.
3. Brignoli, C. A., J. E. Kensella and J. L. Weihrauch. 1976. Comprehensive evaluation of fatty acids in foods. V. Unhydrogenated fats and oils. J. Amer. Dietetic Assn. 68:224-229.
4. Cobia, D. W. and D. E. Zimmer. 1978. Sunflower production and marketing. N.D. State Univ., Fargo. Ext. Bull. 25, pp. 7-8.
5. Dorrell, D. G. 1978. Processing and utilization of oilseed sunflower, pp. 407-440. In J. F. Carter (Ed.) Sunflower Science and Technology, Am. Soc. Agron., Madison, WI.
6. Doty, H. O., Jr. 1978. Future of sunflower as an economic crop in North America and the world, pp. 457-488. In J. F. Cater (Ed.) Sunflower Science and Technology, Am. Soc. Agron., Madison, WI.
7. Leverton, R. M. 1974. Fats in food and diet. ARS, USDA Agr. Inf. Bull. 361.
8. Smith, D. 1978. In Ron Smith, Enterprise budget recommended for producers before year's end, Southeast Farm Press, November 29, p. 14.

EXPORTING SOYBEANS THROUGH THE MISSISSIPPI RIVER GULF PORT

W. M. Gauthier, G. R. Hadder, and H. D. Traylor

Soybean exports accounted for 209.44 million hectoliters, or over one-third of the 528 million hectoliters of grain exported through nine export elevators located on the lower Mississippi River between Baton Rouge and the mouth of the Mississippi River in 1977 (8,2). The ability to export soybeans is dependent upon the ability to export all grains; consequently, reference to grain export capability is synonymous with soybean export capability.

The export elevator complex consists of export elevator facilities and varying combinations of barges, rail units, trucks, and ships. Export elevators differ from each other in export potential due to differences in facility design and storage capacity. The size of the mobile fleet available for grain transport varies by mode with prevailing economic and climatic conditions. Barges are the principal mode of in-bound movement. Rail units are also quite active during the winter months when the upper Mississippi River is frozen. Trucks also provide some in-bound movement and serve to tie country elevators within a 100 to 150 mile radius into two of the export elevators. For all practical purposes, ships are the mode of outbound movement.

At the first World Soybean Research Conference, various transportation and storage problems were identified as major contributors to bottlenecks in the movement of export grains (4,5). The relationships between risk, uncertainty, and speculation in soybean marketing outlined at the conference offer useful background for formulating hypotheses about bottlenecks that hamper the system's ability to export grain (3).

PROBLEM

The individual export elevator exists to support a grain merchandising program that seeks maximum profits. It does this by receiving, storing, conditioning, and shipping out grains brought to the elevator by barges, rail units, and trucks. Ships provide for movement of the conditioned grain out of the system to make room for other incoming grains. Conceptually, a coordinated operation among the elevators and the mobile modes that maximizes turnover at every elevator in turn maximizes profits for grain merchandisers and minimizes bottlenecks for the system. Profit maximization and bottleneck minimization, however, are not necessarily complementary goals.

Barge and rail units are performing complementary transport and mobile storage functions while moving grain into Mississippi River Gulf elevators. Between arrival and unloading, these complementary functions become competitive insofar as a loaded mobile mode awaiting cargo discharge is substituting for fixed storage which makes it unavailable for transport. Even if demurrage charges make the owners of the mobile modes indifferent as to whether their equipment is used for storage or transport, producers and other elements of society bear the social costs of bottlenecks created due to the denial of transportation capability into and/or out of the port.

Conceptually, demurrage charges act as incentives to encourage the users of the mode to minimize their use for storage and to release them for transport. The persistence of bottlenecks despite increases in both absolute numbers of elevators and increased storage turnovers for all elevators over time, however, suggests that there are economic incentives for bottlenecks that outweigh demurrage charges as deterrents to bottlenecks (2).

The dynamic interactions that are associated with export grain handling provide few, if any, *ceteris paribus* conditions that make it possible to explain bottlenecks with a given set of variables. The uncertainties created by the dynamics of world grain export demands and alternative uses for the mobile modes make it difficult to identify the optimum combinations of export elevator storage and handling capacities and sizes of mobile fleets needed with any degree of statistical reliability. These limitations to conclusive analyses intensify the need for an assessment of the grain export capability of the Mississippi River Gulf Port system.

The objectives of this chapter are to: (a) describe a simulation model of the Mississippi River Gulf grain export system; (b) describe a procedure using model output to project maximum export grain handling capability for the system; and (c) compare projections of grain handling capability for the system against projections of 1985 grain export requirements.

THE SIMULATION MODEL

The Fortran-based simulation language GASP-IV, was selected as the tool for analysis (6). It protrays the Mississippi River Gulf grain export system as a multi-server queueing model with finite queues. Grain inputs at a particular service facility (elevator) can be implemented by three modes of arrival: barge, rail unit, and truck. Grain removal is accomplished by ship. All grain transfer activities at an elevator are constrained by grain inventory at that site. Ship loading is restrained by weather conditions. It is assumed that the transfer rate distribution properly reflects the impact of variables such as differences in grain type, grain blending, sampling conditions, and daily fluctuations in crew performance and machinery efficiency.

The forms and parameters of the statistical distributions for modal service are derived from primary data for a base elevator. Service rates among elevators are relative to a base elevator. Other input reflects unique storage capacities, number of servicing berths, and initial inventories for each elevator.

Model output used in the development of this paper consisted of service facility utilization measurement for each mode at each elevator. Other model outputs include measures of modal queues in terms of physical units and time, total and per bushel demurrage costs, export volumes handled by each mode, storage turnovers, and percentages of elevator blockage caused by too much or too little grain available in the elevator for the unloading of inbound modes or loading of ships, respectively.

Simulation is not an analytical tool that provides a unique optimum. It provides for the replication of a model under stochastic conditions. These stochastic conditions, in turn, yield unique measure for each set of conditions. In any given run, the points and severities of bottlenecks vary with the random numbers chosen to initiate the simulation. Multiple runs are performed to enhance the reliability of the measurements. The projections of grain handling capability in this paper are based upon five multiple executions of the model.

PROCEDURE

In 1977, 528 million hectoliters of grain were exported through the system of nine export elevators. The percentage distribution of both volume and mode of arrival varied among the elevators in the system. The first major assumption is that the maximum export handling capability of an elevator can be projected given the intensity measurement of utilization at its busiest facility.

A measure of utilization expresses the fractional intensity of use from 0 to 1. Its inverse yields a factor of potential volume expansion at an elevator given a static environment. The product of the potential volume expansion factor and the 1977 base volume is the projected export elevator handling capability. The capability of the elevator to handle the projected volume needs to be verified by the model, however, as expansions at other elevators in the system could conceivably limit the expansion at any given elevator.

The second major assumption is that the system's grain export potential is equal to the sum of the nine individual export elevators' export potential. The export handling potential for the system was measured under most limiting and under average limiting criteria at each elevator. The range in grain handling capability for the system under those two criteria varies from 602 to 689 million hectoliters.

COMPARATIVE ANALYSES

The reasonableness of the model based grain handling capability projections for the system can be evaluated in light of two earlier works. One of these works provides estimates of grain export handling capability required for 1985 while the other provides actual peak monthly volumes handled between 1973 and March 1978. Model based estimates serve to synthesize all three pieces of work for purposes of assessing the capability of the Mississippi River Gulf Port system to handled projected 1985 volumes of grain. The low and high estimates of 1985 grain exports for the entire U.S. range from 1,091 to 1,760 million hectoliters (7). Between 1970 to 1977, approximately 39% of all U.S. grain exports passed through the Mississippi River Gulf Port system (2). Assuming that this percentage remains constant, 1985 grain exports through the Mississippi River Gulf Port system are expected to range from 426 to 686 million hectoliters.

A 1978 study reported that the Mississippi River Gulf Port system experienced monthly peak levels of 50.69 million hectoliters three times between 1973 and March 1978 (1). In a footnote, an actual maximum of 51.11 million hectoliters was inspected in one month for export. If these volumes were to be sustained over time, annual export handling capacity would range between 608 and 613 million hectoliters.

Table 1 summarizes grain handling capability required to handle the 1985 projected volume of grain exports as well as estimated grain handling capability available. One elevator lost to a recent explosion was included in these estimates. It, however, has been replaced by another elevator so that the net number in operation within the system remains constant. These estimates suggest that the existing system has the capacity to handle the low estimated 1985 volume of 426 million hectoliters, but that the high estimated 1985 volume of 686 million hectoliters would tax the system to its limits.

Projections of the range in grain export handling capability available from the existing system illustrates one of many types of information available from the model. Individual elevator capability can also be addressed with the model. General findings surrounding the projects of grain export handling capability for the system suggest the most likely facilities at which to target investment dollars for enhancing the grain export capability of the system.

Table 1. **A comparison of grain handling capability and requirement estimates for the Mississippi River Gulf Port System (1,7).**

Estimates of Grain Handling	Million Hectoliters Estimates	
	Low	High
Requirement for 1985	426	686
Capability from peak performance data	608	613
Capability from model	602	689

Assessing the Lower Limit of the Range

Barge facilities were limiting at two elevators and ship facilities were limiting at the other seven elevators under the most limiting criteria. When limiting, utilization levels tended to be higher on ships than barges. The potential expansion factors, however, were below 10%. Two of the nine elevators had no expansion potential. The existing system of elevators would be capable of expanding its 1977 volume of 528 million to a minimum of 602 million hectoliters.

Assessing the Upper Limit of the Range

The ship facility is the most limiting at eight of the nine elevators under average limiting criteria. The potential expansion factor of increase for the elevators ranged from 1.09 to 1.86. Two elevators had an expansion factor of less than 20%. Seven elevators had expansion factors greater than 20%. The existing system of elevators appears to be capable of expanding its 1977 volume of 528 million to 689 million hectoliters.

SUMMARY AND IMPLICATIONS

The ability to handle increased volumes of grain exists at seven of the nine export elevators in the system. Analyses based on average limiting criteria indicate the potential for increases at all nine elevators. The magnitude of the increases at each elevator requires model verification. The lower limit of expansion to 602 million represents a 14% increase over the 1977 volume of 528 million hectoliters. The upper limit of 689 million hectoliters represents a 30.5% increase over the 1977 volume. Comparisons of high projected requirements and capability suggest that the existing system will be taxed to its limit. The analyses suggest that the ship facility is the most critical at most elevators. Present restrictions on ship loading during rain were incorporated into the model. The suspension of this rule as well as additional investments in facilities could be investigated to note the effect upon grain handling capability within the system.

The projections of grain handling capability for the system based on the model appear reasonable in light of earlier work. The explosion in December of 1977 that destroyed one of the export elevators does not invalidate the general conclusions as a new elevator became operational in the latter half of 1978. Other dynamic impacts resulting from changes in either number(s) and/or capacities of existing elevators will affect the grain handling capability of the Mississippi River Gulf Port system.

NOTES

Wayne M. Gauthier, Gerald R. Hadder, and Harlon D. Traylor, Department of Agricultural Economics and Agribusiness, Louisiana Agricultural Experiment Station, Louisiana State University, Baton Rouge, Louisiana 70803.

REFERENCES

1. Gaibler, F. D. 1978. "The Transportation System's Capacity to Meet Agriculture's Short-Term Export Demand." In Foreign Agricultural Trade of the United States. Economics, Statistics, and Cooperative Service, U.S.D.A., Washington, D.C., pp. 5-10.
2. Gauthier, W. M. and H. D. Traylor. 1978. "Bottleneck Problems at Mississippi River Gulf Port Grain Export Elevators." Louisiana Rural Economist, 4:2-4.
3. Hieronymous, T. A. 1976. "Speculative Pricing Problems." In Lowell D. Hill (ed.) World Soybean Research, Proceedings of the World Soybean Research Conference. The Interstate Printers and Publishers, Inc., Danville, Illinois, pp. 692-698.
4. Hill, L. D. 1976. "Changing Transportation Modes and Marketing Channels for Illinois Soybeans." In Lowell D. Hill (ed.) World Soybean Research, Proceedings of the World Soybean Research Conference. The Interstate Printers and Publishers, Inc., Danville, Illinois, pp. 750-761.
5. Nicholas, C. J. and W. A. Bailey. 1976. "Transport Problems in Marketing Soybeans Overseas." In Lowell D. Hill (ed.) World Soybean Research, Proceedings of the World Soybean Research Conference. The Interstate Printers and Publishers, Inc., Danville, Illinois, pp. 692-698.
6. Pritsker, A. B. 1974. The GASP-IV Simulation Language. John Wiley & Sons, Inc., New York.
7. Thurston, K., M. J. Phillips, J. E. Haskell, and D. Volkin. 1976. "Improving the Export Capability of Grain Cooperatives." FCS Research Report 34, Farmer Cooperative Service, U.S.D.A., Washington, D.C., p. 29.
8. U.S.D.A., U.S. Foreign Agricultural Trade Statistical Report, Calender Year 1977. A Supplement to the Monthly Foreign Agriculture Trade of the United States, June, 1978, Economics, Statistics, and Cooperative Service, Washington, D.C.

SOYBEAN PROCESSING INDUSTRY – U.S. AND FOREIGN

J. J. Mogush

The soybean processing industry has expanded at a rapid rate in the past 10 yr. Soybeans today are processed for meal and oil on all of the continents of the world with the exception of Antarctica. Total value of products produced in 1978 is between 13 and 14 billion dollars.

Increasing world population and increasing per capita consumption of protein meals and edible oils seem certain to require further expansion of oilseed production and processing capacity. Various sources of meals and oils have competed for a share of this worldwide market. Different forms of business enterprise have sought to hold or increase their participation in this growing industry. Political entities have maneuvered to aid, protect, and expand industry within their territory. Economic realities have played their role in plant locations and size. The processing industry has struggled to hold down inflationary pressures on costs through improved technology and economies of size.

The objective of this chapter is to identify recent trends in oilseed processing which might be helpful in determining the future course of development of the soybean processing industry in the U.S. and the rest of the world.

WORLD OILSEED PRODUCTION

Strong demand for protein meals and edible vegetable oils has stimulated production of oilseeds in all producing countries of the world. In the past decade, growth has been at a compound annual rate of close to 5% (Table 1). Production of every single oilseed has increased in terms of gross tons produced, with the exception of groundnuts. The largest increase by far has

Table 1. World oilseed production. [a]

Oilseed	1967-68 Tons	1967-68 % of Total	1977-78 Tons	1977-78 % of Total
		– 000 M.T. –		
Soybeans	39,105	43.1	77,510	54.2
Cottonseed	17,870	19.7	24,730	17.3
Groundnut	11,605	12.8	11,170	7.8
Sunflower	9,790	10.8	12,660	8.8
Rapeseed	5,715	6.3	7,970	5.6
Copra	3,310	3.6	4,800	3.4
Palm Kernel	785	.9	1,100	.8
Linseed	2,545	2.8	3,170	2.2
TOTAL	90,725	100.0	143,110	100.0

[a]Sources: 1967–68, "Oil World Semi-Annual," December 1973, ISTA, Mielke & Company; and 1977-78, "Oil World," Special Issue 4, November 10, 1978, ISTA, Mielke & Company.

occurred in soybean production. The soybean is the only oilseed which increased its share of the world market—growing from 43% in 1967 to 54% of total world production of oilseeds in 1977. This represents a gain of 1%/yr in market share. Predictions of growth for this crop consistently have been on the low side. The principal reason for this is that the soybean has two high-quality products: soybean meal and soybean oil. Also, the combination of roughly 80% protein meal and 20% oil has satisfied the increasing needs of the world better than any other oilseed.

LOCATION OF SOYBEAN PROCESSING FACILITIES

The geographic location of soybean processing plants in the world has been influenced by both economic and political factors. In countries which are producers of soybeans, such as the U.S., sites for processing facilities have been selected after studying and weighing three primary considerations. In order of importance these are (a) areas of heavy production of soybeans, (b) areas of heavy consumption of soybean meal, and (c) sites which combine flexibility, efficiency, and comparative cost advantage in transportation. These include sites on major inland waterways, at major railroad and highway intersections, and at seaports.

Secondary considerations have been (a) proximity to edible oil refineries, (b) cost, availability, and quality of labor, and (c) cost and availability of energy. In addition, the outlook for further development of each of these factors needs to be favorable. Fortunately for most processors, soybean production and soybean meal demand have tended to increase in areas where processing plants have been constructed.

The industry has pursued a policy of expanding capacity ahead of demand for products (Table 2). Currently in the U.S., all areas are adequately served by modern, efficient processing plants. Selecting a new site today is a difficult problem. It is cheaper and less complicated to expand capacity at an existing facility. Much of this type of expansion has occurred. Size of the market to be served sets a limit on plant capacity, but no doubt we will see more of this type expansion.

The substantial increase in production of soybeans in South America, starting in 1972, has changed the outlook at many U.S. processing locations. The government of Brazil aggressively encouraged the building of a modern, efficient soybean processing industry. It has developed a system of export subsidies and taxes which favors the export of soybean meal and soybean oil over the export of soybeans. An export quota system is also used to hold back soybeans for domestic processing rather than allowing them to be exported.

In locating plants in Brazil, processors basically pondered the same considerations applicable to any producing country. However, Brazil currently consumes domestically only about 20% of its soybean meal production and 60% of its soybean oil production. The remainder has to move for export and, hence, this is a major factor in siting plants. This heavy marketing of soybean meal and oil into world markets out of Brazil has adversely affected U.S. plant locations which serve the same markets.

Soybean processing plants are also located in countries which are not producers of soybeans but which are sizable consumers of both soybean meal and soybean oil. Sites in these countries are selected using similar considerations to those used in producing countries. The emphasis is different. Deep water ports capable of accommodating very large ocean-growing vessels with incoming soybeans have cost advantages over shallow water ports and inland sites. It is more economic to handle and transport soybeans than soybean meal and oil. In addition, some of these countries maintain import duties and use other trade restrictions to increase the advantage of their processing industry over the same industry located in producing countries. Naturally, this does hinder the expansion of soybean processing in the U.S.

SIZE AND NUMBER OF SOYBEAN PROCESSING PLANTS

U.S. soybean processors have provided needed processing capacity ahead of demand for products. While accomplishing this, they have abandoned inefficient, poorly located plants, expanded and modernized existing plants, and constructed new, larger mills at better locations.

Table 2. Estimated number of soybean plants and processing capacity in the U.S., 1968-78.

Year	Processing Mills[a]	Annual Processing Capacity: Total[b]	Utilized[c]	Excess[d]	Ratio Utilized to Total	Average/Mill: Processing Capacity	Capacity Utilized[c]
	– number –		*– mil bu –*		*– % –*	*– mil bu –*	
1968	134	750	606	144	81	5.6	4.5
1969	132	800	737	63	92	6.1	5.6
1970	130	875	760	115	87	6.7	5.8
1971	123	900	720	180	80	7.3	5.9
1972	117	925	722	203	78	7.9	6.2
1973	113	1,000	821	179	82	8.8	7.3
1974	108	1,050	701	349	67	9.7	6.5
1975	103	1,100	865	235	79	10.7	8.4
1976	103	1,200	790	410	66	11.7	7.7
1977	99	1,250	845	405	68	12.6	8.5
1978[e]	100	1,265	985	280	78	12.7	9.9

[a]Estimates developed from census data and trade directories. Sources: 1968-77 (1); and 1978, author's estimates. Includes cottonseed and other oilseed mills that process significant quantities of soybeans.
[b]Estimates shown here are approximations of capacity at the beginning of the marketing year.
[c]Soybeans actually crushed.
[d]Difference between total capacity and soybeans utilized (crushed).
[e]Forecast.

In the past 10 yr, the number of soybean plants declined from 134 to 100. Any plant with daily capacity of less than 6,000 bu/day is not included in these numbers. The average plant capacity increased from 5.6 million bu/yr to 12.7 million bu/yr. Total U.S. capacity increased from 750 million bu/yr to 1,265 million bu/yr. Actual bushels processed increased from 606 million in 1968 to an estimated 995 million in 1978.

Ownership of processing plants in the industry has changed somewhat. The reasons behind the changes are varied and complex. One thing is certain, the business is not profitable all of the time. At some locations it is not profitable much of the time.

Cooperative owned processing plants accounted for an estimate 5% of industry capacity in 1946, 21% in 1971, and an estimated 24% in 1978. Six different cooperatives now are multi-plant processors with 13 plants under their management. Eight additional cooperatives are single-plant processors. Eleven corporations are multi-plant processors. They operate a total of 65 plants. Twenty-two additional corporations are single-plant processors. The total number of firms which processed a significant volume of soybeans in the U.S. in 1978 is 47.

Efforts to capitilize on economies of size have resulted in larger sized plants. In 1964 only 9% of U.S. processing plants could process 1,000 tons of soybeans/day. In 1974 this increased to 26%, and today, 50%, or one-half of the units can process at least 1,000 tons of soybeans/day (Table 3).

Brazil and other countries of the world have many more small plants than the U.S. A small plant now can be defined as one having a daily capacity of less than 1,000 tons of beans/day. Fifty percent of U.S. plants are large, 13% of Brazil's plants are large, and 29% of the plants in other countries of the world are large. Also, 29% of all soybean processing plants in the world are large; that is, they can process one thousand tons or more of soybeans/day (Table 4).

A headstart, a large domestic soybean meal market, and a market oriented economy are the principal reasons for the U.S. lead over other countries. It is generally assumed that in addition to size of plant, the U.S. industry is more efficient in other respects such as unit costs and yields.

Table 3. Size distribution of soybean processing plants in the U.S.[a]

Size	1964		1974		1978	
Tons/Day	Dist.	Accum.	Dist.	Accum.	Dist.	Accum.
			– % –			
Switch or <400	19	59	14	44	25	25
400 - 799	23	82	17	61	16	41
800 - 999	9	91	13	74	9	50
1000 - 1199	3	94	5	79	9	59
1200 - 1599	5	99	16	95	19	78
>1600	1	100	5	100	22	100

[a]Sources: 1964-74, (1); and 1978, author's estimate.

Table 4. Current size distribution of soybean processing plants in the U.S., Brazil and other countries of the world.[a]

Size	U.S.		Brazil		Other Countries		All Countries	
Tons/Day	Dist.	Accum.	Dist.	Accum.	Dist.	Accum	Dist.	Accum.
				– % –				
Switch or <400	25	25	66	66	38	38	44	44
400 - 799	16	41	17	83	25	63	20	64
800 - 999	9	50	4	87	8	71	7	71
1000 - 1199	9	59	4	91	5	76	6	77
1200 - 1599	19	78	2	93	7	83	9	85
>1600	22	100	7	100	17	100	15	100

[a]Author's estimate.

Excluding the East Bloc, USSR and China, a total of 337 plants operated in the world this past year. One hundred of these plants are located in the U.S., 117 in Brazil, and 120 in other countries of the world.The world now has a capacity to process 3 billion, 219 million bu of soybeans/yr. The U.S. has 39.3% of that capacity (Table 5).

TECHNOLOGY

Basic processing procedures have not been changed a great deal. Plant capacities have been increased by installing new equipment designed to handle larger volumes or by adding additional units of the same size as the old. Over the years, some improvement has been achieved in extractors and other processing equipment.

The most progress has been achieved in the field of energy conservation. Additional heat exchangers have been installed. Economizers have been installed in boilers to conserve waste heat. The number of BTU's required to process a bushel of soybeans has been reduced. Some plants have converted to coal boilers from oil/gas boilers in an effort to reduce costs and to implement national policy.

INTEGRATION

U.S. soybean processors are involved in most of the many different businesses which use the products they produce. Practically every processor is connected corporately with the feed business. Some go beyond this to the broiler, egg and hog producing businesses. Many have an edible protein business. Many processors also have integrated vertically into the vegetable

Table 5. Number and size of soybean processing plants, U.S., Brazil and other countries.[a]

Tons/Day	U.S.	Brazil	Other Countries	World Total[b]
Switch or <400	25	77	46	148
400 - 799	16	20	30	66
800 - 999	9	5	10	24
1000 - 1199	9	5	6	20
1200 - 1599	19	2	8	29
>1600	22	8	20	50
TOTAL	100	117	120	337
Annual Capacity/ Million Bu.	1265	656	1298	3219
% of World	29.3	20.4	40.3	100

[a]Author's estimate.

[b]The East Bloc, USSR, and China are not included.

oil refining business. A few go so far as to produce and market margarine and shortening. Economies in rapidly escalating costs of handling and transporting goods have stimulated this trend towards integration.

TRANSPORTATION

Some weaknesses in the U.S. transportation system have become apparent in the past few years. Inflation has increased costs of moving raw material and products. Periodic shortages of hopper cars, boxcars, and tankcars have not been resolved. Utilization of equipment in the nation's railroads has declined. In the past 10 yr, annual number of trips/yr for hoppers declined from 17.6 to 15.2; for boxcars from 17.4 to 12.2. The deterioration of Lock and Dam 26 has been well publicized. As of this writing, construction of a new lock and dam two miles downstream has not yet begun. Extremely cold weather the past three winters has accentuated transportation problems and costs. These factors have caused an increase in the percentage of truck transportation of soybeans and products. At times, processing plants have not been able to move enough product out of plants to run at capacity. This has been a surprising development which raises interesting questions as to the optimum size of processing plants. Some deregulation of transportation is now viewed as a possibility for strengthening the nation's transportation system.

GOVERNMENT REGULATION

The cost of complying with constantly increasing government regulations has surged. The two federal agencies which have had the greatest impact on the soybean processing industry in the past decade are the Occupational Safety and Health Administration (OSHA) and the Environmental Protection Agency (EPA). Real benefits have been gained for society from some of the changes required by these agencies. Some adverse consequences of regulation are becoming more visible. In too many cases, the costs of regulation are outweighing the benefits. The productivity of the nation has been affected. A Brookings Institution economist has calculated that OSHA and EPA regulations cut 1.4 points/yr from U.S. productivity growth between 1967 and 1975. Increased per bushel costs in processing soybeans have been unavoidable. Additional personnel are required to staff new departments which study new regulations and implement procedures to assure compliance.

FUTURE DIRECTIONS IN THE INDUSTRY

The soybean industry has expanded steadily for the past 40 yr. All things must come to an end sometime, yet no one can yet see an end to continued growth in demand for soybean products. Much of the world's population cannot afford even a minimum caloric diet. Another large segment does consume the minimum necessary calories, but does not enjoy proper nutrition which can be provided by oilseed products and the industries which use them to produce consumer foods. World population will continue to grow and a better standard of living will be demanded and achieved. This will require increased production of protein meals and edible oils. Soybean meal and soybean oil are excellent products. Producers will be challenged to produce more soybeans every year. Plant breeders will continue their efforts to increase yields per acre. Producers will continue to upgrade and improve farming practices. Through a combination of higher yields per acre and increased demand, soybeans will effectively be able to compete with other crops for the additional acres needed to meet total demand.

The U.S. soybean processing industry is not likely to expand proportionately with soybean production. A very low population growth rate in this country and an already high standard of living leave little potential for increased domestic demand. Processing capacity is more likely to be increased in those other countries which have economic and non-economic advantages as well as a more rapidly increasing domestic demand. Average plant capacity worldwide is likely to increase some more. On the other hand, one cannot be impressed with the possibilities of further increasing the size of the world's largest soybean plants.

U.S. soybean exports will gain on U.S. exports of soybean meal and soybean oil. The trend towards vertical integration will continue in the U.S. and other countries of the world. Lack of horizontal expansion opportunities in soybean processing in the U.S. will further this trend. Some deregulation of the transportation system is likely. This should be helpful to the market oriented soybean processing industry. Overall government regulation of industry will increase. Single-plant owners will find it increasing difficult to cope. Competition between corporate and cooperative multi-plant owners will intensify and will benefit both producers and consumers.

NOTES

John J. Mogush, Domestic Soybean Crushing Division, Cargill, Inc., Minneapolis, Minnesota 55402.

REFERENCES

1. American Soybean Association, "Soybean Digest Blue Book," Volume 38, June 1978.

AUTHOR INDEX*

Abruña, F., 35
Ahmad, R. A., 129
Allwood, J. K., 831

Baker, D. N., 533
Barfield, C. S., 249
Bell, R. E., 763
Bhattacharya, A. K., 291
Boswell, F. C., 417
Bowen, H. D., 451, 647
Bradley, J. R., Jr., 343
Brun, W. A., 153
Burton, J. C., 89

Campbell, M. F., 713
Curry, R. B., 557

Dina, S. O., 275
Doty, H. O., Jr., 871

Egli, D. B., 19, 501
Emken, E. A., 667
Esmay, M. L. 471
Eldridge, A. C., 721
Erickson, D. R., 851
Ezueh, M. I., 275

Falb, R. A., 851
Findley, P., 11
Frans, R., 399
Fukushima, D., 729

Gauthier, W. M., 883
Guarisma, R., 637

Hadder, G. R., 883
Hanway, J. J., 409
Hardy, R. W. F., 57
Hashimoto, H., 729
Hauck, S. J., 801
Havelka, U. D., 57
Heytler, P. G., 57
Hill, L. D., 823
Hoki, M., 471
Housley, T. L., 101
Hummel, J. W., 451
Huynh, V. M., 601

Ignoffo, C. M., 327
Israel, D. W., 111

Jaafari, H., 647
Johnson, V. A., 179
Jones, J. W., 571

Keeling, B. L., 367
Kenworthy, W. J., 217
Kenyon, D. E., 859
Kmetz, K. T., 355
Knipscheer, H. C., 807
Kogan, M., 303
Korsakov, N. I., 241

Lamm, R. M., Jr., 839
Leggett, J. E., 19
Liener, I. E., 703

Matzinger, D. F., 191
McWhorter, C. G., 371
McClure, P. R., 111
Mederski, H. J., 129, 557
Meyer, G. E., 557
Misra, R. N., 623
Mogush, J. J., 889
Moss, D. N., 225
Mounts, T. L., 659
Munns, D. N., 47

Nakamura, H., 1
Nave, W. R., 433
Nelson, A. I., 745
Newsom, L. D. 261
Nichols, T. E., Jr., 781
Noodén, L. D., 139

Patterson, D. T., 371
Paulsen, M. R., 493
Powell, T. E., 601
Rathore, Y. S., 291
Ravalo, E. J., 519
Rawlings, J. O., 201
Rodda, E. D., 519
Ross, I. J., 501
Rudd, W. G., 547

Schmidt, J. W., 179
Schmitthenner, A. F., 355
Schrader, L. E., 101
Schwarz, F. H., 831
Servaites, J. C., 101
Setter, T. L., 153
Siddall, J. N., 601
Siemens, J. C., 477
Sinclair, J. B., 519
Slife, E. W., 393
Smajstrla, A. G., 571
Sprague, G. F., 233
Steinberg, M. P., 745
Stimac, J. L., 249
Streeter, J. G., 129, 557
Szuhaj, B. F., 681

Taylor, H. M., 161
Tenne, F. D., 519
Throckmorton, R. I., 485
Traylor, H. D., 883
Turnipseed, S. G., 285

Updaw, N. J., 781

Van Duyn, J. W., 343

Wei, L. S., 745
Wernsman, E. A., 191
White, G. M., 501
Wolf, W. J., 721, 767

Young, J. H., 623

Zilch, K. T., 693
Zobel, R. W., 73

*A complete list of all conference contributors is printed in the *World Soybean Research Conference II: Abstracts,* Westview Press, Boulder, Colorado.

For Product Safety Concerns and Information please contact our EU representative GPSR@taylorandfrancis.com Taylor & Francis Verlag GmbH, Kaufingerstraße 24, 80331 München, Germany

Batch number: 10725929

Printed by Printforce, the Netherlands